S0-BHV-463

The Bacteria

VOLUME VII: MECHANISMS OF ADAPTATION

THE BACTERIA

A TREATISE ON STRUCTURE AND FUNCTION

Volume I: Structure (I. C. Gunsalus/R. Y. Stanier, eds.)

Volume II: Metabolism (I. C. Gunsalus/R. Y. Stanier, eds.)

Volume III: Biosynthesis (I. C. Gunsalus/R. Y. Stanier, eds.)

Volume IV: The Physiology of Growth (I. C. Gunsalus/R. Y. Stanier, eds.)

Volume V: Heredity (I. C. Gunsalus/R. Y. Stanier, eds.)

Volume VI: Bacterial Diversity (L. N. Ornston/J. R. Sokatch, eds.)

Volume VII: Mechanisms of Adaptation (J. R. Sokatch/L. N. Ornston, eds.)

The Bacteria

A TREATISE ON STRUCTURE AND FUNCTION

EDITOR-IN-CHIEF

I. C. Gunsalus

Department of Chemistry
School of Chemical Sciences
University of Illinois at Urbana-Champaign
Urbana, Illinois

VOLUME VII: MECHANISMS OF ADAPTATION

VOLUME EDITORS

J. R. Sokatch

University of Oklahoma
Health Sciences Center
Oklahoma City, Oklahoma

L. Nicholas Ornston

Department of Biology
Yale University
New Haven, Connecticut

1979

ACADEMIC PRESS • NEW YORK SAN FRANCISCO LONDON

A Subsidiary of Harcourt Brace Jovanovich, Publishers

ACADEMIC PRESS, INC.
111 Fifth Avenue, New York, New York 10003

United Kingdom Edition published by
ACADEMIC PRESS, INC. (LONDON) LTD.
24/28 Oval Road, London NW1 7DX

Library of Congress Cataloging in Publication Data

Gunsalus, Irwin Clyde, 1912– ed.
The bacteria.

Vol. 6– edited by L. N. Ornston and J. R. Sokatch.
Includes bibliographies.
Contents.––v. 1. Structure.––v. 2. Metabolism.––v. 3. Biosynthesis.––[etc]––v. 7. Mechanisms of adaptation.
1. Bacteriology. I. Stanier, Roger Y. II. Sokatch, John Robert, III. Ornston, L. Nicholas.
QR41.G78 589.9 59–13831

PRINTED IN THE UNITED STATES OF AMERICA

79 80 81 82 9 8 7 6 5 4 3 2 1

CONTENTS OF VOLUME VII

CONTRIBUTORS TO VOLUME VII

WINSTON J. BRILL, *Department of Bacteriology, and Center for Studies of Nitrogen Fixation, University of Wisconsin, Madison, Wisconsin 53706*

MARTIN DWORKIN, *Department of Microbiology, University of Minnesota, Minneapolis, Minnesota 55455*

HARRISON ECHOLS, *Department of Molecular Biology, University of California at Berkeley, Berkeley, California 94720*

CHARLES E. HELMSTETTER, *Department of Experimental Biology, Roswell Park Memorial Institute, Buffalo, New York 14263*

MARGARET HOLMES, *Department of Radiation Medicine, Roswell Park Memorial Institute, Buffalo, New York 14263*

D. E. KOSHLAND, JR., *Department of Biochemistry, University of California at Berkeley, Berkeley, California 94720*

OLGA PIERUCCI, *Department of Radiation Medicine, Roswell Park Memorial Institute, Buffalo, New York 14263*

MILTON H. SAIER, JR., *Department of Biology, John Muir College, University of California at San Diego, La Jolla, California 92093*

J. R. SOKATCH, *University of Oklahoma, Health Sciences Center, Oklahoma City, Oklahoma 73190*

MOON-SHONG TANG, *Department of Experimental Biology, Roswell Park Memorial Institute, Buffalo, New York 14263*

D. J. TIPPER, *Department of Microbiology, University of Massachusetts Medical School, Worcester, Massachusetts 01605*

MARTIN WEINBERGER, *Department of Experimental Biology, Roswell Park Memorial Institute, Buffalo, New York 14263*

A. WRIGHT, *Department of Molecular Biology and Microbiology, Tufts University Medical School, Boston, Massachusetts 02111*

FOREWORD

THE BACTERIA was conceived as a comprehensive source on the general biological properties of these organisms viewed from structure to heredity in five volumes. The initial aim was to include, in addition to the areas well advanced, an understanding of those needed for current advances whose attention had been infrequent or perhaps neglected.

The reception of THE BACTERIA, as indicated from comments and total copies in circulation, far exceeded the primary audience visualized as students of the bacteria per se. In fact, it included many chemists and general biologists seeking the familiarity required to use the bacteria as a tool for the elucidation of molecular and cellular processes. The initial plan had entertained the possibility of a detailed analysis of the general biology of many of the principal or perhaps more important groups of bacteria. This concept was, however, discarded for space considerations and in deference to what apparently had proven to be a more useful format.

The vastly extended knowledge of biological processes in which bacteria, their viruses, plasmids, and processes have taken a leading role since the appearance of THE BACTERIA as a basic sourcebook had led to questions of update from colleagues directly and through the publishers. Perusal of the volumes and reflection of scientific advances has served to emphasize the steepness of the climb, sharp as well as gentle turns, and fruitful as well as false branchings since the initial effort. Equally, those interested had indicated in the Prefaces of both the first and fifth volumes that the general, basic structure of understanding of the bacteria had occurred in the 25 years before the appearance of the first set of sourcebooks. Since projection of the future seemed at least as fragile as past experience, the decision was reached to leave the sourcebooks intact and supplement those areas of prime development or currently in need of clear definition.

This Foreword announces, therefore, the decision to provide two additional volumes, VI and VII, under vigorous, young editors, L. N. Ornston, Yale University and J. R. Sokatch, University of Oklahoma Medical School. Happily, their planning and compatibility has prompted them into joint planning and editorship of the two volumes. These volumes aim to supply comprehensive statements from young authors of advances and understanding of diverse and adaptive mechanisms.

Again, the general properties of the bacteria are stressed rather than specific properties of individual species and genera which peculiarly adapt to nature's ecological niches. With the vastly increased attention to the development of the single cells of eucaryotes as well as procaryotes, an extension in the use of the methods of microbiologists, increased emphasis on cell surfaces and membrane transport, and the drastically sharpened methods of the geneticists are understandable. They may again be anticipated to broaden the community of scholars devoted to cellular systems to whom we hope these volumes will be helpful as sourcebooks. Many of the problems which only a decade ago were considered biology are today increasingly expressed in the molecular terms of primary processes in the language of chemistry and physics. In fact, this may alter emphasis in other volumes which may be added as warranted under these or other editors.

Comments, suggestions, and constructive criticisms to the editors, authors, or publisher are encouraged. This opportunity to express appreciation to the editors for the quality of their creative effort and their successful planning, author selection, and management is indeed a pleasure. The professional expertise and many courtesies as again unfailingly tenured by the publisher and staff have been a rewarding and enjoyable experience.

I. C. Gunsalus

PREFACE

Bacteria are descendents of forms of life that existed in the anaerobic atmosphere of earth about 4 billion years ago. Much of the evolutionary success of bacteria has been due to the development of simple, effective mechanisms for fulfilling their needs. The present microbial biosphere is rich in diversity, and the solutions that bacteria have devised to deal with the problems presented by their environment make them inviting targets for study. Similar adaptations need not reflect a single evolutionary path, and comparison of seemingly similar bacterial substructures suggests that our appreciation of bacterial diversity still is limited by our knowledge. In some cases an adaptation is unique to procaryotes, as in the ability to fix elemental nitrogen. In other instances, for example sensory systems, bacteria continue to supply models that may increase understanding of complex eucaryotic processes. The small size of procaryotes makes the constancy of their intracellular environment all the more remarkable and demands analysis of the surface layers that separate the inside of bacterial cells from fluctuating, sometimes hostile, external environments. The triumph of pure culture technique should not distract attention from the complexity of bacterial habitats, as suggested by sophisticated interactions between bacteria and their phage. Despite environmental fluctuations, the synthesis of cellular materials and cell division remain orderly processes that have been revealed by the experimental analysis of procaryotes. Bacterial evolution was discussed in Volume VI of this treatise, and mechanisms of bacterial adaptations are discussed in this volume.

J. R. Sokatch
L. N. Ornston

The Bacteria

A TREATISE ON STRUCTURE AND FUNCTION

VOLUME I: STRUCTURE

VOLUME II: METABOLISM

VOLUME III: BIOSYNTHESIS

VOLUME IV: THE PHYSIOLOGY OF GROWTH

VOLUME V: HEREDITY

VOLUME VI: BACTERIAL DIVERSITY

CHAPTER 1

Spores, Cysts, and Stalks

MARTIN DWORKIN

ISBN 0-12-307207-7

I. Introduction

Interest in the physiological activities of bacteria has traditionally focused on the process of growth and its regulation. This certainly reflects Francois Jacob's (1970)* aphorism, "one bacterium, one amoeba, . . . ; what destiny could they have other than to form two bacteria, two amoebae, . . . ?" It seems clear that another destiny is indeed available for some bacteria; that, in addition to maximizing growth during those brief moments of nutritional plenty, these bacteria have evolved mechanisms for persisting during periods of suboptimal circumstances. One of these mechanisms involves the formation of specialized cells such as spores, cysts, and akinetes. Such cells have a reduced level of metabolic activity and are usually resistant to various environmental pressures. Their development is separate from and alternative to growth and the process is reversible or cyclic.

An additional adaptive device has been the development of functionally differentiated populations of cells. This developmental strategy is far more limited among prokaryotes than among eukaryotes and includes heterocyst formation by the blue-green bacteria and formation of stalked cells by *Caulobacter* and *Asticacaulis*. It is possible that the myxobacteria also form functionally differentiated cells, although this has not yet been clearly demonstrated. In these processes, in contrast to those resulting in resistant resting cells, development serves to improve growth efficiency and is coordinate with and parallel to growth rather than alternative to it. Nevertheless, in the case of the heterocyst, the differentiated cell itself is a developmental deadend that cannot, as a rule, either divide or germinate. Somewhat analogously, the stalked cell of *Caulobacter*, once formed, persists as a sort of primitive stem cell that produces but never itself reverts to a stalkless cell. One would indeed expect that these different developmental strategies, either complexly involved with growth or independent and alternative to it, would entail fundamentally different regulatory processes. Unfortunately, the understanding of developmental regulation in prokaryotes does not yet allow us to decide whether or not this is indeed the case.

This chapter will discuss the structure, properties, formation, and regulation of spores, cysts, and stalks among the following groups: *Bacillus*, actinomycetes, *Azotobacter*, blue-green bacteria, myxobacteria, and *Caulobacter*. While this coverage is not intended to be comprehensive, either with regard to organism or to structure, it will, however, attempt to be representative.

The primary emphasis will be to describe the individual developmental

* The references cited in this chapter may either refer to the original work, or in those cases where an assortment of references is necessary, to a review article.

patterns that each of these groups has evolved. However, an attempt will also be made to point out where, on the one hand, these strategies overlap and, on the other, where the patterns are unique or unusual.

II. The *Bacillus* Endospore

Members of the family Bacillaceae form a resting cell called an endospore. Formation of the endospore is an alternative to vegetative growth, and may occur under laboratory conditions when a culture is in transition from exponential to stationary phase growth. The necessary environmental stimulus seems to be a nutritional deprivation, and it has been suggested that catabolite repression is the controlling regulatory mechanism (Schaeffer *et al.*, 1965). The spore is formed within the parent vegetative cell and is a round to ovoid, optically refractile, resistant, metabolically quiescent cell. It can remain in a dormant or crytobiotic state essentially indefinitely until environmental circumstances are appropriate for its germination. Under these conditions it gives rise to the vegetative cell capable of growth and division.

There are sound reasons why the bacterial endospore continues to occupy the attention of biologists and biochemists. (a) The dramatic resistance of endospores is of practical concern to those seeking to achieve sterility, whether in food, laboratory media, or space ships. Further, the biochemical basis of this resistance has continued to elude investigators. (b) The condition of cryptobiosis or metabolic dormancy is one extreme form of a widespread biological phenomenon that allows organisms to cope with broadly fluctuating environmental circumstances. (c) The occurrence of a well-defined cellular morphogenesis in a prokaryote offers an attractive model system for investigating the biochemical and regulatory features of a developmental process.

While endospores are formed by a number of genera (*Bacillus, Thermoactinomyces, Clostridium, Sporosarcina, Desulfotomaculum* and *Sporolactobacillus*) (Buchanan and Gibbons, 1974), the available evidence suggests that the events are essentially similar. Thus, this section shall be exclusively concerned with the genus *Bacillus*.

A. Endospore Morphology

Figure 1 is a phase-contrast micrograph of *B. fastidiosus* illustrating the intracellular and optically refractile nature of the endospore. Figure 2 is an electron micrograph of a thin section of a *B. megaterium* endospore. The

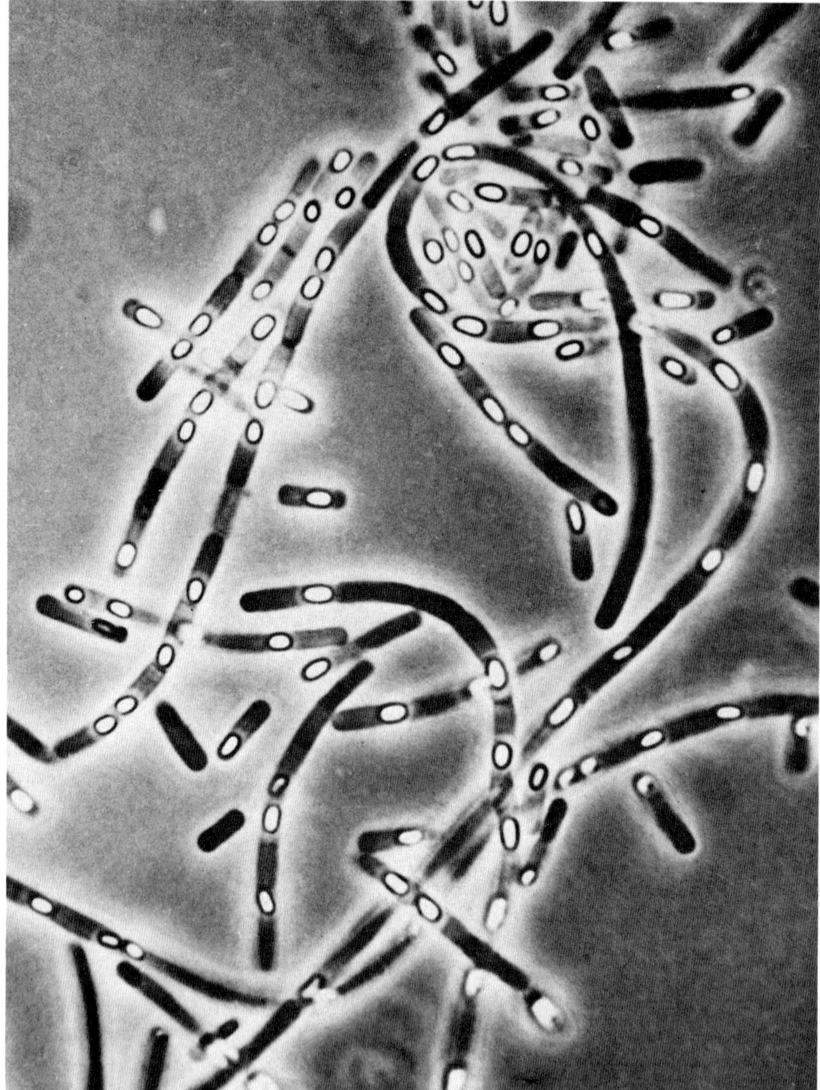

FIG. 1. Phase-contrast photomicrograph of spores and vegetative cells of *Bacillus fastidiosus*. Courtesy of Dr. S. C. Holt.

unique features of spore morphology are the cortex (immediately external to the protoplast membrane), the spore coat, and the exosporium present in some, but not all, types of spores. The various layers of the endospore are illustrated diagramatically in Fig. 3.

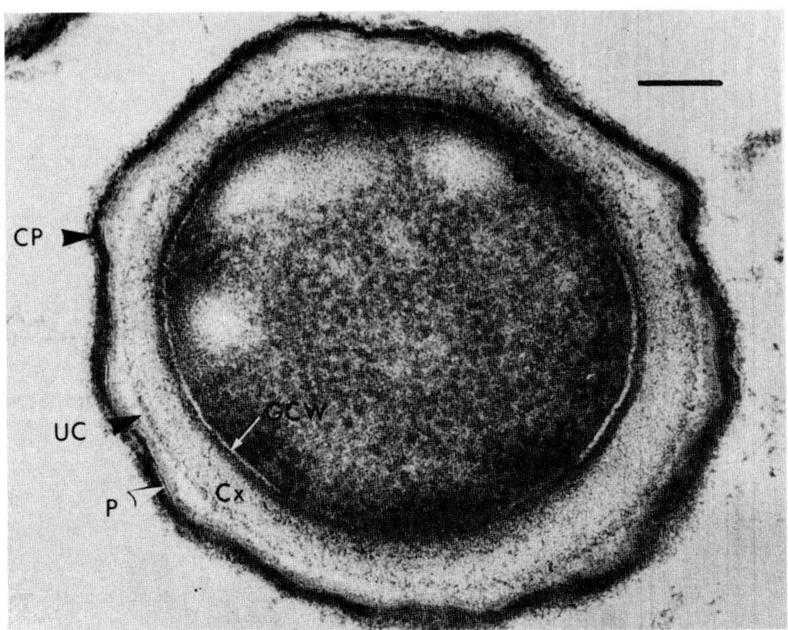

FIG. 2. Electron micrograph of a thin section of a spore of *Bacillus megaterium*. CP, Cross patched layer; P, pitted layer; UC, undercoat; Cx, cortex; GCW, germ cell wall. Scale, 100 nm. (From Aronson and Fitz-James, 1976.)

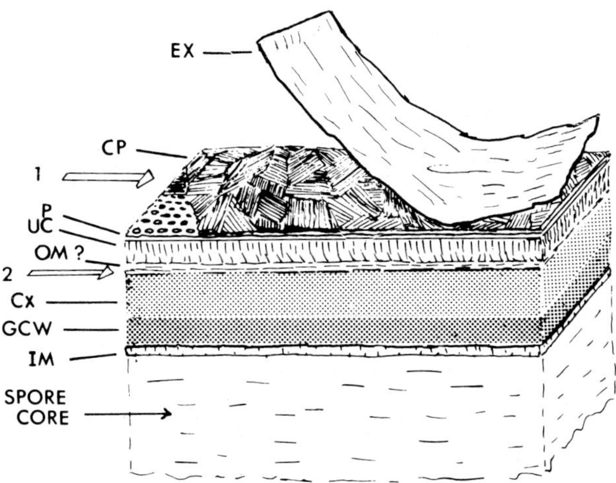

FIG. 3. Diagrammatic representation of the various layers of the mature spore of *Bacillus cereus*. Ex, exosporium; CP, cross patched layer; P, pitted layer; UC, undercoat; OM, outer forespore membrane; Cx, cortex; GCW, germ cell wall; IM, inner forespore membrane; The flat arrows (1 and 2) indicate the frequent cleavage planes in freeze-etch preparations. (From Aronson and Fitz-James, 1976.)

B. Composition of the Endospore

1. DNA

While it is not clear that it is indeed a necessary and bona fide part of the sporulation process (Balassa, 1971), the earliest morphological event during sporulation concerns the organization and partition of the DNA. When the parent vegetative cell contains two nucleoids, these are present as compact spherical bodies that fuse to a single axial filament within minutes after sporulation is initiated (Mandelstam *et al.*, 1975). Subsequently, the axial filament is partitioned between a forespore nucleoid containing roughly 50% of the parent DNA and the mother cell DNA (Young and Fitz-James, 1959a,b). In the case of *B. subtilis* the DNA content of the spore is 5.4×10^{-15} gm with a molecular weight of 3.3×10^9 (Doi, 1969), although this figure varies from species to species (Fitz-James and Young, 1959). The DNA in the *B. subtilis* spore is a double-stranded helix (Mandel and Rowley, 1963), and while the evidence suggests that the secondary structure and conformation of spore DNA is similar to that of the vegetative cells, there seem to be differences in their physical state. These are indicated by the differing nature of photoproducts produced by uv light irradiation of DNA (Donnellan and Setlow, 1965) and intact cells (Smith and Yoshikawa, 1966) and on the different banding properties of the DNA in density gradients (Douthit and Halvorson, 1966). Doi (1969) has suggested that these differences in the physical state of the DNA may be functionally related to the differential transcription of spore and vegetative cell DNA.

2. Spore Cortex

During formation of the forespore, the spore DNA becomes enveloped in a double membrane. [While both of these are essentially derived from the vegetative cell membrane, one is the forespore membrane proper while the other is from the membrane of the mother cell that has surrounded and engulfed the forespore protoplast (see Figs. 2 and 3)]. The spore cortex which develops between these two concentric membranes and consists of modified peptidoglycan (Warth and Strominger, 1969) is illustrated in Fig. 4 (Tipper *et al.*, 1977b). In *B. subtilis*, disaccharide subunits of alternating β-1,4-linked *N*-acetylglucosamine and *N*-acetylmuramic acid are variously substituted with (a) the tetrapeptide L-alanyl → α-D-glutamyl → (L)-*meso*-diaminopimelyl → (L)-D-alanine (30%), 20% of which are cross-linked; (b) 15% of the *N*-acetylmuramic acid residues are substituted at the D-lactyl with C-terminal L-alanine; (c) the remaining 55% of the *N*-acetylmuramic acid residues exist as the lactam between the carboxyl group of the lactate moiety and the amino group on C-2 of muramic acid. Cleveland and Gilvarg (1975) have suggested

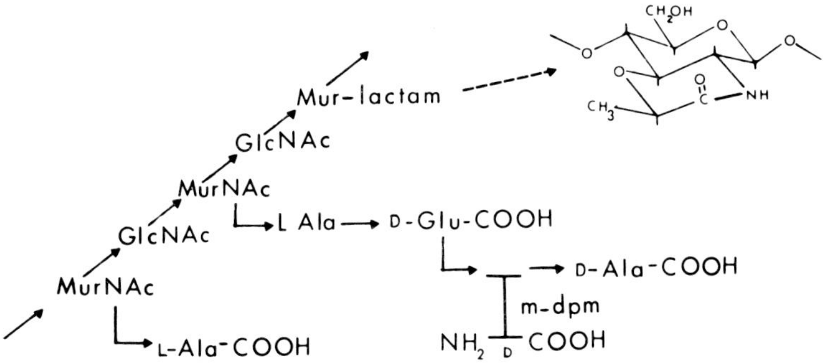

FIG. 4. Peptidoglycan of the cortex of spores of *Bacillus subtilis*. (From Tipper *et al.*, 1977b.)

that the cortex peptidoglycan contains not only the unique spore peptidoglycan described above, but that 20% of the cortex peptidoglycan is of the vegetative type. This portion of the cortex peptidoglycan is resistant to the autolytic enzymes activated during spore germination and may serve as a temporary source of osmotic protection as well as a primer for vegetative peptidoglycan biosynthesis (Cleveland and Gilvarg, 1975).

3. Spore Coat

The spore cortex of *B. cereus* is surrounded by a multiple-layered structure called the spore coat that comprises up to 40% of the total spore protein. It is a highly resistant structure composed essentially of protein, that contains four to five times the amount of cysteine as does the vegetative cell protein. The various layers of the spore coat are schematically illustrated in Fig. 3. Kondo and Foster (1967) originally recognized that the spore coat contained multiple structural components and were able to fractionate the spore coat into an alkali-stable fraction, an insoluble, pronase-resistant fraction, and a fraction solubilized by sonic oscillation. The multiple layers are evident morphologically using freeze-etch preparations (Aronson and Fitz-James, 1976). Figures 5–7 illustrate the layers of the spore coat of *B. cereus*. The outer layer, first described by Holt and Leadbetter (1969), consists of cross patches (CP) of 50 nm made up of parallel fibers (Fig. 5). This is followed by an underlying pitted (P) layer (Fig. 6) and the innermost undercoat (UC) layer (Fig. 7). Aronson and Fitz-James (1976) have examined the nature of the spore coat protein of *B. cereus* and, using dithioerythritol and SDS, at pH 9.5 were able to solubilize 95% of the spore coat protein. While other investigators have described two (Spudich and Kornberg, 1968) and three major protein components of the spore coat of other species of *Bacillus* the data of Aronson and Fitz-James suggest that the spore coat

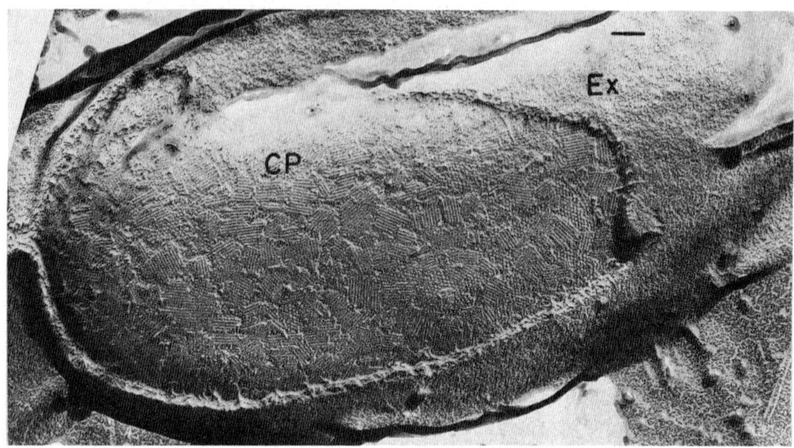

FIG. 5. Electron micrograph of a spore of *Bacillus cereus* prepared by freeze-etching. The photograph shows the cross-patched layer (CP) and the exosporium (Ex). Scale, 100 nm. (From Aronson and Fitz-James, 1976.)

of *B. cereus* consists of a single major protein with a MW of 12,000. They suggest further that the three morphologically distinct components of the spore coat reflect three different tertiary structures of a single polypeptide. While it is not yet possible to assign a specific function to the spore coat, its considerable resistance to physical and enzymatic degradation suggests that it plays a role in spore resistance.

4. Dipicolinic Acid

Dipicolinic acid (DPA) is absent from vegetative cells, is an invariable constituent of all endospores, and represents as much as 10% of the spore dry weight (Murrell and Warth, 1965). It has been demonstrated that Ca^{2+}, which represents 1–3% of the dry weight of spores (Gould and Dring, 1974) exists in a chelate complex with DPA in the spores (Bailey *et al.*, 1965). Both electron probe x-ray microanalysis of intact spores (Scherrer and Gerhardt, 1972) and β-attenuation studies with tritium-labeled DPA (Leanz and Gilvarg, 1973) strongly suggest that DPA is localized in the core portion of the cell. It has for many years been sporulation dogma that DPA is causally associated with heat resistance. This has been based on a number of lines of evidence. The appearance of DPA during sporulation coincides closely with the acquisition of heat resistance (Hashimoto *et al.*, 1960; Vinter, 1969), and DPA-less mutants of *B. cereus* (Wise *et al.*, 1967) also became heat sensitive. However, some heat-resistant revertants of these

FIG. 6. Electron micrograph of a spore of *Bacillus cereus* prepared by freeze-etching. P, pitted layer is indicated. Scale, 100 nm. (From Aronson and Fitz-James, 1976.)

mutants remained DPA-less (Hanson *et al.*, 1972) even though their heat resistance and dormancy was not maintained as well during storage. Similarly, heat-resistant DPA-less mutants of *B. subtilis* have also been isolated (Zytkovicz and Halvorson, 1972). Gould and Dring (1974) have recently suggested that DPA acts as a Ca^{2+} buffer, the cellular level of Ca^{2+} being necessary to maintain dormancy and heat resistance. Presumably, DPA-less mutants are unable to maintain a necessary level of Ca^{2+} and the spores gradually lose their heat resistance and dormancy.

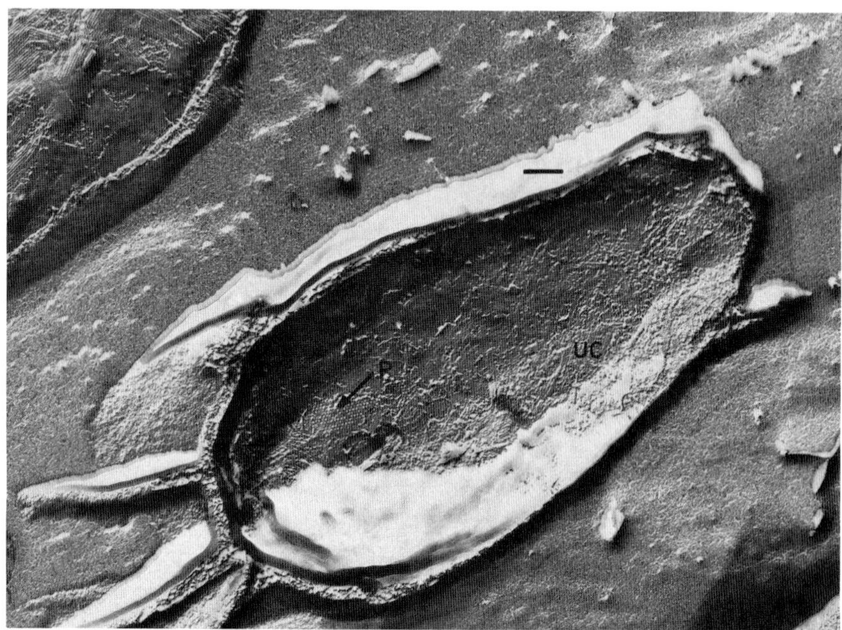

FIG. 7. Electron micrograph of the concave undersurface of the coat of a spore of *B. cereus* prepared by freeze-etching. The undercoat layer (UC) and the undersurface of the pitted layer (P) are indicated. Scale, 100 nm. (From Aronson and Fitz-James, 1976.)

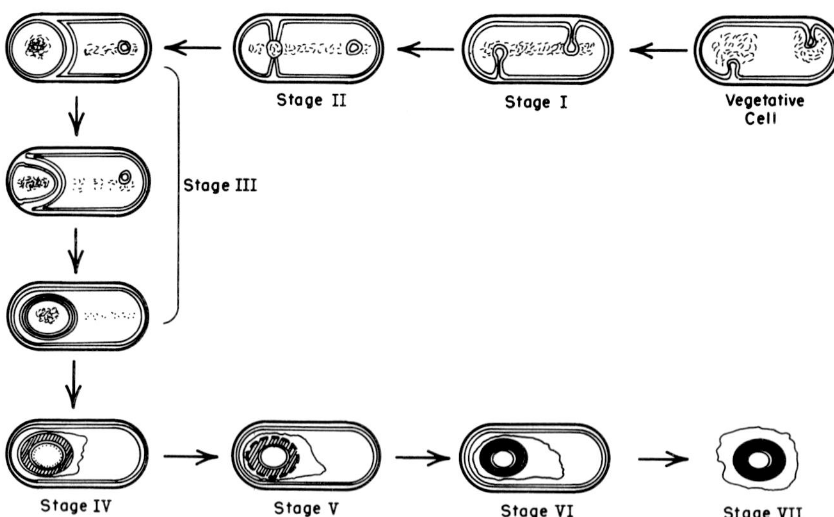

FIG. 8. Diagrammatic representation of the seven stages of endospore formation.

C. Morphology of Sporulation

The sequence of morphological events leading to spore formation is diagramatically illustrated in Fig. 8. Based on cytological observations of a variety of sporulation mutants, the process has been divided by Schaeffer (1969) into seven stages. Stage I involves the axial elongation of the nuclear material. Stage II is characterized by the separation of the spore DNA by the membrane-derived forespore septum. In Stage III the membrane of the adjacent sporangial mother cell engulfs the enclosed forespore resulting in an outer membrane with reversed polarity. In Stage IV the space between the inner and outer spore membranes is filled with spore cortex and before this process is complete, Stage V begins, involving the deposition of spore coat on the outer surface of the outer membrane. During Stage VI the processes of spore coat synthesis are completed, and in Stage VII, the mature spore is released.

D. Properties of the Spore

1. Resistance

The resistance of the endospore surpasses that of any other organism. It is resistant to temperature extremes, desiccation, ionizing and nonionizing radiation, chemicals, metabolic inhibition, and physical disruption (Roberts and Hitchins, 1969).

The heat resistance of *Bacillus* endospores varies from *B. megaterium* with a $D_{100^\circ} = 1$ min (Briggs, 1966), to the thermophile *B. stearothermophilus* with a $D_{115^\circ} = 22.6$ min (Briggs, 1966).* Such figures should be viewed with the understanding that they depend on conditions of sporulation as well as conditions during treatment and recovery (Roberts and Hitchins, 1969).

Theories explaining spore heat resistance fall into three categories: (a) synthesis during sporulation of intrinsically resistant macromolecules; (b) stabilization of spore components by modification of the components; (c) protoplast dehydration resulting from cortex contraction (Lewis *et al.*, 1960) or expansion (Alderton and Snell, 1963; Gould and Dring, 1975).

There is no evidence that would support the notion that purified spore macromolecules are intrinsically more stable at elevated temperature than are the analogous vegetative macromolecules. Alternatively stated, a thorough comparison of the purified proteins has in all cases led to the conclusion that both spore and vegetative enzymes shared the same primary structure

* The "D" value of a spore is the time required for a tenfold killing at the temperature specified by the subscript.

and were thus coded for by the same genetic determinants (Sadoff, 1969). Furthermore, alkaline phosphatase (Glenn and Mandelstam, 1971), DNA polymerase (Falaschi and Kornberg, 1966), adenylate kinases (Spudich and Kornberg, 1969), purine nucleoside phosphorylase (Gardner and Kornberg, 1967), pyruvate kinase, RNA polymerase (Chambon *et al.*, 1968), and inorganic pyrophosphatases (Tono and Kornberg, 1967) from spores and vegetative cells were indistinguishable by a variety of physical parameters. On the other hand, a considerable body of evidence now indicates that modifications of the enzymatic environment [e.g., protonation of purified glucose dehydrogenase and increasing the ionic strength (Sadoff *et al.*, 1965)] substantially increased the heat resistance of the enzyme, that Ca^{2+} considerably increased the heat resistance of purified aldolase of *B. cereus* spores (Sadoff *et al.*, 1969), that proteolytic modification of vegetative cell enzymes may result in smaller more heat-stable proteins (Sadoff *et al.*, 1970), and that binding to subcellular particles substantially increased the heat resistance of ribosidase (Nakata, 1957) and alanine racemase (Stewart and Halvorson, 1954) from spores of *B. cereus*.

Many of the enzymes involved in DNA synthesis have been examined in spores, and in no case have any significant differences been found between the spore and vegetative cell enzymes (Doi, 1969).

While the spore has been shown to be permeable to water (Black and Gerhardt, 1962), its content of total water (w/w) is somewhat lower than that found in vegetative cells (about 65 versus 77%) (Black and Gerhardt, 1962). This assumes a uniform distribution throughout the cell; the water content of the core may, of course, be substantially lower. The heat resistance of a variety of spore types was shown to respond dramatically to changes in the relative humidity of the environment (Murrell and Scott, 1966). This was consistent with the hypothesis that the lower water content of the spore was in some way responsible for heat resistance. Lewis *et al.* (1960) proposed what has come to be called the contractile cortex hypothesis, viz. that the compressive contraction of the cortex onto the core results in a state of relative dehydration. The mechanism that has been suggested by Alderton and Snell (1963) is that replacement of the protons of carboxyl groups by other cations could result in electrostatic repulsion that would in effect cause a swelling of the cortex against the core. Gould and Dring (1975) have also proposed that cortical expansion rather than contraction is responsible for the compressive effect, and have provided evidence that cortex peptidoglycan is the macromolecule whose expansion and contraction regulates the state of hydration of the core. They propose that Ca^{2+} acts as the ion-exchange cation and its delivery to the cortex is accomplished by DPA acting as a calcium ionophore. Warth (1977) has suggested how the oriented synthesis of the lactam-containing cortical peptidoglycan could result in an inner-directed swelling pressure exerted on the spore core. There seems to be general agreement that a reduction in the volume and water content of the core plays a major role in the heat resistance of the spore; and while there are a number of ingenious theories to explain how this reduction in water content may be accomplished, none of them is compelling or supported by sufficient data to consider the problem resolved. Gerhardt and Murrell (1978) have tabulated the various theories to explain heat resistance of the encospore.

2. Dormancy

Two aspects of the problem of spore dormancy should be separated—the maintenance of the dormant state and the trigger for germination. With regard to the former, explanations fall into one or more of three categories (Keynan, 1972): (a) Theories based on the presence of self-inhibitors that maintain spore enzymes in an inactive state; (b) maintenance of dormancy by structural alterations of the spore, e.g., dehydration by means of a contractile or expansive cortex; and (c) physical or chemical alterations of spore enzymes resulting in reversible inactivation.

Generally, the picture that has emerged is that the spore is in a state of latent self-sufficiency, that is, the essential enzyme systems necessary for germination and outgrowth are present in the spore, albeit in an inactive state. This is not meant to imply that all vegetative cell enzymes are present in the spore. Setlow and Kornberg (1970b) have shown that the ability to carry out *de novo* purine or pyrimidine nucleotide biosynthesis disappears during sporulation and does not reappear in germinating spores until protein synthesis has occurred. Switzer *et al.* (1975) have shown that sporulation in *B. subtilis* is accompanied by the loss of aspartate transcarbamylase activity as well as the loss of immunologically cross-reacting protein.

One of the earliest investigations of the metabolic activities of spores was concerned with terminal electron transport. Based on the work of Doi and Halvorson (1961) the generalization emerged that spores, unlike the parent vegetative cells, were devoid of cytochrome pigments. More recent work has indicated, however, that sporulating cells (Felix and Lundgren, 1972) as well as mature spores contain a complete electron transport pathway including cytochromes b, c, $a + a_3$, menaquinone, cytochrome-linked NADH oxidase, NADH-dichlorophenolindophenol reductase, and succinate dehydrogenase (Weber and Broadbent, 1975). More than 60 enzymes have now been shown to be common to vegetative cells and spores (Sadoff, 1969). It thus seems clear that, with regard to dormancy, the focus of the problem must be on regulation of enzyme activity rather than on the elimination of enzyme protein.

In this regard, Keynan (1972) has emphasized the possible role of Ca-DPA as a self-inhibitor. The evidence for this is indirect and consists essentially of the demonstration by Halvorson and Swanson (1969) that the glucose oxidase activity of spores of DPA-less mutants was inhibited if the cells were grown in the presence of DPA and that this inhibition was released by heat shock. While the evidence referred to in Section II.B.4 identifying the core as the site of most of the cellular DPA is consistent with this theory, the properties of the DPA-less mutants indicate that DPA is not indispensible for either heat resistance or dormancy. The notion that the spore protoplast exists in a dehydrated state has been used to explain the cryptobiotic as well as the resistance properties of the spore. Murrell and Scott (1966) have shown that the heat resistance of spores varies considerably with their water actvity, and Dring and Gould (1975) were able to reestablish the heat resistance of germinated spores by osmotically dehydrating the cells. Again, as in the case of resistance, the function of the cortex in maintaining the core in a state of dehydration is emphasized.

In vitro examination of the metabolic activities of spores is characterized by what is almost a technical "Catch 22." The act of opening the cell so as to obtain cell-free enzyme activities, unless done under circumstances that clearly inhibit all metabolic activities, may in itself activate the spore. A number of investigators (Deutscher *et al.*, 1968) have been subject to this criticism (Idriss and Halvorson, 1969). On the other hand, the physical durability of the spore requires breakage techniques that may be sufficiently drastic to disrupt or deter the activity one is seeking. For example, the difficulty in demonstrating polyribosomes in spores by some investigators (Idriss and Halvorson, 1969) but not by others (Douthit and Kieras, 1972) may reflect this difficulty. With these caveats in mind, what can be said about spore enzymes *vis-à-vis* dormancy? Generally speaking, three mechanisms can be envisaged that would modify the activity of the enzyme within the spore: (a) Binding to some structural component of the spore; (b) modification of the immediate intracellular environment of the enzyme, e.g., pH; (c) modification of the enzyme itself. The effect of enzyme binding to structural components of the spore has already been discussed with regard to heat resistance. While it is clear that the activity of spore enzymes, varies in response to pH, there is no evidence that this plays a role *in vivo*. Finally, whereas there is considerable evidence (Sadoff, 1969) that the physical properties of many spore enzymes differ from those of the equivalent vegetative cell enzyme, it has not been possible to relate these differences to *in vivo* activity. Wright (1973) has discussed the problem in relation to developmental changes in polysaccharide synthesis in *Dictyostelium discoideum*. She has emphasized that enzyme activity or amount of enzyme proteins is not the sole critical variable in development, but rather that substrate pool levels and metabolic fluxes of intermediates may critically determine the rate of metabolic reactions. It is not unlikely that this view pertains equally well to the regulation of spore metabolism.

E. Synthesis of Organelles during Sporulation

1. DNA

Stage I of sporulation is characterized by the conversion of the circular nuclear bodies to an elongated axial filament. While recognizing that this may not be an indispensible part of the sporulation process (Balassa, 1971), Mandelstam *et al.* (1975) have suggested that it functions to reorient the mesosome prior to formation of the forespore septum.

There is little or no DNA synthesis during sporulation proper (Ryter and Aubert, 1969), however, prior to sporulation, chromosome completion usually takes place (Doi, 1969) and is indeed required. In fact, Piggot and Coote (1976) have suggested, based on the work of Mandelstam's group (e.g., Mandelstam and Higgs, 1974), that DNA replication in the sporulation medium is obligatory for induction of sporulation. More specifically, Dunn *et al.* (1976) have suggested that during DNA replication there may be an early signal that prevents the subsequent formation of a vegetative division septum and a later signal to form a sporulation septum at the cell pole.

2. Spore Membranes

The membrane of the forespore consists of two layers, the inner forespore membrane (IFSM) and the outer forespore membrane (OFSM). The IFSM is derived from the polar portion of the vegetative cell membrane and the distal half of the sporulation septum. Upon spore germination it will once again become the vegetative cell membrane and must also contain the enzymes for biosynthesis of the germ cell wall. Following the formation of the IFSM, the proximal half of the spore septum expands and engulfs the forespore, resulting in the OFSM. Consequently, the inner–outer polarity of the OFSM is reversed, so that the inner space enclosed by the two membranes is bounded by what was originally the two outer surfaces of the vegetative cell membrane. This inner space will become the spore cortex, the enzymes for whose synthesis are contained in the OFSM (Murrell *et al.*, 1969). Biochemical evidence of the reversed polarity of the OFSM as been presented by Wilkinson *et al.* (1975).

The kinetics of lipid synthesis during sporulation of *B. cereus* (Fitz-James and Young, 1969) and *B. megaterium* (Pitel and Gilvarg, 1971) indicated that sporulation was accompanied by an accelerated rate of lipid synthesis which then declined during spore maturation. Peptidoglycan synthesis ceased upon initiation of sporulation and during the period of lipid synthesis (Pitel and Gilvarg, 1971). The authors suggested that in the absence of cell wall synthesis, the growing membrane buckles and thus envelopes the forespore.

The phospholipids of the IFSM of *B. cereus* have been shown to be the same as those in the vegetative cell membrane and contain an alanine ester of phosphatidylglycerol, lipophosphatidylethanolamine, a glycolipid of unknown structure, phosphatidylglycerol, phosphatidylethanolamine, diphosphatidylglycerol, and one unidentified phospholipid (Felix and Lundgren, 1972). Similarly, no substantial differences between spores and vegetative cells were found in the phospholipids of *B. popilliae* and *B. thuringiensis* (Bulla and St. Julian, 1972). However, the proportion, amount, turnover rate, and timing during growth and sporulation were shown to differ (Bulla and St. Julian, 1972).

Methods are now available for the isolation of forespores containing the IFSM (Felix and Lundgren, 1972) and both membranes (Eaton and Ellar, 1974; Andreoli *et al.*, 1975). This plus the possibility of generating membrane vesicles offers the opportunity of examining in detail the specific properties of developing spore membranes. Questions such as the differences between membranes of forespores and mother cells, kinetics of membrane changes during sporulation, temporal regulation of peptidoglycan and lipid synthesis, specific properties of IFSM as compared to OFSM, and the inside–outside properties of the two forespore membranes are of obvious interest.

3. Cortex

Cortex formation results from the biosynthesis and gradual deposition of spore peptidoglycan between the two membrane layers of the forespore. This biosynthesis can be inhibited by penicillin and cycloserine added during the period of cortex synthesis. Spores thus formed are unstable and leaky for Ca-DPA (Pearce and Fitz-James, 1971). Tipper and Gauthier (1972) have suggested that the spore peptidoglycan in *B. sphaericus* is derived from a nascent polymer of GlcNAc→L-Ala→γ-D-Glu→*meso*-DAP→D-Ala→D-Ala. This may be subsequently cross-linked by transpeptidation or, if uncross-linked, modified by hydrolysis of the L-Ala→D-Glu bond or by complete removal together with lactam synthesis. It has been suggested that some of the enzymes involved in synthesis of cortex peptidoglycan are peculiar to spore formation and thus of particular interest from the point of view of understanding the control of unique sporulation events (Tipper *et al.*, 1977b).

4. Spore Coat

Visible evidence of coat formation appears during Stage IV of sporulation.

The work of Aronson and Fitz-James (1976), however, indicates that in *B. cereus* the synthesis of spore coat protein begins at about Stage II and continues throughout sporulation. Incorporation of labeled precursors into the insoluble portion of the coat increased threefold late in sporulation and then decreased at the time of spore brightening. At the time of increase of incorporation, the rate of cysteine incorporation increased twofold, and a variety of evidence suggested that this increase was correlated with the appearance of the outer spore coat. Aronson and Fitz-James (1976) have suggested that a disulfide interchange reaction occurs at this time and is essential for the reorganization and self-assembly of the coat polypeptides into the fibrillar matrix of the coat.

The successful pursuit of explanations for the regulation of sporulation depend on being able to define the sporulation process (and sporulation mutants) in terms of specific macromolecules. The characterization of spore coat protein and specifically the disulfide exchange reaction represent an important approach in this direction.

F. Spore Enzymes

As indicated earlier, the spore contains a considerable number of enzyme activities. Two broad questions one wishes to ask about them are (a) what role do they play in sporulation, and (b) how is the expression of their activity regulated?

Hanson *et al.* (1970) have divided spore enzymes into four categories: (a) vegetative cell enzymes whose synthesis (and presumably function) persists during sporulation. Enzymes in this category would include those involved in energy metabolism, amino acid biosynthesis, macromolecular synthesis, etc. (b) Enzymes whose activities appear or increase during sporulation but that seem to play no role in the process. It should be emphasized that conclusions that enzyme changes during sporulation are indeed sporulation-specific events must be arrived at cautiously. It is clear that the nutritional history of the cells may determine events occurring during the transition from growth to sporulation that are essentially irrelevant to sporulation. For example, both arginase (Laishley and Bernlohr, 1966) and amylase (Schaeffer, 1969) are normally repressed during growth on glucose but appear during postlogarithmic growth. Neither of these enzymes seems essential for sporulation. Information for their biosynthesis may be coderepressed along with that of sporulation genes as a result of nutrient depletion. (c) Enzymes that appear or increase during sporulation, whose activities are vital to sporulation, but are not unique to the spore. The enzymes of the tricarboxylic acid (TCA) cycle fall in this category. Hanson *et al.* (1964a) has shown that the TCA enzymes in *B. subtilis* and *B. licheniformis* are repressed during vegetative growth on a complex glucose-containing medium but derepressed during sporulation. However, if the cells are grown on a minimal salts–glucose medium, the TCA enzymes are present during vegetative growth (Hanson *et al.*, 1964b). Also included in this category are some of the extracellular proteases (Schaeffer, 1969), NADH oxidase (Szulmajster and Schaeffer, 1961), and enzymes involved in protein and RNA turnover (Kornberg *et al.*, 1968). (d) Enzymes in this category appear only during sporulation, seem to have a unique sporulation function, and are indispensable for sporulation. This includes those enzymes shown to be required for DPA and cortical peptidoglycan and those postulated for spore coat protein biosynthesis.

Attempts to understand the regulation of so-called sporulation enzymes are complicated by the fact that the transition between growth and sporulation is usually not a sharp one. Freese and Fujita (1976) have emphasized this by pointing out that it is usually impossible to define t_0, considered to be the beginning of the sporulation-specific changes. If the period of post-logarithmic growth involves a gradual shift from exponential to linear growth or stationary phase, there will be a continued series of adaptive regulatory modifications of the cell's metabolism, and it is not at all clear at which point these modifications become part of the sporulation processes. With the exception of those truly sporulation-specific processes (e.g., synthesis of DPA, cortex, and coat), the nature and extent of enzyme change is likely to reflect the previous growth medium as well as the sporulation medium.

There are a number of enzymes of *Bacillus* that are inducible by their substrate and that appear during postlogarithmic growth. There are inducible enzymes that are catabolite repressed during growth, and there are enzymes that appear during postlogarithmic growth due to a release of end-product repression. Freese and Fujita (1976) have presented a detailed list of such enzymes.

While there have been no reports of classical end-product inhibition by allosteric control during sporulation, several authors have described catabolic (Bernlohr and Leitzman, 1969) and biosynthetic (Staley and Bernlohr, 1967) enzymes whose activities are rapidly lost prior to sporulation. While proteolytic inactivation has been implicated in some cases (Diesterhaft and Freese, 1973), it seems not to play a role in others (Bernlohr and Gray, 1969). The mechanism of inactivation remains unknown.

G. Antibiotic Formation

Sporulating bacteria produce a variety of peptide antibiotics. These are often cyclic or partially cyclic, contain D-amino acids, and are produced prior to sporulation. There is a strong correlation between antibiotic formation and sporulation that suggests a causal relationship. Inhibition of sporulation either by mutational blocks (Schaeffer, 1969) or by specific sporulation inhibitors (Bernlohr and Novelli, 1959) also inhibits antibiotic formation. Conversely, mutants selected for inability to produce antibiotics are blocked for sporulation (Schmitt and Freese, 1968), and restoration of antibiotic production in mutants by reversion, transformation, or transduction results also in restoration of spore-forming ability (Hanson *et al.*, 1970). There has been a great deal of discussion of this relationship and, in general, of the relationship between antibiotic formation and development in microbes. While the complete explanation as to the developmental function of antibiotic synthesis is not available, Paulus and Sarkar (1976) have proposed that the peptide antibiotics of *Bacillus* function in the regulation of RNA synthesis during the early steps of sporulation by selectively inhibiting the transcription of vegetative cell genes. Specifically, Mukherjee and Paulus (1977) have shown that a number of mutants of *B. brevis* that do not produce gramicidin form spores that have an increased heat sensitivity and a reduced DPA content. The deficiency could be overcome by adding gramicidin to presporulation cells at the end of exponential growth, and gramicidin-producing revertants also regained the ability to produce heat-resistant spores. Hanson *et al.* (1970) Schaeffer (1969), Sadoff (1972), and Katz and Demain (1977) have reviewed the earlier data on this subject in detail.

H. Regulation of Sporulation

An aspect of the regulation of sporulation concerns the initial "trigger" of the process. The transition between growth and sporulation is characterized by a point after which the addition of nutrient no longer will reverse the course of sporulation. This has been referred to as "commitment" (see Piggot and Coote, 1976). While it is experimentally and conceptually simpler to think in terms of a single "trigger" and its subsequent commitment point, the evidence suggests that the situation is, indeed, more complex and subtle. Freese and Fujita (1976) have suggested the "multiple-step hypothesis" instead of a "trigger" hypothesis. According to this view, there occur during the growth–sporulation transition a series of sequential individual changes in enzyme activity, metabolite pools, and structural proteins, each of which may be reversible but in concert become increasingly difficult to reverse. The nature of the prior growth and sporulation medium determine the "point" at which "commitment" is recognized (Cooney *et al.*, 1975). Sterlini and Mandelstam (1969) using the addition of casein hydrolysate to reverse sporulation, have demonstrated that there are a number of "commitments," each corresponding to a different portion of the sporulation process.

A goal of developmental biologists is to understand the regulation of developmental events at a molecular level. One approach to this understanding is to define the events themselves in biochemical terms so that one may investigate the biosynthesis of specific, developmentally relevant gene products. Generally speaking, the process of sporulation has not yet been sufficiently well-defined to permit this sort of approach. Instead, a somewhat more indirect strategy has often been employed wherein the more general nature of protein and RNA synthesis during sporulation have been examined. It has thus been possible to examine the relationship between sporulation and changes in the synthesis (Doi, 1969) and half-life of mRNA (see Piggot and Coote, 1976), patterns of isoaccepting tRNA (Vold, 1973), RNA polymerase (Losick and Sonenshein, 1969) and components of cell-free protein synthesizing systems (Chambliss and Legault-Demare, 1975).

The repression and derepression of enzyme synthesis during sporulation leads to the question of whether these changes are controlled at a transcriptional or translational level. Using DNA–RNA hybridization studies it has been shown that new RNA sequences appear during sporulation, and certain of the vegetative cell genes are no longer transcribed (Doi, 1969). [Dawes and Hansen (1972) have pointed out that the presence of relatively large numbers of nonsporulating cells in sporulating cultures could modify interpretations as to the extent of vegetative gene transcription during sporulation.] Recently, Losick's group has shown with *B. subtilis*, using hybridization competition experiments, that the pattern of RNA synthesis in sporulating cells (t_2 and t_4) was essentially the same as that from an asporogenous mutant in stationary phase (Pero *et al.*, 1975). These data emphasize once again the difficulty in distinguishing between regulatory events occurring during postlogarithmic growth and those associated with sporulation.

Losick and Sonenshein (1969) made the important discovery that template specificity of *B. subtilis* RNA polymerase apparently changed during sporulation. They suggested that the change in specificity was due to an alteration, during sporulation, of one of the RNA polymerase core enzyme β-subunits that altered the interaction between the core enzyme and its complementary σ factor. More recent evidence has attributed the change in specificity to a sporulation-specific polypeptide that interferes with the σ–core interaction and is associated with the RNA polymerase core (Segall *et al.*, 1974). Fukuda and Doi (1977) have recently defined two modifications that occur to the RNA polymerase core during Stages III and IV of sporulation by *B. subtilis*. The core enzyme acquires a δ^1 factor during Stage III and a δ^2 factor during Stage IV. The presence of these new polypeptides alters the specific activity of the enzyme and interferes with the association between the vegetative holoenzyme and σ. δ may be the same polypeptide described by Segall *et al.* (1974).

A second approach to understanding the role of transcriptional control in sporulation has been to select RNA polymerase mutants that are defective for sporulation. Sumida-Yatsumoto and Doi (1977) have refined this approach by isolating a variety of rifampin-resistant mutants of *B. subtilis* that are temperature sensitive for various stages of sporulation. The behavior of these mutants is consistent with the notion that the RNA polymerase from sporulating cells is different than that from vegetative cells and that the conformation of the polymerase core plays an important role in the control of transcription during sporulation. This is consistent with earlier work that showed that certain rifampin-resistant mutants also lost the ability to sporulate and upon reversion to rifampin sensitivity regained the ability to sporulate (Sonenshein and Losick, 1970). Finally, as a word of caution, Piggot and Coote (1976) have described a number of experiments which strongly suggest that the asporogeny that sometimes results from changes in the RNA polymerase, may be an indirect consequence of the mutation rather than a direct result of altered specificity toward sporulation genes.

The role of translational levels of regulation of sporulation remains unresolved. Such levels could include the production of relatively stable species of sporulation-specific mRNA's, modified ribosomes, tRNA's, or combinations of the above. Edgell *et al.* (1975) were unable to detect differences in hybridization between vegetative cell DNA and ribosomal RNA from 23 S subunits of vegetative or sporulating cells. It has, however, been demonstrated that spore ribosomes are deficient in their ability to bind aminoacyl-tRNA and thus are unable to participate in protein synthesis (Kobayashi and Halvorson, 1968). This defect has been shown to be due to the absence of some of the normal ribosomal protein (Kobayashi, 1972). Tipper *et al.* (1977a) have isolated several hundred erythromycin-resistant single-site mutants of *B. subtilis*, all of which are temperature sensitive for sporulation but are otherwise identical. The erythromycin-resistant and sporulation-deficient phenotypes cotransduce and cotransform 100%, and revertants to normal sporulation simultaneously regain erythromycin sensitivity. The data further indicate that the L17 protein of the 50 S ribosomal subunit is altered in the mutant and may thus participate specifically in the sporulation process.

Vold (1975) has described changes in the relative concentration of isoaccepting species of tRNA during sporulation; however, it is not clear that these are indeed sporulation-specific events. While there have been reports of stable mRNA in sporulation (DelValle and Aronson, 1962; Sterlini and Mandelstam, 1969) other investigators, using rifampin as a specific inhibitor of RNA synthesis (Leighton, 1974) and a temperature-sensitive rifampin-resistant RNA polymerase mutant have been unable to demonstrate any role for stable mRNA.

Rhaese *et al.* (1975) have reported a series of highly phosphorylated nucleotides (HPN) that are apparently synthesized by sporulating *B. subtilis* but not by vegetative cells. These have been identified as ppApp(HPNI) and ppAppp(HPNII) and tentatively pppAppp(HPNIV) and ppZpUp(HPNIII) (Z is an unidentified sugar (Rhaese *et al.*, 1975). They have also shown that asporogenous mutants blocked early in sporulation fail to synthesize HPN, whereas revertants reacquire the ability to sporulate and synthesize HPN. (Rhaese *et al.*, 1977). While the function of these molecules is still unknown, and the events are not yet causally related, the finding is especially interesting because of the analogy with the regulatory role of nucleotides such as ppGpp and pppGpp.

I. Germination

Under appropriate circumstances the dormant spore undergoes a series of physiological and morphological changes resulting in its conversion to the vegetative cell. This process has been divided into three stages (Keynan and Halvorson, 1965): (a) activation, (b) initiation, and (c) outgrowth. Activation has been described as rapid aging resulting in the release of spore enzymes from the cryptobiotic state (Keynan, 1972). Fresh spores exposed to optimal germination conditions will usually not germinate unless they have been activated. Activating treatments include sublethal heating, exposure to low pH, thiol agents, or strong oxidizing agents (Keynan and Evenchik, 1969). Initiation involves the irreversible conversion of the activated spore to a cell that has lost all the physiological properties of the spore. Outgrowth is the actual cell morphogenesis requiring growth conditions and resulting in the conversion of the germinated spore to a vegetative rod.

There are a variety of treatments that have been shown to initiate germination. These include nutritional, chemical, enzymatic, and physical factors (Gould and Dring, 1972). Among nutritional initiators are L-alanine or certain other amino acids, glucose or other sugars, adenine or other free bases, and other assorted compounds (Gould, 1969). Among nonnutrient chemical initiators, Ca-DPA (Riemann and Ordal, 1961) and long-chain alkyl amines (Rode and Foster, 1961) have been shown to be effective. Lysozyme has been shown to have initiation activity after pretreatment of the cells with chemicals that reduce disulfide bonds (Gould and Hitchins, 1963). The same result could be achieved using a spore-lytic enzyme extracted from the spores themselves (Gould and Hitchins, 1965). Finally, certain physical treatments, such as hydrostatic pressure (Gould and Sale, 1970) or abrasion (Rode and Foster, 1960), have initiator effects. The mechanism of initiation is not understood, and it has generally not been possible to detect metabolic products of initiators; thus there are no connecting reactions to the germination events that follow. Spore DNA clearly plays a role in germination as indicated by its activity as an initiator (Riemann and Ordal, 1961), its excretion as one of the earlier germination events (Powell and Strange, 1953), and by the difficulty in obtaining germination of DPA-less mutants (Halvorson and Swanson, 1969). In addition, its ability to chelate Ca^{2+} and thus regulate intracellular levels of Ca^{2+} seems relevant. However, its role remains unclear. An interesting direction concerns the role of ions, with particular empnasis on K^+. A variety of data point to the role of cation movement in germination (Gould and Dring, 1972) that may play a role either in relieving the cortex contractility, activating enzymes, altering membrane properties, or creating an energy-yielding cation flux. The recent emphasis on the role of proton and K^+ gradients in transport and phosphorylation (see Chapter 7, Vol. VI) suggest that this may be a useful approach to understanding the early events in germination.

Table I lists the physiological events that occur during germination. The early stage of germination in a single spore may take less than 1 minute. During this brief moment the resistance properties are lost, a number of enzymes are activated, and up to 30% of the dry weight of the spore is lost.

TABLE I

SEQUENCE OF EVENTS DURING *Bacillus* SPORE GERMINATION[a]

	Time[b] for 50% completion of event			
Event	*B. megaterium*	*B. cereus*	*B. cereus*	*B. subtilis*
Heat sensitivity	41	54	61	
Chemical sensitivity	41			
Release of K^+				50
Release of Ca^{2+}			86	
Release of DPA	64	65	86	
Stainability	77	70		
Phase-darkening	95	60		
Release of hexosamine		92		
Release of DAP			97	
Fall in extinction	100	100	100	100

[a] From Gould and Dring (1972).

[b] Time relative to fall in extinction, taken as 100.

Most of the weight loss is due to the excretion of DPA, roughly equivalent amounts of Ca^{2+} and fragments of cortex peptidoglycan. It should be noted that the loss of heat resistance is the first observable physiological change and does not seem to be preceded by any measurable biochemical change.

Although the energy charge of the dormant spore is about 11% that of exponentially growing cells, the spore contains substantially more 3-phosphoglyceric acid than does the vegetative cell (Setlow and Kornberg, 1970a). Even though the initial events in germination do not require ATP, there is a rapid burst of ATP synthesis via the conversion of phosphoglyceric acid to ATP and pyruvate (Setlow and Kornberg, 1970a). The amount of ATP thus generated is easily sufficient to convert the pool of spore ribonucleotides to triphosphates for nucleic acid biosynthesis. This is followed by ATP synthesis via oxidative phosphorylation, although the exact pathways of energy metabolism at this stage are unclear.

RNA biosynthesis in *B. megaterium* begins by the second minute of germination and continues until the pool of stored nucleotides is exhausted. This pool is replenished by the breakdown of about 5% of the spore RNA (Setlow and Kornberg, 1970b) but not by any *de novo* nucleotide biosynthesis. The precise nature of the RNA synthesized or degraded has not been determined, although it has been suggested that it is either ribosomal (Deutscher *et al.*, 1968) or messenger (Setlow, 1975) RNA present in the spore.

RNA polymerase is present in the dormant spore (Chambon *et al.*, 1968). The first period of nucleic acid biosynthesis lasts about 15 minutes and is referred to as the "turnover stage" (Setlow and Kornberg, 1970b). This is followed by a "biosynthetic stage" during which *de novo* synthesis of ribonucleotides takes place. This period coincides with the beginning of the second stage of ATP synthesis during which oxidation of exogenous energy sources take place (Setlow and Kornberg, 1970a).

Net DNA synthesis indicative of chromosome replication does not occur to any sufficient extent until 30 to 90 minutes of germination, even though there seems to be some repair synthesis at about the fifth minute of germination (Rana and Halvorson, 1972). In contrast with RNA biosynthesis, the enzymes for *de novo* synthesis of deoxynucleotides are present in the dormant spore, even though the nucleotides themselves are not. In addition, deoxynucleotides are generated by the reduction of the ribonucleotides produced by RNA breakdown.

As with RNA and repair DNA synthesis, protein synthesis begins during the first 2 or 3 minutes of germination and utilizes endogenous amino acids during a turnover stage. The events of macromolecular synthesis and nucleotide metabolism during germination are summarized in Fig. 9.

Enzymes for most of the pathways of *de novo* amino acid synthesis are

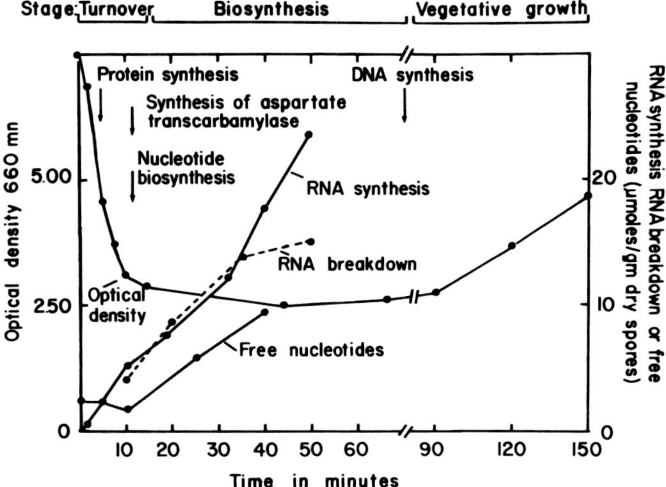

FIG. 9. Macromolecular synthesis and nucleotide metabolism during spore germination. [From Dawes, I. W., and Hansen, J. N. (1972). Morphogenesis in sporulating bacilli. *Crit. Rev. Microbiol.* **1**(4), 479–520. Chem. Rubber Publ. Co., Cleveland, Ohio. Used by permission of the Chemical Rubber Publishing Co.]

absent until quite late in germination (Setlow and Primus, 1975a), and the amino acids are generated by the proteolytic breakdown of approximately 20% of the spore proteins (Setlow and Primus, 1975b). Germination proteolysis is limited to a few distinct proteins located in the spore protoplast and absent from vegetative cells, stationary phase cells, and germinated spores. These proteins comprise 40–50% of the spore core protein, and it is not clear whether or not they also serve a structural function in the spore (Setlow and Primus, 1975b).

Using hydridization competition experiments, it has been shown that synthesis of mRNA is not random but that transcriptional controls are operating during outgrowth (Hansen *et al.*, 1970). Furthermore, it has been possible to isolate and map a series of mutants of *B. subtilis* blocked at various stages of outgrowth. The mutant blocks can be located on the chromosome in an order that corresponds to the order of the associated morphological changes (Nukushina and Ikeda, 1969), and Kennett and Sueoka (1971) have shown that the sequence of cistrons on the chromosome directly controls the order of expression of five enzymes during outgrowth. However, in contrast with these results, Galizzi *et al.* (1975) have isolated temperature-sensitive mutants of *B. subtilis* conditionally blocked in outgrowth. Genetic analysis of the mutants revealed no correlation between map location and temporal expression of the gene. With regard to germination, the striking self-sufficiency of spores, the rapidity with which germination occurs, and their insensitivity to metabolic inhibitors all suggest an initial trigger reaction perhaps involving an allosteric change in one or more cellular components.

J. Genetics

One aspect of the genetic analysis of *Bacillus* has involved attempts to gain insights into the regulation of spore development. This has mainly concerned the characterization of a large number of sporulation mutants and the mapping of sporulation genes by means of transformation and transduction. There have been two approaches to the developmental use of mutants. One has involved the isolation of asporogenous and oligosporogenous mutants. This has provided information as to the various developmental stages at which the process can be genetically blocked and has allowed the correlation of physiological and biochemical events with the morphological stages. However, the pleiotropic nature of sporulation mutants and/or the dependent sequence of developmental events has made it difficult, if not impossible, to assign a specific biochemical defect to the constellation of properties usually associated with the sporulation mutants. To compound the difficulty, it has not been possible to rescue the mutants, i.e., to reverse the block by cross-feeding between asporogenous mutants (Mandelstam, 1969). A second approach has been to select for mutants with known biochemical defects in the hope of simultaneously interfering with a developmental step. When possible, this has been a successful albeit limited maneuver. For example, it has helped to clarify the relationship between RNA polymerase and sporulation (Sumida-Yasumoto and Doi, 1977). The isolation of DPA-less mutants has been useful in investigating the developmental role of DPA (Halvorson and Swanson, 1969). The use of mutants temperature sensitive for DNA replication has been useful in relating DNA replication to sporulation (Dworkin *et al.*, 1972), and the isolation of mutants with altered transcriptional properties (e.g., rifamicin-, streptolydigin-, and streptovaricin-resistant mutants) has provided strong evidence for the regulatory role of RNA polymerase in sporulation (Piggot and Coote, 1976).

The mapping of sporulation genes has revealed two interesting and important features of sporulation. One is that there are approximately 40–50 loci specifically concerned with sporulation (Hranueli *et al.*, 1974), and secondly, that these loci are widely dispersed along the chromosome (Piggot and Coote, 1976). The genetic linkage map of *B. subtilis* is illustrated in Fig. 10 (Piggot and Coote, 1976).

Piggot and Coote (1976) have pointed out that the strategy of genetic analysis of sporulation has largely followed the model for analysis of regulation of metabolic pathways. There are a variety of reasons why this has often

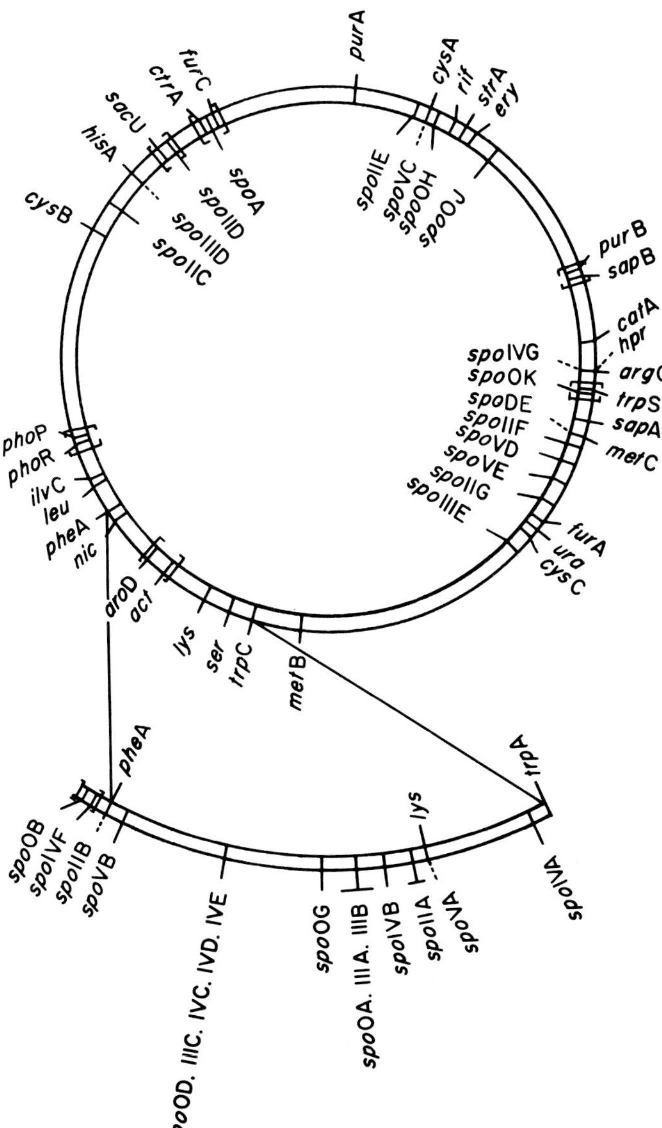

FIG. 10. Genetic map of the sporulation loci of *B. subtilis* (Piggot and Coote, 1976). The format is based on published genetic maps of *B. subtilis* and the total circumference of the circle is approximately 22PBS1 transduction units. The position of loci on the map may not be exactly to scale [precise linkage data together with the appropriate references may be found in Piggot and Coote (1976)]. In the expanded map of the *phe*A to *trp*C region, the groups of *spo* loci where orientation relative to each other has not been determined are shown as mapping at a single site where no orientation is known. The established linkage is indicated by a dotted line. Map positions placed in parentheses have not been ordered relative to outside markers.

been inappropriate. Whereas metabolic pathways are usually well defined biochemically, the blocks imposed by sporulation mutations are not; they are pleiotropic, more delicately controlled, and not amenable to cross-feeding experiments. Furthermore, experiments that have examined the epistatic relationship among pairs of sporulation mutants in *B. subtilis* have shown that for late stage mutations there is no single linear dependent sequence of gene activation; rather, the results are compatible with parallel paths of gene activation (Coote and Mandelstam, 1973). Finally, no primary products of the locus of a sporulation mutant have been isolated or quantitated, and even though it is now possible to construct merodiploids of *B. subtilis* (Audit and Anagnostopoulos, 1972), their analysis is severely limited. It is clear that sporulation involves a complex, nonlinear network of regulatory events and that genetic analysis of developing microbes must take a new and different direction from that which has proved so successful with other systems.

III. Actinospores

Despite the rich variety of developmental events among the actinomycetes they have not received the same attention from developmental microbiologists as have other microbial groups. To some extent this reflects the fact that most of these developmental events only take place on the surface of a solid medium; they are thus less amenable to experimental manipulation than those cells that can be induced to sporulate and germinate in liquid media (e.g., *Bacillus* endospores, myxospores, and *Azotobacter* cysts). While there have been some claims of actinospore formation in liquid (Carvajal, 1947), Kalakoutskii and Agre (1976) have emphasized that these differ from spores formed on aerial mycelia.

There are essentially three types of spores formed by different members of the actinomycetes: hyphal spores (arthrospores) produced at the tips of aerial mycelia, endospores similar in many respects to those formed by the Bacillaceae, and sporangial spores formed by the Actinoplanaceae (Cross and Atwell, 1975). A number of other "spores" have been described [e.g., chlamydospores, blastospores, microcysts (Kalakoutskii and Agre, 1976)] but will not be discussed here.

A. Sporangiospores

Members of the Actinoplanaceae form spores contained within a sac, superficially analogous to the sporangiospores formed by fungi. While there

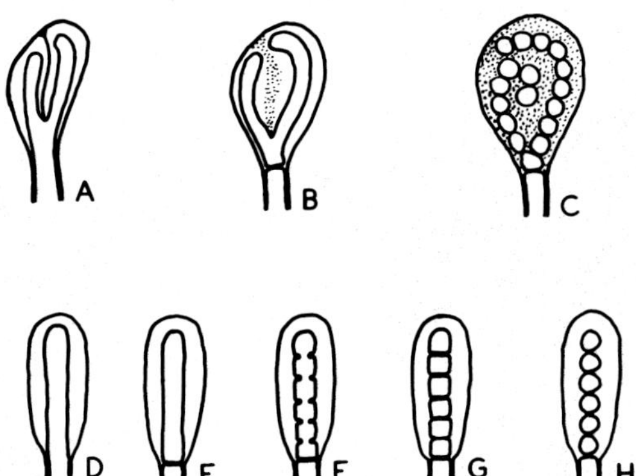

FIG. 11. Sporangial development (A)–(C) and spore development (D)–(H) in *Actinoplanes*. (From Lechevalier and Holbert, 1965.)

is relatively little information as to the nature and developmental biology of these spores, somewhat more information is available concerning the so-called zoospores of the genus *Actinoplanes*. These are formed by the synchronous septation and division of a segment of branched and coiled hypha within a sporangial envelope (Lechevalier and Holbert, 1965). The process is illustrated diagrammatically in Fig. 11 and with electron micrographs in Figs. 12 and 13. During the process leading to the release of the spores (dehiscence), the spores within the sporangium acquire flagella, and upon release are motile. The mature spores, within the sporangium are not flagellated (Higgins, 1967), and thus the term zoospore seems inappropriate. Higgins (1967) has shown that spore release is triggered by wetting the hydrophobic sporangium, whereupon the spores swell, split the sporangial wall, and are released as motile cells. While the precise timing of the acquisition of flagella has not been determined, the process requires a source of carbon and is insensitive to inhibitors of protein and nucleic acid synthesis. Conversion of the motile cells to hyphae (this process has been referred to as germination) requires a source of nutrient and is sensitive to chloramphenicol and actinomycin D (Higgins, 1967).

B. ENDOSPORES

Members of the thermophilic genus of actinomycetes, *Thermoactinomyces*, form endospores that are essentially identical to those formed by the Bacillaceae. They are formed within the hyphal cell, have a high heat resistance, a fine structure similar to that of *Bacillus* spores, and contain DPA (Kalakoutskii and Agre, 1976). A unique feature of the spores, however, is that they are polyhedral with 12 pentagonal and 12 hexagonal faces, each slightly

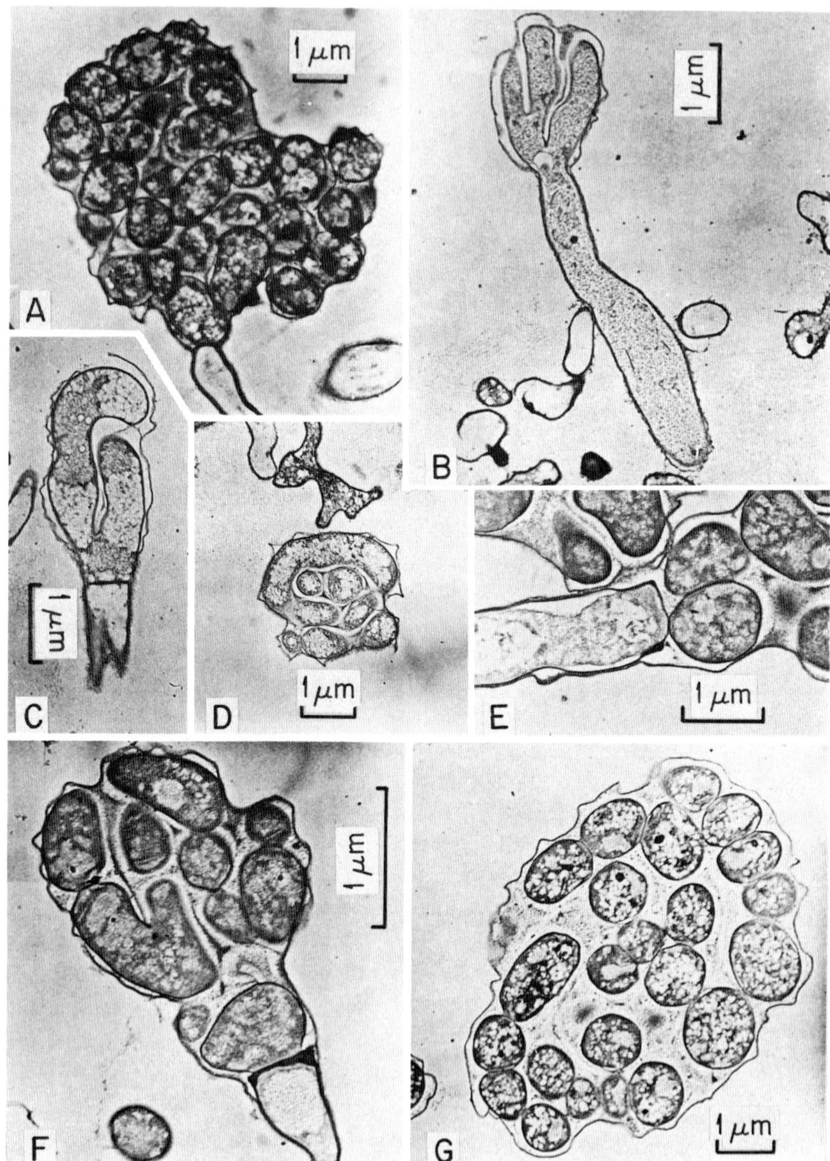

FIG. 12. Electron micrographs of sections through a sporangium of *Actinoplanes* at various stages of development. (A) Mature sporangium (4 days old). (B), (C), and (D), Immature sporangium (2 days old). (E) and (F), Mature sporangium (4 days old) note that the sporangial wall is continuous with the sporangiophore sheath. (G), Mature sporangium (4 days old). (From Lechevalier and Holbert, 1965.)

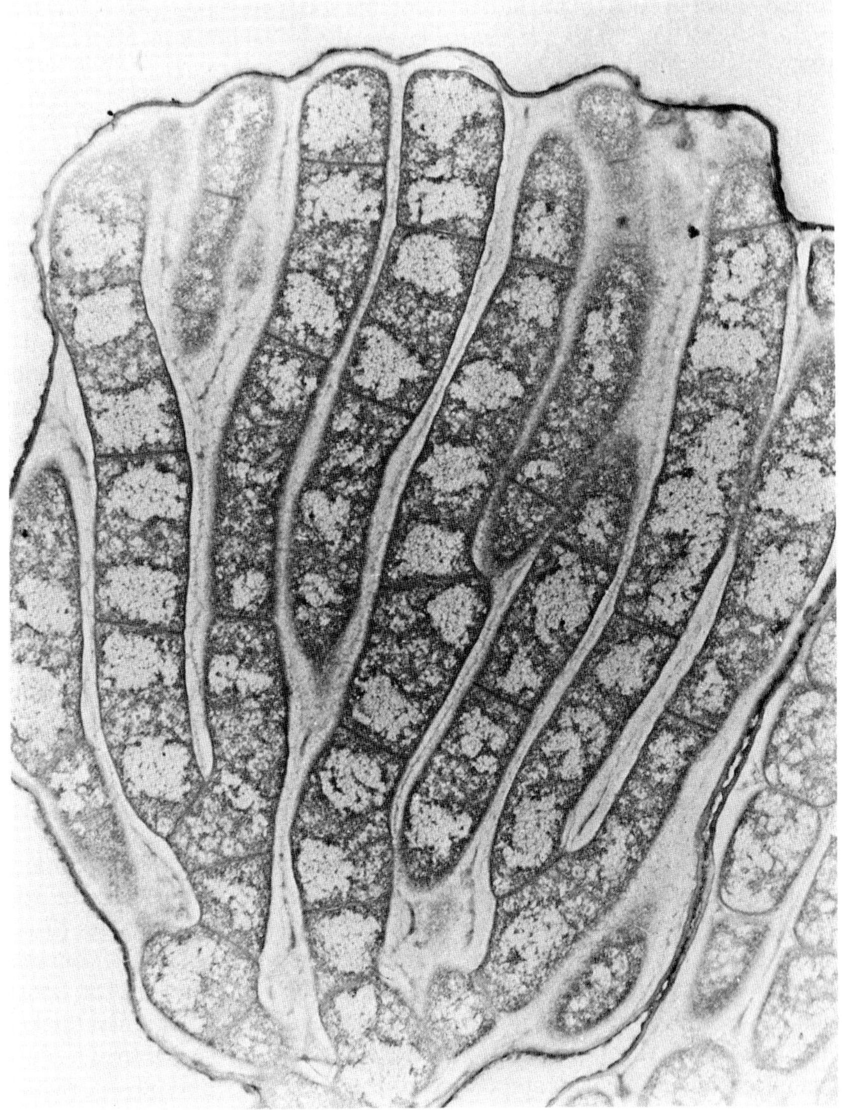

FIG. 13. Electron micrograph of a section through an immature sporangium of *Actinoplanes*. Courtesy of Dr. H. Lechevalier.

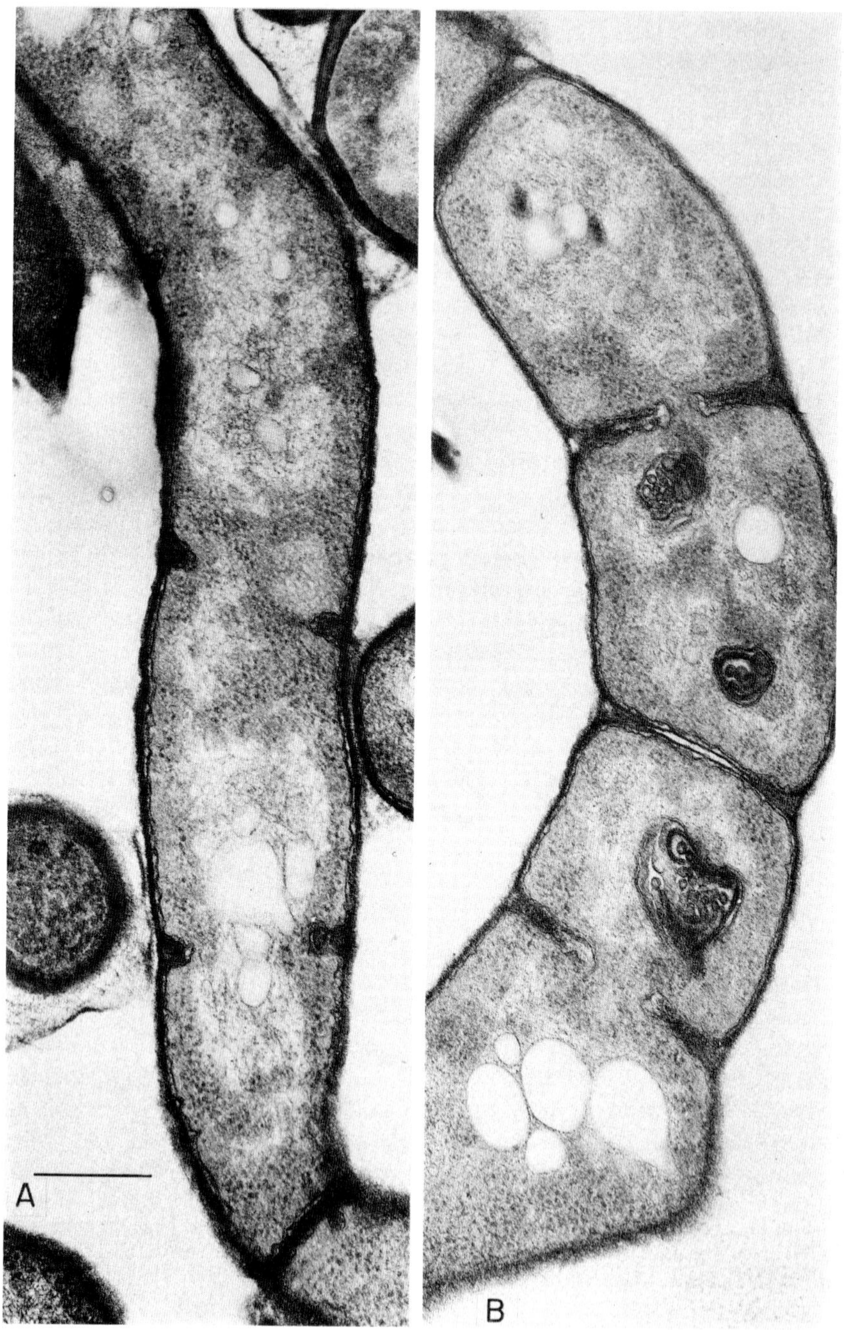

FIG. 15. Electron micrograph of thin section of *Streptomyces coelicolor* aerial mycelium during (**A**) early stage of synchronous sporulation septum and (**B**) late stage of sporulation septation. Scale, 0.5 μm. (From Chater and Hopwood, 1973.)

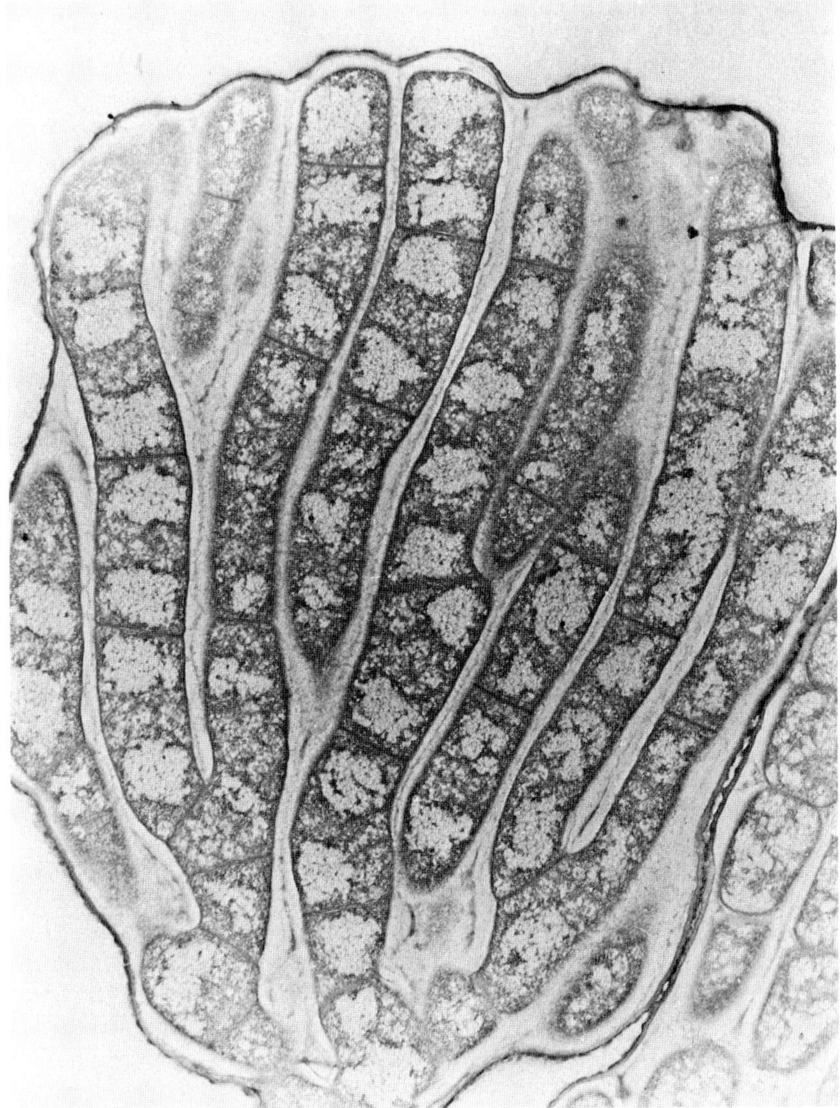

FIG. 13. Electron micrograph of a section through an immature sporangium of *Actinoplanes*. Courtesy of Dr. H. Lechevalier.

concave. The fibrous sheath surrounding the spore which may be equivalent to the exosporium consists of several layers of parallel fibers (McVittie *et al.*, 1972).

C. Hyphal Spores

Most of the work on hyphal spores has been done with *Streptomyces*. Figure 14 illustrates the vegetative hyphae and immature and mature spores of *Streptomyces coelicolor*. Spore germination on agar is followed by the development of a highly branched network of hyphae in and on the agar. A small, smooth, "hairless" colony is formed that subsequently develops a fuzzy appearance, as the aerial hyphae are formed. Wildermuth and Hopwood (1970) have divided the sporulation process into four morphological stages. (1) The long, coenocytic cells of the aerial mycelium become coiled. (2) A series of septa 1–2 μm apart, often accompanied by mesosomes, are synchronously formed along the length of the cell. The septa are formed by the ingrowth of the cell wall and cell membrane, and the process is morphologically distinct from the process of septation during normal vegetative growth. (3) A thick wall is laid down surrounding the spore. (4) The spores become ellipsoidal and the cell wall material connecting the spores disintegrates, the spores remaining connected only by a small fibrous sheath. These stages are illustrated by Figs. 15 and 16. The entire colony is a heterogeneous collection of substrate and aerial mycelia, spores, and ghosts of lysed cells as illustrated by Fig. 17.

1. Properties of Hyphal Spores

From a morphological point of view the *Streptomyces* spore is essentially similar to the vegetative cell of the aerial hypha. Both share the characteristic hydrophobic exterior and, in some cases, hairy or spiny appendages (Rancourt and Lechevalier, 1964) whose function is unknown. The spore contains no characteristic organelles or layers, such as the endospore cortex, spore coat, or exosporium. The wall of the spore is, however, 1.5–2 times thicker than that of the vegetative wall. This does not seem to be reflected by any qualitative change in the components of the cell wall peptidoglycan (Kalakoutskii and Agre, 1976). Figure 16 is an electron micrograph of a thin section of spores of *S. coelicolor*. Note the thick wall surrounding the inner cell.

There are conflicting reports as to the number of nucleoids per spore, with some investigators reporting one and others two (Kalakoutskii and Pozharitskaja, 1973).

A characteristic feature of *Streptomyces* is the pigmentation of sporulating colonies. The pigment appears to be associated with the spores and is lost

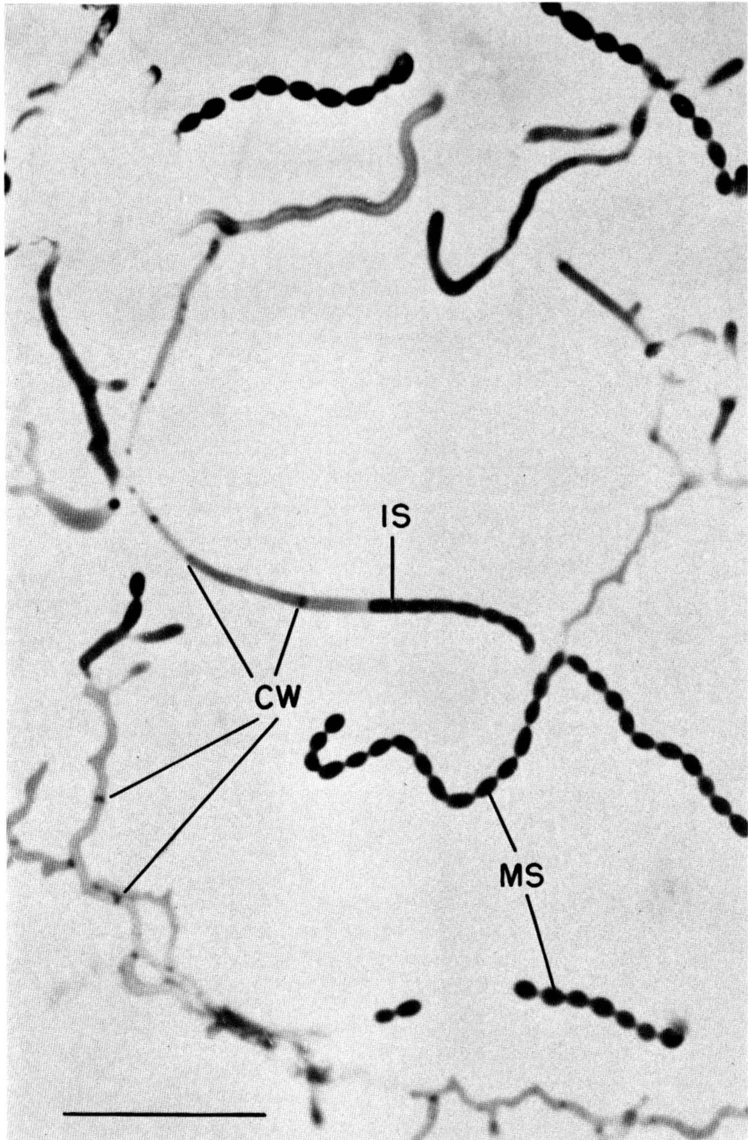

FIG. 14. Phase contrast photomicrograph of *Streptomyces coelicolor*. Note the cross walls (CW) in the vegetative hyphae, rod-shaped immature spores (IS), and the helical chain of mature spores (MS). Scale, 10 μm. (From Chater and Hopwood, 1973.)

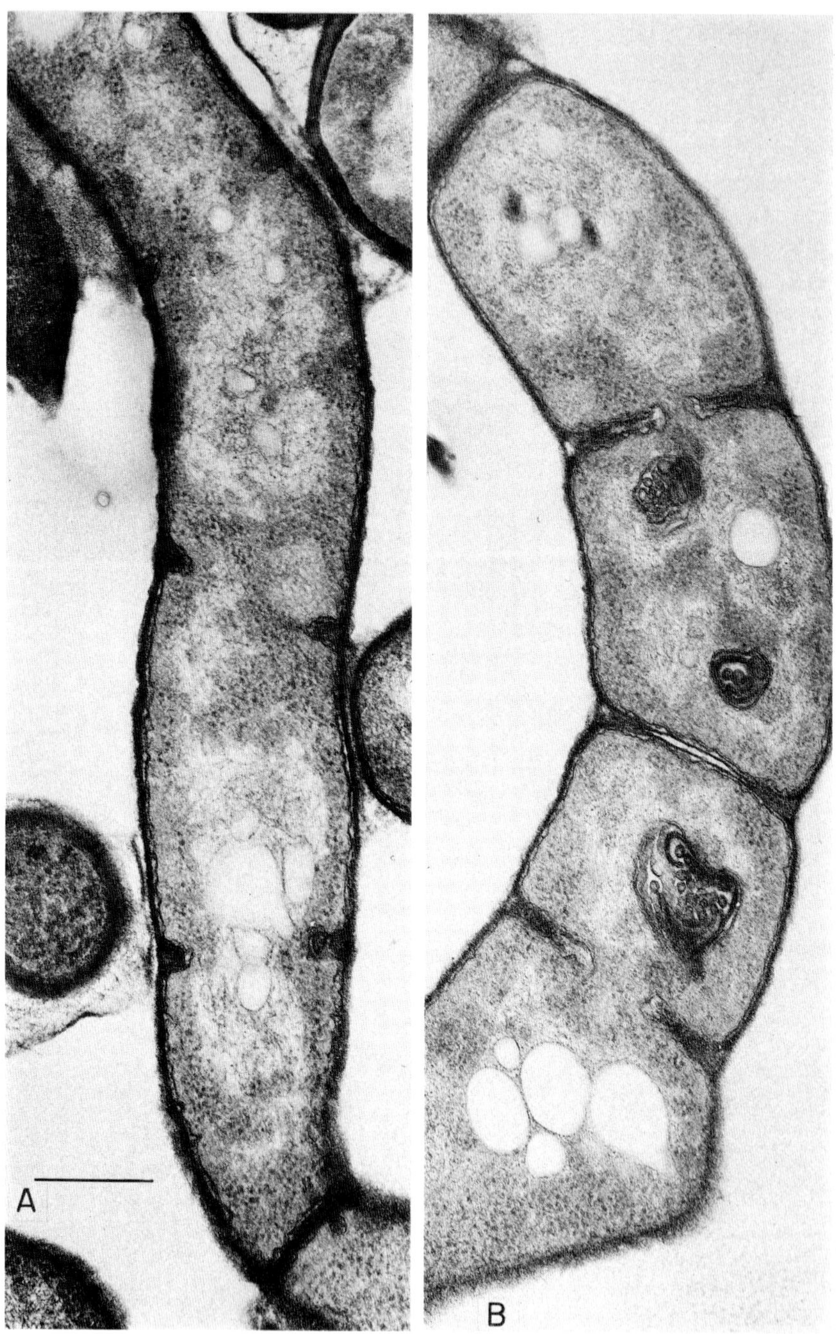

FIG. 15. Electron micrograph of thin section of *Streptomyces coelicolor* aerial mycelium during (**A**) early stage of synchronous sporulation septum and (**B**) late stage of sporulation septation. Scale, 0.5 μm. (From Chater and Hopwood, 1973.)

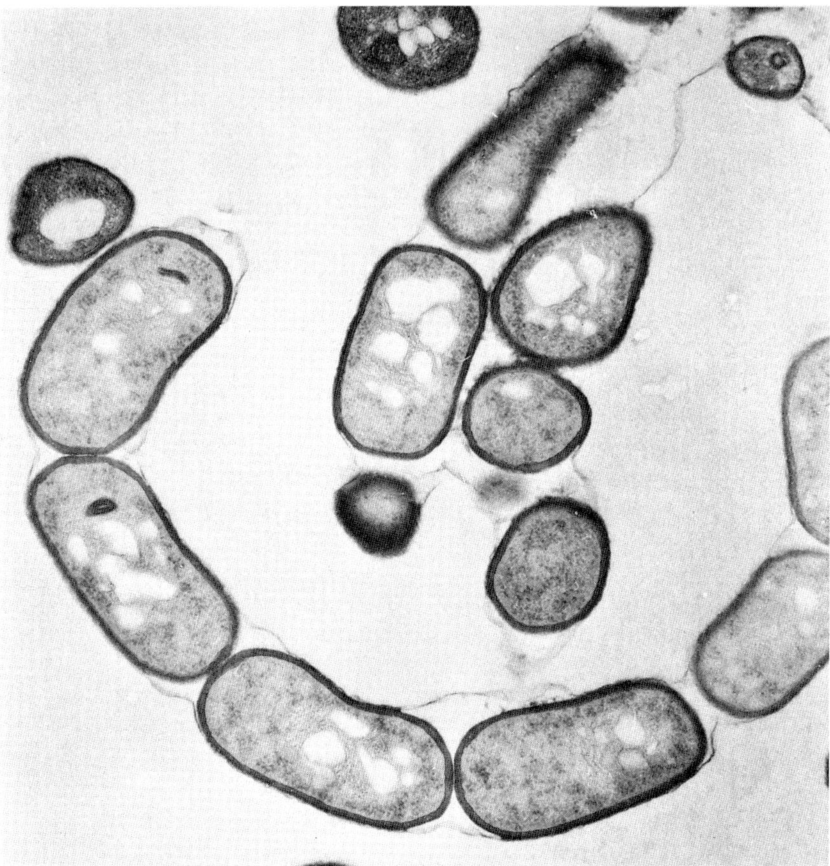

FIG. 16. Electron micrograph of thin section of mature spores of *Streptomyces coelicolor*. Note the somewhat thickened walls. (From McVittie, 1974.)

upon spore germination (Kalakoutskii and Agre, 1976). Scribner *et al.* (1973) have reported that the melting temperature (T_m) and the buoyant density of spore DNA differs from that of vegetative cell DNA and have attributed this difference to a pigment bound to spore DNA. Upon acetone extraction of the spore DNA its properties match those of vegetative cell DNA. While this suggests the interesting possibility that the pigment has a regulatory function in the spore, at the level of the DNA, it is not yet possible to exclude preparational artifacts.

Streptomyces spores do not contain DPA (Kalakoutskii *et al.*, 1969), indicating yet another substantial difference between them and endospores.

Mikulík *et al.* (1975) have examined the ribosomes of *Streptomyces* and despite rather careful preparation were unable to demonstrate polysomes in

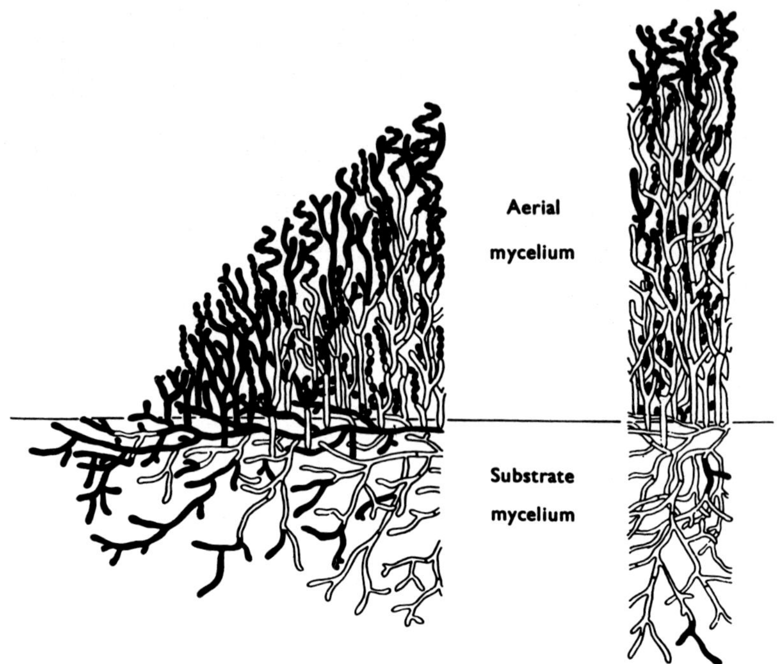

FIG. 17. Idealized diagram of a vertical section through the center of a sporulating colony of *Streptomyces coelicolor*. Black represents intact cells and white disintegrating or completely lysed cells. (From Chater and Hopwood, 1973.)

mature spores. The spores did, however, contain free ribosomes whose properties and response to monovalent cations differed from that of vegetative cell ribosomes. While conventional isolation techniques resulted in the demonstration of 16 and 23 S RNA from vegetative cell ribosomes, the spore rRNA was unstable, yielding 16 S RNA and degraded 23 S RNA. Furthermore, polyacrylamide gell electrophoresis revealed that spore ribosomes contained two extra proteins undetected in vegetative cell ribosomes. The spore ribosomes also contained a dark pigment that could not be removed by extensive washing or dialysis. (Once again, a regulatory role for the sporulation pigment is suggested but not demonstrated.) It seems clear, in general, that some degree of biosynthetic regulation is occurring at the ribosomal level.

Little work has been done on the metabolic activities of *Streptomyces* spores, but the generalization has been made that their metabolic rate is low (Kalakoutskii and Agre, 1976). On the other hand, radiorespirometric examination of dry spores revealed no measurable endogenous respiration

(Hirsch and Ensign, 1975). Despite this, a number of enzyme activities have been demonstrated in mature spores (Kalakoutskii and Agre, 1976). Calorimetric determinations (Kalakoutskii and Pozharitskaja, 1968) and electron spin resonance signals in spores (Pozharitskaja *et al.*, 1972) are characteristic of metabolically active cells, and ^{32}P is incorporated into RNA (Kalakoutskii *et al.*, 1966). The examination of dry versus wet spores may determine whether or not a metabolic activity is detected, and it is possible that wetting the spores is sufficient to trigger an immediate activation of endogenous metabolic activity. In any case, the *Streptomyces* spores seem to fall between extremely dormant endospores and normally active vegetative cells.

Spores of *Streptomyces* seem to be able to survive long periods of storage (Kalakoutskii and Pozharitskaja, 1973), desiccation (Aslanjan *et al.*, 1971), and toxic chemicals (Aslanjan *et al.*, 1972). They are somewhat more resistant than vegetative cells to heat and radiation (Kalakoutskii and Pozharitskaja, 1973). There have been no studies that relate the increased resistance of the spore to any structural or chemical properties of the spore.

2. Spore Germination

There has been some lack of uniformity as to the term used to describe the transition of a *Streptomyces* spore to the vegetative cell. Atwell and Cross (1973) have suggested a standard terminology for endospores wherein the entire process is referred to as "germination" and is divided into activation, initiation, and outgrowth. They suggest that, in *Streptomyces*, the term "germination" shall likewise refer to the entire process, while recognizing that the stage between activation and outgrowth ("emergence") is not sharply delineated. The recent work of Hirsch and Ensign (1975) more clearly defines this intermediate stage, and we shall thus use the endospore terminology without implying that the mechanisms are similar.

In the past, collection of spores was usually done in an aqueous milieu, thus possibly obscuring the early events in germination. Hirsch and Ensign (1975) have collected spores of *S. viridochromogenes* with dry, glass beads; in addition they have facilitated initiation by wetting the hydrophobic spores with detergent. Using this system they have begun to define activation and initiation somewhat more stringently than has been done previously.

Initiation lasted about 80 minutes and was coincident with an approximately 15% decrease in the optical density of the suspension. At the end of the 80 minutes the spores had lost their phase brightness and outgrowth began. Using the rate of optical density drop as a parameter of initiation, it was shown (Hirsch and Ensign, 1975) that the spores are subject to heat activation. Heating at 55°C for 10 minutes was the optimal treatment, and the effect was reversible. There is no information available as to the mechanism of activation.

The minimal nutritional requirements for initiation were defined and shown to be L-glutamate, L-alanine, adenosine, *p*-aminobenzoic acid, Ca^{2+}, Mg^{2+}, and CO_2. (It is interesting that these organic molecules are not required for vegetative growth of the organism.) Oxidation of alanine and adenosine after the optical density drop was essentially complete, while glutamate oxidation began when the spores had lost their optical refractility. Endogeneous metabolism was low during the first 20 to 40 minutes, then increased sharply and remained high throughout germination. Whereas dried spores had a low ATP content, that level rose sharply during germination after about a 10-minute lag. The concentration increased about twentyfold by 50 minutes of germination and remained at that level (Hirsch and Ensign, 1975).

At the very beginning of germination adenosine was incorporated into spore RNA. While a small amount of the adenosine carbon was oxidized to CO_2, by 90 minutes 40% of it had been incorporated into protein and the remainder into RNA. As expected, nearly all the alanine and glutamate were incorporated into protein starting at about 15 minutes (Hirsch and Ensign, 1975).

The effect of inhibitors on macromolecular synthesis was examined, and while rifampin and chloramphenicol completely inhibited protein synthesis, germination as measured by optical density drop was only partially inhibited (Hirsch and Ensign, 1975). The inhibition of protein synthesis by rifampin suggests the absence of stable mRNA, and the increase in ATP in the absence of protein synthesis suggests that the cells are enzymatically self-sufficient, at least for the early stages of germination.

It is interesting that germination between 10 and 80 minutes was accompanied by the excretion of large amounts of spore carbon into the medium (Hirsch and Ensign, 1975). Whether this reflects the degradation and excretion of spore components that are simply not reused by the cells, or some kind of specific signal is not known.

Figure 18 illustrates some of the physiological and biochemical changes as a function of the morphological progress of germination (Kalakoutskii and Agre, 1976).

3. Genetics

Hopwood's group has developed the genetics of *S. coelicolor*, and there is now available (1) a system for genetic recombination, (2) an extensive linkage map, and (3) a large selection of stable developmental mutants. Figure 19 indicates the genes thought to be implicated in the various developmental steps. Thus, the potential for a genetic approach to development in *Streptomyces* is available and is being extensively exploited (Hopwood *et al.*, 1973).

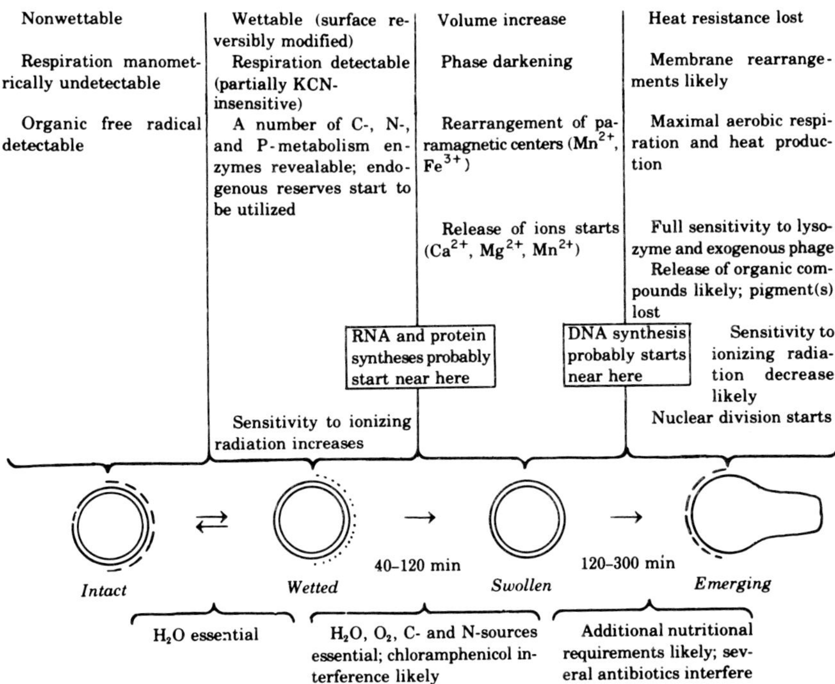

FIG. 18. Events accompanying the germination of *Streptomyces* spores. (From Kalakoutskii and Agre, 1976.)

In summary then, the developmental biology of actinospores is a diverse, relatively unexplored and exciting subject. While some actinospores (e.g., *Thermoactinomyces*) are essentially identical in their properties with endospores of the Bacillaceae, others (e.g., arthrospores of *Streptomyces*) are unique. They share with endospores some features such as dormancy, certain kinds of resistance, and metabolic self-sufficiency; on the other hand, they are structurally dissimilar, the events leading to their formation and germination differ, and there are certain to be different regulatory networks involved in their development.

IV. Myxospores

A. LIFE CYCLE

The myxobacteria are gram-negative bacteria that are characterized by a unique developmental cycle which includes both colonial and cellular

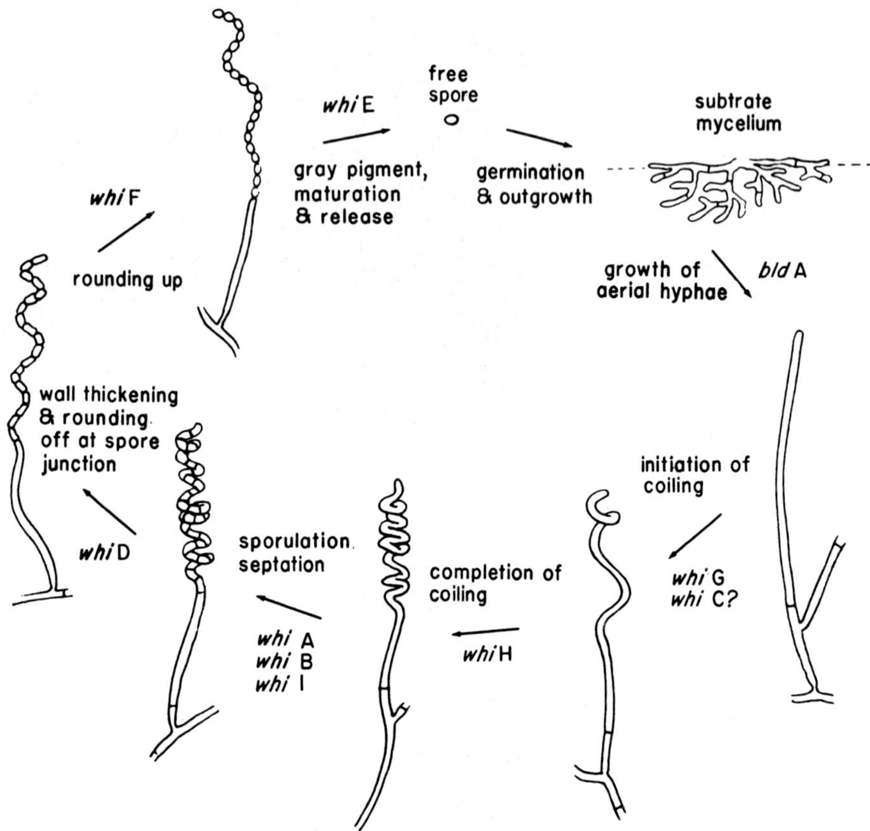

FIG. 19. Diagram of the genetic control of morphogenesis in *Streptomyces coelicolor*. The designation of genes thought to be implicated in each step are given outside the arrows. (From Hopwood *et al.*, 1973.)

morphogenesis. The colonial phase takes place on a solid surface in response to specific nutritional depletion of the medium. Under these conditions, as an alternative to vegetative growth, the cells aggregate—perhaps in response to a chemotactic signal. This is followed by the construction of multicellular fruiting bodies, within which the cells undergo a cellular morphogenesis to myxospores. [This section shall be concerned only with cellular morphogenesis of myxospores; a discussion of the colonial aspects of development can be found in two recent reviews (Dworkin, 1973; Wireman and Dworkin, 1975).] In the case of genera such as *Stigmatella* and *Chondromyces*, the resting cells are enclosed in a cyst, and germination involves their release and activation. While the physiological properties of such myxospores have not been determined, the fine structure of *Stigmatella aurantiaca* has been characterized (Reichenbach *et al.*, 1969), and in this organism it is possible to induce myxospore formation artificially (Reichenbach and Dworkin, 1970). In other myxobacteria (i.e., *Myxococcus*) the individual cells within the fruiting body convert to round, optically refractile, metabolically quiescent, resistant myxospores. Essentially all of the recent work done on myxospores has been done with *Myxococcus xanthus*, and unless otherwise noted it shall be the subject of this discussion. Figure 20 illustrates the life cycle of *M. xanthus*. An important factor that focused experimental attention on *M. xanthus* was the development of a system for inducing the cellular morphogenesis of vegetative rods to myxospores without the intervening process of fruiting body formation (Dworkin and Gibson, 1964). The addition of a relatively high concentration of glycerol or a variety of other inducers, to a liquid suspension of vegetative cells results in the rapid, synchronous, and complete conversion of these cells to myxospores. While there are certain important differences between the properties of these myxospores and those formed on solid media within fruiting bodies, they are essentially similar in morphology, the sequence of events leading to their formation, resistance, and ability to germinate.

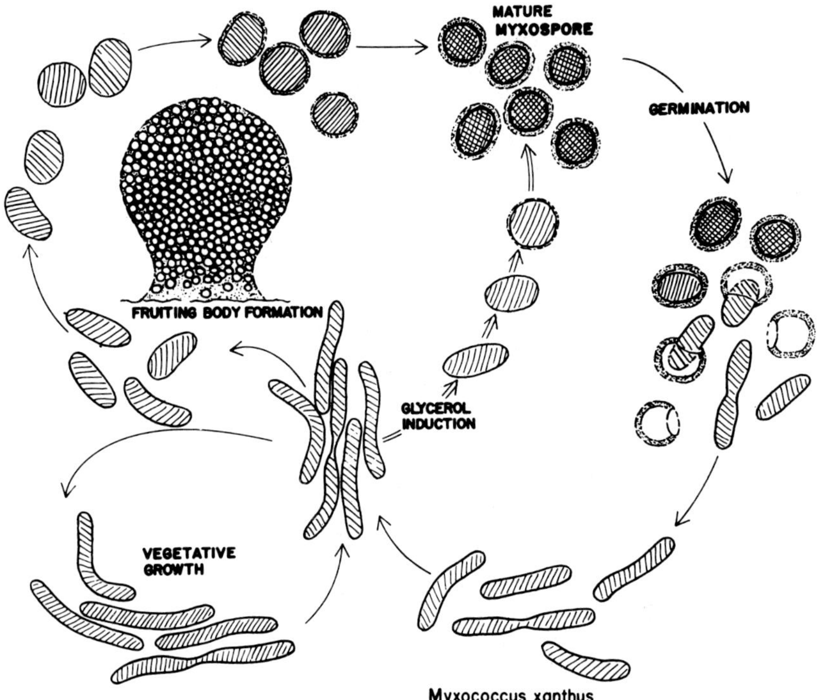

FIG. 20. Life cycle of *Myxococcus xanthus*. (From Dworkin, 1973.)

B. Morphology

Figure 21 is a series of phase-contrast micrographs illustrating the sequence of morphological events and optical density changes during glycerol induction of myxospores of *M. xanthus*. At approximately 35 minutes there is the first visible morphological change, the cells becoming slightly shortened. At 50 minutes most of the cells have converted to short rods. At 60 minutes the cells are prolate spheroids, and at 70 minutes are round, phase dark spheres. At about 85 minutes the spheres have acquired phase refractility which gradually increases in intensity until about 4 or 5 hours. While the myxospores at this stage are morphologically indistinguishable from their mature counterparts formed within the fruiting body, certain metabolic changes continue beyond this point.

Figure 22 is an electron micrograph of a thin section of glycerol-induced myxospores. The distinguishing features are the thick myxospore coat, the vesicularlike bodies between the spore coat and cell envelope, and the cell envelope invaginations. Contrast this with Fig. 23 which shows a fruiting body myxospore.

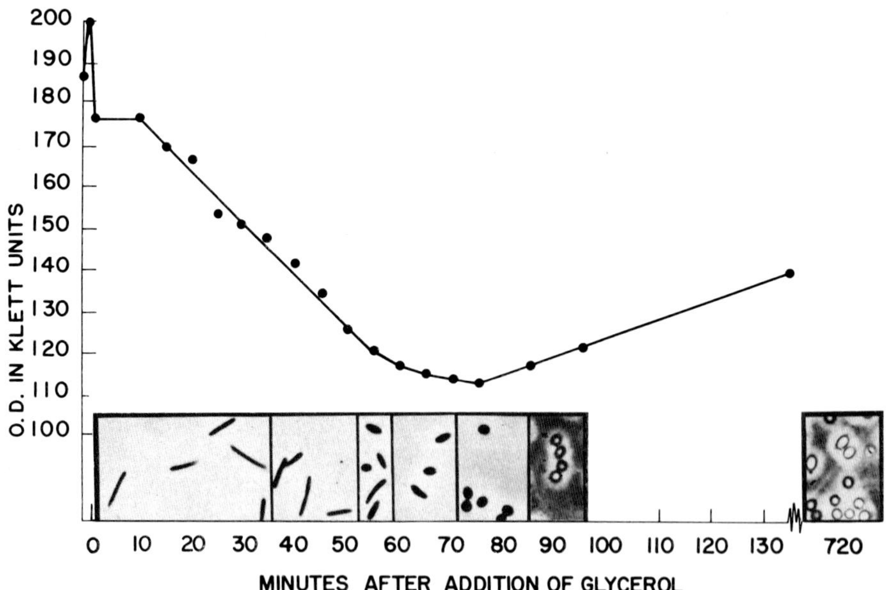

Fig. 21. Glycerol-induced myxospore formation in *Myxococcus xanthus*. (From Dworkin and Gibson, 1964; copyright 1964 by the American Association for The Advancement of Science.)

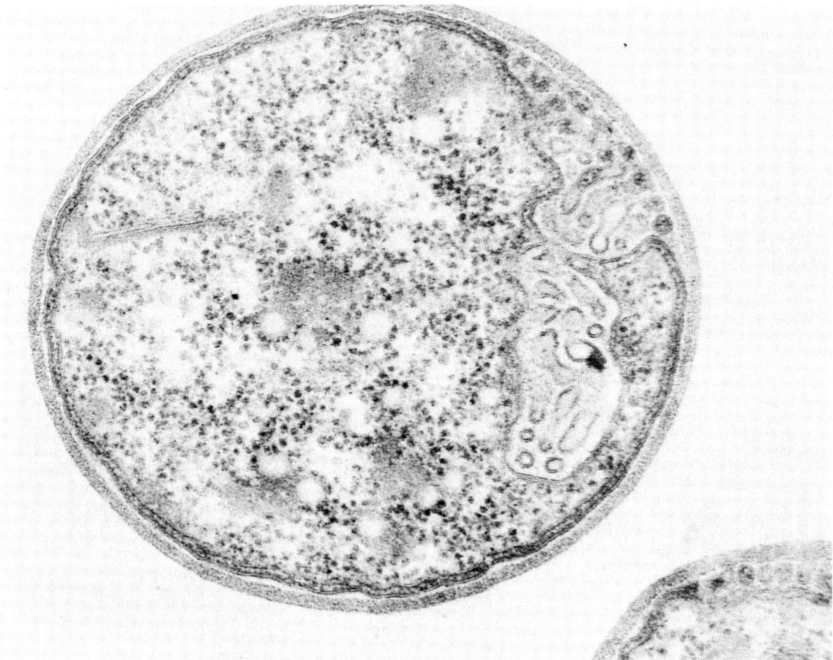

FIG. 22. Electron micrograph of a thin section of a glycerol-induced myxospore of *Myxococcus xanthus* (induced for 2 hours). ×90,000. Courtesy of Dr. Herbert Voelz.

C. PROPERTIES OF MYXOSPORES

The myxospores are more resistant than the vegetative cells to a variety of treatments. These include heat, uv light irradiation, desiccation, and physical and chemical disruption. Whereas 90% of a population of vegetative cells of *M. xanthus* were killed by 1 minute at 50°C, the myxospores were completely resistant to this temperature. However treatment at 60°C resulted in a slow but steady rate of killing (Sudo and Dworkin, 1969). Unlike vegetative cells, the myxospores are substantially resistant to such treatments as sonic oscillation, explosive decompression by the French pressure cell, and lysis by sodium dodecyl sulfate.

A discussion of the physiological properties of myxospores must emphasize that the properties of glycerol-induced spores differ somewhat from those of myxospores formed in fruiting bodies. For example, while it was not possible to detect any endogenous uptake of O_2 in fruiting body myxospores (measured manometrically) (Dworkin and Niederpruem, 1964), a low but measurable rate was detected in spores induced by glycerol (Bacon *et al.*,

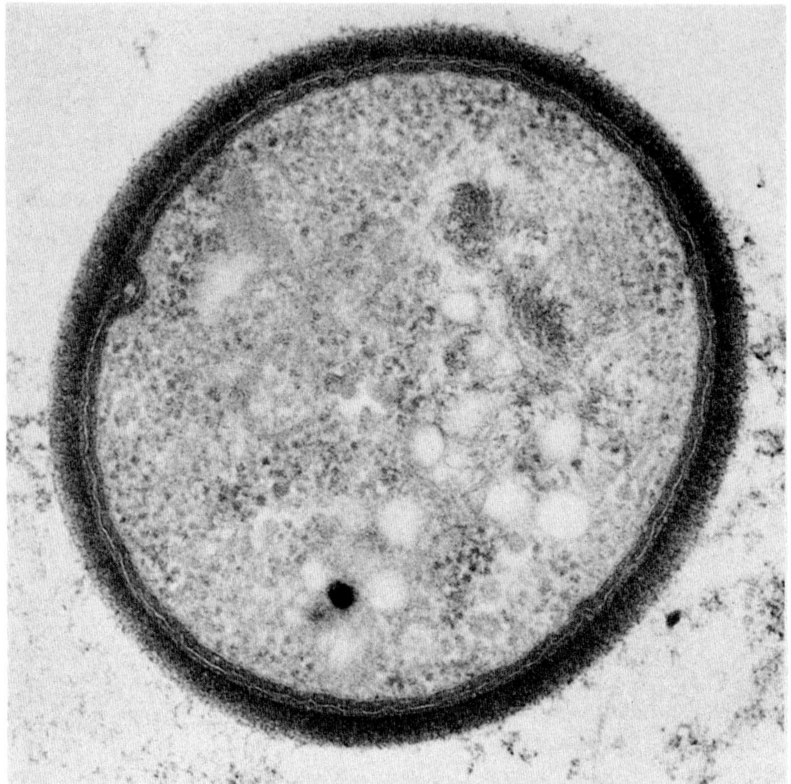

FIG. 23. Electron micrograph of a thin section of a myxospore of *Myxococcus xanthus* formed within a fruiting body. ×75,000. Courtesy of Dr. Herbert Voelz.

1975). During glycerol induction the rate of O_2 consumption gradually decreased and levelled off at 8–10 hours at a value about 20% of that in vegetative cells. It is interesting that the residual O_2 consumption was resistant to sodium azide. Keynan (1972) has pointed out that electron transport in dormant cells often differs from that found in vegetative cells and is frequently resistant to carbon monoxide and cyanide. An examination of the electron transport components of vegetative cells and fruiting body myxospores revealed some differences, but indicated that the major components of the system were present in both types of cells (Dworkin and Niederpruem, 1964).

Glycerol-induced myxospores have been examined for a variety of catabolic enzymes known to be present in vegetative cells, including glycolytic and Krebs cycle enzymes; glucose-6-phosphate dehydrogenase; 6-phophogluconate dehydrogenase; UDP-glucose pyrophosphorylase (Watson and

Dworkin, 1968); alanine, glutamate, and glycine dehydrogenases, and glutamate and alanine-glyoxylate aminotransferases (Kottel *et al.*, 1974). The results indicated that all the above enzymes present in vegetative cells were also present at roughly equivalent amounts in the glycerol-induced myxospores. However, alanine and aspartate aminotransferases and the enzymes of the glyoxylate pathway increased sharply during myxospore induction (Hanson and Andreoli, 1973). The function of these changes is unknown, but the glyoxylate pathway often serves for the gluconeogenic metabolism of fatty acids via acetate. This suggestion is consistent with the observation (Bacon *et al.*, 1975) that during myxospore formation there is a 150–200% increase in neutral polysaccharide. Thus, it seems that gluconeogenesis becomes a major biosynthetic activity during myxospore induction. Part of the increased polysaccharide synthesis results in an α-1,3-linked glucan (Sutherland and Mackenzie, 1977) probably located on the surface of the cell. The remainder of the polysaccharide synthesis in myxospores is directed toward the myxospore-specific polysaccharide in the spore coat consisting of glucose (about 20%), galactosamine (about 55%), glycine (about 7%), and protein (about 14%) (Kottel *et al.*, 1975). The large amount of glycine in the spore coat may explain the increases in the glyoxylate pathway activities during myxospore induction (Orlowski *et al.*, 1972) and the presence of L-alanine glyoxylate aminotransferase (Kottel *et al.*, 1974) in the cells.

Recently, the activity of six enzymes responsible for the synthesis of UDP-*N*-acetylgalactosamine from fructose 6-phosphate have been examined and shown to increase during myxospore induction prior to the appearance of the glucosamine-containing spore coat polysaccharide (Filer *et al.*, 1977). Furthermore, these activities decreased later in myxospore development to the level found in vegetative cells. A similar decrease was found to occur with aspartokinase (Rosenberg *et al.*, 1973) and isocitrate lyase (Orlowski and White, 1974b) during myxospore induction. Orlowski and White (1974a) have suggested that enzyme inactivation during later stages of myxospore formation may be attributed to intracellular proteases whose activity fluctuates during myxospore induction. There is obviously a network of interacting enzymatic changes related to the synthesis of spore coat polysaccharide during myxospore formation, and investigation of spore coat biosynthesis may offer a useful approach for examining the regulation of a specific and biochemically definable developmental process.

The peptidoglycan of myxospores has been examined and compared to that of vegetative cells (White *et al.*, 1968). While there are approximately equal amounts of peptidoglycan in both cell types, the amount of cross-linking increased substantially during myxospore formation. The peptidoglycan of vegetative cells was suggested to be a discontinuous layer with patches of peptidoglycan connected by a trypsin–SDS-sensitive material (White *et al.*, 1968). Substantial amounts of glucose and glycine are also associated with the peptidoglycan, (White, 1975) and these may be part of the nonpeptidoglycan material. It is interesting that glucose and glycine disappear from this layer during myxospore formation, and concommitantly large amounts of glucose and glycine appear as components of the spore coat (Kottel *et al.*, 1975). Furthermore, the spore peptidoglycan layer can no longer be disrupted with trypsin and SDS and has thus presumably lost its interpatch connecting material.

Recently, Inouye *et al.* (1979) have described a small protein synthesized by cells undergoing fruiting body formation. The protein (called protein S) is initially found in the soluble fraction of developing cells, but eventually appears as an insoluble component of the surface of mature myxospores. It represents an extremely high proportion of the total protein of the myxospore and is absent from vegetatively growing cells and glycerol-induced myxospores. Its function is unknown; myxospores from which protein S has been removed retain their viability and resistance to heat and sonication. The authors suggest that protein S is a surface adhesive that holds the myxospores together in the fruiting body.

In *Bacillus* endospores, the state of metabolic dormancy has been correlated with the absence of nucleoside triphosphate pools and an extremely low adenylate energy charge (Setlow and Kornberg, 1970a). In contrast, during glycerol-induced myxospore formation in *M. xanthus*, pools of nucleoside triphosphates increased slightly and the adenylate energy charge of the cells (0.85) remained constant (Hanson and Dworkin, 1974). Thus the glycerol-induced myxospore differs fundamentally from the *Bacillus* endospore in that its dormancy persists in a state of energetic self-sufficiency. This is consistent with the observation that under appropriate conditions, glycerol-induced myxospores (unlike endospores) will germinate in distilled water.

It has recently been shown (Westby and Tsai, 1977) that vegetative cells and glycerol-induced myxospores undergo active glycine and purine uptake as well as purine synthesis, salvage, and interconversion. This is yet another indication either that the myxospores are less dormant than *Bacillus* endospores, which do not carry out purine *de novo* synthesis (Switzer *et al.*, 1975), or that 8-hour-old myxospores have not yet reached a state of complete maturity and dormancy.

D. Mechanism of Myxospore Induction

The discovery that it was possible to induce myxospore formation artificially (Dworkin and Gibson, 1964) represented a substantial improvement in studies of myxospore morphogenesis. Induction was accomplished by the addition of 0.5 *M* glycerol, or one of a variety of other inducers [e.g. dimethyl sulfoxide (Bacon and Rosenberg, 1967) or other molecules with primary or secondary hydroxyl groups (Sadler and Dworkin, 1966)] to a suspension of vegetative cells. The morphological conversion proceeded rapidly (about 80–100 minutes), synchronously, and essentially quantitatively. This induction has been shown to be effective with other species of *Myxococcus* as well as with the genus *Stigmatella* (Reichenbach and Dworkin, 1970). In the latter case, unlike *Myxococcus*, monovalent cations will also induce myxospore conversion, suggesting a different mechanism of induction.

Rosenberg's group has demonstrated that in one strain of *M. xanthus* (FBmp) starvation for methionine induced myxospore formation and that this process was inhibited by spermidine, a metabolic product of methionine (Witkin and Rosenberg, 1970). They suggested that spermidine was an endogenous inhibitor of myxospore formation and that its levels were reduced by methionine starvation. [It is interesting, by the way, that the protein associated with the myxospore coat is completely devoid of methionine (Kottel *et al.*, 1975)]. Recently, it has been demonstrated (Filer *et al.*, 1973) that aspartokinase activity in *M. xanthus* is stimulated *in vitro* by spermidine and methionine. The authors suggested that reduced levels of methionine and spermidine decrease the activity of aspartokinase and that this may be an additional link in the regulatory process.

There are no convincing hypotheses that explain the mechanism of myxospore induction. Furthermore, it is not clear how the aspartokinase data relate to the mechanism of glycerol induction, since it has been demonstrated that the inducer need not be metabolized in order to be effective (Sadler and Dworkin, 1966). Since it is possible to obtain mutants that cannot be induced with glycerol but that can form myxospores normally in fruiting bodies (Burchard and Parish, 1975), it is likely that the glycerol technique represents an override mechanism that bypasses the normal regulatory control of myxospore formation.

E. Macromolecular Synthesis during Myxospore Formation

Recent experiments on the relation between myxospore induction and DNA synthesis indicated that during induction there is about a 40% net increase in DNA per cell. This is consistent with the value calculated for chromosome completion in a culture with an average of 1.4 chromosomes per cell (Kimchi and Rosenberg, 1976). Thus, the mature myxospore contains two completed chromosomes per cell. In addition, these experiments have suggested that for vegetative cell division to occur there must not only be chromosome completion but that replication at or near the origin of the chromosome triggers the formation of a protein required for subsequent division. The data also indicated that during myxospore formation this protein is destroyed and must be reestablished during germination for subsequent division to occur. There is thus an interesting analogy with the suggestion (Dunn *et al.*, 1976) that signals during DNA replication in *Bacillus* control the subsequent formation of vegetative and sporulation septa.

Upon myxospore induction, net RNA synthesis stops almost immediately (Bacon and Rosenberg, 1967). Turnover of all classes of RNA continues, however, throughout myxospore formation, and RNA–DNA hybridization competition studies have shown that a small fraction of total mRNA synthesized during myxospore formation consists of new species of mRNA (Okano *et al.*, 1970). This suggests, of course, that transcriptional changes occur during the developmental process. Turnover of ribosomal RNA may be related to the alteration in the physical properties and protein composition of the 30 S subunit reported to occur during myxospore formation (Foster and Parish, 1973). This would then suggest that there is also a translational control of protein synthesis during myxospore formation.

During the first 2 hours of myxospore induction there is a net increase of 35% in protein per cell (Sadler and Dworkin, 1966). Using incorporation of amino acids as a parameter of protein synthesis, however, the rate of incorporation continues unchanged for an additional 3 hours, suggesting that the later stages of induction include a period of substantial protein turnover. This is consistent with measurements of protein turnover (Orlowski and White, 1974b) and changes in protease activity (Orlowski and White, 1974a) during this period.

F. Myxospore Germination

When myxospores are placed in growth medium they germinate and outgrow as indicated in Fig. 24. At about 40 minutes the cells begin to lose phase refractility; this process is essentially complete by 60 minutes. Under appropriate circumstances the cells then begin outgrowth which is complete by about 120 minutes. During outgrowth the cells elongate and emerge from the ruptured spore coat which may be left behind or adhere to the cell (Fig. 25). The physiological optima and nutritional requirements for germination have been determined (Ramsey and Dworkin, 1968). The cells show no heat activation, and germination can be induced by complete growth medium or by a number of individual amino acids. Glycerol-induced myxospores (but not fruiting body myxospores) showed germination and outgrowth in dis-

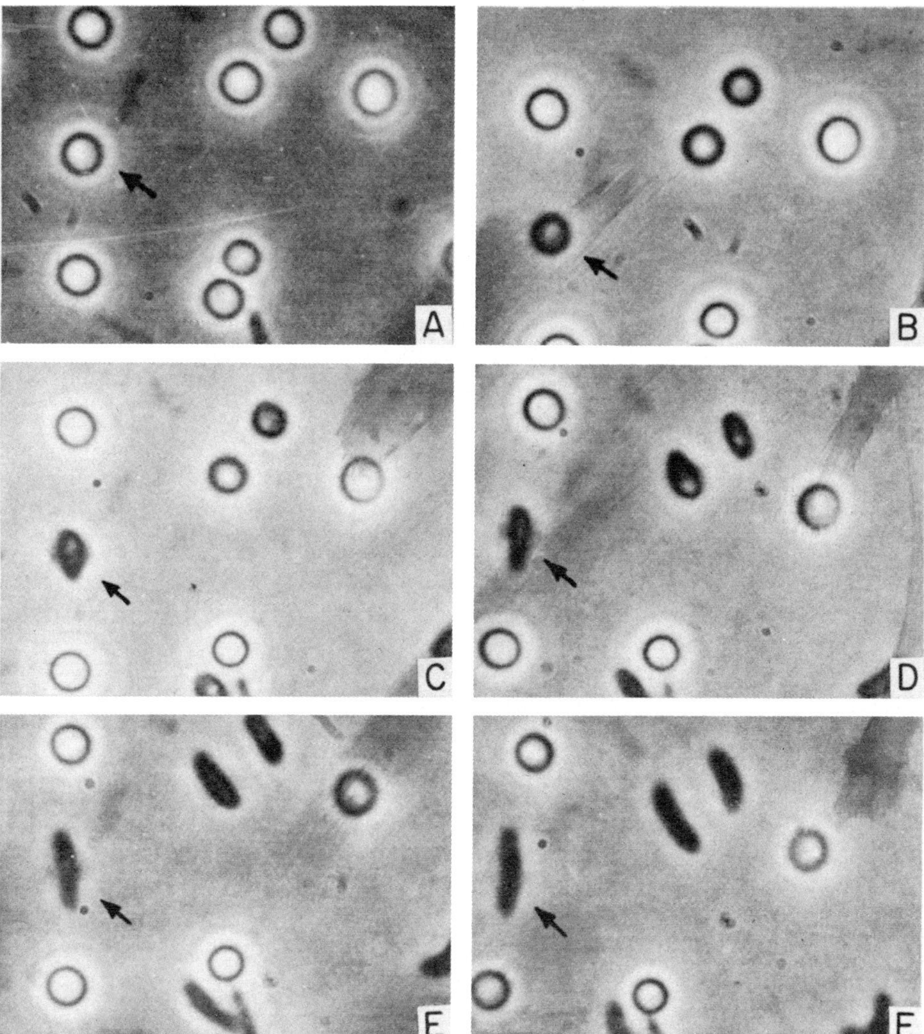

FIG. 24. Phase contrast micrograph of germinating myxospores of *Myxococcus xanthus*. (From Dworkin and Voelz, 1962.)

tilled water but only if the cell concentration was greater than about 10^9/ml. The mechanism whereby the spores determined their own cell density was by the excretion of orthophosphate, the external concentration of which served as a parameter of cell density (Ramsey and Dworkin, 1968).

During myxospore germination, RNA and protein synthesis take place (Ramsey and Dworkin, 1970) and seem to be necessary for germination

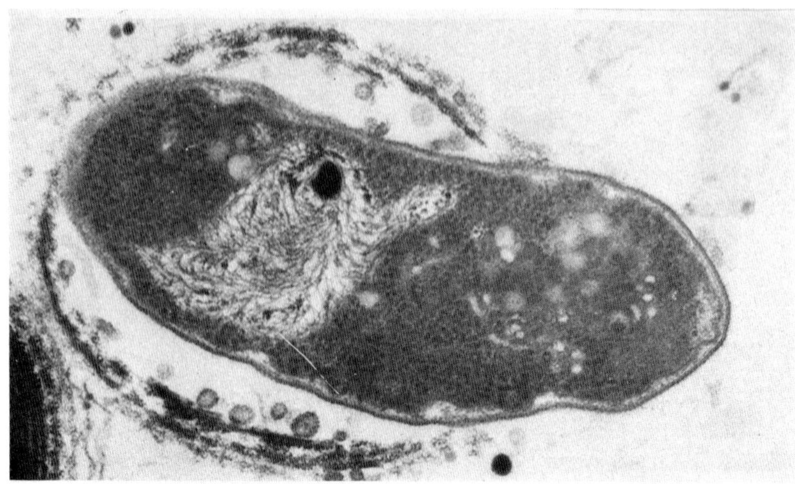

FIG. 25. Beginning of outgrowth of a myxospore of *Myxococcus xanthus*. ×80,000. (From Voelz, 1966.)

(Juengst and Dworkin, 1973). Even though the myxospore contains complete chromosomes (Rosenberg *et al.*, 1967), DNA synthesis of the first 5% of the chromosome (followed by transcription and translation of this portion of the genome) seems to be required for subsequent division of the germinated cells (Kimchi and Rosenberg, 1976). Other than this, there is no requirement for net synthesis of DNA.

While most of the work on myxospore morphogenesis has been done with glycerol-induced myxospores, the myxobacteria afford the rather unique opportunity to study myxospore morphogenesis under conditions that closely approximate the natural ones, i.e., as a part of fruiting body formation. When this has been done, certain important differences between glycerol-induced and fruiting body spores have emerged. The fine structures of the spores while similar are not identical. The rate of endogenous metabolism of glycerol-induced spores is measurably greater than that of spores from fruiting bodies (Bacon *et al.*, 1975). Finally the ability to obtain non-glycerol-inducible mutants that can nevertheless form spores in fruiting bodies indicates at least a separate induction mechanism. While the utility of the glycerol induction system for investigating morphogenesis under controlled and reproducible conditions is considerable, it must be kept in mind when comparing spores formed by the two routes that the differences in the process as well as the final development product may be significant.

G. Genetics

While there is not yet available any well-developed system for classical genetic analysis in *Myxococcus*, there are a number of interesting possibilities: (a) Campos *et al.* (1977) and Martin *et al.* (1978) have isolated a number of transducing phages for *M. xanthus* including at least one high frequency generalized transducer, (b) Kaiser and Dworkin (1975) have described the transfer of drug resistance to *M. xanthus* from the enteric phage P1, (c) Parish (1975) has reported the intergeneric transfer of drug resistance carried by enteric R factors, and (d) the intergenetic chromosomal mobilization by these R factors.

Two classes of sporulation mutants have been obtained. There are cells that are resistant to glycerol induction but will sporulate within fruiting bodies (Burchard and Parish, 1975), and cells that are unable to sporulate either in response to glycerol or as a part of the normal fruiting process (Dworkin and Sadler, 1966). In addition, a variety of fruiting body mutants (Hagen *et al.*, 1978) and motility mutants (Burchard, 1970; Hodgkin and Kaiser, 1977) have been isolated. Other mutants to auxotrophy, phage resistance, antibiotic resistance, etc., may be routinely obtained.

H. Bacteriophage

Burchard and Dworkin (1966) described a virulent DNA phage (MX-1) for *M. xanthus* that related in an interesting way to myxospore development. If cells were infected simultaneously with the initiation of myxospore morphogenesis, the phage infection preempted the myxospore induction, and a cycle of virulent phage development occurred. During normal development, however, the myxospore became resistant to phage infection at about 60–75 minutes [the time during which the cells converted from ovoids to spheres (see Fig. 21)]. While the evidence is not conclusive, the loss of sensitivity to the phage is probably due to alterations of the cell surface that preclude phage adsorption. When phage were added to developing cells that had begun the morphological change, but had not yet lost their receptors (between about 20–50 minutes), the phage were adsorbed and the cells became infected. Nevertheless, morphogenesis was completed, resulting in an apparently normal myxospore. Upon germination, however, some of the outgrown, latently infected cells completed the cycle of phage infection, resulting in lysis and release of phage (Burchard and Voelz, 1972). The situation is somewhat analogous to that subsequently described in *B. subtilis*

(Sonenshein and Roscoe, 1969). It is not known whether the control of phage replication that is interrupted in *M. xanthus* is at the level of the RNA polymerase as is the case of *B. subtilis* (Losick and Sonenshein, 1969).

As indicated earlier, Campos *et al.* (1977) and Martin *et al.* (1978) have isolated and described a number of additional phages for *M. xanthus*, including at least one high frequency generalized transducing phage.

The myxobacteria offer a number of unique advantages for examining prokaryotic development. The glycerol induction system is probably the most convenient, rapid, and reproducible system for inducing spore formation and as such has been a useful experimental system. Furthermore, spore formation is only part of a complex life cycle involving developmental chemotaxis, aggregation, fruiting body formation, and a variety of intriguing cell interactions.

V. *Azotobacter* Cysts

Azotobacter is a genus of large, ovoid, gram-negative bacteria that are distinguished by their ability to fix nitrogen and to form resistant resting cells called cysts.

A. Morphology and Life Cycle

Figure 26 is a phase-contrast micrograph of an intact cyst. An electron micrograph of a thin section of a cyst is shown in Fig. 27. The distinctive morphological features of the cyst are the exine, intine, and prominent poly-β-hydroxybutyrate granules.

If cells are grown under nutritional conditions that lead to a maximal growth rate (e.g., glucose or other carbohydrates as a source of carbon and energy) cysts appear at the end of the growth cycle, and the yield of cysts is poor (less than 0.1%) (Lin and Sadoff, 1968). Winogradsky (1938) noted that if the cells were grown on a carbon source such as *n*-butanol or ethanol, growth was poor but the cells converted completely to cysts in about 5 days. During an attempt to identify a catabolite of *n*-butanol that was responsible for inducing encystment, Lin and Sadoff (1968) discovered that replacing glucose-grown cells in a medium containing β-hydroxybutyrate or crotonate as carbon sources induced encystment. The induction was repressed by glucose and by ammonia, and, unlike glycerol-induction of myxospores, required the metabolism of the inducer. In addition, the inducer may have a direct effect on the cells, in that replacement into the β-hydroxybutyrate medium resulted in the immediate decrease in the rate of nitrogen fixation (Hitchens and Sadoff, 1973). The mechanism of induction is unknown.

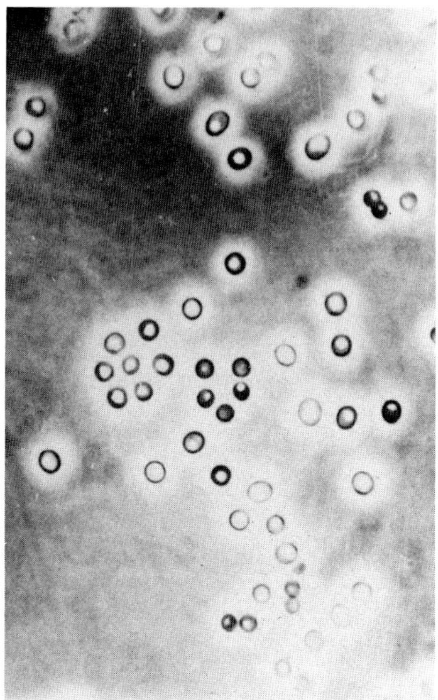

FIG. 26. Phase-contrast photomicrograph of cysts of *Azotobacter vinelandii.* Courtesy of Dr. H. L. Sadoff.

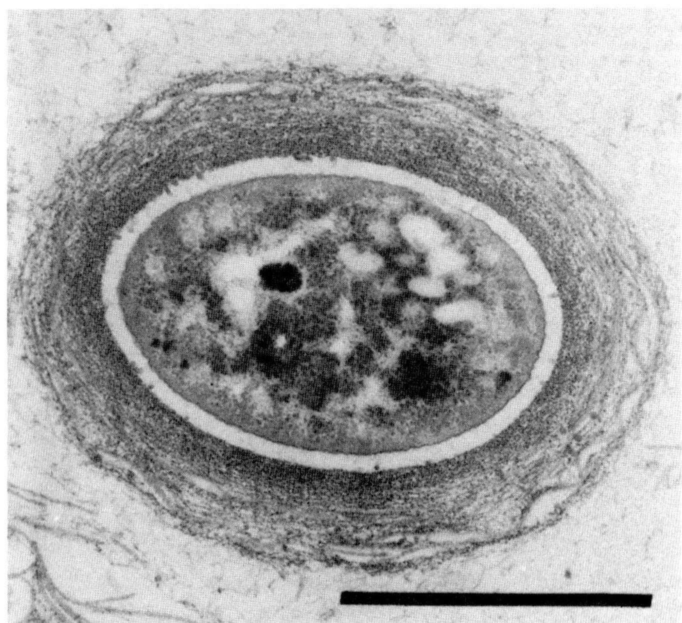

FIG. 27. Thin section of a mature cyst of *Azotobacter vinelandii.* The multilayered exine surrounds the cell. Scale, 1.0 μm. (From Hitchins and Sadoff, 1970.)

Figure 28 illustrates the sequence of morphological changes that occur during cyst formation and germination. The cells lose their flagella, undergo a final division resulting in two nonmotile, heavily encapsulated cells referred to as precysts. These assume a spherical shape and accumulate heavy deposits of granules of poly-β-hydroxybutyric acid. By about 12 hours numerous membranous blebs are formed on the surface of the central cell, protruding into the electron-transparent space between the central cell and its surrounding capsule. Over a period of about 18 hours these blebs continuously break free, coalesce to form the fragmented layers that comprise the exine and give it its characteristic shingled appearance. At about 36 hours the exine is present as a distinct layer and there is a reduction in the number of poly-β-hydroxybutyric acid granules. By 4 to 5 days the electron-transparent, structureless intine is present and the typical triple-layered wall and membranes have been formed.

It is possible by the use of appropriate chelating agents to fractionate the cyst into a viable central body, a solubilized intine and a disrupted exine (Parker and Socolofsky, 1966). Such central bodies can be germinated, but until germination show the same metabolic dormancy as the cyst (Loperfido and Sadoff, 1973). They have, however, lost their resistance to uv light irradiation and sonic oscillation. The solubilized intine consists mainly of lipid and carbohydrate, 72% of which consists of uronic acids (Lin and

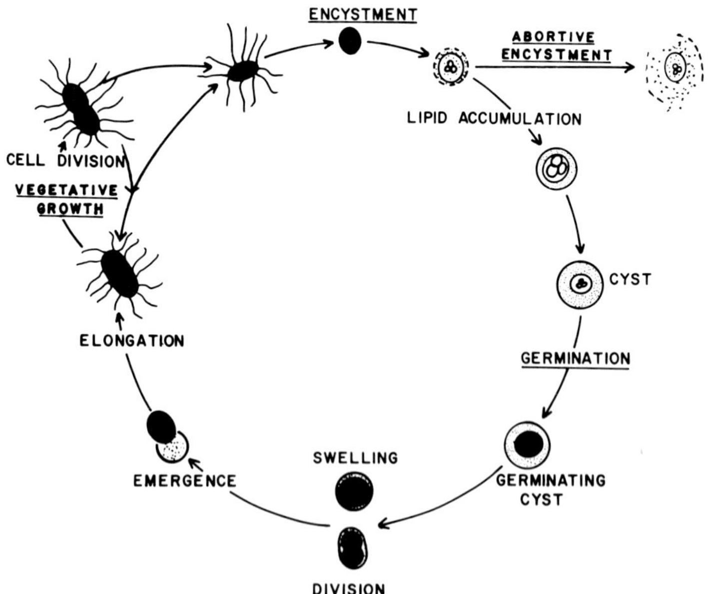

FIG. 28. Schematic diagram of the life cycle of *Azotobacter vinelandii*. (From Sadoff, 1975.)

Sadoff, 1969). The exine contains approximately equal amounts (on a dry weight basis) of lipid, protein, and carbohydrate. Forty percent of the exine carbohydrate is uronic acids, mainly polyguluronic acid. Whereas the exine polysaccharide is peculiar to that structure, the intine polysaccharides are similar to those found in the vegetative cell capsule. The physical and chemical properties of the exine and intine polysaccharides have been examined (Sadoff, 1975), and the rigid nature of the exine structure, in contrast to the more flexible quality of the inner intine, is consistent with the physical properties of the respective polymers. The cysts contain twice as much lipid as do vegetative cells and a far greater variety of fatty acids. Likewise they contain a substantially higher level of Ca^{2+}, most of which is localized in the central body. Much of the Ca^{2+} is also associated with the uronic acids of the exine (Sadoff, 1975). The ability to separate the various structures of the cyst while retaining central body viability has offered the opportunity to examine structure–function relationships in *Azotobacter* cysts to a far greater extent than is the case with any of the other prokaryotic resting cells. This approach has in fact indicated that cysts stripped of their exine layer, while still viable and dormant, have lost the characteristic cyst resistance to desiccation, radiation, and sonic oscillation (Parker and Socolofsky, 1966).

Cyst induction is accompanied by distinct shifts in the metabolic pathways of the organism. Vegetative cells growing in glucose primarily utilize the Entner–Doudoroff pathway for glucose breakdown. During encystment the cells shift to a metabolism geared to the oxidation of acetate and to gluconeogenesis. Glucose-6-phosphate dehydrogenase disappears as does nitrogen fixation. Conversely, glyoxylate shunt and gluconeogenic enzyme (fructose-1,6-diphosphate aldolase and fructose-1,6-diphosphatase) activities appear during encystment (Hitchins and Sadoff, 1973). It should be emphasized that none of these changes are unique to cyst formation [e.g., aldolase can be induced in cells growing in glucose by increasing aeration of the culture (Nagai *et al.*, 1971)]. As a result of the loss of N-fixing ability, subsequent biosynthesis of N-containing spore constituents requires that the developing cells utilize existing nitrogenous pools or protein turnover (Sadoff, 1975).

As in the case of myxospores, encystment involves the completion of chromosome replication but no initiation of new replication cycles. Again, similar to the general situation during myxospore induction, net RNA synthesis decreases and then stops, while RNA turnover of an undefined nature probably continues. Protein synthesis continues at a high rate during the early stages of encystment and then continues at a reduced rate until cyst formation is complete (Hitchins and Sadoff, 1973). There is no information available concerning possible regulatory mechanisms of macromolecular synthesis during cyst formation. Similarly no information is yet available defining the regulatory nature of the metabolic shifts during cyst formation.

B. Germination

Germination, here defined as the entire conversion of a cyst to a vegetative cell, occurs when cysts are placed under conditions suitable for aerobic growth. Germination does not require activation, occurs over an 8-hour period and can be divided into initiation and outgrowth. Cells lose their optical refractility; the central body swells, occupies the intine volume, and finally bursts through the exine, releasing a nonmotile dividing cell. The cells regain motility after the first postgermination division.

Germination can be blocked by inhibition of protein, RNA, or ATP synthesis. Upon placing the cysts in the presence of glucose, respiration and CO_2 production begin immediately, followed shortly by RNA and protein synthesis. DNA synthesis and N_2 fixation do not occur until outgrowth begins, and are accompanied by increases in the rate of RNA and protein synthesis and respiration to values characteristic of vegetative cells (Loperfido and Sadoff, 1973).

C. Genetics

The isolation of auxotrophic or developmental mutants of *Azotobacter* has been relatively unsuccessful. *Azotobacter* contains 1.6×10^{-13} gm of DNA/cell, an amount sufficient for approximately 50 genomes/cell (Sadoff *et al.*, 1977). The ensuing problems of segregation have been suggested as the basis for the organism's resistance to mutation. A transformation system for genetic exchange has been demonstrated (Page and Sàdoff, 1976), but at the moment, the lack of developmental mutants limits the utility of such a system.

VI. Heterocysts and Akinetes of Blue-Green Bacteria

In addition to the classical developmental questions one may ask about cellular morphogenesis, the blue-green bacteria offer a unique opportunity to examine population differentiation, cell interactions, and pattern formation in a prokaryote. The blue-green bacteria contain an extraordinary variety of cellular types and life cycles, and, until recently, work on these organisms has been largely morphological and descriptive. Most blue-green bacteria have not been grown in pure culture, grow slowly, and are often difficult to cultivate. As a result, recent developmental studies have tended to be limited to the *Hormogonales*, a group of filamentous blue-green bacteria that produce two differentiated cell types, heterocysts and akinetes.

It should be pointed out, however, that despite these difficulties Stanier's recent studies on the nutrition and metabolism of the *Chroococcales* (Stanier *et al.*, 1971), Wolk's (1973) and Carr's work on physiology and metabolism of filamentous blue-greens (Carr and Bradley, 1973) as well as that of a number of other groups has now firmly established a modern analytical approach to studying the blue-green bacteria.

This section will deal with the properties and development of heterocysts and akinetes and with their various interactions with the vegetative cells.

A. Heterocysts

Heterocysts are produced by some members of the filamentous blue-green bacteria and are distinguished by the following morphological features (Fay, 1973).

1. They are somewhat larger and more rounded than the adjacent vegetative cells (Fig. 29 and Table II).
2. They are attached to adjacent vegetative cells by a narrow constricted connection (Fig. 30).

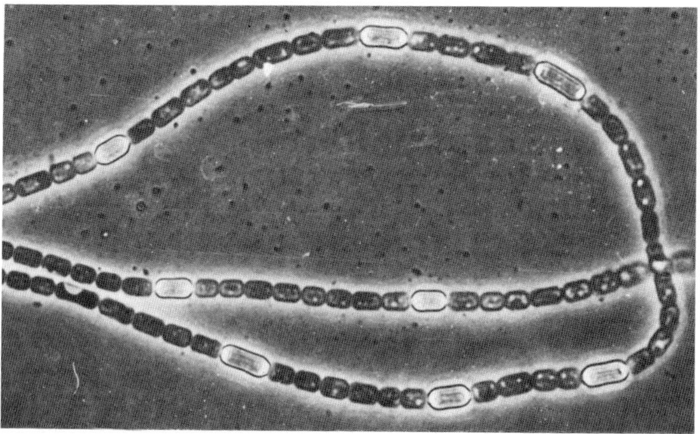

FIG. 29. Vegetative cells and heterocysts of *Anabaena cylindrica* showing the regular heterocyst spacing. ×600. (From Wilcox *et al.*, 1975.)

TABLE II

AVERAGE CELL SIZE OF THE VARIOUS CELLS OF *Anabaena cylindrica*[a]

	Cell size (μm)	Cell volume (μm^3)
Vegetative cells	3.2 × 4.8	38.6
Heterocysts	4.2 × 6.0	83.0
Spores	6.3 × 16.4	527.0

[a] From Fay (1969).

3. They are surrounded by a conspicuous thickened coat or capsule.
4. They lack the granular cytoplasmic inclusions characteristic of the vegetative cells.
5. They contain at the polar ends strongly refractile structures called "polar nodules."
6. The absence of phycocyanin results in a yellow-green appearance rather than the typical blue-green color.

Physiologically, heterocysts of *Anabaena* are distinguished from the vegetative cells by the absence of photosystem II (Donze *et al.*, 1972) and of phycocyanin (Thomas, 1970), increased respiratory activity (Fay and Walsby, 1966), an altered lipid content (Walsby and Nichols, 1969), increased capability for N_2 fixation (see Stewart, 1973) and increased levels of glyoxylate cycle enzymes (Carr and Bradley, 1973).

Figure 31 is an electron micrograph of a thin section of a heterocyst of *Anabaena*. This is diagrammatically represented in Fig. 32. The heterocyst is surrounded by a thick complex of layers including an external fibrous region, a thick homogeneous layer, and a laminated layer immediately external to the cell wall proper. The two most external layers are apparently composed largely of carbohydrates, comprising 73% glucose, 21% mannose, 4% xylose, and 3% galactose (Dunn and Wolk, 1970). The overall carbohydrate composition of the entire envelope was 62%. Most of the lipid is contained in the laminar layer and consists entirely of four compounds unique to the heterocysts, 25-hydroxyhexacosanoic acid (1α-D-glycopyranose) ester, 1-(*O*-α-D-glycopyranosyl)-3,25-hexacosanediol, 1-(*O*-α-D-glycopyranosyl)-3,25,27-octacosanetriol, and 25,27-dihydroxyoctacosanoic and (1α-D-glucopyranose) ester (Lambein and Wolk, 1973). In these lipids, glucose comprises 90% of the carbohydrate and galactose the remainder. Wolk (1973) has suggested that this hydrophobic layer constitutes a barrier to the escape of products of nitrogen fixation which can then be transmitted to the adjacent vegetative cells only via the polar regions of the heterocyst. The heterocysts

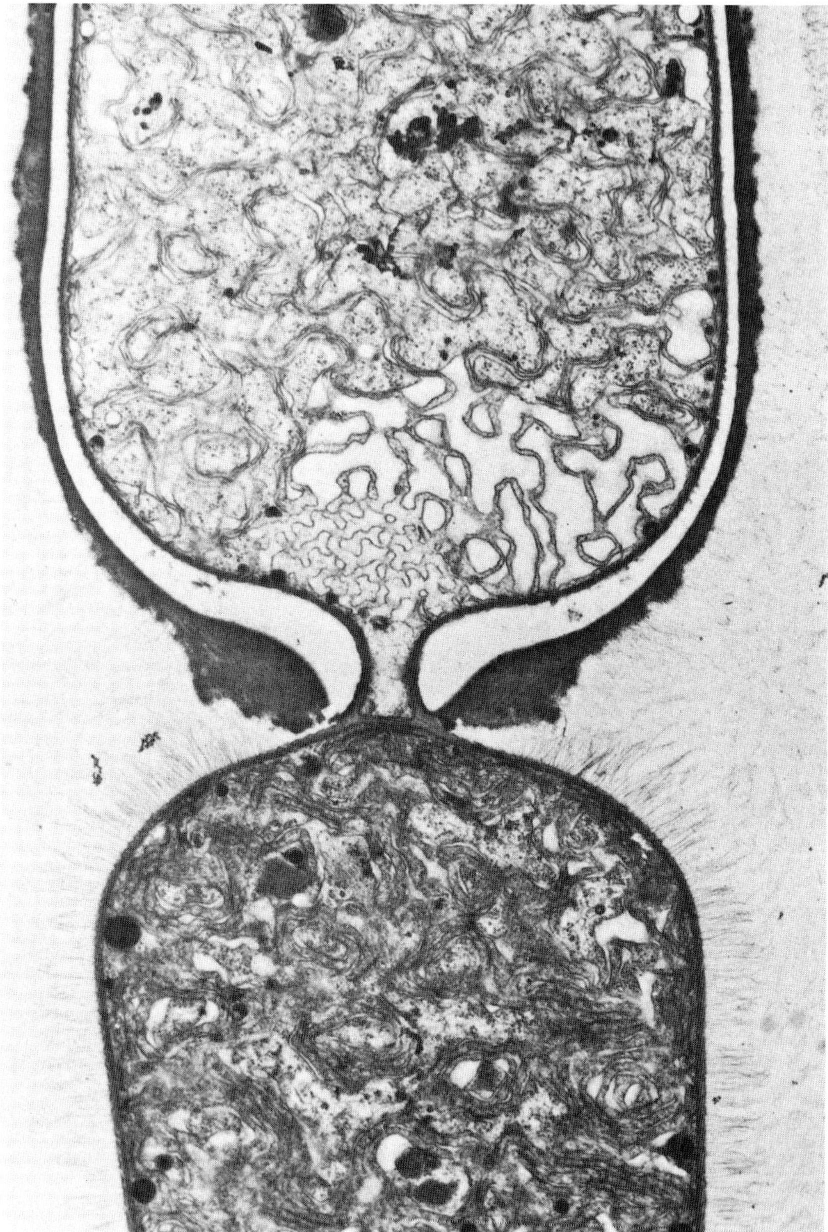

FIG. 30. Vegetative cell (bottom) and heterocyst (top) of *Anabaena cylindrica*. ×27,000. (From Lang, 1968.)

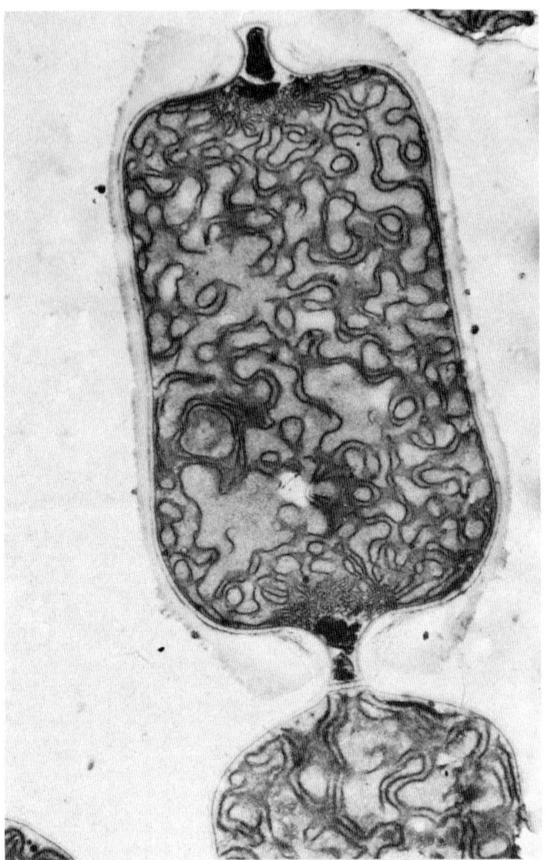

FIG. 31. Electron micrograph of a heterocyst of *Anabaena cylindrica*. × 14,000. (From Fay and Lang, 1971.)

contain somewhat reduced amounts of chlorophyll, carotenoids, and phycocyanin (Fay, 1969) (Table III). Other reports, however, indicate that phycocyanin may be present under certain conditions in the reduced bleached state (see Wolk, 1973). While photosystem II, responsible for the photolytic generation of reductant from water, is absent, the heterocyst is capable of photophosphorylation via photosystem I (Donze *et al.*, 1972). Since nitrogen fixation requires considerable amounts of reducing power, it is not clear how, in the absence of photosystem II, the heterocyst is able to generate this reductant. The heterocysts are capable of respiration (Fay and Walsby, 1966), but the absence of ribulosediphosphate carboxylase (Winkenbach and Wolk, 1973) results in their inability to fix CO_2.

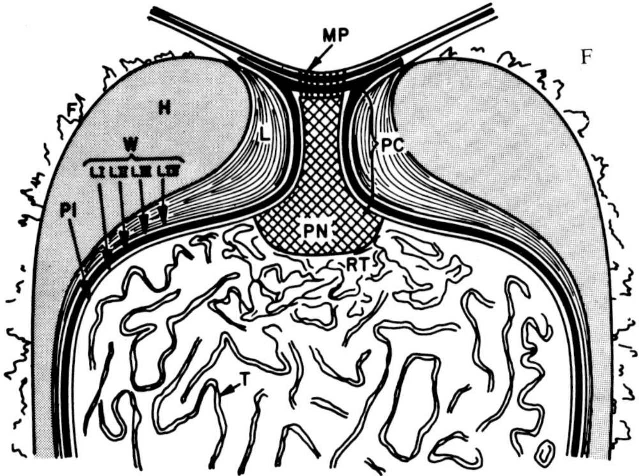

FIG. 32. Diagram of part of a heterocyst showing the envelope consisting of a fibrous layer (F), a homogeneous layer (H), and a laminated layer (L); the subjacent wall (W) with its four layers (LI–LIV) the plasmalemma (Pl); microplasmodesmata (MP); a polar nodule (PN) in the pore channel (PC); and thylakoids appearing relatively "normal" near the center of the cell (T) and reticulated (RT) near the ends of the cell. The end wall (top) of the adjacent vegetative cell is also shown. The thickness of the wall layers LI–LIV are exaggerated for clarity. (From Wolk, 1973.)

TABLE III

AVERAGE PHOTOSYNTHETIC PIGMENT CONTENT OF THE VARIOUS CELL TYPES OF *Anabaena cylindrica*[a,b]

	Chlorophyll		Carotenoid		Phycocyanin	
	Cells (μg/10^7)	Dry wt (%)	Cells (μg/10^7)	Dry wt (%)	Cells (μg/10^7)	Dry wt (%)
Filaments	7.72	1.79	0.75	0.17	49.8	11.53
Heterocysts	2.20	0.92	0.14	0.06	7.2	3.12
		(49)		(65)		(73)
Spores	6.20	0.47	0.46	0.03	18.5	1.39
		(74)		(82)		(88)

[a] From Fay (1969).
[b] Data in parentheses are the percent decreases with reference to the filaments.

A considerable amount of evidence has accumulated relating heterocysts and nitrogen fixation (Fay *et al.*, 1968). Evidence supporting this idea is (1) the inhibition of nitrogen fixation and heterocyst formation by the products of nitrogen fixation, e.g., NH_3 (Fogg, 1949); (2) the ability to demonstrate nitrogen fixation in pure suspensions of heterocysts (Stewart *et al.*, 1969); and (3) the observation that the appearance of nitrogenase in *Anabaena* is preceded by the formation of heterocysts (Carr and Bradley, 1973). This does not preclude the possibility that some level of nitrogen fixation occurs in vegetative cells; in fact, such is indeed the case (see Wolk, 1973). It does seem, however, that in those filamentous blue-green bacteria that do form heterocysts, the heterocyst is the primary and perhaps under some conditions, the sole site of nitrogen fixation. If so, this is a unique example of a functional, organized population differentiation in prokaryotes. The idea is consistent with the observed sensitivity of nitrogenase to oxygen. Unlike the vegetative cells that are producing oxygen as a by-product of the photolysis of water, the heterocyst is essentially an anaerobic cell; both the absence of an oxygen generating photosystem and the oxygen-scavenging effect of respiration would tend to maintain this condition and thus contribute to the optimal conditions for nitrogen fixation.

Heterocyst formation is induced by the absence of a readily assimilable source of fixed nitrogen and stimulated by the presence of organic carbon (Fogg, 1949). While heterocyst frequency within the filament varies greatly with different species and under different environmental conditions (see Carr and Bradley, 1973), there is a striking regularity of the spacing of the heterocysts along the filament (Fig. 29). This had led to the suggestion that the linear diffusion of an inhibitor or stimulator of heterocyst formation occurs along the filament, representing a cellular interaction between vegetative cells and heterocysts. Fogg (1949) originally suggested that gradients of NH_4^+ arising as a result of nitrogen fixation in heterocysts could control heterocyst spacing. Recent evidence has discounted this possibility but has remained focused on the idea of an inhibitor produced by and diffusing from the heterocyst. Wolk (1975) has presented evidence that supports the notion of heterocysts arising randomly along the filament and producing a diffusible inhibitor that is destroyed at a rate proportional to its concentration. Wilcox *et al.* (1975) have obtained data that, while supporting the idea of a diffusible inhibitor produced by heterocysts, suggests in addition a predetermination of presumptive heterocysts. He has observed that growing cells of *Anabaena* show a recursive asymmetric division illustrated in Fig. 33. The division of each cell gives rise to a larger and a smaller daughter cell; if the parent cell was itself the left (or right) member of a previous division its left (or right) daughter cell will be the smaller of the pair. Further, heterocysts arise subsequently only from the smaller daughter cells. Wilcox, thus, suggests that the

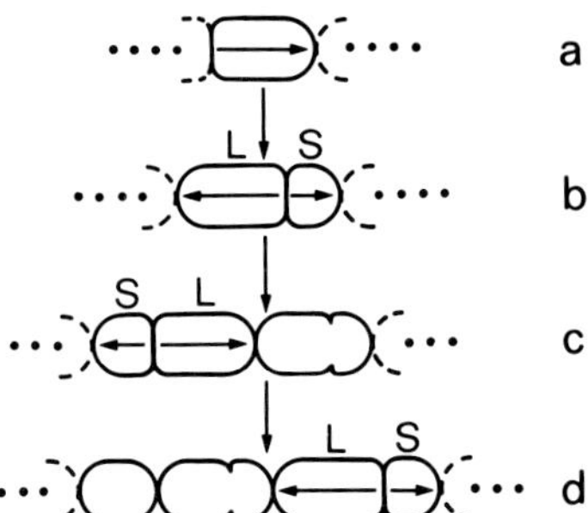

FIG. 33. Division rule. (a) A cell is represented with an arrow pointing away for the newly formed septum. The rest of the filament is indicated by dotted outlines. (b) The two daughter cells are shown. Smaller daughters take longer than larger daughters (by some 40%) to reach a subsequent division; this introduces an asychrony which is indicated here. Thus, the division of the first L cell is complete in (c) but that of the first S cell only in (d). (From Wilcox *et al.*, 1975.)

combination of the inhibitory zone–threshold mechanism and the competitive interaction between larger and smaller daughter cells combine to provide a finely tuned spacing mechanism. The regulation of the asymmetric division, the mechanism of asymmetric heterocyst predetermination, the nature of the diffusible inhibitor, and the mechanism of inhibition remain undetermined and represent exciting problems in heterocyst development.

B. AKINETES

Under certain cultural conditions vegetative cells convert to akinetes. In comparison to the vegetative cells the akinetes have an altered form; are somewhat more resistant to desiccation, heat (see Fritsch, 1945), and physical breakage (Fay, 1970) are metabolically altered and are capable of germination. While their function has not been clearly defined, they are in many ways similar to the spores formed by other prokaryotes.

Figure 34 illustrates the appearance and location of akinetes along a filament of vegetative cells. Notice that the akinetes characteristically flank a heterocyst. Figure 35 illustrates the ultrastructural details of an akinete. The cell is characteristically larger than the vegetative cells and contains a thick outer coat and large numbers of cyanophycin granules. The chemical composition of the coat is indicated in Table IV and, like that of the heterocyst, consists largely of polysaccharide (Dunn and Wolk, 1970). An examination of the pigment content of akinetes of *Anabaena* (Fay, 1969) revealed that on a per cell basis, chlorophyll, carotenoids, and phycocyanin were present in reduced amounts. Since the akinete contains over ten times the volume of

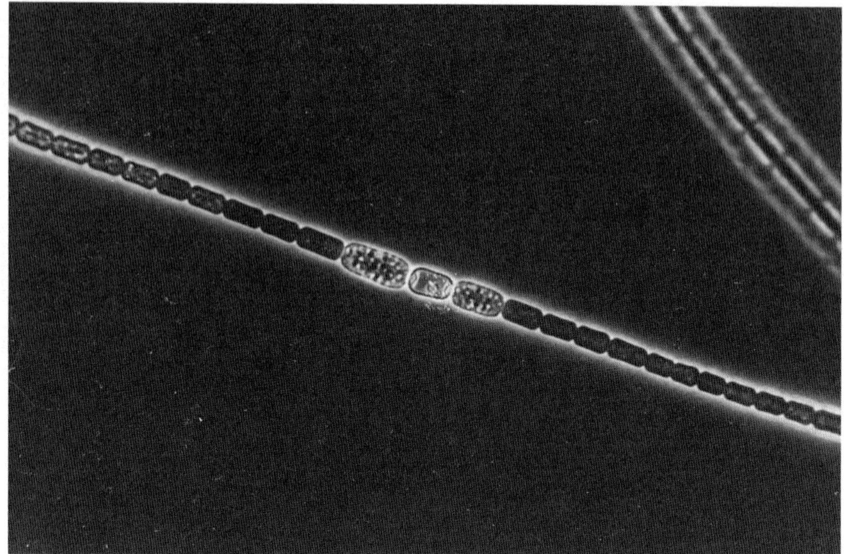

FIG. 34. Chain of vegetative cells of *Anabaena cylindrica* containing a heterocyst flanked by a pair of akinetes. ×750. Courtesy of Dr. N. G. Carr.

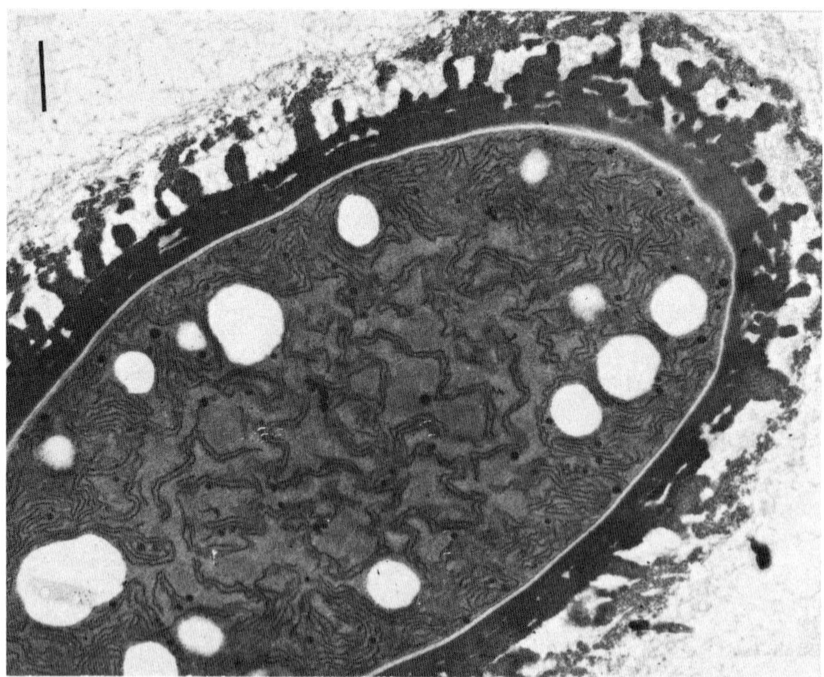

FIG. 35. Akinete of *Cylindrospermum*. Note the dense fibrillar layer and the large cyanophycin granules. (From Miller and Lang, 1968.)

TABLE IV

COMPOSITION OF THE WALLS OF VEGETATIVE CELLS, HETEROCYSTS, AND AKINETES OF *Anabaena cylindrica*[a]

Cell type	Vegetative	Heterocysts	Akinete
Total analysis			
Carbohydrate	18	62	41
Lipid	3	15	11
Amino compounds	65	4	24
Polysaccharide constituents (%)			
Glucose	35	73	76
Mannose	50	21	17
Galactose	5	3	3
Xylose	8	4	4
Fucose	2	0	0

[a] From Dunn and Wolk (1970).

the vegetative cell (Fay, 1969), the pigment content as a percentage of dry weight is reduced even further (Table III). Isolated akinetes of *Anabaena* were able to fix CO_2 in the light, albeit at a reduced rate, were able to produce respiratory CO_2 in the dark at a rate greater than that of vegetative cells, and were unable to fix N_2 (Fay, 1969). Akinetes contain larger amounts of cyanophycin, a copolymer of arginine and aspartic acid, with a molecular weight of 30,000 daltons (Simon, 1971). These are also present in vegetative cells (Miller and Lang, 1968) but not in fully developed heterocysts (see Fay, 1973). These granules are absent from freshly germinated akinetes (Miller and Lang, 1968) and may thus serve as a reserve of carbon, nitrogen, and energy for the akinete.

Unlike the heterocysts, which are formed during the period of active vegetative growth, akinete formation seems to occur after the period of active growth. Wolk has studied the optimal conditions for akinete formation in *Anabaena*, and these conditions include the absence of phosphate, the presence of nontoxic buffer (e.g., alanylglycine), the presence of Ca^{2+} and acetate, optimal light intensity, and a heavy inoculum (Wolk, 1965). In addition, the observation has been frequently made that akinete formation generally occurs adjacent to the heterocyst (Fig. 34). Wolk (1973) has shown that this association is not random and has suggested that the heterocysts may be creating zones of influence for akinete formation as they do to regulate heterocyst spacing.

Akinete germination can be induced by transferring mature akinetes to fresh culture medium. In *Cylindrospermum*, the most common mode of germination involved the emergence of the germ cell from a pore dissolved at one end of the cell. This was preceded by one or two cell divisions (Miller and Lang, 1968).

While cellular morphogenesis in the blue-green bacteria has not yet been examined as extensively as in other prokaryotes, an exciting start has been made on the problems of cell interactions and pattern formation. In this sense they represent an excellent system for examining the nature, role, and regulation of diffusible morphogenetic gradients. It is quite clear, however, that any substantial developmental insights must be preceded by an accumulation of fundamental information describing the organism itself. It is encouraging that such information is now forthcoming.

VII. *Caulobacter*

Most of the developmental events discussed thus far occur as an alternative to vegetative growth. Even though growth and development, as in the case of endospore formation in *Bacillus*, may not be sharply demarcated, the two processes occur essentially independently of each other. In *Caulobacter* this is not the case; growth and development occur concomitantly and are complexly related. Thus, *Caulobacter* development is rather unusual and the necessity of weaving together cell division, DNA replication, and development may require regulatory mechanisms distinct from those involved in cyst or spore formation.

A. Life Cycle

Figure 36 illustrates the life cycle of *Caulobacter* and the relative timing of the various growth and development events. In general terms, the flagellated swarmer cell loses its flagellum and synthesizes in its place the characteristic *Caulobacter* stalk. During the period of stalk elongation an asymmetric cell division occurs resulting in a stalked cell and a smaller swarmer cell. While each swarmer cell that is formed goes through the same cycle, the stalked cell repeatedly divides giving rise to new swarmer cells, in effect, acting as a primitive stem cell. There are a number of interesting developmental features of the cycle that focus about spatial and temporal asymmetries. The cell division is asymmetric in terms both of the size of the daughter cells (Terrana and Newton, 1975) as well as the nature of their polar structure. This asymmetry is expressed prior to the actual cell division and consists of the simultaneous polar appearance of pili, flagella, and DNA phage receptors. DNA synthesis in the daughter cells is likewise

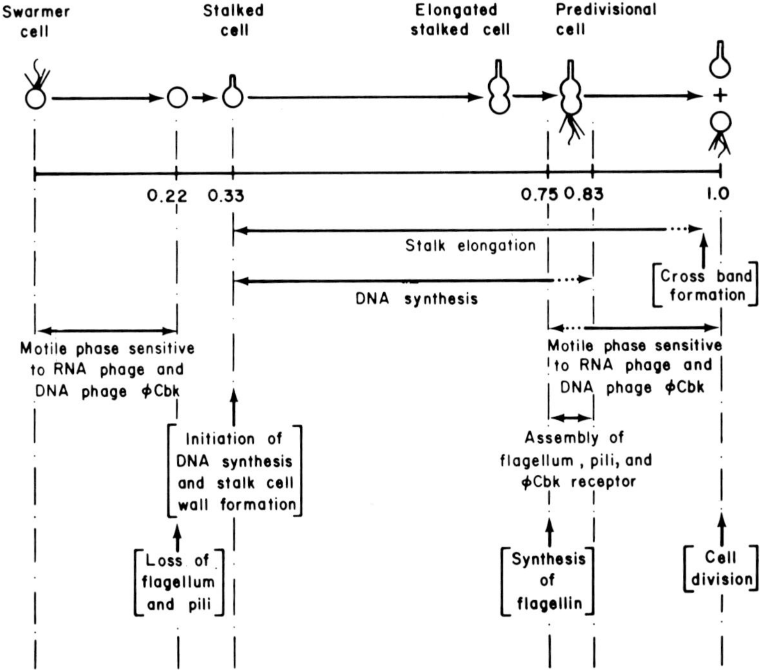

FIG. 36. Diagram of the life cycle events of *Caulobacter crescentus*. (From Shapiro, 1976.)

asymmetric; upon cell division or the completion of chromosome replication a regulatory event occurs that delays DNA replication in the progeny swarmer cell until it has formed a stalk. The progeny stalk cell, on the other hand, begins DNA replication immediately (Degnan and Newton, 1972). These asymmetric events represent a differentiation of the population into functionally distinct coexisting cell types. In this sense the situation is analogous to that occurring during heterocyst formation in the blue-green bacteria and it is interesting that in both cases cell differentiation is preceded by an asymmetric cell division.

DNA synthesis occupies the period between 0.33 and 0.83 division units (Degnan and Newton, 1972). This implies that the rate of DNA synthesis in *Caulobacter* is a function of the growth rate and thus may differ from the enteric bacteria where, within a broad range, it is independent of growth rate.

B. THE STALK

The most distinctive morphological feature of *Caulobacter* is its stalk. This polar appendage (that may be located laterally in another genus,

Asticacaulis) is relatively long and slender (about 0.1 × 3.0 μm) and is bounded by a continuation of the normal cell membrane and envelope (Fig. 37). The stalk contains a series of annular cross-bands (Fig. 37) comprised of concentric rings (Jones and Schmidt, 1973) and containing peptidoglycan (Schmidt, 1973). It has been suggested that these reflect the age of the stalked cells (Staley and Jordan, 1973). While little is known about stalk function, it has been suggested that the stalk serves either to increase the flotation properties of the cell (Poindexter, 1964) or to increase the surface area available for respiration and permeability (Pate and Ordal, 1965). Bearing in mind that the stalked cell acts as a stem cell that repeatedly produces swarmer cells, there have been no determinations of its longevity.

C. Regulation

Different species of proteins appear at defined stages of the life cycle of *Caulobacter*. Perhaps the best example is the appearance of flagellin monomers at 0.75 division units (Shapiro and Maizel, 1973). This is a fortunate occurrence, since it is thus possible to describe a morphogenetic event by the appearance of a well-defined product at a specific point in development. While it has been shown that RNA synthesis is required throughout the developmental cycle (Newton, 1972), it has not been possible to locate specific transcriptional controls. For example, an examination of purified RNA polymerase from stalked, swarmer, and predivisional cells has not revealed any differences (Bendis and Shapiro, 1973).

The addition of cyclic GMP to actively growing cultures caused the loss of flagella, pili, and DNA phage receptors but had no effect on the growth rate of the cells (Kurn and Shapiro, 1976). This reflected a more general block of development resulting in the accumulation of cells at the elongated stalk stage. The addition of dibutyryl cyclic AMP reversed the effect and released the block. The effect of cyclic GMP was shown to be the result of the inhibition of synthesis of specific gene products, i.e., flagellin and three outer membrane proteins (Kurn and Shapiro, 1976). The level at which these cyclic nucleotides exert their effect is not known.

An examination of *Caulobacter* DNA failed to reveal any plasmids (Wood *et al.*, 1976). However, reassociation kinetics of denatured and reannealed DNA indicated that 4% of the DNA was present as a rapidly renaturing component. On the basis of a number of criteria this was attributed to the presence of approximately 350 inverted repeat sequences per chromosome with an average size of 300 nucleotide pairs (Wood *et al.*, 1976). The recent work of Zeig *et al.* (1977) which showed that an invertible sequence regulates phase conversion between the two flagellar antigens of *Salmonella* suggests

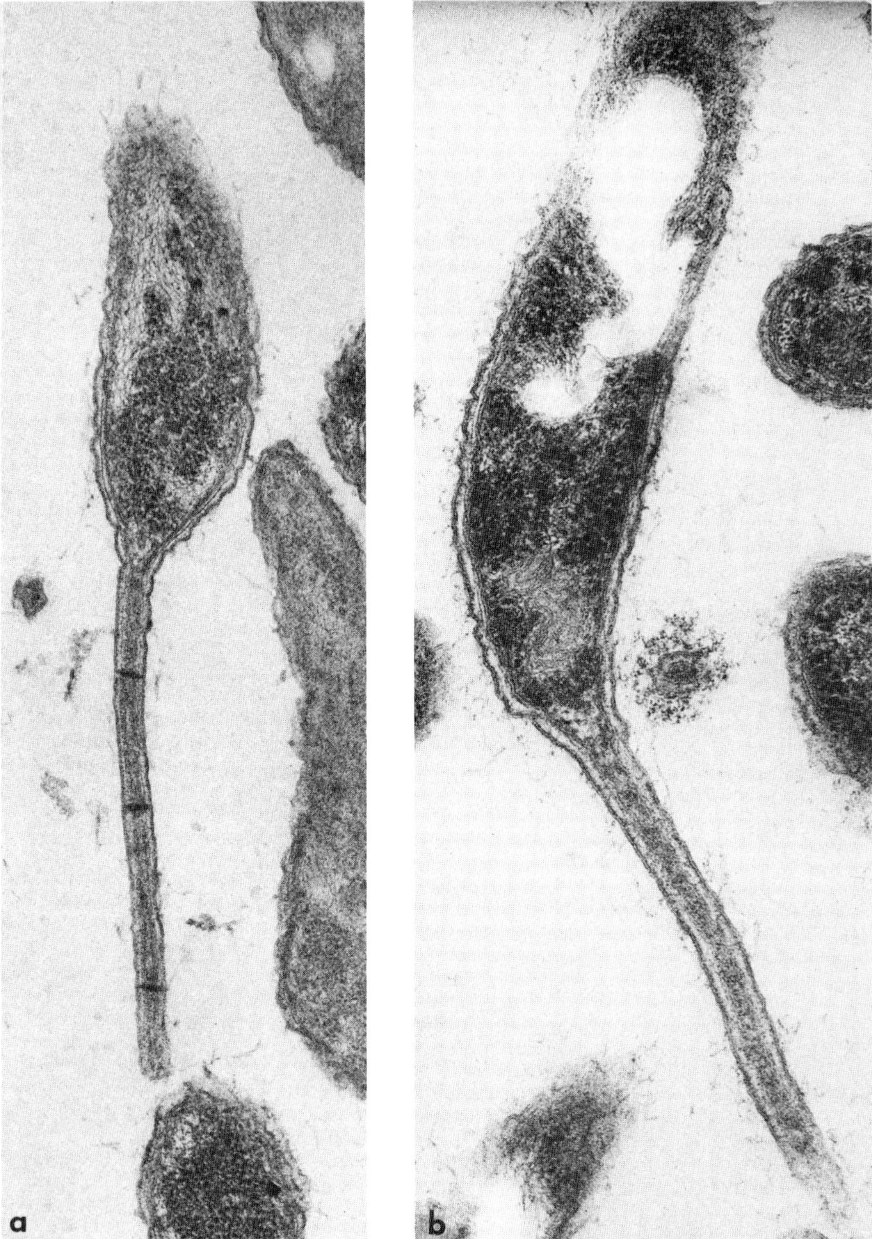

FIG. 37. Electron micrographs of a thin section of *Caulobacter crescentus* showing the stalk and the annular cross-bands. (a) ×60,000; (b) ×84,000. (From Schmidt and Stanier, 1966.)

that this type of sequence or the inverted repeats in *Caulobacter* may play a role in the sort of metastable regulation required for development.

D. Genetics

While there have been no specialized transducing phages isolated for *Caulobacter*, the nonlysogenic phage ϕCr30 has been shown to carry out high frequency generalized transduction (Ely and Johnson, 1975). A conjugational system has been described for *Caulobacter* (Jollick and Wright, 1974) and thus the techniques for genetic mapping are now available. In addition, a variety of nutritional (Johnson and Ely, 1976) and developmental (Kurn *et al.*, 1974) mutants are now at hand.

VIII. Summary and Conclusions

Bacteria have evolved two developmental strategies for expanding their ability to deal with an environment that is systematically changeable and that often fluctuates between extremes exceeding the normal conditions for growth. One of these involves the ability to form differentiated populations containing coexisting cells that are functionally distinct and specialized. The other involves the formation of resistant resting cells as an alternative to growth.

While there is no evidence to suggest that prokaryotic differentiation represents an evolutionary precursor to any existing multicellular organisms, it is clearly a more sophisticated evolutionary stage than the homogeneous populations of most eubacteria. Among the blue-green bacteria, the ability to fix nitrogen greatly expands the biological versatility of the organism, and it is thus not surprising that heterocyst differentiation focuses around this property. The well-known sensitivity of nitrogenase to oxygen must always pose a dilemma to an aerobic organism, and in the case of *Azotobacter* it is not unreasonable to assume that the extremely high respiratory rate of the organism is, in part, a strategy for maintaining relatively low levels of intracellular oxygen. In the blue-green bacteria the very process of the generation of reducing power via the oxidation of water produces rather than uses oxygen. Thus, the tactic of separating the photogeneration of reducing power and nitrogen fixation by the formation of a specialized cell lacking photosystem II is a simple and elegant biological solution. However, this particular solution creates other problems. Namely, the determination and regulation of optimal spacing of the heterocyst and the need for intercellular communication among the differentiated cells. The spacing phenomenon has created the unusual and valuable opportunity to examine the control of pattern formation in its simplest sense, i.e., in one dimension. It is possible to ask questions of perennial concern to developmental biologists, e.g., is the spacing by means of physical or chemical signals? If it is the latter, how is the gradient generated, regulated, maintained, and perceived? These questions are of course in addition to the more strictly morphogenetic questions of heterocyst formation. An interesting beginning has been made with the blue-green bacteria and one looks forward to new areas of developmental inquiry in a prokaryote.

In the case of *Caulobacter* the spatial–temporal development is visible in a simpler and more obvious fashion than in most other developing systems. Unfortunately it is not yet possible to understand the function of the stalked cell–swarmer differentiation. But it occurs in a highly regulated and systematic fashion so one must have faith that it does indeed have a function and wait patiently for it to be revealed. Perhaps the most attractive feature of *Caulobacter* development is that the various developmental processes can be described in terms of gene products. From the spatial point of view development involves the oriented assembly of pilin and flagellins into pili and flagella, respectively. Furthermore, the developmental appearance of phage receptor sites, while not yet biochemically characterized, is certainly amenable to such characterization. The regular appearance of pili, flagella, phage receptor sites, and stalks at fixed and predetermined poles of the cell represents a most useful system for examining the mechanisms whereby the cell recognizes and remembers which end is which.

In addition, the different temporal patterns of DNA synthesis in stalked and swarmer cells provides an opportunity to examine the factors involved in regulating the timing of DNA synthesis.

The second developmental strategy involves the formation of resistant resting cells as an alternative to growth. Here, development is cyclic and in a sense reversible, unlike the unidirectional formation of heterocyst or stalk cells. There are two broad categories of resistant resting cells formed by prokaryotes. These include the endospores formed by the Bacillaceae as well as by *Sporosarcina*, *Desulfotomaculum*, and *Thermoactinomyces* and spores or cysts formed by myxobacteria, *Azotobacter*, blue-green bacteria, and various actinomycetes.

For obvious practical and historical reasons most research has centered on the endospores and, unlike the other types of resting cells, a tremendous amount of information is available. Attempts to synthesize a comparative biology of prokaryotic resting cells (Sudo and Dworkin, 1973; Sadoff, 1973) have been hindered by the unequal amounts of information available. Furthermore, those clues that are now available indicate that there are as many fundamental differences between endospore development and that of other resting cells as there are interesting similarities. Initiation of spore or cyst formation seems in all cases to be in response to a nutritional depletion. In *Bacillus*, a few lines of evidence point to the release of catabolite repression as the regulatory mechanism even though cAMP does not seem to be involved. The facts that nutrient repression of fruiting body formation in *Myxococcus xanthus* can be overridden by the addition of cAMP, and that glucose will inhibit cyst formation in *Azotobacter* suggest that catabolite repression may play a regulatory role in both these organisms. Since bona fide sporulation in *Streptomyces* occurs only on the surface of solid media concomitant with growth, lysis, and germination, it has been impossible, thus far, to obtain any valid generalizations about the process in this organism.

Finally, while akinete formation in *Anabaena* occurs after the period of active growth, there is insufficient information to suggest any regulatory mechanism.

The evidence now seems convincing that in *Bacillus* the nutritional signals for sporulation must occur while a specific portion of the chromosome is being replicated. This coupling of nutrient derepression and DNA synthesis has not yet been demonstrated in the other sporulation systems.

The morphological difference between endospores and all other bacterial spores is fundamental and is reflected in the nature of the presporulation events as well as the final state of the DNA and the cell walls and membranes. It must be no accident that almost all endospore formers are gram-positive and may indeed reflect some peculiar mechanistic constraints imposed by the nature of the cell wall.*

In endospores there is an equal partition of the DNA between two unequal cells. This raises of course the question of a differential regulation between the two separate genomes—one to form the spore and the other to fade gracefully away. In the case of other spore-forming systems, sporulation includes the completion of a round of replication with the mature spore then containing two full complements of DNA in contrast to the single chromosome in the endospore.

In the case of endospores, the peculiar envelopment of the forespore by a double membrane results in an outer membrane with a reversed polarity located between the cortex and the spore coat. Whether or not this membrane has any regulatory function is unknown, but if so, the topological problems necessary for a reorganization of permeability control, for example, are intriguing. In the case of myxospores and *Azotobacter* cysts, the reorganization of the cell surface seems essentially to involve the addition of new

* *Desulfotomaculum*, a sulfate-reducing spore-former is described as gram-negative, however, its fine structure has not been examined so the true nature of its cell wall remains undetermined.

layers—the intine and exine in *Azotobacter* and the spore coat in *Myxococcus*.

Net RNA synthesis in both *Bacillus* and in *Myxococcus* ceases during sporulation. In both cases, however, turnover continues with the formation of sporulation-specific messages. Similar experiments have not been done with the other sporulating systems, although during encystment in *Azotobacter*, the rate of net RNA synthesis decreases and finally stops at a relatively early stage in cyst formation.

Extensive protein turnover occurs during endospore formation, although there is little if any net synthesis. This may reflect the accessibility of a considerable amount of protein precursors available as a result of the degradation of the mother cell protein. In *Myxococcus xanthus* and *Azotobacter*, where the entire cell is converted to the spore, no such source or protein is available, and the cell must rely on net synthesis to form such structures as the spore coat, exine, and intine. This independence of an external source of protein precursors may represent one advantage of the endospore strategy.

In all of the sporulation systems that have been examined, there is a shift of metabolic emphasis during sporulation. In *Bacillus*, the cells shift from a glycolytic to an oxidative direction, the TCA cycle appears, presumably for energetic reasons, poly-β-hydroxybutyric acid is utilized rather than formed, and there is a transient increase in glyoxylate cycle activity. While the metabolic flow has not been examined in sporulating *Myxococcus xanthus* there are sharp increases both in the activities of enzymes required for synthesis of spore coat polysaccharides and in glyoxylate pathway enzymes. In a similar fashion, the requirement for coat polysaccharide synthesis in *Azotobacter* cysts is satisfied by a decrease in Entner–Doudoroff metabolism and the appearance of gluconeogenic and glyoxylate pathway enzymes.

While all prokaryotic spores manifest some degree of increased resistance to temperature, desiccation, radiation, and physical disruption, only endospores are resistant to temperature extremes. The unique features of the endospore that might play a role are the core Ca-DPA and the intracellular cortex. The existence of DAP-less heat-resistant mutants excludes the possibility that DPA is solely responsible for heat resistance, and one is then left with the possibility that the cortex does indeed play a critical role, perhaps, as has been suggested, by maintaining the core in a state of relative dehydration.

With regard to spore dormancy, three possibilities have been mentioned: (a) the enzymes are absent, (b) they are in an altered and inactive conformation, (c) the spore is depleted of and impermeable to substrates and metabolites. In those spores that have been examined, it is possible to exclude the first possibility; and it is not unlikely that the complete description of dormancy will include a combination of altered conformation of macromolecules and restricted flux of substrates and metabolites.

The energy charge of endospores and the ATP content of *Streptomyces* spores was considerably lower than in corresponding vegetative cells. In glycerol-induced myxospores, however, there was no reduction in energy charge, suggesting that the spores had not sufficiently matured or reduction in energy charge is not a part of the state of dormancy. It is interesting that these myxospores, whose endogenous respiratory rate is substantially lower than that of vegetative cells, are also still actively undergoing *de novo* and salvage purine biosynthesis.

While germination in endospores and *Streptomyces* spores required activation, such is not the case with myxospores or *Azotobacter* cysts. Again, this may reflect either that the processes are fundamentally different in this respect or that the spores that do require activation have not sufficiently aged.

In those spores that have been examined for patterns of macromolecular synthesis during germination, the sequence of events is essentially similar; RNA and protein synthesis occur relatively early in the process and are followed by DNA synthesis at the beginning of outgrowth.

One is tempted to try to arrange the various processes of spore morphogenesis into some all-inclusive unified model. However, the facts (albeit far from complete) do not allow it. There are important differences between endospores and all other spores, between spores formed by gram-positive and gram-negative organisms, and even between such similar spores as those of *Azotobacter* and *Myxococcus*. The morphological, macromolecular, and metabolic differences emphasize the diversity of mechanisms used to arrive at common solutions.

ACKNOWLEDGMENT

The author wishes to acknowledge the valuable comments and criticism of Dr. David White.

IX. Selected Reviews

A. BACILLUS ENDOSPORES

1. Aronson, A. I., and Fitz-James, P. (1976). Structure and morphogenesis of the bacterial spore coat. *Bacteriol. Rev.* **40,** 360–420.
2. "The Bacterial Spore." (1969). (G. W. Gould and A. Hurst, eds.). Academic Press, New York.
3. Dawes, I. W., and Hansen, J. N. (1972). Morphogenesis in sporulating bacilli. *Crit. Rev. Microbiol.* **1,** 479–520.
4. Doi, R., and Sanchez-Anzaldo, F. S. (1976). Complexity of protein and nucleic acid synthesis during sporulation of bacilli. *In* "Microbiology—1976" (D. Schlessinger, ed.), pp. 145–163. Am. Soc. Microbiol., Washington, D.C.

5. Freese, E., and Fujita, Y. (1976). Control of enzyme synthesis during growth and sporulation. *In* "Microbiology—1976" (D. Schlessinger, ed.), pp. 164–184. Am. Soc. Microbiol., Washington, D.C.
6. Gould, G. W., and Dring, G. J. (1972). Biochemical mechanisms of spore germination. *In* "Spores V" (H. O. Halvorson, R. Hanson, and L. L. Campbell, eds.), pp. 401–408. Am. Soc. Microbiol., Washington, D.C.
7. Gould, G. W., and Dring, G. J. (1974). Mechanisms of spore heat resistance. *Adv. Microb. Physiol.* **11,** 137–164.
8. Hanson, R. S. (1975). Role of small molecules in regulation of gene expression and sporogenesis in Bacilli. *In* "Spores VI" (P. Gerhardt, R. N. Costilow, and H. L. Sadoff, eds.), pp. 318–326. Am. Soc. Microbiol., Washington, D.C.
9. Hanson, R. S., Peterson, J. A., and Yousten, A. A. (1970). Unique biochemical events in bacterial sporulation. *Annu. Rev. Microbiol.* **24,** 53–90.
10. Katz, E., and Demain, A. L. (1977). The peptide antibiotics of *Bacillus*; chemistry, biogenesis and possible functions. *Bacteriol. Rev.* **41,** 449–494.
11. Keynan, A. (1972). Cryptobiosis: A review of the mechanisms of the ametabolic state in bacterial spores. *In* "Spores V" (H. O. Halvorson, R. Hanson, and L. L. Campbell, eds.), pp. 355–362. Am. Soc. Microbiol., Washington, D.C.
12. Keynan, A. (1978). "Spore structure and its relation to resistance, dormancy, and germination." *In* "Spores VII" (G. Chambliss and J. C. Vary, eds.), pp. 43–53. Am. Soc. Microbiol., Washington, D.C.
13. Kornberg, A., Spudich, J. A., Nelson, D. L., and Deutscher, M. P. (1968). Origin of proteins in sporulation. *Annu. Rev. Biochem.* **37,** 51–78.
14. Piggot, P. J., and Coote, J. G. (1976). Genetic aspects of bacterial endospore formation. *Bacteriol. Rev.* **40,** 908–962.
15. Schaeffer, P. (1969). Sporulation and the production of antibiotics, exoenzymes and exotoxins. *Bacteriol. Rev.* **33,** 48–71.
16. Setlow, P. (1975). Energy and small-molecule metabolism during germination of bacterial spores. *In* "Spores VI" (P. Gerhardt, R. N. Costilow, and H. L. Sadoff, eds.), pp. 443–450. Am. Soc. Microbiol., Washington, D.C.
17. Sonenshein, A. L., and Campbell, K. M. (1978). "Control of gene expression during sporulation." *In* "Spores VII" (G. Chambliss and J. C. Vary, eds.), pp. 179–192. Am. Soc. Microbiol., Washington, D.C.
18. Tipper, D. J., and Gauthier, J. J. (1972). Structure of the bacterial endospore. *In* "Spores V" (H. O. Halvorson, R. Hanson, and L. L. Campbell, eds.), pp. 3–12. Am. Soc. Microbiol., Washington, D.C.

B. Actinomycetes

1. "Actinomycetales: Characteristics and Practical Importance." (1973). (G. Sykes and F. A. Skinner, eds.). Academic Press, New York.
2. Chater, K. F., and Hopwood, D. A. (1973). Differentiation in actinomycetes. *In* "Microbial Differentiation" (J. M. Ashworth and J. E. Smith, eds.), 23rd Symposium of the Society for General Microbiology, pp. 143–160. Cambridge Univ. Press, London and New York.
3. Hopwood, D. A., Chater, K. F., Dowding, J. E., and Vivian, A. (1973). Advances in *Streptomyces coelicolor* genetics. *Bacteriol. Rev.* **37,** 371–405.
4. Kalakoutskii, L. V., and Agre, N. (1976). Comparative aspects of development and differentiation in actinomycetes. *Bacteriol. Rev.* **40,** 469–524.

C. Myxobacteria

1. Dworkin, M. (1965). Biology of the myxobacteria. *Annu. Rev. Microbiol.* **20,** 75–106.
2. Dworkin, M. (1972). The myxobacteria: New directions in studies of procaryotic development. *Crit. Rev. Microbiol.* **1,** 435–452.
3. Dworkin, M. (1973). Cell–cell interactions in the myxobacteria. *In* "Microbial Differentiation" (J. M. Ashworth and J. E. Smith, eds.), 23rd Symposium of the Society for General Microbiology, pp. 125–142. Cambridge Univ. Press, London and New York.
4. White, D. (1975). Myxospores of *Myxococcus xanthus. In* "Spores VI" (P. Gerhardt, R. N. Costilow, and H. L. Sadoff, eds.), pp. 44–51. Am. Soc. Microbiol., Washington, D.C.
5. Wireman, J. W., and Dworkin, M. (1975). Morphogenesis and developmental interactions in myxobacteria. *Science* **189,** 516–523.

D. Azotobacter

1. Sadoff, H. L. (1975). Encystment and germination in *Azotobacter vinelandii. Bacteriol. Rev.* **39,** 516–539.
2. Sadoff, H. L., Page, W. J., and Reusch, R. N. (1975). The cyst of *Azotobacter vinelandii:* Comparative view of morphogenesis. *In* "Spores VI" (P. Gerhardt, R. H. Costilow, and H. L. Sadoff, eds.), pp. 52–60. Am. Soc. Microbiol., Washington, D.C.

E. Caulobacter

1. Newton, A., Osley, M. A., and Terrana, B. (1975). *Caulobacter crescentus:* A model for the temporal and spatial control of development. *In* "Microbiology—1975" (D. Schlessinger, ed.), pp. 442–452. Am. Soc. Microbiol., Washington, D.C.
2. Poindexter, J. S. (1964). Biological properties and classification of the *Caulobacter* group. *Bacteriol. Rev.* **28,** 231–295.
3. Shapiro, L. (1976). Differentiation in the *Caulobacter* cell cycle. *Annu. Rev. Microbiol.* **30,** 377–408.

F. Blue-Green Bacteria

1. Carr, N. G., and Bradley, S. (1973). Aspects of development in blue-green algae. *In* "Microbial Differentiation" (J. M. Ashworth and J. E. Smith, eds.), 23rd Symposium of the Society for General Microbiology, pp. 161–188. Cambridge Univ. Press, London and New York.
2. Carr, N. G., and Whitton, B. A., eds. (1973). "The Biology of Blue-Green Algae." Univ. of California Press, Berkeley.
3. Nichols, J. M., and Carr, N. G. (1978). "Akinetes of cyanobacteria." *In* "Spores VII" (G. Chambliss and J. C. Vary, eds.), pp. 335–343. Am. Soc. Microbiol., Washington, D.C.
4. Wilcox, M., Mitchison, G. J., and Smith, R. J. (1975). Spatial control of differentiation in the blue-green alga *Anabaena. In* "Microbiology—1975" (D. Schlessinger, ed.), pp. 453–463. Am. Soc. Microbiol., Washington, D.C.
5. Wolk, C. P. (1973). Physiology and cytological chemistry of blue-green algae. *Bacteriol. Rev.* **37,** 32–101.

6. Wolk, C. P. (1975). Differentiation and pattern formation in blue-green algae. *In* "Spores VI" (P. Gerhardt, R. N. Costilow, and H. L. Sadoff, eds.), pp. 85–96. Am. Soc. Microbiol., Washington, D.C.

G. Comparative

1. Sadoff, H. L. (1973). Comparative aspects of morphogenesis in three prokaryotic genera. *Annu. Rev. Microbiol.* **27,** 133–153.
2. Sudo, S. Z., and Dworkin, M. (1973). Comparative biology of procaryotic resting cells. *Adv. Microb. Physiol.* **9,** 153–224.

References

Alderton, G., and Snell, N. (1963). *Biochem. Biophys. Res. Commun.* **10,** 139–143.

Andreoli, A. J., Saranto, J., Baeker, P. A., Svehiro, S., Escamilla, E., and Steiner, A. (1975). *In* Spores VI" (P. Gerhardt, R. N. Costilow, and H. L. Sadoff, eds.), pp. 418–424. Am. Soc. Microbiol., Washington, D.C.

Aronson, A. I., and Fitz-James, P. C. (1976). *Bacteriol. Rev.* **40,** 360–402.

Aslanjan, R. R., Agre, N. S., Kalakoutskii, L. V., and Kirillova, I. P. (1971). *Mikrobiologija* **40,** 293–296.

Aslanjan, R. R., Agre, N. S., and Tartakovski, I. S. (1972). *Mikrobiologija* **41,** 746–747.

Atwell, R. W., and Cross, T. (1973). *In* "Actinomycetales: Characteristics and Practical Importance" (G. Sykes and F. A. Skinner, eds.), pp. 197–207. Academic Press, New York.

Audit, C., and Anagnostopoulos, C. (1972). *In* "Spores V" (H. O. Halvorson, R. Hanson, and L. L. Campbell, eds.), pp. 117–125. Am. Soc. Microbiol., Washington, D.C.

Bacon, K., and Rosenberg, E. (1967). *J. Bacteriol.* **94,** 1883–1889.

Bacon, K., Clutter, D., Kottel, R. H., Orlowski, M., and White, D. (1975). *J. Bacteriol.* **124,** 1635–1636.

Bailey, G. F., Karp, S., and Sacks, L. E. (1965). *J. Bacteriol.* **89,** 984–987.

Balassa, G. (1971). *Curr. Top. Microbiol. Immunol.* **56,** 99–192.

Bendis, I. K., and Shapiro, L. (1973). *J. Bacteriol.* **115,** 848–857.

Bernlohr, R. W., and Gray, B. H. (1969). *In* "Spores IV" (L. L. Campbell, ed.), pp. 186–195. Am. Soc. Microbiol., Bethesda, Maryland.

Bernlohr, R. W., and Leitzman, C. (1969). *In* "The Bacterial Spore" (G. W. Gould and A. Hurst, eds.), pp. 183–213. Academic Press, New York.

Bernlohr, R. W., and Novelli, G. D. (1959). *Nature (London)* **184,** 1256–1257.

Black, S. H., and Gerhardt, P. (1962). *J. Bacteriol.* **83,** 960–967.

Briggs, A. (1966). *J. Appl. Bacteriol.* **29,** 490–504.

Buchanan, R. E., and Gibbons, N. E. (1974). "Bergey's Manual of Determinative Bacteriology," 8th Ed. Williams & Wilkins, Baltimore, Maryland.

Bulla, L. A., Jr., and St. Julian, G. (1972). *In* "Spores V" (H. O. Halvorson, R. Hanson, and L. L. Campbell, eds.), pp. 191–196. Am. Soc. Microbiol., Washington, D.C.

Burchard, R. P. (1970). *J. Bacteriol.* **104,** 940–947.

Burchard, R. P., and Dworkin, M. (1966). *J. Bacteriol.* **91,** 1305–1313.

Burchard, R. P., and Parish, J. H. (1975). *Arch. Mikrobiol.* **104,** 289–292.

Burchard, R. P., and Voelz, H. (1972). *Virology* **48,** 555–566.

Campos, J. M., Geisselsoder, J., and Zusman, D. R. (1977). *J. Mol. Biol.* **119,** 167–178.
Carr, N. G., and Bradley, S. (1973). *In* "Microbial Differentiation" (J. M. Ashworth and J. E. Smith, eds.), pp. 161–188. Cambridge Univ. Press, London and New York.
Carvajal, F. (1947). *Mycologia* **39,** 425–440.
Chambliss, G. H., and Legault-Demare, L. (1975). *In* "Spores VI" (P. Gerhardt, R. N. Costilow, and H. L. Sadoff, eds.), pp. 314–317. Am. Soc. Microbiol., Washington, D.C.
Chambon, P. E., DuPraw, E. J., and Kornberg, A. (1968). *J. Biol. Chem.* **243,** 5101–5109.
Chater, K. F., and Hopwood, D. A. (1973). *Symp. Soc. Gen. Microbiol.* **23,** 143–160.
Cleveland, E. F., and Gilvarg, C. (1975). *In* "Spores VI" (P. Gerhardt, R. N. Costilow, and H. L. Sadoff, eds.), pp. 458–464. Am. Soc. Microbiol., Washington, D.C.
Cooney, P. H., Freese, E. B., and Freese, E. (1975). *In* "Spores VI" (P. Gerhardt, R. N. Costilow, and H. L. Sadoff, eds.), pp. 187–194. Am. Soc. Microbiol., Washington, D.C.
Coote, J. G., and Mandelstam, J. (1973). *J. Bacteriol.* **114,** 1254–1263.
Cross, T., and Attwell, R. W. (1975). *In* "Spores VI" (P. Gerhardt, R. N. Costilow, and H. L. Sadoff, eds.), pp. 3–14. Am. Soc. Microbiol., Washington, D.C.
Dawes, I. W., and Hansen, J. N. (1972). *Crit. Rev. Microbiol.* **1,** 479–520.
Degnan, S. T., and Newton, A. (1972). *J. Mol. Biol.* **64,** 671–680.
DelValle, M. R., and Aronson, A. I. (1962). *Biochem. Biophys. Res. Commun.* **9,** 421–425.
Deutscher, M. P., Chambon, P., and Kornberg, A. (1968). *J. Biol. Chem.* **245,** 5117–5125.
Diesterhaft, M. D., and Freese, E. (1973). *J. Biol. Chem.* **248,** 6062–6070.
Doi, R. H. (1969). *In* "The Bacterial Spore" (G. W. Gould and A. Hurst, eds.), pp. 125–166. Academic Press, New York.
Doi, R. H., and Halvorson, H. (1961). *J. Bacteriol.* **81,** 51–58.
Donnellan, J. E., Jr., and Setlow, R. B. (1965). *Science* **149,** 308–310.
Donze, M., Haveman, J., and Schiereck, P. (1972). *Biochim. Biophys. Acta* **256,** 157–161.
Douthit, H. A., and Halvorson, H. O. (1966). *Science* **153,** 182–183.
Douthit, H. A., and Kieras, R. A. (1972). *In* "Spores V" (H. O. Halvorson, R. Hanson, and L. L. Campbell, eds.), pp. 264–268. Am. Soc. Microbiol., Washington, D.C.
Dring, G. J., and Gould, G. W. (1975). *Biochem. Biophys. Res. Commun.* **66,** 202–208.
Dunn, G., Torgerson, D. M., and Mandelstam, J. (1976). *J. Bacteriol.* **125,** 766–779.
Dunn, J. H., and Wolk, C. P. (1970). *J. Bacteriol.* **103,** 153–158.
Dworkin, M. (1973). *Symp. Soc. Gen. Microbiol.* **23,** 125–142.
Dworkin, M., and Gibson, S. M. (1964). *Science* **146,** 243–244.
Dworkin, M., and Niederpruem, D. J. (1964). *J. Bacteriol.* **87,** 316–322.
Dworkin, M., and Sadler, W. (1966). *J. Bacteriol.* **91,** 1516–1519.
Dworkin, M., and Voelz, H. (1962). *J. Gen. Microbiol.* **28,** 81–85.
Dworkin, M., Higgins, J., Glenn, A., and Mandelstam, J. (1972). *In* "Spores V" (H. O. Halvorson, R. Hanson, and L. L. Campbell, eds.), pp. 233–247. Am Soc. Microbiol., Washington, D.C.
Eaton, M. W., and Ellar, D. J. (1974). *Biochem. J.* **144,** 327–337.
Edgell, M. H., Hutchinson, C. A., III, and Bott, K. F. (1975). *In* "Spores VI" (P. Gerhardt, R. N. Costilow, and H. L. Sadoff, eds.), pp. 195–201. Am. Soc. Microbiol., Washington, D.C.
Ely, B., and Johnson, R. (1975). *Abstr. Annu. Meet. Am. Soc. Microbiol.* S-22, p. 217.
Falaschi, A., and Kornberg, A. (1966). *J. Biol. Chem.* **241,** 1478–1482.
Fay, P. (1969). *Arch. Mikrobiol.* **67,** 62–70.
Fay, P. (1970). *J. Exp. Bot.* **20,** 100–109.
Fay, P. (1973). *In* "The Biology of Blue-Green Algae" (N. G. Carr and B. A. Whitton, eds.), pp. 238–259. Univ. of California Press, Berkeley.
Fay, P., and Lang, N. J. (1971). *Proc. R. Soc. London, Ser. B* **178,** 185–192.

Fay, P., and Walsby, A. E. (1966). *Nature (London)* **209,** 94–95.
Fay, P., Stewart, W. D. P., Walsby, A. E., and Fogg, G. E. (1968). *Nature (London)* **220,** 810–812.
Felix, J., and Lundgren, D. G. (1972). *In* "Spores V" (H. O. Halvorson, R. Hanson, and L. L. Campbell, eds,), pp. 35–44. Am. Soc. Microbiol., Washington, D. C.
Filer, D., Rosenberg, E., and Kindler, S. H. (1973). *J. Bacteriol.* **115,** 23–28.
Filer, D., Kindler, S. H., and Rosenberg, E. (1977). *J. Bacteriol.* **131,** 745–750.
Fitz-James, P. C., and Young, I. E. (1959). *J. Bacteriol.* **78,** 743–754.
Fitz-James, P. C., and Young, I. E. (1969). *In* "The Bacterial Spore" (G. W. Gould and A. Hurst, eds.), pp. 39–72. Academic Press, New York.
Fogg, G. E. (1949). *Ann. Bot. (London)* **15,** 23–35.
Foster, H. A., and Parish, J. H. (1973). *J. Gen. Microbiol.* **75,** 391–400.
Freese, E., and Fujita, Y. (1976). *In* "Microbiology—1976" (D. Schlessinger, ed.), pp. 164–184. Am. Soc. Microbiol., Washington, D.C.
Fritsch, F. E. (1945). *In* "The Structure and Reproduction of Algae," Vol. 2. Cambridge Univ. Press, London.
Fukuda, R., and Doi, R. H. (1977). *J. Bacteriol.* **129,** 422–432.
Galizzi, A., Siccardi, A. G., Albertini, A. M., Amileni, A. R., Meneguzzi, G., and Polsinelli, M. (1975). *J. Bacteriol.* **121,** 450–454.
Gardner, R., and Kornberg, A. (1967). *J. Biol. Chem.* **242,** 2383–2388.
Gerhardt, P., and Murrell, W. G. (1978). *In* "Spores VII" (G. Chambliss and J. C. Vary, eds.), pp. 18–20. Am. Soc. Microbiol., Washington, D.C.
Glenn, A. R., and Mandelstam, J. (1971). *Biochem. J.* **123,** 129–138.
Gould, G. W. (1969). *In* "The Bacterial Spore" (G. W. Gould and A. Hurst, eds.), pp. 397–444. Academic Press, New York.
Gould, G. W., and Dring, G. J. (1972). *In* "Spores V" (H. O. Halvorson, R. Hanson, and L. L. Campbell, eds.), pp. 401–408. Am. Soc. Microbiol., Washington, D.C.
Gould, G. W., and Dring, G. J. (1974). *Adv. Microb. Physiol.* **11,** 137–164.
Gould, G. W., and Dring, G. J. (1975). *In* "Spores VI" (P. Gerhardt, R. N. Costilow, and H. L. Sadoff, eds.), pp. 541–546. Am. Soc. Microbiol., Washington, D.C.
Gould, G. W., and Hitchins, A. D. (1963). *J. Gen. Microbiol.* **33,** 413–423.
Gould, G. W., and Hitchins, A. D. (1965). *In* "Spores III" (L. L. Campbell and H. O. Halvorson, eds.), pp. 213–221. Am. Soc. Microbiol., Ann Arbor, Michigan.
Gould, G. W., and Sale, A. J. H. (1970). *J. Gen. Microbiol.* **60,** 335–346.
Hagen, D. C., Bretscher, A. P., and Kaiser, A. D. (1978). *Developmental Biology* **64,** 284–296.
Halvorson, H. O., and Swanson, A. (1969). *In* "Spores IV" (L. L. Campbell, ed.), pp. 121–132. Am. Soc. Microbiol., Bethesda, Maryland.
Hansen, J. N., Spiegelman, G., and Halvorson, H. O. (1970). *Science* **168,** 1291–1298.
Hanson, C. W., and Andreoli, A. J. (1973). *Arch. Mikrobiol.* **92,** 1–10.
Hanson, C. W., and Dworkin, M. (1974). *J. Bacteriol.* **118,** 486–496.
Hanson, R. S., Blicharska, J., and Szulmajster, J. (1964a). *Biochem. Biophys. Res. Commun.* **17,** 1–7.
Hanson, R. S., Blicharska, J., Arnaud, M., and Szulmajster, J. (1964b). *Biochem. Biophys. Res. Commun.* **17,** 690–695.
Hanson, R. S., Peterson, J. A., and Yousten, A. A. (1970). *Annu. Rev. Microbiol.* **24,** 53–90.
Hanson, R. S., Curry, M. V., Garner, J. V., and Halvorson, H. O. (1972). *Can. J. Microbiol.* **18,** 1139–1143.
Hashimoto, T., Black, S. H., and Gerhardt, P. (1960). *Can. J. Microbiol.* **6,** 203–212.
Higgins, M. L. (1967). *J. Bacteriol.* **94,** 495–498.
Hirsch, C. F., and Ensign, J. C. (1975). *In* "Spores VI" (P. Gerhardt, R. N. Costilow, and H. L. Sadoff, eds.), pp. 28–35. Am. Soc. Microbiol., Washington, D.C.

Hitchens, V. M., and Sadoff, H. L. (1970). *J. Bacteriol.* **104,** 492–498.
Hitchens, V. M., and Sadoff, H. L. (1973). *J. Bacteriol.* **113,** 1273–1279.
Hodgkin, J., and Kaiser, A. D. (1977). *Proc. Natl. Acad. Sci. U.S.A.* **74,** 2938–2942.
Holt, S. C., and Leadbetter, E. R. (1969). *Bacteriol. Rev.* **33,** 346–378.
Hopwood, D. A., Chater, K. F., Dowding, J. E., and Vivian, A. (1973). *Bacteriol. Rev.* **37,** 371–405.
Hranueli, D., Piggot, P. J., and Mandelstam, J. (1974). *J. Bacteriol.* **119,** 684–690.
Idriss, J. M., and Halvorson, H. O. (1969). *Arch. Biochem. Biophys.* **133,** 442–453.
Inouye, M., Inouye, S., and Zusman, D. R. (1979). *Proc. Natl. Acad. Sci. U.S.A.* in press.
Jacob, F. (1970). "La Logique Du Vivant: Une Histoire De L'Hérédite." Gallimard, Paris.
Johnson, R. C., and Ely, B. (1976). *Abstr. Annu. Meet. Am. Soc. Microbiol.* H-79, p. 92.
Jollick, J. D., and Wright, B. (1974). *J. Gen. Virol.* **22,** 197–205.
Jones, H. C., and Schmidt, J. M. (1973). *J. Bacteriol.* **116,** 466–470.
Juengst, F. W., Jr., and Dworkin, M. (1973). *J. Bacteriol.* **113,** 786–797.
Kaiser, D., and Dworkin, M. (1975). *Science* **187,** 653–654.
Kalakoutskii, L. V., and Agre, N. (1976). *Bacteriol. Rev.* **40,** 469–524.
Kalakoutskii, L. V., and Pozharitskaja, L. M. (1968). *J. Gen. Appl. Microbiol.* **14,** 209–212.
Kalakoutskii, L. V., and Pozharitskaja, L. M. (1973). *In* "Actinomycetales: Characteristics and Practical Importance" (G. Sykes and F. A. Skinner, eds.), pp. 155–178. Academic Press, New York.
Kalakoutskii, L. V., Bobkova, E. A., and Krassinlnikov, N. A. (1966). *Dokl. Akad. Nauk. SSSR* **170,** 705–707.
Kalakoutskii, L. V., Agre, N. S., and Aslanjan, R. R. (1969). *Dokl. Akad. Nauk SSSR* **184,** 1214–1216.
Katz, E., and Demain, A. L. (1977). *Bacteriol. Rev.* **41,** 449–494.
Kennett, R. H., and Sueoka, N. (1971). *J. Mol. Biol.* **60,** 31–44.
Keynan, A. (1972). *In* "Spores V" (H. O. Halvorson, R. Hanson, and L. L. Campbell, eds.), pp. 255–362. Am. Soc. Microbiol., Washington, D.C.
Keynan, A., and Evenchik, Z. (1969). *In* "The Bacterial Spore" (G. W. Gould and A. Hurst, eds.), pp. 359–396. Academic Press, New York.
Keynan, A., and Halvorson, H. (1965). *In* "Spores III" (L. L. Campbell and H. O. Halvorson, eds.), pp. 174–179. Am. Soc. Microbiol., Ann Arbor, Michigan.
Kimchi, A., and Rosenberg, E. (1976). *J. Bacteriol.* **128,** 69–79.
Kobayashi, Y. (1972). *In* "Spores V" (H. O. Halvorson, R. Hanson, and L. L. Campbell, eds.), pp. 269–276. Am. Soc. Microbiol., Washington, D.C.
Kobayashi, Y., and Halvorson, H. O. (1968). *Biochim. Biophys. Acta* **123,** 622–632.
Kondo, M., and Foster, J. W. (1967). *J. Gen. Microbiol.* **47,** 257–271.
Kornberg, A., Spudich, J. A., Nelson, D. L., and Deutscher, M. P. (1968). *Annu. Rev. Biochem.* **37,** 51–78.
Kottel, R. H., Orlowski, M., White, D., and Grutsch, J. (1974). *J. Bacteriol.* **119,** 650–651.
Kottel, R. H., Bacon, K., Clutter, D., and White, D. (1975). *J. Bacteriol.* **124,** 550–557.
Kurn, N., and Shapiro, L. (1976). *Proc. Natl. Acad. Sci. U.S.A.* **73,** 3303–3307.
Kurn, N., Ammer, S., and Shapiro, L. (1974). *Proc. Natl. Acad. Sci. U.S.A.* **71,** 3157–3161.
Laishley, E. J., and Bernlohr, R. W. (1966). *Biochem. Biophys. Res. Commun.* **24,** 85–90.
Lambein, F., and Wolk, C. P. (1973). *Biochemistry* **12,** 791–798.
Lang, N. J. (1968). *In* "Algae, Man and the Environment" (D. F. Jackson, ed.), pp. 235–248. Syracuse Univ. Press, Syracuse, New York.
Leanz, G., and Gilvarg, C. (1973). *J. Bacteriol.* **114,** 455–456.
Lechevalier, H., and Holbert, P. E. (1965). *J. Bacteriol.* **89,** 217–222.
Leighton, T. J. (1974). *J. Biol. Chem.* **249,** 7808–7812.

Lewis, J. C., Snell, N. S., and Burr, H. K. (1960). *Science* **132,** 544–545.
Lin, L. P., and Sadoff, H. L. (1968). *J. Bacteriol.* **95,** 2336–2343.
Lin, L. P., and Sadoff, H. L. (1969). *J. Bacteriol.* **100,** 480–486.
Loperfido, B., and Sadoff, H. L. (1973). *J. Bacteriol.* **113,** 841–846.
Losick, R., and Sonenshein, A. L. (1969). *Nature* (*London*) **224,** 35–37.
McVittie, A., (1974). *J. Gen. Microbiol.* **81,** 291–302.
McVittie, A., Wildermuth, H., and Hopwood, D. A. (1972). *J. Gen. Microbiol.* **71,** 367–381.
Mandel, M., and Rowley, D. B. (1963). *J. Bacteriol.* **85,** 1445–1446.
Mandelstam, J. (1969). *Symp. Soc. Gen. Microbiol.* **19,** 377–404.
Mandelstam, J., and Higgs, S. A. (1974). *J. Bacteriol.* **120,** 38–42.
Mandelstam, J., Kay, D., and Hranueli, D. (1975). *In* "Spores VI" (P. Gerhardt, R. N. Costilow, and H. L. Sadoff, eds.), pp. 181–186. Am. Soc. Microbiol., Washington, D.C.
Martin, S., Sodergren, E., Masuda, T., and Kaiser, D. (1978). *Virol.* **88,** 44–53.
Mikulík, K., Janda, I., Řičicová, A., and Vinter, V. (1975). *In* "Spores VI" (P. Gerhardt, R. N. Costilow, and H. L. Sadoff, eds.), pp. 15–27. Am. Soc. Microbiol., Washington, D.C.
Miller, M. M., and Lang, N. J. (1968). *Arch. Mikrobiol.* **60,** 303–313.
Mukherjee, P. K., and Paulus, H. (1977). *Proc. Natl. Acad. Sci. U.S.A.* **74,** 780–784.
Murrell, W. G., and Scott, W. J. (1966). *J. Gen. Microbiol.* **43,** 411–425.
Murrell, W. G., and Warth, A. D. (1965). *In* "Spores III" (L. L. Campbell and H. O. Halvorson, eds.), pp. 1–24. Am. Soc. Microbiol., Ann Arbor, Michigan.
Murrell, W. G., Ohye, D. F., and Gordon, R. A. (1969). *In* "Spores IV" (L. L. Campbell, ed.), pp. 1–19. Am. Soc. Microbiol., Bethesda, Maryland.
Nagai, S., Nishizawa, Y., Onodera, M., and Aiba, S. (1971). *J. Gen. Microbiol.* **66,** 197–203.
Nakata, H. (1957). *In* "Spores I" (H. O. Halvorson, ed.), pp. 97–104. Am. Inst. Biol. Sci., Washington, D.C.
Newton, A. (1972). *Proc. Natl. Acad. Sci. U.S.A.* **69,** 447–451.
Nukushina, J., and Ikeda, Y. (1969). *Genetics* **63,** 63–74.
Okano, P., Bacon, K., and Rosenberg, E. (1970). *J. Bacteriol.* **104,** 275–282.
Orlowski, M., and White, D. (1974a). *Arch. Microbiol.* **97,** 347–357.
Orlowski, M., and White, D. (1974b). *J. Bacteriol.* **118,** 103–111.
Orlowski, M., Martin, P., White, D., and Wong, M. C. (1972). *J. Bacteriol.* **111,** 784–790.
Page, W. J., and Sadoff, H. L. (1976). *J. Bacteriol.* **125,** 1080–1087.
Parish, J. H. (1975). *J. Gen. Microbiol.* **87,** 198–210.
Parker, L. T., and Socolofsky, M. D. (1966). *J. Bacteriol.* **91,** 297–303.
Pate, J. L., and Ordal, E. J. (1965). *J. Cell Biol.* **27,** 133–150.
Paulus, H., and Sarkar, N. (1976). *In* "Molecular Mechanisms in the Control of Gene Expression" (D. P. Nierlich, W. J. Rutter, and C. F. Fox, eds.), pp. 177–194. Academic Press, New York.
Pearce, S. M., and Fitz-James, P. C. (1971). *J. Bacteriol.* **105,** 339–348.
Pero, J., Nelson, J., and Losick, R. (1975). *In* "Spores VI" (P. Gerhardt, R. N. Costilow, and H. L. Sadoff, eds.), pp. 202–212. Am. Soc. Microbiol., Washington, D.C.
Piggot, P. J., and Coote, J. G. (1976). *Bacteriol. Rev.* **40,** 908–962.
Pitel, D. W., and Gilvarg, C. (1971). *J. Biol. Chem.* **246,** 3720–3724.
Poindexter, J. S. (1964). *Bacteriol. Rev.* **28,** 231–295.
Powell, J. F., and Strange, R. E. (1953). *Biochem. J.* **54,** 205–209.
Pozharitskaja, L. M., Vanin, A. F., and Kalakoutskii, L. V. (1972). *Dokl. Akad. Nauk SSSR* **202,** 1429–1431.
Ramsey, W. S., and Dworkin, M. (1968). *J. Bacteriol.* **95,** 2249–2257.
Ramsey, W. S., and Dworkin, M. (1970). *J. Bacteriol.* **101,** 531–540.
Rana, R. S., and Halvorson, H. O. (1972). *J. Bacteriol.* **109,** 606–615.

Rancourt, M. W., and Lechevalier, H. A. (1964). *Can. J. Microbiol.* **10,** 311–316.
Reichenbach, H., and Dworkin, M. (1970). *J. Bacteriol.* **101,** 325–326.
Reichenbach, H., Voelz, H., and Dworkin, M. (1969). *J. Bacteriol.* **97,** 905–911.
Rhaese, H. J., Dichtelmüller, H., Grade, R., and Groscurth, R. (1975). *In* "Spores VI" (P. Gerhardt, R. N., Costilow, and H. L. Sadoff, eds.), pp. 335–340. Am. Soc. Microbiol., Washington, D.C.
Rhaese, H. J., Hoch, J. A., and Groscurth, R. (1977). *Proc. Natl. Acad. Sci. U.S.A.* **74,** 1125–1129.
Riemann, H., and Ordal, Z. J. (1961). *Science* **133,** 1703–1704.
Roberts, T. A., and Hitchens, A. D. (1969). *In* "The Bacterial Spore" (G. W. Gould and A. Hurst, eds.), pp. 611–670. Academic Press, New York.
Rode, L. J., and Foster, J. W. (1960). *Proc. Natl. Acad. Sci. U.S.A.* **46,** 118–128.
Rode, L. J., and Foster, J. W. (1961). *J. Bacteriol.* **81,** 768–779.
Rosenberg, E., Katarski, M., and Gottlieb, P. (1967). *J. Bacteriol.* **93,** 1402–1408.
Rosenberg, E., Filer, D., Zafriti, D., and Kindler, S. H. (1973). *J. Bacteriol.* **115,** 29–34.
Ryter, A., and Aubert, J. P. (1969). *Ann. Inst. Pasteur, Paris* **117,** 601–611.
Sadler, W., and Dworkin, M. (1966). *J. Bacteriol.* **93,** 1402–1408.
Sadoff, H. L. (1969). *In* "The Bacterial Spore" (G. W. Gould and A. Hurst, eds.), pp. 275–299. Academic Press, New York.
Sadoff, H. L. (1972). *In* "Spores V" (H. O. Halvorson, R. Hanson, and L. L. Campbell, eds.), pp. 157–166. Am. Soc. Microbiol., Washington, D.C.
Sadoff, H. L. (1973). *Annu. Rev. Microbiol.* **27,** 133–153.
Sadoff, H. L. (1975). *Bacteriol. Rev.* **39,** 516–539.
Sadoff, H. L., Bach, J. A., and Kools, J. W. (1965). *In* "Spores III" (L. L. Campbell and H. O. Halvorson, eds.), pp. 97–110. Am. Soc. Microbiol., Ann Arbor, Michigan.
Sadoff, H. L., Hitchens, A. D., and Celikkol, E. (1969). *J. Bacteriol.* **98,** 1208–1218.
Sadoff, H. L., Celikkol, E., and Engelbrecht, H. L. (1970). *Proc. Natl. Acad. Sci. U.S.A.* **66,** 844–849.
Sadoff, H. L., Shimei, B., and Ellis, S. (1977). *Abstr. Annu. Meet. Am. Soc. Microbiol.* H-99, p. 151.
Schaeffer, P. (1969). *Bacteriol. Rev.* **33,** 48–71.
Schaeffer, P., Millet, J., and Aubert, J. P. (1965). *Proc. Natl. Acad. Sci. U.S.A.* **54,** 704–711.
Scherrer, R., and Gerhardt, P. (1972). *J. Bacteriol.* **112,** 559–568.
Schmidt, J. M. (1973), *Arch. Mikrobiol.* **89,** 33–40.
Schmidt, J. M., and Stanier, R. Y. (1966). *J. Cell. Biol.* **28,** 423–436.
Schmitt, R., and Freese, E. (1968). *J. Bacteriol.* **96,** 1255–1265.
Scribner, H. E., III, Tang, T., and Bradley, S. G. (1973). *Appl. Microbiol.* **25,** 873–879.
Segall, J., Tijian, R., Pero, J., and Losick, R. (1974). *Proc. Natl. Acad. Sci. U.S.A.* **71,** 4860–4863.
Setlow, P. (1975). *In* "Spores VI" (P. Gerhardt, R. N. Costilow, and H. L. Sadoff, eds.), pp. 443–450. Am. Soc. Microbiol., Washington, D.C.
Setlow, P., and Kornberg, A. (1970a). *J. Biol. Chem.* **245,** 3637–3644.
Setlow, P., and Kornberg, A. (1970b). *J. Biol. Chem.* **245,** 3645–3652.
Setlow, P., and Primus, G. (1975a). *J. Biol. Chem.* **250,** 623–630).
Setlow, P., and Primus, G. (1975b). *In* "Spores VI" (P. Gerhardt, R. N. Costilow, and H. L. Sadoff, eds.), pp. 451–457. Am. Soc. Microbiol., Washington, D.C.
Shapiro, L. (1976). *Annu. Rev. Microbiol.* **30,** 377–407.
Shapiro, L., and Maizel, J. V. (1973). *J. Bacteriol.* **113,** 478–485.
Simon, R. D. (1971). *Proc. Natl. Acad. Sci. U.S.A.* **68,** 265–267.
Smith, K. C., and Yoshikawa, H. (1966). *Photochem. Photobiol.* **5,** 777–786.
Sonenshein, A. L., and Losick, R. (1970). *Nature* (*London*) **227,** 906–909.

Sonenshein, A. L., and Roscoe, D. H. (1969). *Virology* **39,** 265–276.
Spudich, J. A., and Kornberg, A. (1968). *J. Biol. Chem.* **243,** 4588–4599.
Spudich, J. A., and Kornberg, A. (1969). *J. Bacteriol.* **98,** 69–74.
Staley, D. P., and Bernlohr, R. W. (1967). *Biochim. Biophys. Acta* **146,** 467–476.
Staley, J. R., and Jordan, T. L. (1973). *Nature (London)* **246,** 155–156.
Stanier, R. Y., Kunisawa, R., Mandel, M., and Cohen-Bazire, G. (1971). *Bacteriol. Rev.* **35,** 171–205.
Sterlini, J. M., and Mandelstam, J. (1969). *Biochem. J.* **113,** 29–37.
Stewart, B. T., and Halvorson, H. O. (1954). *Arch. Biochem. Biophys.* **49,** 167–178.
Stewart, W. D. P. (1973). *In* "The Biology of Blue-Green Algae" (N. G. Carr and B. A. Whitton, eds.), pp. 260–278. Univ. of California Press, Berkeley.
Stewart, W. D. P., Haystead, A., and Pearson, H. W. (1969). *Nature (London)* **224,** 226–228.
Sudo, S. Z., and Dworkin, M. (1969). *J. Bacteriol.* **98,** 883–887.
Sudo, S. Z., and Dworkin, M. (1973). *Adv. Microb. Physiol.* **9,** 153–224.
Sumida-Yasumoto, C., and Doi, R. H. (1977). *J. Bacteriol.* **129,** 433–444.
Sutherland, I. W., and MacKenzie, C. L. (1977). *J. Bacteriol.* **129,** 599–605.
Switzer, R. L., Turnbough, C. L., Jr., Brabson, J. S., and Waindle, L. M. (1975). *In* "Spores VI" (P. Gerhardt, R. N. Costilow, and H. L. Sadoff, eds.), pp. 327–334. Am. Soc. Microbiol., Washington, D.C.
Szulmajster, J., and Schaeffer, P. (1961). *Biochem. Biophys. Res. Commun.* **6,** 217–223.
Terrana, B., and Newton, A. (1975). *Dev. Biol.* **44,** 380–385.
Thomas, J. (1970). *Nature (London)* **228,** 181–183.
Tipper, D. J., and Gauthier, J. J. (1972). *In* "Spores V" (H. O. Halvorson, R. Hanson, and L. L. Campbell, eds.), pp. 3–12. Am. Soc. Microbiol., Washington, D.C.
Tipper, D. J., Johnson, C. W., Ginther, C. L., Leighton, T., and Wittman, H. G. (1977a). *Mol. Gen. Genet.* **150,** 147–159.
Tipper, D. J., Pratt, I., Guinand, M., Holt, S. C., and Linnett, P. E. (1977b). *In* "Microbiology—1977" (D. Schlessinger, ed.), pp. 50–68. Am. Soc. Microbiol., Washington, D.C.
Tono, H., and Kornberg, A. (1967). *J. Biol. Chem.* **248,** 2375–2382.
Vinter, V. (1969). *In* "The Bacterial Spore" (G. W. Gould and A. Hurst, eds.), pp. 73–123. Academic Press, New York.
Voelz, H. (1966). *Arch. f. Mikrobiol.* **55,** 110–115.
Vold, B. S. (1973). *J. Bacteriol.* **114,** 178–182.
Vold, B. S. (1975). *In* "Spores VI" (P. Gerhardt, R. N. Costilow, and H. L. Sadoff, eds.), pp. 282–289. Am. Soc. Microbiol., Washington, D.C.
Walsby, A. E., and Nichols, B. W. (1969). *Nature (London)* **221,** 673–674.
Warth, A. D. (1977). *Adv. Microbiol. Physiol.* **15,** 1–45.
Warth, A. D., and Strominger, J. L. (1969). *Proc. Natl. Acad. Sci. U.S.A.* **64,** 528–535.
Watson, B. F., and Dworkin, M. (1968). *J. Bacteriol.* **96,** 1465–1473.
Weber, M. M., and Broadbent, D. A. (1975). *In* "Spores VI" (P. Gerhardt, R. N. Costilow, and H. L. Sadoff, eds.), pp. 411–417. Am. Soc. Microbiol., Washington, D.C.
Westby, C. A., and Tsai, W. C. (1977). *Abstr. Annu. Meet. Am. Soc. Microbiol.* I-77, p. 167.
White, D. (1975). *In* "Spores VI" (P. Gerhardt, R. N. Costilow, and H. L. Sadoff, eds.), pp. 44–51. Am. Soc. Microbiol., Washington, D.C.
White, D., Dworkin, M., and Tipper, D. J. (1968). *J. Bacteriol.* **95,** 2186–2197.
Wilcox, M., Mitchison, G. J., and Smith, R. J. (1975). *In* "Microbiology—1975" (D. Schlessinger, ed.), pp. 453–463. Am. Soc. Microbiol., Washington, D.C.
Wildermuth, H., and Hopwood, D. A. (1970). *J. Gen. Microbiol.* **60,** 51–59.
Wilkinson, B. J., Deans, J. A., and Ellar, D. J. (1975). *Biochem. J.* **152,** 561–569.
Winkenbach, F., and Wolk, C. P. (1973). *Plant Physiol.* **52,** 480–483.

Winogradsky, S. (1938). *Ann. Inst. Pasteur, Paris* **60,** 351–400.
Wireman, J. W., and Dworkin, M. (1975). *Science* **189,** 516–523.
Wise, J., Swanson, A., and Halvorson, H. O. (1967). *J. Bacteriol.* **94,** 2075–2076.
Witkin, S., and Rosenberg, E. (1970). *J. Bacteriol.* **103,** 641–649.
Wolk, C. P. (1965). *Dev. Biol.* **12,** 15–35.
Wolk, C. P. (1973). *Bacteriol. Rev.* **37,** 32–101.
Wolk, C. P. (1975). *In* "Spores VI" (P. Gerhardt, R. N. Costilow, and H. L. Sadoff, eds.), pp. 85–96. Am. Soc. Microbiol., Washington, D.C.
Wood, N. B., Rake, A., and Shapiro, L. (1976). *J. Bacteriol.* **126,** 1305–1315.
Wright, B. E. (1973). "Critical Variables in Differentiation." Prentice-Hall, Englewood Cliffs, New Jersey.
Young, I. E., and Fitz-James, P. C. (1959a). *J. Cell Biol.* **6,** 467–481.
Young, I. E., and Fitz-James, P. C. (1959b). *J. Cell Biol.* **6,** 483–498.
Zeig, J., Silverman, M., Hilmen, M., and Simon, M. (1977). *Science* **196,** 170–172.
Zytkovicz, T. H., and Halvorson, H. O. (1972). *In* "Spores V" (H. O. Halvorson, R. Hanson, and L. L. Campbell, eds.), pp. 49–52. Am. Soc. Microbiol., Washington, D.C.

CHAPTER 2

Nitrogen Fixation

WINSTON J. BRILL

I. Introduction

Most organisms are limited in growth because of N limitation. This is apparent from the mere observation that many plants increase in yield when nitrogenous fertilizer is added, and many animals (including humans) are lacking in protein, the major nitrogenous constituent in cells. This nitrogen deprivation seems paradoxical when it is realized that the gas, N_2, is ubiquitous and accounts for 80% of the air we breathe. The paradox is explained by the fact that no animal or plant can use N_2 as a nitrogen source. The ability to break the extremely stable triple bond between the two N atoms in N_2 is restricted to a few species of bacteria, the N_2-fixing bacteria.

Why is it that plants, animals, and most bacteria are unable to fix N_2? There certainly seems to be selective pressure for such a metabolic system. What special adaptations have to be made in order for an organism to fix N_2? These will be the major questions that will be discussed in this chapter.

Nitrogen fixation plays an integral role in the nitrogen cycle, because N_2 fixation is required to recycle fixed nitrogen that is lost to the atmosphere during denitrification. Nitrogen-fixing organisms are best known for their ability to supply fixed nitrogen to plants, such as legumes, allowing these plants to grow without nitrogenous fertilizer. Bacteria that fix N_2 represent

ISBN 0-12-307207-7

a minority of species, but exist within a wide range of genera. These bacteria frequent diverse habitats. Physiological, biochemical, ecological, and genetic studies have been performed with many of these bacteria. Examples of N_2-fixing bacteria include *Azotobacter vinelandii* (strict aerobe), *Clostridium pasteurianum* (strict anaerobe), *Klebsiella pneumoniae* (facultative aerobe), *Rhodospirillum rubrum* (photosynthetic), *Anabaena cylindrica* (a cyanobacterium), and *Rhizobium japonicum* (nodulates soybean roots).

If the ability to fix N_2 appeared early in the evolutionary development of prokaryotes, then we must attempt to understand why most organisms are unable to fix N_2. However, if N_2 fixation is a reaction that has been acquired relatively recently (on an evolutionary scale), then we should try to understand the barriers that keep the N_2-fixing ability within restricted species of prokaryotes and we should determine what must be overcome to extend this ability to organisms that cannot use N_2. These considerations will be developed after the properties of nitrogenase and N_2-fixing organisms have been discussed.

Insight into N_2 fixation has developed rapidly within the last decade. More detailed information on this process can be obtained from recently published books and reviews that are available on general topics (Quispel, 1974; Burns and Hardy, 1975), regulation (Brill, 1975), genetics (Streicher and Valentine, 1974; Brill, 1975), biochemistry (Burris and Orme-Johnson, 1974; Eady and Postgate, 1974; Zumft and Mortenson, 1975; Newton and Nyman, 1976; Winter and Burris, 1976), the *Rhizobium*–legume symbiosis (Quispel, 1974; Nutman, 1976), and cyanobacteria (Stewart, 1973; Wolk, 1973).

II. Biochemistry

A. Nitrogenase

It will be apparent that an understanding of the biochemistry of N_2 fixation is necessary to appreciate the demands specifically made for an organism to fix N_2 and thus adapt and compete successfully in a nitrogen-deficient environment. Nitrogenase catalyzes the reduction of N_2 to ammonium. Because nitrogenases from a variety of bacteria share many common properties, generalizations can be made about the enzyme even though nitrogenases that have been studied in greatest detail come from only a few organisms—*C. pasteurianum* (Tso *et al.*, 1972; Tso, 1974; Nakos and Mortenson, 1971a), *A. vinelandii* (Burns *et al.*, 1970; Kleiner and Chen, 1974), and *K. pneumoniae* (Eady *et al.*, 1972).

The enzyme is composed of two proteins (reviewed in Winter and Burris, 1976) that can readily be separated by chromatography on a DEAE-cellulose column (Bulen and LeComte, 1966). A common nomenclature for these proteins has not yet been agreed upon. The first protein to be eluted from the column has been called component I, Mo-Fe protein, or molybdoferredoxin. The protein that binds most tightly to the column is known as component II, Fe-protein, or azoferredoxin. Component I has a molecular weight of approximately 220,000 daltons and is composed of a pair each of two dissimilar subunits (reviewed in Eady and Postgate, 1974). There are about 32 nonheme Fe atoms, 26 acid-labile sulfide atoms, and 2 atoms of molybdenum in each molecule of component I. Component II has a molecular weight of approximately 65,000 daltons and is comprised of two identical subunits which share 4 nonheme Fe atoms and 4 acid-labile sulfide atoms. Studies with electron paramagnetic resonance spectroscopy showed that reduced component II is responsible for reducing component I (Orme-Johnson *et al.*, 1972; Smith *et al.*, 1972). Presumably, the active site for reduction of N_2 is located on component I (Bui and Mortenson, 1969).

A variety of Mo-containing proteins (e.g., nitrate reductase, xanthine oxidase, sulfite oxidase, aldehyde oxidase, and component I of nitrogenase) seem to contain a common cofactor, since acidification of these proteins yields a factor that is able to reconstitute activity *in vitro* in an extract of a mutant of *Neurospora crassa* that has an inactive nitrate reductase (Ketchum *et al.*, 1970; Nason *et al.*, 1971). A Mo-containing cofactor, isolated from nitrogenase component I, is a small peptide and contains 1 Mo : 8 Fe as well as 6 acid-labile sulfides (Shah and Brill, 1977). This cofactor (FeMo-co) is capable of activating extracts of mutant strains of *A. vinelandii* (Nagatani *et al.*, 1974) and *K. pneumoniae* (St. John *et al.*, 1975) that produce component I but are unable to produce an active FeMo-co. Spectral data indicates that FeMo-co is an active site of nitrogenase (Orme-Johnson, Shah, and Brill, unpublished observations).

FeMo-co from component I is not identical to the Mo cofactors from the other Mo-containing enzymes. None of these Mo cofactors has yet been characterized, but the Mo cofactor from nitrate reductase, at least, cannot contain any Fe, since the purified enzyme has no nonheme Fe (Jacob, 1976). Presumably, the FeMo-co must share some common features with the Mo cofactor of nitrate reductase, since acid-treated component I will activate nitrate reductase that lacks the Mo cofactor (Nason *et al.*, 1971).

When a cell uses N_2 as its sole nitrogen source, approximately 1–2% of the total cell protein is nitrogenase (V. K. Shah, unpublished results). This high level of enzyme is required because nitrogenase has a substrate turnover number that is quite low: 50–100 molecules of N_2 reduced per minute per nitrogenase molecule (Burns and Hardy, 1975). This turnover number should be compared with the turnover number for nitrate reductase, which is approximately 100 times greater (Jacob, 1976).

B. Energy Requirement

Since nitrogenase is a major protein in a N_2-fixing cell, it is apparent that a great deal of energy is required to generate nitrogen-fixing capabilities. Besides protein synthesis, energy also is needed to transport and metabolize exogenously supplied Fe, Mo, and S. The synthesis of FeMo-co may be an additional energy drain on the cell.

The most significant energy drain, however, is due to actual catalysis of the reaction. For each N_2 fixed, six electrons are required. These electrons are derived from a variety of carbon sources; the final reduction seems to be mediated through ferredoxins and/or flavoproteins (Mortenson, 1964; Yates, 1972; Yoch, 1974). In cyanobacteria, the electrons are derived from water via photosynthesis. A most striking phenomenon, common to all nitrogenases, is that 12–30 ATP's are hydrolyzed to ADP and phosphate for each molecule of N_2 that is reduced (Hardy and Knight, 1966; Winter and Burris, 1968; Hadfield and Bulen, 1969). This is surprisingly high, since the overall reaction of N_2 reduction to 2 NH_3 by H_2 is exergonic (Bayliss, 1956). The molybdenum cofactor (FeMo-co) is readily inactivated by water (Shah and Brill, 1977). A possible role for ATP might be to remove water from FeMo-co through ATP hydrolysis.

The energy drain on a N_2-fixing cell is best demonstrated by the observation that a N_2-fixing bacterium generally grows much more rapidly in media containing ammonium than in nitrogen-free media in which growth depends on N_2 fixation. For instance, *A. vinelandii* doubles every 2 hours in ammonium, but doubles every 3 hours on N_2 (Gordon and Brill, 1972).

C. Oxygen Sensitivity

Both components of nitrogenase are extremely O_2 labile. This is true for nitrogenases isolated from aerobes (Shah and Brill, 1973) as well as from anaerobes (Zumft and Mortenson, 1975). Mere addition of reducing agents or removal of O_2 are not sufficient to reverse this inactivation. All biochemical manipulations with purified components, therefore, are performed under strictly anaerobic conditions. The FeMo-co also is very sensitive to O_2—total inactivation results from less than 1-minute exposure to air (Shah and Brill, 1977). The component I that lacks FeMo-co in a mutant strain still is sensitive to O_2; therefore, FeMo-co is not the only site in component I that is inactivated by O_2 (Shah, unpublished observations).

Nitrogen-fixing cells obviously have to keep O_2 away from their nitrogenase. For strict anaerobes there is no problem. However, facultative anaerobes such as *K. pneumoniae* fix N_2 only under anaerobic conditions, although they can grow aerobically or anaerobically on fixed N compounds (Pengra and Wilson, 1958). These organisms seems to be unable to protect nitrogenase from being O_2 inactivated. *Spirillum lipoferum* (Okon *et al.*, 1976a) grows on fixed N compounds aerobically, but requires microaerophilic conditions for N_2 fixation. Enough O_2 must be available for ATP synthesis, but too much O_2 destroys nitrogenase. The most remarkable situation is found in strict aerobes, such as *Azotobacter*. This organism has an extremely high respiratory quotient (Phillips and Johnson, 1961) which probably plays a role in reducing free O_2 before it reaches nitrogenase. Many of the aerobic N_2-fixing organisms have a thick slimy capsule which may help to keep high levels of O_2 from reaching the cell (Hill, 1971). It should be noted, however, that a nongummy mutant strain of *A. vinelandii* fixes N_2 and grows very well in air (Bush and Wilson, 1959). *Azotobacter* cells are much larger than most bacteria, therefore the surface–volume ratio is relatively low. This also should impede O_2 uptake (Mulder and Brotonegoro, 1974).

Many cyanobacteria grow in chains of cells which form filaments. When some of these species are starved of fixed N, the filaments develop a low percentage of specialized cells known as heterocysts. These cells can easily be distinguished by light microscopy due to their thick wall. It seems that N_2 fixation occurs exclusively in these heterocysts (Stewart *et al.*, 1969; Wolk and Wojciuch, 1971; Fleming and Haselkorn, 1973). Unlike other cells of the filament, these specialized cells lack photosystem II (the O_2-evolving photosystem) and thus do not generate O_2 in the vicinity of nitrogenase (Fay, 1970; Lyne and Stewart, 1973). Heterocysts, therefore, protect nitrogenase from O_2. The thick wall of the heterocyst may also act as a barrier to O_2. There are some unicellular cyanobacteria that fix N_2 without producing a heterocyst-like cell. An example is *Gleocapsa*, which fixes N_2 most efficiently under microaerophilic conditions (Stewart and Lex, 1970).

In the case of the *Rhizobium*–legume symbiosis, a myoglobin-like protein, leghemoglobin, surrounds N_2-fixing *Rhizobium* cells in the nodule. This protein apparently distributes O_2 to *Rhizobium* for oxidative phosphorylation, thus supplying ATP for nitrogenase activity. At the same time it keeps free O_2 from inactivating nitrogenase (Bergersen and Briggs, 1958a).

In facultative aerobes, such as *K. pneumoniae*, the ATP required for N_2 fixation cannot come from the optimum ATP source, oxidative phosphorylation, because O_2 prevents N_2 fixation. Instead, these bacteria have to sacrifice ATP production merely to keep nitrogenase from being O_2 inactivated. Since there is continued selective pressure to produce an O_2-stable nitrogenase in nature, why has no mutant strain been found that produces an O_2-insensitive nitrogenase? First of all, at least three O_2-labile sites must be modified—apocomponent I, FeMo-co, and component II. Second, O_2 is an inhibitor of nitrogenase activity (Wong and Burris, 1972). It seems, therefore, that the nitrogenase active sites (containing reduced metals and sulfide) can only be formed in a structure that cannot exclude O_2.

D. Hydrogen Evolution

Another problem faced by N_2-fixing cells is that nitrogenase catalyzes the apparently wasteful process of ATP-dependent H_2 evolution (Bulen *et al.*, 1965). In the absence of the substrate, N_2, electron donors still reduce the enzyme, and these electrons combine with protons to yield H_2. This reaction wastes both ATP and electrons. A natural situation, in which N_2 is not available to the enzyme, probably is unusual. However, even in the presence of N_2, nitrogenase distributes about half of the incoming electrons to N_2 and the other half to H_2 (Bulen *et al.*, 1965). This is not an artifact caused by *in vitro* assays, but occurs *in vivo* as well (Lee and Wilson, 1943; Hamilton *et al.*, 1964; Smith *et al.*, 1976).

Of what possible value is ATP-dependent H_2 evolution to the organism? It does not seem that protection from O_2 might be achieved by using these electrons to reduce O_2. Anaerobically grown *K. pneumoniae* shunts approximately the same high percentage of electrons from nitrogenase to H_2 as does the aerobe, *A. vinelandii* (Smith *et al.*, 1976). There seems to be no good reason why these cells need this ATP-requiring electron loss. It is possible that the active sites of the various nitrogenases are made in such a manner that H_2 formation is a secondary reaction and that these active sites (at this stage in their evolution) have no means of preventing this wasteful reaction.

Electron loss and ATP hydrolysis, by nitrogenase-catalyzed H_2 evolution, has not yet been overcome by modification of the nitrogenase structure, but several organisms are able to recoup the electrons and/or ATP by oxidizing H_2 via a hydrogenase that is not a part of nitrogenase (Hyndman *et al.*, 1953; Dixon, 1972). The H_2-oxidizing hydrogenase recycles H_2 lost by nitrogenase and resupplies electrons to nitrogenase. Hydrogen oxidation seems to cause a net synthesis of ATP which also can be used for N_2 fixation. This might be one of the reasons that *Azotobacter* fixes N_2 more efficiently than *C. pasteurianum in vivo* (see Smith *et al.*, 1976). Some *Rhizobium*–legume symbioses may be more efficient than others because the nodules contain a H_2-oxidizing hydrogenase (Schubert and Evans, 1976).

III. Regulation

A. Inhibition of Nitrogenase Activity

The previous section focused on the special requirements that an organism has to satisfy if it is to fix N_2. The demands for high levels of ATP and electrons certainly tax a N_2-fixing bacterium. When such an organism finds itself in a medium containing sufficient fixed N, an obvious regulatory mechanism is feedback inhibition. This should prevent valuable ATP and electron waste under conditions where N_2 fixation is not necessary. Strangely, many N_2-fixing bacteria do not have a mechanism for feedback inhibition of nitrogenase. When N_2-fixing cultures are placed in a medium containing a very good N source, such as ammonium, nitrogenase activity remains high. Examples of nitrogenases that exhibit no feedback inhibition include those from *A. vinelandii*, *K. pneumoniae*, and *C. pasteurianum* (Gordon, Shah, and Brill, 1976; Gordon, unpublished observations). Perhaps these organisms have an environment that is constantly deficient in available fixed N, and thus the bacteria have not evolved a mechanism to inactivate nitrogenase. This seems unlikely since there is strong control of nitrogenase synthesis (discussed in Section III,B). Another possibility, although unlikely, is that the nitrogenase proteins are unable to tolerate the addition of a site that could be used for feedback inhibition.

When ammonium is added to a culture of *R. rubrum*, on the other hand, nitrogenase is immediately inactivated (Schick, 1971). When this organism is exposed to air, nitrogenase also is inactivated but can be reactivated when O_2 is removed (Neilson and Nordlund, 1975). Reactivation in *R. rubrum* requires a constitutively made protein (activating factor) which interacts with component II, but not component I, of nitrogenase (Ludden and Burris, 1976). The same type of control mechanism has been detected in *S. lipoferum* (Ludden, Okon, and Burris, unpublished observations). Activation by the activating factor requires ATP, but the exact mechanism of activation and inactivation is not yet understood.

B. Regulation of Nitrogenase Synthesis

1. Introduction

Since N_2 is ubiquitous, control by induction would not be useful to a cell so it is not surprising that N_2 is not required for nitrogenase synthesis (Parejko and Wilson, 1970; Daesch and Mortenson, 1972). Rather, nitrogenase synthesis is under repression–derepression control (Shah *et al.*, 1972; Tubb and Postgate, 1973; Collmer and Lamborg, 1976).

2. Repression by Fixed N

In all N_2-fixing systems that have been examined, utilizable fixed nitrogen decreases the N_2-fixing ability of the system. The faster a cell will grow on a particular nitrogen source, the greater the repression of nitrogenase synthesis is caused by that source (Wilson *et al.*, 1943). Ammonium is the best nitrogen source for *A. vinelandii*, *C. pasteurianum*, and *K. pneumoniae*. No trace of nitrogenase activity can be detected when these bacteria are grown for many generations in media containing excess ammonium (Parejko and Wilson, 1970; Daesch and Mortenson, 1972; Shah *et al.*, 1972). Nitrate represses nitrogenase synthesis approximately 50% in *A. vinelandii*, as long as nitrate reductase is active. A mutant strain lacking nitrate reductase is not repressed for nitrogenase synthesis by nitrate (Sorger, 1969); therefore, ammonium formation is required for repression.

In *K. pneumoniae*, the mechanism for repression of nitrogenase synthesis seems to involve the enzyme glutamine synthetase. Many species of bacteria use glutamine synthetase and glutamate synthase for production of glutamine when cells are limited for growth by fixed nitrogen (Meers *et al.*, 1970). When cells are grown in media containing excess ammonium, then glutamate dehydrogenase is responsible for glutamate synthesis (Fig. 1). High glutamate dehydrogenase and low glutamine synthetase activities are found in ammonium-grown cells; low glutamate dehydrogenase and high glutamine synthetase activities are in cells grown on poor nitrogen sources. It seems that N_2-fixing bacteria also use glutamate dehydrogenase when they grow with excess ammonium, but use glutamine synthetase for glutamate biosynthesis when they fix N_2 (Nagatani *et al.*, 1971). As discussed above, ammonium is a much better nitrogen source than N_2 for many N_2-fixing bacteria.

In enteric bacteria, at least, glutamine synthetase plays not only a catalytic role but also plays a direct regulatory role in the synthesis of enzymes involved with the degradation of compounds that are poor N sources (see Magasanik *et al.*, 1974). Glutamine synthetase regulates transcription of genes coding for enzymes of certain catabolic pathways, such as histidine and proline degradation. Glutamine synthetase also regulates the synthesis of both itself and glutamate dehydrogenase. The active deadenylylated form of glutamine synthetase activates transcription of genes such as the *hut* (histidine catabolic) genes of *K. aerogenes* and *S. typhimurium* (Tyler *et al.*, 1974).

Glutamate synthesis

In excess NH_3

$$\alpha\text{-Ketoglutarate} + NH_3 + NADPH \xrightarrow[\text{dehydrogenase}]{\text{glutamate}} \text{glutamate} + NADP + H_2O$$

In limiting NH_3

$$\text{Glutamate} + NH_3 + ATP \xrightarrow[\text{synthetase}]{\text{glutamine}} \text{glutamine} + ADP + P_i + H_2O$$

$$\alpha\text{-Ketoglutarate} + \text{glutamine} + NADPH \xrightarrow[\text{synthase}]{\text{glutamate}} 2\ \text{glutamate} + NADP$$

FIG. 1. Pathways of glutamate biosynthesis.

Evidence that glutamine synthetase plays a similar regulatory role in the synthesis of nitrogenase in *K. pneumoniae* comes from the observation that glutamine auxotrophs, with defects in the gene specifying glutamine synthetase, are unable to synthesize nitrogenase (Streicher *et al.*, 1974; Tubb, 1974). In addition, certain mutations in the gene that codes for glutamine synthetase allow nitrogenase to be synthesized, even in the presence of excess ammonium (Streicher *et al.*, 1974; Tubb, 1974).

It is not known whether glutamine synthetase plays the same regulatory role in N_2-fixing bacteria other than *K. pneumoniae*. An analog of glutamate, *L*-methionine-DL-sulfoximine, inactivates glutamine synthetase while causing derepression of nitrogenase synthesis in *K. pneumoniae* (Gordon and Brill, 1974). Nitrogenase synthesis also is derepressed when this analog is added to *A. vinelandii* (Gordon and Brill, 1974), *Anabaena cylindrica* (Stewart and Rowell, 1975), *S. lipoferum* (Okon *et al.*, 1976b), or *R. rubrum* (Weare and Shanmugam, 1976) growing in the presence of excess ammonium. In the case of *A. cylindrica*, heterocyst formation also is derepressed when methionine sulfoximine is added to ammonium-grown cells (Stewart and Rowell, 1975).

The analog may interact with glutamine synthetase in such a manner that glutamine synthetase is in a conformation that will allow transcription of the *nif* (nitrogen fixation) genes. Another explanation, not involking a regulatory role for glutamine synthetase, is that glutamine (or some metabolite of glutamine) is responsible for repression and that methionine sulfoximine stops glutamine production by inactivating glutamine synthetase.

3. Regulation by Molybdenum

Since nitrogenase activity depends on the presence of Mo in the enzyme, it is critical that sufficient Mo becomes available when cells need to fix N_2. Natural Mo-deficient environments have been shown to limit N_2 fixation (Anderson, 1946; Evans *et al.*, 1951; Lobb, 1953). When a cell needs to fix N_2 but is unable to find Mo, no N_2 fixation is expected, since any component I synthesized would be inactive. Therefore, synthesis of component I should not occur because it would waste energy and amino acids. In fact, *K. pneumoniae* and *C. pasteurianum* (Brill *et al.*, 1974; Cardenas and Mortenson, 1975) do not produce either nitrogenase component when Mo is lacking. The fact that component II is not made is evidence that some control system, sensitive to the internal Mo concentration, affects nitrogenase synthesis.

Azotobacter vinelandii (Nagatani and Brill, 1974) and the non-heterocyst-forming cyanobacterium, *Plectonema boryanum* (Nagatani and Haselkorn, 1977), do synthesize component II when no Mo is available. In the case of *A. vinelandii*, a Mo-storage protein is made that is capable of binding several hundred times the amount of Mo than is required for maximum growth on N_2 (Pienkos and Brill, 1977). This uptake and storage system is found in ammonium-grown cells as well, even though the organism grows as well on ammonium with or without any Mo. *Azotobacter vinelandii*, therefore, takes up large amounts of Mo any time that Mo is available and this intracellular Mo can support N_2 fixation for many generations in an environment that contains no Mo.

4. Regulation by Oxygen

What happens when *K. pneumoniae* is starved for N in an aerobic environment? Nitrogenase might be synthesized since excess fixed N is not available to repress nitrogenase synthesis, but this nitrogenase would presumably be immediately inactivated by O_2. To prevent this, *K. pneumoniae* does not synthesize any nitrogenase in the presence of O_2 (St. John *et al.*, 1974). Degradation of O_2-inactivated nitrogenase is not an explanation for this result since it seems to be diluted out of the cell during cell division. Oxygen also seems to regulate the synthesis of nitrogenase in *P. boryanum* as well (Nagatani and Haselkorn, 1977). The mechanism of this control is not understood, but is may share similar regulatory units that are required for repression of other enzymes that only are needed when cells grow anaerobically.

IV. Genetics

Most of the genetic work with respect to N_2 fixation has been done with *K. pneumoniae*. This organism is closely related to *Escherichia coli* and has a chromosome map order that is similar to that of *E. coli* and *S. typhimurium* (Matsumoto and Tazaki, 1971). Phage P1, which is usually used for transduction crosses in *E. coli*, can be used for genetic mapping in *K. pneumoniae* as well (Streicher *et al.*, 1971). Mu, a phage that has been valuable (Howe and Bade, 1975) for inactivating genes and for selecting deletions in *E. coli*, will also infect a genetically modified *K. pneumoniae* (Bachhuber *et al.*, 1976; Rao, 1976). The mutation presumably affects a gene involved with lipopolysaccharide biosynthesis. These tools and techniques, commonly used for studying *E. coli* genetics, have been invaluable for studying the *nif* genes in *K. pneumoniae*.

The *nif* genes all seem to be clustered in a region that is cotransducible (Streicher *et al.*, 1971; St. John *et al.*, 1975; Bachhuber *et al.*, 1976) with the genes involved with histidine biosynthesis (*his*). There is evidence that the cluster of *nif* genes is made up of at least three operons and that DNA, not required for expression of *nif* genes, separates these operons (Kennedy and Dixon, 1977; T. MacNeil, unpublished observations). Complementation analyses have indicated that there are at least eight *nif* genes (Kennedy and Dixon, 1977; T. MacNeil, unpublished observations). The products of all of these genes have not yet been determined. However, at least two genes are required for the structure of component I apoprotein, one gene is necessary to code for component II, one gene is required for the synthesis of the iron–molybdenum cofactor, and one gene seems to be involved with electron transport to nitrogenase (St. John *et al.*, 1975; G. Roberts, unpublished observations).

Transformation is the only technique (Page and Sadoff, 1976) that has been useful for mapping *nif* mutations in *A. vinelandii.* The *nif* genes of *A. vinelandii* seem to be scattered around the chromosome (Bishop and Brill, 1977) rather than being clustered as in *K. pneumoniae. Nif* mutations in *Rhodospeudomonas capsulata* have been transferred (Wall *et al.*, 1975) between mutant strains, but nothing is known about the arrangement of *nif* genes in this organism. Nif^- mutant strains have also been isolated in *Clostridium pasteurianum* (Simon and Brill, 1971), *Anabaena variabilis* (Currier *et al.*, 1977), and *Rhizobium japonicum* (Maier and Brill, 1976), but no gene transfer experiments with *nif* mutations have yet been performed in these systems.

The cluster of *nif* genes in *K. pneumoniae* was transferred from the chromosome onto an F factor as well as to a drug-resistance transfer factor (Dixon and Postgate, 1971; Dixon *et al.*, 1976; Cannon *et al.*, 1976). A plasmid containing the *K. pneumoniae nif* genes was then transferred by conjugation to bacteria that normally do not fix N_2 (Dixon and Postgate, 1972; Dixon *et al.*, 1976). Examples of such recipients are *E. coli, S. typhimurium* (Postgate and Krishnapillai, 1977), and *Agrobacterium tumefaciens.* In the first two species, the exconjugants fixed N_2 only under anaerobic conditions. *Agrobacterium*, however, is a strict aerobe, and the exconjugants produced antigenically detectable nitrogenase but were unable to fix N_2 (Dixon *et al.*, 1976). Presumably, O_2 rapidly inactivated nitrogenase when it was synthesized in *A. tumefaciens.*

V. N_2-Fixing Symbioses

Various symbiotic relationships are known to involve a N_2-fixing prokaryote with a eukaryote. While the *Rhizobium*–legume root system symbiosis is most important in agriculture, other N_2-fixing plant symbioses also involve N_2-fixing bacteria in root nodules. Nodulated N_2-fixing trees include *Alnus* (alder), *Myrica, Casuarina,* and *Elaeagnus;* nodulated shrubs include *Purshia* and *Ceanothus* (see Bond, 1974; Quispel, 1974; Becking,

1976b). All of these plants seem to contain N_2-fixing species of *Frankia* in the nodules. The water fern, *Azolla*, has pores in its leaves that contain the N_2-fixing cyanobacterium, *Anabaena* (Peters, 1976). *Gunnera* is a stout herb that contains glands on the stem which are invaded by species of cyanobacteria (*Nostoc*) that form N_2-fixing nodules within the stem (see Millbank, 1974; Becking, 1976a).

The plants mentioned above can obtain all of their N from their N_2-fixing symbiont; therefore, such plants usually dominate in environments that are deficient in fixed nitrogen. The prokaryote, in return, receives photosynthate and a protected environment from the plant. After repeated growth of these plants in a nitrogen-limited environment, the soil or water becomes enriched with fixed nitrogen and other plants tend to overtake the N_2-fixing ones. The enrichment of the soil is taken advantage of by crop rotation between legumes and nonlegumes.

Many lichens harbor cyanobacteria that fix N_2 for the fungus (Bond and Scott, 1955; Hitch and Stewart, 1973; Millbank and Kershaw, 1969). Nitrogen fixation can continue in a lichen during several weeks of desiccation (Henriksson and Simu, 1971). In lichens, fungi presumably supply protection and minerals to the cyanobacteria.

These symbiotic relationships are generally disrupted when sufficient fixed nitrogen is available for the eukaryote. For example, no nodules are formed on legumes that grow in the presence of suitable quantities of fixed nitrogen (Fred *et al.*, 1932; Maier and Brill, 1976).

The best-studied N_2-fixing symbiosis is the *Rhizobium*–legume system. Legumes are pod-bearing plants such as soybean, clover, alfalfa, peanut, bean, and pea. There are several species of *Rhizobium*, each somewhat specific for the range of plants that it can nodulate (Fred *et al.*, 1932). For instance, *R. japonicum* nodulates soybean, *R. meliloti* nodulates alfalfa, *R. trifolii* nodulates clover, and *R. leguminosarum* nodulates pea. Some *Rhizobium* species are less specific. For instance, *Rhizobium* of the "cowpea" type may nodulate both soybeans and cowpeas.

It seems that the plants produce proteins (lectins) that recognize the cell surface of the specific *Rhizobium* symbiont (Hamblin and Kent, 1973; Bohlool and Schmidt, 1974; Dazzo and Hubbell, 1975; Wolpert and Albersheim, 1976). Capsular polysaccharide, possibly the O antigen of the bacterium (Dazzo and Hubbell, 1975; Wolpert and Albersheim, 1976; Dazzo and R. M. Maier, unpublished observations), is the site that interacts with the plant lectin. The capsular polysaccharide from *R. trifolii* binds specifically to clover root hairs (Dazzo and Brill, 1977). Most of the lectin is found on the surface of root hair tips (Dazzo, unpublished observations). During the initial stages of infection, *Rhizobium* cells attach to the tips of root hairs. The binding of both the capsular polysaccharide to clover and the clover lectin (called trifoliin) to *R. trifolii* cells is inhibited by 2-deoxyglucose (Dazzo and Hubbell, 1975; Dazzo and Brill, 1977).

A model proposes that trifoliin recognizes both a site on the capsular polysaccharide and a site on the root hair surface. The two sites share common features in that both specific recognition reactions are inhibited by 2-deoxyglucose and that the two sites share common antigenic determinants (Dazzo and Hubbell, 1975).

After the recognition step, the root hair wall invaginates from the tip and follows a path into the body of the plant cell (Fred *et al.*, 1932; Napoli and Hubbell, 1975). *Rhizobium* cells follow the invagination (known as the infection thread), which sometimes passes through many plant cells. The plant cells enlarge and multiply at the site which will become a developing nodule. Nodule cells become almost completely filled with *Rhizobium*, which are contained within packets of plant-derived membrane (Bergersen and Briggs, 1958b; Tu, 1974).

Rhizobium contains the genetic information for the structural genes of nitrogenase (Phillips *et al.*, 1973; Bishop *et al.*, 1975; Keister, 1975; Kurz and LaRue, 1975; Maier and Brill, 1976; McComb *et al.*, 1975; Pagan *et al.*, 1975; Tjepkema and Evans, 1975; Page, 1977), and active nitrogenase is found within bacteria isolated from the nodule (Koch *et al.*, 1967; Bergersen and Turner, 1970). Plant nodule cells that contain the bacteria are filled with the O_2-binding protein, leghemoglobin (Bergersen and Briggs, 1958a). This protein is coded for by the plant (Cutting and Schulman, 1968; Dilworth, 1969; Verma *et al.*, 1974) and presumably delivers O_2 for respiration, but prevents free O_2 from inactivating nitrogenase (Wittenberg *et al.*, 1974) in the *Rhizobium* (see Section II,C). It is not clear whether the leghemoglobin (the only known hemoglobin in the plant kingdom) surrounds the membrane-bound packets (Verma and Bal, 1976) or whether the leghemoglobin is enclosed within the packets (Bergensen and Goodchild, 1973). Leghemoglobin is only found in legume nodule tissue. Nodules from other N_2-fixing plants do not contain leghemoglobin, but might have other O_2-scavenging proteins or may have alternate mechanisms for keeping O_2 away from nitrogenase (see Bond, 1974).

The control of nodulation must be quite complex since millions of bacteria produce only tens of nodules and fully fertilized plants do not produce any nodules at all. Presumably the control of infection is the key step that prevents N_2 fixation when it is not needed.

No one has yet conclusively demonstrated that any *Rhizobium* strain is capable of growing on N_2 asymbiotically. There are certain strains, however, that fix N_2 in the absence of the plant; but these strains need a fixed N source for growth and thus are not able to grow solely on the N_2 that is fixed (Keister, 1975; Kurz and LaRue, 1975; Tjepkema and Evans, 1975). There does not seem to be any significant repression by fixed N of this activity (Keister, 1975; Kurz and LaRue, 1975). However, there is some indication that glutamine synthetase plays some role in the regulation of nitrogenase synthesis (Kondorosi *et al.*, 1977; Ludwig and Signer, 1977). It seems that most of the N_2 that is fixed by the *Rhizobium* is excreted (Bergersen and Turner, 1967; O'Gara and Shanmugan, 1976). Furthermore, the only way asymbiotic N_2-fixing activity can be produced is under very stringent conditions of low partial pressures of O_2 (Tjepkema and Evans, 1975; Keister and Evans, 1976). Too much O_2 inactivates nitrogenase, and too little O_2 prevents sufficient ATP from being made through respiration.

The legume N_2-fixing systems obviously are much more difficult to work with than the free-living N_2-fixing bacteria causing progress in understanding the biochemistry, genetics, and regulation of these symbiotic systems to be quite slow. Mutant strains of *Rhizobium* have recently been obtained with defects in genes specifically required for N_2 fixation (Maier and Brill, 1976; Beringer *et al.*, 1977; W. Leps, unpublished observations) and nodulation (Maier and Brill, 1976; W. Leps, unpublished observations). Gene transfer systems are now available, and chromosomes of several *Rhizobium* species are being mapped with respect to antibiotic resistance and auxotrophic markers (Beringer and Hopwood, 1976; Meade and Signer, 1977). It has been possible to transfer *nif* genes (Page, 1977) as well as genes involved in the infection process (Bishop *et al.*, 1977) from *Rhizobium* to Nif$^-$ mutant strains of *A. vinelandii*.

VI. Nitrogen-Fixing Associations

A. Bacteria–Plant Associations

Nitrogen-fixing nodules are the most efficient systems for supplying a plant with fixed nitrogen produced by bacterial N_2 fixation. Inside the nodule, the bacteria are given photosynthate and are protected from competitive bacteria and from environmental stresses such as desiccation. There are reports of looser associations between N_2-fixing bacteria and plants that do not result in the formation of nodules or other specialized structures. For instance, *Beijerinckia* is associated with the rhizosphere of sugar cane (Döbereiner, 1961; Döbereiner *et al.*, 1972a). Perhaps the bacterium supports growth and N_2 fixation by catabolizing sucrose from the root exudate. The grass, *Paspalum notatum* (bahiagrass), has *Azotobacter paspali* on its roots, (Döbereiner *et al.*, 1972b; Kass *et al.*, 1971), but it is not clear whether the stimulation of plant growth by *A. paspali* is due to N_2 fixation or to plant growth hormones produced by the bacteria (Brown, 1976).

Digitaria decumbens is a tropical grass that has the N_2-fixing bacterium, *Spirillum lipoferum*, associated with its roots. Claims have been made that *S. lipoferum* supplies substantial fixed N to the plant (Von Bülow and Döbereiner, 1975; Döbereiner and Day, 1976). Grasses, such as *D. decumbens*, are C-4 plants which have malate as a primary photosynthetic product (Hatch and Slack, 1970). Interestingly, *S. lipoferum* grows best on N_2 when malate is the carbon source (Okon *et al.*, 1976b). Since *S. lipoferum* fixes N_2 only microaerophilically, very specific conditions must be met for N_2 fixation to proceed. It is possible that the bacteria are within root cortical cells (Döbereiner and Day, 1976) or in clumps which create a microaerophilic environment for the bacteria. Since the highest reported N_2-fixing activities for the *S. lipoferum*–root association seem to be in wet tropical climates, it may be that N_2 fixation only occurs under certain soil conditions having high metabolic activity depleting most of the O_2. The amount of fixed nitrogen given to the plant through this association, however, actually seems to be quite low (Tjepkema and van Berkum, 1977). A survey of grasses in Wisconsin showed that *S. lipoferum* can be isolated from most of these grasses (S. L. Albrecht, Okon, and Burris, unpublished observations), but the bacteria do not seem to supply significant amounts of fixed N to the plant.

It has been observed that roots from unusually green corn plants have higher N_2-fixing activity than roots of normal plants (Raju *et al.*, 1972). The organism that might be responsible for this activity is *Enterobacter cloacae*. It has not been determined, however, whether the association causes the superior greening or whether some other factor is responsible for the greening and the *Enterobacter* merely are selected for by healthier plants. Several non-*Rhizobium* microbial inoculants, consisting of N_2-fixing bacteria, are currently sold commercially. Many more studies need to be made to determine whether some of these N_2-fixing associations with crop plants actually contribute significant amounts of fixed N to the plant. Such associations would certainly be a major contribution to agriculture.

Another important association is the relationship of wood-rotting fungi with N_2-fixing bacteria, mostly gram-negative fermentative bacteria (Aho *et al.*, 1976). These bacteria are probably supplied soluble carbohydrate by the fungi in return for fixed N. In a northern hardwood forest, approximately 68% of the nitrogen entering the ecosystem each year comes from N_2 fixation (Bormann *et al.*, 1977). The sources of this fixation could include the fungal associations, lichens, as well as free-living N_2-fixing bacteria. Legumes do not seem to be a factor.

B. Bacteria–Animal Associations

Termites and shipworms (a bivalve mollusk) live on food, such as wood, that is very high in carbon but low in nitrogen. These organisms have the unusual property of harboring significant numbers of N_2-fixing bacteria in their intestine (Benemann, 1973; Breznak *et al.*, 1973; Carpenter and Culliney, 1975). The termite gut fixes N_2 by means of bacteria, presumably *Enterobacter agglomerans* (Potrikus and Breznak, 1976) and/or *Citrobacter freundii* (French *et al.*, 1976). All termites tested had this activity and no N_2-fixing activity was found in a wide variety of other insects. When termites were fed substrates with high fixed N levels, N_2 fixation was depressed (Breznak *et al.*, 1973).

Nitrogen-fixing bacteria have been isolated from intestinal contents of animals, including humans (Bergersen and Hipsley, 1970). Species capable of N_2 fixation have been isolated from human feces, and they include *K. aerogenes*, *Enterobacter cloacae*, and *E. coli* (Bergerson and Hipsley, 1970). These bacteria probably do not contribute significant amounts of fixed N to the human diet.

Klebsiella pneumoniae is an organism that is frequently found in the soil, on plant material, in water, and also as a pathogen of man. One might suspect that strains isolated from plants would fix N_2, whereas pathogenic strains would be unable to fix N_2. The surprising result is that many clinical isolates were capable of fixing N_2 (Chambers and Silver, 1977). It is not known whether such strains fix N_2 during infection. Perhaps these disease-forming organisms actually were introduced by plant material with which the patient had contact. *Klebsiella pneumoniae* is known to be an opportunistic pathogen and thus may be around all the time, causing disease when an individual is already weakened by another disease. If this is true, then it would be very interesting to examine adaptations that *K. pneumoniae* has to make to survive in such different environments as the lung and the surface of a plant.

VII. Evolution of N_2 Fixation

Is the ability of fix N_2 a relatively recent acquisition, or has this ability been with some bacteria before the development of eukaryotes? There have been several discussions on the evolution of N_2 fixation, and it certainly is not clear when the first N_2-fixing organisms developed (Silver and Postgate, 1973; Postgate, 1974; Burns and Hardy, 1975). It is still debatable whether substantial amounts of ammonium were present in the prebiotic environment (Bada and Miller, 1968; Ferris and Nicodem, 1972). if we can understand how N_2 fixation was acquired and the particular selective pressures that allowed this system to develop, then this information should be useful for manipulating present systems by such techniques as plant breeding, genetic engineering, or enzyme engineering to produce new more efficient N_2-fixing systems.

Points that must be considered to rationalize an evolutionary scheme include: (a) nitrogenases, from bacteria having widely different physiological properties, are very similar (see Eady and Postgate, 1974); (b) a nitrogenase component from one bacterium usually complements *in vitro* the other component from a different bacterium (Detroy *et al.*, 1968); (c) even components from noncomplementing combinations (e.g., *C. pasteurianum* and *A. vinelandii*) interact with each other (Emerich and Burris, 1976); (d) corresponding components from different organisms cross-react when tested by serological techniques (Maier and Brill, 1976; Brill, unpublished observations); (e) there are a variety of substrates (e.g., acetylene, cyanide, azide, nitrous oxide) other than N_2 that all nitrogenases are capable of reducing (reviewed in Zumft and Mortenson, 1975; Burns and Hardy, 1975); (f) the active centers (FeMo-co) from different nitrogenases are identical when analyzed by electron paramagnetic resonance and Mössbauer spectroscopy (Shah, Orme-Johnson, E. Münck, and Brill, unpublished observations). Did these striking similarities arise from an ancestral N_2-fixing organism, or are these common properties merely due to convergent evolution?

If nitrogenase was acquired early in evolution, then it would seem that some plants should have taken this important enzyme. There is no apparent reason why a legume could not have evolved to integrate the *nif* genes into its chromosome or mitochondrial DNA. These genes, together with the leghemoglobin-coding genes, might be controlled so that they are only expressed when the plant is starved of fixed N.

If a functional nitrogenase evolved recently, on an evolutionary scale, then the barriers already set in established eukaryotes might be too great for a stable N_2-fixing system to develop in plant cells. As discussed in Section II, O_2 toxicity, high metal sequestering ability, low turnover number, and the accumulation of a potentially toxic product (ammonium) might be too many barriers for an established eukaryote to overcome.

The similarities between nitrogenases are evidence that a functional nitrogenase can only be made with a very restricted conformation and with the metals, iron and molybdenum, in a unique array. This would be true whether convergent or divergent evolution occurred. The large amount of nitrogenase that needs to be made in order for an organism to be able to grow on N_2 might be taken as evidence that nitrogenase has not yet been developed to be a very efficient enzyme. ATP-driven H_2 evolution, the high ATP requirement, and the wide range of reducible substrates are other examples of the inefficiency of nitrogenase. On the other hand, biological N_2 fixation may only be able to occur with a very specific tertiary structure. Convergent evolution, therefore, could have been the cause for the similarities among nitrogenases.

Nothing is known about the amino acid sequences of the nitrogenase components, but a statistical method has been developed that compares relatedness between proteins based on data of the amino acid contents (Marchalonis and Weltman, 1971). A term, $S\Delta Q$, is calculated to compare relatedness between two proteins. The lower the $S\Delta Q$ value, the greater the relatedness is presumed for two proteins. Among 100 unrelated proteins, the $S\Delta Q$ values were greater than 50. All but two proteins, thought to be unrelated, had values lower than 100. Relatedness between nitrogenase components from different organisms was demonstrated (Dilworth, 1974). The $S\Delta Q$ value between component I from *C. pasteurianum* and *K. pneumoniae* is 43, and the $S\Delta Q$ value between component II from these two organisms is 51.

If nitrogenase is a recent acquisition, what genes were modified to make this enzyme? In other words, what is the progenitor protein (or proteins) of nitrogenase? No obvious relationships were detected between nitrogenase components and other Fe–S proteins (Dilworth, 1974). Some relatedness, however, was observed between the components and proteins that had some type of interaction with ATP as well as with some other Mo-containing proteins (Kleiner *et al.*, 1976). It was previously shown that many ATP-requiring enzymes show relatedness (Weltman and Dowben, 1973). Xanthine oxidase, a Mo-containing protein, has the smallest $S\Delta Q$ value (26) when compared to component I (Kleiner *et al.*, 1976).

In one organism, *C. pasteurianum*, amino acid contents are available for the two subunits of component I (Chen *et al.*, 1973), component II (Chen *et al.*, 1973), hydrogenase (Nakos and Mortenson, 1971b), rubredoxin (Lovenberg, 1974), ferredoxin (Tanaka *et al.*, 1966), and flavodoxin (Knight and Hardy, 1967). This organism is assumed to be similar to cells that appeared early in evolution (Hall, 1971). The $S\Delta Q$ value (including cysteines) for the two subunits of 60,000 and 50,000 daltons is 11 (Fig. 2). This indicates that the two subunits (I_α and I_β, respectively) are very closely related. The same degree of relatedness is seen between the two component I subunits from *K. pneumoniae* (Kennedy *et al.*, 1976). The $S\Delta Q$ value for *C. pasteurianum* components I and II is 94; thus, they do not show an obvious relationship to each other.

An enzyme in *C. pasteurianum* that bears some similarity to nitrogenase is hydrogenase. This hydrogenase evolves H_2 and is a Fe–S protein with Fe–S clusters similar to those found in components I and II (Chen *et al.*, 1976). Ferredoxin can reduce both hydrogenase and nitrogenase (see Lovenberg, 1974). Hydrogenase (Bothe *et al.*, 1977) as well as nitrogenase (Schöll-

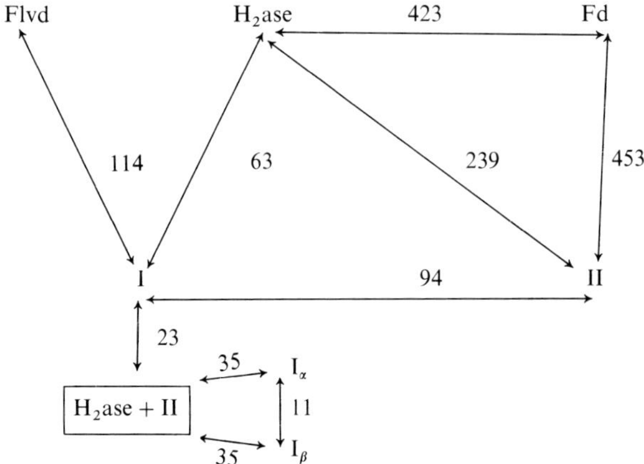

FIG. 2. $S\Delta Q$ relationships among proteins from *Clostridium pasteurianum*. Flvd = flavodoxin; H_2ase = hydrogenase; Fd = ferredoxin; I_α, I_β = two different subunits of component I.

horn and Burris, 1967) is inhibited by acetylene. Hydrogenase from *C. pasteurianum* is a single polypeptide with a molecular weight of 60,500 daltons (Chen *et al.*, 1973). A speculation is that component I from *C. pasteurianum* evolved from a fusion of genes coding for hydrogenase and component II.

The $S\Delta Q$ value for component I and hydrogenase is 63 (Fig. 2). The $S\Delta Q$ value for component I and the sum of hydrogenase and a single subunit of component II (component II has two identical subunits) is calculated to be 23, which implies a very close relatedness. In this model, we assume that hydrogenase and component II (or closely related proteins) were present before nitrogenase appeared. Component II may have played a role in some aspect of electron transport. Evolution of new functions has been predicted to occur through duplication of existing genes (Watts and Watts, 1968). It is not too difficult to envision a rare situation by which a fusion occurs with concomitant duplication (Jackson and Yanofsky, 1973; Beeftinck *et al.*, 1974; Anderson and Roth, 1977). The resultant protein, proto-I, might react with component II and the molybdenum cofactor to produce a nitrogenase with very little activity. Carbon dioxide reductase in *C. pasteurianum* seems to contain Mo (Thauer *et al.*, 1973); therefore, a molybdenum cofactor probably already is present in the cytoplasm. Selective pressure could force a stable duplication of the gene, and then mutations with further selection could yield the subunits known today, I_α and I_β. The latter step will account for the fact that $S\Delta Q$ values are higher between proto-I and each subunit than between proto-I and the sum of the two subunits (I_α and I_β).

Comparison of amino acid sequences of hydrogenase and both components of nitrogenase should support or refute this model.

No information is available to test this model similarly for the evolution of nitrogenase in other organisms. However, *Azotobacter*, *K. pneumoniae* (Smith *et al.*, 1976), and *Anabaena cylindrica* (Bothe *et al.*, 1977) have H_2-oxidizing hydrogenases that are inhibited by acetylene. Nothing is known about the amino acid contents of these hydrogenases. Perhaps some, or all, of the N_2-fixing bacteria obtained their nitrogenases (or proto-nitrogenases) from organisms such as *C. pasteurianum*.

Is it possible for nitrogenase to become a more efficient enzyme during further evolution? Or, more to the point, is it possible for man to genetically modify the enzyme so that it will have less of an ATP requirement, be unable to evolve H_2, not be inactivated by O_2, or have a higher substrate turnover number? Will it be possible perhaps through protoplast fusion and other techniques, to give an agronomically important cereal plant the capability to fix its own N_2? What would the agronomic and ecological effect of such a breakthrough be?

Less speculative ways to increase N_2 fixation might include plant breeding methods to improve symbioses already established. Plant–bacterium associations might be useful for supplying microbially fixed N to cereal plants such as corn, wheat, or rice. Either photosynthesis by cyanobacteria or carbon compounds from plant exudates might be used as the energy source for N_2 fixation. However, in any of these associations, some mechanisms of adaptation must be available to allow the associative N_2-fixing bacterium to successfully compete with other bacteria in the environment.

References

Aho, P. E., Seidler, R. J., Evans, H. J., and Nelson, A. D. (1976). *Proc. Int. Symp. Nitrogen Fixation, 1st* **2,** 629–640.

Anderson, A. J. (1946). *J. Counc. Sci. Ind. Res.* **19,** 1–18.

Anderson, R. P., and Roth, J. R. (1977). *Annu. Rev. Microbiol.* **31,** 473–505.

Bachhuber, M., Brill, W. J., and Howe, M. M. (1976). *J. Bacteriol.* **128,** 749–753.

Bada, J. L., and Miller, S. L. (1968). *Science* **159,** 423–425.

Bayliss, N. S. (1956). *Aust. J. Biol. Sci.* **9,** 364–370.

Becking, J. H. (1976a). *Proc. Int. Symp. Nitrogen Fixation, 1st* **2,** 556–580.

Becking, J. H. (1976b). *Proc. Int. Symp. Nitrogen Fixation, 1st* **2,** 581–591.

Beeftinck, F., Cunin, R., and Glansdorff, N. (1974). *Mol. Gen. Genet.* **132,** 241–253.

Benemann, J. R. (1973). *Science* **181,** 164.

Bergersen, F. J., and Briggs, M. J. (1958a). *J. Gen. Microbiol.* **19,** 345–354.

Bergersen, F. J., and Briggs, M. J. (1958b). *J. Gen. Microbiol.* **19,** 482–490.

Bergersen, F. J., and Goodchild, D. J. (1973). *Aust. J. Biol. Sci.* **26,** 741–756.

Bergersen, F. J., and Hipsley, E. H. (1970). *J. Gen. Microbiol.* **60,** 61–65.

Bergersen, F. J., and Turner, G. L. (1967). *Biochim. Biophys. Acta* **141,** 507–515,

Bergersen, F. J., and Turner, G. L. (1970). *Biochim. Biophys. Acta* **214,** 28–36.
Beringer, J. E., and Hopwood, D. A. (1976). *Nature* (*London*) **264,** 291–293.
Beringer, J. E., Johnston, A. W. B., and Wells, B. (1977). *J. Gen. Microbiol.* **98,** 339–343.
Bishop, P. E., and Brill, W. J. (1977). *J. Bacteriol.* **130,** 954–956.
Bishop, P. E., Evans, H. J., Daniel, R. M., and Hampton, R. O. (1975). *Biochim. Biophys. Acta* **381,** 248–256.
Bishop, P. E., Dazzo, F. B., Appelbaum, E. R., Maier, R. J. and Brill, W. J. (1977). *Science* **198,** 938–940.
Bohlool, B. B., and Schmidt, E. L. (1974). *Science* **185,** 269–271.
Bond, G. (1974). *In* "The Biology of Nitrogen Fixation" (A. Quispel, ed.), pp. 342–378. North-Holland Publ., Amsterdam.
Bond, G., and Scott, G. D. (1955). *Ann. Bot.* (*London*) **19,** 67–77.
Bormann, F. H., Likens, G. E., and Melillo, J. M. (1977). *Science* **196,** 981–983.
Bothe, H., Tennigkeit, J., Eisbrenner, G., and Yates, M. G. (1977). *Planta* **133,** 237–242.
Breznak, J. A., Brill, W. J., Mertins, J. W., and Coppel, H. C. (1973). *Nature* (*London*) **244,** 577–580.
Brill, W. J. (1975). *Annu. Rev. Microbiol.* **29,** 109–129.
Brill, W. J., Steiner, A. L., and Shah, V. K. (1974). *J. Bacteriol.* **118,** 986–989.
Brown, M. E. (1976). *J. Appl. Bacteriol.* **40,** 341–346.
Bui, P. T., and Mortenson, L. E. (1969). *Biochemistry* **8,** 2462–2465.
Bulen, W. A., and LeComte, J. R. (1966). *Proc. Natl. Acad. Sci. U.S.A.* **56,** 979–986.
Bulen, W. A., Burns, R. C., and LeComte, J. R. (1965). *Proc. Natl. Acad. Sci. U.S.A.* **53,** 532–539.
Burns, R. C., and Hardy, R. W. F. (1975). "Nitrogen Fixation in Bacteria and Higher Plants." Springer-Verlag, Berlin and New York.
Burns, R. C., Holsten, R. D., and Hardy, R. W. F. (1970). *Biochem. Biophys. Res. Commun.* **39,** 90–99.
Burris, R. H., and Orme-Johnson, W. H. (1974). *In* "Microbial Iron Metabolism" (J. B. Neilands, ed.), pp. 187–209. Academic Press, New York.
Bush, J. A., and Wilson, P. W. (1959). *Nature* (*London*) **184,** 381.
Cannon, F. C., Dixon, R. A., and Postgate, J. R. (1976). *J. Gen. Microbiol.* **80,** 227–234.
Cardenas, J., and Mortenson, L. E. (1975). *J. Bacteriol.* **123,** 978–984.
Carpenter, E. J., and Culliney, J. L. (1975). *Science* **187,** 551–552.
Chambers, C. A., and Silver, W. S. (1977). *Abstr. Annu. Meet. Am. Soc. Microbiol.* No. K80, p. 199.
Chen, J. S., Multani, J. S., and Mortenson, L. E. (1973). *Biochim. Biophys. Acta* **310,** 51–59.
Chen, J. S., Mortenson, L. E., and Palmer, G. (1976). *In* "Iron and Copper Proteins" (K. T. Yasunobu, H. F. Mower, and O. Hayaishi, eds.), pp. 68–82. Plenum, New York.
Collmer, A., and Lamborg, M. (1976). *J. Bacteriol.* **126,** 806–813.
Currier, T. C., Haury, J. F., and Wolk, C. P. (1977). *J. Bacteriol.* **129,** 1556–1562.
Cutting, J. A., and Schulman, H. M. (1968). *Fed. Proc., Fed. Am. Soc. Exp. Biol.* **27,** 768.
Daesch, G., and Mortenson, L. E. (1972). *J. Bacteriol.* **110,** 103–109.
Dazzo, F. B., and Brill, W. J. (1977). *Appl. Environ. Microbiol.* **33,** 132–136.
Dazzo, F. B., and Hubbell, D. H. (1975). *Appl. Microbiol.* **30,** 1017–1033.
Detroy, R. W., Witz, D. F., Parejko, R. A., and Wilson, P. W. (1968). *Proc. Natl. Acad. Sci. U.S.A.* **61,** 537–541.
Dilworth, M. J. (1969). *Biochim. Biophys. Acta* **184,** 432–441.
Dilworth, M. J. (1974). *Annu. Rev. Plant Physiol.* **25,** 81–114.
Dixon, R., Cannon, F., and Kondorosi, A. (1976). *Nature* (*London*) **260,** 268–271.
Dixon, R. A., and Postgate, J. R. (1971). *Nature* (*London*) **234,** 47–48.

Dixon, R. A., and Postgate, J. R. (1972). *Nature* (*London*) **237,** 102–103.
Dixon, R. O. D. (1972). *Arch. Mikrobiol.* **85,** 193–201.
Döbereiner, J. (1961). *Plant Soil.* **15,** 211–216.
Döbereiner, J., and Day, J. M. (1976). *Proc. Int. Symp. Nitrogen Fixation, 1st* **2,** 518–536.
Döbereiner, J., Day, J. M., and Dart, P. J. (1972a). *Plant Soil* **37,** 191–196.
Döbereiner, J., Day, J. M., and Dart, P. J. (1972b). *J. Gen. Microbiol.* **71,** 103–116.
Eady, R. R., and Postgate, J. R. (1974). *Nature* (*London*) **249,** 805–810.
Eady, R. R., Smith, B. E., Cook, K. A., and Postgate, J. R. (1972). *Biochem. J.* **128,** 655–675.
Emerich, D. W., and Burris, R. H. (1976). *Proc. Natl. Acad. Sci. U.S.A.* **73,** 4369–4373.
Evans, H. J., Purvis, E. R., and Bear, F. E. (1951). *Soil Sci.* **71,** 117–124.
Fay, P. (1970). *Biochim. Biophys. Acta* **216,** 353–356.
Ferris, J. P., and Nicodem, D. E. (1972). *Nature* (*London*) **238,** 268–269.
Fleming, H., and Haselkorn, R. (1973). *Proc. Natl. Acad. Sci. U.S.A.* **70,** 2727–2731.
Fred, E. B., Baldwin, I. L., and McCoy, E. (1932). "Root Nodule Bacteria and Leguminous Plants" Univ. of Wisconsin Press, Madison.
French, J. R. J., Turner, G. L., and Bradbury, J. F. (1976). *J. Gen. Microbiol.* **95,** 202–205.
Gordon, J. K., and Brill, W. J. (1972). *Proc. Natl. Acad. Sci. U.S.A.* **69,** 3501–3503.
Gordon, J. K., and Brill, W. J. (1974). *Biochem. Biophys. Res. Commun.* **59,** 967–971.
Gordon, J. K., Shah, V. K., and Brill, W. J. (1976). *Abstr. Annu. Meet. Am. Soc. Microbiol.* No. K164, p. 163.
Hadfield, K. L., and Bulen, W. A. (1969). *Biochemistry* **8,** 5103–5108.
Hall, J. B. (1971). *J. Theor. Biol.* **30,** 429–454.
Hamblin, J., and Kent, S. P. (1973). *Nature* (*London*), *New Biol.* **245,** 28–30.
Hamilton, I. R., Burris, R. H., and Wilson, P. W. (1964). *Proc. Natl. Acad. Sci. U.S.A.* **52,** 637–641.
Hardy, R. W. F., and Knight, E., Jr. (1966). *Biochim. Biophys. Acta* **122,** 520–531.
Hatch, M. D., and Slack, G. R. (1970). *Annu. Rev. Plant Physiol.* **21,** 141–162.
Henriksson, E., and Simu, B. (1971). *Oikos* **22,** 119–121.
Hill, S. (1971). *J. Gen. Microbiol.* **67,** 77–83.
Hitch, C. J. B., and Stewart, W. D. P. (1973). *New Phytol.* **72,** 509–524.
Howe, M. M., and Bade, E. G. (1975). *Science* **190,** 624–632.
Hyndman, L. A., Burris, R. H., and Wilson, P. W. (1953). *J. Bacteriol.* **65,** 522–531.
Jackson, E. N., and Yanofsky, C. (1973). *J. Bacteriol.* **116,** 33–40.
Jacob, G. S. (1976). Ph.D. Thesis, Univ. of Wisconsin, Madison.
Kass, D. L., Drosdoff, M., and Alexander, M. (1971). *Soil Sci. Soc. Am., Proc.* **35,** 286–289.
Keister, D. L. (1975). *J. Bacteriol.* **123,** 1265–1268.
Keister, D. L., and Evans, W. R. (1976). *J. Bacteriol.* **129,** 149–153.
Kennedy, C., and Dixon, R. A. (1977). *Abstr. Annu. Meet. Am. Soc. Microbiol.* No. K81, p. 199.
Kennedy, C., Eady, R. R., Kondorosi, E., and Rekosh, D. K. (1976). *Biochem. J.* **155,** 383–389.
Ketchum, P. A., Cambier, H. Y., Frazier, W. A., III, Madansky, C. H., and Nason, A. (1970). *Proc. Natl. Acad. Sci. U.S.A.* **66,** 1016–1023.
Kleiner, D., and Chen, C. H. (1974). *Arch. Mikrobiol.* **98,** 93–100.
Kleiner, D., Littke, W., Bender, H., and Walenfels, K. (1976). *J. Mol. Evol.* **7,** 159–165.
Knight, E., Jr., and Hardy, R. W. F. (1967). *J. Biol. Chem.* **242,** 1370–1374.
Koch, B., Evans, H. J., and Russell, S. (1967). *Plant Physiol.* **42,** 466–468.
Kondorosi, A., Svab, Z., Kiss, G. B., and Dixon, R. A. (1977). *Mol. Gen. Genet.* **151,** 221–226.
Kurz, W. G. W., and LaRue, T. A. G. (1975). *Nature* (*London*) **256,** 407–409.
Lee, S. B., and Wilson, P. W. (1943). *J. Biol. Chem.* **151,** 377–385.
Lobb, W. R. (1953). *N.Z. Soil News* **3,** 9–16.

Lovenberg, W. (1974). *In* "Microbial Iron Metabolism" (J. B. Neilands, ed.), pp. 161–185. Academic Press, New York.
Ludden, P. W., and Burris, R. H. (1976). *Science* **194,** 424–426.
Ludwig, R. A., and Signer, E. R. (1977). *Nature (London)* **267,** 245–248.
Lyne, R., and Stewart, W. D. P. (1973). *Planta* **109,** 27.
McComb, J. A., Elliot, J., and Dilworth, M. J. (1975). *Nature (London)* **256,** 409–410.
Magasanik, B., Prival, M. J., Brenchley, J. E., Tyler, B. M., DeLeo, A. B., Streicher, S. L., Bender, R. A., and Paris, C. G. (1974). *Top. Cell. Regul.* **8,** 119–138.
Maier, R. J., and Brill, W. J. (1976). *J. Bacteriol.* **127,** 763–769.
Marchalonis, J. J., and Weltman, J. K. (1971). *Comp. Biochem. Physiol. B* **38,** 609–616.
Matsumoto, H., and Tazaki, T. (1971). *Jpn. J. Microbiol.* **15,** 11–20.
Meade, H. M., and Signer, E. R. (1977). *Proc. Natl. Acad. Sci. U.S.A.* **74,** 2076–2078.
Meers, J. L., Tempest, D. W., and Brown, C. M. (1970). *J. Gen. Microbiol.* **64,** 187–194.
Millbank, J. W. (1974). *In* "The Biology of Nitrogen Fixation" (A. Quispel, ed.), pp. 238–264. North-Holland Publ., Amsterdam.
Millbank, J. W., and Kershaw, K. A. (1969). *New Phytol.* **68,** 721–729.
Mortenson, L. E. (1964). *Proc. Natl. Acad. Sci. U.S.A.* **52,** 272–279.
Mulder, E. G., and Brotonegoro, S. (1974). *In* "The Biology of Nitrogen Fixation" (A. Quispel, ed.), pp. 37–85. North-Holland Publ., Amsterdam.
Nagatani, H. H., and Brill, W. J. (1974). *Biochim. Biophys. Acta* **362,** 160–166.
Nagatani, H. H., and Haselkorn, R. (1977). *Abstr. Annu. Meet. Am. Soc. Microbiol.* No. K124, p. 206.
Nagatani, H. H., Shimizu, M., and Valentine, R. C. (1971). *Arch. Mikrobiol.* **79,** 164–175.
Nagatani, H. H., Shah, V. K., and Brill, W. J. (1974). *J. Bacteriol.* **120,** 697–701.
Nakos, G., and Mortenson, L. E. (1971a). *Biochemistry* **10,** 455–458.
Nakos, G., and Mortenson, L. E. (1971b). *Biochemistry* **10,** 2442–2449.
Napoli, C. A., and Hubbell, D. H. (1975). *Appl. Microbiol.* **30,** 1003–1009.
Nason, A., Lee, K. Y., Pan, S. S., Ketchum, P. A., Lamberti, A., and DeVries, J. (1971). *Proc. Natl. Acad. Sci. U.S.A.* **68,** 3242–3246.
Neilson, A. H., and Nordlund, S. (1975). *J. Gen. Microbiol.* **91,** 53–62.
Newton, W. E., and Nyman, C. J. (1976). *Proc. Int. Symp. Nitrogen Fixation, 1st* **1,** pp. 311.
Nutman, P. S. (1976). "Symbiotic Nitrogen Fixation in Plants." Cambridge Univ. Press, London and New York.
O'Gara, F., and Shanmugam, K. T. (1976). *Biochim. Biophys. Acta* **437,** 313–321.
Okon, Y., Albrecht, S. L., and Burris, R. H. (1976a). *J. Bacteriol.* **127,** 1248–1254.
Okon, Y., Albrecht, S. L., and Burris, R. H. (1976b). *J. Bacteriol.* **128,** 592–597.
Orme-Johnson, W. H., Hamilton, W. D., Ljones, T., Tso, M.-Y. W., Burris, R. H., Shah, V. K., and Brill, W. J. (1972). *Proc. Natl. Acad. Sci. U.S.A.* **69,** 3142–3145.
Pagan, J. D., Child, J. J., Scowcroft, W. R., and Gibson, A. H. (1975). *Nature (London)* **256,** 406–407.
Page, W. J. (1977). *Abstr. Annu. Meet. Am. Soc. Microbiol.* No. K82, p. 199.
Page, W. J., and Sadoff, H. L. (1976). *J. Bacteriol.* **125,** 1080–1087.
Parejko, R. A., and Wilson, P. W. (1970). *Can. J. Microbiol.* **16,** 681–685.
Pengra, R. M., and Wilson, P. W. (1958). *J. Bacteriol.* **75,** 21–25.
Peters, G. A. (1976). *Proc. Int. Symp. Nitrogen Fixation, 1st* **2,** 592–610.
Phillips, D. A., and Johnson, M. J. (1961). *J. Biochem. Microbiol. Technol. Eng.* **111,** 277–309.
Phillips, D. A., Howard, R. L., and Evans, H. J. (1973). *Physiol. Plant.* **28,** 248–253.
Pienkos, P. T., and Brill, W. J. (1977). *Abstr. Annu. Meet. Am. Soc. Microbiol.* No. K77, p. 199.
Postgate, J. R. (1974). *Symp. Soc. Gen. Microbiol.* **24,** 263–292.

Postgate, J. R., and Krishnapillai, V. (1977). *J. Gen. Microbiol.* **98,** 379–385.
Potrikus, C. J., and Breznak, J. A. (1976). *Abstr. Annu. Meet. Am. Soc. Microbiol.* No. I64, p. 122.
Quispel, A. (1974). *In* "The Biology of Nitrogen Fixation" (A. Quispel, ed.), pp. 499–520. North-Holland Publ., Amsterdam.
Raju, P. N., Evans, H. J., and Seidler, R. (1972). *Proc. Natl. Acad. Sci. U.S.A.* **69,** 3474–3478,
Rao, R. N. (1976). *J. Bacteriol.* **128,** 356–362.
St. John, R. T., Shah, V. K., and Brill, W. J. (1974). *J. Bacteriol.* **119,** 266–269.
St. John, R. T., Johnston, H. M., Seidman, C., Garfinkel, D., Gordon, J. K., Shah, V. K., and Brill, W. J. (1975). *J. Bacteriol.* **121,** 759–765.
Schick, H. J. (1971). *Arch. Mikrobiol.* **75,** 89–101.
Schöllhorn, R., and Burris, R. H. (1967). *Proc. Natl. Acad. Sci. U.S.A.* **58,** 213–216.
Schubert, K. R., and Evans, H. J. (1976). *Proc. Natl. Acad. Sci. U.S.A.* **73,** 1207–1211.
Shah, V. K., and Brill, W. J. (1973). *Biochim. Biophys. Acta* **305,** 445–454.
Shah, V. K., and Brill, W. J. (1977). *Proc. Natl. Acad. Sci. U.S.A.* **74,** 3249–3253.
Shah, V. K., Davis, L. C., and Brill, W. J. (1972). *Biochim. Biophys. Acta* **256,** 498–511.
Silver, W. S., and Postgate, J. R. (1973). *J. Theor. Biol.* **40,** 1–10.
Simon, M. A., and Brill, W. J. (1971). *J. Bacteriol.* **105,** 65–69.
Smith, B. E., Lowe, D. J., and Bray, R. C. (1972). *Biochem. J.* **130,** 641–643.
Smith, L. A., Hill, S., and Yates, M. G. (1976). *Nature (London)* **262,** 209–210.
Sorger, G. J. (1969). *J. Bacteriol.* **98,** 56–61.
Stewart, W. D. P. (1973). *Annu. Rev. Microbiol.* **27,** 283–316.
Stewart, W. D. P., and Lex, M. (1970). *Arch. Mikrobiol.* **73,** 250–260.
Stewart, W. D. P., and Rowell, P. (1975). *Biochem. Biophys. Res. Commun.* **65,** 846–856.
Stewart, W. D. P., Haystead, A., and Pearson, H. W. (1969). *Nature (London)* **224,** 226–228.
Streicher, S. L., and Valentine, R. C. (1974). *In* "Microbial Iron Metabolism" (J. B. Neilands, ed.), pp. 211–229. Academic Press, New York.
Streicher, S. L., Gurney, E. G., and Valentine, R. C. (1971). *Proc. Natl. Acad. Sci. U.S.A.* **68,** 1174–1177.
Streicher, S. L., Shanmugam, K. T., Ausubel, F., Morandi, C., and Goldberg, R. B. (1974). *J. Bacteriol.* **120,** 815–821.
Tanaka, M., Nakashima, T., Benson, A., Mower, H., and Yasonubu, K. T. (1966). *Biochemistry* **5,** 1666–1681.
Thauer, R. K., Fuchs, G., Schnitker, U., and Jungermann, K. (1973). *FEBS Lett.* **38,** 45–48.
Tjepkema, J., and Evans, H. J. (1975). *Biochem. Biophys. Res. Commun.* **65,** 625–628.
Tjepkema, J., and van Berkum, P. (1977). *Appl. Environ. Microbiol.* **33,** 626–629.
Trinick, M. J. (1976). *Proc. Int. Symp. Nitrogen Fixation, 1st* **2,** 507–517.
Tso, M.-Y. W. (1974). *Arch. Mikrobiol.* **99,** 71–80.
Tso, M.-Y. W., Ljones, T., and Burris, R. H. (1972). *Biochim. Biophys. Acta* **267,** 600–604.
Tu, J. C. (1974). *J. Bacteriol.* **119,** 986–991.
Tubb, R. S. (1974). *Nature (London)* **251,** 481–485.
Tubb, R. S., and Postgate, J. R. (1973). *J. Gen. Microbiol.* **79,** 103–117.
Tyler, B., DeLeo, A., and Magasanik, B. (1974). *Proc. Natl. Acad. Sci. U.S.A.* **71,** 225–229.
Verma, D. P. S., and Bal, A. K. (1976). *Proc. Natl. Acad. Sci. U.S.A.* **73,** 3843–3847.
Verma, D. P. S., Nash, D. T., and Schulman, H. M. (1974). *Nature (London)* **251,** 74–77.
Von Bülow, J. F. W., and Döbereiner, J. (1975). *Proc. Natl. Acad. Sci. U.S.A.* **72,** 2389–2393.
Wall, J. D., Weaver, P. F., and Gest, H. (1975). *Nature (London)* **258,** 630–631.
Watts, R. L., and Watts, D. C. (1968). *Nature (London)* **217,** 1125–1130.
Weare, N. M., and Shanmugam, K. T. (1976). *Arch. Mikrobiol.* **110,** 207–213.
Weltman, J. K., and Dowben, R. M. (1973). *Proc. Natl. Acad. Sci. U.S.A.* **70,** 3230–3234.

Wilson, P. W., Hull, J. F., and Burris, R. H. (1943). *Proc. Natl. Acad. Sci. U.S.A.* **29,** 289–294.
Winter, H. C., and Burris, R. H. (1968). *J. Biol. Chem.* **243,** 940–944.
Winter, H. C., and Burris, R. H. (1976). *Annu. Rev. Biochem.* **45,** 409–426.
Wittenberg, J. B., Bergersen, F. J., Appleby, C. A., and Turner, G. L. (1974). *J. Biol. Chem.* **249,** 4027–4066.
Wolk, C. P. (1973). *Bacteriol. Rev.* **37,** 32–101.
Wolk, C. P., and Wojciuch, E. (1971). *Planta* **97,** 126–134.
Wolpert, J. S., and Albersheim, P. (1976). *Biochem. Biophys. Res. Commun.* **70,** 729–737.
Wong, P. P., and Burris, R. H. (1972). *Proc. Natl. Acad. Sci. U.S.A.* **69,** 672–675.
Yates, M. G. (1972). *FEBS Lett.* **27,** 63–67.
Yoch, D. C. (1974). *J. Gen. Microbiol.* **83,** 153–164.
Zumft, W. G., and Mortenson, L. E. (1975). *Biochim. Biophys. Acta* **416,** 1–52.

CHAPTER 3

Bacterial Chemotaxis

D. E. KOSHLAND, JR.

I. Introduction

Bacteria are able to control their movements to optimize their physiological health. They are able to swim toward compounds, such as nutrients, that are favorable for their survival and away from compounds, such as toxic substances, which are deleterious. By doing this they are able to con-

ISBN 0-12-307207-7

trol their environmental conditions in a way which is possibly analogous to pain and pleasure in higher species (Koshland, 1977a). Moreover, this ability to respond to chemicals in the environment is widespread in living species and in wide varieties of bacteria.

The phenomenon of chemotaxis was discovered by Pfeffer (1883) and Englemann (1881) in the 1880's and for a period of time was studied vigorously. After an extended period of little activity, renewed interest in this area was generated by Adler and his group in the 1960's. At present there has been resurgence of interest in other laboratories, and, as a result, a number of reviews have been recently written (Adler, 1969, 1975; Berg, 1975; Clayton, 1964; Koshland, 1974, 1977a,b; Macnab, 1978a; Metzner, 1920; Nultsch, 1970; Seymour and Doetsch, 1973; Weibull, 1960; Ziegler, 1962). This chapter will emphasize work done on *Salmonella typhimurium* and *Escherichia coli*, the two organisms in which most of the experimentation in regard to the mechanism of bacterial chemotaxis has been performed. Reference to other organisms will be made where relevant with the hope that the interested reader will consult other reviews for more extensive discussions of some of the classic studies on other species.

Our current understanding of bacterial chemotaxis can be summarized briefly. There are receptors on the periphery of the bacterium which receive signals from the environment. These signals are then processed by chemotactic machinery involving not less than nine genes. The information obtained from this signaling system is then transmitted to the flagella, causing reversal in flagella rotation and an abrupt turn (called a tumble) in the bacterial trajectory. It is the modulation of the tumbling frequency which is responsible for the net bacterial motion.

The chemotactic system offers a particularly attractive phenomenon for studying a regulatory system. In more complex multicellular systems, feedback from other organs, hormonal messages, etc., make it difficult to interpret all of the factors in the system. The moderate complexity of the system and the presence of all components in one cell make the bacteria an ideal organism for comprehensive study. Moreover, the availability of mutants provides a particular experimental advantage.

The nomenclature in this area is complex since Pfeffer (1883), Englemann (1881) and others (Weibull, 1960) used the term "chemotaxis" to describe this ability to move toward or away from chemicals. Later Fraenkel and Gunn (1961) developed their nomenclature based on the mechanism of the response. In this view, taxis refers to directed orientation reaction and kineses to orientation achieved by altering the speed of movement or the frequency of turning. Since the mechanism by which the bacteria migrated was unknown, it was referred to in the literature as "chemotaxis" following historical tradition. Recently, however, it has been shown that chemotaxis is indeed klinokinesis (Macnab and Koshland, 1972; Berg and Brown, 1972) and is interesting in this regard, since it is the first demonstration of klinokinesis in chemical sensing. Since the term "chemotaxis" has been widely accepted by biochemists and others not directly concerned with mechanistic analogies and historical literature, the term "chemotaxis" will be used here in the historical sense to refer to the generalized capacity to alter migration toward chemicals without any implications in regard to mechanism.

This chapter will consider the bacterial receptors which receive the signals from the environment, the nature of the sensing mechanism in broad general terms, and the current biochemical information in regard to the sensing mechanism. The various methods that have been used to develop this information will be described as will the information which chemotaxis studies have revealed in regard to bacterial individuality.

II. The Chemotactic Receptors

A. Numbers of Receptors and Their Specificity

The ability of bacteria to respond to chemicals is dependent on a repertoire of receptors of varying specificities. Most of the bacterial receptors appear to be highly specific in responding to one or two chemicals at high affinity, but respond to others at lower affinities. In the best studied example of *E. coli*, a pattern of approximately 15 potential receptors for attractants (Adler, 1975) and approximately 8 potential receptors for repellents are now known (Tso and Adler, 1974a). Studies of other organisms, such as *Salmonella typhimurium* and *Bacillus subtilis*, indicate similar patterns, although they have certainly not been studied nearly as intensively. Each organism has its own mixture of attractants and repellents. For example, *Salmonella typhimurium* is sensitive to phenol, while *Escherichia coli* is not (Lederberg, 1956). Cyclic AMP is an attractant for slime mold, but does not seem to serve as an attractant for *E. coli* or *S. typhimurium* (Gerisch and Hess, 1974). The repertoire of receptors for *E. coli* and *Salmonella* is shown in Table I, and the responses for a wide variety of other bacteria subjected to gradients of attractants and repellents is shown in Table II.

The receptors listed do not show absolute specificity for the compounds indicated. For example, the affinities of sugars for the galactose receptor of *Salmonella typhimurium* (Zukin *et al.*, 1977a) are shown in Table III. It is clear in this case that galactose and glucose bind very strongly to the receptor and are undoubtedly the major physiological attractants. However,

TABLE I

LIST OF RECEPTORS FOR CHEMOTAXIS OF *E. coli* AND *S. typhimurium*[a]

Potential receptor for chemoeffector tested	Other compounds which interact with receptor	Shown to be present in: *E. coli*	*S. typhimurium*	Attractant (A) or repellent (R)
N-Acetylglucosamine		X		A
D-Fructose		X		A
D-Galactose	D-Glucose, D-Fucose	X	X	A
D-Glucose		X		A
D-Mannose	D-Glucose	X		A
Maltose		X		A
Mannitol		X		A
Ribose		X	X	A
D-Sorbitol		X		A
Trihalose				A
Aspartate	Glutamate	X	X	A
Serine	Cysteine, Alanine, Glycine	X	X	A
Acetate	Propionate, butyrate, valerate	X	X	R
Isopropanol	Ethanol, isobutanol	X		R
Leucine	Isoleucine, Valine	X	X	R
Indole	Skatole	X	X	R
Phenol		X		R
Benzoate	Salicylate	X		R
H^+, OH^-		X	X	A, R
Mg^{2+}	Ca^{2+}, Sr^{2+}, Ni^{2+}	X	X	A
Citrate			X	A
O_2		X	X	A
S^{2-}		X		R

[a] Data from Adler (1969), Lederberg (1956), Weibull (1960), Tsang *et al.* (1973), Macnab and Koshland (1974), Tso and Adler (1974a), Koshland (1977b), and Miller *et al.* (1978).

arabinose, fucose, and lactose bind at higher concentrations and thus would be detected by the organism in the absence of the other sugars. Finally, compounds such as ribose and allose do not bind at all to this receptor. The receptor for serine has also been found to bind cysteine, alanine, and glycine with comparable affinities (Mesibov and Adler, 1972). Thus, the specificity of bacterial receptors is similar to enzymes in not being absolutely specific for a single compound but also not being the low specificity associated with bitter or sweet tastes in man.

Of particular interest is the fact that certain bacteria can respond to electric fields and others to the magnetic field of the earth (Blakemore, 1975). Thus, receptors for these particular physical conditions must exist. In addition, Englemann discovered in the early studies that bacteria can respond to temperature (Englemann, 1883), and adaptive responses of temperature have been observed more recently (Miller and Koshland, 1977a).

The evidence for the number of specific receptors has been obtained by competition studies which are performed in the following manner. Competition studies using the capillary assay of Adler (1973) involve placing compound A, a known attractant, inside the capillary and the bacteria in the outside medium and then measuring the bacteria which accumulate in the capillary. The experiment is repeated, but compound B is then added in high uniform concentration in both the capillary and the outside medium; if compound B prevents the normal accumulation of bacteria in the capillary then presumptive evidence of competition has been obtained. If it fails to alter the normal accumulation, it is presumed to act at some other receptor or not be active at all. In the temporal gradient assay of Tsang *et al.* (1973), bacteria are subjected to a temporal gradient with compound A alone and the time to adapt to normal tumbling patterns is recorded. The experiment is then repeated with compound B in both the bacterial and the external medium solutions and the new time for return to normal tumbling patterns recorded. If the chemotactic response is altered or reduced, compound B is presumed to compete with compound A for its receptor; if no alteration in the time dependence is observed, compound B is presumed to act with other receptors or not at all. The two assays have different advantages. The temporal gradient response is very rapid and simple and can be performed on a microscopic slide by simply adding a drop of attractant or repellent to the bacteria and observing their response. As such, it is now the method of choice for screening a wide variety of compounds. On the other hand, the compounds which give a very short time response, for example, galactose in *Salmonella typhimurium*, are difficult to measure by this assay, and in such cases the capillary response appears to be more sensitive. In those cases in which purified receptors have been isolated, the results have been consistent with both the capillary and temporal gradient analyses.

While it is likely that the total list of receptors for any organism is incomplete, it is unlikely, in the case of *E. coli* and *S. typhimurium*, finding a large number of new receptors. Hence, it is pertinent to ask if any simple pattern has emerged which could explain the repertoire of receptor species in any particular bacterium. An easy generalization would be that bacteria swim toward compounds needed for their survival and away from compounds detrimental to it. Buried in this generaliztion is the fact that a number of compounds do not fall in those categories. Some deviations are trivial. For example, some compounds, such as fucose, are not metabolized but undoubtedly serve as attractants merely because they are bound by the galactose receptor. Other compounds, such as tryptophan, act as repellents

TABLE II

CHEMOTACTIC RESPONSES TO CARBOHYDRATES BY TEN SPECIES OF MOTILE BACTERIA

Carbohydrates in Ionagar plug[a]	Final pH in Ionagar plug	Responses of[c]									
		Alcaligenes faecalis	*Bacillus licheniformis*	*Erwinia carotovora*	*Escherichia coli* K-12	*Proteus morganii*	*Pseudomonas aeruginosa*	*Pseudomonas fluoresceus*	*Sarcina ureae*	*Serratia marcescens*	*Spirillum serpens*
Monosaccharides											
Glycerol	6.6	0	+v	+	0	+	0	0	0	0	+v
meso-Erythritol	6.3	0	+v	0	0	+v	0	0	0	0	0
Adonitol	6.2	0	0	+v	+v	+	+	+v	0	0	+
D-Arabinose	6.1	0	0	+	+v	0	0	0	0	0	+
L-Arabinose	6.1	0	0	+v	+v	0	0	0	0	0	+
D-Arabitol	6.2	0	0	0	0	+v	+v	+v	0	0	+
L-Arabitol	6.2	0	0	0	0	+v	+v	+v	0	0	+
D-Lyxose	6.4	0	0	+v	0	+	+v	+	0	0	+
L-Lyxose	6.3	0	0	+	0	+	+v	0	0	0	+
D-Ribose	6.0	0	0	0	+v	0	+v	+	0	0	0
Xylitol[b]	6.3	0	0	+v	0	0	+v	0	0	+	+
α-D-Xylose[b]	6.3	0	+v	+	0	+v	0	0	0	0	+v
β-D-Fructose[b]	6.3	0	0	+	nd	0	+v	0	0	0	+v
D-Fucose	6.0	0	+v	0	0	+v	+v	0	0	+v	+
L-Fucose	6.3	0	0	0	0	0	0	0	nd	0	0
Galacitol	6.8	0	0	0	0	0	0	0	0	0	0
α-D-Galactose[b]	6.2	0	0	+	0	0	0	0	0	+v	+v
D-Glucose	6.2	0	0	+	+v	0	0	0	0	+v	+v
D-Glucose[b]	6.2	0	0	+v	nd	0	+	0	0	0	0
2-Deoxy-D-glucose[b]	6.5	0	0	nd	0	0	0	0	nd	+	0
iso-Inositol	6.4	0	0	+v	0	0	0	0	0	0	+v
D-Mannose	6.2	0	+	+	+v	0	0	0	0	+v	+
D-Mannitol[b]	6.3	0	0	+	nd	0	0	0	0	+v	+
Potassium gluconate	6.6	0	0	+v	+v	0	0	0	0	0	+v
L-Rhamnose	6.3	0	+	+	0	+	0	0	0	+v	+
D-Sorbitol	6.3	0	0	+	+v	+v	+v	+	0	0	+
L-Sorbose	6.6	0	0	+	nd	0	0	0	0	0	+
Sedoheptulose	6.7	0	+v	+v	0	+v	+	+	0	0	+v
Disaccharides											
D-Cellobiose	6.3	0	0	0	0	0	0	0	0	nd	0
Cellobiose octa-acetate	6.6	0	0	0	0	0	0	0	0	0	0
D-Lactose · H_2O	6.4	0	0	+	0	0	0	0	0	+	+v
Maltose	6.1	0	0	+v	+v	0	0	0	nd	0	0
Melibiose[b]	6.5	0	0	0	0	0	0	0	0	0	0
Sucrose[b]	6.4	0	0	+v	0	0	0	0	0	0	0
Trehalose	6.6	0	0	+v	0	0	0	0	nd	nd	0
Turanose	5.2	0	0	0	0	+v	0	0	nd	0	+v

(*Continued*)

TABLE II (*Continued*)

Carbohydrates in Ionagar plug[a]	Final pH in Ionagar plug	Responses of[c]									
		Alcaligenes faecalis	*Bacillus licheniformis*	*Erwinia carotovora*	*Escherichia coli* K-12	*Proteus morganii*	*Pseudomonas aeruginosa*	*Pseudomonas fluoresceus*	*Sarcina ureae*	*Serratia marcescens*	*Spirillum serpens*
Trisaccharides											
Melezitose	6.4	0	0	+v	0	0	+v	0	nd	0	0
Raffinose	6.6	0	0	0	0	nd	0	0	nd	0	0
Polysaccharides											
Arabic acid	2.5	–	–	–	–	–	–	–	–	–	–
β-Cyclodextrin	6.6	0	0	0	0	0	0	nd	0	0	0
Dextrin	6.3	0	0	0	0	+v	0	+v	0	0	0
Heparin	6.6	0	0	0	0	0	0	nd	0	0	nd
Amino sugars											
N-Acetylglucosamine	5.6	0	0	0	0	0	nd	+v	nd	0	nd
N-Acetylneuraminic acid	4.3	0	0	+v	0	nd	0	nd	+v	0	0
Glycosides											
Aesculin	5.8	0	0	0	0	0	0	0	0	p	nd
Aloin	6.1	0	0	0	0	0	nd	nd	0	0	nd
Arbutin	6.2	0	0	0	0	nd	0	+	0	0	nd
αMethyl-D-glucoside[b]	6.3	0	0	+	nd	0	0	0	0	+	0
α-Methyl-D-mannoside[b]	5.5	0	0	+	0	0	0	+	0	0	+v
Salicin	6.4	0	0	0	+v	0	0	0	0	0	0

[a] Arranged alphabetically in order of increasing number of carbon atoms in the basic carbohydrate unit.

[b] PMBN used instead of PMBA.

[c] +, Positive response; –, negative response; 0, no response; v, variable response; nd, not determined. Molar concentrations were used except for: (1) aloin, dextrin and heparin, which were incorporated in 1.5% (w/v) Ionagar as 10% solutions; and (2) aesculin, arbutin, cellobiose, cellobiose octaacetate, β-cyclodextrin and raffinose, which were incorporated in 1.5% (w/v) Ionagar in 10% (v/v) amounts of saturated solutions. PMBA was used to suspend bacteria and as solvent for the carbohydrates.

Data from Seymour and Doetsch (1973).

TABLE III

Binding of Saccharides to the Purified Galactose Receptor of *S. typhimurium*

Substrate	Dissociation constant (*M*)
Galactose	2×10^{-7}
Glucose	10^{-7}
Arabinose	4×10^{-5}
Lactose	6×10^{-4}
Fucose	6×10^{-3}
Methyl galactoside	No binding
Ribose	No binding
Allose	No binding

From Zukin *et al.* (1977a).

to *Salmonella typhimurium*, an apparent anomaly which is easily explained when it is seen that tryptophan is such a weak repellent that its effect would easily be nullified by the presence of even trace amounts of serine and other amino acid attractants. Since the likelihood of a source of protein having both serine and tryptophan is very great, bacteria would inevitably swim toward the source of amino acids even though tryptophan had slight repellent action. The existence of repellent compounds, such as acetic acid, seems anomalous, since it is a known excretion product, but these excretion products would be an indicator of crowded conditions where food supply might become limiting and hence a signal for the bacterium to migrate elsewhere.

It is not clear why different species have different repertoires of receptors, although it might be expected that a change in receptors could be explained ultimately by the varying metabolic needs of different organisms.

B. Metabolism and Transport

The chemotactic response does not depend on the metabolism of the compound. This has been demonstrated in a variety of ways (Adler, 1969, 1975). Some compounds which are extensively metabolized are not attractants, e.g., histidine. Some chemicals which are not metabolized serve as attractants (e.g., D-fucose, a 6-deoxy analog of D-galactose, which binds to the galactose receptor). Similarly α-methyl aspartate can bind to the aspartate receptor. Finally, it is possible to block normal metabolism so that a compound which is a normal nutrient does not metabolize. Thus mutants of *S. typhimurium* blocked in ribose metabolism, are still capable of responding chemotactically to ribose (Adler, 1969; Aksamit and Koshland, 1974).

In the case of repellents many of these compounds are clearly not metabolized. Sometimes toxic compounds, such as phenol (Lederberg, 1956), serve as repellents in the chemotactic responses observed at levels well below those which are toxic to the organism. This makes sense since the organism is induced to swim out of an area containing phenol long before the concentration becomes lethal.

The fact that metabolism or transport is not necessary for chemotaxis should not obscure the fact that many of the compounds for which chemotaxis is designed are ultimately transported and metabolized. Since they are nutrients, this is the purpose of the chemotactic response. The separation of the two functions simply means that bacteria, have two separate systems, one to generate a sensory response and the other to metabolize the compound for their energy needs. The evidence that transport can be separated from the sensory response has also been obtained by mutant studies (Adler, 1969; Hazelbauer and Adler, 1971; Aksamit and Koshland, 1974; Ordal and Adler, 1974).

C. Purified Receptors

Receptors identified with chemotaxis have been purified and characterized in the case of the periplasmic proteins. The galactose-binding protein involved in transport was originally isolated by Anraku (1968) and further studied by Silhavy *et al.* (1974). This protein was shown to be involved in chemotaxis by Hazelbauer and Adler (1971) by mutation studies. A ribose response in *E. coli* was detected by Adler and the ribose-binding protein from *S. typhimurium* was isolated by Aksamit and Koshland (1974) and shown by similar mutation studies to be responsible for the chemotactic response in their organisms. In both cases there are approximately 10^4 molecules of periplasmic receptor protein per bacterium. Both proteins have molecular weights of around 30,000 and have been investigated for their similarities and differences (Zukin *et al.*, 1977a). Table IV presents a comparison of the properties of the periplasmic chemoreceptor proteins isolated from *Salmonella* and from *E. coli* together with the amino acid composition and properties of some other proteins. There is obviously a similarity in amino acid composition and in some of their physical properties.

In addition to these proteins, the arabinose-binding protein, also identified with chemotaxis, has been idolated, purified, and preliminary X-ray crystallographic studies have been made (Parsons and Hogg, 1974). The maltose receptor has also been purified (Kellerman and Szmelcman, 1974).

So far no integral membrane-bound receptor has been purified. Studies using osmotic shock have revealed that the serine receptor of both *E. coli*

TABLE IV

AMINO ACID COMPOSITIONS OF SOME TRANSPORT BINDING PROTEINS AND CHEMOTACTIC RECEPTORS

	S. typhimurium		*E. coli*	
Amino acid	Ribose-binding protein (mole/29,000 gm)	Galactose-binding protein (mole/33,000 gm)	Galactose-binding protein (mole/35,000 gm)	Arabinose-binding protein (mole/38,000 gm)
Lys	26.0	31.7	27.3	29.6
His	3.0	3.1	3.0	2.7
Arg	5.8	6.0	5.7	8.4
Asp	35.6	50.0	43.4	34.2
Thr	12.2	12.9	12.2	17.9
Ser	8.0	13.3	11.0	14.4
Glu	26.1	29.0	25.2	38.4
Pro	8.2	10.3	8.0	17.9
Gly	23.7	22.0	19.4	33.1
Ala	36.7	39.0	37.7	36.1
Val	25.5	25.9	26.1	33.8
Met	4.9	5.7	5.2	11.4
Ile	12.0	13.3	13.2	16.0
Leu	24.0	22.7	21.2	23.6
Tyr	2.7	7.6	5.0	7.6
Phe	6.7	6.0	5.8	16.0
Cys	0	0	0	1.7
Trp	0	1.0	3.5	7.0

From Zukin *et al.* (1977a).

and *Salmonella* is not diminished by the shock procedures under circumstances in which all of the ribose-binding activity is removed (Anderson and Koshland, 1978). Shock procedure has eliminated 50% of the aspartate activity. These results suggest that receptors are bound to the inner membrane with different affinities, some floating freely in the periplasmic space and others firmly anchored in the membrane.

D. CONFORMATIONAL CHANGE

The presence of purified proteins has made it possible to provide evidence for what has in the past been an act of faith on the part of most sensory physiologists. It has widely been assumed that conformational change is induced in a receptor molecule on binding of the chemical stimulant. However, the difficulty in purifying receptors and the need to have large amounts of purified protein has hindered any direct demonstration of such an effect. In the case of the galactose-binding protein evidence of fluorescent changes and absorption changes on binding of galactose was obtained (Boos *et al.*, 1972), but these suffered from the ambiguity of whether the effect was direct or indirect. By the method of attaching a fluorescent reporter group at a distance from the active site and demonstrating that only one galactose molecule was bound per mole of protein, it was demonstrated (Zukin *et al.*, 1977b) that a dramatic conformation change occurs in both the galactose and the ribose receptors on binding of their respective sugars. Moreover, from the energy transfer experiments, it was possible to show that this conformation change extended at least 30 Å through the molecule. Thus, an extensive conformation change is induced on binding of sugar to the purified protein, and this undoubtedly explains the first stage in the signaling process.

Moreover, the conformation change explains the absence of any need for metabolism in the initial sensing response. The chemical induces a conformation change in the protein receptor which then transmits the signal to the next part of the processing system, while metabolism of the initial sugar is not needed. Since the same molecule is involved as a receptor for chemotaxis and transport, it is possible that this initial binding then triggers both chemotaxis and transport. The evidence cited below for a free-floating receptor and the mutation studies which separate transport and chemotaxis suggest that some molecules of the receptor–chemoeffector complex in the periplasmic space bind to the components of the signaling system for chemotaxis, and other molecules bind to the transport machinery which then translocates the sugar into the cell. In any case, the demonstration of a conformation change readily explains the mechanism by which the two separate systems are activated by binding of attractant to the receptor molecule.

E. Competition Model

The manner in which the receptor transmits its information to the next component in the signaling system is of great interest. A variety of interactions can be envisaged. One possibility is that the receptor is free-floating; a conformation change is induced which generates a binding site, thus allowing the receptor to interact with the next component of the signaling system and trigger the transmitting process (cf. Fig. 1). A second alternative is that the receptor is already attached to the next component of the signaling system but must be triggered by the induced conformational change of the chemoeffector binding. A third mechanism involves the receptor–signal–

LIGAND-INDUCED RECEPTOR ASSOCIATION WITH SIGNALING SYSTEM

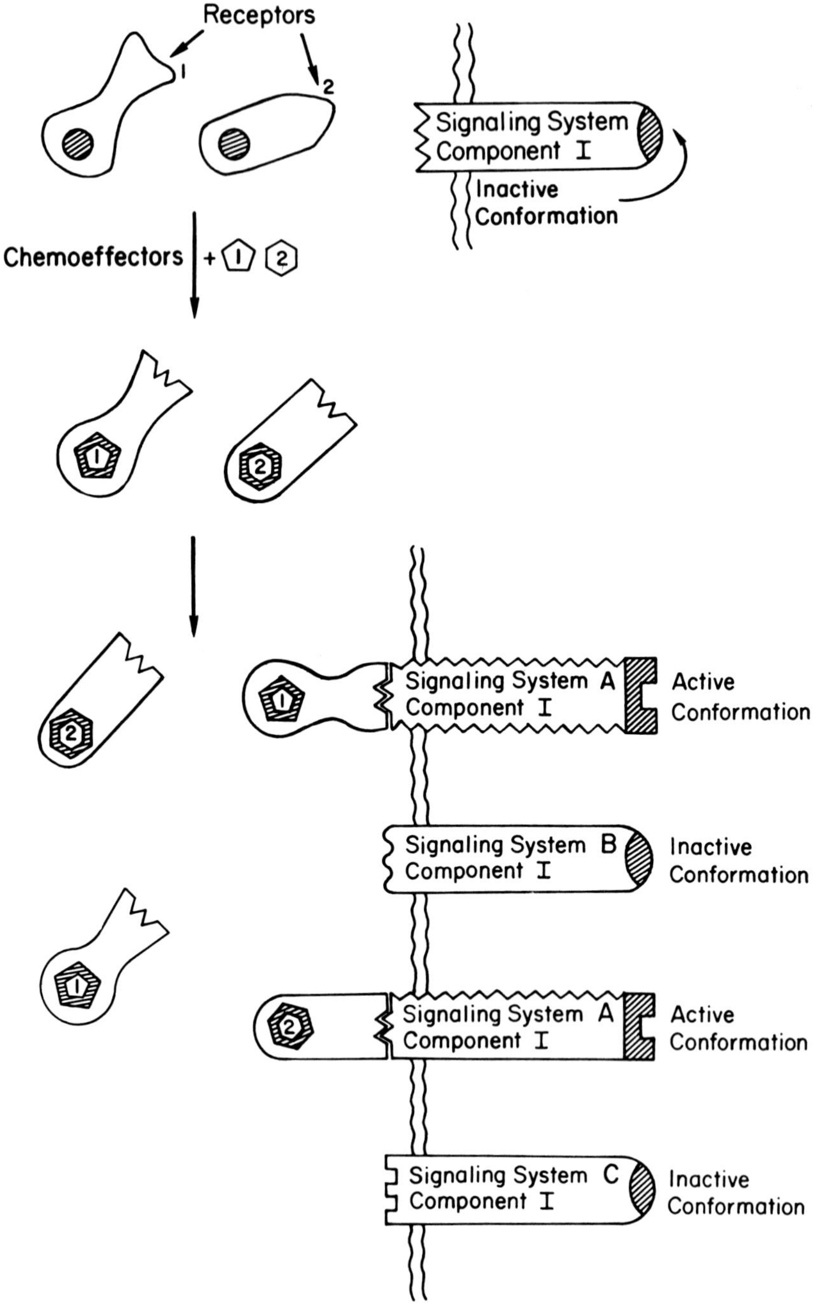

component complex in which the receptor dissociates from the signal component on binding chemoeffector. The signal component is activated by the dissociation of the receptor.

The first of these alternatives appears to be used in the ribose–galactose signaling pathway as demonstrated by competition studies in *Salmonella typhimurium* (Strange and Koshland, 1976).

The competition model was established by a series of studies. First, the purified galactose-binding protein in *Salmonella* was found to bind galactose tightly but not ribose (Strange and Koshland, 1976). Likewise, the ribose-binding protein, was purified and found to bind ribose but not galactose (Aksamit and Koshland, 1974). If the competition occurred between the two receptors and the concentrations of the receptors were appropriate, galactose would inhibit ribose in a competition study of the type mentioned above. This, in fact, was found to be the case (Strange and Koshland, 1976). Ribose completely inhibited galactose taxis as shown in Fig. 2. Such inhibition could, however, be explained by a difference in specificity between the purified protein and that existing in the membrane. The second piece of evidence was obtained by isolating a mutant lacking the ribose-binding protein and finding that it did not show any ribose inhibition of galactose taxis. Third, the mechanism described in Fig. 1 would require an induced conformational change on galactose binding to the galactose receptor and ribose binding to the ribose receptor, and these were also established (Zukin *et al.*, 1977b). Fourth, a mutant (*trg*) was found in *E. coli* (Ordal and Adler, 1974) which failed to respond to either ribose or galactose. A similar mutant in *Salmonella* was shown to contain both the ribose- and the galactose-binding proteins (Strange and Koshland, 1976). Thus, the possibility of a double mutation turning off both responses was eliminated. The mechanism of Fig. 1, therefore, appears to be quite solidly established.

This finding has a number of implications. First, it suggests an easy rationalization of the observation that the same receptor protein is used for transport and chemotaxis. The receptor is freely floating in the periplasmic space. On binding of the chemoeffector, a conformation change then occurs in some fraction of the periplasmic binding protein molecules and triggers the sensory system, whereas the vast majority of molecules are involved in the transport of the sugar to the interior of the cell. Second, the mechanism

---

FIG. 1. Floating receptor model. Receptors are initially in conformations that are not attracted to component I, but are induced into new conformations by the chemoeffectors. As a result, individual chemoeffector–receptor complexes are induced to encounter and associate with the first component of the signaling system. If one binds there, it induces a conformation change, which activates the signaling system and begins a signal that can be amplified in a cascade process. If two receptor–chemoeffector complexes compete for the same site, one stimulus can diminish or completely block another.

raises a caveat in regard to the receptor competition experiments described in Section II,A. Although those experiments have been successful in identifying receptors, this new type of competition can obscure identification of receptors by competition studies alone. Obviously, the kind of experiments described in Section II,A would suggest that ribose and galactose were bound by the same receptor which, in fact, is not the case. Third, the mechanism suggests a processing system in which certain stimuli are focused through a specialized system prior to central processing as compared to a parallel system in which all stimuli proceed directly to the central system (cf. Fig. 3).

It might logically be asked what advantage such a mechanism would have, particularly in a sensory system. First, there is an economy and simplicity in a common response. In the case of chemotaxis, a rough additivity in the

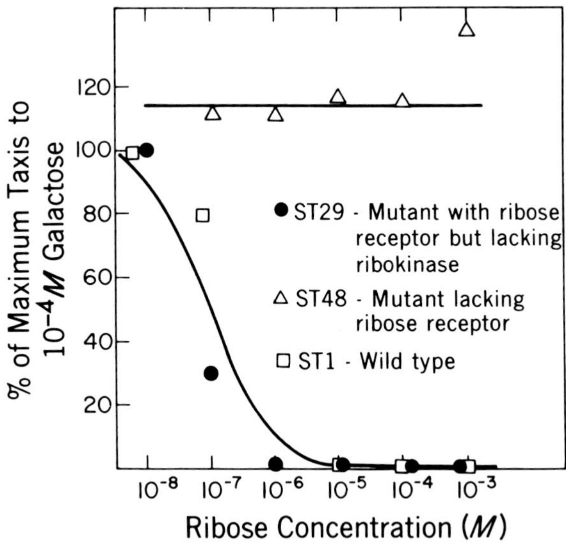

FIG. 2. Interaction of galactose and ribose receptors as indicated by competition experiments. In experiments indicated by open squares, ribose at the concentration indicated on the abscissa is placed both inside the capillary and in the outside solution, whereas galactose is placed only inside the capillary. Increasing concentrations of ribose inhibit the galactose response until it is completely suppressed at concentrations above 10^{-4} *M*. The open triangles indicate a similar experiment with the mutant ST48, which lacks the ribose-binding protein. As can be seen there is no inhibition of the galactose response by ribose over the entire concentration range. The closed circles represent the same experiment performed with ST29, a mutant that blocks metabolism of ribose but contains the ribose-binding protein. Ribose is seen to inhibit in this case as in the wild type. The shift to lower concentration occurs because of the block on metabolism in the capillary assay in the ribokinase mutant.

signals generated by attractants and repellents indicates a common pool of a response regulator that can integrate a variety of stimuli. In nerves, synapses providing excitation and inhibition are utilized. Adenylate cyclase obviously provides a similar integrating function in hormonal systems. Second, the mechanism provides a focusing of stimuli, which has added control benefits. In this case, an organism saturated with a good carbon source, such as ribose, would not respond to an added superfluous carbon source, such as galactose, but it could still respond to a nitrogen source. In higher species, such as man, a number of different sensory phenomena feed into a common brain, but separate focused pathways prevent saturation of one system (e.g., the visual) from desensitizing a second (e.g., the auditory system). (If we are deafened we can still see, blinded we can still taste.) Third, an interaction system of the type shown in Fig. 3 provides a maximum sensitivity with a maximum of control. In most cases an organism is subject to one stimulus at a time and at low levels. Hence, maximum sensitivity is achieved by a tight binding of chemoeffector to an excess of receptor with subsequent attraction of the chemoeffector to the signaling component molecules. If sufficient component I molecules (cf. Fig. 1) were present, however, the occasional situation in which the organism is bombarded by many stimuli could result in overstimulation and metabolic breakdown. The focused pathways limit maximum response, while maintaining sensitivity to small amounts of stimuli.

The advantage of using a protein for focusing such a system can be explained in terms of specificity. The receptors that have been isolated so far

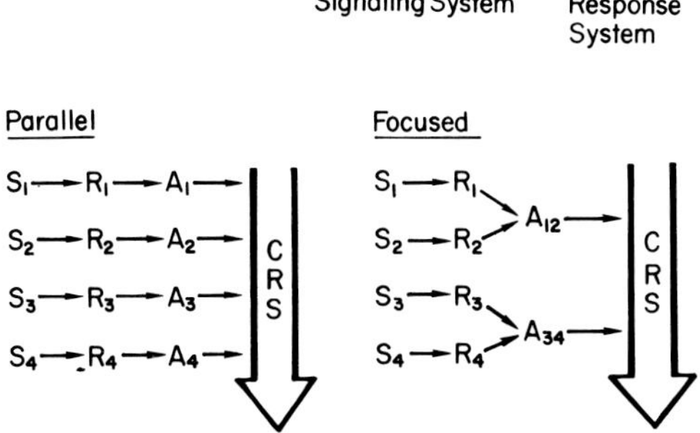

FIG. 3. Possible signaling systems. Stimulus activates a receptor that interacts with component I which feeds into a central signaling system. In a "parallel" system, the receptors and component I molecules work independently, and integration in terms of additions or subtractions occurs in the central signaling system. In a focused system, the receptors compete for common component I molecules, which then feed signals into the central response system (CRS). In this mechanism some integration of signals would occur prior to the main stream of the central response system.

are highly specific. Each has a very limited range of compounds which bind to it. The galactose receptor mentioned above binds glucose and galactose tightly; arabinose, lactose, and fucose weakly; and ribose and allose not at all. Since this is a typical protein specificity one might ask how a sensory system could obtain the advantages of focusing while still achieving discrimination between similarly structured chemical compounds. In the case of galactose and glucose, conventional competition at a single receptor site is possible because these two compounds differ only by inversion at a single carbon atom. Thus, the active site can be made to bind both galactose and glucose and exclude many saccharides which differ more significantly in structure. However, it would be extremely difficult to design a site which bound ribose and galactose but excluded compounds such as fucose, arabinose, etc. In that case, receptor competiton provides an answer. The binding sites are tailored for the chemoeffector, and the "adapter end" of the receptor is tailored to bind with the next component of the signaling system. In that way "focused" specialized pathways can be devised for chemicals which are structurally diverse. A mutant is known which eliminates taxis toward serine and repellents (Hazelbauer *et al.*, 1969), which are two very different structural classes.

F. Multiple Functions of Receptors

As more and more receptor proteins are identified, an interesting generalization appears to be emerging, i.e., that the receptor proteins have more than one function. The galactose- and ribose-binding proteins have been identified as receptors in chemotaxis (Aksamit and Koshland, 1974; Hazelbauer and Adler, 1971) and also serve as part of the transport system for those compounds (Anraku, 1968; Boos, 1974; Hazelbauer and Adler, 1971; Oxender, 1972). The glucose receptor for chemotaxis (Adler and Epstein, 1974) is also part of the phosphotransferase system (Kundig and Roseman, 1971). A blue-light effect which disturbed the sensory system could also be identified with a perturbation of the electron-transport system (Macnab and Koshland, 1974; Taylor and Koshland, 1975). The oxygen, nitrate, and fumarate receptors are identified with the enzymes involved in electron transport (Taylor *et al.*, 1978).

These results would seem to suggest an economy of function and an evolutionary explanation for some subunit interaction in regulatory enzymes. It seems likely that most enzymes undergo induced conformational changes, that the conformational changes can be transmitted from one subunit to another, and that the evolutionary selection for interactions between subunit binding sites are as selected as the interaction of substrate and enzyme at active sites (Cook and Koshland, 1969). A mutation in a subunit designed initially for catalytic purposes might allow that subunit to bind to a totally different protein. The induced conformational change of the new association, both in the presence and absence of ligand, could then have regulatory functions in the new system. The initial function might be lost by mutation as the subunit becomes specialized in its new task, e.g., the regulatory subunit of aspartyl transcarbamylase might have been an enzyme initially or the subunit could serve different functions in different aggregates. The noncatalytic lactabumin subunit of galactose transferase (Hill and Brew, 1975), and the subunit of tryptophan synthetase (Yanofsky and Crawford, 1972) might be such proteins. Possibly the specialization of higher species is such that the initial enzymatic nature of the receptor has been lost over evolutionary time, but it may also be true that higher species, such as bacteria, preserve structure so that one peptide chain can serve several functions.

G. Location of Receptors

The evidence accumulated so far does not allow an easy generalization as to the location of receptors. The galactose, ribose, arabinose, and maltose receptors mentioned above are clearly in the periplasmic space of the gram-negative bacteria. The serine and aspartate receptors are solidly attached to the inner membrane. An indication that cytochrome oxidase and nitrate reductase might be receptors for these specific chemoeffectors is intriguing, since these enzymes are in the cytoplasmic membrane. Thus, evidence for both periplasmic and inner surface membrane-bound receptors has been obtained. In other species, such as gram-positive organisms, there is no periplasmic space, and the similar behavior of *Bacillus subtilis* in many reactions suggests that the receptors can function equally well being bound into the membrane itself (Boos, 1974). In fact, there does not seem to be any discrepancy between the mechanism now adduced for the action of the periplasmic protein and the membrane-bound receptor, since the periplasmic receptor induces a conformational change in a membrane-bound component as shown in Fig. 2. The only difference then between the ribose receptor and the serine receptor would be that all of the serine-receptor components would be in the membrane fraction. Thus, signals both from the inner surface of the membrane, the outer surface, and within the membrane itself would be received by the sensory system.

H. Induced Proteins

An important feature of the receptors are that their numbers can vary from species to species and also with various growth conditions. The ribose receptor and the galactose receptor are induced (Adler, 1969; Wilson, 1974) under appropriate growth conditions, whereas the serine receptor appears to be constant and constitutive. The nitrate receptor (Showe and DeMoss, 1968) is induced by growth in nitrate, whereas the oxygen receptor, if it is cytochrome oxidase, appears to be constitutive. Thus, each species appears to have a mixture of receptors which are induced under some growth conditions and others which are present in every bacterium.

III. The Sensing Mechanism

The ability of bacteria to sense a gradient at all is incredible when it is realized that the length of most bacterial bodies are about 2 μm in length. Although Pfeffer had indicated that bacteria sense concentration ratios (Pfeffer, 1888), the experiments could be rationalized on the basis of some general allosteric effect which suppressed tumbling as higher concentrations of attractant were achieved. This alternative was eliminated by studies in defined gradients, studying migrations of colonies of bacteria (Dahlquist *et al.*, 1972). These studies confirmed that bacteria sensed ratios of concentration and, therefore, had some comparative analytical device.

A. The Biased Random Walk

Bacteria observed in a microscope appear to swim in straight lines interrupted by abrupt changes in direction. Some of these changes appear to involve the bacterium tumbling "head over heels," and then heading off randomly in a new direction. In other cases, a dramatic change in direction can be observed. These observations were placed on a quantitative basis with the aid of the elegant tracking device by Berg and Brown (1972) who showed that indeed the bacteria followed a classic random walk pattern. Not only were the lengths of the runs (the more or less straight, smooth swimming part of the trajectory) distributed in a Poissonian manner but the angles of the turns were also in a Poissonian distribution averaging an angle of 62°. When Berg and Brown subjected their bacteria to a gradient situation, they observed an alteration in the tumbling frequency. The bacteria proceeding up a gradient of attractant tumbled less frequently. Bacteria moving downward in the gradient, tumbled with normal frequency. As a result, a biased selective movement up the gradient of attractant was observed.

While these experiments were proceeding, Macnab and Koshland (1972) performed their experiments using temporal gradients and observed a suppression of tumbling frequency when attractant concentration was increased and an increase in tumbling frequency when attractant concentration was decreased. Thus, these independent studies agreed in the general mechanism of the chemotactic response, i.e., that migration was achieved by alteration of tumbling frequency, but disagreed in the quantitative distribution of the response. The discrepancy between these two initial conclusions was quickly resolved. It was found that bacteria indeed do increase their tumbling frequency in going down a gradient of attractant, but that the quantitative effect is far smaller than the suppression of tumbling on going favorable directions. Thus, in the shallow gradients used in the experiments of Berg and Brown, the increase in tumbling frequency in going down the gradient was barely above experimental error, whereas in the steeper gradients of Macnab and Koshland both suppression and the generation of tumbling were evident.

In studies with the same temporal gradient apparatus, repellent gradients were found to be the same as attractant gradients except for an inverted algebraic sign (Tsang *et al.*, 1973). The bacteria tumble more frequently when traveling up gradients of repellents.

These studies, thus, lead to a remarkably consistent overall picture. Bacteria move through space in a random walk pattern which becomes biased in a gradient. The bias is created by suppression of tumbling on movement in a favorable direction, and generation of tumbling on movement in an unfavorable direction. The suppression of tumbling is quantitatively more significant in the gradients normally seen by the bacteria. Unlike some higher species, the bacteria do not turn in a favorable direction. They simply use an on–off switching device which increases or decreases the frequency of tumbling to achieve a net migration.

At first glance, it would appear that such a movement is inefficient. This qualitative impression can be easily dismissed by observing bacteria under the microscope if a capillary containing an attractant is placed in the solution. Bacteria seem to move rapidly into the capillary. Considering the above mechanism, this qualitative observation is readily confirmed. If the bacteria proceed for very short distances in the wrong direction before tumbling, and go long distances in the right direction before tumbling, the random walk is so biased in the favorable direction that it is highly efficient. In fact, a theoretical calculation by Dahlquist *et al.* (1976) showed that in a maximal gradient, bacteria proceed in a favorable direction at approximately one-half their maximal velocity.

B. The Temporal Sensing Mechanism

If bacteria direct migration by modulating tumbling frequency, the question arises as to how the sensing system monitors the outside environment. The two mechanisms are general in nature: (a) an instantaneous comparison of receptors located at different topological positions on the surface of the organism, and (b) the temporal sensing over time. That the latter is the mechanism utilized by bacteria was demonstrated by temporal gradient experiments utilizing an apparatus similar to the stopped flow apparatus of enzyme kinetics (Macnab and Koshland, 1972).

First, *Salmonella typhimurium* bacteria were shown to swim normally in the absence of a gradient at different absolute concentrations of attractant, e.g., no differences in motility behavior were observed between bacteria swimming in 10^{-4} *M* serine, 10^{-5} *M* serine, or in the absence of serine (Macnab and Koshland, 1972). By normal swimming we mean a random walk-type behavior in which bacteria swim in approximately straight lines interrupted by random tumbles. The bacteria were then subjected to a sudden change in concentration and examined only after the mixing process was complete. If the sensing mechanism utilized the instantaneous comparison between concentrations at the head and tail, the bacteria in the observation chamber would sense only a uniform distribution of attractant and would therefore behave as though they were in a "no gradient" situation. If, on the other hand, they were comparing concentrations over time, a sudden decrease in concentration would cause them to behave as if they were swimming down a gradient and, an increase as if they were swimming up a gradient. The latter was precisely what occurred (Macnab and Koshland, 1972). The bacterial tumbling increased dramatically if the concentration of attractant was decreased suddenly (simulating swimming down a gradient). Bacterial tumbling was suppressed ("smooth swimming") if the concentration was increased. No change in tumbling pattern was observed if the concentration was kept constant (the control). When the swimming pattern was recorded over a time interval, the patterns observed immediately after mixing gradually relaxed back to normal, precisely what would be expected in a temporal process as the "memory" of the previous environment faded over time. These experiments eliminated the instantaneous comparison alternative and indicated that bacteria had a rudimentary "memory" which was utilized to direct migration by suppressing tumbling as the bacteria moved in favorable directions and increasing tumbling on traveling in unfavorable directions.

The term bacterial memory is used to describe the behavior pattern of the bacteria because it involves a time-dependent comparison of past and present. It is certainly not "long-term memory" in the sense of higher species, but the time span involved can be shown to be selected to optimize its biological function (Macnab and Koshland, 1973).

To decide how long a "useful memory" of a bacterium should be (Macnab and Koshland, 1973), it is important to remember that bacteria swim in a random walk manner, a series of straight runs and tumbles distributed in a Poissonian manner. A very short memory would mean the comparison would have to be made over a very short interval of time. In that case, the amount of the gradient recorded by the bacterium would be barely longer than the interval of gradient seen from the head to the tail of the bacterium. It would then lose the advantage of comparing over many body lengths. On the other hand, a very long memory might mean that directional information might not be processed until after a random tumble had been made and the bacterium was traveling in a different direction. A brief memory span, therefore, would offer a high correlation between information received and current direction of motion, but little reduction in analytical accuracy. A long memory span would offer an advantage in analytical accuracy, but a lower correlation with direction of motion. To optimize their sensory system, bacteria might be expected to have an intermediate length memory span, and that is indeed what is observed.

The bacteria can sense a gradient which gives a change of only one part in 10^4 over its body length. The effective memory span for the bacterium swimming in a gradient is approximately 20 to 100 body lengths. The "memory" thereby reduces the analytical problem from detecting one part in 10^4 to one part in 10^2–10^3, still a formidable analytical challenge (Macnab and Koshland, 1973).

It might well be asked whether the temporal gradient studies which subject the bacterium to larger stimuli than they normally confront in swimming through solutions are not a measure of an artificial condition and whether a totally different mechanism might operate under true physiological conditions. To test this possibility, experiments have been done with shallow gradients using both the enzyme degradation procedure (Brown and Berg, 1974) or the temporal gradient method (Macnab and Koshland, unpublished observations). The results provide a smooth extrapolation up to the point at which no detectable change can be observed. There does not appear to be any disontinuity between the shallow gradients in which a small difference in concentration is experienced and the steeper gradients where the responses are easier to measure. It appears, therefore, that a single memory device is operating over the entire concentration range.

This result can be confirmed in a second way. Dahlquist *et al.* (1976) have taken the results from the population–migration apparatus and analyzed them in terms of the prevailing theory making a number of *ad hoc*, but reasonable, assumptions. They assume, for example, that the chemotactic potential is caused by suppression of tumbling when the bacteria are moving in a favorable direction and involve essentially no alteration in mean free paths when the bacteria are moving in the unfavorable direction. This is in rough agreement with the observation described above in which the quantitative contribution to the favorable direction to chemotaxis is certainly far greater. With these and other assumptions, the calculated movement of the bacteria in three-dimensional space agrees with observations. This agreement means that the temporal gradient and tracking studies are indeed studying the mechanism by which bacterial behavior is controlled in its real world.

C. The Tumble Regulator

How can the bacterial memory be explained in chemical terms? A rudimentary model is shown in Eq. (1) and described further in Fig. 4.

$$\begin{array}{c} W \xrightarrow{V_f} X \xrightarrow{V_d} Y \\ \text{(tumble regulator)} \\ \downarrow \\ \text{control of flagellar function} \end{array} \tag{1}$$

In this model X represents a tumble regulator which operates somewhat like a thermostat to activate or suppress tumbling. It is formed from W at a rate designated by V_f and decomposed at a rate indicated by V_d. Either or both of these steps can be modified by signals from receptors. For illustrative purposes, we assume that the tumble regulator suppresses tumbling when it rises above the threshold level and increases tumbling when it falls below the threshold (Koshland, 1977a; Macnab and Koshland, 1972).

A bacterium moving up a gradient of attractant would initially increase V_f more than V_d (because V_f responds more rapidly to the change in chemoeffector concentration) and thus lead to an increased level of the tumble regulator. This would suppress tumbling for an interval, but if no further stimulus were encountered, the decomposition rate (V_d) which would be a function of (X) would increase until X decreases to its former level. On going down a gradient the inverse would occur, the level of X would be momentarily depressed and tumbling would be increased. Although the details of the system are not yet known, the features that emerge from the experiments are (a) there is some parameter whose level regulates tumbling, (b) the formation and decomposition of this parameter is under the influence of stimuli so that the level can be perturbed in gradients, and (c) the bacterial "memory" is a function of the time-dependent characteristics of this tumble regulator.

This simple model allowed an explanation of a number of observed phenomena and led to several predictions which have been verified. Some of these are illustrated in Fig. 5. For example, the explanation of non-chemotactic mutants became clear. If modification tumbling frequency was essential to the behavior pattern, mutants which tumbled all the time or

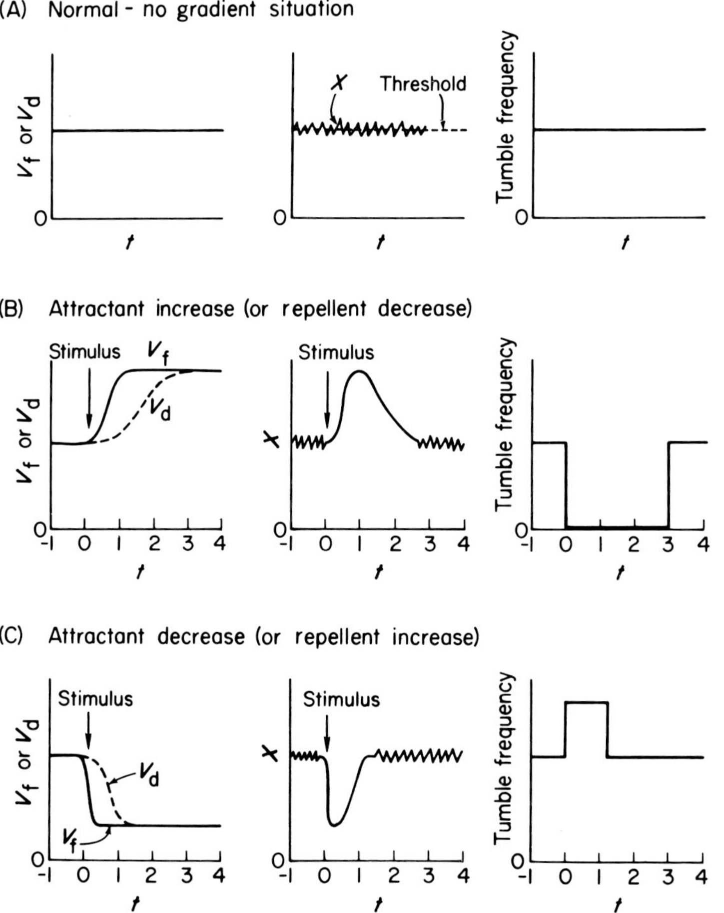

FIG. 4. Response of wild-type bacterium to attractants and repellents, as explained by a response (tumble) regulator model. The variation over time for the enzyme activities, the level of tumble regulator and the tumbling frequency is shown for three situations. (A) In absence of a gradient, $V_f = V_d$ are constant over time, and X (the tumble regulator) concentration varies around threshold in a Poissonian manner. The tumble frequency remains essentially constant. (B) Sudden increase in attractant increases rate of V_f faster than V_d leading to transient increase in concentration of X and transient decrease in tumbling frequency. Repellent decrease gives the same effect. (C) Sudden decrease in repellent decreases rate of V_f more rapidly than V_d leading to decrease in concentration of X and transient increase in tumbling frequency.

none of the time could obviously not follow a gradient. The nonchemotactic mutants observed so far all fall in one of these two patterns. Moreover, constantly tumbling mutants might be able to exhibit smooth swimming if given a large enough attractant stimulus, whereas smooth-swimming mutants would be unaffected. This was demonstrated by Aswad and Koshland (1974). Likewise, smooth-swimming mutants could be induced to tumble by a strong repellent stimulus, and this also has been shown (Fig. 5) (see also Koshland *et al.*, 1976).

A model with some variable level of a tumble regulator readily explains how the bacterium optimizes its "useful memory." The instant the bacterium starts up a gradient of attractant, it begins producing higher levels of tumble regulator and hence increases its probability of traveling a longer than normal distance in that run. The instant a bacterium heads in the wrong direction, it starts reducing the level of tuble regulator and shortening the average path length in that direction. The optimal compromise to give a "useful memory" span occurs by a balance between the rates of change in formation and decomposition of the tumble regulator.

The "memory span" is embodied in the pool level of the tumble regulator and normally does not vary far from the threshold levels for a bacterium

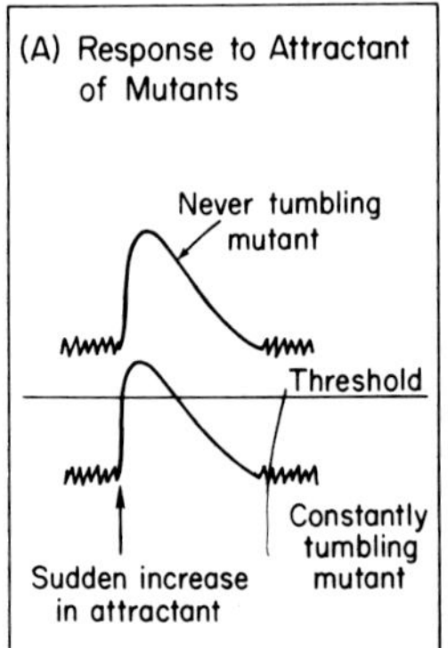

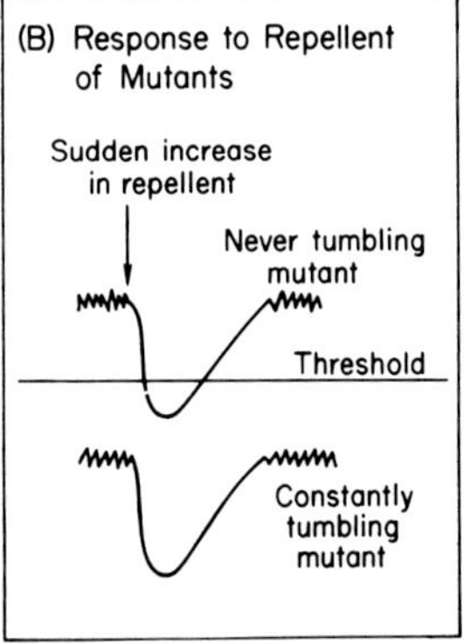

FIG. 5. The tumble regulator model applied to mutants undergoing stimuli. (A) A constantly tumbling mutant has the level of tumble regulator below the threshold in the absence of a gradient, leading to constant tumbling. Stimulus with an attractant can raise the tumble regulator for a brief period above the threshold, after which it returns, in the absence of a further stimulus, to constant tumbling patterns. A smooth-swimming nonchemotactic mutant has a tumble regulator level above the threshold at all times. An increase in attractant can further increase the level of X, but does not change the behavioral pattern. (B) A smooth-swimming nonchemotactic mutant has a tumble regulator level above threshold. When treated with a sudden increase in repellent, the tumble regulator level is lowered to give a momentary tumbling response. Same treatment of a constantly tumbling mutant would have no observable behavioral effect.

swimming in usual gradients, thus maximizing sensitivity to a change in direction. It can, however, be extended for longer intervals artificially in the laboratory by intense temporal stimuli, and this has been of great experimental value.

D. Additivity of Stimuli

The relationship of stimuli to each other could be studied using various quantitative devices developed for such studies. The first such additivity relationship involved the use of the temporal gradient method for comparing attractants and repellents to demonstrate that they contribute to a common temporal system. It was found that repellents gave the same type of temporal responses as attractants, but opposite in algebraic sign, i.e., an increase in phenol, a repellent, generated tumbling whereas an increase in serine, an attractant, suppressed tumbling. A mixture of these two led to neutralization of the effects (Tsang *et al.*, 1973). A similar finding was obtained using the capillary assay (Tso and Adler, 1974b). Thus, it was apparent that whatever the initial receptors, the ultimate response fed into a common sensory system.

The availability of a quantitative analytical tool for studying an instantaneous response made it possible to study the additivity of stimuli more directly (Spudich and Koshland, 1975). One study of this sort is shown in Fig. 6.

A given stimulus, from 0 to 0.5 m*M* L-serine, was broken up into two smaller stimuli, from 0 to 0.02 m*M* and from 0.02 to 0.5 m*M* serine. What effect would this have on the response? As seen in Fig. 6, the results are additive. The areas under the curves of the two small stimuli add up to the area under the curve of the one large one. Since the areas under the curves represent the average recovery times in the experiment, it means that the $\tau_{13} = \tau_{12} + \tau_{23}$, where τ_{13} is the average time of recovery to the overall stimulus and τ_{12} and τ_{23} are the average times of recovery to the fractional stimuli. This same experiment was performed with the attractant aspartate as well as serine and also with different varieties of fractional increments. In all cases, the results were the same, additivity between fractional increments and tumbles suppressed as long as the stimulant was a single chemoeffector. This is quite consistent with the evidence in regard to the time independence of the delivery of chemoeffector.

The quantitative method allowed a comparison between responses of the whole organism and the physical parameters of a purified receptor. In Fig. 7 are shown the data obtained for the responses to a D-ribose stimulus of a living bacterial cell of *S. typhimurium*. The purified receptor utilized was the

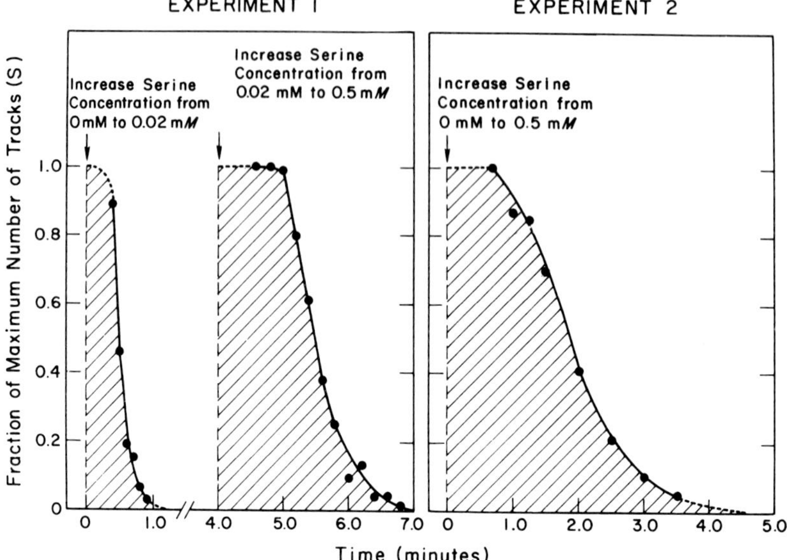

FIG. 6. Experiment to test additivity of stimuli and responses. In experiment 1 response of ST171, a constantly tumbling mutant, to a 0.02 m*M* serine stimulus was assayed as described in the text. After 4 minutes of incubation on a gyratory shaker in the 0.02 m*M* serine, the bacteria were rapidly mixed with a concentration of serine that yielded a final concentration of 0.5 m*M*, and the assay was initiated. In experiment 2, the response to a single stimulus of 0.5 m*M* was assayed (see text, Section VII). The mean recovery time, t_R, for each stimulus is the area under the recovery curve. The areas in experiment 1 are 0.5 and 1.5 minutes. In experiment 2 the area is 2.0 minutes.

purified ribose-binding protein isolated from *Salmonella* (Aksamit and Koshland, 1974). The theoretical line is calculated from the dissociation constants (K_d) of sugar from the ribose-binding protein which had a K_d for ribose of 3×10^{-7} and for allose of 3×10^{-4}. The change in the receptor–chemoeffector complex can be deduced from the mass action law [Eq. (2)] and the change in receptor occupancy [Eq. (3)].

$$K_d = [R][C]/[RC] \tag{2}$$

where R is protein receptor, C is the chemoeffector, and RC is the receptor–chemoeffector complex. The solid lines are based on the calculation of the change in receptor occupancy, i.e., Δ[RC] of Eq. (3) using the K_d's from the purified protein, where C_f is the final concentration after mixing and C_i is the initial concentration in the medium.

$$\Delta[RC] = [R(C_f - C_i)]/K_d \tag{3}$$

The points are the recovery times obtained from the tumble frequency assay on the whole living bacterium. The agreement is excellent for ribose and almost as good for allose (Fig. 7).

These experiments together with the previous ones show that the tumbling suppression is proportional to Δ[RC], the change in receptor occupancy, and is not proportional to $d(\Delta\,\mathrm{RC})/dt$. Mesibov *et al.* (1973) have shown that the sensitivity curves based on the capillary response are correlated with the change in receptor occupancy.

When the quantitative method is applied to different types of stimuli, the results are not quite as simplistic. It was found in the early experiments described above that it was not possible to calculate from the separate responses of repellents and attractants exactly what concentration would lead to a nullification of the response. When mixtures of attractants and repellents were mixed or when two different attractants were added to each other, the additivity of the individual responses was not as neatly summed as that shown in Fig. 6. Some particular results obtained by Rubik and Koshland (1977) are shown in Table V. Thus, it is clear that the processing

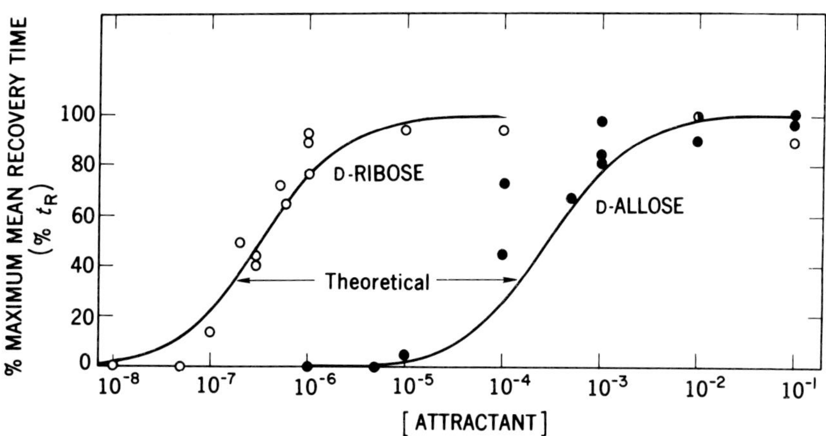

FIG. 7. Comparison of response with receptor occupancy. Points are percent maximum mean recovery times (t_R) for D-ribose and D-allose concentration increases from 0 to the concentration shown on the abscissa. Each point represents the average of three consecutive assays. Results are for three experiments performed on different days. Maximum t_R were 0.56, 0.58, and 0.62 minute for ribose experiments and 0.54, 0.55, and 0.58 minute for allose experiments. Theoretical curves were calculated assuming that noncooperative chemoreceptor binding constants equal 3.3×10^{-7} for ribose and 3.0×10^{-4} for allose (as determined *in vitro*) and that the response t_R is proportional to the change in fraction of binding protein occupied. The half-filled circle represents coincidence of the points ○ and ●.

TABLE V

ADDITIVITY RELATIONSHIPS IN MULTIPLE STIMULI IN DIFFERENT RECEPTOR CLASSES[a]

Stimulus No. 1		Stimulus No. 2		Combined stimulus		Difference of sum of individual stimuli and combined stimulus
Gradient	Time	Gradient	Time	Gradient	Time	
Ser (0.01)	0.58	Asp (0.01)	0.40	Ser (0.01) + Asp (0.01)	0.54	0.44
Rib (0.01)	0.35	Ser (0.001)	0.26	Rib (0.01) + Ser (0.001)	0.58	+0.03
Rib (0.01)	0.36	Asp (0.01)	0.46	Rib (0.01 + Asp (0.01)	+0.48	+0.34
Asp (0.2)	1.2	Rib (0.2)	0.95	Asp (0.2) + Rib (0.2)	2.27	−0.13

[a] In each case stimulus involves mixing bacteria with compound indicated so that final concentration (m*M*) is reached after mixing period. Recovery times for each stimulus are recorded in minutes. Combined stimulus is achieved by simultaneously adding compounds for stimulus No. 1 and stimulus No. 2 previously studied in separate experiments.

From Rubik and Koshland (1978).

system acquires additional complexity when more than one type of attractant or repellent is providing a stimulus than if one is analyzing simply one type of stimulant. Nevertheless, there is a basic summation of responses, and, in fact, it has been possible to show that there is additivity of stimuli in a rough qualitative sense for metal ion and attractants and repellents, for light responses and attractants and repellents, for temperature effects, and for mixtures of attractants and repellents. Therefore, the detailed mathematics of the processing system remains to be evaluated, but the fact that all of these attractants and repellents ultimately go through a central processing system appears quite clear.

E. WEBER–FECHNER LAW

In the 1880's, Pfeffer performed ingenious experiments with meat extracts and showed that to a first approximation the bacteria followed Weber's law. Weber's law defines a response in terms of a just noticeable stimulus, leading to the conclusion that $\Delta\Psi/\Psi$ is a constant where Ψ is the stimulus and $\Delta\Psi$ the just noticeable change in the stimulus (Stevens, 1970). The advent of a quantitative method for measuring bacterial population migration made it possible for this phenomenon to be examined more closely. Using the population–migration apparatus mentioned above, Dahlquist *et al.* (1972) investigated the movement of bacteria under the influence of various gradients of attractant. In a linear gradient of attractant (ΔC per unit distance equals a constant with distance) the bacteria showed a peculiar pattern which was difficult to analyze quantitatively. However, when the same experiment was performed with a logarithmic gradient ($\Delta C/C$ per unit distance equals a constant) the bacteria gave a uniform migrational velocity over the entire range of the logarithmic gradient. This result can only be explained by the assumption that the bacteria are migrating with a velocity following $d(\ln c)$. From this result alone, one would conclude that bacterial migration obeys the Weber–Fechner law. However, these data were put to a further test because the apparatus allowed study of a wide range of concentrations. A horizontal straight line would be observed if the Weber–Fechner law were obeyed at all concentrations. The results showed clearly that it is not (Dahlquist *et al.*, 1972).

Similar conclusions were reached by Mesibov *et al.* (1973) by studying capillary tests. By ingenious methods of correction, Mesibov *et al* (1973) were able to factor out a roughly quantitative Weber–Fechner law relationship over a narrow range and found deviations occurred over a wide range. A theoretical curve which gave rather good agreement was obtained for α-methyl-L-aspartate, an extremely strong attractant, but agreement was not nearly as good in the case of poor attractants such as D-galactose. In the process of this analysis, they found that the needed threshold for the bacteria to detect a gradient was a 12% change in the occupancy of the receptor for galactose but only a 0.4% change for α-methyl-L-aspartate.

If we assess these conclusions in terms of the excellent correspondence between change in receptor occupancy and the dissociation constant of the pure protein (Fig. 7), it is clear why the capillary assay and Weber's law are correct only over a narrow range. The plot of percentage saturation versus the log of the concentration of the free ligand is as shown in the theoretical curves of Figs. 6 and 7 (cf. also Stevens, 1970; Beidler, 1954, 1962). This is the curve one obtains if one plots a biological response against stimulus, provided the stimulus obeys a simple receptor occupancy law. It is quite clear though that this curve is not linear over its total concentration range, but is approximately linear only over small regions, in particular over its middle portion. Complex behavioral phenomena subjected to numerous inputs will certainly not *a priori* be proportional to receptor occupancy alone. However, it is quite possible that the initial trigger, whether it be a chemical occupying a receptor or a photon activating a photon receptor, can induce a chain of very complex events which produces a response that is proportional to the initial stimulus over a limited range. Therefore, if the biological response in a complex assay, such as the capillary assay, or a direct assay, such as the tumble frequency assay, is studied over a limited range, it may obey the Weber–Fechner law. This is probably also the case in more complex phenomena in higher species (Stevens, 1970). As the range is increased, however, both assays should deviate and that is indeed observed.

The results of Section IV,C also throw light on statements about "just noticeable response" and "thresholds" of stimuli. According to the receptor occupancy curves of Fig. 7, there is complete correspondence with a theoretical curve of binding to a pure isolated receptor. This means that there is no threshold as far as the stimulus is concerned. The smallest amount of attractant changes the probability of tumbling immediately. Whatever threshold seems to be present exists because the analytical techniques for measuring the behavioral responses have different sensitivities. Thus, when capillary assays assign thresholds above those of the tumble frequency assay, it simply means that one assay is more sensitive than another. Thresholds, therefore, are a practical matter but not a theoretical one.

It is quite straightforward to explain these results in terms of the tumble regulator model outlined above. Since the memory span of the bacterium depends on the level of the tumble regulator parameter, it may be a small molecule or a membrane potential or the lifetime of some chemoeffector–receptor protein complex. It is readily seen that the additivity of this response would depend on the inputs and rates of degradation of this tumble regulator intermediate. As different stimuli contribute to the concentration of this parameter, it would lead to simple additive responses only under restrictive circumstances where the formation and degradation of the intermediate follow simple laws. As more complex relationships between the formation and degradation of the intermediates arise (analogous to a tank of water with many faucets leading into the tank, many drains leading out), the process of presenting or removing any one stimulus would produce complex kinetics in relation to the other stimuli. Hence, it is not surprising that under limited circumstances simple relationships, such as the Weber–Fechner law and additivity, are observed whereas over the entire response range the simplistic relationships break down. Nevertheless, the conclusion of a central processing system into which all stimuli proceed and from which a final response is generated is consistent with all of the data.

It should be emphasized that the tumble regulator model represents a general type rather than a specific prescription. Thus, there is no reason to select a model in which the response regulator turns off tumbling when it is above the threshold vis-a-vis a model in which tumbling is generated above the threshold and suppressed below it. Also, it is equally acceptable to assume step V_d responds rapidly to changes in effector and V_f responds slowly. Moreover, it is equally possible that the effector alters the rate of enzyme 1 and the threshold detector as compared to enzyme 1 and enzyme 2. Those alternatives reflect detailed molecular assignments to the mathematical framework and must await further experimentation. An analogy might be to the use of *d* and *l* nomenclature for stereoisomers before absolute configurations were known. A self-consistent set of principles and relationships could be established and then placed on an absolute basis once a single stereoisomer structure was known. The interrelationships of the response regulator and stimuli can similarly be established even though the absolute relationships are yet to be delineated.

IV. Biochemistry of the Sensing System

The biochemistry of the sensing system is as yet largely unexplored. Nevertheless, some important clues as to the complexity of the system and some of the biochemical intermediates have been obtained. A summary of this work is given below.

A. Mutants in the Central Processing System

In the bacterial system, the use of genetic mutation is a classic tool, and it has been applied in a helpful way in regard to bacterial chemotaxis. In initial studies Armstrong *et al.* (1967) used the swarm plate method to select generally nonchemotactic mutants from other types of mutations. By generally nonchemotactic, they meant mutants which were defective in the sensing system and not in motility. These mutants could not respond to gradients of any attractant and hence appeared defective in the transmission system and not in a specific receptor. Forty independent mutants were isolated which fell in three complementation groups called *cheA*, *cheB*, and *cheC*. Armstrong and Adler (1969) subsequently mapped these mutants in three groups which they placed in between the *his* and *trp* regions in the *E. coli* chromosomes. Using the same techniques with the addition of an *F′* episome, Parkinson (1974) has added a fourth gene which he called *cheD*.

Aswad and Koshland (1975a) used the preformed liquid gradient technique for selection of mutants from *Salmonella typhimurium*. This technique which separated nonmotile from nonchemotactic mutants as well as from wild types produced a total of 72 mutants, of which 58 were identified and separated into 6 complementation groups. This work was then carried further by Warrick *et al.* (1977) who separated 110 additional generally nonchemotactic mutants and proceeded to characterize them by abortive transduction, complementation, episome, and deletion techniques. A total of nine different complementation groups were found, of which eight have now been mapped in specific regions of the chromosome of the *Salmonella;* one in a separate location is yet not mapped. These were named *che* genes *P, Q, R, S, T, U, V, W*, and *X*.

Two of the mutational classes, *cheU* and *cheV*, map precisely in regions previously identified with *fla* genes (Warrick *et al.*, 1977). Silverman and Simon (1973) also observed overlap between a chemotactic gene in *E. coli*, and similar results have been obtained by Collins and Stocker (1976). The *fla* genes are identified on the basis of the flagella fail to assemble. The non-chemotactic mutants are motile but cannot receive signals. The most simple explanation of these facts is that these genes produce a protein which is part of the flagellar machinery. Hence some mutants prevent flagellar assembly, and others are defective in that portion which receives signals from the sensory apparatus.

The discrepancy between the number of genes found in *E. coli* (Parkinson, 1976) and *Salmonella* (Warrick *et al.*, 1977) was quickly resolved by the studies of Silverman and Simon (1977) who cloned the bacterial genes of *E. coli* onto a lambda phage and identified the peptide fragment from the various gene loci. They identified eight genes, which they placed in order in the region between the *mot* and the *fla* genes, very similar to the regions identified in *Salmonella*. Since only one gene is coincident with a *fla* gene in *E. coli*, this evidence strongly supports the minimum of nine genes observed in the *Salmonella* analysis. The two gene maps are shown in Fig. 8, although the minutes in the corresponding maps do not correspond precisely. The

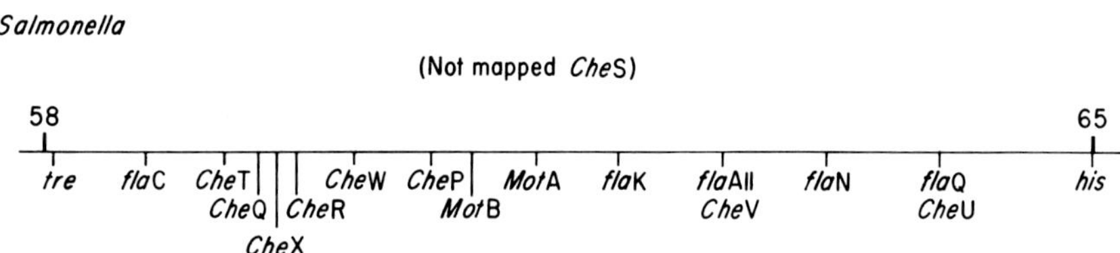

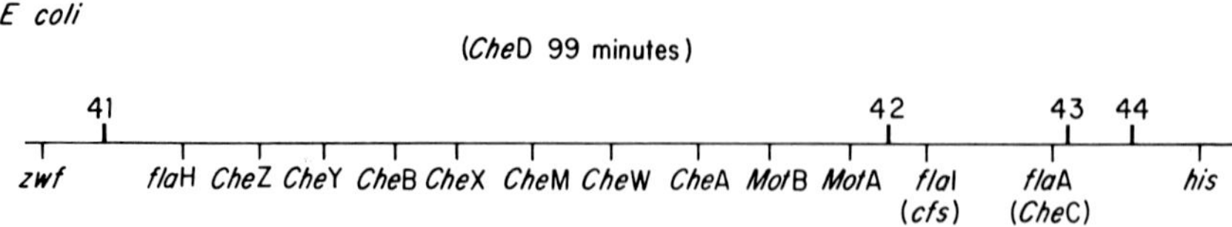

FIG. 8. Genetic maps of the chemotactic region from *Salmonella typhimurium* and *Escherichia coli*. Map positions in minutes are used to give order but precise distances between genes are not known.

cluster of chemotactic genes is not only oriented similarly in the two bacteria, but they are nestled in the center of the flagella genes in a very similar manner.

The total number of these mutant classes gives some idea of the complexity of the central processing machinery of the sensory system. There appear to be a minimum of nine gene products identified with the chemotactic machinery. Since an exhaustive search for mutants has not been completed, additional genes are still possible. Nevertheless, it is interesting that the study of a number of different mutants in two different organizers by four different laboratories has produced so far only nine genes. This would suggest that the total number is not far different from this, and hence that some idea of the total complexity of the system is available.

It should be clear that the entire chemotactic response cannot be identified with just nine genes. There are approximately twenty or more genes identified with individual receptors, and twenty or more genes identified with the flagellum itself. All these are essential for the chemotactic response. However, there are nine genes which are central to all responses to outside stimuli, and these genes are related to the complexity of the processing system to interpret the stimuli received by the receptors. They are the "brain" of the bacterium; the receptors are its eyes and ears, and the flagella, its arms and legs.

B. Role of Methionine and Methylation

The identification of genes coding for central processing functions inevitably raises questions in regard to the biochemical function of the individual genes. One clue came from an early observation that a methionine auxotroph of *E. coli* was nonchemotactic toward all attractants when starved for methionine (Adler and Dahl, 1967a). The effect of methionine starvation was not caused by lack of protein synthesis because starvation for other amino acids, such as thereonine or leucine, had no effect on chemotaxis. Moreover, methionine starvation does not lead to loss of motility, but rather an alteration in the type of motility displayed, i.e., the bacteria swam smoothly without tumbling.

This observation raised the question of whether methionine was involved simply in the mechanical ability of the flagella to tumble or in the sensory apparatus. This was resolved by transducing the block in methionine metabolism to a constantly tumbling mutant of *Salmonella typhimurium* ST4 (Aswad and Koshland, 1974). The bacteria continued to tumble in the absence of methionine. Hence the methionine is not needed specifically for tumbling. The mutant was then subjected to a sudden increase in the concentration of serine. Smooth swimming resulted followed by a return to the normal tumbling pattern of the mutant. The rate of return to the constantly tumbling state was appreciably longer in the methionine-depleted bacteria than in the bacteria to which methionine had been added. These indicated that methionine was involved in maintaining the relative levels of the tumble regulator and the threshold for detection of the tumble regulator, and not in the mechanics of the tumbling process (Aswad and Koshland, 1974).

It then became of some interest to find whether or not methionine was itself involved or whether it was a precursor for some other compound. Experiments on *E. coli* (Armstrong, 1972) and on *S. typhimurium* (Aswad and Koshland, 1975b) indicated that *S*-adenosylmethionine (SAM) had to be synthesized in order to obtain the chemotactic response. These results were obtained in several ways. Methionine analogs which replaced methionine as substrates in the SAM synthetase reaction replaced methionine in restoring chemotaxis, while those which were nonsubstrates showed no replacement ability. Cycloleucine, a methionine analog which inhibits SAM synthesis, produces the same behavioral changes as methionine starvation. SAM is only one of two metabolites derived from methionine which is formed rapidly and present in high concentrations during incubation in the chemotaxis medium. Concentration of the other metabolite, spermidine, does not correlate with chemotactic ability. When a methionine-starved chemotactically wild-type cell is given an exogenous supply of methionine, the SAM pool is replaced at the same time that spontaneous tumbling reappears. Likewise, the rapid turnover of SAM correlates well with the rapid loss of tumbling, which occurs when the exogenous supply of methionine is removed. These results provide evidence that SAM is involved in the signaling mechanism of the chemotactic machinery.

The next clue was provided by Kort *et al.* (1975) who found that a protein in the membrane of bacteria with a molecular weight of approximately 65,000 was methylated. There was a relationship, but a somewhat indistinct one, between methylation and the chemotactic response. Springer and Koshland (1977) isolated a mutant of *Salmonella typhimurium* which lacked the *cheR* gene product, and found that no methylated protein was formed under these circumstances. Further studies showed that this gene produced a protein methylase which transferred the methyl group from *S*-adensylmethionine to a 65,000 molecular weight protein of *Salmonella*. It was shown that the *Salmonella* protein was similar to the same methyl-accepting chemotactic protein (MCP) that had been demonstrated (Kort *et al.*, 1975) in *E. coli*.

Recently a demethylating enzyme, a carboxymethyl esterase has been identified by Stock and Koshland (1978) as the *cheX* gene product of *Salmonella typhimurium.* Moreover Springer *et al.* (1977) have shown that the methylation levels are altered by attractants and repellents *in vivo.* Hence a reversible methylation-demethylation system is central for the chemotactic response of these bacteria.

If a bacterium encounters slightly unpleasant or deleterious surroundings it migrates chemotactically to a new better environment. If the medium is extremely deleterious so that its survival is threatened it would be best for it to travel rapidly in a straight line to a new locale. Thus an override mechanism similar to the role of adrenalin in higher systems would be useful to respond to extreme danger. Such an override mechanism appears in the chemotactic system whenever the level of S-adenosyl methionine is depleted. The bacterium then swims in straight lines and is incapable of responding to stimuli which might make it turn and tumble. Once the level of S-adenosyl methionine is restored the bacterium resumes its chemotactic migrations. Obviously the chemotactic mechanisms are more productive of directed migration to a specific environment. However, the availability of an override mechanism to respond rapidly to a survival threatening environment provides the bacterium with a device similar to that utilized by higher species.

Studies of the methyltransferase specificity of *Salmonella* (Van Der Werf and Koshland, 1977) and *E. coli* (Kleene *et al.*, 1977) showed that a glutamic acid residue was methylated. The *Salmonella* studies resulted in a purification of the enzyme and demonstration that it can modify membrane-bound proteins *in vitro* (Springer and Koshland, 1977). Interestingly, this methyltransferase enzyme is present in the cytoplasm as is the demethylene (Stock and Koshland, 1978).

The *cheR* mutant which is deficient in methyltransferase activity has a smooth swimming phenotype, which is consistent with the finding that prevention of SAM synthesis produced smooth swimming. A strong phenol gradient was found to cause tumbling in any of these nonmethylating mutants (Springer and Koshland, 1977). Thus, the tumble regulator and the ability to tumble are not eliminated in a bacterium lacking the methylating ability. This tends to reinforce again the earlier conclusion that methylation is involved in fine tuning of the system to provide proper levels of tumble regulator intermediate, but it is not essential for the mechanism of tumbling or for the generation of a sensory response that stimulates tumbling. This leads to a modification of the tumble regulator model of the type shown in Fig. 9, where again the indicated modifications are schematic in nature. A definitive model will not be possible until more of the methylating properties are known.

C. Light Effect

Exposure of free-swimming *S. typhimurium* to a high intensity light caused tumbling (Macnab and Koshland, 1974). The response was found to be reversible provided short light pulses were used. The optimal wavelength for inducing tumbling was found to be between 360 and 450 nm. It had previously been reported that photoinactivation of bacteria was sensitized by photoinducible dyes and molecular oxygen, a phenomenon which became to be known as the "photodynamic effect" (Raab, 1900). However, the light-induced tumbling effect is quite clearly different from the photodynamic effect and involves a direct perturbation of the transmission machinery. It was subsequently found that there were three distinct light effects (Taylor and Koshland, 1975). One of these was a light-induced tumbling which occurred immediately. A second involved a smooth response, generated on somewhat more prolonged exposures to light. Finally, a killing or

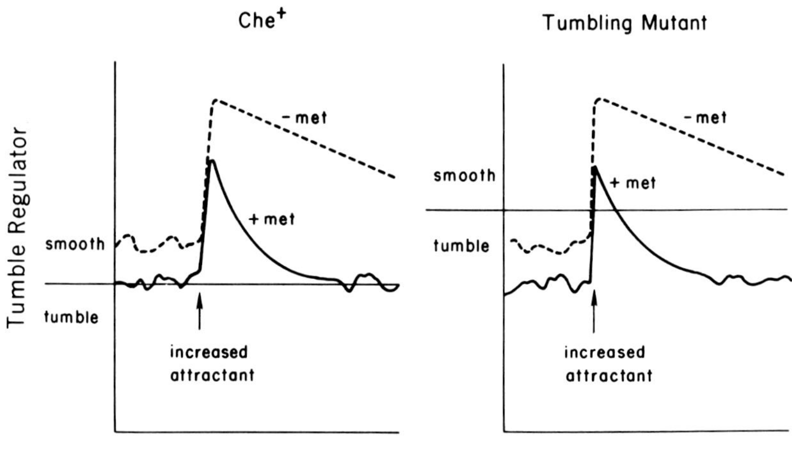

FIG. 9. A model for a possible role of AdoMet in regulating tumbles. When the tumble regulator (wavy line) rises above some threshold value (horizontal solid line) tumbling is suppressed; when it falls below, tumbling is increased. In the absence of a gradient, the regulator level fluctuates randomly around the threshold, but a rapid temporal increase in attractant makes it rise way above. The regulator is maintained at a steady-state level by a delicate balance between its synthesis and degradation. It is postulated that AdoMet is necessary for the degradation reaction. Thus, during methionine starvation, the level of regulator is higher than normal, and when its level is further perturbed by a temporal gradient, it takes longer than normal to fall to its steady-state level. By assuming that the uncoordinated mutant has an abnormally high threshold, one can easily explain its behavior in both the presence and absence of methionine.

paralyzing effect was found due to very long exposures to light. The paralyzing effect, the so-called photodynamic effect observed earlier, is of relatively little importance to this particular analysis, although it may indeed involve destruction of the chemotactic machinery. However, such extensive changes occur that it woulf be difficult to disentangle the specific causes.

In the case of the light-induced tumbling, the effect was found to be involved in the chemotactic signaling in several ways. First of all, light-induced tumbling was additive with chemical-induced tumbling. Hence, there is an additive interrelationship between the light-induced tumbling and the normal response to chemoattractants. Second, the mutants which had defective transmission machinery (*cheU* and *cheV*) failed to show the light effect. These mutants were motile, but were swimming without tumbling, and the light could not generate tumbling in them. Moreover, methionine auxotrophs did not exhibit the light effect unless methionine was added to the culture.

The action spectrum of the light effect was studied and appeared to give the action spectrum of a flavin (Macnab and Koshland, 1974). This strongly suggests that the initial absorption of the light occurs in some flavinlike molecule. The light-induced tumbling effect could be generated by externally added dyes (called the extrinsic effect) as well as by internal compounds (the intrinsic effect) (Taylor and Koshland, 1975). For the extrinsic effect to operate, the electron transport machinery had to be operating, thus suggesting that the light effect causes some interruption in the electron transport chain or modifies an intermediate of the electron transport chain which in turn induces the tumbling. The light effect was also found in *E. coli*, but seemed to be more sluggish in response. The observation of such a light effect provides a handle in probing the transmission apparatus more deeply, and possibly the flavin action spectrum will help in identifying individual proteins involved in the chemotactic sensing process.

The studies of this light effect add one further complexity and additional insight into the chemotactic process. The light intensities required to generate tumbling are far higher than the bacterium normally experiences, and hence the light effect is not a normal physiological response. Nevertheless, the identification of the tumbling response with some part of the electron transport chain occurring during oxidation suggests that the light is perturbing some essential part of the system. Similar conclusions are reached from the finding of the divalent metal ion effect in the Mg^{2+},Ca^{2+}-ATPase. The levels of metal ions required to obtain a chemotactic response are far higher than the bacterium needs to obtain its essential requirements, and this response appears to be nonphysiological. Thus it appears that chemotactic responses can be generated in the laboratory by perturbing the level of tumble regulator without necessarily utilizing a receptor selected by evolution for its physiological importance. Whether some of the receptors now listed as natural receptors fall in the same category is too early to say, although it is apparent that they work at levels which from teleological reasoning would be important to survival of the organism. These effects, even though nonphysiological, may be very helpful as clues to understanding the biochemical processes.

D. Membrane Potential

One of the intriguing ideas for the control of tumble frequency is that the membrane potential is the tumble regulator. Many cells identified with sensory systems respond to transient or permanent changes in membrane potential when confronted with specific stimuli. Behavior of this sort is seen in cells as diverse as sensory neurons and in unicellular protozoa. That membrane potential can affect the swimming pattern was indicated by the finding that a number of membrane-active drugs as well as inhibitors and uncouplers of oxidative phosphorylation can influence motility (Caraway and Krieg, 1972; Faust and Doetsch, 1971; Ordal and Goldman, 1975, 1976). Moreover, it was known that the energized membrane state was essential for the motility itself (Larsen *et al.*, 1974a). Ordal and Goldman (1975, 1976) showed in *Bacillus subtilis* that uncouplers of oxidative phosphorylation cause behavioral changes analogous to those caused by repellents in the gram-negative bacteria. DeJong *et al.* (1976) also concluded that transient changes in potential cause swimming behavior alterations. Szmelcman and Adler (1976) studied permeant cation distribution using triphenylmethylphosphonium ion and found a change in membrane potential as measured by this cation correlated with additions of attractants or repellents. Since the blue light effect caused changes in the oxidation machinery (Macnab and Koshland, 1974), it seemed logical that it also would cause changes in membrane potential which were important in the alterations of tumbling response.

Studies of cyanine dyes as measures of membrane potential were made in *Bacillus subtilis* as a means of detecting alteration in membrane potential for correlation with changes in tumbling frequency (Miller and Koshland, 1977b). The net result of these studies was to demonstrate that the changes in the membrane potential produced changes in the behavioral response very similar to that of attractants and repellents. Increases in membrane potential cause suppression of tumbling, while decreases in membrane potential cause generation of tumbling. However, additions of attractants and repellents did not cause detectable changes in membrane potential as measured by these cyanine dyes. It was concluded (Miller and Koshland, 1977b) that the alteration in membrane potential was detected by the chemotactic system in the same way that alteration in levels of nutrients or toxic substances was detected. The above results indicate quite strongly, however, that a change in the total potential cannot be directly correlated with the change in rotation. A specific potential, such as a particular ion or a particular combination of ions, is not excluded. In that case, the change in overall potential would alter the specialized potential and that would explain the responsiveness of the system to change in overall potential. It would also mean that one could alter attractants and repellents and change some small potential which only made a minor contribution to the overall potential, or, for example, an Na^+–K^+ antiport system could be activated without any change in overall potential i.e., Na^+ goes in one direction wheras K^+ goes in the other. It is also known that there is a compensatory effect between protons and potassium ions, and, therefore, a change in attractants and repellents could cause a change in one potential which would be offset by a different potential change leading to no change in the overall potential. Future developments in these studies should be of great interest.

V. The Mechanism of Tumbling

The motility of the bacteria is intimately related to the chemotactic response, since it is the manner by which the response is expressed. Some of the proteins involved are common to both sensing and motility. Since the mechanism of motility is described in another chapter of this volume (Sokatch, Chapter 5, this volume) and in a recent review by Macnab (1977b), it will not be discussed in detail here. The energy to drive the flagella is identified with the energized membrane state and not with ATP directly (Larsen *et al.*, 1974a). This is in contrast to skeletal muscle which is driven by ATP. The mechanism of tumbling has been shown by Silverman and Simon (1974) to be a reversal of the flagella rotation by means of a study of tethered bacteria. Furthermore, it was established that reversal of flagella rotation can be identified with tumbling (Larsen *et al.*, 1974b). When the bacterium is swimming smoothly, the flagella rotate clockwise when viewed looking backward from the body of the bacterium. Tumbling is generated by reversal of the flagella rotation to a counterclockwise rotation. However, Macnab has demonstrated that simple reversal of rotation should not lead to tumbling but rather to jamming of the flagella bundle, and therefore has shown that the flagellar helix undergoes a change to the right-handed helical form (Macnab, 1977c; Macnab and Ornston, 1977). As a result of the large resistance of the "dog leg" form of the flagella in transition between left- and right-handed pitches, the two structures are presumed to make a major contribution to the tumbling phenomenon. It has been observed, in addition, that the flagella fly apart during the tumbling process (Macnab and Koshland, 1974). The detailed mechanism of tumbling, has now been clarified by the use of mutants (Rubik & Koshland, 1978) and the observations on flagellar structure (Macnab & Ornston, 1977). The theory is quite simple (Khan *et al.*, 1978). Prolonged rotation in either direction produces smooth swimming. In the counterclockwise direction the flagellar helix is left-handed (CCW-LH) and is normal smooth swimming of wild type bacteria. In the clockwise direction the helix is right-handed (CW-RH), and prolonged rotation in this direction in mutants can also give smooth swimming. Tumbling is caused by reversals of rotation either from CCW to CW or CW to CCW' The random swimming in normal wild-type is caused by largely CCW rotation punctuated by CW reversals. However, random swimming can also be due to predominantly CW rotation punctuated by CW reversals. Continual reversals do not allow flagellar bundle formation to occur at all and hence cause constant tumbling. The way in which tumbling can be used to analyze mutant and wild-type behavior has been described by Khan *et al.* (1978).

VI. Bacterial Individuality

Studies on chemotactic properties have led to interesting evidence in relation to bacterial individuality. The problem of nongenetic variation of cells has been one of central interest to biochemists. Studies on bacteria, in particular, have led to consideration of this phenomenon in relation to bacteriophage burst sizes, cell division time (Delbruck, 1945; Powell, 1958), cell flagella phases (Lederberg and Iino, 1956; Stocker, 1949) and β-galactosidase concentrations (Benzer, 1953; Maloney and Rotman, 1973; Novick and Weiner, 1957). The spatial heterogeneities which inevitably develop in stationary surface growth are continuously randomized in swirled liquid cultures of bacteria. The further development of the techniques for chemotaxis made available an assay which could be used to study properties over a very small portion of the bacteria's life cycle, and moreover could be used to study an individual bacterium by tethering. Such instantaneous measures of the individual cell, therefore, allowed studies of an individual bacterium throughout its life cycle. Moreover, the property that controls tumbling frequency is clearly an essential property of the bacterial cell, one that has been selected for its contribution to the survival of the organism. For all of these reasons, a study of the chemotactic response could provide useful clues in regard to bacterial individuality.

Studies of this phenomenon were made on the populations as a whole and on individual bacterium (Spudich and Koshland, 1976). Briefly, it was found that there was a Poissonian distribution of properties relating bacterial responses to chemotactic stimuli. It could be shown that the bacterial response to a stimulus followed Eq. (4):

$$1 - F(t) = \frac{S(t) - S_{\infty}}{1 - S_{\infty}} \tag{4}$$

where $F(t)$ is defined as the fraction of the population that has recovered at time T, $S(t)$ is the probability of producing a smooth-swimming track, and S_{∞} is the number of smooth tracks at infinite time. Thus individual bacteria of identical genetic constitution presented with a stimulus responded differently. Experiments to test various alternative hypotheses for this phenomenon established that the differences could not be caused by genetic variation or by the presence of the bacteria in different phases of the cell cycle. It could readily be shown that a single bacterium gave the same responses within experimental error over its entire lifetime. Furthermore, the differences in responses between bacteria are not due to individual fluctuations of their nutritive pool during various phases of their life cycle.

The conclusion reached by Spudich and Koshland (1976) in this case was that the bacteria produced some enzymes or parts of bacterial systems in amounts which varied in a Poissonian manner at some crucial point in the bacterial development. If this molecule is present in small amounts at crucial stages in development, then Poissonian variation in the behavior of the "adult" bacterium would be established by this variation. Hence, bacteria would exist as individuals with respect to this property. The analysis indicated that certain properties might be distributed in a Poissonian manner and others might be identical for all bacteria.

The existence of such variability leads one to ask whether it has some advantage to the organism or whether it is a disadvantage which must be surmounted in the development of the species. Although there may be no significant selective pressure against nongenetic variability, one can certainly make an argument that nongenetic variability aids in the survival of a population subjected to widely varying conditions during its lifetime. If the bacterium has a single or monolithic reaction to a particular chemical gradient, it might migrate as a colony into a toxic situation or fail as a colony to be sufficiently sensitive to an essential new nutrient. If there were individual bacterial variations in a genetically homogeneous population, a few fractions of the bacteria might either be supersensitive or insensitive. These bacteria would normally not survive as effectively as the bulk of the more average optimized population. They would then have lower survival probabilities in most circumstances, an unimportant loss if the main body of the colony survived. If, on the other hand, the toxic and lethal situations described above arose, a small fraction, on the wings of the probability distribution, might constitute the only bacteria to survive and would then reproduce a mutant poorly adapted to the more common conditions of the environment. The wild type, however, was selected over evolutionary time for survival in all of the widely varying conditions of the environment. Thus, nongenetic variability would be a preferred mechanism for accommodation to random short-term fluctuations in the environment and genetic variability the pre-preferred mechanism for accommodation to long-lasting environmental changes. Induction and repression of enzyme synthesis reflect intermediate accommodations to fluctuations of a lengthy but impermanent type.

VII. Methods

The early studies of Pfeffer and Englemann utilized microscopic observation almost entirely. Watching *E. coli* or *S. typhimurium* for even a brief period leaves a clear impression of bacteria swimming in straight lines, tumbling, and heading in a new direction. Schematic representation of this observation in a two-dimensional plane shows the "runs" of roughly straight lines followed by abrupt changes in direction called "tumbles," "twiddles," or "turns." This movement can be recorded permanently by open shutter photography with steady (Dryl, 1958; Vaituzis and Doetsch, 1969) or stroboscopic light (Macnab and Koshland, 1972).

Recently the use of high intensity light (Macnab and Koshland, 1974) has revealed that individual flagella can be observed, and this technique has now been elegantly exploited by Macnab (1977c; Macnab and Ornston, 1977) and Asakura (Shimada *et al.*, 1975, 1976) to elucidate the helical turns in the flagellas, the change in phase on reversal of rotation, and many other features involved in their locomotion. The subject of motility is discussed in Chapter 5 by Sokatch in this volume and will not be repeated here. However, in this section, some of the methods which have been particularly useful in the development of our understanding of bacterial chemotaxis will be described.

A. The Capillary Assay

In the 1880's Pfeffer demonstrated chemotaxis by placing attractant in a capillary tube and observing microscopically the accumulation of bacteria near the mouth and inside the capillary. In the 1960's Adler made the method quantitative by transferring the bacteria from the capillary to agar plates and counting the colonies. One form of Adler's method (Adler and Dahl,

1967b) involves a small chamber formed by laying a glass tube from a 5 cm length of melting point capillary between a microscopic slide and a cover slip. The chamber is filled with about 0.2 ml of the bacterial suspension. After incubation for 1 hour, the capillary is removed and the exterior is rinsed with a thin stream of water from a wash bottle. The sealed end is then broken and the contents delivered into a test tube containing tryptone broth at 0°C. Suitable dilutions are made and a sample is mixed with 2 ml soft tryptone agar at 45°C and then poured into a tryptone agar plate which has a high plating efficiency. After incubation at 37°C, the colonies are counted. Averages of several runs are used to reduce variability due to convection currents, temperature variations, etc. Variations are in the range of 27% between two values.

The number of bacteria in the capillary will somewhat depend on the conditions of the experiment. If the experiment is allowed to proceed too long, the nutrient will diffuse out of the capillary tube completely. If it is carried out for too short a period the attractant will not have time to diffuse throughout the solution and only a few bacteria will be attracted inside the capillary. Thus, a compromise between too short or too long periods of incubation is needed, which is usually about 1 hour. The decreased response of high concentrations is not due to a repellent action but rather is a function of the nature of the capillary assay. At the highest levels of a sugar or amino acid, the diffusion of the attractant will create a zone just outside the capillary mouth which saturates the bacterial receptor. Therefore, no further reason exists for swimming into the capillary and the number inside starts to decrease.

Capillary curves can also be affected by metabolism of the attractant, since it is quite possible for the bacteria to digest a significant fraction of the attractant when its concentration is as low as 10^{-6} *M*. This alters the concentration in the neighborhood of the mouth of the capillary and hence displaces the capillary curve. This situation does not occur with nonmetabolizable compounds or mutants defective in metabolic pathways.

These complications make certain theoretical interpretation of the capillary assay results difficult, but nevertheless it has been an extraordinarily effective tool for rapid screening of compounds and the investigation of simple responses. Moreover, it can be used for delineating quantitative relationships by careful analysis of sensitivity curves and dose–response relationships (Mesibov, Ordal, & Adler, 1973).

B. Trackers

The first tracker to be designed was that of Berg (1971) which utilizes fiber optics, photomultiplier pairs, and an automatic feedback device to keep the bacterium in a fixed position in laboratory space. The optical fibers are arranged in the x, y, and z directions so that movement of the bacteria in any direction is recorded in the automatic machinery of the tracker, allowing it to follow the bacterium through three-dimensional space. As the bacterium moves on the stage of the microscope, its positions are recorded in a computer for subsequent analysis.

The tracking device can follow the movement of *E. coli* within the very small chamber quite accurately. The center of focus floats within the body length of the bacterium which is 1 μm wide and approximately 2–5 μm long. The optical systems do not allow the focus to stay on one fixed point within the bacterium, and hence the center of focus tends to slide from one edge of the bacterium to another in the process of the tracking. To follow movements in a gradient a capillary is introduced into the side of the vessel (Berg and Brown, 1972). Since the attractant or repellent is diffusing from the capillary into the bulk of the vessel, the gradient obviously varies with time, and therefore calculation of the variation of the gradient over time must be allowed for in the analysis of the movement of the bacteria over time (Futrelle and Berg, 1972). Brown and Berg (1974) have recently developed a method to vary the gradient of concentration in the tracker over time by enzymatic depletion of a chemical. This appreciably simplifies the mathematical analysis as compared to the complexities of diffusion from a capillary and makes the instrument more accurate.

The second type of tracker operating on somewhat different principles was developed by Lovely *et al.* (1974). A stable gradient of attractant was produced in a relatively large reaction vessel (1 cm × 1 m × 11 cm) by utilizing glycerol (a nonattractant) to stabilize the system against convection currents. An observer, looking through a long-range microscope keeps the bacterium on a cross-hair and in focus by moving a pedal which controls movement of the microscope stage in the z direction and by operating a "joy-stick" which controls movement in the x and y directions. The stage of the microscope is connected to a computer which records the three-dimensional movements of the bacterium.

The two instruments have complementary advantages and disadvantages. The tracker of Berg obtains more accurate data because the automatic recording device responds more rapidly to sudden changes in direction than a human observer. The instrument, however, has inertial characteristics which have limited it to a bacterium with movements similar to those of *E. coli*, and the accuracy of the data decreased when attempts were made to follow movements such as those of *Salmonella*, which swim appreciably faster.

The tracker of Lovely *et al.* (1974) is less accurate in regard to the individual movements of the bacteria, but can follow *Salmonella* and has the advantage that the bacteria are moving in a stable gradient which can be observed over long periods of time. This type of tracker is particularly useful for examining "persistence time," i.e., the analysis of the movement of the bacteria in terms of the component in the direction of the gradient (Macnab and Koshland, 1973). It should be noted that this instrument utilizes a one-dimensional gradient only, an advantage which simplifies theoretical interpretation of data.

C. Temporal Gradient Methods

The temporal gradient apparatus of Macnab and Koshland (1972) was introduced to establish whether bacteria utilize a temporal sensing system, but the approach has also become very useful for chemotaxis assays. The apparatus is described in Fig. 10. Bacteria in one bottle which may or may not contain attractant (or repellent) are rapidly mixed with a solution containing the same ingredients but a different concentration of attractant (or repellent). The two solutions are pushed rapidly through the mixing chamber and into an observation chamber on the stage of a microscope. When bacteria are observed the attractant (or repellent) is uniformly (isotropically) distributed. Since the mixing time is rapid (on the order of 200 milliseconds), time in the mixing chamber is less than the memory time of the bacteria, and

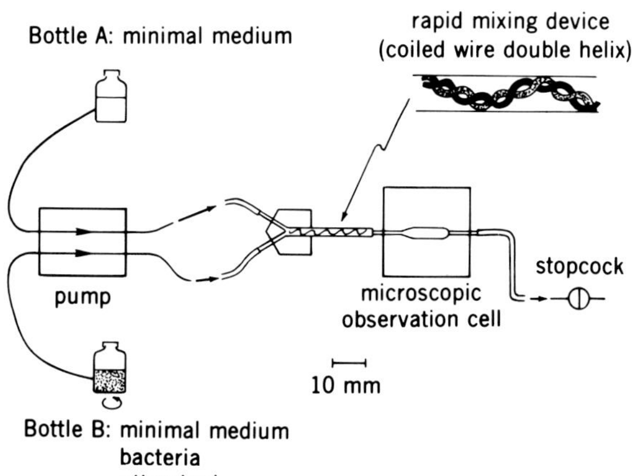

FIG. 10. Schematic illustration of temporal gradient apparatus. Attractant concentrations are: (i) Bottle B, C_i (≥ 0); (ii) bottle A, C_i' ($>$, $=$, or $< C_i$); (iii) observation cell (as a result of stream mixing) C_f ($>$, $=$, or $< C_i$). Bacteria experience $C_i \rightarrow C_f$, and thus can be subjected to positive, zero, or negative temporal gradients as desired. Gradient is given by $\Delta C/\Delta t$, where $\Delta C = C_f - C_i$ and Δt is mixing time.

the bacteria receive a stimulus equivalent to swimming up or down a gradient rapidly.

The bacterial motility pattern is modified dramatically when subjected to the stimulus (Fig. 5A). Increased tumbling is observed as a result of increasing repellent or decreasing attractant concentration. Decreased tumbling frequency is observed as a result of increased attractant or decreased repellent. The altered swimming pattern then returns to normal motility over an interval of time. By timing the return to the normal swimming patterns a qualitative picture of the influence of the attractant or repellent is easily obtained (Fig. 5B). The instrument, thus, provides a very easy and rapid screening method for attractants and repellents.

For even more rapid screening the bacteria can also be placed on a microscope slide and a small amount of attractant or repellent added to the side of the cover glass with a micropipette. As the chemical diffuses into the liquid on the slide, the alteration in motility pattern is easily followed. The effect of methionine, dyes, uncouplers of oxidative phosphorylation, pH, etc., have been studied in this way (Ordal and Goldman, 1975; Taylor and Koshland, 1975; Tsang *et al.*, 1973). The time it takes a bacterium to return to normal is a function of the strength of the stimulus and the nature of the stimulant. Although not quantitative as in the procedure described below, these semiquantitative assays are frequently of great value in rapidly assessing mechanisms or testing unknown chemicals.

D. Tumble Frequency Assay

The temporal gradient approach described above was made quantitative by Spudich and Koshland (1975). In this method the responses of bacteria gradients are recorded photographically. By leaving a camera shutter open for 0.8 seconds, during which stroboscopic light at 5 flashes per second falls on the bacteria, photographs such as those shown in Fig. 11 are obtained. In this photograph, the mutant in the absence of a stimulus tumbles incessantly, so no smooth tracks are observed prior to addition of attractant. Right after addition ($t = 0.4$ minutes), all the bacteria are smooth swimming and the stroboscopic light shows them at four separate positions appearing roughly in a straight line track. Later ($t = 0.6$ seconds), the bacteria that tumble produce successive images near the same position, and hence are pictured as splotches of light. Sometimes these turn perpendicular to the plane of focus, and disappear from observation. By counting the tracks in such a sequence of pictures, it is possible to determine quantitatively the number of bacteria which are swimming smoothly and the number which are tumbling in any interval of time. The raw data can be plotted to give an appropriate recovery curve which serves as a quantitative assay.

For routine use it is convenient to use a tumbling mutant of the type shown in Fig. 11, since the number of smooth tracks is zero initially and again after complete recovery. However, the method is applicable to wild-type populations, but in that case the base is not zero.

E. Tethered Bacteria

Silverman and Simon (1974) recently showed that flagella rotate like propellers rather than propagate conformation waves. They did this by tethering bacterial cells to the walls of vessels by their flagella causing the bacteria to rotate clockwise or counterclockwise. Tethered bacteria can therefore be subjected to additions of repellents or attractants, and their changes in rotation observed under the microscope (Larsen *et al.*, 1974b). In order to observe such a bacterium in the case of *E. coli* and *Salmonella typhimurium*, which have many flagella, it is necessary to grow the bacteria under nutritive conditions which suppress production of flagella. If this is not done, a second flagellum also sticks to the glass and prevents rotation. This has the difficulty of forcing study with bacteria under appreciably altered nutritive states, but the assay has been useful in studying the bacterial responses. Recently, a mutant which produces only a few flagella (usually only one) has been particularly useful in tethered assays (Spudich and Koshland, 1976).

F. The Population-Migration Assay

The population-migration apparatus was devised to measure the movement of a population of bacteria in a defined gradient of attractant or repellent (Dahlquist *et al.*, 1972). An observation cell with optical bottom and sides is filled with bacteria, usually uniformly distributed over the entire cell. A gradient of attractant or repellent is stabilized by means of glycerol concentration which maintain the gradient against thermal convection and mechanical vibrations. The gradient of attractant or repellent can be altered in various ways by appropriate mixing devices, and some of the more useful gradients are linear, step, and exponential gradients. A laser beam is directed through the bottom of the vessel, and a recording device reads the optical density of the scattered light at various positions in the observation vessel. From the intensity of the scattered light, the concentration of bacteria at any position can then be determined. Various gradients, mixtures of attractants, and repellents or repellents plus attractants can be used.

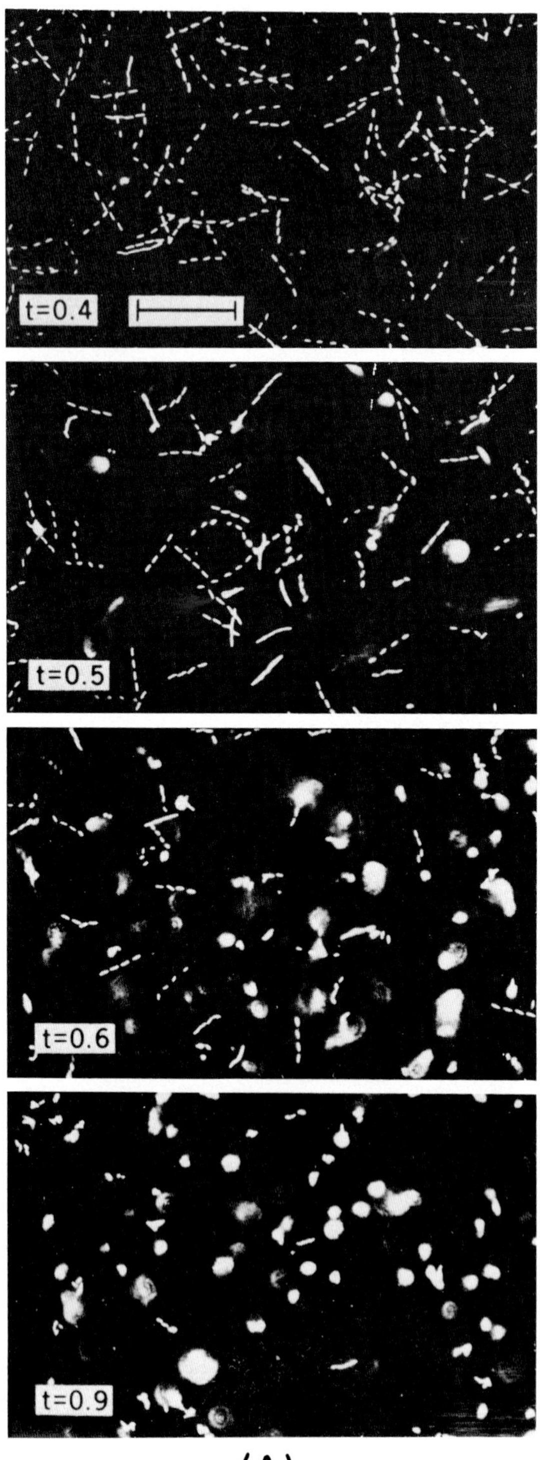

(A)

The population migration apparatus is appreciably more cumbersome to set up than the capillary assay and hence should not be used for routine screening of attractants or repellents. On the other hand, it provides a stable gradient which can be defined accurately and maintained for long intervals and measures bacterial movement quantitatively. Hence the movement of the bacteria in such a gradient is not only quantitatively determinable but can be analyzed theoretically in ways which are not available in the more empirical assays.

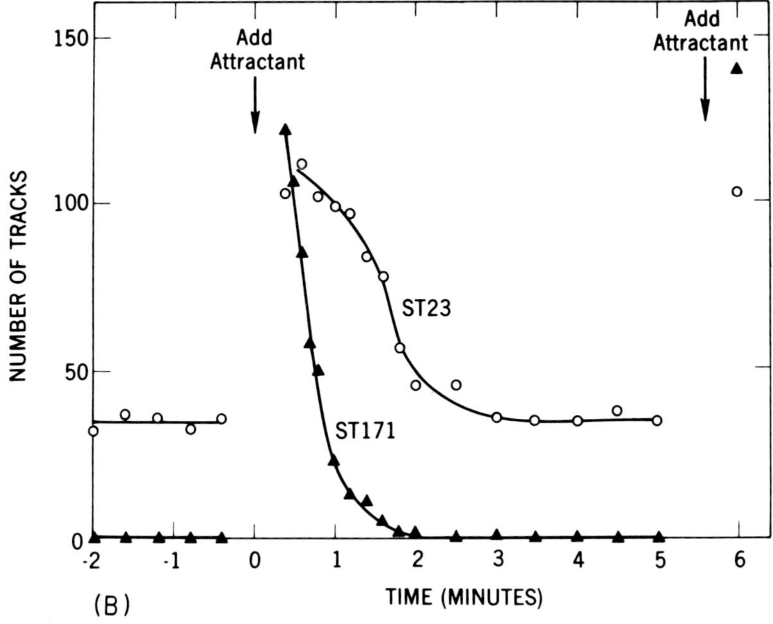

FIG. 11. Illustration of the quantitative tumble frequency assay. In (A) are shown the tracks of a constantly tumbling bacterium that has been subjected to a sudden temporal gradient of L-serine (0.02 m*M*). At the indicated number of minutes after mixing, photographs were taken by opening the shutter of a camera for 0.8 seconds while four stroboscopic flashes occurred at constant intervals. The length of the bar in the first photograph is 50 mm, and the bacteria are all seen swimming smoothly at 0.4 minutes after mixing. At 0.5 minutes after mixing, some bacteria have started tumbling, as indicated by splotches of light caused by bacteria turning end over end in the same position. Some bacteria are still smooth swimming, as indicated by the four separate positions in the bacterial trajectory. At 0.6 minutes very few bacteria are swimming smoothly, and at 0.9 minutes practically none are. In (B) these data are plotted together with other points for the constantly tumbling mutant ST171 and for wild type. Before mixing the number of smooth-swimming tracks is finite for the wild type and 0 for the constantly tumbling mutant. After stimulus both bacteria return to their prestimulus condition. The number of tracks counted is, of course, an arbitrary number based on the number of bacteria in the photographic field.

G. Selection of Nonchemotactic Mutants

Mutants which have lost the ability to sense a gradient but are motile are defective in the sensing system. One type of mutant to be discussed below is defective in a specific receptor, but a second type, called generally nonchemotactic, is defective in the more general transmission system. Two methods for selection of such mutants have been devised.

Armstrong *et al.* (1967) have used bacterial swimming on soft agar plates to select generally nonchemotactic mutants and the technique has been

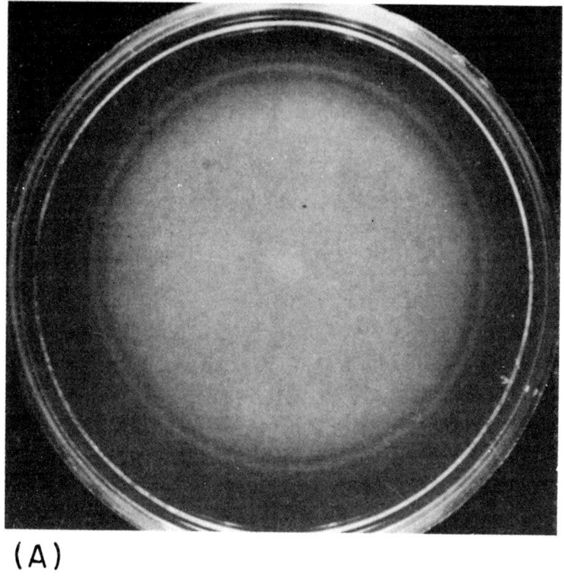

Fig. 12. (A) Illustration of Adler swarm assay technique using semi-solid gel. (B) Preformed liquid gradient used for isolation of chemotaxis mutants. The drawing on the left indicates how the gradient was constructed. The drawing on the right shows the distribution of the selection tube contents at the beginning of the enrichment period. All solutions were made up in minimal medium and pumped through a stainless steel inlet pipe (No. 20 gauge hypodermic needle) that extended to the bottom of the 18 × 150 mm glass tube. First 2.0 ml of 1.0 m*M* serine (attractant) was pumped in, followed by 2.5 ml of 1.0 m*M* serine containing glycerol. The glycerol concentration was initially 0% (w/v) and increased linearly to 1.0% at the end of the 2.5 ml pumping segment. Next, 0.2 ml of bacterial suspension (4×10^7 bacteria/ml in 0.5 ml of a solution of 0.5 m*M* serine, 1.5% glycerol) was pumped in, followed by 2.5 ml glycerol, which increased linearly from 2.0 to 3.0%. The glycerol gradient was necessary to stabilize the attractant gradient against convection. (C) Behavior of wild type (w^+, strain ST1), nonmotile (mot^-, strain TA1859), and motile but nonchemotactic (che^-, strain ST20) bacteria in a vertical step gradient of attractant. The selection apparatus is described in (B) above. The distribution of the bacteria was measured by light scattering, using the apparatus described by Dahlquist *et al.* (1972). The distributions were recorded at 6 and 30 minutes after introducing the bacteria.

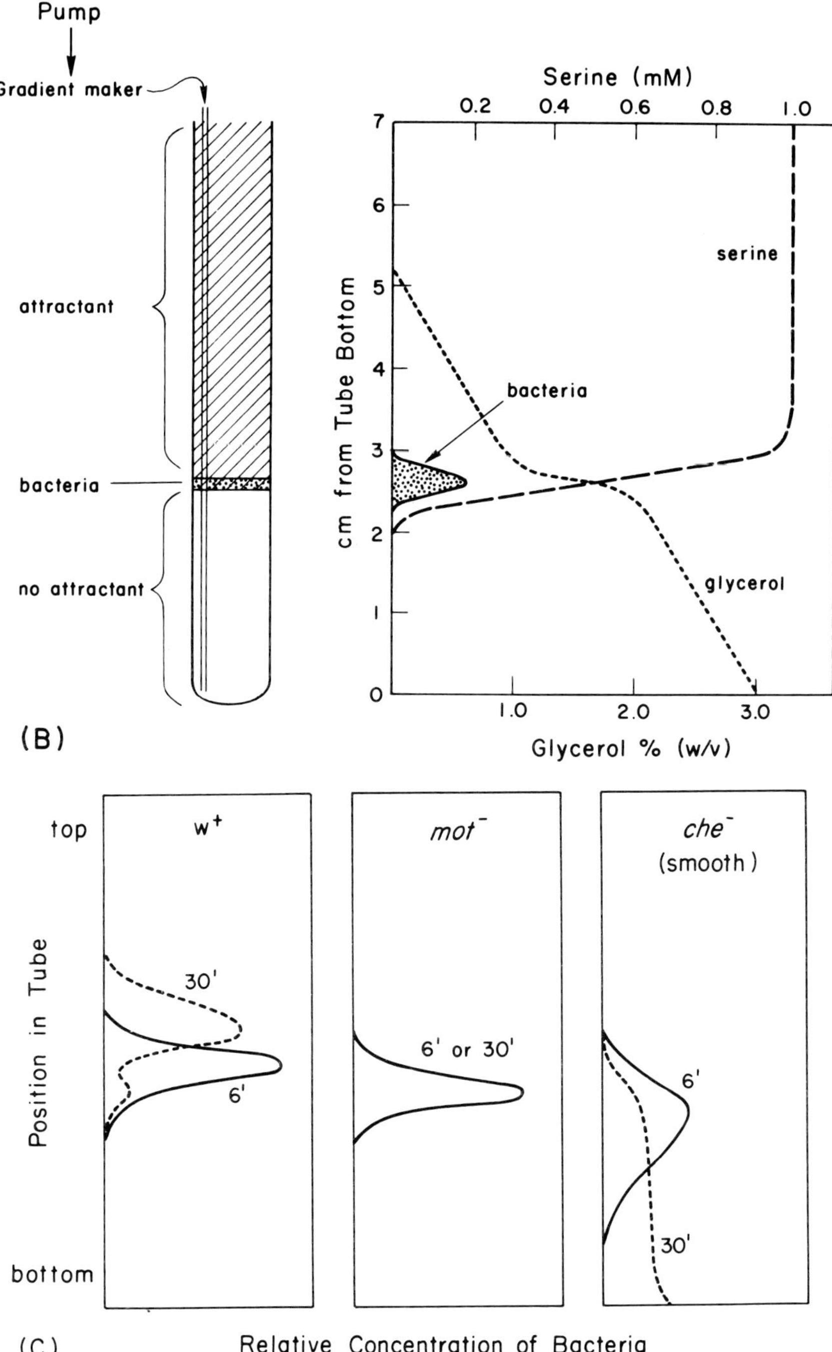
Pump
Gradient maker
attractant
bacteria
no attractant
(B)
Serine (mM)
0.2 0.4 0.6 0.8 1.0
cm from Tube Bottom
serine
bacteria
glycerol
1.0 2.0 3.0
Glycerol % (w/v)
top
bottom
Position in Tube
w+
30'
6'
mot⁻
6' or 30'
che⁻
(smooth)
6'
30'
(C)
Relative Concentration of Bacteria

called the swarm plate method. This involves spotting a colony on the center of a semisolid tryptone agar plate and incubating for 16 to 18 hours at 35°C in a water-saturated incubator. After this time the wild-type bacteria have swarmed out in several rings. Samples of the cells remaining at the origin are removed and suspended in 1 ml of tryptone broth. After sedimentation at 4600 *g*, a loopful of the pellet is deposited on the center of a second plate and the procedure is repeated. After ten such transfers swarming is considerably reduced and the origin has become dense with nonmotile and nonchemotactic bacteria. A final plug is removed from the origin. This contains a very few residual wild-type bacteria, some nonchemotactic mutants, and various paralyzed mutants with abnormal flagella. To get rid of the large number of nonmotile cells, the culture is suspended in suitably diluted antiflagella antibody serum which had been previously preabsorbed with nonflagellated mutant to eliminate nonspecific antibodies. Only colonies which do not swarm normally, but react fully with antiserum are further characterized. This procedure has been used both for observing natural mutations and specifically mutagenized cultures. Parkinson (1974) has applied similar methods to isolate mutants adding the use of an F′ epizome.

Aswad and Koshland (1975a) have used a "preformed" liquid gradient technique as shown in Fig. 12. The bacteria are placed in the center of the column and the wild-type bacteria are attracted to the top of the column by an attractant in that region. The nonmotile mutants remain at the center position at which the bacteria were introduced, but the motile nonchemotactic mutants diffuse away from the center and therefore occupy the bottom of the tube essentially free of wild-type and nonmotile bacteria. After 60 minutes of incubation, all but the bottom 1 ml of the tube contents are aspirated off, and the walls of the tube are rubbed with ethanol-soaked cotton to kill bacteria which might be adhering. A 2.0-ml volume of sterile VBC is added to the tube which is then capped and incubated for 24 hours. Appearance of colonies on a plate allows final separation of nonchemotactic from nonmotile or wild-type bacteria. The method has the advantage of separating nonmotile as well as wild-type bacteria from the motile but nonchemotactic mutants. It is, furthermore, amenable to the selection of temperature-sensitive mutants.

VIII. Conclusion

A great deal is now known about bacterial chemotaxis. Some of these phenomena appear quite similar to those of other organisms, such as phototaxis in phototactic bacteria and chemotaxis in molds, spirilla, etc. However, it certainly could not be said that all unicellular organisms have the same chemotactic mechanism. It does appear that larger organisms, such as protozoa, can move by reorienting themselves relative to a direction of the gradient. Nevertheless, there are indications that the unitary processes are not dissimilar from those of the bacterium. Thus, a bacterium responds by a simple reversal of rotation on receiving signals; protozoa reverse the motion of flagella on receiving the signal, indicating a striking similarity. The correlation of the response time and the mathematics of the responses in bacteria and neurons is also striking in many ways. Thus, it would appear likely that basic biochemical mechanisms are operating in all cells in a roughly similar manner.

In the bacterial system, a generalized summary of the information available so far would be as follows. There are approximately 25 to 30 receptors on the surface of the bacteria. These receptors undergo a conformational change on binding of the chemoeffector, and it is this conformation change which transmits the signal to the remaining parts of the flagella apparatus. As a result the chemoeffector need not be metabolized to generate the signal, although it may be subsequently metabolized to satisfy the nutrient requirement of the cell. The signal from the receptor is processed through a signaling system which is composed of at least nine gene products and which processes all of the signals from the many diverse stimuli which can affect bacterial behavior. This processing apparatus can provide a roughly additive response of the various stimuli after the algebraic sign has been considered. However, the additive relationship is not a simple one and will require more detailed computer analysis. The signal processed in this manner can be interpreted in terms of changing levels of some response regulator which then alters the gear mechanism of the flagella motor. If the motor is reversed from its normal swimming mode, tumble is generated. It is this on–off switching device which converts the sensory stimuli into a behavioral response. Tumbling is suppressed when the bacterium is proceeding in a favorable direction; tumbling occurs when it proceeds in an unfavorable direction. A combination of these two processes leads to a biased random walk, which is quite efficient in producing a net bacterial motion.

Sometimes bacteria are referred to as a lowly species, presumably because the complexity of their sensory apparatus is low compared to that of higher species. Certainly a sensory system composed of nine components is in striking contrast to a mammalian brain composed of 10^{12} neurons. When one considers that this simple apparatus has many of the features we identify with higher species, such as "choice," "adaptation," "discrimination," "memory," etc., it can perhaps be marveled that such a simple organism provides these properties with such elegant simplicity. That the relation of these words is not simply semantic but relates to basic biochemical events has been argued elsewhere (Koshland, 1977a) and will not be repeated here.

What is important from the point of view of the bacteria is that processes which are important to the survival of higher species are stripped down to a system of simplicity to provide the same kind of optimization of environmental conditions for the bacteria that pain and pleasure provides in higher species. The complete elucidation of chemotactic behavior, now in the state of a healthy infant, should afford a rewarding chapter to bacteriologists and indeed to all biologists.

References

Adler, J. (1969). *Science* **166,** 1588–1597.
Adler, J. (1973). *J. Gen. Microbiol.* **74,** 77–91.
Adler, J. (1975). *Annu. Rev. Biochem.* **44,** 341–356.
Adler, J., and Dahl, M. (1967a). *J. Gen. Microbiol.* **46,** 161–173.
Adler, J., and Dahl, M. (1967b). *J. Bacteriol.* **93,** 390–398.
Adler, J., and Epstein, W. (1974). *Proc. Natl. Acad. Sci. U.S.A.* **71,** 2895–2899.
Aksamit, R., and Koshland, D. E., Jr. (1974). *Biochemistry* **13,** 4473–4478.
Anderson, M., and Koshland, D. E., Jr. (1978). In preparation.
Anraku, Y. (1968). *J. Biol. Chem.* **243,** 3116–3135.
Armstrong, J. B. (1972). *Can. J. Microbiol.* **18,** 1695–1701.
Armstrong, J. B., and Adler, J. (1969). *J. Bacteriol.* **97,** 156–161.
Armstrong, J. B., Adler, J., and Dahl, M. M. (1967). *J. Bacteriol.* **93,** 390–398.
Aswad, D., and Koshland, D. E., Jr. (1974). *J. Bacteriol.* **118,** 640–645.
Aswad, D., and Koshland, D. E., Jr. (1975a). *J. Mol. Biol.* **97,** 225–235.
Aswad, D., and Koshland, D. E., Jr. (1975b). *J. Mol. Biol.* **97,** 207–223.
Beidler, L. M. (1954). *J. Gen. Physiol.* **38,** 133–139.
Beidler, L. M. (1962). *Prog. Biophys. Biophys. Chem.* **12,** 107–151.
Benzer, S. (1953). *Biochim. Biophys. Acta* **11,** 383–395.
Berg, H. C. (1971). *Rev. Sci. Instrum.* **42,** 868–871.
Berg, H. C. (1975). *Annu. Rev. Biophys. Bioeng.* **4,** 119–136.
Berg, H. C., and Brown, D. A. (1972). *Nature (London)* **239,** 500–504.
Blakemore, R. (1975). *Science* **190,** 377–379.
Boos, W. (1974). *Annu. Rev. Biochem.* **43,** 123–146.
Boos, W., Gordon, A. S., Hall, R. E., and Price, H. D. (1972). *J. Biol. Chem.* **247,** 917–924.
Brown, D. A., and Berg, H. C. (1974). *Proc. Natl. Acad. Sci. U.S.A.* **71,** 1388–1392.
Caraway, B. H., and Krieg, N. R. (1972). *Can. J. Microbiol.* **18,** 1749–1759.
Clayton, R. K. (1964). *Photophysiology* **2,** 51–77.
Collins, A. L. J., and Stocker, B. (1976). *J. Bacteriol.* **128,** 754–765.
Cook, R. A., and Koshland, D. E., Jr. (1969). *Proc. Natl. Acad. Sci. U.S.A.* **64,** 247–254.
Dahlquist, F. W., Lovely, P., and Koshland, D. E., Jr. (1972). *Nature (London)* **236,** 120–123.
Dahlquist, F. W., Elwell, R. A., and Lovely, P. (1976). *J. Supramol. Struct.* **4,** 329–342.
DeJong, M. H., van der Drift, C., and Vogels, G. D. (1976). *Arch. Mikrobiol.* **111,** 7–11.
Delbruck, M. (1945). *J. Bacteriol.* **50,** 131–135.
Dryl, S. (1958). *Bull. Acad. Phys. Sci.* **4,** 429.
Englemann, T. W. (1881). *Pfluegers Arch. Gesammte Physiol.* **25,** 285–292.
Faust, M. A., and Doetsch, R. N. (1971). *Can. J. Microbiol.* **17,** 191–196.
Fraenkel, G. S., and Gunn, D. L. (1961). "The Orientation of Animals." Dover, New York.
Futrelle, R. P., and Berg, H. C. (1972). *Nature (London)* **239,** 517–518.

Gerisch, G., and Hess, B. (1974). *Proc. Natl. Acad. Sci. U.S.A.* **71,** 2118–2122.

Hazelbauer, G. J., and Adler, J. (1971). *Nature (London) New Biol.* **30,** 101–104.

Hazelbauer, G. L., Mesibov, R. E., and Adler, J. (1969). *Proc. Natl. Acad. Sci. U.S.A.* **64,** 1300–1307.

Hill, R. L., and Brew, K. (1975). *Adv. Enzymol. Relat. Areas Mol. Biol.* **43,** 411–484.

Kellerman, O., and Szmelcman, S. (1974). *Eur. J. Biochem.* **47,** 139–149.

Khan, S., Macnab, R., DeFranco, A., and Koshland, D. E., Jr. (1978). *Proc. Natl. Acad. Sci. U.S.A.* In Press.

Kleene, S. J., Toews, M. L., and Adler, J. (1977). *J. Biol. Chem.* **252,** 3214–3218.

Kort, E. N., Goz, M. F., Larsen, S. H., and Adler, J. (1975). *Proc. Natl. Acad. Sci. U.S.A.* **72,** 3939–3943.

Koshland, D. E., Jr. (1974). *FEBS Lett.* **40,** S3–S9.

Koshland, D. E., Jr. (1977a). *Science* **196,** 1055–1063.

Koshland, D. E., Jr. (1977b). *In* "Advances in Neurochemistry" (B. W. Agranoff and M. H. Aprison, eds.), Vol. 2, pp. 277–341. Plenum, New York.

Koshland, D. E., Jr., Warrick, H., Taylor, B., and Spudich, J. (1976). *Cold Spring Harbor Symp. Quant. Biol.* , 1976 "Cell Motility" Vol. *1* 57–69.

Kundig, W., and Roseman, S. (1971). *J. Biol. Chem.* **246,** 1393–1406.

Larsen, S. H., Adler, J., Gargus, J. J., and Hogg, R. W. (1974a). *Proc. Natl. Acad. Sci. U.S.A.* **71,** 1239–1243.

Larsen, S. H., Reader, R. W., Kort, E. N., Tso, W.-W., and Adler, J. (1974b). *Nature (London)* **249,** 74–77.

Lederberg, J. (1956). *Genetics* **41,** 845–871.

Lederberg, J., and Iino, T. (1956). *Genetics* **41,** 743–757.

Lovely, P., Dahlquist, F. W., Macnab, R. M., and Koshland, D. E., Jr. (1974). *Rev. Sci. Instrum.* **45,** 683–686.

Macnab, R. M. (1978a). *Crit. Rev. in Biochem.*. In Press.

Macnab, R. M. (1978b). *In* "Encyclopedia of Plant Physiology" (W. Haupt and M. E. Feinleib, eds.), pp. 291–341. Springer-Verlag, Berlin and New York. Vol. 7.

Macnab, R. M. (1977c). *Proc. Natl. Acad. Sci. U.S.A.* **74,** 221–225.

Macnab, R. M., and Koshland, D. E., Jr. (1972). *Proc. Natl. Acad. Sci. U.S.A.* **69,** 2509–2512.

Macnab, R. M., and Koshland, D. E., Jr. (1973). *J. Mechanochem. Cell Motil.* **2,** 141–148.

Macnab, R. M., and Koshland, D. E., Jr. (1974). *J. Mol. Biol.* **84,** 399–406.

Macnab, R. M., and Ornston, M. K. (1977). *J. Mol. Biol.* **112,** 1–30.

Maloney, P. C., and Rotman, B. (1973). *J. Mol. Biol.* **73,** 77–91.

Mesibov, R., and Adler, J. (1972). *J. Bacteriol.* **112,** 315–326.

Mesibov, R., Ordal, G. W., and Adler, J. (1973). *J. Gen. Physiol.* **62,** 203–223.

Metzner, P. (1920). *Jahrb. Wiss. Bot.* **59,** 325–412.

Miller, J. B., and Koshland, D. E., Jr. (1977a). *J. Mol. Biol.* **111,** 183–201.

Miller, J. B., and Koshland, D. E., Jr. (1977b). *Proc. Natl. Acad. Sci. U.S.A.* **74,** 4752–4756.

Novick, A., and Weiner, M. (1957). *Proc. Natl. Acad. Sci. U.S.A.* **43,** 553–566.

Nultsch, W. (1970). *In* "Photobiology in Microorganisms" (P. Halldal, ed.), pp. 213–251. Wiley, New York.

Ordal, G. W., and Adler, J. (1974). *J. Bacteriol.* **117,** 517–526.

Ordal, G. W., and Goldman, D. J. (1975). *Science* **189,** 802–804.

Ordal, G. W., and Goldman, D. J. (1976). *J. Mol. Biol.* **100,** 103–108.

Oxender, D. L. (1972). *Annu. Rev. Biochem.* **41,** 777–814.

Parkinson, J. F. (1974). *Nature (London)* **252,** 317–319.

Parkinson, J. F. (1976). *J. Bacteriol.* **126,** 758–770.

Parsons, R. G., and Hogg, R. W. (1974). *J. Biol. Chem.* **249,** 3602–3607.

Pfeffer, W. (1883). *Ber. Dtsch. Bot. Ges.* **1,** 524–533.
Pfeffer, W. (1888). *Unters. Bot. Inst. Tubrigen.* **2,** 582–663.
Powell, E. O. (1958). *J. Gen. Microbiol.* **18,** 382–417.
Raab, O. (1900). *Z. Biol.* (*Munich*) **39,** 524.
Rubik, B., and Koshland, D. E., Jr. (1978). *Proc. Natl. Acad. Sci. U.S.A.* **75,** 2820–2824.
Seymour, F. W. K., and Doetsch, R. N. (1973). *J. Gen. Microbiol.* **78,** 287–296.
Shimada, K., Kamiya, R., and Asakura, S. (1975). *Nature* (*London*) **254,** 332–334.
Shimada, K., Ikkai, T., Yoshida, T., and Asakura, S. (1976). *J. Mechanochem. Cell Motil.* **3,** 185–193.
Showe, M. K., and DeMoss, J. A. (1968). *J. Bacteriol.* **95,** 1305–1313.
Silhavy, T. J., Boos, W., and Kalckar, H. M. (1974). *In* "Biochemistry of Sensory Functions" (L. Gasincke, ed.), pp. 165–205. Springer-Verlag, Berlin and New York.
Silverman, M. R., and Simon, M. I. (1973). *J. Bacteriol.* **116,** 114–122.
Silverman, M. R., and Simon, M. I. (1974). *Nature* (*London*) **249,** 73–74.
Silverman, M. R., and Simon, M. I. (1977). *J. Bacteriol.* **130,** 1317–1325.
Springer, W. R., and Koshland, D. E., Jr. (1977). *Proc. Natl. Acad. Sci. U.S.A.* **74,** 533–537.
Springer, M. S., Gay, M. F., and Adler, J. (1977). *Proc. Natl. Acad. Sci. U.S.A.* **74,** 3312–3316.
Spudich, J., and Koshland, D. E., Jr. (1975). *Proc. Natl. Acad. Sci. U.S.A.* **72,** 710–713.
Spudich, J., and Koshland, D. E., Jr. (1976). *Nature* (*London*) **262,** 467–471.
Stevens, S. S. (1970). *Science* **170,** 1043–1050.
Stock, J. B., and Koshland, D. E., Jr. (1978). *Proc. Natl. Acad. Sci. U.S.A.* In Press.
Stocker, B. A. D. (1949). *J. Hyg.* **47,** 398–413.
Strange, P. G., and Koshland, D. E., Jr. (1976). *Proc. Natl. Acad. Sci. U.S.A.* **73,** 762–766.
Szmelcman, S., and Adler, J. (1976). *Proc. Natl. Acad. Sci. U.S.A.* **73,** 4387–4391.
Taylor, B. L., and Koshland, D. E., Jr. (1975). *J. Bacteriol.* **123,** 557–569.
Taylor, B. L., Miller, J. B., Warrick, H. M., and Koshland, D. E., Jr. In preparation.
Tsang, N., Macnab, R. M., and Koshland, D. E., Jr. (1973). *Science* **181,** 60–63.
Tso, W.-W., and Adler, J. (1974a). *J. Bacteriol.* **118,** 560–576.
Tso, W.-W., and Adler, J. (1974b). *Science* **184,** 1292–1294.
Vaituzis, A., and Doetsch, R. N. (1969). *Appl. Microbiol.* **17,** 584–588.
Van Der Werf, P., and Koshland, D. E., Jr. (1977). *J. Biol. Chem.* **252,** 2793–2795.
Warrick, H. M., Taylor, B. J., and Koshland, D. E., Jr. (1977). *J. Bacteriol.* **130,** 223–231.
Weibull, C. (1960). *In* "The Bacteria" (I. C. Gunsalus and R. Y. Stanier, eds.), Vol. 1, pp. 153–205. Academic Press, New York.
Wilson, D. B. (1974). *J. Biol. Chem.* **249,** 553–558.
Yanofsky, C., and Crawford, I. P. (1972). *In* "The Enzymes" (P. Boyer, ed.), 3rd Ed., Vol. 7, pp. 1–32. Academic Press, New York.
Ziegler, H. (1962). *Handb. Pflanzenphysiol.* **17,** Part II, 484–532.
Zukin, R. S., and Koshland, D. E., Jr. (1976). *Science* **193,** 405–408.
Zukin, R. S., Strange, P. G., Heavey, L. R., and Koshland, D. E., Jr. (1977a). *Biochemistry* **16,** 381–386.
Zukin, R. S., Hartig, P. R., and Koshland, D. E., Jr. (1977b). *Proc. Natl. Acad. Sci. U.S.A.* **74,** 1932–1936.

CHAPTER 4

The Role of the Cell Surface in Regulating the Internal Environment

MILTON H. SAIER, JR.

A finely balanced network of biological regulatory interactions must simultaneously incorporate conditions for homeostasis and responsiveness to change.

"Le roi est mort; vive le roi"

ISBN 0-12-307207-7

I. Introduction

The intracellular concentration of any metabolite is determined largely by the rates at which the molecule is synthesized and consumed and the rates at which it enters and exits from the cytoplasmic compartment. Chemical interconversions are generally catalyzed by enzymes, and the rates of these reactions are dependent on the concentrations of substrates, products, coenzymes, and regulatory molecules which interact with and influence the activities of the enzymes. Enzyme concentration is also a determinative factor so that the rates of enzyme synthesis and degradation become important. For most intrabacterial enzymes, the rates of turnover appear to be low relative to the generation time of the organism (Nath and Koch, 1970). Consequently, the biosynthetic processes (DNA transcription and RNA translation) must be subject to regulation if enzyme concentrations are to influence the cellular levels of substrates or products. For catabolic enzymes involved in carbon and energy metabolism, intracellular levels of inducer and cyclic adenosine 3′,5′-monophosphate (cyclic AMP) determine the rates of transcription. By contrast, it has been suggested that the glutamine : α-ketoglutarate ratio and appropriate inducers influence the synthesis of enzyme systems involved in nitrogen metabolism.

Of primary importance in nutrient acquisition is the process of transmembrane solute translocation. In the case of the gram-negative bacteria, with which we will be primarily concerned in this chapter, there are two membranes which must be traversed by an incoming solute molecule (Fig. 1). The outer membrane consists of a phospholipid bilayer with lipopolysaccharide and proteins embedded in it as integral constituents. This membrane (but not the inner membrane) apparently contains nonspecific protein "pores" through which small neutral solute molecules (MW <800) can

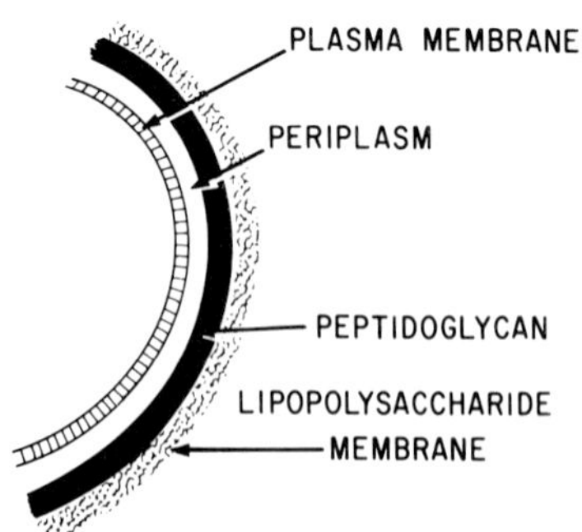

FIG. 1. Anatomy of the gram-negative bacterial cell.

readily pass (see Wright and Tipper, this volume, Chapter 7). Anionic molecules (such as glucose 6-phosphate, α-glycerophosphate, and succinate), large oligosaccharides, and peptides may cross the outer membrane via inducible solute-specific permeases localized within this phospholipid bilayer (Boos *et al.*, 1977). Small nutrients, therefore, enter the periplasmic space [the region of the bacterial cell delineated by the inner and outer membranes (Saier and Stiles, 1975)] by passive or facilitated diffusion.

Nutrients which gain access to the periplasm can pass through the inner cytoplasmic membrane only if a transport system specific for that solute is present. As in the case of an intracellular enzyme, both the activity and the synthesis of a transport system may be subject to regulation. Some permease systems catalyze exchange transport (which allows exchange of one substrate for another), but frequently energy is used to render the translocation process essentially irreversible. In the latter case, "leakage" of the accumulated intracellular metabolite into the extracellular fluid is minimized. Another mechanism which prevents loss of accumulated metabolites involves modification of the solute during entry. A neutral molecule may, for example, be converted to a charged membrane-impermeable species. Clearly, the processes responsible for the synthesis, functioning, and regulation of the membrane gates are of great importance in determining the composition of the intracellular environment.

In order for a nutrient to enter the cytoplasm rapidly, it must be present in the extracellular medium at a sufficiently high concentration. With this in mind, bacteria have evolved peripheral organelles and functions involved in nutrient detection and acquisition. Among these structures and functions are the organelles of motility and adhesion, the bioluminescent systems of marine bacteria, and the extracellular and periplasmic enzymes, toxins, and lysins which allow for preliminary extracellular digestion of large food stuffs. As noted above for the permeases and metabolic enzymes, both the activities and the syntheses of the proteins which perform these functions may be subject to regulation. Because the processes of nutrient detection and extracellular digestion may be the first steps in a chain of events leading to nutrient acquisition in the natural environment, we will begin with this subject.

II. Bacterial Functions Involved in Nutrient Detection and Acquisition

A. Motility and Chemotaxis

Many bacterial species are endowed with flagella, the organelle of motility, as well as with membrane associated chemoreceptors (Berg, 1975; Koshland, this volume, Chapter 3). Consequently, the cells recognize and respond to environmental stimuli. Bacterial chemoreception elicits a response that causes the organism to swim up a gradient of an attractant or down a gradient of a repellent. In *E. coli*, a signal is apparently transmitted from the cell surface receptor molecules to the basal portion of the flagellum. This signal in some unexplained way determines the direction of flagellar rotation. The rotational direction of the flagellar filament, clockwise or counterclockwise, in turn, determines whether the organism "tumbles" or swims in a straight trajectory, and tumbling, in general, results in a change of direction. Hence, the frequency of tumbling determines how long a cell will proceed in a single direction.

When an *E. coli* cell is suspended in a medium of uniform composition, it changes direction with low frequency. But when it swims up a concentration gradient of an attractant, the tumbling motion is suppressed so that the organism swims without interruption for an increased period of time. If, on the other hand, the organism swims down the attractant concentration gradient, the frequency of tumbling is enhanced. Hence, modulation of the tumbling frequency determines the chemotactic response and allows the cell to seek optimal nutrient concentrations.

Recent studies have revealed that both the activities and the syntheses of chemotactic systems are subject to regulation. Strange and Koshland (1976) provided evidence that galactose and ribose chemoreceptors compete in eliciting a chemotactic response. It was suggested that the chemoeffector–receptor complexes in some way compete with each other for a common component molecule in the signaling system (Strange and Koshland, 1976).

A possible mechanism, is discussed in Chapter 3 by Koshland in this volume and is shown in his Fig. 3. According to this model, the receptors compete for common component I molecules which then feed signals into the central response system. Integration of signals presumably occurs prior to signal input into the main stream of the central response system. In support of this proposal, a mutant of *S. typhimurium* was isolated which failed to respond to either ribose or galactose but did respond to serine and other amino acid attractants. Since the mutant contained both the ribose and the galactose receptors, it was presumed to be defective for component I.

The regulatory interactions discussed above appear to depend upon the simultaneous presence of an extracellular attractant and the membrane receptor specific for that attractant. Synthesis of the latter must presumably be induced to a high level before inhibition becomes appreciable. Mutual inhibitory interactions of this type provide the organism with a mechanism for eliminating the chemotactic response to one source of carbon when another is present in the medium in excess. Such a mechanism prevents the wasteful expenditure of energy for chemotaxis when a sufficient nutrient source is available.

Genetic experiments in several laboratories have shown that the chemotactic systems for sugars and amino acids are multicomponent systems coded for by several distinct genes (Berg, 1975). One of the components of each of the maltose, galactose, and ribose chemotactic systems appears to be a high affinity sugar-specific binding protein which also functions in sugar transport. In addition, Adler and Epstein (1974) have suggested that the sugar-activated membrane-bound components of the phosphoenolpyruvate: sugar phosphotransferase system in *E. coli* participate in the chemotactic response. These proteins are known to bind and transport the sugar substrates of this system (Postma and Roseman, 1976). Thus, the sugar-specific binding proteins of several transport systems appear to serve as chemoreceptors.

Since syntheses of the transport and corresponding catabolic enzyme systems are subject to inductive and repressive control, similar regulatory constraints might be expected to exist for the chemoreceptor complexes. That this is, in fact, the case has been substantiated by several investigations (Berg, 1975; Adler and Epstein, 1974). Recently, it was shown that the receptors which mediate taxis toward glucose are not synthesized in *Salmonella* strains which lack adenylate cyclase or the cyclic AMP receptor protein (Rephaeli and Saier, 1976). Since synthesis of the flagellar system is itself under cyclic AMP control, this demonstration required that synthesis of the flagellar system occurred in the absence of a functional cyclic AMP–cyclic AMP receptor protein complex. Pursuant of this goal, Silverman and Simon isolated mutants in which a specific flagellar gene, the *cfs* (constitutive flagellar synthesis) gene was defective (Silverman and Simon, 1974). The genetic lesion allowed flagellar synthesis to occur in the absence of cyclic AMP or its receptor protein. An episome from *E. coli* bearing the flagellar genes and a mutated *cfs* gene was transferred to *S. typhimurium* strains which lacked either adenylate cyclase or the cyclic AMP receptor protein due to nonsense mutations in the structural genes for these proteins. These *Salmonella* strains synthesized *E. coli* flagella, were fully motile, and exhibited normal chemotactic behavior in a gradient of aspartate. Chemotaxis toward glucose, however, was impaired, and glucose phosphotransferase activity (measured *in vitro*) and glucose transport activity (measured *in vivo*) were not inducible (Rephaeli and Saier, 1976). These observations provided evidence for the conclusion (Adler and Epstein, 1974) that the sugar-specific proteins of the phosphotransferase system function as chemoreceptors and suggested that synthesis of the glucoreceptors was subject to cyclic AMP control.

B. Bacterial Adhesion—Fimbriae

A second bacterial appendage which may function in nutrient acquisition is the bacterial fimbrium. Fimbriae are hairlike structures which project peritrichously from the surfaces of gram-negative bacteria in large numbers. Electron microscopic and biochemical analyses have shown that fimbriae are less rigid and in most cases straighter and thinner than the organelles of motility. They are also structurally and functionally distinct from pili which play a role in sexual conjugation. The fiber of group 1 subtype 1 fimbriae, which appears to consist of a single helically arranged protein (fimbrilin, MW = 16,600) is apparently anchored to the cell envelope by the fimbrial basal body (Ottow, 1975).

Fimbriated bacteria are far more adhesive than are nonfimbriated strains because fimbrilin, the structural protein of the fimbrial filament, is a carbohydrate-binding protein—a lectin protein with a high degree of specificity for mannosyl residues in glycoproteins, glycolipids, and polysaccharides. By virtue of the presence of these structures, gram-negative bacteria can adhere to a variety of eukaryotic cells (those of plants, animals, and fungi) as well as to other bacteria. These cells and the macromolecules to which the fimbriae adhere may serve as a useful source of carbohydrate for growth. Thus, bacterial fimbriae may function in nutrient acquisition when the available nutrients are in a particulate form or are associated with solid matter.

C. Bioluminescence

Bioluminescence is a third bacterial function which indirectly allows for nutrient acquisition and is apparently under cyclic AMP control (Hastings and Nealson, 1977). Many marine bacteria are bioluminescent, and of the four known luminous species three have been shown to form symbiotic relationships with appropriate host fish. Within the luminous gland of the fish the bacteria produce vast amounts of light. Field studies have shown that a primary function of bioluminescence is in nutrient acquisition: the light given off by the bacteria in the luminous organ of the fish attract phototactic copopods and other small marine organisms. Consequently, the bioluminescent bacteria facilitate acquisition of nutrients by the host, and the host presumably supplies the symbiotic bacteria with a fraction of the resultant blood glucose.

A primary function of cyclic AMP in regulating bioluminescence is to control synthesis of bacterial luciferase, the enzyme responsible for light emission (Hastings and Nealson, 1977). The biochemistry of this enzyme has been extensively studied. Bacterial luciferase is an α–β dimer of approximately 80,000 MW. It is a mixed-function oxidase which simultaneously catalyzes the oxidation of reduced flavin mononucleotide and a long-chain aldehyde with the resultant emission of light. Studies with mutant enzymes and chemically modified luciferases indicated that although both subunits are required for activity, the active site is located on the α subunit.

Synthesis of bacterial luciferase is controlled by an autoinductive mechanism as well as by cyclic AMP. The bacteria produce a small diffusible molecule at a constant rate, and this molecule accumulates in the medium. When the autoinducer concentrations becomes sufficient, the components of the luminous system are induced. Although the exact structure of the inducer is not known with certainty, NMR and mass spectral data indicate that the molecule from *Photobacterium fischerii* has a molecular weight of 159 and is probably an epoxide of indole acetaldehyde (Nealson and Fenical, unpublished observations).

When autoinducer is added to growing bacterial cells, luciferase is synthesized. This response is rapid and is blocked by either rifampicin or chloramphenicol. Since the molecule is small, acts from without, and is effective at extremely low concentrations (10^{-13} M), it can be thought of as a bacterial pheromone (Hastings and Nealson, 1977).

The phenomenon of catabolite repression is superimposed on autoinduction. In *Beneckea harveyi* glucose strongly represses synthesis of the luminous system, even in the presence of excess autoinducer. Cyclic AMP reverses this repression (Nealson *et al.*, 1972). Mutants which are altered with respect to catabolite repression have been isolated and studied. They fall into three groups: those which lack cyclic AMP or the cyclic AMP receptor protein, those which are defective for components of the phosphoenolpyruvate : sugar phosphotransferase system, and those which are resistant to glucose repression (Lin *et al.*, 1976). Adenylate cyclase negative mutants synthesized five to six orders of magnitude less luciferase than the wild-type strain, and exogenous cyclic AMP, in concentrations ranging from 10^{-6} to 10^{-3} M, restored bioluminescence. Cyclic AMP receptor mutants showed a similar phenotype, but exogenous cyclic AMP did not restore light emission.

Mutants which lacked either enzyme I or HPr of the phosphotransferase system (*pts* mutants) had the interesting characteristic of being dim or dark (Lin *et al.*, 1976). On the basis of preliminary analyses, the relationship of the phosphotransferase system to cyclic nucleotide regulation appeared to be similar to that in *Salmonella* (Saier and Feucht, 1975). The repressed state of the bioluminescent system in *pts* mutants may, therefore, reflect the depressed rate of cyclic AMP synthesis in these strains.

Mutants resistant to catabolite repression (bright on glucose) appeared to gain this resistance through complex and varied means (Hastings and

Nealson, 1977). Some such mutants were found to lack the enzyme phosphoenolpyruvate carboxylase, although the physiological basis for derepression of the bioluminescent system was not determined. Mutant analyses should be of utility in understanding the mechanistic details of catabolite repression.

D. Extracellular Degradative Enzymes

Serratia marcescens is a gram-negative bacterium, closely related to *E. coli*, which secretes macromolecular catabolic enzymes into the extracellular fluids. The exoenzymes which have been studied include a nuclease, a protease, and a lipase (Winkler *et al.*, 1975). Of these three enzymes, the synthesis of the lipase was found to be sensitive to glucose repression. This was shown in experiments in which the wild-type cells were grown in minimal glucose or glycerol medium. Glucose-grown cells produced much less lipase than the glycerol-grown cells, but excess cyclic AMP in the growth medium restored lipase synthesis. Additionally, a mutant which lacked adenylate cyclase, and therefore could not synthesize cyclic AMP, produced lipase only in response to exogenous cyclic AMP. In this mutant, extracellular lipase activity was detected 5 minutes after addition of cyclic AMP to the culture medium. Since the appearance of lipase activity in the presence of cyclic AMP was sensitive to low concentrations of chloramphenicol (5 μg/ml), it was concluded that cyclic AMP was required for the *de novo* synthesis of lipase and not for the release of preformed enzyme from the cell.

It is interesting to note that in this organism the pigment, prodigiosin, which is responsible for the bright red color of the bacterium, is also under cyclic AMP control. Although the physiological function of prodigiosin in *Serratia marcescens* is not known, this result suggests that it may function in a nutritive capacity.

In addition to the extracellular macromolecular degradative enzymes, several periplasmic enzymes function in nutrient acquisition. Periplasmic hydrolases of *E. coli* and *Salmonella typhimurium*, for example, cleave organic phosphate esters and small peptides into their constituents. At least some of these enzymes are under cyclic AMP control (Kier *et al.*, 1977).

III. Transmembrane Solute Transport

The intracellular environment is controlled in part by the molecular gates which allow solutes to pass into and out of the cell. In some cases the solute of interest is equilibrated across the membrane in an energy-independent process termed *facilitated diffusion.* In other cases, the solute may be accumulated intracellularly or actively pumped out of the cell by an energy-requiring process which does not modify the substrate molecule. These mechanisms are grouped together under the term *active transport.* Still other solutes are acted upon by transport systems which modify the transported species during passage through the membrane. These transport systems are said to catalyze the process of *group translocation.* All three types of transport systems appear capable of catalyzing an additional transmembrane solute translocation process in which one solute molecule, initially present in the cytoplasm, is exchanged for another molecule, initially present in the extracellular medium. This process has been termed *exchange transport,* or in the case of group translocating systems, *exchange group translocation.* In this section, these processes will be discussed from mechanistic and physiological standpoints, and the relationships between the different systems will be briefly considered.

A. Facilitated Diffusion

The simplest of the protein-mediated solute transport mechanisms is an energy-independent process which may or may not involve stereo-specific solute recognition. In eukaryotic cells numerous solutes cross the plasma membrane by facilitated diffusion. Among the best characterized animal cell permease systems of this type are the anion (Rothstein *et al.*, 1976) and glucose (Mawe and Hempling, 1965) transport systems in the human erythrocyte, and the ion-specific channels found in muscle and nerve cells (Saier and Stiles, 1975). The latter can be nonspecific with respect to the ion transported, as in the case of the tetrodotoxin-sensitive sodium channel.

Recent work has revealed the presence of analogous transport proteins in gram-negative bacteria, such as *E. coli* and *Salmonella typhimurium.* As already discussed and illustrated in Fig. 1, these bacteria possess two membranes, an inner cytoplasmic membrane and an outer lipopolysaccharide-containing membrane. The space between these two membranes, the periplasm, is a distinct cellular compartment which possesses a complement of soluble proteins which differ from those found in the cytoplasm (Saier and Stiles, 1975). Very little is known about the biogenic mechanisms by which subcellular protein compartmentation is brought about.

Each of the two gram-negative bacterial membranes appear to provide a barrier function. The outer membrane allows transmembrane passage of small oligosaccharides and peptides, but it is essentially impermeable to hydrophilic macromolecular species with molecular weights in excess of 800. Some hydrophobic antibiotics of considerably larger size can penetrate this membrane, presumably as a consequence of their solubility in the nonpolar phase of the phospholipid bilayer (Nikaido, 1976). It has been postulated that the permeability of the outer membrane to these hydrophobic drugs is restricted in wild-type bacteria by the core region of the lipopolysaccharide, which limits the extent of exposure of the phospholipid bilayer regions to the external aqueous environment (Nikaido, 1976).

Studies on the permeability of the outer bacterial membrane to small hydrophilic solutes and antibiotics led to the possibility that these molecules diffuse through the outer membrane via transmembrane aqueous pores (Decad and Nikaido, 1976). Preliminary evidence in support of this postulate included the demonstration that penetration of these molecules into the periplasmic space was marginally affected by temperature. Unambiguous evidence for a nonspecific pore in the outer membrane resulted from membrane reconstitution studies. Membrane vesicles reconstituted from phospholipids and lipopolysaccharide isolated from the outer membrane were impermeable to sucrose. Addition of a mixture of outer membrane proteins to the reconstitution system produced sucrose-permeable vesicles, and the active protein constituent was shown to be associated with the insoluble residue remaining after extraction of the outer membrane with 2% sodium dodecyl sulfate. A single polypeptide chain (MW = 36,500) apparently formed the oligoprotein aggregates which constituted the outer membrane diffusion channels (Nakae, 1976). These studies provided evidence for a transmembrane channel which allowed small hydrophilic molecules to permeate passively, in a nonstereospecific fashion, through the outer membrane of the gram-negative bacterial cell.

Analogous nonspecific channels are thought to be absent from the inner cytoplasmic membrane. Instead, solute-specific transport systems are generally responsible for the uptake of nutrients into the cytoplasm and the expulsion of deleterious compounds and end products of metabolism into the periplasmic space. Most of these appear to function by energy-linked permeases, but a few apparently catalyze facilitated diffusion. Thus, glycine may enter the *E. coli* cell by a substrate-saturable energy-independent mechanism which merely allows equilibration of the solute between the intracellular and extracellular compartments (Kaback and Stadtman, 1968). Since phospholipid bilayers are reasonably permeable to glycerol, the absence of an energy-dependent active transport system for this solute can be rationalized. If glycerol were accumulated against a concentration gradient, passive diffusion across the membrane into the extracellular medium would prevent solute retention, and, hence, the energy expended for its accumulation would be wasted (Lin, 1970).

B. Active Transport

It is now recognized that several different forms of energy can be coupled to the active transport of solute molecules. Light energy can be utilized to pump protons against a concentration gradient, both in bacteriochlorophyll-containing photosynthetic bacteria and in halophilic bacteria which possess bacteriorhodopsin (Bogomolni and Stoeckenius, 1974). Electrical energy (utilized during the passage of electrons from an appropriate electron donor, such as NADH, down the electron transfer chain to molecular oxygen in *E. coli*) is apparently coupled to the transmembrane extrusion of protons from the cytoplasm (Harold, 1974, 1977). In addition, the chemical energy stored in the terminal pyrophosphate bond of ATP can be utilized to extrude protons via the bacterial proton translocating ATPase (Harold, 1977). In each of these three cases a *primary* source of energy is utilized to drive the translocation process, and, hence, it is referred to as *primary active transport*. If an electrochemical ion gradient is generated by primary active transport, and the dissipation of this electrochemical gradient is secondarily utilized for the accumulation of a second solute molecule, the coupled process by which a gradient of the latter solute is generated is referred to as *secondary active transport*. In this case, the energy source directly responsible for solute accumulation is chemiosmotic in nature (Harold, 1974, 1977). While energy clearly functions in active transport to allow the accumulation of some solute molecules within the cell and the extrusion of other solutes from the cytoplasm, it is possible that energy also functions to regulate the activities of the permease systems. This possibility will be considered after a discussion of selected active transport systems.

1. Primary Active Transport of Protons in Bacteria

The source of energy utilized for the uptake or extrusion of numerous cations, anions, sugars, amino acids, vitamins, and other nutrients, all of which are transported by secondary active transport, is the proton electrochemical gradient or "proton motive force" (Harold, 1977). In photosynthetic bacteria, light can drive the extrusion of protons from the cell, resulting in a pH gradient (basic inside) and a transmembrane electrical potential (negative inside). Many of the details by which a single rhodopsin-like protein in the purple membrane of *Halobacterium halobium* catalyzes this process have been elucidated (Stoeckenius and Lozier, 1974).

The protein apparently spans the membrane in a nonsymmetrical fashion. Based on electron microscopic and X-ray diffraction studies, a three-dimensional model for the purple membrane has been proposed (Henderson and Unwin, 1975). Each subunit (MW 28,000) appears to consist of seven closely packed α-helical segments (35–40 Å in length) which extend perpendicular to the plane of the membrane. Three identical polypeptide chains comprise the unit cell, and the trimeric molecules are arranged in the latticelike planar membrane in a hexagonal array. Protein–protein interactions must be primarily responsible for the biogenesis and maintenance of this regular molecular arrangement. The spaces between the helical segments and the individual protein molecules do not contain solvent but appear to be filled with regions of phospholipid bilayer. Thus, the protein (75% by weight) and lipid (25% by weight) form a two-dimensional hydrophobic barrier through which hydrophilic species cannot passively diffuse.

Bacteriorhodopsin is a chromophoric protein containing 11-*trans*-retinal in Schiff's base linkage to a lysyl residue in the protein. Incident light induces a cyclic series of rapid conformational changes in the protein. One of these conformational transitions is apparently accompanied by proton release, while another depends upon proton uptake. The entire cycle takes 1–10 milliseconds under ordinary conditions (Stoeckenius and Lozier, 1974). These observations are presumably related to the proton pumping activity of the molecule in intact cells and membrane vesicles. If, during the rapid cycling process, protons are taken up from the cytoplasm and released on the opposite side of the membrane, net translocation of protons across the membrane should result.

A variety of spectroscopic studies indicated that the molecule undergoes

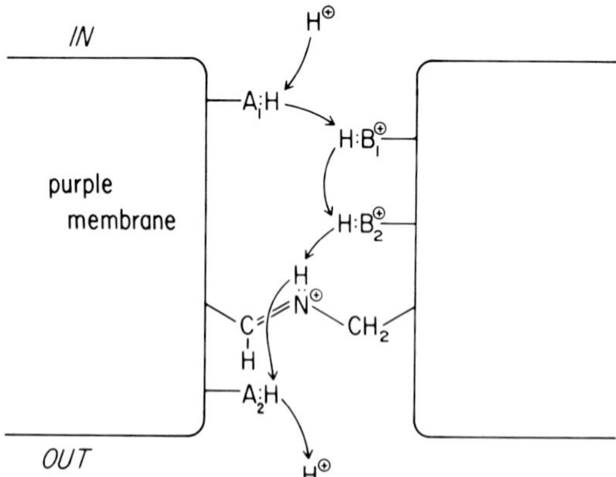

FIG. 2. Hypothetical model depicting the flow of protons through bacteriorhodopsin in a light driven process.

little movement during the photoreaction cycle. Therefore, it is unlikely that proton transport is accompanied by rotation or translocation of all or part of the bacteriorhodopsin molecule. Instead, a mechanism has been proposed in which photon absorption causes a p*K* change in a group in the protein which is linked to one residue on each surface through a proton transfer chain of suitable acidic and basic groups in the protein. This model is depicted in Fig. 2. The following evidence suggests that the group which changes p*K* upon excitation is the Schiff's base between retinal and protein and that it receives and passes protons from and to other amino acid residues in the protein. First, the H^+ of the protonated Schiff's base exchanges slowly with deuterium ions when the purple membrane is suspended in deuterium oxide in the dark. Upon illumination of the sample, more rapid replacement occurs. Second, the Schiff's base has been shown to be protonated in one of the bacteriorhodopsin photointermediates, but not in another. This fact leads to the suggestion that functional conformational changes in the protein influence the acidity of the Schiff's base. Third, in isolated purple membranes, the spectral shift between the unprotonated photointermediate and the protonated intermediate is insensitive to pH between 4.0 and 8.0. This observation suggests that the proton appearing on the Schiff's base is not directly derived from the medium (Stoeckenius and Lozier, 1974).

The results summarized above suggest that light-driven proton translocation in *H. halobium* is mediated by a "channel" mechanism involving the participation of a single polypeptide chain. Several other proteins or protein complexes have also been shown to catalyze proton translocation.

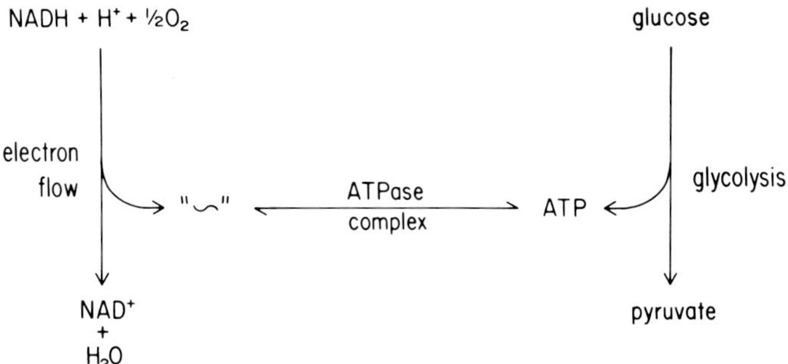

FIG. 3. Interconversion of various forms of energy in *Escherichia coli.* Electron transfer via the electron transport chain is illustrated on the left, while the process of glycolysis is indicated on the right. Because of the proton translocation function of the electron carriers, electron flow generates a proton electrochemical gradient, here indicated as "∼". Chemical energy, in the form of ATP, is generated by anaerobic sugar metabolism via the glycolytic pathway. Chemiosmotic and chemical energy are interconverted as a result of the action of the proton translocating ATPase, which functions as the central link in this scheme.

Thus, the mitochondrial, chloroplast, and bacterial ATPase complexes apparently couple proton translocation to ATP hydrolysis in a reversible reaction. Genetic studies provide evidence that, as in the case of bacteriorhodopsin, a channel mechanism may be operative (Senior, 1973). Reconstitution studies have also established that the cytochrome oxidase complex and other components of the electron transfer chain catalyze transmembrane proton translocation concomitant with changes in redox state. These facts provide a molecular basis for the coupling of various primary sources of energy to the generation or dissipation of chemiosmotic energy. The essential elements of the cellular constituents involved in these energy interconversions in *E. coli* are schematically shown in Fig. 3. As illustrated in the figure, chemical, electrical, and chemiosmotic forms of energy are readily interconverted by proton translocating transport systems which are integrated in the cytoplasmic membrane.

2. Secondary Active Transport of Cations

Both monovalent and divalent cations are actively transported by bacterial cells. The polarity of the pumping activity depends on the cation. Thus, K^+ and Mg^{2+} are accumulated against large concentration gradients within the cell, while Na^+ and Ca^{2+} are actively extruded. Some evidence suggests that the proton electrochemical gradient provides the source of energy for transport of at least some of these cations. Sodium, for example, appears to be extruded at the expense of an incoming proton. One H^+ probably enters the cytoplasm for every Na^+ which exits. This coupled process, in which the two species are transported with opposite polarity, is referred to as *countertransport* or *antiport*. If the two molecules are translocated with the same polarity, the process is called *cotransport* or *symport*. The processes of Ca^{2+} efflux and K^+ uptake will be discussed below.

Extensive studies have served to characterize the process by which Ca^{2+} is extruded from bacterial cells. This became possible when methods for the preparation of inverted (wrong side out) membrane vesicles became available (Futai, 1974). Inverted membrane vesicles of *E. coli* accumulated Ca^{2+} in the presence of electron donors, such as D-lactate and NADH, or in the presence of ATP. Uptake driven by electron donors was specifically blocked by inhibitors of electron transport, while dicyclohexylcarbodiimide, an inhibitor of the proton translocating ATPase, specifically abolished ATP-dependent uptake. Antisera prepared against the ATPase also effectively inhibited Ca^{2+} uptake in the presence of ATP. Uncouplers of oxidative phosphorylation (proton conductors) blocked uptake by all energy sources tested (Tsuchiya and Rosen, 1975).

Calcium transport into inverted membrane vesicles of *E. coli* was observed in the absence of exogenous energy when an artificial proton gradient (acidic inside) was generated (Tsuchiya and Rosen, 1976). The characteristics of the Ca^{2+} uptake processes, driven by exogenous energy sources and artificial proton gradients, were similar, suggesting that the same system was involved. The results were interpreted in terms of a calcium–proton countertransport mechanism.

Many solutes are transported across bacterial membranes by more than a single permease system. Kinetic evidence suggests that two transport systems, both with outwardly directed polarity, may be responsible for Ca^{2+} efflux. One of these systems is a low-affinity high-efficiency system, while the other functions with low efficiency, but binds Ca^{2+} with high affinity. Similarly, K^+ is acted upon in the *E. coli* cell by more than one transport process (Rhoads and Epstein, 1977). In this case, three or possibly four transport systems have been identified. One of the systems is a high rate K^+ repressible system which binds its substrate with high affinity (K_m = 2 μM). This system appears to depend on chemical energy (a high-energy phosphorylated compound such as ATP) for activity, but not on chemiosmotic energy. The system is probably coded for by four linked *kdp* genes and is sensitive to osmotic shock, suggesting that one component of the system might be an osmotic-shock-releasable K^+-binding protein.

A second, nonsaturable, low rate K^+ uptake system (*TrkF*) is resistant to osmotic shock and appears to be driven exclusively by chemiosmotic energy. This system functions in the total absence of high-energy phosphorylated compounds. A third very high rate K^+ transport system (*TrkA*) exhibits moderate affinity for K^+ (K_m = 1.5 mM) and is synthesized constitutively. The system appears to be coded for by a single gene. It is not sensitive to osmotic shock, but apparently depends on both forms of energy. In the absence of either chemiosmotic or chemical energy this system does not exhibit appreciable activity. It appears, therefore, that uptake of a particular solute (such as K^+) can occur either by primary or secondary active transport, or uptake may depend both on a primary and a secondary source of energy. A transport system which exhibits a dual energy requirement may utilize one form of energy for solute accumulation and another for maintenance of a functional permease conformation. A transport system which requires only one form of energy may utilize that energy source in more than a single capacity (see below).

3. Secondary Active Transport of Organic Nutrients

Numerous amino acids, carboxylic acids, sugars, and vitamins cross the cytoplasmic membrane of bacteria by secondary active transport. There appear to be at least three distinct mechanisms by which metabolic energy is coupled to the uptake process: H^+ coupling, Na^+ coupling, and chemical coupling (Harold, 1977). Proline, lactose, and many other low molecular weight solutes cross the cytoplasmic membranes of *E. coli* and *Salmonella typhimurium* via permease systems which operate by solute–H^+ cotransport. In these same cells, melibiose and tricarboxylic acids, such as citrate and isocitrate, apparently cross the membrane by Na^+ cotransport mechanisms. By contrast, the uptake of maltose and leucine appears to depend on chemical energy. A single example of each of these transport processes will be discussed.

The energization of lactose uptake in *E. coli* has been extensively studied (Harold, 1977; Wilson *et al.*, 1972). In intact *E. coli* cells lactose accumulation can occur either aerobically in the presence of an oxidizable substrate, such as succinate, or anaerobically in the presence of a sugar, such as glucose. In order to determine the primary source of energy for lactose accumulation, various inhibitors of energy metabolism were studied (Harold, 1977). Table I summarizes the effects of these inhibitors on lactose uptake under both aerobic and anaerobic conditions. The sites of action of the various inhibitors are revealed by reexamination of Fig. 3. Aerobically driven lactose uptake into *E. coli* cells or membrane vesicles was strongly inhibited by inhibitors of electron flow and by uncouplers of oxidative phosphorylation. However, inhibitors of the proton translocating ATPase, arsenate (which abolishes ATP synthesis), and sodium fluoride (which blocks glycolysis) had no effect. These observations suggested that ATP was not required for active lactose uptake and that the normal functioning of the glycolytic scheme and the proton translocating ATPase could be dispensed with. Under aerobic conditions, only electron flow and the maintenance of the proton electrochemical potential ("$\sim$") appeared to be required. By contrast, when cells were maintained anaerobically, lactose uptake required that ATP be synthesized (i.e., by glycolysis) and that the proton translocating ATPase be present in a functional state. While uncouplers of oxidative phosphorylation, which dissipated the proton electrochemical gradient, blocked anaerobic lactose uptake, inhibitors of electron flow were without effect. Evaluation of the data in Table I reveals that the only agents which inhibited lactose uptake under both aerobic and anaerobic conditions were the compounds which dissipated the proton electrochemical gradient. This source of energy was, therefore, the most likely candidate for the driving force responsible for lactose accumulation.

How lactose uptake is energetically coupled to the proton electrochemical gradient was revealed by experiments in which the uptake of lactose and the disappearance of protons from the extracellular medium were simultaneously measured in wild-type *E. coli* cells (Wilson *et al.*, 1972). Using a cell

TABLE I

EFFECTS OF METABOLIC INHIBITORS ON LACTOSE ACCUMULATION IN *E. coli* CELLS UNDER AEROBIC AND ANAEROBIC CONDITIONS

Inhibitor	Inhibits or acts on	Effect on lactose uptake under: Aerobic conditions	Anaerobic conditions
Potassium cyanide	Electron flow	−[a]	+
Dinitrophenol	Proton electro-chemical gradient	−	−
Dicyclohexyl-carbodimide	ATPase complex	+	−
Sodium arsenate	ATP synthesis	+	−
Sodium fluoride	Glycolysis	+	−

[a] −, agent inhibits lactose accumulation; +, lactose accumulation is observed in the presence of this agent under the specified conditions.

suspension which was incapable of both glycolysis and electron flow, and which was maintained in an anaerobic temperature-controlled electrode vessel, changes in the pH of the medium were measured after the addition of lactose. Under these conditions, lactose could enter the cell but could not be accumulated against a concentration gradient. The results suggested that for every molecule of lactose that entered the cell, a proton disappeared from the extracellular compartment. Uncouplers of oxidative phosphorylation did not inhibit lactose entry, but they abolished the pH changes which normally accompanied lactose entry. The measurements showed that a single proton disappeared from the extracellular medium for every lactose molecule which entered the cytoplasmic space. A lactose–proton cotransport mechanism was inferred.

Genetic studies served to confirm this conclusion and to provide evidence that proton cotransport was the mechanism of energy coupling for lactose accumulation. In these studies, a mutant of *E. coli* was isolated in which lactose transport was partially uncoupled from metabolic energy. Facilitated diffusion of galactosides occurred normally in this mutant, but galactoside accumulation was depressed (Wilson *et al.*, 1972). The mutation was shown to map in the *lacY* (permease) gene. Using the procedures described above, it was further shown that the genetic defect uncoupled lactose transport from proton uptake. This observation could be readily explained if proton co-transport provided the energy for lactose accumulation.

Correlative studies provided evidence that in *E. coli* membrane vesicles, lactose uptake was primarily energized by the transmembrane electrical potential (Harold, 1974, 1977). Large lipophilic cations which readily penetrate the hydrophobic matrix of the phospholipid bilayer distribute across the membrane in accordance with the membrane potential (a component of the proton electrochemical gradient). The extent of lactose accumulation, mediated by the lactose permease, was proportional to the extent of uptake of the lipophilic cation. Since uptake of the latter was an approximate measure of the membrane potential, the results suggested that the membrane potential served to energize lactose uptake.

Confirmation of this conclusion resulted from studies in which an electrical potential was artificially generated across the vesicle membrane in the absence of an exogenous source of energy. Vesicles were loaded with a K^+ salt and were then suspended in a medium in which Na^+ was the predominant cation. Because the bacterial cytoplasmic membrane is not selectively permeable to either Na^+ or K^+, these manipulations did not result in the generation of a transmembrane electrical potential. Introduction of the K^+-specific ionophore, valinomycin, into the membrane rendered the membrane selectively permeable to K^+. As a consequence, K^+ flowed down its concentration gradient, out of the cell, leaving a deficiency of positive charge in the vesicles. This manipulation resulted in the transient generation of a membrane potential—negative inside (Harold, 1977; Saier and Stiles, 1975). In response to this potential, either lactose or a lipophilic cation could be accumulated against a concentration gradient in the intravesicular compartment. These results provided substantiation for the suggestion that the membrane potential energizes lactose accumulation. Since lactose is a neutral molecule, these results can be explained only if lactose uptake is coupled to the uptake of a positively charged species or the extrusion of a negatively charged species. Proton cotransport, therefore, provides an explanation for the observations cited above.

While the membrane potential clearly energizes lactose accumulation by a proton cotransport mechanism, it is not clear that this is its sole function. Recent studies have provided convincing evidence for the possibility that energy is required for the binding of lactose analogs to the external surface of the lactose permease protein (Rudnick *et al.*, 1976; Belaich *et al.*, 1976). In these studies, the binding of a fluorescent galactoside or an azidophenyl galactoside (a photoinactivator of the lactose permease) was measured. Although these compounds could not penetrate the membrane, they were bound to the permease protein and competitively inhibited lactose uptake. Binding of either compound to the external surface of the permease required the generation of a membrane potential—negative inside. It has, therefore, been proposed that the membrane potential influences the conformation of the permease protein so that it gains affinity for its substrates on the external face of the membrane (Rudnick *et al.*, 1976). Whether or not the lactose permease catalyzes facilitated diffusion in the complete absence of energy is still debatable (see below).

While many solutes appear to cross the cytoplasmic membranes of enteric bacteria by proton cotransport, a few appear to be accumulated by Na^+ cotransport. Moreover, the latter mechanism may predominate in marine bacteria. The first unambiguous demonstration of sugar–Na^+ cotransport in bacteria resulted from studies on melibiose uptake in *Salmonella typhimurium* (Stock and Roseman, 1971). In this organism, melibiose uptake is strictly dependent on the extracellular Na^+ concentration. Kinetic analysis of sugar uptake via this transport system revealed hyperbolic kinetics as a function of both the sugar and the extracellular Na^+ concentrations. Moreover, uptake of a sugar substrate via the melibiose permease was accompanied by the transient appearance of intracellular $^{22}Na^+$ when this radioactive ion was present in the extracellular medium. Since Na^+ is apparently normally extruded from the cell by Na^+–H^+ countertransport, the Na^+ electrochemical gradient apparently served as the energy source for melibiose accumulation.

Recent studies on the uptake of the [^{14}C]tricarboxylic acids, into *S. typhimurium* cells provided convincing evidence for the suggestion (noted above) that energy may be required even for the facilitated translocation of some solutes across the bacterial cell membrane (M. H. Saier, Jr., unpublished observations). Uptake of these substrates could be observed when *S. typhimurium* cells were grown in minimal medium containing citrate, isocitrate, or *cis*-aconitate as the sole source of carbon for growth. The three tricarboxylic acids all served as transport substrates, although the order of transport efficiency and binding affinity decreased in the order: citrate > isocitrate > *cis*-aconitate. Uptake of these rapidly metabolized substrates was dependent on Na^+ ions, and the activation of tricarboxylic acid uptake by Na^+ followed hyperbolic kinetics. Sodium ions could not be replaced by Li^+, K^+, or Rb^+. A tricarboxylic acid–Na^+ cotransport mechanism was inferred.

Tricarboxylic acid uptake in *Salmonella* was strictly dependent on energy, and uptake of these acids was completely blocked by energy uncouplers. As shown in Fig. 4, carbonylcyanide *m*-chlorophenylhydrazone (CCCP) was more effective than tetrachlorosalicyl anilide (TCS) in inhibiting citrate and isocitrate uptake. Both uncouplers were effective at high concentrations although neither inhibited oxygen consumption. Similar inhibitory effects have been noted for the uptake of [^{14}C]3-phosphoglycerate which is also metabolized upon entry into the cytoplasm (Saier *et al.*, 1975b). These results would not have been anticipated if the transport systems were capable of catalyzing rapid facilitated diffusion. They lead to the suggestion that the

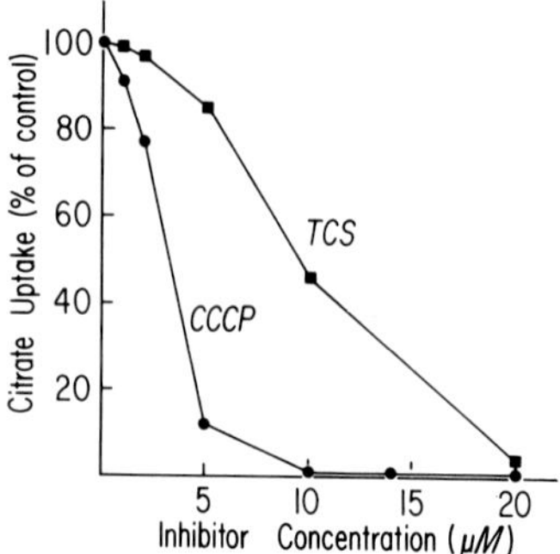

FIG. 4. Inhibition of citrate uptake by uncouplers of oxidative phosphorylation in *Salmonella typhimurium* strain LT-2. Cells were grown in medium 63 containing 0.4% sodium citrate as the sole source of carbon, harvested during logarithmic growth, washed three times, and resuspended in medium 63 containing 10 μM NaCl to a cell density of 0.16 mg dry cells per ml (Saier *et al.*, 1970). The cell suspension was equilibrated at 37°C, and the inhibitor was added to the cell suspension 30 seconds before initiation of the uptake experiment by addition of the radioactive substrate (specific activity, 10^6 cpm per μmole) to a final concentration of 0.2 μM. The uninhibited rate of [^{14}C]citrate uptake was 8 μmoles per minute per gram dry weight of cells. The uninhibited rate of [^{14}C]isocitrate uptake was 3 μmoles per minute by the same cell suspension. CCCP, carbonylcyanide *m*-chlorophenylhydrazone; TCS, tetrachlorosalicylanilide.

membrane potential may function in two unrelated capacities: first, to allow solute accumulation against a concentration gradient, and, second, to maintain the transport protein in a functional conformation. This important hypothesis is illustrated schematically in Fig. 5 and helps to explain a large body of previously anomalous observations (Kaback, 1972).

Uptake of the solutes discussed above is catalyzed by permease systems

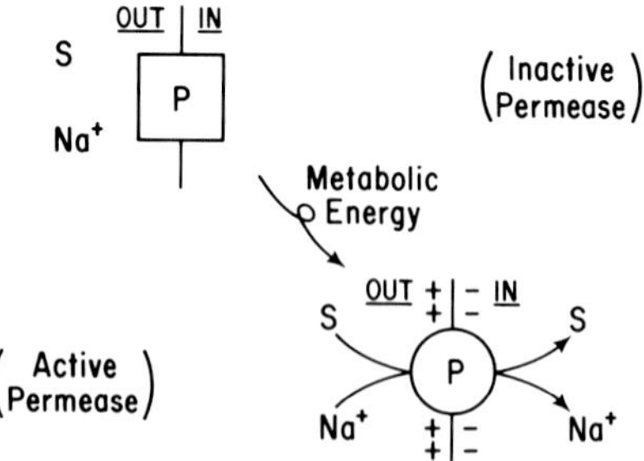

FIG. 5. Schematic representation of a proposal suggesting that energy regulates the activity of the citrate permease in two capacities: by allowing solute accumulation by Na^+ cotransport and by influencing the conformational state of the permease protein. This suggestion may be applicable to numerous bacterial permeases.

which are tightly associated with the cytoplasmic membrane. The activities of these transport systems are not sensitive to osmotic shock and can be demonstrated in isolated bacterial membrane vesicles. However, many transport systems require the participation of solute-binding proteins which are loosely associated with the external surface of the cytoplasmic membrane (Berger and Heppel, 1974). Osmotic shock treatments release these proteins from the periplasmic space into the extracellular medium, and transport activities are simultaneously lost. Consequently, the uptake of these solutes into bacterial membrane vesicles has not been extensively studied.

Recent investigations with intact *E. coli* have provided evidence for a second reason why uptake of solutes via these transport systems is not observed in membrane vesicles: their uptake apparently depends upon a chemical energy source such as ATP (Berger and Heppel, 1974). Energy coupling to at least some of these solutes occurs by a mechanism which is independent of the chemiosmotic state of the cell. Nevertheless, it is possible that solute accumulation via these transport systems will be found to exhibit a dual energy requirement: normal rates of transmembrane translocation may require chemiosmotic as well as chemical energy where one of these sources functions in a regulatory capacity.

4. Excretion of Waste Products and the Active Extrusion of Cellular Metabolites

Little information is available concerning the mechanisms responsible for metabolite excretion in bacteria. However, under many conditions of bacterial growth, normal cellular metabolites accumulate at rates which exceed their rates of utilization. When this situation exists, these metabolites are excreted into the extracellular medium. The extracellular end products of sugar metabolism may be, for example, ethanol, acetate, lactate, and/or pyruvate. There is evidence that excretion of at least some of these compounds occurs by reversal of the activities of transport systems which normally function with inwardly directed polarity (Harold and Levin, 1974). Similarly, when nonmetabolizable sugars or sugar phosphates accumulate intracellularly, their excretion may be effected by permeases which normally exhibit inwardly directed polarity (Dietz, 1976). These observations suggest that metabolite excretion may occur as a result of reversal of the functioning of the normal metabolite uptake systems.

It is unlikely that this picture is sufficient to account for metabolite excretion in general for several reasons: (a) Some transport systems appear to catalyze essentially irreversible vectorial processes (Parnes and Boos, 1973; Wilson, 1976). (b) Some transport systems are induced when the substrate inducer is present in the extracellular medium but not if it is present only in the cytoplasm (see Section IV,B). (c) Various nucleotides and charged cellular metabolites have been identified in the extracellular medium of bacterial cultures although transport systems responsible for the uptake of these compounds have not been demonstrated. (d) Specific transport systems with outwardly directed polarity have been identified for Na^+ and Ca^{2+} (see Section III,B,2). It seems likely that transport systems for the extrusion of other deleterious compounds will also be found.

In light of these considerations, it is of interest that several years ago Makman and Sutherland (1965) demonstrated efflux of adenosine 3′,5′-cyclic phosphate (cyclic AMP) from *E. coli* cells. They showed that the intracellular concentration of this nucleotide rose when cells were depleted of a carbon source but that readdition of the carbon source simultaneously inhibited synthesis and stimulated efflux of the compound. This discovery later led to a realization of the involvement of cyclic AMP as a key regulatory nucleotide involved in the control of carbon catabolic enzyme synthesis (Pastan and Adhya, 1976).

A recent report has dealt with the nature of the process responsible for the transport of cyclic AMP (Saier *et al.*, 1975a). Sugars and other utilizable sources of carbon and energy were found to lower intracellular concentrations of cyclic AMP by promoting the release of the nucleotide into the extracellular medium. The stimulatory effects of the carbohydrates were diminished when their metabolism was inhibited, and nonmetabolizable sugar analogs were not effective. These results suggested that the outwardly directed transport of cyclic AMP was an energy-requiring process.

Confirmation of this conclusion resulted from studies with intact cells and bacterial membrane vesicles. In the vesicle preparations, sugars did not stimulate efflux of the cyclic nucleotide, but various electron donors, including D-lactate, were effective. Uncouplers of oxidative phosphorylation, which inhibited proline uptake, exerted corresponding inhibitory effects on cyclic AMP transport. However, arsenate, which reduced ATP levels to nondetectable levels, had no effect. These results led to the suggestion that the energy source for cyclic AMP transport was chemiosmotic in nature and that chemical energy was not involved. Since cyclic AMP is negatively charged, and the membrane potential is normally negative inside, efflux of the compound corresponds to movement down its electrochemical gradient. Cation cotransport or countertransport need not be postulated in order to provide the driving force for the extrusion process.

C. Group Translocation

Several solutes appear to cross the bacterial cytoplasmic membrane by mechanisms which involve chemical modification of the molecule during its passage through the membrane (Postma and Roseman, 1976). Evidence has been presented which suggests that membrane-associated nucleoside phosphorylases catalyze the uptake of the ribose moiety of nucleosides into *Salmonella typhimurium* cells and membrane vesicles (Rader and Hochstadt, 1976). Adenosine phosphorylase, for example, apparently binds adenosine on the outer membrane surface, splits the compound into the constituent base and sugar, releases adenine into the external medium, and translocates the ribose moiety across the membrane. The product released in the cytoplasm is ribose 1-phosphate where the phosphoryl moiety is derived from inorganic phosphate (Rader and Hochstadt, 1976). A second enzyme (adenine phosphoribosyl pyrophosphate phosphoribosyltransferase) apparently catalyzes the uptake of the base (Hochstadt-Ozer, 1972). This enzyme is reported to bind adenine on the external surface of the membrane and phosphoribosyl pyrophosphate on the inner surface of the membrane, and following catalysis, releases its products, AMP and inorganic pyrophosphate, in the cytoplasm. Two sequential group translocation processes are, therefore, responsible for the uptake of the two moieties of an extracellular nucleoside (Hochstadt-Ozer, 1972).

Another group translocation system, responsible for the coupled transmembrane transport and phosphorylation of several sugars (glucose, mannose, fructose, mannitol, etc.) is the phosphoenolpyruvate-dependent phosphotransferase system. Extensive studies have revealed the nature of the structural constituents of the system and many of the mechanistic details by which energy coupling occurs (Postma and Roseman, 1976). Phosphorylation of a particular sugar (such as glucose in *E. coli* or lactose in *Staphylococcus aureus*) requires the participation of four distinct proteins of the phosphotransferase system. Two of these proteins, enzyme I and the small heat-stable protein, HPr, function as general energy-coupling proteins and are required for the phosphorylation of all of the sugar substrates of the system. The other two proteins, the enzymes II and III, exhibit sugar specificity and function in the transport and phosphorylation of only one or a few sugars. The bacterial cell apparently possesses several pairs of these sugar-specific proteins, accounting for the broad substrate specificity of the system.

Sugar phosphorylation does not involve the direct transfer of the phosphoryl moiety of phosphoenolpyruvate to sugar, but requires the sequential phosphorylation of enzyme I, HPr, and a sugar-specific enzyme III. The final phosphoryl transfer reaction, catalyzed by the corresponding enzyme II, involves transfer of the phosphoryl moiety of phosphoenzyme III to sugar (Postma and Roseman, 1976). Because this last reaction is presumably the one which is coupled to transmembrane translocation, it is of particular interest.

Of the four proteins discussed above, only one, the enzyme II, is an integral membrane protein. The other proteins are either soluble or releasable from the membrane in a water-soluble form after appropriate treatments. The integral membrane constituent is also the protein which functions in sugar recognition. Binding studies and kinetic analyses suggest that the sugar and phosphoenzyme III simultaneously bind to the enzyme II before phosphoryl transfer occurs. A concerted mechanism which does not involve phosphorylation of the enzyme II has been suggested (Postma and Roseman, 1976). This process is illustrated schematically in Fig. 6A.

Recently, the membrane-associated enzyme II complexes have been shown to catalyze chemical reactions in the absence of the soluble proteins of the system (Saier and Newman, 1976). These reactions are simple sugar phosphate: sugar phosphoryl exchange, or transphosphorylation reactions:

$$\text{Sugar-P} + {}^*\text{sugar} \rightarrow {}^*\text{sugar-P} + \text{sugar} \qquad (\Delta G^\circ = 0)$$

The process proved to be of considerable interest when it was found that the reaction could occur in a vectorial fashion in whole cells and membrane vesicles. In this process, intracellular sugar phosphate can drive the uptake of extracellular sugar. When this occurs, however, the phosphoryl moiety of

A. Active Group Translocation (PTS)

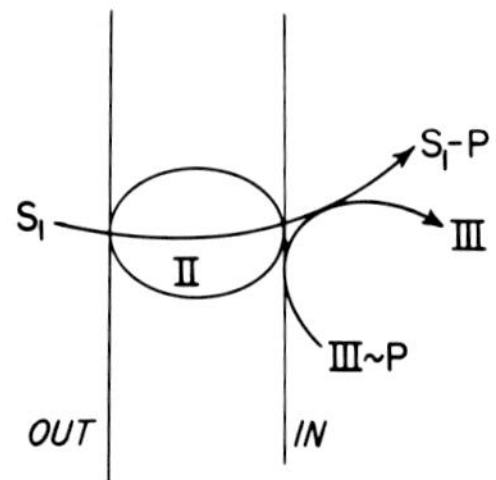

B. Exchange Group Translocation (PTS)

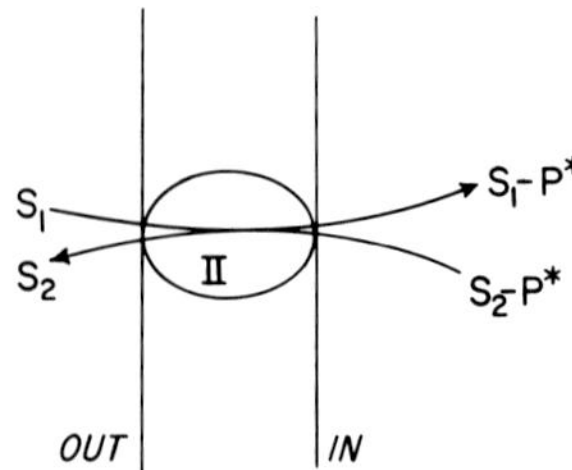

FIG. 6. Schematic representation of (A) active and (B) exchange group translocation catalyzed by the bacterial phosphotransferase system.

the intracellular sugar phosphate is transferred to the incoming sugar, while the sugar moiety of the sugar phosphate is apparently expelled from the cell. This process, termed exchange group translocation, is illustrated in Fig. 6B. It apparently involves no net change in the intracellular sugar phosphate pool. Several mechanistic aspects of the process have been elucidated (Rephaeli *et al.*, 1977).

Although the net concentration of intracellular sugar-P is not altered by the vectorial transphosphorylation process described above, the reaction may be of physiological significance. Thus, if a bacterium finds itself in a medium containing both a metabolizable sugar substrate of enzyme II (such as glucose) and a nonmetabolizable sugar substrate of the same enzyme II (such as methyl α-glucoside), then both sugars should be taken up into the intracellular compartment and phosphorylated at the expense of cellular phosphoenolpyruvate. The relative rates of uptake of the two sugars will depend on the ratio of their extracellular concentrations. However, regardless of the ratio, only the former sugar can be metabolized so that in the absence of a mechanism for its removal, the latter sugar-P will accumulate in the cytoplasm until toxic concentrations are attained. Exchange group translocation provides a mechanism for the elimination of the toxic sugar-P.

In the presence of an extracellular metabolizable sugar, the nonmetabolizable sugar-P should serve as an effective phosphoryl donor for the uptake of the metabolizable carbohydrate. Because the nonmetabolizable sugar must exit from the cell at the time its phosphoryl moiety is transferred to incoming sugar (Fig. 6B), expulsion of the toxic substance from the cell is accomplished. Moreover, the phosphoryl moiety of the nonmetabolizable sugar is quantitatively transferred to the incoming sugar, and energy wastage is avoided. Precisely this process has been demonstrated experimentally (Saier *et al.*, 1977). Exchange group translocation therefore allows the utilization of intracellular phosphoenolpyruvate via the phosphotransferase system for the exclusive acquisition of metabolizable carbohydrates. Worthy of note is the fact that facilitated diffusion and active transport systems catalyze exchange transport by a mechanism which may be related to exchange group translocation and which may serve a similar physiological function (Mawe and Hempling, 1965; Rothstein *et al.*, 1976).

D. Relatedness of Carbohydrate Transport Systems

Very little information is available concerning the evolutionary origins of membrane transport systems. However, as discussed in detail elsewhere (Saier, 1977), it is likely that many of these proteins have a common origin and share certain structural features. The establishment of these relationships will require that the transport proteins be purified and their structures analyzed. However, it can be anticipated that detailed studies will reveal that permeases which function via the same energy coupling mechanism will have diverged relatively recently. For example, some evidence suggests that the three hexitol–enzyme II complexes of the phosphotransferase system which exhibit specificity toward glucitol, mannitol, and galactitol may possess similar structures. There is also some evidence that the glucose– and mannose–enzyme II complexes diverged relatively recently. Possibly all of the genes which code for the enzyme II complexes of the phosphotransferase system will eventually be traced back to a single ancestral gene.

It is also reasonable to suppose that transport proteins which catalyze facilitated diffusion or active transport may share features in common with the group translocating permeases. For example, all three types of permease systems catalyze transmembrane solute exchange transport processes (Mawe and Hempling, 1965; Rothstein *et al.*, 1976). The possibility (discussed in Section IV) that the glycerol-specific and the disaccharide-specific transport systems (which, respectively, catalyze facilitated diffusion and active transport) recognize a protein constituent of the phosphotransferase system (Saier and Feucht, 1975; Saier and Stiles, 1975) also argues in favor of a common ancestry. It is possible that the processes by which permeases catalyze passage of solute molecules through the bacterial membrane will prove to be mechanistically similar, regardless of the energy coupling mechanisms (Saier, 1977).

IV. Regulation of the Synthesis of Carbohydrate Permeases and Catabolic Enzymes

Synthesis of a specific carbohydrate-metabolizing enzyme system, such as that involved in lactose breakdown in *E. coli*, has been shown in extensive studies to rely on the presence of two small cytoplasmic molecules, inducer and cyclic AMP. Many details of the regulatory process have been elucidated. While the principles established from studies of this system have proved applicable to numerous systems, both in prokaryotic and in eukaryotic cells, some evidence for the existence of alternative and quite different modes of regulation has appeared. Specifically, the induced synthesis of some permeases apparently requires the presence of extracellular inducers, and the synthesis of certain catabolic enzyme systems appears to depend on the integrity of specific permease proteins. In this section both the established and the less well understood more recently documented examples of transcriptional regulation will be discussed.

A. Control of Carbohydrate Catabolic Enzyme Synthesis by Intracellular Inducer and Cyclic AMP

1. Repression of Enzyme Synthesis

Even before the turn of the century, it was known that when microorganisms were grown in the presence of two different carbon sources, one of these was frequently utilized preferentially. For example, when *E. coli* cells are grown in the presence of both glucose and lactose, the glucose is metabolized first, and lactose catabolism begins only after the glucose has been completely removed from the medium (Saier and Moczydlowski, 1978). Subsequent investigations revealed that glucose functioned in this capacity by reducing the concentrations of the enzymes involved in lactose utilization. This phenomenon became known as the *glucose effect*, and it was shown, particularly by Gale and co-workers and by Monod, that it was limited to certain carbohydrate degradative enzymes. In 1961 and again in 1970, Magasanik published reviews in which he attributed the glucose effect to three distinct processes, termed catabolite repression, transient repression, and inducer exclusion (Magasanik, 1970). *Catabolite repression* referred to the inhibition of enzyme synthesis, caused by catabolites of the repressing sugar present within the cell. Actually, the causal catabolites were never identified. *Transient repression* was distinguished kinetically. It was more intense than catabolite repression but of short duration, and data were available showing that the repressing sugar did not have to be metabolized. *Inducer exclusion* was thought to occur when the inducer for a given catabolic enzyme system was forced out of the cell upon addition to the inducing medium of a second sugar. Although the molecular mechanisms responsible for these three processes were unknown, the permeases for the respective sugars were implicated both in transient repression and in inducer exclusion. This fact, however, provided little information about the mechanism of enzyme repression until the phosphotransferase system was discovered and the constituents of this system were characterized both structurally and functionally (see Section III,C).

It is worth reemphasizing, that the phosphotransferase system consists, in essence, of a phosphate transfer chain; the phosphoryl moiety of phosphoenolpyruvate is first transferred to enzyme I with the formation of a high energy phosphoryl protein with the phosphate group covalently linked to a histidyl residue in the protein. Second, the phosphoryl moiety is transferred from phosphoenzyme I to the heat-stable protein, HPr, with the formation of a second high-energy phosphoprotein. As in the case of enzyme I, a histidyl residue in HPr is phosphorylated. The third phosphoryl transfer reaction in this sequential progression of events involves transfer of the phosphoryl moiety from the general, non-sugar-specific protein, HPr, to one of the several sugar-specific enzymes III. Again, a high-energy phosphoryl protein is formed, but in this case, either a histidyl or a glutamyl residue may be phosphorylated, depending on the enzyme III. Finally, in the presence of the membrane-associated enzyme II, the only protein component of the system for which phosphorylation has not been demonstrated, phosphate is transferred to sugar. This last reaction is presumably coupled to sugar transport (Section III,C).

In 1965, Makman and Sutherland identified cyclic AMP in *E. coli* and determined the conditions which gave rise to altered intracellular concentrations of the compound (see Section III,B,4). As shown in several laboratories, the cyclic nucleotide is synthesized by a membrane-associated adenylate cyclase, degraded to 5′-adenosine monophosphate by a cytoplasmic cyclic nucleotide phosphodiesterase, and transported out of the cell by a membrane-associated permease system (Fig. 7) (Pastan and Adhya, 1976; Saier *et al.*, 1975a). Since the original studies were conducted with an *E. coli* strain which lacked cyclic nucleotide phosphodiesterase, variations in intracellular cyclic AMP concentrations had to reflect the differential rates of cyclic AMP synthesis and excretion. When rapidly growing cells were harvested, washed free of carbohydrate, and resuspended in a salt medium in the absence of a carbon source, intracellular concentrations of cyclic AMP rose to very high values, approaching 1 m*M*. Under these conditions a significant fraction of the cellular adenine nucleotide pool was in the form of cyclic AMP. If glucose was then added to such a cell suspension, the intracellular concentration of cyclic AMP was observed to drop to less

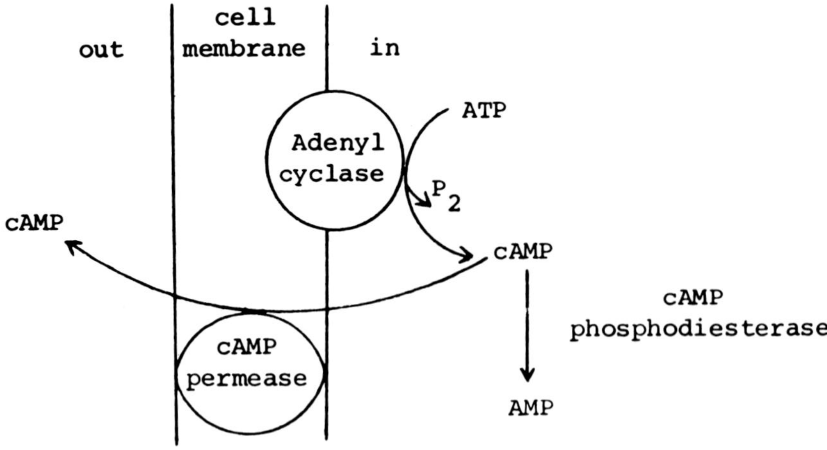

FIG. 7. Metabolism of cyclic AMP in *E. coli*.

than 1% of its original value within a few minutes. This was due to the fact that glucose stimulated the efflux of cyclic AMP from the bacterial cell while strongly inhibiting the synthetic reaction. How glucose and other energy sources stimulated cyclic AMP efflux was discussed in Section III,B,4. The regulation of adenylate cyclase will be considered below.

2. Coordinate Regulation of Adenylate Cyclase and Inducer Uptake

The studies of Makman and Sutherland prompted Pastan and Perlman to consider the possibility that cyclic AMP was a key metabolite in the regulation of catabolic enzyme synthesis (Pastan and Adhya, 1976). As a consequence of their investigations and of those in several other laboratories, the general model for transcriptional regulation of the lactose (*lac*) operon shown in Fig. 8 was proposed. Three proteins can bind specifically to the controller region of the *lac* operon: RNA polymerase which transcribes the DNA sequence of the operon into messenger RNA, and two regulatory proteins which determine the frequency with which messenger RNA synthesis is initiated, the *lac* repressor and the cyclic AMP receptor protein (CR protein). The latter two proteins can each exist in either of two conformations, only one of which can bind to the DNA. The conformations of these proteins are controlled by two small cytoplasmic ligands, inducer and cyclic AMP, respectively. The free form of the *lac* repressor binds to the *lac* operator region of the operon to prevent transcription, but inducer

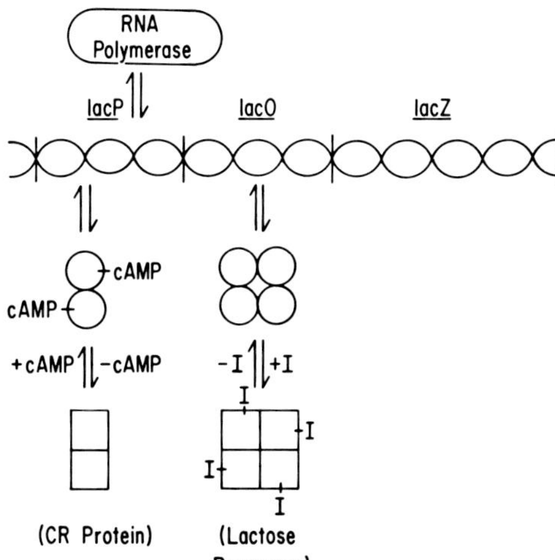

FIG. 8. Proposed mechanism of transcriptional regulation of the lactose operon in *E. coli*.

molecules bind to the repressor protein, converting it to a form which possesses low affinity for the DNA. Consequently, an elevated intracellular concentration of inducer causes the *lac* repressor protein to dissociate from the *lac* operon. By contrast, the free form of the cyclic AMP receptor protein possesses low affinity for the controller region of the operon, but the binding of cyclic AMP to the protein converts it to a conformation with high affinity for the *lac* promoter. Thus, an enhanced intracellular concentration of cyclic AMP causes the CR protein to associate with the DNA. But since the binding of the cyclic AMP–CR protein complex to the promoter promotes transcription (positive control) while the binding of the *lac* repressor to the operator inhibits transcription (negative control), an elevation in the concentration of either cyclic AMP or inducer should enhance the rate of *lac*-specific messenger RNA synthesis. Expression of the operon is clearly under dual control by the two small cytoplasmic molecules: inducer and cyclic AMP.

If carbohydrate catabolic enzyme systems are generally under dual control by inducer and cyclic AMP (and extensive evidence supporting this notion is now available), then the bacterium might benefit from a mechanism which results in the coordinate regulation of the intracellular levels of these two molecules. Recent studies have shown that the transport systems responsible for the uptake of several inducers and the cyclic AMP synthetic enzyme, adenylate cyclase, are, in fact, subject to coordinate regulation in a variety of gram-negative bacterial strains. As a consequence of extensive genetic, physiological, and biochemical studies, a specific but highly speculative regulatory mechanism involving the proteins of the bacterial phosphotransferase system has been proposed (Fig. 9) (Castro *et al.*, 1976; Peterkofsky, 1976; Postma and Roseman, 1976; Saier and Feucht, 1975; Saier and Moczydlowski, 1978; Saier and Stiles, 1975).

According to this proposal, both the permease proteins and adenylate cyclase are subject to allosteric regulation, and the allosteric effector molecule is a central regulatory protein, termed RPr. Furthermore, RPr can be phosphorylated by a mechanism involving enzyme I and HPr as outlined above. Phosphate is first transferred from phosphoenolpyruvate to enzyme I, then to HPr, and finally to RPr. In all cases, the resultant phosphoryl proteins are of high energy. Phosphorylation of RPr is thought to be a key step in the regulatory sequence. A carbohydrate permease system which is sensitive to phosphotransferase system (PTS)-mediated regulation is thought to normally exist in an active form, but interaction of the free form of RPr with the allosteric binding site of the permease protein would inhibit its activity. In other words, the permeases are thought to be subject to negative control by free RPr. By contrast, adenylate cyclase is thought to normally exist in a conformation of low activity, and the enzyme may be activated by the binding of phospho-RPr to the adenylate cyclase allosteric binding site. In this case, the enzyme is thought to be subject to positive control by the phosphorylated form of RPr.

It is instructive at this point to reconsider the experiment of Makman and Sutherland (described above). Glucose-grown bacteria were washed free of carbohydrate and resuspended in sugar-free salt medium. Under these conditions intracellular phosphoenolpyruvate, derived from endogenous energy reserves, should phosphorylate enzyme I, HPr, and RPr. Consequently, only RPr-P (not free RPr) will be present in the cell. In response to the cellular pool of RPr-P, adenylate cyclase will be activated. Because there is little or no free RPr in the cell, the permeases will operate at the uninhibited rates when inducer molecules are present in the extracellular medium. Thus, the rates of cyclic AMP synthesis and inducer uptake will be maximal.

If, now, glucose is added to the cell suspension, phosphate will be drained off of the phosphorylated phosphotransferase proteins including RPr because the phosphoryl moieties of all of these proteins are of high energy and are, therefore, in equilibrium with each other. This should increase the cellular level of free RPr, while decreasing the concentration of phospho-RPr. As a consequence, adenylate cyclase should become deactivated while

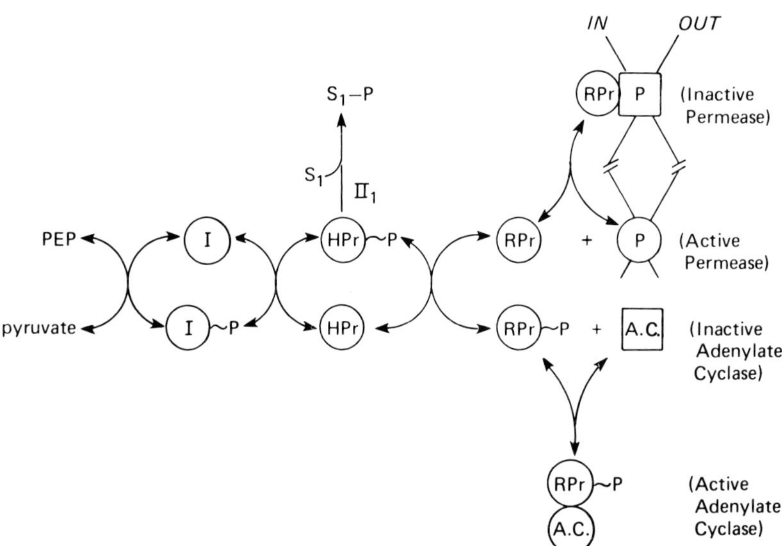

FIG. 9. Proposed mechanism for the coordinate regulation of carbohydrate permeases and adenylate cyclase in *E. coli* and *S. typhimurium*. Ac, adenylate cyclose; RPr, regulatory protein; S, sugar substrate of PTS.

the permeases become inhibited. The model appears to account for coordinate regulation of these systems and for a large body of experimental evidence bearing on the regulation of cyclic AMP synthesis and inducer uptake in bacteria (Postma and Roseman, 1976; Saier and Moczydlowski, 1978).

As discussed in section V,A, carbohydrate permeases appear to be subject to regulation by at least three distinct mechanisms. One mechanism involves the direct participation of the enzymes of the phosphotransferase system. Other proposed mechanisms require the participation of intracellular metabolites or the membrane electrical potential (Saier and Moczydlowski, 1978). As for the permeases, there is evidence that adenylate cyclase is subject to regulation by more than a single mechanism (Saier *et al.*, 1976). Several metabolizable carbohydrates which are not substrates of the phosphotransferase system were found strongly to inhibit adenylate cyclase activity in *E. coli* provided that the corresponding catabolic enzyme systems had been induced to high levels. Inhibition by lactose was abolished by the genetic loss of the lactose permease and greatly diminished by loss of β-galactosidase. By contrast, a mutation which reduced the cellular concentration of enzyme I had no effect on this inhibition. These results suggested that lactose metabolism contributed to the regulatory effects observed with this sugar and that a mechanism was operative which was distinct from that by which sugar substrates of the phosphotransferase system exert regulatory control over adenylate cyclase.

3. Desensitization to Regulation by the Phosphotransferase System

Regardless of the mechanism by which the phosphotransferase system regulates the activities of various permease systems and adenylate cyclase, the fact that these activities are subject to coordinate control raises another important question. Is the inhibition of cyclic AMP synthesis, or that of inducer uptake primarily responsible for the intense repression of catabolic enzyme synthesis observed when bacteria are simultaneously exposed to two carbon sources, such as glucose and lactose? Alternatively, are both regulatory responses of equal physiological significance? Extensive studies have led to the conclusion that while both phenomena contribute to the repression of carbohydrate catabolic enzyme synthesis, inducer exclusion is normally of greater quantitative importance (Saier and Moczydlowski, 1978). This conclusion, however, led to an apparent anomaly. Several investigators had observed that cyclic AMP, in the presence of normal concentrations of an exogenous metabolizable source of inducer (such as lactose), could completely overcome the repression of catabolic enzyme synthesis (Pastan and Adhya, 1976). This would not be expected if the regulation of inducer uptake were largely responsible for enzyme repression.

Recent experiments have resolved this apparent inconsistency (Keeler *et al.*, 1977). When bacteria are incubated with inducer under appropriate experimental conditions, and when by any means the intracellular cyclic AMP levels are allowed to increase to high values, then the transport systems responsible for inducer uptake become desensitized to PTS-mediated inhibition. Addition of glucose or another sugar substrate of the PTS to a desensitized cell suspension does not inhibit the uptake of inducer. Under the same conditions, inducer uptake will be inhibited more than 90% by glucose in a normally sensitive cell suspension. Although the mechanism of this type of physiological desensitization has not yet been elucidated, the process seems to require protein synthesis. Since desensitization only occurs when the cellular levels of cyclic AMP are high, it appears that cyclic AMP plays a direct role in this process and, therefore, functions in the bacterial cell in two related capacities: it controls the synthesis of catabolic enzyme synthesis and it determines the sensitivity of corresponding carbohydrate transport systems to regulation by sugar substrates of the phosphotransferase system. Both of these functions, in turn, determine the rate at which a particular carbohydrate will be taken up and metabolized by a bacterial cell suspension.

4. Additional Regulatory Constraints Influencing Sensitivity of Catabolic Enzyme Synthesis to Repression

Considerable published evidence suggests that catabolic enzyme synthesis may be subject to regulatory constraints in addition to those described above. Some of these interactions are still poorly understood. For example, permeases and adenylate cyclase are apparently subject to inhibition by intracellular metabolites, and they may be responsive to the membrane potential (Saier and Moczydlowski, 1978). In addition, the magnitude of catabolite repression depends not only on the nature of the repressing sugar but also on the availability of utilizable nitrogen. For example, in certain *E. coli* strains, maximal sensitivity of β-galactosidase synthesis to repression is achieved in the presence of a rapidly metabolizable source of carbon and a slowly utilizable source of nitrogen (Contesse *et al.*, 1960). Maximal derepression of carbon catabolic enzyme synthesis occurs when the nitrogen source is rapidly assimilated and the carbon source is utilized slowly. The mechanism by which nitrogen influences the degree of catabolite repression is not known, but it is worth noting in this regard that the unusual nucleotide, guanosine 5′-diphosphate 3′-diphosphate, has been shown to influence expression of the lactose operon *in vitro* (Aboud and Pastan, 1973). Moreover, preferential stimulation of *lac*-specific messenger RNA synthesis by this compound was observed only if cyclic AMP was present (Smolin and Umbarger, 1975). The mechanistic and physiological bases for this observation have yet to be established.

Still another type of regulatory mechanism, superimposed on those described above, has been discovered. Some catabolic enzymes in *E. coli* are apparently subject to regulation by a repressor protein which coordinates synthesis of a galactose-containing capsular polysaccharide (Hua and Markovitz, 1974). A mutation in the *capR* gene, which codes for this repressor, results in the overproduction and secretion of the polysaccharide. A number of enzymes involved in the synthesis of the precursors of the secreted polymer are synthesized at increased rates in *capR* mutants. Among these enzymes are those coded for by the galactose operon. It has been postulated that the repressor protein (the product of the *capR* gene) binds to a site within the operator region of the galactose operon and exerts a repressing effect upon the rate of transcription. Loss of the *capR* repressor by mutation, therefore, results in derepression of all of the genes normally regulated by this protein. Available evidence suggests that regulation by the *capR* repressor occurs independently of other regulatory phenomena, such as inducer promoted and cyclic AMP promoted synthesis of galactose messenger RNA (Hua and Markovitz, 1974; Mackie and Wilson, 1972). It appears that expression of an operon may be subject to three or more independent forms of regulation due to the presence of multiple protein recognition sites within the regulatory region of the operon.

B. Exogenous Inducer Requirements for Permease Synthesis

1. The Hexose Phosphate Transport System

Gram-negative enteric bacteria possess a transport system which allows a variety of hexose phosphates including the 1- and 6-phosphate esters of glucose, mannose, fructose, and 2-deoxyglucose to cross the cytoplasmic membrane without hydrolysis. This hexose phosphate transport system functions by active transport and can accumulate appropriate hexose phosphates at least tenfold against a concentration gradient (Dietz, 1976). Synthesis of the system depends upon cyclic AMP and an appropriate inducer. Of the various substrates, only glucose 6-phosphate (glucose-6-P) and 2-deoxyglucose 6-phosphate (2-deoxyglucose-6-P) appear to function as inducers. Other hexose phosphates are inactive unless they are converted enzymatically to glucose-6-P. However, even when this interconversion occurs rapidly within the cytoplasm, induction may not occur. This is due to the fact that the inducer must be present in the extracellular medium to enhance the rate of synthesis of the hexose phosphate transport system.

When *E. coli* cells are exposed to glucose or 2-deoxyglucose, these sugars are transported across the membrane and phosphorylated to the corresponding hexose 6-phosphates via the phosphotransferase system. Mutants which lack glucose-6-P dehydrogenase and phosphoglucoisomerase cannot metabolize intracellular glucose 6-phosphate, and the ester accumulates in the cytoplasm to concentrations in excess of 50 m*M*. Under these conditions, synthesis of the hexose phosphate transport system is not induced (Dietz, (1976). However, if a low concentration of glucose-6-P or 2-deoxyglucose-6-P (0.1 m*M*) is added exogenously to the cell suspension, rapid induction of the transport system is observed. Since the system accumulates its substrate only tenfold against a concentration gradient, and since it catalyzes a reversible process, the sugar phosphate which accumulated in the cytoplasm by means of the action of the phosphotransferase system is released into the extracellular medium. These observations clearly show that exogenous (but not cytoplasmic) inducer influences the synthesis of the transport system.

2. Other Permease Systems Requiring Exogenous Inducer for Induction

Evidence is accumulating that a requirement for exogenous inducer may be a property of several transport systems. For example, the phosphoglycerate transport system in *Salmonella typhimurium* is readily induced by 0.1 m*M* exogenous 3-phosphoglycerate, 2-phosphoglycerate, or phosphoenolpyruvate, but cells growing exponentially with glucose, glycerol, or glycerate as the sole source of carbon did not induce for the phosphoglycerate transport system (Saier *et al.*, 1975b). Under these conditions, the intracellular concentration of each of the inducers of the system is probably in excess of 1 m*M*. Still a third transport system which appears to be subject to regulation by extracellular inducers is the tricarboxylic acid transport system in *S. typhimurium*, discussed in Section III,B,3. Although all of the three tricarboxylic acid inducers of the system are normally present in the cytoplasm, induced synthesis is observed only when one of these compounds is added to the extracellular medium (Saier, unpublished results).

It is worthy of note that in each of the three examples cited above the substrates of the transport systems are compounds which are normally generated as a result of the metabolism of other carbon sources. In contrast, transport systems which act on carbohydrates that are not normal cellular constituents generally utilize intracellular inducers. In view of these observations, a generalization appears justified: When the substrate of a permease system is generated internally as a result of normal metabolic activity, then the inducer of the system must be present in the extracellular (or periplasmic) compartment in order for induction to occur. This requirement clearly prevents the wasteful induction of metabolite transport systems and prevents leakage of metabolites from the cell.

3. Possible Mechanisms of "Induction from Without"

The mechanisms by which exogenous inducers exert their effects are not known. However, it is worthy of note that synthesis of all three of the systems discussed above requires intracellular cyclic AMP and the cyclic AMP receptor protein, since mutants which lack adenylate cyclase or the receptor protein cannot be induced for these activities (Saier *et al.*, 1975b). In terms of the classic operon model (Pastan and Adhya, 1976), these observations might suggest that a membrane-associated repressor (or activator) with an inducer-binding site on the external side of the membrane might regulate transcription at the operator site, while the intracellular cyclic AMP receptor protein controls transcription at the promoter site. This hypothesis requires that the regulatory region of the operon be in association with the membrane. Further, the operator of each operon which responds to an exogenous inducer would have to be in association with a specific membrane protein. These considerations render the postulate unlikely. Other possibilities, such as the generation of a labile intracellular inducer from an exogenous "pre-inducer" or an involvement of the uptake process per se in induction, are also difficult to accept.

We shall consider below a simple mechanism which appears to be consistent with the available evidence and accounts for induction by exogenous inducers. Two types of permease-specific mutants, altered with respect to either the hexose phosphate transport system (Dietz, 1976) or the phosphoglycerate transport system (Saier *et al.*, 1975b), can be easily isolated. One class of mutants does not synthesize functional permease, while the other class synthesizes the permease constitutively in the absence of inducer. Among mutants of the former class are some which appear to be structural gene mutants, but others appear to be regulatory in nature. These observations suggest that synthesis of the two permeases is subject to both positive and negative control and that two distinct proteins may, therefore, mediate the response to exogenous inducers. One of these proteins might be a cytoplasmic repressor (or activator) which interacts with the operator region of the operon coding for the permease proteins. The other protein is assumed to be an integral transmembrane regulatory protein, hereafter designated the MR protein, which possesses a binding site for the inducer molecule on the external surface of the cytoplasmic membrane, and a binding site for the repressor (or activator) protein on the inner surface of the membrane (Fig. 10). The affinity of the MR protein for the repressor (or activator) on the inner surface of the membrane is postulated to be allosterically controlled by the binding of inducer to its external ligand binding site.

If the soluble regulatory protein functions as a repressor, then the MR protein must possess low affinity for the repressor when not complexed with inducer. The binding of inducer at the external surface would enhance the

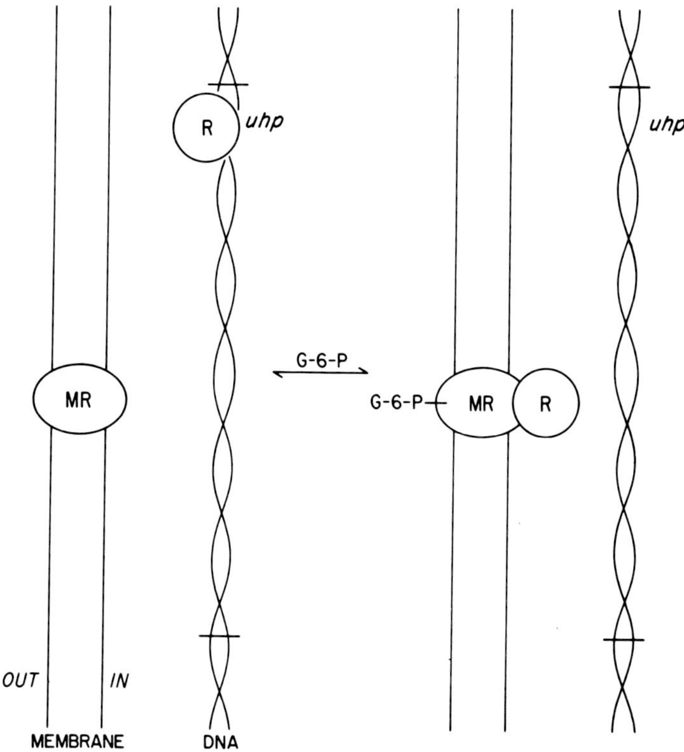

FIG. 10. A possible mechanism for the regulation of gene expression by exogenous inducer. The model suggests an explanation for transcriptional regulation of the hexose phosphate transport system in *E. coli* as well as that of the phosphoglycerate transport system in *S. typhimurium*. Two regulatory proteins are proposed. One is a repressor protein (R) which possesses affinity for the controller region of the operon coding for the transport protein. It also binds to a transmembrane regulatory protein (MR) which possesses a glucose-6-P binding site on the external surface of the cytoplasmic membrane. The binding of glucose-6-P to the MR protein determines its affinity for the R protein. Data are not available which distinguish this model from one in which a cytoplasmic activator protein (instead of a repressor protein) controls expression of the operon (see text).

affinity of the MR protein for the repressor so that in an equilibrium situation the DNA–repressor complex would dissociate. In this model it is assumed that the MR protein is present in a quantity which exceeds that of the soluble regulatory protein. On the other hand, if the soluble protein promotes transcription of the operon, the MR protein should possess high affinity for the activator when extracellular inducer is not present. Inducer binding would presumably alter the conformation of the MR protein with a resultant decrease in affinity for the activator.

One virtue of these models arises from the fact that they can be tested and distinguished experimentally. For example, the repressor-mediated and activator-mediated transcriptional regulation can be distinguished genetically as follows: If a repressor protein were involved, then the constitutive mutants (defective for the soluble repressor protein) would be epistatic over the permease negative regulatory mutants (defective for the MR protein). By contrast, if an activator protein determines sensitivity of the operon to transcriptional regulation by the mechanism described above, then the permease-negative phenotype, which would result from a defective activator, would be epistatic over the constitutive phenotype, which would result from a defective MR protein. The use of nonsense mutants would facilitate interpretation of these experiments.

It is interesting to note that the proposed mechanism involving transmembrane control of events occurring at the cytoplasmic surface of the cell membrane may be analogous to the regulation of catalytic functions by extracellular hormones in higher eukaryotic systems. It is even possible that an evolutionary relationship exists between these two types of membrane receptors.

C. Direct Involvement of Permease Proteins in the Regulation of Enzyme Synthesis

In several instances, a protein which is coded for by a gene which comprises part of an operon apparently functions as a regulatory constituent controlling expression of that operon (Calhoun, 1975). In some cases, the regulatory protein may be a catabolic or anabolic enzyme with a well-defined catalytic function. This type of induction has been termed autogenous induction. Evidence is beginning to emerge which suggests that certain transport proteins may also play dual catalytic and regulatory roles. A few examples are cited below.

1. The Lactose Regulon in *Staphylococcus aureus*

Lactose is transported across the cytoplasmic membrane of *Staphylococcus aureus* by the lactose phosphotransferase system (see Section III,C). The uptake and phosphorylation of lactose via this system requires the functional integrity of enzyme I, HPr, enzyme II^{lac} and enzyme III^{lac}. The product, cytoplasmic lactose 6-phosphate (lactose-6-P), is then cleaved to glucose and galactose 6-phosphate by a phospho-β-galactosidase. The latter compound is the true inducer of the regulon which codes for the lactose-specific constituents of the system, enzyme II^{lac}, enzyme III^{lac}, and the phospho-β-galactosidase. This phosphate ester can induce the operon when added to a cell suspension at high concentration, even when the phosphotransferase system is nonfunctional (Postma and Roseman, 1976; Saier and Stiles, 1975).

Mutants defective for lactose utilization were isolated from two *S. aureus* parental strains (Simoni and Roseman, 1973). One was the lactose inducible wild-type strain, and the other was a lactose constitutive strain derived from the inducible parent. The two strains each yielded three identical classes of lactose-negative mutants defective in either enzyme I, enzyme II^{lac}, or phospho-β-galactosidase. When the inducible parent was studied, a fourth class of mutants lacked all three of the proteins of the lactose regulon, enzyme II^{lac}, enzyme III^{lac}, and phospho-β-galactosidase. Extensive attempts to isolate mutants lacking only enzyme III^{lac} from the inducible parent gave negative results. By contrast, the constitutive parental strain did not yield mutants lacking the three proteins of the lactose regulon, but yielded a major class lacking only enzyme III^{lac}. The difference in behavior of the inducible and constitutive strains suggested that enzyme III^{lac} may play a role in regulating transcription or translation (or both) of the lactose regulon in *S. aureus* in addition to its role in the transport of lactose.

2. The Maltose Regulon in *Escherichia coli*

Extensive genetic studies on maltose utilization in *E. coli* provided evidence for a second example in which a transport protein possibly functions directly in the regulation of gene expression. The maltose catabolic enzyme system is coded for by genes which are included within at least three operons (Table II). These operons comprise the maltose regulon and are subject to control by the product of the *malT* gene. In the absence of the *malT* protein, expression of the maltose regulon is suppressed. The structural genes, *malP* and *malQ* in the *malA* region (at 74 minutes on the *E. coli* chromosome) code for maltodextrin phosphorylase and amylomaltase, respectively (Hofnung *et al.*, 1974). The structural genes *malE*, *malF*, *malK*, and *lamB* in the *malB* region (at 90 minutes on the *E. coli* chromosome) code for the protein constituents of the maltose transport system. The *malE* gene product is a periplasmic maltose-binding protein (see Section III,B,3), while the *lamB* gene product is the lambda phage receptor whose biological function in the outer lipopolysaccharide membrane is to facilitate the transport of

TABLE II
GENES COMPRISING THE MALTOSE REGULON IN *E. coli* AND THEIR PROTEIN PRODUCTS[a]

malA region	(74 minutes)
malT	Positive regulatory protein (activator)
operon 1	
↓ *malP*	Maltodextrin phosphorylase
↓ *malQ*	Amylomaltase
malB region	(90 minutes)
operon 2	
↓ *malE*	Maltose binding protein
↓ *malF*	Essential constituent of maltose permease
operon 3	
↓ *malK*	Essential constituent of maltose permease
↓ *lamB*	Phage lambda receptor protein; a nonessential constituent of the maltose permease; probably a protein in the outer membrane which facilitates the permeation of maltooligosaccharides across this structure

[a] The arrows indicate direction of transcription.

maltooligosaccharides across this membrane. Of these four genes only the *lamB* gene is nonessential for transport of maltose across the cytoplasmic membrane.

Mutations in the *malB* region have startling effects on the expression of the *malPQ* operon. Loss of *malE* or *malF* function renders the catabolic enzymes noninducible or poorly inducible, but mutation of the *malK* gene resulted in a level of constitutive expression of the *malPQ* operon equal to about 50% of the maximally induced wild-type level (Hofnung *et al.*, 1974). Although other interpretations can be entertained, this unexpected result suggests that the protein product of the *malK* gene may play a direct role in regulating the expression of the maltose regulon.

3. THE *pts* OPERON IN *Salmonella typhimurium*

In *S. typhimurium* and *E. coli* the genes which code for enzyme I and HPr of the phosphotransferase system comprise one operon (Postma and Roseman, 1976). The expression of these two genes is inducible to the extent of about three-fold by any sugar substrate of the phosphotransferase system (Saier *et al.*, 1970). However, the genetic loss of the enzyme II complex specific for a sugar substrate of the system renders that sugar noninductive. Moreover, disaccharides (such as maltose and melibiose) which are hydrolyzed intracellularly to the constituent free sugars, and sugar phosphates which are transported across the membrane via the hexose phosphate transport system are noninductive (Saier and Roseman, unpublished observations). These observations suggest that glucose and other sugar substrates of the phosphotransferase system enhance the expression of the *pts* operon as a consequence of an interaction with the sugar-specific membrane constituents of the system. Possibly, passage of the sugar through the membrane by group translocation is in some unknown way coupled to gene expression.

Preliminary evidence suggests that similar interactions may control synthesis of some of the sugar-specific proteins of the phosphotransferase system (Rephaeli and Saier, unpublished observations). Thus, the enzyme II^{Glc} and enzyme II^{Man} complexes in the membrane are inducible by extracellular glucose to an extent of about fourfold. Glucose, generated intracellularly, and glucose 6-phosphate, transported into the cell via the hexose phosphate transport system, are noninductive. Moreover, the genetic loss of enzyme I or HPr renders synthesis of both proteins constitutive. These results suggest that while the expression of the *pts* operon is dependent on enzyme II function, induction of the synthesis of some of the enzyme II complexes may depend on the activities of the general energy coupling proteins of the phosphotransferase system.

V. Regulation of the Activities of Carbohydrate Permeases and Catabolic Enzymes

A. Regulation of Sugar Permease Activities

The subject of bacterial carbohydrate transport regulation has recently been reviewed (Saier and Moczydlowski, 1978) and will be dealt with only briefly here. Just as a diversity of transport mechanisms have evolved in bacteria (see Section III), so have a number of mechanisms controlling their activities. In Section IV,A,2 one such regulatory mechanism was discussed. It was suggested that the proteins of the phosphotransferase system act directly in the allosteric regulation of several permeases which catalyze facilitated diffusion or active transport of carbohydrates (see Fig. 9).

A second process in which carbohydrate permeases are inhibited apparently results from the accumulation of intracellular sugar phosphates (Saier and Moczydlowski, 1978). Many transport systems appear to be subject to this type of negative regulation, and genetic loss of the phosphotransferase system-mediated regulatory mechanism does not interfere with this second mechanism (Saier, unpublished observations).

A third mechanism of regulation was referred to briefly in Section III,B,3. The membrane potential may control the conformation of a permease protein and thereby regulate its activity. This can occur in a positive sense (as discussed above for the citrate permease) (see Fig. 5), or in a negative sense, as has been suggested for the glucose permease in *E. coli* (Saier and Moczydlowski, 1978). An inhibitory effect of chemiosmotic energy on the activity of the glucose enzyme II of the phosphotransferase system may result if the conformation of this protein is responsive to the transmembrane electric potential.

It is worthy of note that the three recognized mechanisms for the regulation of carbohydrate permeases each involves a distinct form of energy. Thus, regulation by the phosphotransferase system is responsive to the extracellular availability of the sugar substrates of this system; inhibition by intracellular sugar phosphate reflects the cellular levels of metabolites which normally can be utilized as sources of carbon and energy, and inhibition by the membrane potential results whenever the level of cellular chemiosmotic energy builds up. All three mechanisms, therefore, appear to guard against the uptake of energy sources in excess of the needs of the cell. The permeases are apparently subject to novel types of feedback regulation where the ultimate allosteric effector is an energy source rather than a chemical species.

B. Allosteric Regulation of Metabolic Pathways in Bacteria

Numerous biosynthetic (anabolic) and degradative (catabolic) pathways are subject to regulation due to sensitivity of one or more of the constituent enzymes to inhibition and/or stimulation by cellular metabolites (Stouthamer, 1977). In the case of an anabolic sequence, the regulatory metabolite is frequently the end product of the pathway, and the enzyme which is sensitive to feedback control is the one which initiates the chemical transformation sequence. A variety of basic homeostatic negative regulatory mechanisms have been demonstrated, including single metabolite feedback, cooperative feedback, concerted feedback, and cumulative feedback (Sanwal, 1970). By contrast, strictly degradative sequences are generally not subject to end product inhibition, but exhibit sensitivity to regulation by compounds which serve as indicators of the energy state of the cell. Such indicators are frequently inorganic phosphate, pyrophosphate, or adenine nucleotides. These compounds signal the availability of the ultimate products of energy metabolism common to the diverse catabolic channels.

These same regulatory constraints appear to be superimposed on "amphibolic" pathways, metabolic sequences which fulfill both anabolic and catabolic functions in the cell. In the present abbreviated discussion of this subject, we will consider the enzymes of glycolysis and the citric acid cycle in *E. coli.* These pathways furnish carbon skeletons for biosynthetic purposes as well as sources of energy to be utilized through terminal oxidative pathways. They are, therefore, representative of amphibolic pathways. *Escherichia coli* has evolved complex regulatory mechanisms which ensure the coordinated flow of appropriate carbon sources into biosynthetic channels and energy generating pathways. More detailed discussions of these regulatory interactions and of the integration of these enzymatic sequences with other pathways can be found elsewhere (Sanwal, 1970).

The glycolytic scheme and the citric acid cycle can be thought of as a continuous sequence of metabolic reactions consisting of pathway segments in which a physiologically irreversible reaction is followed by one or more reversible reactions (Fig. 11). Each of the enzymes catalyzing an irreversible

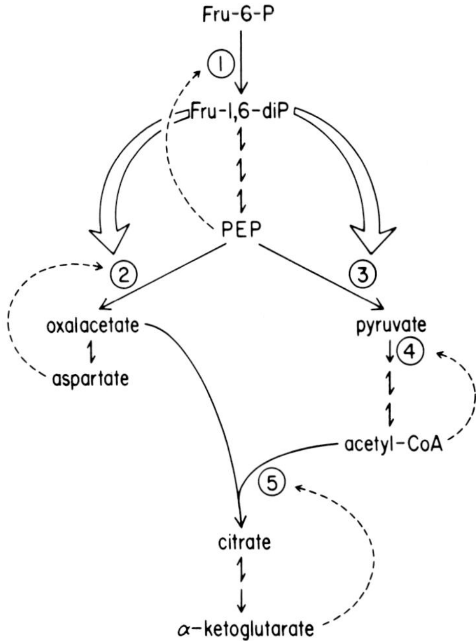

FIG. 11. Partial reaction sequences for glycolysis and the citric acid cycle. Single headed solid arrows (→) indicate reactions which are essentially irreversible under physiological conditions. Reversible reactions are indicated by double headed arrows (⇌). End product feedback inhibition is indicated by dashed arrows (---→), while precursor activation is indicated by solid double lined arrows (⇒). The enzymes indicated by number are (1) phosphofructokinase (EC 2.7.1.11), (2) phosphoenolpyruvate carboxylase (EC 4.1.1.31); (3) pyruvate kinase (EC 2.7.1.40), (4) pyruvate decarboxylase (EC 1.2.4.1), and (5) citrate synthetase (EC 4.1.3.7). Abbreviations: Fru-6-P, fructose 6-phosphate; Fru-1,6-P, fructose 1,6-diphosphate; PEP, phosphoenolpyruvate.

reaction is generally subject to negative allosteric feedback regulation by a product generated later in the reaction scheme. Four documented examples of this behavior are phosphofructokinase which is inhibited by phosphoenolpyruvate, phosphoenolpyruvate carboxylase which is inhibited by aspartate and other dicarboxylic acids, pyruvate decarboxylase (one of the three constituent enzymes of the pyruvate dehydrogenase–multienzyme complex) which is subject to inhibition by acetyl-CoA, and citrate synthetase which is inhibited by α-ketoglutarate (Fig. 11). In each case, the inhibitory response can be considered analogous to classic end product control, even though the products are utilized further for catabolic as well as anabolic purposes.

In dealing with the same metabolic pathway, it has been found that a metabolite generated early in the sequence may activate an enzyme which catalyzes a reaction further along in the sequence. As for the enzymes which are subject to end product inhibition, precursor-activated enzymes usually catalyze essentially irreversible reactions. For example, in *E. coli* both phosphoenolpyruvate carboxylase and pyruvate kinase are activated by fructose 1, 6-diphosphate (Fig. 11). Precursor activation may be restricted to amphibolic pathways, since the phenomenon has not been reported for strictly anabolic or catabolic pathways (Sanwal, 1970).

In addition to end product and precursor control, allosteric enzymes which catalyze irreversible reactions in amphibolic pathways may be subject to control by compounds which serve as indicators of the cellular energy level. As for catabolic pathways, the compounds which usually function in this capacity are AMP, ADP, ATP, pyrophosphate, and inorganic phosphate. However, since other purine and pyrimidine nucleotides may be in equilibrium with the adenine nucleotides by means of nucleoside kinases, the former compounds may also serve as indicators of cellular energy availability. Examples of energy-sensitive enzymes represented in Fig. 11 include phosphofructokinase which is activated by ADP and GDP, one of the two pyruvate kinases in *E. coli* which is activated by AMP, and the pyruvate decarboxylase component of the pyruvate dehydrogenase complex which is activated by AMP and GDP. In view of these and other observations, it has been suggested that the ratio of the cellular concentrations of purine nucleotides may be an important parameter in determining the activities of enzymes which consume or produce ATP (Atkinson, 1968).

The above discussion illustrates the intricacies of regulatory interactions which control metabolic pathways in bacteria. However, the situation is known to be far more complex; interacting feedback and feed-forward loops of diverging and converging channels require the operation of additional allosteric controls. It has been shown, for example, that phosphoenolpyruvate carboxylase is activated by acetyl-CoA as well as by fructose

1, 6-diphosphate and that enzymes of the citric acid cycle are subject to negative allosteric control by NADH (Sanwal, 1970). These interactions serve to integrate each of the segments of amphibolic pathways with the various catabolic and anabolic sequences. They prevent the wasteful production of excessive amounts of specific metabolites, while ensuring that these compounds will be supplied in amounts sufficient to supply the needs of the cell. They allow the maintenance of homeostasis under widely varying physiological conditions.

C. Control of Cellular Metabolite Concentrations by Multienzyme Systems and Multifunctional Enzymes

In the preceding section it was noted that the intracellular concentrations of metabolites can be controlled, in part, by elaborate regulatory interactions which influence the rates at which individual enzymes catalyze the chemical modification of their substrates. In general, allosteric enzymes possess multiple semiautonomous globular structures or domains. These globular domains may represent contiguous stretches of a single polypeptide chain, or they may be localized to distinct polypeptide chains present in close association with one another in an oligomeric complex. The binding of a ligand to the allosteric site influences the conformation and, hence, the catalytic activity of a neighboring protein domain. In this relatively simple situation, one globular structure possesses an active center and therefore a catalytic function, while the other is associated with a ligand-binding site and a consequential regulatory function.

Multifunctional proteins are known which resemble single polypeptide allosteric enzymes in that the polypeptide chain consists of more than a single functional unit. At least two of the globular domains associated with the protein possess catalytic activity. In all known multifunctional enzymes the different functions of the native protein are related. Moreover, the different active sites usually catalyze sequential metabolic steps, $A \rightarrow B \rightarrow C$. The intermediate, B, sometimes reaches the second active site by diffusion, but more often it is channeled from one site to the other or covalently bound to a mobile arm of the enzyme. In the latter cases the intermediate need not accumulate in the cytoplasm and does not equilibrate with the cytoplasmic metabolite pools generated via other pathways. The net result is compartmentation of metabolic processes; one metabolite pool may be channeled strictly for biosynthetic purposes while another is utilized exclusively for catabolic purposes. While most well-characterized multifunctional proteins have biosynthetic functions in the cell, some play catabolic or oxidative roles (Kirschner and Bisswanger, 1976).

A further degree of complexity in macromolecular organization is introduced by the multienzyme complex (Ginsburg and Stadtman, 1970). These large protein aggregates may include, in addition to simple catalytic proteins, multifunctional and/or classic allosteric enzymes [i.e., the pyruvate dehydrogenase complex discussed in Section V,B or the tryptophan synthetase complex in various bacterial and eukaryotic species (Crawford, 1975)]. In some cases, such as the fatty acid synthetase of yeast, catalytic function may depend upon the integrity of the complex. In such complexes, the dissociated components lack enzymatic activity, presumably because specific aggregative associations stabilize the active configurations of the multiple enzymes. As noted above for multifunctional proteins, channeling of metabolic intermediates through a multienzyme complex allows efficient utilization of the primary substrate for one purpose while eliminating competition with other anabolic or catabolic pathways. Metabolic compartmentation results from the physical separation of catalytic aggregates.

In some multienzyme complexes the enzyme aggregate shows no enhancement of activity over that of the constituent proteins (Ginsburg and Stadtman, 1970). Instead, complexation confers regulatory properties on the individual enzymes. In such instances, regulation of one enzyme might occur through the binding of a metabolite to the catalytic site of another subunit in the complex. Ligand-induced changes in the conformation of the latter enzyme might be transmitted to the former by heterologous interactions. Thus, effector molecules have been shown to have both stimulatory and inhibitory effects on reactions catalyzed by the aggregate, even though the separated components are insensitive to regulation.

Finally, intracellular multienzyme systems may be still more highly organized, being integrated with other parts of the cellular metabolic machinery in the membrane. Experimental evidence is available to suggest that in some eukaryotic cells, very few of the proteins which comprise the complement of cellular catalytic macromolecules are free in the cytoplasm (Kempner and Miller, 1968). The same may be true in bacteria. A largely unprobed field for future investigation deals with the understanding of the biogenic process and physical forces responsible for the assembly of enzymes in multienzyme complexes in tight or loose association with membraneous structures.

VI. Regulation of Nitrogen, Phosphorus, and Sulfur Metabolism in Bacteria

In Section IV,A, it was noted that catabolic enzyme systems which function in the acquisition of carbon and energy are, in general, subject to control by cyclic AMP. Cyclic AMP is the cytoplasmic indicator of carbon insufficiency. In a similar fashion it has been suggested that distinct second messengers, or "alarmones" may exist to signal a deficiency of utilizable nitrogen, phosphorus, or sulfur (B. N. Ames, personal communication). These cytoplasmic messengers could either be small molecules, such as cyclic AMP, or they could be proteins which can exist in alternate states. In the latter case, the two states of the protein might exert opposing regulatory effects. Alternatively, only one of the two forms of the protein might function as a regulatory agent, the other (modified) form being inactive. It has been suggested that the regulation of nitrogen metabolism occurs by just such a mechanism where the central regulatory protein is glutamine synthetase. This enzyme can exist in two states, one of which is catalytically active under physiological conditions and may function in the regulation of enzyme synthesis; the alternative state, in which the enzyme is adenylated, is viewed as being inactive in both capacities.

A. Possible Involvement of Glutamine Synthetase in the Control of Nitrogen Metabolism

Work in several laboratories has established that the activity of glutamine synthetase is subject to regulation by several mechanisms, one of which involves enzymatic adenylation of the protein (Stadtman and Ginsburg, 1974). Recent studies, particularly in Magasanik's laboratory, have provided evidence for a direct involvement of glutamine synthetase in the transcriptional regulation of operons involved in nitrogen metabolism in species of *Klebsiella*. A specific formulation of the mechanism by which synthesis of the nitrogen metabolic enzymes is controlled by ammonium availability has been advanced, and substantial experimental evidence exists to support this postulate (Magasanik *et al.*, 1974). However, other evidence suggests that the proposed mechanism is either incorrect (in part or in entirety) or that the situation is more complex than proposed. In the paragraphs below, the properties of the proteins involved in glutamine synthetase regulation will be discussed and the proposed mechanism for their involvement in transcriptional regulation will be briefly presented. Subsequently, the data which bring this proposal into question or suggest a higher degree of complexity will be considered.

Glutamine synthetase catalyzes the reaction which is responsible for the fixation of ammonium ions as organic nitrogen. Evidence has been presented suggesting that in *Klebsiella*, it also functions as a positive control element for the transcription of genes which code for enzymes catalyzing the utilization of nitrogen-containing compounds. Among the enzymes which may be subject to direct regulation by glutamine synthetase are (a) histidase and the other enzymes involved in histidine utilization, (b) proline oxidase which initiates the sequence of reactions which permits proline degradation, (c) glutamine synthetase itself, and (d) nitrogenase which is the enzyme complex in *Klebsiella* and other nitrogen-fixing bacteria which catalyzes the reduction of molecular nitrogen to ammonium ions. Some of the enzymes which appear to be subject to glutamine synthetase regulation in *Klebsiella* are also under the control of the cyclic AMP receptor protein. Conditions of either carbon or nitrogen starvation allows synthesis of these enzymes in the presence of an appropriate inducer molecule. The simultaneous but alternate induction of the expression of these enzyme systems by either nitrogen or carbon limitation can be easily rationalized since metabolism of the corresponding compounds (i.e., histidine and proline) yields both nitrogen and carbon in a utilizable form. Syntheses of enzymes which are repressed by NH_4^+ but which do not act on carbon-containing compounds are not subject to cyclic AMP control. Nitrogenase is an example of an enzyme whose expression is subject to NH_4^+ repression but not to cyclic AMP control.

When NH_4^+ is present in the culture medium in quantities which exceed the needs of the cell, α-ketoglutarate (αKG) is converted to glutamine (Gln) with glutamic acid (Glu) as an intermediate as shown in Eqs. (1–4). When

$$\text{Glu} + \text{NH}_4^+ + \text{ATP} \xrightarrow[\text{synthetase}]{\text{glutamine}} \text{Gln} + \text{ADP} + \text{P}_i \tag{1}$$

$$\alpha\text{KG} + \text{Gln} + \text{NADPH} \xrightarrow[\text{synthase}]{\text{glutamate}} 2\ \text{Glu} + \text{NADP}^+ \tag{2}$$

$$\text{Glu} + \text{NH}_4^+ + \text{ATP} \xrightarrow[\text{synthetase}]{\text{glutamine}} \text{Gln} + \text{ADP} + \text{P}_i \tag{3}$$

$$\alpha\text{KG} + 2\ \text{NH}_4^+ + 2\ \text{ATP} + \text{NADPH} \longrightarrow \text{Gln} + 2\ \text{ADP} + 2\ \text{P}_i + \text{NADP}^+ \tag{4}$$

ammonium ions are present in the medium in limiting amounts, glutamine is utilized for biosynthetic purposes. Under these conditions, the glutamate synthase reaction would be expected to reverse to give α-ketoglutarate. Thus, the ratio of α-ketoglutarate to glutamine may be a measure of the availability of cellular ammonium and function as a second messenger or "alarmone" to signal nitrogen depletion. Many of the details determining how the ratio of these compounds controls the physical state of glutamine synthetase have recently come to light (Stadtman and Ginsburg, 1974).

Glutamine synthetase can exist in two forms, one of which is enzymatically active, the other which is not. The ratio of these two forms of the enzyme is responsive to the availability of NH_4^+, i.e., to the ratio of α-ketoglutarate to glutamine. The mechanism by which the relative concentrations of these metabolites control glutamine synthetase depends on the presence of four (or possibly five) proteins, two of which catalyze protein derivatization reactions and two of which are themselves derivatized. One of these proteins may catalyze both the uridylation and deuridylation of a protein, designated P_{II}, which functions as an activator of the second protein modification enzyme. Some evidence suggests that the uridylyltransferase and the uridylyl-removing enzyme are distinct proteins. The enzyme which is regulated by the P_{II} protein catalyzes adenylation and deadenylation of glutamine synthetase. Both the uridylation and adenylation reactions are allosterically controlled by α-ketoglutarate and glutamine. Table III summarizes some of the pertinent information.

The uridylyltransferase–uridylyl-removing enzyme is a protein complex (or possibly two separate enzymes) of 160,000 molecular weight which catalyzes two reactions, the uridylation of the small dimeric protein (P_{II}) with UTP as substrate and the hydrolysis of the derivatized protein to give UMP and the unmodified form of the protein. α-Ketoglutarate allosterically activates and glutamine allosterically inhibits the first reaction, while the hydrolysis reaction has been reported to be insensitive to both compounds (Stadtman and Ginsburg, 1974).

As noted above, the substrate protein of the uridylation reaction, the P_{II} protein, is known to be an activator of the enzyme which catalyzes the derivatization of glutamine synthetase. The free form of P_{II} stimulates the adenylation reaction, while $P_{II}(UMP)_2$ stimulates the deadenylation reaction. In the adenylation reaction, each of the twelve identical subunits of the glutamine synthetase complex can be derivatized with the AMP moiety covalently linked to a tyrosyl residue in the protomer. In the deadenylation reaction, the AMP residues are transferred from the enzyme to inorganic phosphate with the synthesis of ADP as a by-product. As for the uridylation reactions, both adenylation and deadenylation appear to be catalyzed by a single enzyme complex (MW = 130,000) which is subject to allosteric control: the first reaction is stimulated by glutamine and inhibited by α-ketoglutarate, while the reverse is true for the second reaction (Table III). As a result of the allosteric regulatory properties of these enzymes, we can picture the sequence of events which may sensitize the cell to ammonium availability (Fig. 12).

Ammonium depletion causes an increase of the cellular α-ketoglutarate concentration and a concomitant decrease in the concentration of glutamine. The increased concentration of α-ketoglutarate in the cell allosterically activates the uridylyltransferase so that the adenylyltransferase activator protein, P_{II}, is uridylated. This derivatized protein then combines with the adenylyltransferase which also binds α-ketoglutarate, and the binding of these two ligands to allosteric regulatory sites on the enzyme converts it to a form which rapidly deadenylates glutamine synthetase, converting it to

TABLE III

PROPERTIES OF THE PROTEIN CONSTITUENTS OF THE GLUTAMINE SYNTHETASE REGULATORY CASCADE

Enzyme	Abbreviation	Molecular weight	Molecular forms	Reactions catalyzed	Allosteric activators	Allosteric inhibitors
Uridylyltransferase	UTase	160,000 (UTase and UR enzyme)		$P_{II} \xrightarrow{2\ UTP} P_{II}(UMP)_2$ + $2\ P_2$	αKG, ATP	Gln, P_i
Uridylyl-removing enzyme	UR enzyme			$P_{II}(UMP)_2 \xrightarrow{2\ H_2O} P_{II}$ + 2 UMP	—	—
ATase regulatory protein	P_{II}	50,000	P_{II}	Activates adenylation of GS		
			$P_{II}(UMP)_2$	Activates deadenylation of GS	—	—
Adenylyltransferase	ATase	130,000		$GS \xrightarrow{12\ ATP} GS(AMP)_{12}$ + $12\ P_2$	Gln, P_{II}	α-KG
				$GS(AMP)_{12} \xrightarrow{12\ P_i} GS$ + 12 ADP	α-KG	Gln
Glutamine synthetase	GS	600,000	GS $GS(AMP)_{12}$	Active form Inactive form	—	Histidine, tryptophan, AMP, CTP, glucosamine-6-P carbamyl-P

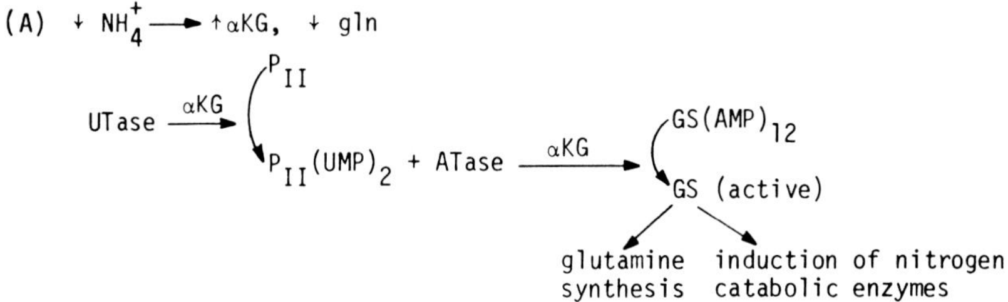

FIG. 12. Proposed mechanism for the control of nitrogen metabolism by glutamine synthetase in bacteria. (A) Ammonium deprivation increases the cellular concentration of α-ketoglutarate (αKG) and decreases the concentration of glutamine (Gln). This leads to a cascade of events which converts glutamine synthetase to the nonadenylated active form. (B) Ammonium availability causes a decrease in the cellular concentration of α-ketoglutarate and a concomitant increase in the concentration of glutamine. The end result is adenylation of glutamine synthetase to the inactive state. The abbreviations used are as indicated in Table III.

the form which is active under physiological conditions with respect to glutamine synthesis. The same form of the enzyme has been postulated to allow induction of the transcription of the nitrogen catabolic enzymes (Magasanik *et al.*, 1974).

Ammonium availability, on the other hand, causes a decrease in the α-ketoglutarate concentration and a simultaneous increase in the glutamine concentration. The sequence of events depicted in Fig. 12B then follows. The uridylyl-removing enzyme hydrolyzes the derivatized form of the adenylyltransferase activator, $P_{II}(UMP)_2$, to the free form of this protein, and the latter, together with glutamine, combines with the adenylyltransferase to stimulate allosterically the adenylation of glutamine synthetase. This cascade type of regulation has several consequences relative to a simpler mode of control: (1) It allows for amplification, i.e., a single uridylyltransferase can control the activities of numerous glutamine synthetase molecules. (2) It increases the number of compounds which can regulate glutamine synthetase. (3) It allows greater sensitivity to a single regulatory agent, such as α-ketoglutarate or glutamine. The net result may be an elaborate mechanism whereby ammonium availability influences the rates of transcription of genes coding for the nitrogen catabolic enzymes. The cytoplasmic messengers are α-ketoglutarate and glutamine. The proteins which sense these compounds are the uridylyl- and adenylyltransferases. The central regulatory protein which is postulated to interact with the promoter region of the sensitive operons is glutamine synthetase (Magasanik *et al.*, 1974).

The mechanism proposed by Magasanik and co-workers, suggesting the involvement of glutamine synthetase in the regulation of nitrogen catabolic enzyme synthesis, does not appear to explain readily all of the pertinent experimental observations. First, mutations in the newly discovered *glnF* gene have been shown to render glutamine synthetase synthesis insensitive to ammonium repression (high level constitutive). The mutations did not map near the structural gene for glutamine synthetase (Garcia *et al.*, 1977). The isolation of these mutants suggests the presence of a positive control element (in addition to or instead of glutamine synthetase), which regulates expression of the glutamine synthetase structural gene. Second, mutations in or near the glutamine synthetase structural gene have been isolated which eliminate glutamine synthetase activity and prevent the synthesis of material which cross-reacts with anti-glutamine synthetase antibody. These mutations frequently render synthesis of nitrogenase and other nitrogen catabolic enzymes constitutive rather than noninducible as expected from the proposed model (Shanmugam *et al.*, 1977). Since no cross-reacting material is detected immunologically in these mutants, it is difficult to argue that an enzymatically inactive polypeptide chain derived from glutamine synthetase is functionally active in promoting transcription. Third, it has been shown that mutants which synthesize nitrogenase constitutively in the presence of excess ammonium ions are nevertheless subject to strong repression by combinations of amino acids (Shanmugam and Morandi, 1976). Finally, it has been shown that mutants which are altered in glutamyl-tRNA synthetase exhibit derepressed levels of glutamate synthase and glutamine synthetase in *E. coli* (La Pointe *et al.*, 1975). These observations must be accounted for in any model which purports to explain transcriptional regulation of the nitrogen catabolic enzymes. They lead to the possibility that the regulatory interactions which control gene expression are more complex than previously supposed (Magasanik *et al.*, 1974).

B. Regulation of Phosphate Metabolism in *Escherichia coli*

A large body of experimental evidence concerning expression of the genes coding for the enzymes involved in carbon and nitrogen metabolism led to very specific postulates regarding the molecular mechanisms responsible for transcriptional regulation. The postulates resulted from a synthesis of genetic, physiological, and biochemical evidence and were confirmed by direct demonstration of the regulatory interactions in cell-free preparations (Magasanik *et al.*, 1974; Pastan and Adhya, 1976). While much information has accumulated regarding the synthesis of enzymes involved in phosphate metabolism, particularly at the genetic level, there is still insufficient information available to allow confident postulation of the molecular details of the regulatory process. In this section, we shall nevertheless consider this evidence which, interestingly, leads to the possibility (discussed in Sections IV,B and IV,C) that permease proteins and extracellular ligands may function directly in the regulation of gene expression. An hypothesis which appears to explain many of the experimental observations will be discussed.

1. Physiological Studies on the Regulation of Alkaline Phosphatase Synthesis

Alkaline phosphatase is localized in the periplasmic space of the *E. coli* cell, between the cytoplasmic and outer membranes (see Fig. 1). It functions to scavenge inorganic phosphate (P_i) from phosphomonoesters external to the cytoplasmic membrane during growth under conditions of limiting exogenous P_i. In wild-type cells, the rate of alkaline phosphatase synthesis is enhanced as much as one thousandfold over the basal repressed rate when exogenous P_i is depleted. Under these conditions, up to 6% of the total cell protein is alkaline phosphatase. In chemostat cultures, the external P_i concentration must drop below 10 μM before appreciable derepression of alkaline phosphatase synthesis occurs.

Exactly how P_i functions in alkaline phosphatase regulation is not known, but several studies have led to the tentative conclusion that intracellular P_i does not function as a corepressor (Wilkins, 1972; Willsky and Malamy, 1976). Some investigators have postulated that low external P_i concentrations as well as defects in P_i transport might result in low internal P_i levels which indirectly (perhaps through the accumulation or depletion of the "true effector") regulate alkaline phosphatase synthesis (Wilkins, 1972). However, there is evidence that internal P_i need not be lowered for induction of alkaline phosphatase synthesis to occur (Willsky and Malamy, 1976).

2. Genetic Studies on the Regulation of Alkaline Phosphatase

Alkaline phosphatase synthesis is apparently subject to both positive and negative control in a complicated fashion. Mutations affecting synthesis of this enzyme are listed in Table IV. These mutations are briefly discussed below.

a. phoA. The *phoA* gene is the structural gene for alkaline phosphatase and, therefore, is of little interest from the standpoint of regulation.

TABLE IV
GENES WHICH AFFECT THE SYNTHESIS OF ALKALINE PHOSPHATASE IN *E. coli*

Gene designation	Map position	Gene product	Properties of the mutants
phoA	9.5	Alkaline phosphatase	Alkaline phosphatase is not active
phoB	9.6	Positive regulatory protein	Alkaline phosphatase, the P_i binding protein, and 2 other periplasmic proteins (which are repressed by P_i) are not synthesized
phoR	9.7	Cytoplasmic regulatory protein	Alkaline phosphatase synthesis is *partially* constitutive (low level)
phoS	73.4	Periplasmic P_i binding protein of phosphate-specific transport system	Alkaline phosphatase synthesis is fully constitutive but can be repressed slightly by high external P_i in most mutants
phoT	73.4	Essential component of phosphate-specific transport system	Regulatory properties are the same as for *phoS*

b. phoB. The *phoB* gene product is necessary for transcription of *phoA* both *in vivo* and *in vitro*. The mutant gene is recessive to *phoB*$^+$ but epistatic over *phoR*, *phoS*, and *phoT*. Alkaline phosphatase expression does not respond to an increase in the dosage of this gene. Mutation in the *phoB* gene prevents synthesis of alkaline phosphatase, the P_i-binding protein, and two periplasmic proteins of unknown function. Mutations which have been designated *phoRcl* give a phenotype which is similar to that of *phoB* mutants except that the *phoS* gene product (the periplasmic P_i-binding protein) is synthesized. A recent study indicates that the *phoB* and *phoRcl* mutations belong to a single complementation group (Pratt and Torriani, 1977). This result leads to the suggestion that there is a single positive control gene for alkaline phosphatase synthesis.

c. phoR. phoR mutants synthesize alkaline phosphatase constitutively at 10–50% of the fully derepressed level. The protein product of this gene, therefore, appears to be required for maximal induction as well as for full repression of the enzyme. *phoR* is epistatic over *phoS* and *phoT*.

d. phoS and *phoT*. These genes code for two constituents of the phosphate-specific transport system, and both types of mutations render alkaline phosphatase synthesis constitutive at a high level. However, repression (0–80%) by high concentrations of external P_i may occur. Both mutant alleles are recessive to the wild-type alleles. It does not appear that transport function alone can account for the involvement of these gene products in the regulation of alkaline phosphatase synthesis.

3. Possible Mechanism of Alkaline Phosphatase Regulation

Based on the physiological and genetic data summarized above, a speculative mechanism involving cytoplasmic and membrane-associated proteins can be entertained (Fig. 13). According to this model, there are two cytoplasmic regulatory proteins, B and R, the products of the *phoB* and *phoR* genes, respectively. These proteins (together as a complex or independently) possess affinity for the promoter regions of operons which are subject to P_i regulation. In addition, these proteins (as a complex) can bind to the T protein, assumed to be an integral membrane constituent of the phosphate-specific transport system. The T protein is presumed to span the cytoplasmic membrane and to interact with the product of the *phoS* gene, the S protein, on the external surface. The affinity of the B–R complex for the T protein is determined by the conformation of the latter, and T protein conformation is presumed to be influenced by the binding of ligand (P_i) to the S protein.

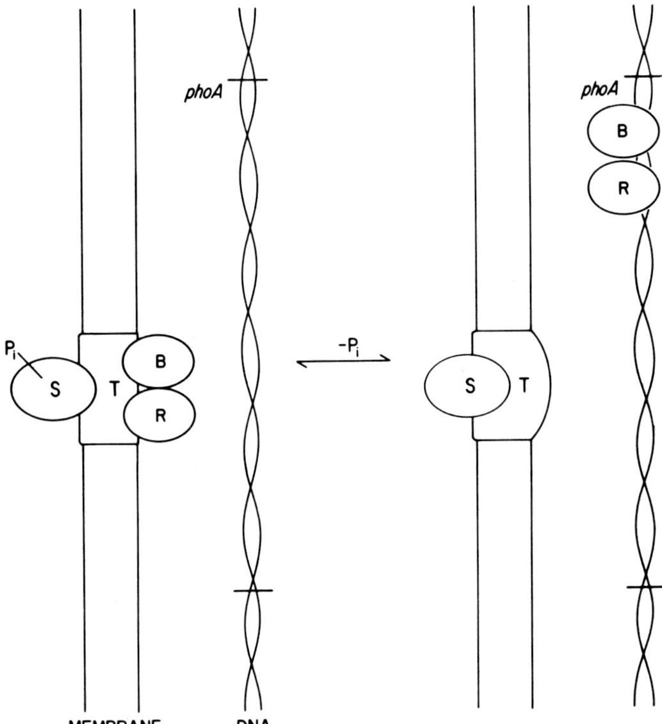

Fig. 13. Postulated mechanism for alkaline phosphatase regulation in wild-type *E. coli* cells. The letters indicated above refer to the protein products of the corresponding genes. S, product of the *phoS* gene; T, product of the *phoT* gene; B, product of the *phoB* gene; R, product of the *phoR* gene.

Thus, when P_i is bound to S, T assumes a conformation which binds B–R with high affinity. However, when extracellular P_i concentrations fall below the value required to saturate the S protein, dissociation of P_i from S occurs, the S protein assumes an altered conformation, and this conformational change induces a change in the T protein which lowers its affinity for the B–R complex. The B and R proteins are then free to bind to appropriate promoter regions on the DNA to promote or otherwise regulate transcription. Such a model requires that the cellular concentrations of the S and T proteins are present in excess of the B and R proteins, even under repressing conditions. It further assumes that the binding of the B–R complex to the phosphate-specific transport system does not inhibit its transport activity. It should be noted that other mechanisms can be proposed which account for most of the published data. However, the model depicted in Fig. 13 is of interest with respect to the possibilities discussed in Sections IV,B and IV,C that permease proteins and their exogenous substrates may play a direct role in the regulation of gene expression. These subjects may prove to be exciting areas for future investigation.

C. Regulation of Sulfur Metabolism in *Salmonella typhimurium*

Too little information is available to warrant the conclusion that sulfur metabolism, like that of carbon and nitrogen, is subject to regulation by internal "alarmones." However, the pathway for sulfate assimilation in *E. coli* and *S. typhimurium* has been shown to involve (a) transmembrane sulfate permeation, (b) two activation steps, the first yielding adenosine 5′-phosphosulfate and the second producing 3′-phosphoadenosine 5′-phosphosulfate, (c) two reductive reactions which convert 3′-phosphoadenosine 5′-phosphosulfate first to sulfite and then to sulfide, and (d) displacement of the acetyl group from *O*-acetyl-L-serine by sulfide to give L-cysteine. The enzymes which catalyze these reactions are all subject to derepression by sulfur starvation and to repression by cysteine or sulfide. Derepression is dependent on internal *O*-acetyl-L-serine and the product of the *cysB* gene, but sulfide and cysteine block the inductive effect of exogenously added *O*-acetylserine (Kredich, 1971). It is, therefore, probable that expression of the *cys* regulon, which consists of at least four operons on the *Salmonella* chromosome, is subject to both positive and negative control. Normally *O*-acetylserine, sulfur starvation, and an intact *cysB* gene are required for derepression of the *cys* regulon.

Genetic and physiological analyses have shown that certain mutations in the *cysB* region lead to partial constitutivity of the cysteine biosynthetic enzymes in the presence or absence of sulfide, cysteine, or *O*-acetylserine

(Cheney and Kredich, 1975). This observation leads to the possibility that the *cysB* gene product may function both in the positive expression of the regulon, controlled by *O*-acetylserine, and in negative control by cysteine or sulfide. Available evidence suggests that the *cysB* region consists of a single cistron which probably codes for an oligomeric protein (Cheney and Kredich, 1975). Assuming this to be the case, the conformation of the protein might be controlled by ligand interactions at multiple sites. Further studies will be required to establish the detailed mechanism by which sulfur assimilation is regulated in bacteria.

VII. Concluding Remarks

This chapter has emphasized those membrane-associated processes which determine the concentrations of the cellular constituents which together comprise the intrabacterial milieu. We have seen that nonmetabolizable ions (protons, alkali metal cations, inorganic anions, etc.) are acted upon by a variety of transmembrane active transport systems. These charged species are either accumulated within the cytoplasm against concentration gradients or are actively extruded from the cell by mechanisms which involve the expenditure of metabolic energy (Section III). The activity gradients generated create electrochemical and osmotic conditions essential to life. The cellular concentrations of metabolizable nutrients are determined, in part, by the activities of elaborate extracellular nutrient detection devices (Section II) and also by the rates of transmembrane solute permeation (Section III) and by the velocities of cytoplasmic metabolite turnover (Sections IV and V). The permeases and metabolic enzymes are, in turn, subject to a variety of regulatory constraints which prevent the useless expenditure of metabolic fuel and ensure that no metabolite will be synthesized or degraded at rates which exceed the needs of the cell (Section V). The synthesis of cellular macromolecular species, particularly of the catalytic elements of the metabolic system, are also subject to stringent control, in part, by virtue of regulatory constraints imposed on the activities of the solute-specific permeases and metabolic enzymes (Sections IV and VI). This elaborate hierarchy of regulatory interactions not only renders the cell an efficient homeostatic system for rapid growth and reproduction under favorable conditions but also confers upon the organism sensitivity to environmental stimuli and potential for change in response to a shift to less favorable conditions.

With regards to this last possibility, it is worth noting that a variety of bacterial species undergo differentiative changes, usually in response to nutrient deprivation. Thus, sporulation in species of *Bacillus* and numerous other bacteria involves a temporal program of events (some reversible, others apparently irreversible) which can be triggered by carbon depletion, phosphate limitation, or amino acid starvation. It has been suggested that the sensing device leading to the initiation of sporulation consists of a membrane-associated enzyme system which synthesizes highly phosphorylated nucleotides (Rhaese and Groseurth, 1976). The synthesis of these cytoplasmic messengers on the membrane surface may be responsive to cytoplasmic as well as extracellular stimuli (Rhaese and Groseurth, 1976).

A much more elaborate type of sporulation process is observed with species of *Myxobacteria*. In these organisms complex developmental sequences involve intercellular communication and ultimately result in the construction of elaborate multicellular fruiting bodies (Wireman and Dworkin, 1975). Both temporal and spatial programs of gene expression must, therefore, be operative. A striking parallel between the progression of events leading to fruiting in *Myxobacteria* and that in the eukaryotic cellular slime molds has been noted (Wireman and Dworkin, 1975).

A very different type of differentiation is observed in photosynthetic blue-green bacteria such as *Anabaena* and *Nostoc*. In response to nitrogen deprivation, strands of these bacteria synthesize, at regular intervals along the strand, nonphotosynthetic nitrogen-fixing cells called heterocysts. Heterocyst development may be the form of prokaryotic differentiation which most resembles differentiation in higher eukaryotes because it involves both temporal and spatial differentiation and appears to be irreversible. A mature heterocyst may be incapable of division and probably cannot dedifferentiate to the vegetative state. Since nitrogen fixation can only occur in an anaerobic environment and since the vegetative cells produce oxygen, a special structure, capable of excluding or rapidly consuming oxygen, must be constructed. The synthesis of this structure and the catalytic components of the heterocyst involves a sequence of programmed events, possibly triggered by glutamine synthetase (see Section VI,A). The program has been "visualized" by electron microscopy, by following the synthesis of individual proteins during heterocyst development by sodium dodecyl sulfate–polyacrylamide gel electrophoresis, and by following the appearance and disappearance of enzymatic activities as a function of time after ammonium removal (Fleming and Haselkorn, 1974).

Single proheterocysts develop within a strand of vegetative *Anabaena* cells with a regular spacing of about one heterocyst per ten vegetative photosynthetic cells. If the proheterocyst is destined to develop in a position near a chain terminus, it arises 70% of the way from the last heterocyst to the end of the strand. In either case, the heterocyst distribution curve is tighter than gaussian, suggesting that spatial differentiation is regulated. To explain these and other observations, it has been postulated that the heterocyst acquires two new traits: first, the ability to produce an inhibitor of heterocyst development, and, second, resistance to the action of the inhibitor. The inhibitor is presumed to diffuse through the strand of vegetative cells and to be lost or inactivated at a constant rate. The new heterocyst will develop from the vegetative cell which is first exposed to a concentration of the inhibitor which is below the threshold for inhibition of proheterocyst formation (Wilcox *et al.*, 1973).

The examples of bacterial differentiation cited above undoubtedly involve the temporally (and spatially) programmed expression and silencing of hundreds of chromosomal genes. While the molecular mechanisms responsible for the switching of genes on and off are only now beginning to be understood (Zieg *et al.*, 1977), one prediction regarding the initiation process can be made with some confidence. The same types of regulatory mechanisms which render gene expression responsive to conditions of nutrient deprivation and sufficiency are likely to be operational in initiating complex programs of development. Internal concentrations of metabolites and external environmental stimuli will frequently be sensed by membrane "receptor" proteins, and the signals generated will be transmitted to the nucleoid as postulated in several specific instances in this chapter. It can be anticipated that at a fundamental level, the principles established with one biological system will be universally applicable across phylogenetic lines, within eukaryotic as well as prokaryotic kingdoms (Saier and Stiles, 1975).

REFERENCES

Aboud, M., and Pastan, I. (1973). *J. Biol. Chem.* **248,** 3356–3358.

Adler, J., and Epstein, W. (1974). *Proc. Natl. Acad. Sci. U.S.A.* **71,** 2895–2899.

Atkinson, D. E. (1968). *Biochemistry* **7,** 4030–4034.

Belaich, A., Simonpietri, P., and Belaich, J.-P. (1976). *J. Biol. Chem.* **251,** 6735–6738.

Berg, H. C. (1975). *Annu. Rev. Biophys. Bioeng.* **4,** 119–136.

Berger, E. A., and Heppel, L. A. (1974). *J. Biol. Chem.* **249,** 7747–7755.

Bogomolni, R. A., and Stoeckenius, W. (1974). *J. Supramol. Struct.* **2,** 775–780.

Boos, W., Hartig-Beecken, I., and Altendorf, K. (1977). *Eur. J. Biochem.* **72,** 571–581.

Calhoun, D. H. (1975). *Annu. Rev. Microbiol.* **29,** 275–299.

Castro, L., Feucht, B. U., Morse, M. L., and Saier, M. H., Jr. (1976). *J. Biol. Chem.* **251,** 5522–5527.

Cheney, R. W., Jr., and Kredich, N. M. (1975). *J. Bacteriol.* **124,** 1273–1281.

Contesse, G., Crépin, M., Gros, F., Ullmann, A., and Monod, J. (1960). *In* "The Lactose Operon" (J. Beckwith and D. Zipser, eds.), pp. 401–415. Cold Spring Harbor Lab., Cold Spring Harbor, New York.

Crawford, I. P. (1975). *Bacteriol. Rev.* **39,** 87–120.

Decad, G. M., and Nikaido, H. (1976). *J. Bacteriol.* **128,** 325–336.

Dietz, G. W. (1976). *Adv. Enzymol. Relat. Areas Mol. Biol.* **44,** 237–259.

Fleming, H., and Haselkorn, R. (1974). *Cell* **3,** 159–170.
Futai, M. (1974). *J. Membr. Biol.* **15,** 15–18.
Garcia, E., Bancroft, S., Rhee, S. G., and Kustu, S. (1977). *Proc. Natl. Acad. Sci. U.S.A.* **74,** 1662–1666.
Ginsburg, A., and Stadtman, E. R. (1970). *Annu. Rev. Biochem.* **39,** 429–472.
Harold, F. M. (1974). *Ann. N. Y. Acad. Sci.* **227,** 297–311.
Harold, F. M. (1977). *In* "The Bacteria" (J. R. Sokatch and L. N. Ornston, eds.), Vol. VI, Ch. 5. Academic Press, New York.
Harold, F. M., and Levin, E. (1974). *J. Bacteriol.* **117,** 1141–1148.
Hasting, J. W., and Nealson, K. E. (1977). *Annu. Rev. Microbiol.* **31,** 549–595.
Henderson, R., and Unwin, P. N. T. (1975). *Nature* (*London*) **257,** 28–32.
Hochstadt-Ozer, J. (1972). *J. Biol. Chem.* **247,** 2419–2426.
Hofnung, M., Hatfield, D., and Schwartz, M. (1974). *J. Bacteriol.* **117,** 40–47.
Hua, S.-S. and Markovitz, A. (1974). *Proc. Natl. Acad. Sci. U.S.A.* **71,** 507–511.
Kaback, H. R. (1972). *Biochim. Biophys. Acta* **265,** 367–416.
Kaback, H. R., and Stadtman, E. R. (1968). *J. Biol. Chem.* **243,** 1390–1400.
Keeler, D. K., Feucht, B. U., and Saier, M. H., Jr. (1977). *Fed. Proc. Fed. Am. Soc. Exp. Biol.* **36,** 685.
Kempner, E. S., and Miller, J. H. (1968). *Exp. Cell Res.* **51,** 141–149.
Kier, L. D., Weppelman, R., and Ames, B. N. (1977). *J. Bacteriol.* **130,** 420–428.
Kirschner, K., and Bisswanger, H. (1976). *Annu. Rev. Biochem.* **45,** 143–166.
Kredich, N. M. (1971). *J. Biol. Chem.* **246,** 3474–3484.
La Pointe, J., Deleuve, G., and Duplain, L. (1975). *J. Bacteriol.* **123,** 843–850.
Lin, E. C. C. (1970). *Annu. Rev. Genet.* **4,** 225–262.
Lin, P., Saier, M. H., Jr., and Nealson, K. E. (1976). *Fed. Proc., Fed. Am. Soc. Exp. Biol.* **35,** 1361.
Mackie, G., and Wilson, D. B. (1972). *J. Biol. Chem.* **247,** 2973–2978.
Magasanik, B. (1970). *In* "The Lactose Operon" (J. Beckwith and D. Zipser, eds.), pp. 189–219. Cold Spring Harbor Lab., Cold Spring Harbor, New York.
Magasanik, B., Prival, M. J., Brenchley, J. E., Tyler, B. M., De Leo, A. B., Streicher, S. L., Bender, R. A., and Paris, C. G. (1974). *Curr. Top. Cell Regul.* **8,** 119–138.
Makman, R. S., and Sutherland, E. W. (1965). *J. Biol. Chem.* **240,** 1309–1314.
Mawe, R. C., and Hempling, H. G. (1965). *J. Cell. Comp. Physiol.* **66,** 95–104.
Nakae, T. (1976). *Biochem. Biophys. Res. Commun.* **71,** 877–884.
Nath, K., and Koch, A. L. (1970). *J. Biol. Chem.* **245,** 2889–2990.
Nealson, K. E., Eberhard, A., and Hastings, J. W. (1972). *Proc. Natl. Acad. Sci. U.S.A.* **69,** 1073–1076.
Nikaido, H. (1976). *Biochim. Biophys. Acta* **433,** 118–132.
Ottow, J. C. G. (1975). *Annu. Rev. Microbiol.* **29,** 79–108.
Parnes, J. R., and Boos, W. (1973). *J. Biol. Chem.* **248,** 4436–4445.
Pastan, I., and Adhya, S. (1976). *Bacteriol. Rev.* **40,** 527–551.
Peterkofsky, A. (1976). *Adv. Cyclic Nucleotide Res.* **7,** 1–48.
Postma, P. W., and Roseman, S. (1976). *Biochim. Biophys. Acta* **457,** 213–257.
Pratt, C., and Torriani, A. (1977). *Genetics* **85,** 203–208.
Rader, R. L., and Hochstadt, J. (1976). *J. Bacteriol.* **128,** 290–301.
Rephaeli, A. W., and Saier, M. H., Jr. (1976). *J. Bacteriol.* **127,** 120–127.
Rephaeli, A. W., Lin, P. H., and Saier, M. H., Jr. (1977). *Fed. Proc., Fed. Am. Soc. Exp. Biol.* **36,** 827.
Rhaese, H.-J., and Groseurth, R. (1976). *Proc. Natl. Acad. Sci. U.S.A.* **73,** 331–335.
Rhoads, D. B., and Epstein, W. (1977). *J. Biol. Chem.* **252,** 1394–1401.

Rothstein, A., Cabantchik, Z. I., and Knauf, P. (1976). *Fed. Proc., Fed. Am. Soc. Exp. Biol.* **35,** 3–10.
Rudnick, G., Schuldiner, S., and Kaback, H. R. (1976). *Biochemistry* **15,** 5126–5131.
Saier, M. H., Jr., (1977). *Bacteriol. Rev.* **41,** 856–871.
Saier, M. H., Jr., and Feucht, B. U. (1975). *J. Biol. Chem.* **250,** 7078–7080.
Saier, M. H., Jr., and Moczydlowski, E. G. (1978). *In* "Bacterial Transport" (B. P. Rosen, ed.), pp. 103–125. Marcel-Dekker, New York.
Saier, M. H., Jr., and Newman, M. J. (1976). *J. Biol. Chem.* **251,** 3834–3837.
Saier, M. H., Jr., and Stiles, C. D. (1975). "Molecular Dynamics in Biological Membranes." Springer-Verlag, Berlin and New York.
Saier, M. H., Jr., Simoni, R. D., and Roseman, S. (1970). *J. Biol. Chem.* **245,** 5870–5873.
Saier, M. H., Jr., Feucht, B. U., and McCaman, M. T. (1975a). *J. Biol. Chem.* **250,** 7593–7601.
Saier, M. H., Jr., Wentzel, D. L., Feucht, B. U., and Judice, J. J. (1975b). *J. Biol. Chem.* **250,** 5089–5096.
Saier, M. H., Jr., Feucht, B. U., and Hofstadter, L. J. (1976). *J. Biol. Chem.* **251,** 883–892.
Saier, M. H., Jr., Feucht, B. U., and Mora, W. K. (1977). *J. Biol. Chem.* **252,** 8899–8907.
Sanwal, B. D. (1970). *Bacteriol. Rev.* **34,** 20–39.
Senior, A. E. (1973). *Biochim. Biophys. Acta* **301,** 249–277.
Shanmugam, K. T., and Morandi, C. (1976). *Biochim. Biophys. Acta* **437,** 322–332.
Shanmugam, K. T., Morandi, C., and Valentine, R. C. (1977). *In* "Iron Sulfur Proteins" (W. Lovenberg, ed.), Vol. 3, pp. 1–14. Academic Press, New York.
Silverman, M., and Simon, M. (1974). *J. Bacteriol.* **117,** 73–79.
Simoni, R. D., and Roseman, S. (1973). *J. Biol. Chem.* **248,** 966–976.
Smolin, D. E., and Umbarger, H. E. (1975). *Mol. Gen. Genet.* **141,** 277–284.
Stadtman, E. R., and Ginsburg, A. (1974). *In* "The Enzymes" (P. D. Boyer, ed.), 3rd Ed., Vol. 10, pp. 755–807. Academic Press, New York.
Stock, J., and Roseman, S. (1971). *Biochem. Biophys. Res. Commun.* **44,** 132–138.
Stoeckenius, W., and Lozier, R. H. (1974). *J. Supramol. Struct.* **2,** 769–774.
Stouthamer, A. (1977). *In* "The Bacteria" (J. R. Sokatch and L. N. Ornston, eds.), Vol. VI, Ch. 8. Academic Press, New York.
Strange, P. G., and Koshland, D. E., Jr. (1976). *Proc. Natl. Acad. Sci. U.S.A.* **73,** 762–766.
Tsuchiya, T., and Rosen, B. P. (1975). *J. Biol. Chem.* **250,** 7687–7692.
Tsuchiya, T., and Rosen, B. P. (1976). *J. Biol. Chem.* **251,** 962–967.
Wilcox, M., Mitchison, G. J., and Smith, R. J. (1973). *J. Cell. Sci.* **12,** 707–723.
Wilkins, A. S. (1972). *J. Bacteriol.* **110,** 616–623.
Willsky, G. R., and Malamy, M. H. (1976). *J. Bacteriol.* **127,** 595–609.
Wilson, D. B. (1976). *J. Bacteriol.* **126,** 1156–1165.
Wilson, T. H., Kashket, E. R., and Kusch, M. (1972). *In* "The Molecular Basis of Biological Transport" (J. F. Woessner, Jr. and F. Huijing, eds.), pp. 219–247. Academic Press, New York.
Winker, U., Schoole, H., and Bohne, L. (1975). *Arch. Mikrobiol.* **104,** 189–196.
Wireman, J. W., and Dworkin, M. (1975). *Science* **189,** 516–523.
Zieg, J., Silverman, M., Hilmen, M., and Simon, M. (1977). *Science* **196,** 170–172.

CHAPTER 5

Roles of Appendages and Surface Layers in Adaptation of Bacteria to Their Environment

J. R. SOKATCH

Nowhere is the adaptation of bacteria to their environment more striking than in the roles played by their organelles and surface layers in dealing with the problems of survival in media which are both life-supporting and hostile. The subjects of motility and chemotaxis have blossomed since the classic review by Weibull in the first volume of this treatise, and much more is known about the several roles of the cell envelope since this treatise first appeared. The structure and functions of bacterial spores and the roles of flagella in chemotaxis, of the cell membrane in transport, and of chromatophores in photosynthesis are covered in chapters by Dworkin, Koshland, Saier, and Dutton and Prince, respectively, in Volumes VI and VII of this series. The chemistry and function of surface layers are discussed in this volume by Tipper and Wright (Chapter 6). This chapter will concentrate on the structure and functions of bacterial appendages and surface layers with as little redundancy as possible with the aforementioned chapters.

ISBN 0-12-307207-7

Mesosomes will not be covered because this subject has been reviewed extensively, most recently by Greenawalt and Whiteside (1975), and also because there is considerable uncertainty about their role and even about their existence (Salton and Owen, 1976).

I. Flagella and Axial Filaments

A. Ultrastructure

Mixing of cell suspensions with a blendor is the oldest method for the preparation of bacterial flagella and axial filaments (Martinez, 1963; see also review in Iino, 1969). Flagellar filaments are sheared off and can be purified by ion exchange chromatography on powdered cellulose, differential centifugation, and density gradient centrifugation. One disadvantage of this procedure is that flagella are broken off at the cell surface leaving part of the structure embedded in the cell. However, ballistic disintegration of *Spirochaeta stenostrepta* by glass beads freed a high proportion of intact axial filaments (Holt and Canale-Parola, 1968). Lysozyme digestion of the cell walls of both gram-positive (Dimmitt and Simon, 1971) and gram-

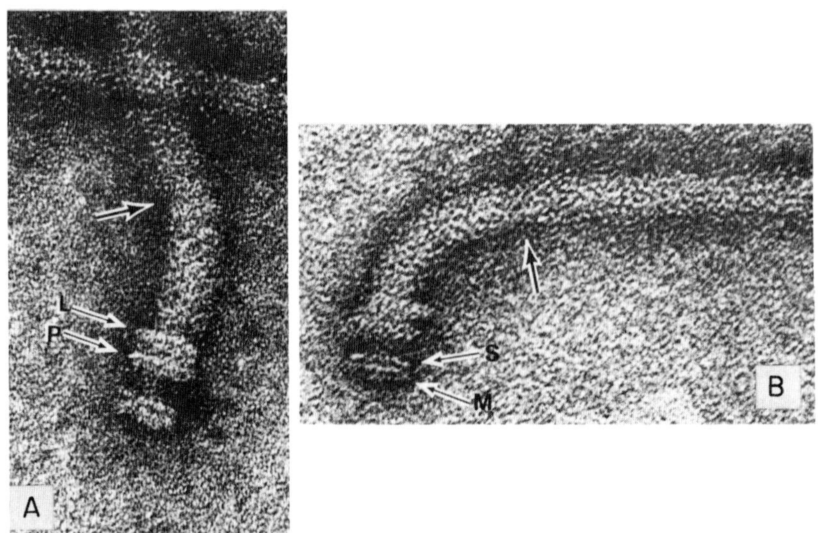

FIG. 1. Filament, hook, and basal body of *Escherichia coli* showing L, P, S, and M rings. In (A) and (B) the arrow marks the juncture between the hook and filament. (From DePamphilis and Adler, 1971b, reproduced with permission of the American Society for Microbiology.)

negative bacteria (DePamphilis and Adler, 1971a) has been used successfully to prepare intact flagella with filaments, hooks, and basal bodies.

Bacterial flagella are composed of three substructures—a helical filament, hook, and a basal body (Fig. 1). DePamphilis and Adler (1971a,b) studied *Escherichia coli* flagella in great detail and similar structures have been described in *Bacillus subtilis* (Dimmitt and Simon, 1971; DePamphilis and Adler, 1971a,b), *Rhodospirillum rubrum* (Cohen-Bazire and London, 1967), *Proteus mirabilis* (Van Iterson *et al.*, 1966), *Vibrio metchnikovii* (Vaituzis and Doetsch, 1969), and the photosynthetic organism *Ectothiorhodospira mobilis* (Remsen *et al.*, 1968). The filament of *E. coli* is approximately 13.5 nm in diameter (Fig. 2) and averages 5 μm in length. The filament of most bacterial flagella is naked, but sheathed flagella occur in *Vibrio* (Glauert *et al.*, 1962; Follett and Gordon, 1963), *Pseudomonas stitzlobii* (Fuerst and Hayward, 1969), and *Bdellovibrio bacteriovorus* (Fig. 3) (Seidler and Starr, 1968). All *Beneckea* species have a sheathed polar flagellum, and several have naked peritrichous flagella in addition, apparently composed of a flagellin distinct from that of the polar flagellum (Baumann and Baumann, 1977; Shinoda *et al.*, 1974).

Flagella are helical with a wavelength of 2–2.5 μm. Pijper *et al.* (1956) first noticed that some bacteria produced flagella with the normal wavelength and also with a wavelength of 1–1.2 μm, or about one-half the normal

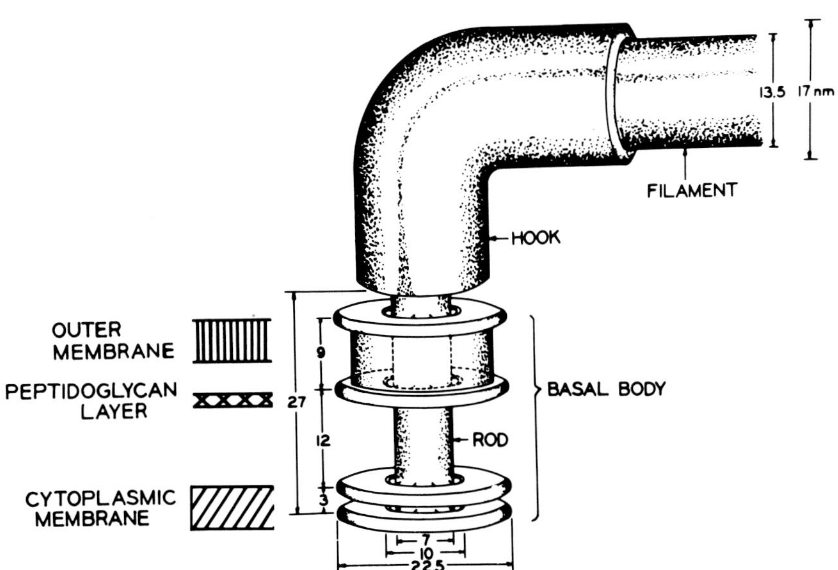

FIG. 2. Diagram of hook and basal body of *Escherichia coli* flagellum. (From DePamphilis and Adler, 1971b, reproduced with permission of the American Society for Microbiology.)

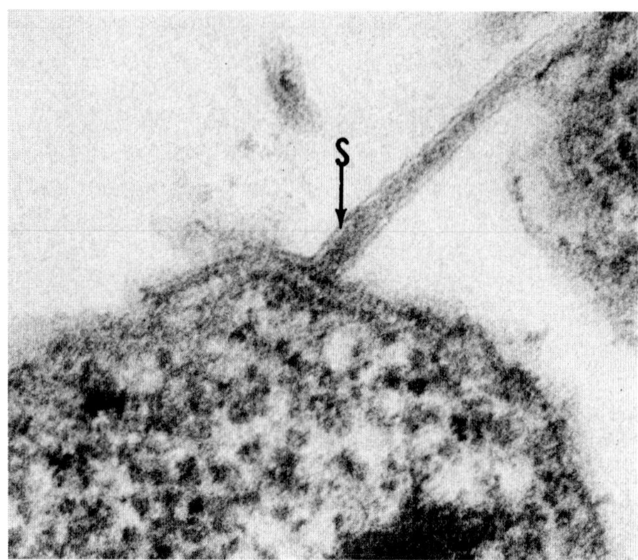

FIG. 3. Sheathed flagellum (S) of *Bdellovibrio bacteriovorus*. This thin section shows that the flagellar sheath is continuous with the cell envelope. (From Seidler and Starr, 1968; reproduced with permission of the American Society for Microbiology.)

pitch, a phenomenon which they named biplicity. Shimada *et al.* (1975) found that isolated flagella of *Salmonella* aggregated into normal bundles with a wavelength of 2.3 μm as well as curly bundles with a wavelength of about 1.1 μm. The normal bundles of flagella were left-handed helices, while the curly bundles were right-handed. A mutant which tumbled consistently had curly flagella, and Shimada *et al.* (1975) suggested that the curly configuration was associated with tumbling.

There have been several studies of flagellar substructure with the objects of comparing them to eukaryotic flagella and of trying to understand how a filament composed of subunits of a single protein can have a helical structure. Lowy and Hanson (1965) studied flagella of several bacteria and suggested that there were two types of flagellar substructurès, one with 8 subunits per turn and a second with 10 subunits per turn, possibly with a central tubule. O'Brien and Bennett (1972) studied substructure of straight flagella produced by a mutant of *S. typhimurium*. Kondoh and Yanagida (1975) did the same with straight flagella from mutants of *E. coli* and similar structures were proposed in both cases. Self-assembly of filaments from flagellin monmers occurs only at one end of the filament, and presumably filaments grow like this in the living cell (Asakura *et al.*, 1968). When the flagellin monomer is incorporated into the growing end of the filament, it undergoes a conformational change which confers polarity to the growing end of the filament. Growth of flagella *in vivo* probably occurs by the same mechanism (Iino, 1974, 1977). Models of flagellar substructure were proposed by Asakura and Iino (1972) and Calladine (1976) which explain the observed substructure and possibly even the helical structure. Czajkowski *et al.* (1974) claimed that the helices were an artifact and that flagella are actually smooth but this, in turn, has been disputed by McCoy *et al.* (1975), who believe they are genuine. The hook is slightly wider than the filament, 17 nm in diameter and about 45 nm long. The curly flagella of an *E. coli* mutant (Silverman and Simon, 1972) were the result of a mutation which failed to terminate assembly of the hook, resulting in flagella that were actually polyhooks. The basal body has four rings (Figs. 1 and 4), the L, P, S, and M rings. DePamphilis and Adler (1971b) believe that the L ring is located in the lipopolysaccharide layer, the P ring in the peptidoglycan layer, the S ring in the area just above the membrane (the supramembrane region), and the M ring in the membrane. Spheroplasts of *E. coli* prepared with lysozyme and EDTA still had flagella attached to the outer membrane by the L ring. Portions of the cytoplasmic membrane were still attached to the M ring, implying that the M ring was embedded in this membrane. Given the dimensions of the basal body and the cell envelope (Section IV,A), they concluded that the S ring lies just above the membrane and that the P ring is embedded in the peptidoglycan layer (Fig. 4).

Bacillus subtilis has only two rings, which apparently correspond to the S and M rings of *Escherichia coli* (DePamphilis and Adler, 1971a). Lysozyme protoplasts of *Bacillus subtilis* show only the bottom ring attached to the

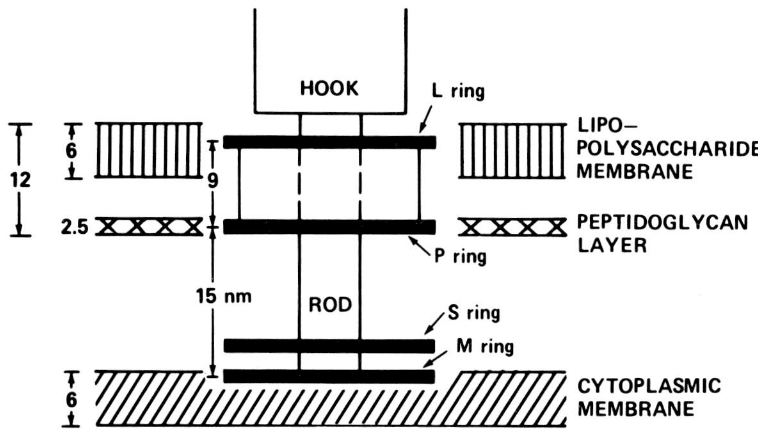

FIG. 4. Diagram of attachment of basal body of *Escherichia coli* flagellum to the cell envelope. (From DePamphilis and Adler, 1971c; reproduced with permission of the American Society for Microbiology.)

cell membrane with the upper ring lying above the membrane. The S ring of *Bacillus subtilis* is 3.5 nm above the M ring, which places it just below or possibly attached to the cell wall. The rings of *Bacillus subtilis* basal body are 21 nm in diameter, close to the dimensions of those of *Escherichia coli.*

At the present time, nearly 30 genes have been identified in *E. coli* and *S. typhimurium* which have some relationship to flagella structure, assembly, chemotaxis, or motility (Silverman and Simon, 1977; Iino, 1977). The *hag*

TABLE I

Amino acid	*Bacillus subtilis* 19[a] 40,000 MW (residues/mole)	*Bacillus subtilis* W23[b] (residues/mole)	*Bacillus subtilis* 168[b] (residues/mole)	*Caulobacter crescentus*[c] 27,900 MW (residues/mole)	24,800 MW (residues/mole)
Lysine	16.9	15	16	16	14
ε-*N*-Methyllysine	—[j]	—	—	—	—
Histidine	4.2	4	4	5	2
Arginine	14.6	17	15	5	5
Aspartic acid	55.4	51	49	22	26
Threonine	20.2	24	18	16	22
Serine	32.3	21	24	34	29
Glutamic acid	46.4	45	42	37	24
Proline	3.3	1	2	6	2
Glycine	22.8	27	19	37	24
Alanine	44.9	42	40	22	29
Valine	14.6	10	14	9	8
Methionine	10.5	11	8	2	1
Isoleucine	28.0	22	24	7	9
Leucine	37.9	33	29	14	22
Tyrosine	2.0	0	1	3	2
Phenylalanine	6.1	7	5	6	6
Cysteine	0	0	0	0	0
Tryptophan	0	0	0	0	0
Total residues	360	330	310	241	225

[a] Martinez *et al.* (1967).
[b] DeLange *et al.* (1973).
[c] Lagenaur and Agabian (1976).
[d] Silverman and Simon (1972).
[e] McDonough (1965).
[f] Chang *et al.* (1969).

gene in *E. coli* and the *H1* gene in *S. typhimurium* are the structural genes for the filament. Genes which control flagellar assembly and structure are designated *fla* in both species. For example, the polypeptide isolated from the polyhook mutant of *E. coli* is known to be the hook protein (Table I) (Silverman and Simon, 1972), and several other polypeptides related to basal body structure have been identified (Silverman and Simon, 1977). Other genes related to flagella function are the *che* genes for chemotaxis

AMINO ACID COMPOSITION OF FLAGELLA AND AXIAL FILAMENTS

Escherichia coli[d]		*Salmonella typhimurium* SW1061[e] 40,000 MW	*Spirillum serpens*[a] 40,000 MW	*Proteus vulgaris*[f] 39,170 MW	*Leptospira* B16[g,i]	*Treponema zuelzerae*[h,i] 37,000 MW
Normal (mole%)	Curly (mole%)	(residues/mole)	(residues/mole)	(residues/mole)	(mole%)	(residues/mole)
7.2	3.1	13.8	14.7	22	6.73	14.9
—	—	10.4	—	—	—	—
0.4	0.6	1.2	0.5	0	1.84	3.1
3.3	2.0	10.4	18.2	12	9.56	20.1
17.0	17.8	61.4	47.9	63	12.64	43.9
9.9	10.0	39.7	34.2	30	4.24	18.7
8.6	9.1	23.4	33.5	31	5.30	23.9
8.5	8.9	37.9	40.2	41	17.25	43.3
1.4	2.6	4.9	1.5	1	1.37	7.2
9.5	10.1	32.6	34.9	30	3.23	27.1
12.1	10.0	59.0	57.9	38	7.54	36.9
5.8	4.4	24.4	21.5	29	4.35	26.2
1.0	2.6	2.2	5.4	2	2.70[k]	12.3
4.2	3.6	19.0	20.7	22	6.62	18.7
7.6	7.7	30.0	28.1	32	9.34	24.3
2.3	2.9	8.3	3.5	3	3.23	8.8
1.4	4.4	5.3	9.7	8	4.17	12.7
0	0	0	0	0	0	0
0	0	0	0	0	0	0
		384	372	364		342

[g] Nauman *et al.* (1969).
[h] Bharier and Rittenberg (1971).
[i] Axial filaments.
[j] Not done.
[k] Reported as methionine sulfone.

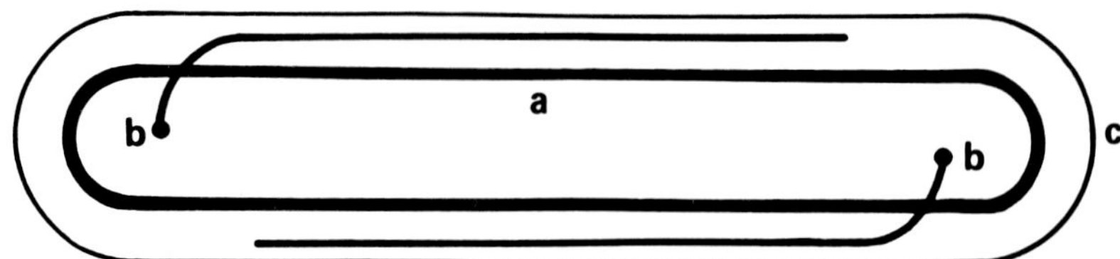

FIG. 5. Diagram of the insertion and arrangement of axial filaments. The protoplast is a, the insertion of axial filaments is at b, and c is the external sheath. Because filaments meet in the center of the cell, a cross section through the center will show twice as many filaments as will a cross section through either end. (From Berg, 1976; reproduced with permission of the *Journal of Theoretical Biology*.)

and the *mot* genes which control motility. One of the very interesting consequences of the genetic studies has been the revelation of the sequence of events in flagellar morphogenesis (Iino, 1977). Assembly begins with the formation of the S and M rings and the rod which holds them, followed by the addition of the P and S rings, hook, and filament.

Considering the similarities in chemical composition and anatomy, it is difficult to avoid the conclusion that flagella and axial filaments share a common heritage. Axial filaments of spirochaetes, leptospires, and borrelia are inserted suberminally at each end of the cell and extend toward the center of the cell (Bharier *et al.*, 1971; Holt and Canale-Parola, 1968; Listgarten and Socransky, 1964; Nauman *et al.*, 1969; Pillot and Ryter, 1965; Joseph and Canale-Parola, 1972). There are usually 2–8 fibrils per cell, but there may be up to several hundred. Axial filaments lay on the cell surface, but below an outer sheath (Fig. 5). Joseph *et al.* (1973) identified peptidoglycan as a constituent of the cell envelope of *Spirochaeta stenostrepta* and believe that the cell surface of this organism is similar to other gram-negative bacteria. If this is true, then the outer sheath of spirochaetes would correspond to the outer membrane of gram-negative bacteria which it resembles in electron micrographs.

The average diameter of axial filaments is 17–18 nm, slightly larger than flagella. Like flagella, axial filaments have a hook and basal body (Fig. 6), but the hook is usually not as pronounced as flagellar hooks. Axial filaments are usually covered by a sheath, which is less common with flagella. The sheath is a complex in *Leptospira* B16 (Fig. 6) (Nauman *et al.*, 1969) but is a single layer in *Treponema zuelzerae* (Bharier *et al.*, 1971).

B. CHEMICAL COMPOSITION

There are at least eleven polypeptides associated with flagella formation. The major protein is named flagellin and is the main component of the

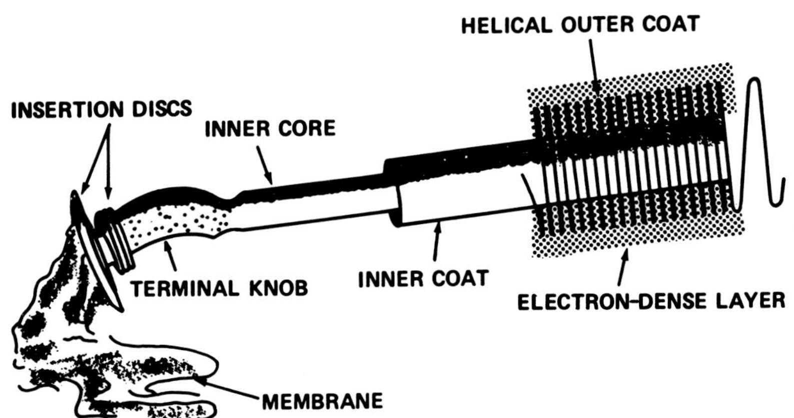

FIG. 6. Diagram of the axial filament of *Leptospira* B16. (From Nauman *et al.*, 1969; reproduced with permission of the American Society for Microbiology.)

flagellar filament (Silverman *et al.*, 1976). One polypeptide is associated with the hook, and nine other polypeptides are located in the basal body structure (Silverman and Simon, 1977). The amino acid compositions of flagellins from both flagella and axial filaments are shown in Table I, and several other examples are given in the review of Smith and Koffler (1971). All flagellins lack cysteine and tryptophan, and are low in phenylalanine, tyrosine, histidine, and proline. *Salmonella* flagellins contain ξ-*N*-methyllysine, which has not been reported in other genera of bacteria. There are strain differences in amino acid composition and molecular weights of flagellins. Table I contains data for three strains of *Bacillus subtilis* which show small, but distinct, differences between the strains. Lagenaur and Agabian (1976) isolated two flagellins from one strain of the stalked bacterium *Caulobacter crescentus* (Table I). The *Caulobacter* flagellins have a similar chemical composition, but the smaller protein has more leucine, alanine, threonine, and aspartic acid which rules out a simple unmasking of the larger protein to produce the smaller one. Nauman *et al.* (1969) detected six proteins in their preparation of purified axial filaments of *Leptospira* B16, but Bharier and Rittenburg found only one protein in purified axial filaments of *Treponema zuelzerae* (Table I). Flagellin from the mutant of *E. coli* with curly flagella (Table I) is probably the hook protein. Molecular weights of flagellin subunits vary from 25,000 for *Caulobacter* (Lagenaur and Agabian, 1976) to 60,000 for *Escherichia coli* K12 (Kondoh and Hotani, 1974). DeLange *et al.* (1976) reported the complete amino acid sequence of *Bacillus subtilis* strain 168 flagellin. These authors also compared the sequence of 27 amino acids at the N-terminal end and 17 amino acids of the C-terminal end of

B. subtilis strains W23 and 168 and found three differences in amino acid sequence (DeLange *et al.*, 1973). This is the first flagellin whose entire amino acid sequence was determined, although portions of the amino acid sequence of *Salmonella typhimurium* (Joys and Rankis, 1972), *S. adelaide* (Davidson, 1971) and *Proteus mirabilis* (Glossman and Bode, 1972) flagellins have been reported. While there appear to be some homologies in amino acid sequence, there are not enough data to draw conclusions about evolutionary processes. There are only a few repeating amino acid sequences in *Bacillus subtilis* 168 flagellin, which means that little gene duplication occurred (DeLange *et al.*, 1976).

C. Flagella as Organs of Motility

The discovery of the mechanism for motility in bacteria is one of the most interesting recent biological discoveries. Direct observation of motile bacteria tethered to glass slides, behavior of mutants which lack filaments, of mutants with polyhook flagella, and of flagella in the chemotactic response (all of which are described in the following sections) provide conclusive proof for the role of bacterial flagella in motility. *Caulobacter* provides a natural example of evidence that flagella are organs of motility (Shapiro *et al.*, 1971). These stalked bacteria divide to produce one stalked cell and one cell with flagella. The flagellated daughter cell is motile but the stalked mother cell is not. The flagellated cell changes to a stalked cell which is nonmotile and then divides again. The subject of bacterial motility has been recently reviewed by Berg (1975).

For a long time it was believed that bacterial flagella functioned by generating a helical wave similar to eukaryotic flagella. However, after Berg and Anderson (1973) reviewed the literature they came to the conclusion that the evidence favored the hypothesis that bacterial flagella actually are rigid and propel the organism by rotation. Silverman and Simon (1974) obtained direct evidence for the concept of rigid rotation using the polyhook mutants of *Escherichia coli.* Antiserum against the hook, layered on the surface of a glass slide, caused the mutant to stick to the slide by its hook; these bacteria rotated counterclockwise when viewed from above. The interpretation of this observation was that the flagellum was stuck to the slide and unable to rotate. This hypothesis required a motor to rotate the flagellum, and the best location for the motor was at the base of the flagellum. The motor continued to rotate and to generate torque and, since the flagellum could not rotate, the cell turned. Silverman and Simon (1974) provided conclusive proof for this idea with a strain of *Escherichia coli* with straight flagella in a medium with anti-flagellar antibody and tiny polystyrene beads,

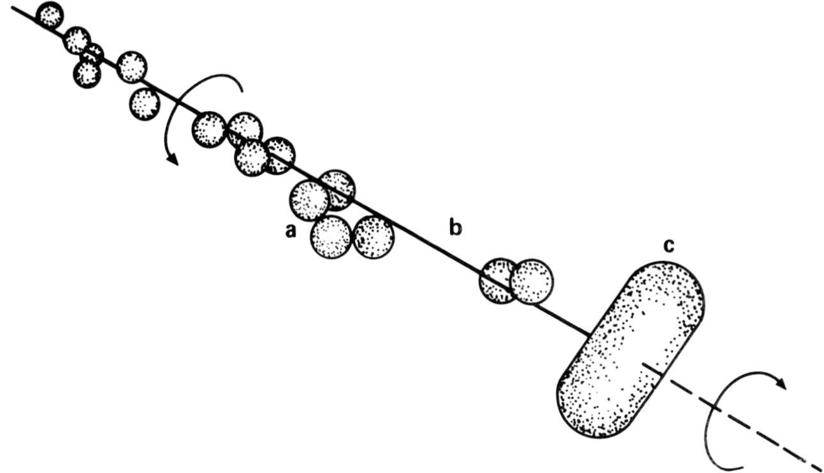

FIG. 7. Silverman and Simon's (1974) experiment with polystyrene beads and a strain of *Escherichia coli* with straight flagella to determine if flagella rotate as rigid bodies. The flagellum (b) cannot be seen with the light microscope, but its direction of rotation can be inferred by the rotation of the beads (a) which are attached to the flagellum by antibody. The cell (c) rotates in the opposite direction. (From Berg, 1975; reproduced with the permission of *Nature*.)

0.7 μm in diameter, stuck to the flagella. The flagella were too fine to be seen but as they rotated the beads turned in the same direction (Fig. 7).

Larsen *et al.* (1974) used the tethered cell technique to study the effect of chemical attractants and repellents on motility. They grew *Escherichia coli* in glucose, which repressed the formation of flagella by catabolic repression, resulting in bacteria with an average of slightly more than one flagellum per cell. Normal cells spent 75–90% of their time rotating counterclockwise when viewed from above, the equivalent of swimming forward. When 0.30 m*M* L-leucine, a repellent, was added to the medium, the percentage of time spent in counterclockwise rotation was decreased to as low as 0.2% with a corresponding increase in clockwise (backward) rotation. When an attractant such as 20 μM L-serine was added to the medium, bacteria rotated counterclockwise 100% of the time. However, mutants which had lost the ability to respond to leucine or serine in the usual chemotaxis assays no longer changed their direction of rotation in the tethered cell assay. For example, a mutant of *Escherichia coli*, which tumbled almost continuously instead of swimming normally rotated clockwise (backward motion) in the tethered cell test 98% of the time. These results fitted neatly with observations of Berg (1975) on the swimming patterns of motile bacteria using a special microscope which tracked their movement in three-dimensions (Fig. 8).

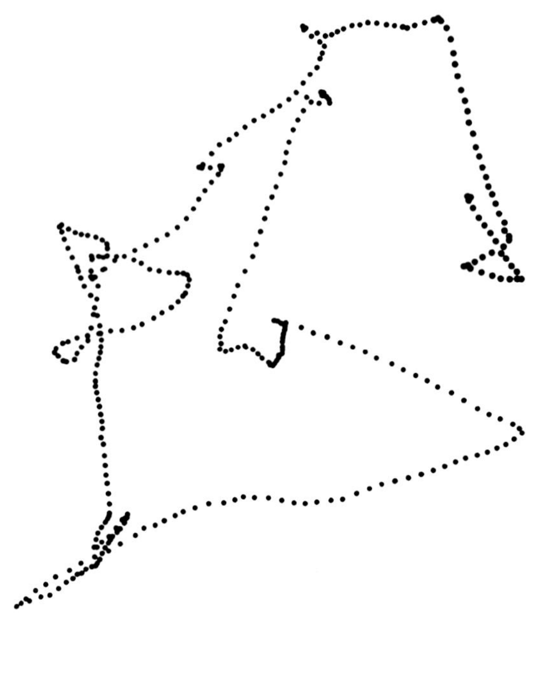

FIG. 8. Track of *Escherichia coli* swimming normally which was made using a special microscope which tracks the organism in three dimensions. The dots are 0.08 seconds apart, and the distance between the dots is a measure of the speed of movement. Dots which are very close together may be also movement at right angles to the plane of the paper. The cell swims forward (a run), tumbles (or twiddles), changes direction, and then swims forward again. Tumbles occur at random but on an average of one per second. (From Berg, 1975; reproduced with the permission of *Nature*.)

Most of the time was spent in forward movement which is counterclockwise rotation. Any change in direction was preceded by a tumble, and then forward swimming again. The implications of these observations on chemotaxis in motile bacteria are discussed in this volume by Koshland (Chapter 3).

When it became certain that flagella rotated as rigid organelles, it followed that there must be a motor at the origin of the flagellum which can apply torque to the filament causing it to rotate. Berg (1974, 1975) proposed that the S and M rings of the flagellum were the motor (Fig. 9). The M ring is attached rigidly to the hook but free to rotate in the membrane. The S ring

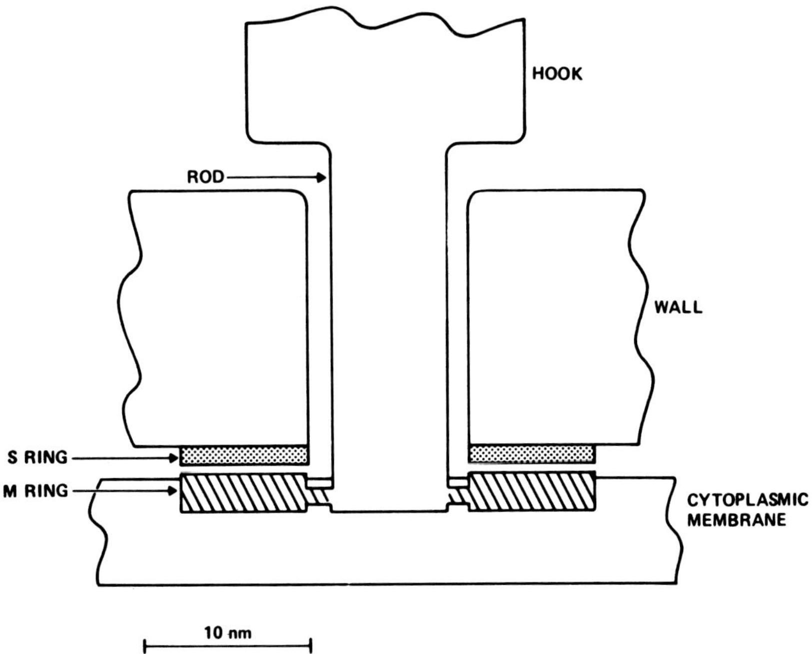

FIG. 9. Berg's (1974) proposal for the flagellar motor of a gram-positive bacterium. The M ring is imbedded in the membrane and fixed to the hook, but free to rotate in the membrane. The S ring is fixed to the peptidoglycan but not to the hook and is not free to rotate. Torque is generated between the S ring and M ring causing the M ring and filament to rotate. The L and P rings of gram-negative bacteria are proposed to act as bushings. (From Berg, 1975; reproduced with the permission of *Nature*.)

is attached rigidly to the peptidoglycan layer but not to the rod. Torque is generated between the M and S rings. The energy source for torque is not ATP (Larsen *et al.*, 1974; Thipayathasana *et al.*, 1974), but is probably proton motive force (see Harold, Volume VI, Chapter 7). Energy for torque could be generated by ion passage through the M ring which interacts with fixed charges on the S ring.

While there is solid experimental evidence for the concepts of bacterial motility by flagella, there are only theories of how spirochaetes move. Because of the chemical and anatomical similarities between flagella and axial filaments, there is good reason to believe that axial filaments take part in motility of spirochaetes, but their location under the outer sheath rules out a propellerlike mechanism. Furthermore, the motion of spirochaetes is very different from bacteria. Spirochaetes propel themselves through the medium by a corkscrew motion and flex or creep over the surface. Berg (1976) proposed a theory for motility in spirochaetes in motion (Jahn and Landman,

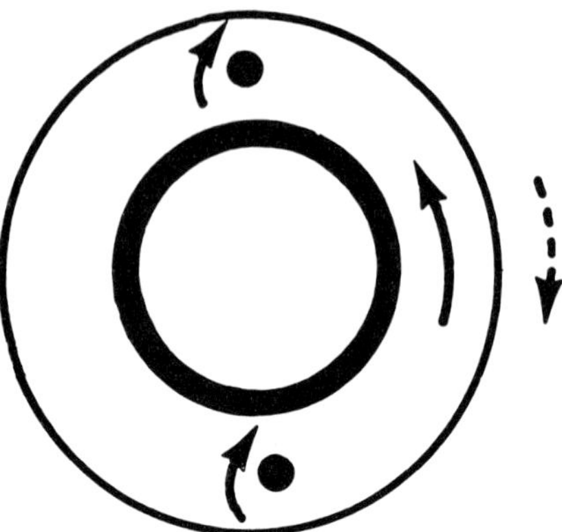

FIG. 10. Cross section of a spriochaete illustrating Berg's (1976) proposal for motility in these organisms. Both axial filaments rotate in a clockwise fashion, rolling on the surface of the cell causing the cell body to rotate also but in the opposite direction. The outer sheath is loose and rotates in the direction opposite to the cell body. The sum of these rotations in a helical organism is forward motion by rotation about the longitudinal axis of the cell. (From Berg, 1976; reproduced with the permission of *The Journal of Theoretical Biology*.)

1965; Wang and Jahn, 1972; Chwang *et al.*, 1974). Berg proposed that axial filaments rotated between the sheath and the semirigid cell body (Fig. 10). The outer sheath must be loose fitting and also free to rotate. Clockwise rotation of the axial filaments causes the cell body and sheath to rotate in opposite directions, and the sum of all these motions is that the helical spirochaete rotates about its longitudinal axis and moves forward. Direction of movement can be changed by a change in rotation of axial filaments. Special cases of this mechanism can be invoked to explain creeping and flexing motility (Berg, 1976).

The swimming of leptospires is different from that of spirochaetes. Cox and Twigg (1974) and Berg *et al.* (1978) showed that motile leptospires usually had the configuration shown in Fig. 11a. There is always a helical wave at the leading edge of the cell which travels part way toward the center of the cell. The trailing edge of the cell may be either hook shaped as shown or straight. One of the configurations of nonmotile cells is shown in Fig. 11b

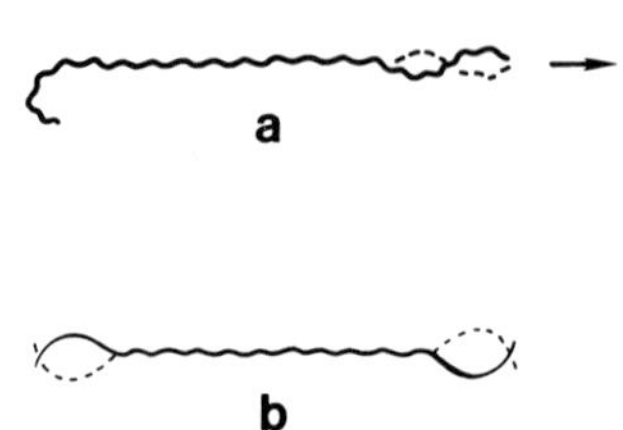

FIG. 11. Configuration of *Leptospira* during forward motion (a) and at rest (b).

with the two ends of the cell flexing in opposite directions. Berg *et al.* (1978) suggested that the axial filaments of leptospires are stiff and helical, but with a wavelength longer than that of the protoplasmic cylinder. The filament is also shorter than that of spirochaetes, so that filaments originating at the ends of the cell do not meet. The protoplasmic cylinder is more flexible than the axial filament. Given these facts, and the assumption that the filament is free to rotate between the outer layer and protoplasmic cylinder, Berg *et al.* (1978) proposed a theory for motility by leptospires. At the leading edge of the cell, the rotation of the axial filament generates a wave in the protoplasmic cylinder which reflects the helical wave of the filament. Thrust is generated by the propagation of the wave from the leading edge of the cell to the center of the cell and by the roll of the protoplasmic cylinder around the filament. During forward motion the filament at the leading edge of the cell rotates in the opposite direction to the filament at the trailing edge. In the resting or nontranslational configuration, the axial filaments rotate in the same direction which causes the two ends of the cell to gyrate in opposite directions but no thrust is generated (Fig. 11b).

II. Sex (Donor) Pili

A. Definition and Distribution

Sex pili are appendages which occur on donor strains of bacteria and whose formation is determined by conjugative plasmids. Novick *et al.* (1976) proposed that these structures be named donor pili. By far, the best characterized sex pili are those of *Escherichia coli* K12, although other enteric bacteria carry the same conjugative plasmids (Brinton, 1965; Valentine *et al.*, 1969). *Pseudomonas aeruginosa* (Bradley, 1972) and *Caulobacter crescentus* also produce structures which may be sex pili, since they are bacteriophage receptors and conjugation occurs in these species (Lagenaur and Agabian, 1977). Some species of soil bacteria form star-shaped clusters of cells and their pili may be responsible for cluster formation as well as play a role in conjugation (Heuman, 1968).

B. Types of Pili

The two major groups of sex pili in *E. coli* are F and I pili whose formation is determined by F and I plasmids, respectively (Table II). Plasmids are classified on the basis of incompatibility with incompatible, and therefore closely related, plasmids in the same group (Novick *et al.*, 1976). At least

TABLE II
CHARACTERISTICS OF DONOR PILI AND PILI WHICH ARE RECEPTORS FOR BACTERIOPHAGES

Incompatibility group	Plasmid determinants	Donor-specific bacteriophages
F pili[a]		
FI	F, R386	RNA: R17 (Crawford and Gesteland, 1964);
FII	R1, R100	MS2 (Meynell *et al.*, 1968); M12 (Brinton *et*
FIII	ColB-K98	*al.*, 1964); F2 (Valentine and Strand, 1965);
FIV	R124	Qβ (Achtman *et al.*, 1971); μ2 (MacFarren
FV	F_0*lac*	and Clowes, 1967).
		DNA: f1 (Caro and Schnös, 1966)
		M13 (Hofschneider, 1963).
I pili[b]		
Iα	ColIb-P9, R144	No RNA phages known
I2	TP114	DNA: If1, If2 (Meynell and Lawn, 1968)
Iγ	R621a	
P pili[c]		
PSA	None known	RNA: PP7 (Bradley, 1966)
		DNA: M6 (Bradley and Pitt, 1974);
		F116 (Pemberton, 1973)
P-1	RP4	RNA: PRR1 (Olsen and Thomas, 1973)
		DNA: pf3 (Stanish, 1974); PR4 (Bradley and Rutherford, 1975); PRDI (Olsen *et al.*, 1974)
P-2	R130, FP39	None yet isolated (Shahrabadi *et al.*, 1975)
W pili[d]		
W	S-a	DNA: PR4 (Bradley and Rutherford, 1975); PRDI (Olsen *et al.*, 1974)
Caulobacter pili[e]		
Swarmer cell pili	None known	RNA: ϕCb5 (Bendis and Shapiro, 1970)
		DNA: ϕCbk (Lagenaur *et al.*, 1977)

[a] Distribution mostly *Escherichia* and other *Enterobacteriaceae* (Lawn *et al.*, 1967; Datta, 1975; Novick *et al.*, 1976).

[b] Distribution mostly *Escherichia* and other *Enterbacteriaceae* (Lawn *et al.*, 1967; Datta, 1975; Novick *et al.*, 1976).

[c] Distribution mostly *Pseudomonas* but occur naturally in a wide variety of species, particularly *Enterobacteriaceae* (Datta, 1975; Novick *et al.*, 1976; Bradley, 1977).

[d] Distribution mostly *Enterobacteriaceae* but can be transferred to *Pseudomonas* (Bradley, 1975).

[e] Schmidt (1966).

two other incompatible groups of plasmids produce pili, P, which are primarily *Pseudomonas* plasmids, but with a wide host range, and W which are plasmids of enteric bacteria, but which can also infect *Pseudomonas.* Considering the burgeoning literature on plasmids, it appears almost a certainty that more pili will be discovered as products of plasmid infection.

Brinton *et al.* (1964) discovered that F plasmids carried genes for the formation of sex pili. However, it was soon apparent that other plasmids, such as R factors and colicin plasmids, also caused the formation of pili and conferred the ability to conjugate on bacteria (Meynell and Datta, 1965). Many of these plasmids also inhibited conjugation in F^+ *E. coli* and were further classified as fi^+ for fertility inhibition. Presumably the inhibition was due to the action of a repressor which affected production of pili by the F plasmid and R factor, signifying a close relationship between these plasmids. F pili are also the receptors for the single-stranded RNA phages R17, M12, MS2, f2, and $\mu 2$ (Table II). In fact, the observation that these donor-specific phages attach to sex pili and not to common pili was the first piece of evidence that linked sex pili to conjugation (Crawford and Gesteland, 1964; Brinton *et al.*, 1964). RNA phages attach to the sides of F pili, while the filamentous single-stranded DNA phages f1 and M13 attached to the tip of the pilus (Fig. 12) (Caro and Schnös, 1966; Meynell *et al.*, 1968).

Plasmids which carry genes for the formation of I pili include ColI and R plasmids (Table II). Most I-like R plasmids do not suppress the action of the F plasmid, however, they produce I pili, which can be identified by the attachment of the I-specific phages If1 and If2 (Meynell and Lawn, 1968). These are filamentous single-stranded DNA phages which attach to the tip of the I pilus just like f1 and M13 described above. Attempts to find I-specific RNA phages so far have been unsuccessful.

It is apparent from the preceding paragraphs that the attachment of single-stranded RNA and DNA phages is a characteristic of F and I pili, and this generalization is true also for *Pseudomonas* and *Caulobacter* pili. However, there are some exceptions. The DNA phages PR3, PR4, and PRD1 are double-stranded DNA phages which contain lipid and which absorb to the cell surface rather than to the pili (Bradley and Rutherford, 1975; Olsen *et al.*, 1974). Another exception occurs in the case of bacteria carrying plasmids of incompatibility group N. These bacteria do not form pili but are sensitive to phage Ike, a single-stranded DNA phage which specifically attaches to the cell surface of bacteria carrying N plasmids (Brodt *et al.*, 1974).

C. Ultrastructure

There are usually one or two sex pili per cell located randomly on the *Escherichia coli* cell surface, although there may be as many as 7–10 and many cells have no pili at all (Fig. 12) (Brinton, 1965; Lawn, 1966). Sex pili are 8.5 (Brinton, 1965) to 9.5 nm (Lawn, 1966) in diameter. Lawn *et al.* (1967) discovered that F pili were usually long, up to 20 μm, while I pili were

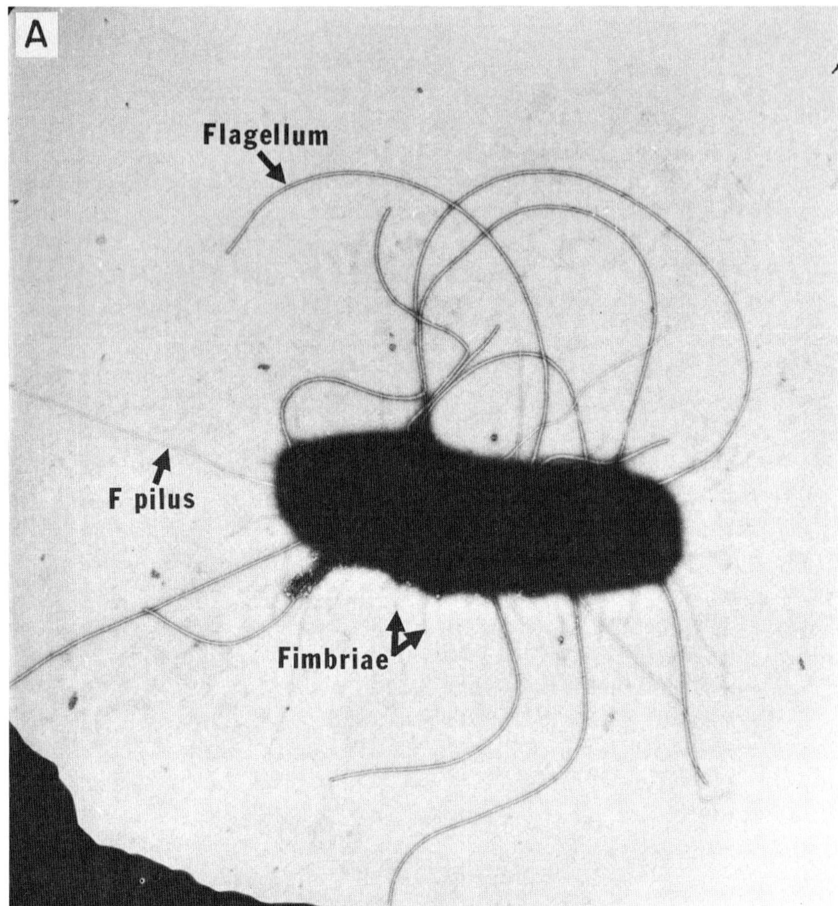

FIG. 12. (A) *Escherichia coli* with flagella, fimbriae, and F pili. (B), F pilus with absorbed MS2 bacteriophage. The other structure is a flagellum. (C), there are three I-like sex pili in this picture, one of which has the filamentous bacteriophage Ifl attached to the top. The phage is labeled with antibody to distinguish it from I pili. (From Meynell *et al.*, 1968; reproduced with the permission of the American Society for Microbiology.)

short, no longer than 2 μm. Sex pili are flexible, but they are not helical as flagella. Several authors have observed an axial dark line in electron micrographs of F pili (but not I pili) which have been interpreted as a tubule, a pair of parallel protein rods, or an artifact (Brinton, 1965; Valentine and Strand, 1965; Lawn, 1966). There does not appear to be a discernible substructure to sex pili comparable to the protein subunits of flagella. Very little is known about the structures that anchor sex pili to the cell, presumably to the cell membrane. W pili, which occur on both *Enterobacteriaceae* and

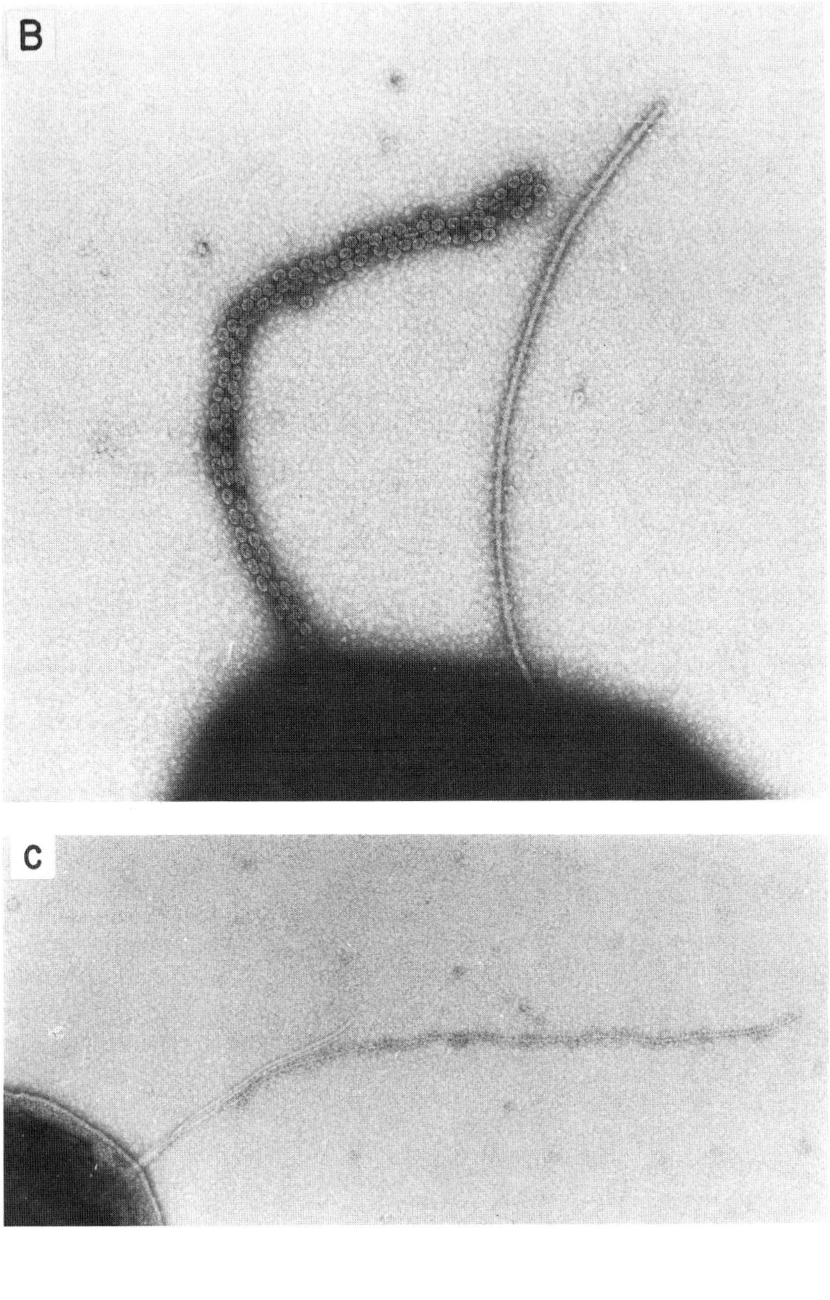

B

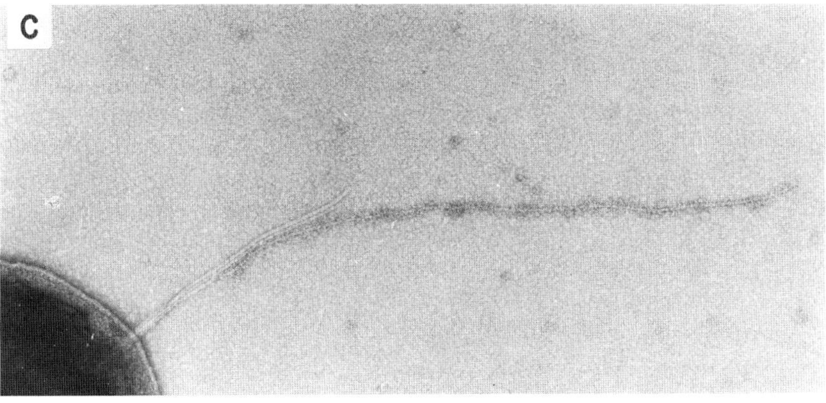

Pseudomonas, are long, flexible, and about 12 nm thick, or somewhat larger than F pili.

There are four types of pili in *Pseudomonas aeruginosa*, including W pili, which could be sex pili (Bradley, 1977). Polar pili, named PSA pili by Bradley are 6 nm thick, flexible, and up to 5 μm long, but with an average length of about 1.3 μm (Bradley, 1972). At the present time, it has not been established that PSA pili are determined by a conjugative plasmid. P-1 pili (Table II) are determined by conjugative plasmids of which RP4 is the type plasmid. P-1 pili are rigid, about 8 nm thick, and about 300 nm long. P-2 pili are determined by plasmids, such as R130, and the sex factor FP39. P-2 pili are flexible, but their dimensions have not been reported. Pili determined by plasmids of the W incompatibility group are long, flexible, and about 12 nm thick. *Caulobacter* pili are formed only at the flagellated pole of swarmer cells (Schmidt, 1966) and are flexible, about 4 nm thick and up to 4.0 μm long.

D. Chemistry of Sex Pili

The presence of other surface appendages makes it difficult to purify pili and there is only one report of the chemical composition *Escherichia coli* sex pili (Brinton, 1971). The details of the purification have not been published but Brinton and his co-workers used a mutant of *Escherichia coli* which produced large amounts of pili which were released into the medium. He reported that F-pilin contained 124 amino acids residues per mole and lacked histidine, arginine, proline and cysteine (Table III). Other unusual features were that F pili contained two phosphates and one glucose per mole. The molecular weight was 11,800, the bouyant density was 1.257 gm/cm^3 and the pI was 4.15.

Beard *et al.* (1972) used another approach to the purification of sex pili. They used a strain of *Escherichia coli* K12 which lacked flagella and which carried a derepressed R factor resulting in a high yield of sex pili. Pili were produced during exponential growth, but disappeared quickly after the culture reached stationary phase. They obtained 5 mg of a preparation which was 98% pure and which retained the ability to bind MS2. Beard *et al.* (1972) reported a bouyant density of 1.296 gm/cm^3, slightly higher than that reported by Brinton (1971). Beard and Connolly (1975) detected a protein similar to F pilin in the outer membrane of *Escherichia coli* but not in the cytoplasm. The subunit of this protein had a molecular weight of 12,500 which was close to the value obtained by Brinton (1971). The presence of pilin subunits in the outer membrane but not in the cytoplasm suggests that assembly of sex pili occurs in the outer membrane.

TABLE III
AMINO ACID COMPOSITION OF PILI AND FIMBRIAE

	Escherichia coli				
Amino acid	Sex pilus[a] (11,800 mw)	Type I fimbriae[b] (16,000 mw)	*Pseudomonas aeruginosa* pili[c] (17,800 mw)	*Caulobacter crescentus* pili[d] (8500 mw)	*Corynebacterium renale* pili[e] (19,400 mw)
Lysine	10	3	14.8	2	11.4
Histidine	0	2	0	1	2.9
Arginine	0	3	4.0	1	6.0
Aspartic acid	8	20	14.9	4	21.9
Threonine	8	20	15.0	10	16.5
Serine	11	10	9.9	6	7.8
Glutamic acid	4	13	14.9	7	22.7
Proline	0	2	9.7	2	8.9
Glycine	15	17	17.7	12	15.3
Alanine	15	34	23.8	13	17.5
Valine	21	13	9.0	5	14.3
Methionine	8	0	2.1	0	1.2
Isoleucine	4	4	11.7	3	5.6
Leucine	9	10	14.0	6	9.7
Tyrosine	2	2	2.2	1	5.8
Phenylalanine	7	8	2.3	2	3.9
Cysteine	0	2	3.9	—	0.78
Tryptophan	2	0	2.0	—	1.0
Density (gm/ml)	1.257		1.221	1.236	

[a] Brinton (1971).
[b] Brinton (1965).
[c] Mutant strain PAK/2PfS, Frost and Paranchych (1977)
[d] Lagenaur and Agabian (1977).
[e] Kumazawa and Yanagawa (1972).

The chemical composition of *Pseudomonas aeruginosa* PAK/2PfS and *Caulobacter crescentus* pili are included in Table II for comparison. Strain PAK/2PfS is a mutant of *Pseudomonas aeruginosa* which produces numerous pili and which is resistant to infection by pilus-specific phages (Frost and Paranchych, 1977). *Caulobacter* strain SW16 is a spontaneous mutant which lacks flagella and a stalk, but which produces pili. The molecular weight of PAK/2PfS pilin is higher than that of F pilin. Furthermore, *Pseudomonas* pili do not contain either phosphate or carbohydrate, and their density is slightly lower than that of F pili. There are neither obvious similarities nor differences between *Pseudomonas* pili, F pili and type I fimbriae. *Caulobacter* pili are more like F pili than type I fimbriae in the proportion of acidic amino acids and very different from their own flagella in this regard (Table I). The molecular weight of *Caulobacter* pili is the lowest of any in Table II, and their density is between that of F pili and *Pseudomonas aeruginosa* pili.

E. The Role of Sex Pili in Phage Infection

The attachment of donor-specific bacteriophages to pili and the sensitivity of piliated bacteria to these viruses immediately raised the question of how viral nucleic acid was passed to the cell. The first suggestion was that nucleic acid passed through a tube in the sex pilus (Brinton, 1965), but it has been difficult to obtain experimental support for this theory. A second possibility was that pili retracted after absorption, placing the virus in contact with the cell surface and allowing viral nucleic acid to pass through the cell wall. A unique opportunity to test this hypothesis presented itself when it was learned that only one PP7 virus attached to each pilus of *Pseudomonas aeruginosa*. The attachment of PP7 was a random rather than to the tip of the pilus as in the case of filamentous DNA viruses. If pili retracted after absorption of viruses, then the average pilus length should be one-half the length before absorption, and this is exactly what Bradley (1972) observed. Furthermore, a mutant of *Pseudomonas aeruginosa* resistant to infection apparently possessed nonretractile pili, since the average length of the pilus before and after infection was the same in this case.

F. The Role of Sex Pili in Conjugation

It is clear that sex pili play an important role in conjugation in *Escherichia coli*, however, their precise function is uncertain. Brinton and his associates studied the kinetics of removal of pili on conjugation (Novotny *et al.*, 1969a). F pili were sheared off by blending, but reappeared rapidly, reaching one-half their full length in 30 seconds and fully grown in 4–5 minutes. Blending greatly reduced the ability of F^+ cells to form mating pairs and the ability of the cell to absorb both R17 and M13 phages (Novotny *et al.*, 1969b). The ability to form mating pairs reappeared in about 5 minutes along with the ability to absorb male phages (Novotny *et al.*, 1969a). Since both RNA and DNA male-specific phages inhibited the formation of mating pairs but had no effect on the preformed mating pairs, both the sides and tip of sex pili appeared to be involved in conjugation. Ou (1973) found that the DNA phage f1 completely prevented the formation of mating pairs, while the RNA phage MS2 inhibited but did not completely prevent pair formation.

All these observations were consistent with an important role for F pili in conjugation. However, other factors must also play a role in conjugation. Falkinham and Curtiss (1976) isolated mutants of *Escherichia coli* with defects in the cell membrane and cell walls which also were impaired in their ability to conjugate, implying that these structures also played a role in conjugation, possibly in cell-to-cell contact.

There is not enough evidence to firmly assign a role to sex pili in conjugation. Both Brinton (1971) and Valentine *et al.* (1969) proposed that sex pili acted as a tubule conveying DNA to the recipient either as pipeline or as a conveyor belt. There is some evidence for an axial canal in F pili, but no evidence for its function as such and little evidence for a canal in I pili. The fact that the DNA phage f1 rapidly inhibited the formation of mating pairs (Ou, 1973) suggested that the tip of the pilus is important in conjugation. Ou proposed that the sides of the pilus were used to contact the recipient cell and that the tip then attached to receptor sites on the F cell, which was followed by the formation of mature mating pairs. Conjugation occurs at higher frequencies in close mating pairs (Ou and Anderson, 1970), and it is possible that pili act as lines to bring the cells into close contact (Curtiss, 1969). However, it should also be pointed out that conjugation can occur in the absence of cell-to-cell contact (Ou and Anderson, 1970) and that the one common feature of all mating pairs is attachment by sex pili. Schweizer and Henning (1977) recently published evidence that highly purified protein II, an outer layer protein, along with lipopolysaccharide specifically inhibited conjugation in *Escherichia coli.* Their interpretation was that protein II acted as a receptor in conjugation, and since mutants of *Escherichia coli* lacking this protein are defective in pair formation and are *con*$^-$; this is a very appealing suggestion. With the preceding information in mind, the role of swarmer cell pili in *Caulobacter* becomes very interesting. Since cell-mediated gene transfer has been demonstrated using piliated strains of *Caulobacter*, does this imply that the piliated swarmer cell is the donor cell?

III. Fimbriae

Short, straight, and hairlike bacterial appendages (Fig. 13) were named fimbriae (Latin, thread) by Duguid *et al.* (1955); later Brinton (1965) called them pili (Latin, hairs), and other names have been proposed. Ottow (1975) suggested that the term fimbriae be used to refer to structures other than flagella and those appendages involved in transfer of bacterial and viral nucleic acids.

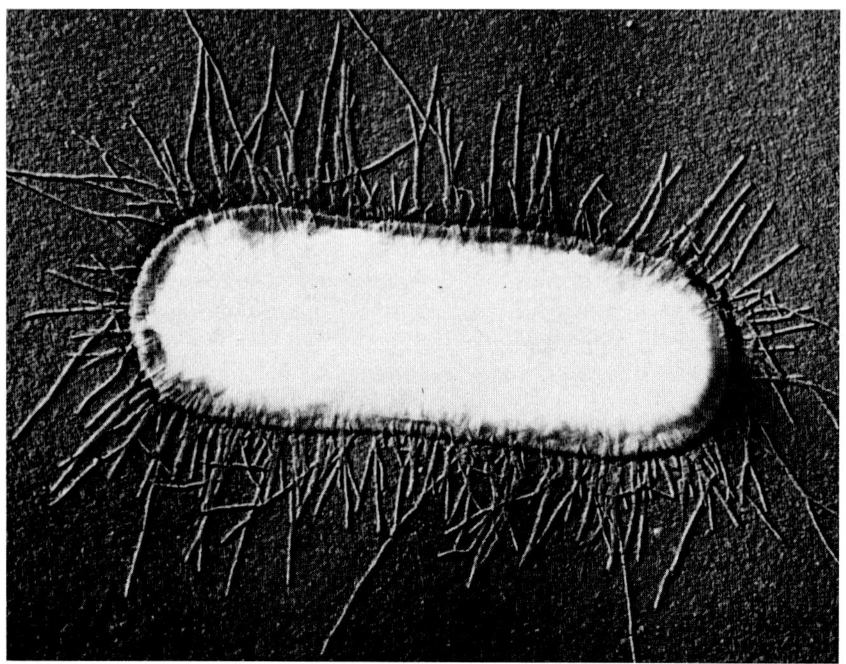

FIG. 13. Fimbriae of *Salmonella anatum*. (From Duguid *et al.*, 1966; reproduced with the permission of *The Journal of Medical Microbiology*.)

A. ULTRASTRUCTURE

Fimbriae are found almost exclusively on gram-negative bacteria (Table IV), although some fimbriae-like appendages have been reported for *Corynebacterium renale* (Yanagawa and Otsuki, 1970). Most fimbriae are narrower than sex pili and flagella (Table IV), although type V fimbriae of *Escherichia coli* are 25 nm in diameter (Brinton, 1965). Brinton (1965) classified fimbriae of *Escherichia coli* into five types, exclusive of sex pilus, which he designated with roman numerals. All fimbriae of *Escherichia coli* were peritrichous. Type I fimbriae were 7.0 nm in diameter with an axial canal of 2 μm and were present on both male and female strains of *Escherichia coli*, making a role in conjugation for these structures unlikely. Purified type I fimbriae formed crystalline aggregates which Brinton studied by X-ray diffraction. Diffraction patterns indicated that the protein subunits were arranged in a helix such that there were 25 protein subunits for every 8 turns of the helix. The molecular weight of the subunits in this arrangement would be approximately 16,000, which is close to the experimentally determined value reported in the next section.

TABLE IV

SPECIES OF BACTERIA WITH FIMBRIAE AND THEIR LOCATION

Bacteria	Diameter (nm)	Location	Reference
Aeromonas liquifaciens	7.9	Peritrichous	Tweedy *et al.* (1968)
Caulobacter crescentus	4.0	Polar	Schmidt (1966)
Caulobacter bacteriodes	—	Polar	Schmidt (1966)
Caulobacter fusiformis	4.0	Polar	Schmidt (1966)
Escherichia coli	3.0–25.0	Peritrichous	Brinton (1965)
Neisseria catarrhalis	4.0–4.5	Peritrichous	Wistreich and Baker (1971)
Neisseria gonorrhoeae	4.5	Septum between cells	Novotny *et al.* (1975)
Neisseria perflava	2.0–3.0	Peritrichous	Wistreich and Baker (1971)
Neisseria subflava	6.0	Peritrichous	Wistreich and Baker (1971)
Pseudomonas aeruginosa	4.5	Polar	Bradley (1966)
	6.0	Polar	Weiss (1971)
	5.2	Polar	Fuerst and Hayward (1969)
Pseudomonas alcaligenes	6.5	Polar	Fuerst and Hayward (1969)
Pseudomonas echinoides	5.0	Monopolar	Heuman and Marx (1964)
Pseudomonas multivorans	6.2	Peritrichous	Fuerst and Hayward (1969)
Pseudomonas multivorans	8.6	Peritrichous	Tweedy *et al.* (1968)
Pseudomonas solanacearum	4.9	Polar	Fuerst and Hayward (1969)
Pseudomonas testosteroni	4.0	Polar	Fuerst and Hayward (1969)
Salmonella sp.	<10	Peritrichous	Duguid *et al.* (1966)
Shigella flexneri	5–10	Peritrichous	Duguid and Wilkinson (1961)
Vibrio cholerae	7.2	Peritrichous	Tweedy *et al.* (1968)
Vibrio eltor	7.9	Peritrichous	Tweedy *et al.* (1968)

Duguid *et al.* (1966) studied 1453 strains of *Salmonella* and found that 1184 were peritrichously fimbriated. The diameter was less than 10 nm, and the length varied from 0.2 to 1.5 μm. Almost all strains (76/78) of *Salmonella paratyphi* A were nonfimbriated, while over 90% of *Salmonella paratyphi* B and over 80% of *Salmonella typhi* strains were fimbriated. Fuerst and Hayward (1969) examined fimbriae of several *Pseudomonas* species by electron microscopy and found that the diameter ranged from 4.0 to 6.5 nm (Table IV) with a maximum length of 1.7 μm. Both polar and peritrichous arrangements occurred, indicating that there was no relationship between arrangement of fimbriae and flagella. *Pseudomonas fluorescens*, *Pseudomonas putida*, and *Pseudomonas oleovorans* were nonfimbriated. Weiss (1971) found that *Pseudomonas aeruginosa* had 15–20 fimbriae per cell, which were both monopolar and bipolar, averaged 6.0 nm in diameter, and were 1 to 2 μm in length. They were flexible and threadlike and, unlike Type I fimbriae of *Escherichia coli*, were not hollow. Fimbriae of *Pseudomonas aeruginosa* were most numerous during logarithmic growth, but declined sharply in number after the culture reached stationary phase. The loss of fimbriae coincided with a change in colony morphology from dry to moist. Fimbriae of *Caulobacter* are located only at the flagellated end of the cell which will develop a stalk (Schmidt, 1966). Fimbriae of *Caulobacter* are also receptors for RNA phages, which suggests a function similar to sex pili of *Escherichia coli*. Fimbriae of *Neisseria gionorrhoeae* are particularly interesting because of the possible relationship of these structures to pathogenicity. They are located in a tuft of up to 100 fibers at the septum between the cell (Fig. 14). Novotny *et al.* (1975) classified fimbriae of *Neisseria gonorrhoeae* into three types: a, b, and c. Type a was the most numerous and was approximately 4.5 nm in diameter and up to 6 μm in length. Type b were narrower in diameter than type a, and types a and c were serologically distinct.

B. Chemistry of Fimbriae

Brinton (1965) purified type I fimbriae of *Escherichia coli* as crystalline aggregates. Both the molecular weight and amino acid composition of *Escherichia coli* fimbriae (Table III) were different from those of the sex pilus and flagella. Table III also includes the amino acid composition of *Corynebacterium renale*, the only gram-positive organism known to have fimbriae.

C. Function

There is much uncertainty about function of the various types of fimbriae, how many types of fimbriae exist, and if there is any commonality of structure and function between species. There is good evidence that fimbriae of enteric bacteria cause aggregation of the bacteria to form a pellicle on the surface of static cultures. It appears that the function of the pellicle is to help fimbriated strains compete better for oxygen, which is in limited supply below the surface of the medium. Brinton (1965) noted that *Escherichia coli* possessing fimbriae which he defined as type I, formed smaller more dense colonies on solid medium than nonfimbriated strains, an observation which was also made of *Neisseria* (Kellogg *et al.*, 1968). Fimbriated cells grew more rapidly than nonfimbriated cells in liquid media when oxygen was limiting. In fact, the number of nonfimbriated cells decreased during the stationary phase of growth, while the number of fimbriated cells increased. Brinton pointed out that this behavior could be an advantage in the environment of the intestine where oxygen is limiting.

Duguid *et al.* (1966) found a high correlation between fimbriae of *Salmonella*, the ability to form a pellicle in static cultures, hemagglutinating

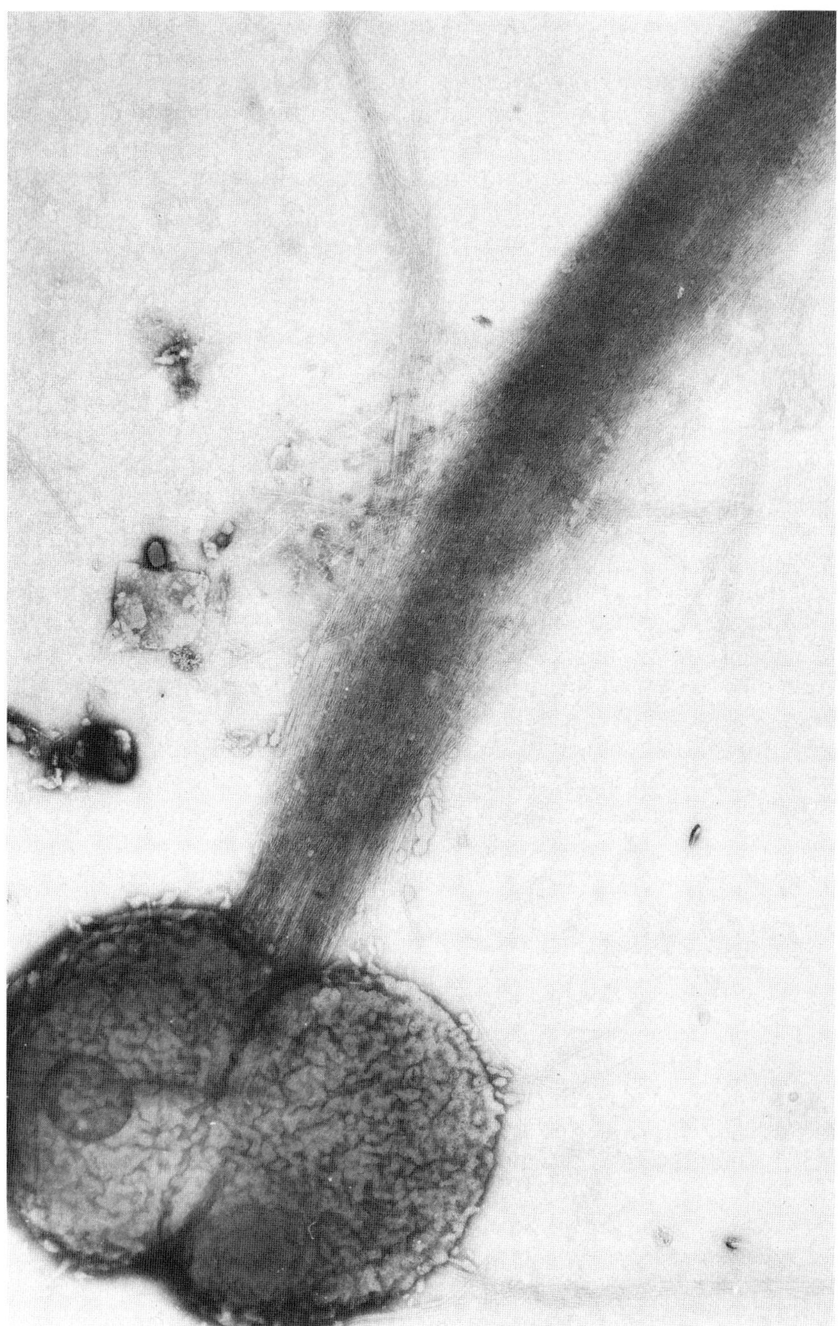

FIG. 14. Fimbriae of *Neisseria gonorrhoeae*. Note the large number of fimbriae and their unusual origin at the intracellular septum. (From Novotny *et al.*, 1975; reproduced with the permission of *The Journal of Medical Microbiology*.)

ability, and adhesiveness. Of 1184 fimbriated strains studied, all but 45 agglutinated guinea pig red blood cells. Only six of the nonfimbriate strains agglutinated red blood cells. The hemagglutinating strains had Duguid's type I fimbriae, while the 45 nonhemagglutinating strains had his type II fimbriae. Hemagglutination by type I fimbriae was inhibited by D-mannose. Fimbriated cells adhered to agglutinated red blood cells as well as to leukocytes, tracheal epithelium, buccal epithelium, and amniotic cells, which implies that these activities are all manifestations of a single property. There is also evidence that pellicle formation is another manifestation of this same property. Old and Duguid (1970) inoculated fimbriated (their type I) and nonfimbrated *Salmonella typhimurium* into broth cultures incubated aerobically but without shaking. Fimbriated bacteria rapidly overgrew the nonfimbriated strains, even when the inoculum had 10^7 nonfimbriated cells to one fimbriated cell. In shake cultures, both types of *Salmonella* grew at about the same rate; an observation also made by Brinton (1965) with *Escherichia coli*. The presence of flagella hastened, but was not essential for, pellicle formation. Pellicle formation did not occur when fimbriated cultures were incubated anaerobically (Old and Duguid, 1970). α-Methylmannoside, which inhibited hemagglutination, also prevented the overgrowth of fimbriated *Salmonella* in mixed cultures of fimbriated and nonfimbriated strains. *Neisseria* fimbriae may have similar functions, since three strains studied by Tweedy *et al.* (1968) also hemagglutinated guinea pig red blood cells and hemagglutination was partially inhibited by D-mannose.

There have been interesting speculations about the function of *Neisseria gonorrhoeae* fimbriae. Kellogg *et al.* (1968) reported that types T1 and T2, which formed small compact colonies, were capable of establishing infection in healthy volunteers. These colony types were also the most frequently isolated from clinical specimens. Types T3 and T4, which produced large granular colonies, caused a mild transitory infection. Jephcott *et al.* (1971) and Swanson *et al.* (1971) almost simultaneously discovered that types T1 and T2 were fimbriated, while Type T3 and T4 were not. Swanson (1973) later showed that T2 cells adhered to human amniotic tissue culture cells, in many cases by their fimbriae, while T4 cells were not nearly so effective in binding to the tissue culture cells. It was tempting to draw the conclusion that T1 and T2 were pathogenic because they were able to attach themselves to the urethral epithelium and establish an infection. However, when Novotny *et al.* (1975) studied the attachment of *Neisseria gonorrhoeae* to cells in exudate from patients, he found that fimbriae were rare and that the *Neisseria* cells seemed to be attached more by cell contact rather than by fimbriae. However, there was a good deal of overlaying matter in these preparations and Novotny did not rule out a role for fimbriae in pathogenicity of *Neisseria gonorrhoeae*.

Pseudomonas echinoides and other species of soil bacteria form star-shaped clusters of cells which are attached to each other by their fimbriae (Heuman and Marx, 1964). This organism was isolated by Heuman, who described its polar fimbriae which attached to and agglutinated red blood cells. Mannose inhibited attachment of bacteria to the red blood cells. Heuman and Marx isolated mutants of *Pseudomonas echinoides* which had lost the ability to form star-shaped clusters. The *sta*$^{-}$ mutants still had fimbriae, but they were longer and narrower than normal fimbriae and had lost the ability to attach to red blood cells. Most bacteria which form star-shaped clusters are also capable of conjugation, suggesting a connection between the ability to form clusters, the possession of fimbriae, and conjugation.

IV. Cell Walls and Envelopes

The bacterial cell wall consists of the peptidoglycan layer and the surface layers of the cell. The cell envelope is the wall plus cell membrane but does not include the capsule or appendages. In order to make this subject manageable, I have restricted this section to a discussion of the structure and functions of the cell wall and cell envelope with an emphasis on function. Cell wall chemistry has been reviewed many times, most recently in this volume by Tipper and Wright (Chapter 6). In addition, the chemistry of the cell wall of gram-negative bacteria has been reviewed by Osborn *et al.* (1974), Osborn (1969), and Lüderitz *et al.* (1971). Braun (1975) reviewed the chemistry of the lipoprotein which connects the peptidoglycan and outer layer of gram-negative bacteria. The chemistry of the cell walls of gram-positive bacteria has been reviewed by Rogers (1974), Schleifer and Kandler (1972), and Ghuysen and Shockman (1973).

A. Cell Envelope of Gram-Negative Bacteria

1. Layers of the Cell Envelope

The early work of the chemistry of the cell wall of *Escherichia coli* by Martin, Frank, and Weidel provided the foundation for understanding the cell wall structure of this organism (Weidel *et al.*, 1960; Martin and Frank, 1962; Weidel and Pelzer, 1964). The surface layers of cell wall preparations can be removed by sodium dodecyl sulfate and phenol, exposing the lipoprotein layer which can be removed with pepsin; the remaining peptidoglycan layer can be dissolved with lysozyme (see Braun, 1975, for electron micrographs of this process). When the pepsin step is omitted, lysozyme

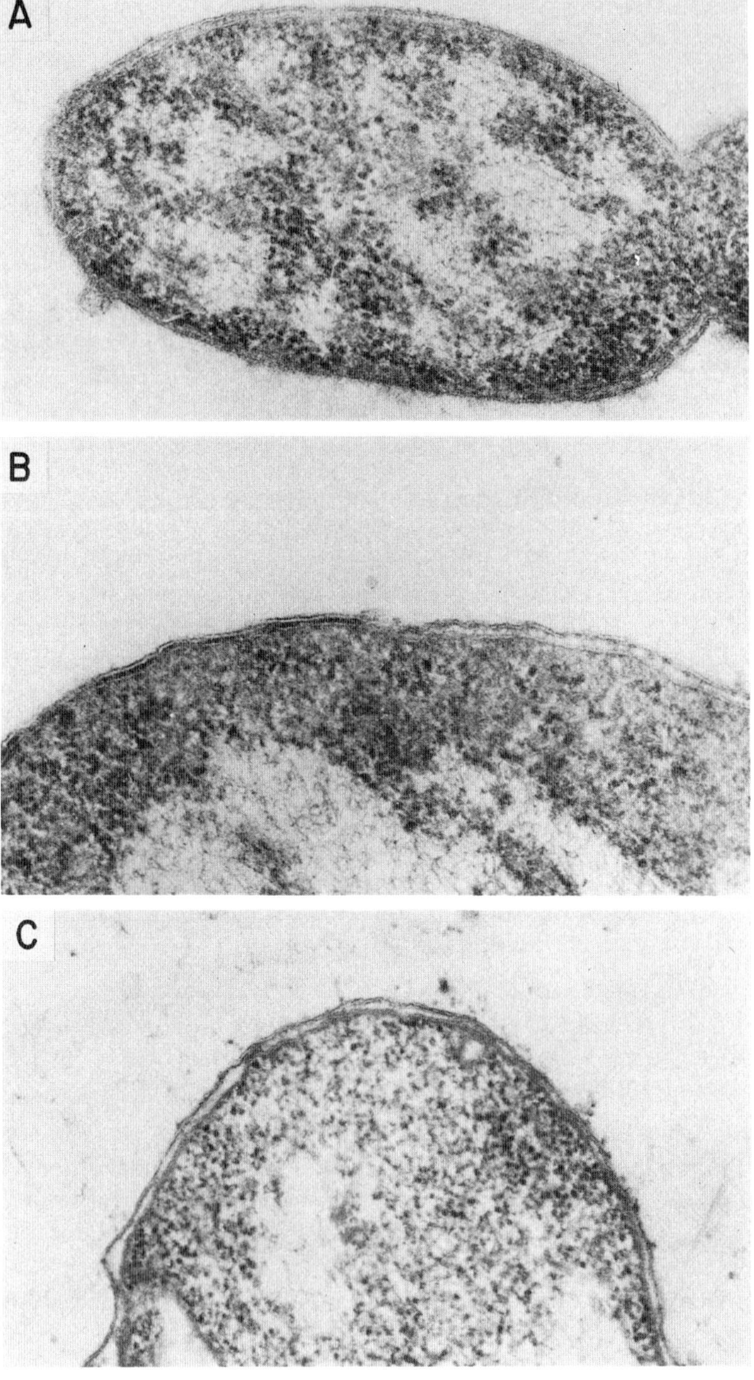
A
B
C

digestion of peptidoglycan leaves behind globules of lipoprotein, mostly the Braun lipoprotein still covalently attached to peptidoglycan fragments.

Murray and DePetris correlated the structure of *Escherichia coli* cell walls with their chemistry. Murray *et al.* (1965) identified the peptidoglycan layer in electron micrographs by its disappearance when the cells were treated with lysozyme or grown in the presence of penicillin (Fig. 15). This placed the cell membrane immediately below the peptidoglycan and lipopolysaccharide in the outer layer immediately above the peptidoglycan. Murray *et al.* (1965) and DePetris (1967) also measured the dimensions of cell wall layers of *Escherichia coli.* The overall thickness of the peptidoglycan and outer layer was approximately 12.0 nm. Both the outer layer and cell membrane had the typical double-track appearance, and both had the dimensions of biological membranes in electron micrographs, both being approximately 6.0 nm in thickness. The peptidoglycan layer was 2.5 to 3.0 nm in thickness, and there was a gap (the g_1 layer of DePetris, 1967) between the peptidoglycan and outer layers of approximately 3.0 nm. A gap of 4.0–4.5 nm was named the M layer by DePetris and was frequently seen between the cell membrane and peptidoglycan; however there is some question if this gap is real. In the living cell, turgor pressure would keep the cell membrane snug against the peptidoglycan layer. Bayer (1974) demonstrated several points of attachment of cell membrane to peptidoglycan and outer layer, which he named adhesion points and which he believed functioned in transport of viral nucleic acid into the cell (see Tipper and Wright, this volume, Chapter 6). For some time it was thought that *Enterobacteriaceae* divided by constriction rather than by septum formation like gram-positive bacteria (Rogers, 1970) and other gram-negative bacteria (Maier and Murray, 1965; Reyn *et al.*, 1970). However, Burdett and Murray (1974) demonstrated the formation of a genuine septum during cell division in *Escherichia coli* and showed that the previous negative results were a matter of sample preparation.

2. The Cell (Inner) Membrane

a. Structure of the Cell Membrane. The double-track appearance of the *Escherichia coli* cell membrane in cross section is typical of most bacteria;

FIG. 15. These photographs are cross sections of *Escherichia coli* viewed with the electron microscope. (A) is an intact cell showing the outer, middle, and inner (cell membrane) layers typical of gram-negative bacteria. Both the outer and inner layers have the double track appearance of cell membranes in electron micrographs. (B) is of *Escherichia coli* growing in the presence of penicillin which results in an incomplete middle layer of peptidoglycan. (C) is a section of *Escherichia coli* treated with lysozyme which completely removes the middle layer of peptidoglycan. (From Murray *et al.*, 1965; reproduced with permission by the *Canadian Journal of Microbiology*.)

however, chromatophores of photosynthetic bacteria appear to be extentions of the cell membrane (Oelze and Drews, 1972; Dutton and Prince, this series, Volume VI, Chapter 8). Recently, a complicated membraneous structure was discovered in *Crenothrix ploysporа*, a sheathed bacterium which is not photosynthetic, but which may be a methane-oxidizing organism (Völker *et al.*, 1977). Valuable information about cell membrane structure comes from freeze-fracture studies where the main fracture plane is through the paraffin interior of the membrane, although other fracture planes occur (Figs. 16–18) (Van Gool and Nanninga, 1971). Particles are usually seen on the interior leaflet (A face) of the cell membrane which is on the cytoplasmic side of the membrane. The particles are 5–10 nm in diameter and have been observed in cell membranes of *Mycoplasma*, gram-negative bacteria, gram-positive bacteria, and eukaryotic cells. They are believed to be protein (Salton and Owen, 1976), and indeed when *Acholeplasma laidlawii* was incubated in media with puromycin or without amino acids the number of particles decreased (Tourtellote and Zupnik, 1973). These particles usually are uniformly distributed, although at low temperatures, there are patches of the *Escherichia coli* membrane inner leaflet devoid of particles as though the transition to the gel state had excluded the particles from the lipid-rich areas (Schechter *et al.*, 1974). Dimples or depressions are seen in the exterior leaflet (B face), which is adjacent to the peptidoglycan layer, and there are usually fewer depressions on the B face than particles in the A face.

The fact that particles are associated with the internal leaflet has been used to determine the orientation of membrane vesicles, which have been used so extensively in transport studies (Kaback, 1974). Altendorf and Staehelin (1974) studied freshly prepared membrane vesicles of *Escherichia coli* by freeze-fracture electron microscopy. The inner (convex) face of the vesicle had particles 6–10 nm in diameter, while the outer (concave) surface was covered with depressions (Fig. 18) leading to the conclusion that vesicles prepared in the usual fashion (Kaback, 1974) have the same orientation as the cell membrane. This point becomes important in the interpretation of transport studies. Finally, there is almost uniform consensus that these particles are protein and that they are enzymes located in the inner leaflet of the cell membrane (see Section IV,2,c).

b. Chemical Properties of Cell Membrane. The cell membrane and outer layer of gram-negative bacteria can be separated from the spheroplasts by density gradient centrifugation (Miura and Mizushima, 1968; Schnaitman, 1970; Osborn *et al.*, 1972). Osborn *et al.* (1972) identified two peaks which sedimented at densities of 1.14 and 1.16 gm/ml as being cell membranes. It is not clear why the cell membrane fraction separated into two bands, although there were minor differences in chemical composition. These fractions contained enzymes of the electron transport system (cytochromes, NADH oxidase, and succinate dehydrogenase), enzymes of active transport (phosphotransferase II and D-lactate dehydrogenase), and enzymes of poly-

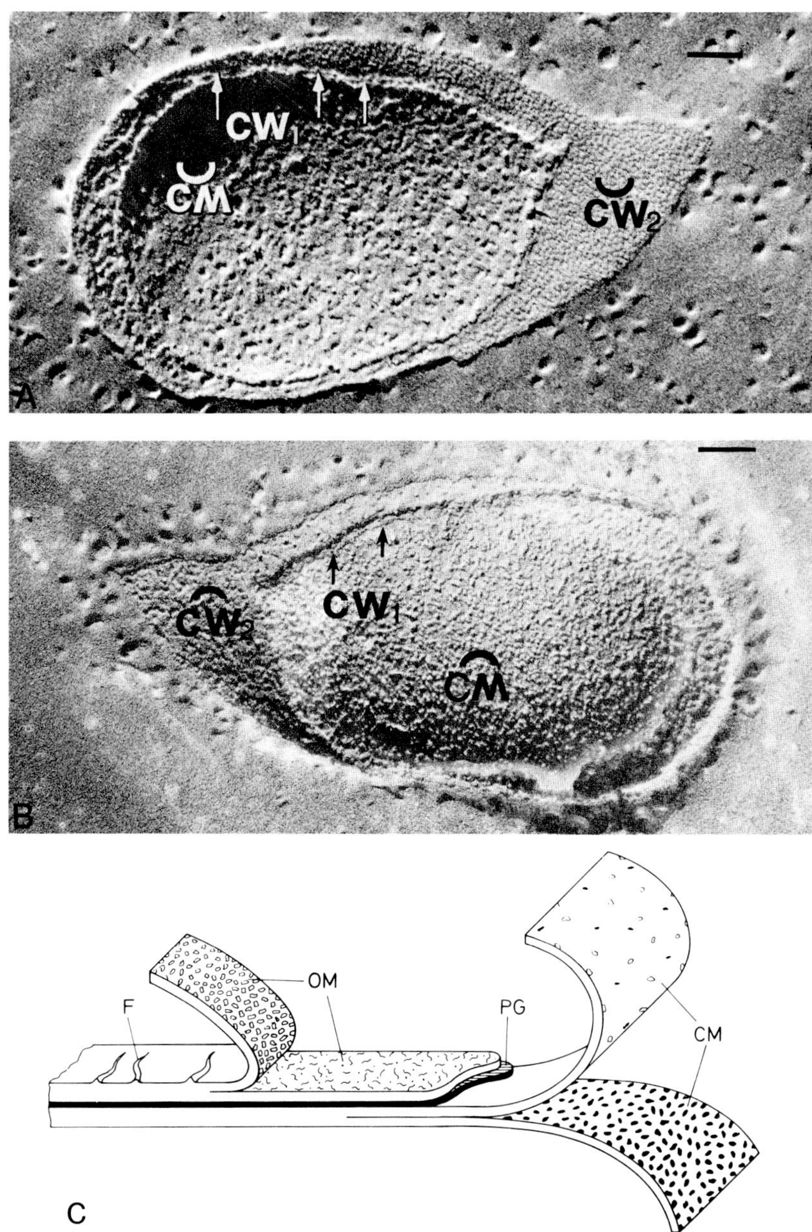

FIG. 16. (A) and (B) are complimentary fracture faces of the *Escherichia coli* cell envelope. The main fracture plane is between the inner (convex) leaf of the cell membrane CM and the outer (concave) leaf of the cell membrane CM. Concave and convex refer to the appearance of these faces after freeze-fracturing and shadowing. There is a less pronounced fracture plane between the corresponding layers of the outer membrane, CW_2 and CW_2. CW_1 is the peptidoglycan layer. (C) shows the probable location of the main fracture planes in *Escherichia coli*. F, flagellum; OM, outer membrane; PG, peptidoglycan; CM, cell membrane. (From Van Gool and Nanninga, 1971; reproduced with permission of the American Society of Microbiology.)

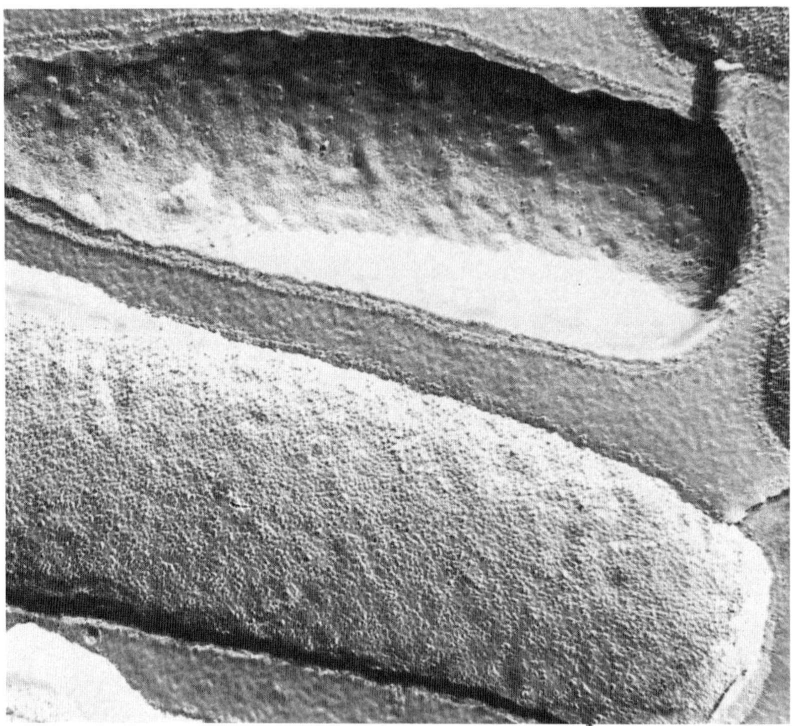

FIG. 17. Electron micrographs of freeze-fractured *Escherichia coli* log phase cells. The bottom structure is the inner (convex) face of the plasma membrane, and the top structure is the outer (concave) face of the plasma membrane. Particles that are suspected to be membrane proteins are evenly distributed over the surface of the inner face. The outer face has some indentations but fewer than the number of particles on the A face. (From Altendorf and Staehelin, 1974; reproduced with permission of the American Society for Microbiology.)

saccharide synthesis (UDPglucose hydrolase). Both membrane fractions were composed of two-thirds protein and one-third phosolipid, with less than 10% of lipopolysaccharide. Osborn *et al.* (1972) resolved the outer and inner fractions into approximately 20 proteins with sodium dodecyl sulfate gel electrophoresis, about equally divided between the inner and outer membranes. This number of proteins is too low, since there are at least 30 proteins in the outer layer alone (Hindennach and Henning, 1975). The lipid fraction was composed of 60% phosphatidylethanolamine, 33% phosphatidylglycerol, and 7% cardiolipin. The combined cell membrane fractions also contained approximately 5% of the total cell peptidolyglycan but did not contain RNA or DNA.

c. Localization of Enzymes in the Cell Membrane. It was pointed out in the preceding paragraph that many enzymes are located in the cell membrane

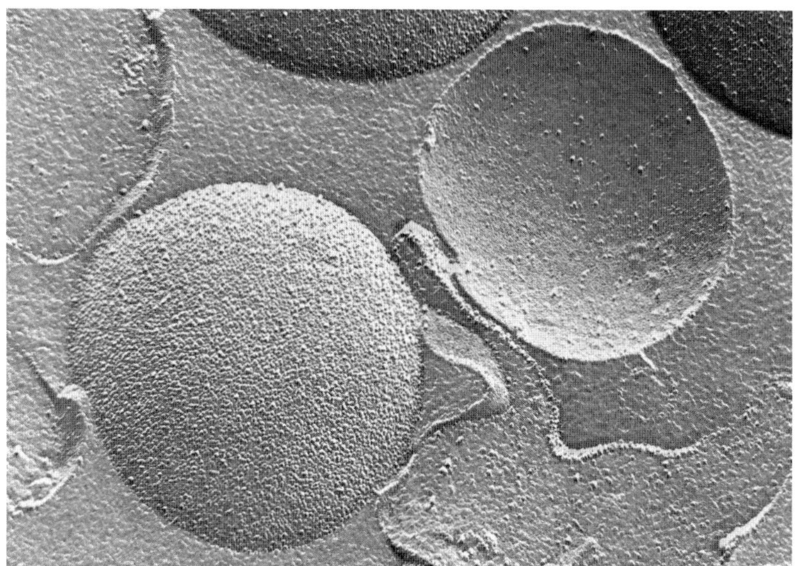

FIG. 18. Inner and outer leaflets of membrane vesicles of *Escherichia coli*. The number of particles on the A face (lower left) identified it as the inner leaflet of the cell membrane, and the lack of particles on the B face (upper right) identified it as the outer leaflet of the cell membrane. Therefore, these vesicles have the same orientation as the membranes of the intact cell. (From Altendorf and Staehelin, 1974; reproduced with permission of the American Society for Microbiology.)

and in Section IV,A,2,a that the inner leaf of the cell membrane contained particles which were probably protein. It is not surprising then that some membrane enzymes are located in the inner leaf. The ATPase of many bacteria is membrane-associated (Salton, 1974a) and is located on the inner leaflet in the case of *Micrococcus lysodeikticus* (See Section IV,B). ATPase of *Escherichia coli* is almost certainly located on the inner leaflet also. Futai (1974) was unable to detect ATPase with intact spheroplasts of *Escherichia coli* until the spheroplasts were treated with toluene to destroy their permability. Membrane vesicles treated with sonic oscillation or passed through the French pressure cell had high ATPase activity, indicating that these vesicles were inside-out. Futai also found NADH oxidase on the inner leaflet of the *Escherichia coli* cell membrane.

α-Glycerol-3-phosphate, succinate, and D-lactate dehydrogenases are also located in the inner leaf of the cell membrane of *Escherichia coli*. Weiner (1974) found very little activity of α-glycerol-phosphate and succinate dehydrogenases with intact spheroplasts, but measured high activity after treatment with toluene to destroy their permeability. Membrane vesicles, however, had about 50% of normal activity of all three dehydrogenases, which was doubled when the vesicles were treated with toluene.

d. Physical Properties of the Cell Membrane. The bacterial cell membrane, as other membranes, is a fluid membrane of phopholipid with protein molecules interspersed (Singer and Nicolson, 1972). At ordinary temperatures, the paraffin chains of phospholipid fatty acids are randomly arranged, and this disorder is responsible for the fluid properties of the membrane (Fig. 19). At low temperatures, the fatty acid side chains become regularly arranged, which transforms cell membrane phospholipids into a gel. Both lipid and protein components of the cell membrane are able to migrate laterally in fluid membranes. Sackmann *et al.* (1973) used spin labelling to measure the rate of lateral disfusion of a probe in the phospholipid layer of *Escherichia coli* membranes and obtained a diffusion coefficient of of 3.25×10^{-8} cm^2/second. Sackmann *et al.* (1973) calculated that a lipid with the above-mentioned diffusion coefficient would allow a probe to travel at a rate of 2.7 μm/second, or about twice the length of an *Escherichia coli* cell in a second. These authors also estimated the diffusion coefficient of a membrane protein with a molecular weight of 100,000 and a diameter of 2.5 nm to be about 3×10^{-10} cm^2/second, which translates to a lateral movement of about 0.3 μm/second. The temperature at which the paraffin side chains of phospholipids change from an ordered to a disordered arrangement (also referred to as crystalline to noncrystalline and gel to liquid) occurs depending on the fatty acid composition of the membrane. Nonlinear fatty acids, such as unsaturated fatty acids, cyclopropane fatty acids, and branched-chain fatty acids, have lower transition temperatures than straight-chain fatty acids. For example, phospholipids containing elaidic acid, an 18-carbon fatty acid with one trans double bond at position 9, has two transition temperatures, one at 12° and one at 26°C. On the other hand, phospholipids with only stearic acid have one transition temperature at 58°C. One reason that *Escherichia coli* is such an attractive organism for the study of phase transitions in cell membranes is that mutants exist which are unable to synthesize unsaturated fatty acids or to oxidize fatty acids. Almost any unsaturated fatty acid will support growth and be incorporated into the phospholipid of *Escherichia coli* (Silbert, 1975). This technique has made it easy to prepare membranes high in elaidic acid, and these membranes have been useful in the study of membrane transition temperatures and the effect of the physical state of the membrane on membrane function.

3. The Outer Layer

The picture which emerges of the outer layer is one where the outer leaf is predominantly lipopolysaccharide and protein, the inner leaf is predom-

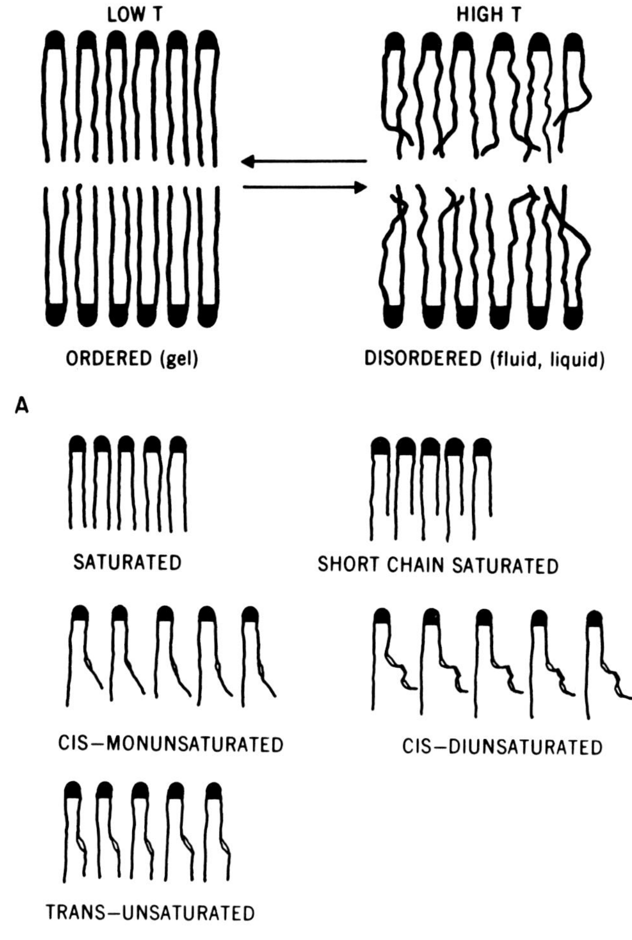

FIG. 19. (A) shows the ordered arrangement of phospholipid fatty acid side chains at temperatures below the transition temperature and the disordered arrangement at temperatures above. (B) shows chemical features of fatty acids that interfere with the ordered arrangement of fatty acid side chains and decrease the transition temperature. The transition temperatures of a homologous series of phospholipids would decrease with the following series of fatty acids: Stearic (58°C), elaidic (12° and 26°C) and linoleic (less than −50°C). (From Cronan and Gelmann, 1975; reproduced with permission of the American Society for Microbiology.)

inantly phospholipid, with proteins such as the matrix protein of Rosenbusch and Braun's lipoprotein being attached to peptidoglycan and extending into the outer layer (Smit *et al.*, 1975). The reader is directed to Chapter 7 by Wright and Tipper in this volume for a detailed discussion of the roles of these proteins in the outer layer. The appearance of the outer layer as a typical biological membrane in cross section was mentioned in Section IV,A,1. It seems almost certain that the outer sheath of spirochaetes corresponds to the outer layer of gram-negative bacteria (Canale-Parola, 1977). Many gram-negative bacteria also present interesting geometrical patterns on their surface (Glauert and Thornley, 1969). Buckmire and Murray (1973) isolated a protein, MW 125–150,000, from *Spirillum serpens* which appeared to be responsible for hexagonal patterns on the cell surface. They were able to reproduce the patterns on cell wall fragments using purified protein and calcium.

The outer layer preparations of Osborn *et al.* (1972) had a density of 1.22 gm/ml and were 40% protein, 40% lipopolysaccharide, and 10–12% phospholipid and contained about 13% of the cell's peptidoglycan. Mühlradt and Golecki (1975) showed that the lipopolysaccharide was located on the outer face of membranes prepared at 0°C using ferritin-labeled antibody and concluded that this was the true location of the lipopolysaccharide. When membranes were prepared at 25°C or higher temperatures, lipopolysaccharide was found on both surfaces of the outer layer, presumably being free to migrate at the higher temperature. The amount of phospholipid in the outer layer is not enough to cover both leaves of the outer layer, and since lipopolysaccharide is concentrated in the outer leaf, phospholipid must be concentrated in the inner leaf (Smit *et al.*, 1975). The lipid composition of the outer layer was significantly different from that of the cell membrane (Section IV,A,2,b). The outer layer phospholipid was 81% phosphatidylethanolamine, 17% phosphatidylglycerol, and 2% cardiolipin.

The outer membrane is enzyme-poor compared to the cell membrane, and the only enzymes detected exclusively in the outer layer were phospholipase A and lysophospholipase. The receptors which participate in transport of vitamin B_{12}, iron, maltose, and nucleosides are discussed by Wright and Tipper (this volume, Chapter 7). UDPsugar hydrolase, ribonuclease I, and endonuclease I were present in both inner and outer membranes. The location of F pilin in the outer layer by Beard and Connolly (1975) was mentioned in Section II.

Although the enzyme content of the outer layer was low compared to the cell membrane, at least 30 polypeptides were identified by Haller *et al.* (1975) using sodium dodecyl sulfate gel electrophoresis. Smit *et al.* (1975) believed that most of the outer layer proteins were concentrated in the outer leaf. This conclusion was based mainly on freeze-etching studies, which showed that there were a large number of particles in the outer leaf of the outer layer. Several of these proteins have been purified and characterized by different investigators, leading to some confusion in terminology (see Wright and Tipper, this volume, Chapter 7). Hindennach and Henning (1975) developed a procedure for isolation of the major outer layer proteins of *Escherichia coli* B/r and confirmed their identity with proteins isolated in other laboratories. They obtained the following yields from 200 gm of cell paste (there are about 10^5 copies of proteins I, II, and IV in *Escherichia coli*, the number of copies of protein III is unknown):

Protein I: They obtained about 120 mg of this protein which has a MW of 38,000. Protein I is identical with Rosenbusch's protein (Rosenbusch, 1974), Schnaitman's protein 1 (Schnaitman, 1970), and Bragg's protein A_1 (Bragg and Hou, 1972). Protein I is a matrix protein for the outer layer and is linked to peptidoglycan.

Protein II: They obtained about 110 mg of protein II. This protein has a MW of 33,000 and is probably the same as the protein purified by Reithmeier and Bragg (1974).

Protein III: Protein III has a MW of 17,000 and they obtained about 50 mg of this protein.

Protein IV: They obtained 30 mg of protein IV with a MW of 7000. This is identical with Braun's lipoprotein (Braun, 1975), but this represents only that fraction which is free in the outer layer. Approximately the same amount is covalently attached to peptidoglycan.

Schnaitman (1974) also separated four major proteins from the outer layer of *Escherichia coli* strain 0111, and, while one protein is known to be different from those listed above the others may be the same proteins. The reader is directed to Chapter 7 by Wright and Tipper in this volume for a detailed discussion of the role of these proteins in the outer layer. Because of the large number of proteins in the outer layer and the relative lack of enzyme activity, one is led to suspect that these are structural proteins. However, the recent findings that mutants of *Escherichia coli* which lack proteins I and II are apparently normal both physiologically and anatomically raises questions about this concept (Henning and Haller, 1975).

4. The Periplasmic Space

The region enclosed by the outer layer and the cell membrane is the periplasmic space, which Mitchell (1961) defined as an enzyme-retaining compartment bounded on the inside by the cell membrane and on the outside by a molecular sieve, now identified as the outer layer. The location of an enzyme in the periplasmic region is established by release of the enzyme into

TABLE V
PERIPLASMIC PROTEINS OF GRAM-NEGATIVE BACTERIA[a]

Class of protein	Approximate molecular weight of substrate	Organism
Hydrolytic enzymes[b]		
Alkaline phosphatase	260	*E. coli*
Acid hexose phosphatase	260	*E. coli*
Nonspecific acid phosphatase	260+	*E. coli*
5′-Nucleotidase	350	*E. coli*
Cyclic phosphodiesterase	330	*E. coli*
ADPglucose pyrophosphatase	590	*E. coli*
β-Lactamase	338	*E. coli*
β-Lactamase	338	*S. typhimurium*
Asparaginase	132	*E. coli*
Nucleases[b]		
Ribonuclease I	Large	*E. coli*
Deoxyribonuclease I	Large	*E. coli*
Enzymes catabolizing deoxyribonucleosides[b]		
Deoxythymidine phosphorylase	242	*E. coli*
Purine deoxyribonucleoside phosphorylase	276	*E. coli*
Deoxyribomutase	214	*E. coli*
Deoxyriboaldolase	214	*E. coli*
Miscellaneous enzymes		
Cytochrome c_{550}[c]	—	*S. itersonii*
Neomycin phosphotransferase[d]	614	*E. coli*
Primary alkyl sulfatase[e]	166	*Pseudomonas* $C_{12}B$
Secondary alkyl sulfatase[e]	211	*Pseudomonas* $C_{12}B$
Cytochrome c_2[f]	—	*Rhodopseudomonas spheroides*
Cytochrome c_2[f]	—	*Rhodopseudomonas capsulata*

the medium either by osmotic shock or by conversion of the cell to a sphereoplast (Heppel, 1971). Enzymes left in solution after either of these two treatments are presumed to be located between the outer layer and the cell membrane (Table V). Direct labeling of periplasmic enzymes, such as alkaline phosphatase, with ferritin-coupled antibodies supports this interpretation (Wetzel *et al.*, 1970; MacAlister *et al.*, 1972).

B. THE STRUCTURE OF CELL WALLS OF GRAM-POSITIVE BACTERIA

Cell walls of gram-positive bacteria usually appear as a single homogenous layer in electron micrographs (Fig. 20), although layers have been observed

TABLE V
(*Continued*)

Class of protein	Approximate molecular weight of substrate	Organism
Binding proteins for[g]		
Sulfate	98	*S. typhimurium*
Phosphate	98	*E. coli*
Galactose	180	*E. coli*
L-Arabinose	150	*E. coli*
Ribose	150	*S. typhimurium*
Thiamine	301	*E. coli*
Cyancobalamine	1580	*E. coli*
Leucine, isoleucine, valine	131, 117	*E. coli*
Glutamine	146	*E. coli*
Glutamic acid	147	*E. coli*
Phenylalanine	165	*Comamonas* sp.
Histidine	155	*S. typhimurium*
Arginine	174	*E. coli*
Lysine, arginine, ornithine	146, 174, 132	*E. coli*

[a] Most of the above data were taken from the reviews by Heppel (1971) and Rosen and Heppel (1973) and are used to illustrate the types of periplasmic enzymes of gram-negative bacteria with a few recent additions. No attempt was made at a complete review of all periplasmic enzymes.

[b] Heppel (1971).

[c] Garrard (1972).

[d] Goldman and Northrup (1976).

[e] Fitzgerald and Laslie (1975).

[f] Prince *et al.*, (1975).

[g] Rosen and Heppel (1973).

in cell walls of gram-positive bacteria (Tipper and Wright, this volume, Chapter 6). Glauert (1962) stated that the cell wall thickness of several gram-positive species ranged from 15 to 35 nm with a maximum of 80 nm for *Lactobacillus acidophilus.* Higgins and Shockman (1970) made careful measurements of the cell wall of *Streptococcus faecalis* and arrived at an average thickness of 27 nm. The narrowest part of the wall was approximately 22 nm at the septum, and the thickest part was 31 nm near the notch (Higgins and Shockman, 1976). Beveridge and Murray (1976) reported that the cell walls of *Bacillus subtilis* varied in thickness from 25 nm at the polar cap to a maximum of 35 nm at the cross wall. Higgins and Shockman (1976) also measured the surface area and volume of cell walls of *Streptococcus faecalis.* Gram-positive bacteria form a prominent septum during cell division (Rogers, 1970) (see also Fig. 1 in Chapter 6 by Tipper and Wright, in this volume).

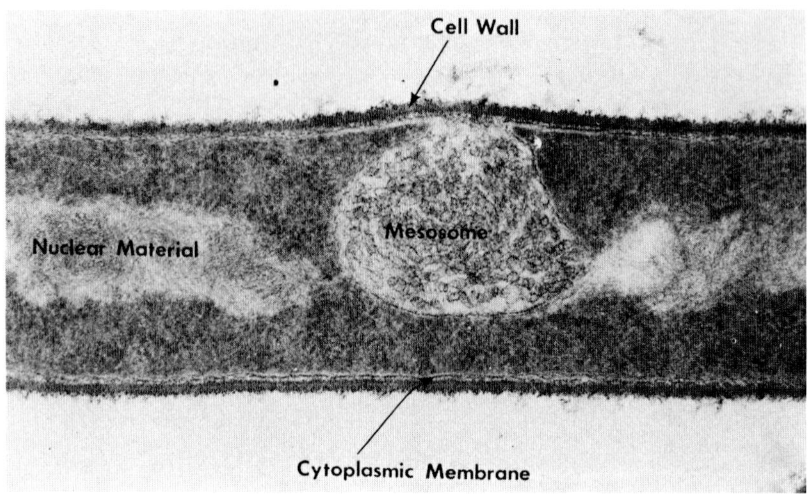

FIG. 20. Electron micrograph of a cross section of *Bacillus licheniformis*. Courtesy of I. Burdett and H. J. Rogers, National Institute for Medical Research London. (From Sokatch and Ferretti, 1976.)

Glauert (1962) gave a thickness of approximately 7.5 nm for the cell membrane of several gram-positive bacteria. Higgins and Shockman (1970) stated that the cell membrane of *Streptococcus faecalis* was 12 nm and gave a figure of 8–10 nm generally for gram-positive bacteria. Cell membranes of gram-positive bacteria contain particles on the inner leaflet just as gram-negative bacteria and mycoplasma. Tsien and Higgins (1974) found particle-free patches on the inner membrane leaflet in *Streptococcus faecalis* chilled to 3°C, while the membrane of *Streptococcus* kept at 25°C was uniformly covered with particles (Fig. 21). Tsien and Higgins (1974) believed that the patches in membranes of chilled cells were caused by gelling of the membrane resulting in lipid-rich areas devoid of protein particles. This same phenomenon was observed by Shechter *et al.* (1974) with *Escherichia coli* and Verkleij *et al.* (1972) with *Acholepasma laidlawii*, and the same explanation was offered. Oppenheim and Salton (1973) clearly demonstrated the localization of ATPase on the inner leaflet of *Micrococcus lysodeikticus* cell membrane with ferritin-labeled antibody against purified *Micrococcus* ATPase. Membrane vesicles in cross section showed stalked particles only on one side of the membrane, and, since intact protoplasts were unlabeled, it was concluded that *Micococcus* ATPase was located on the inner surface of the cell membrane. Membranes are asymmetrical chemically as well as anatomically. Rothman and Kennedy (1977) found that 68% of the phosphatidylethanolamine in membranes of *Bacillus megaterium* was in the inner membrane leaflet and 32% in the outer leaflet.

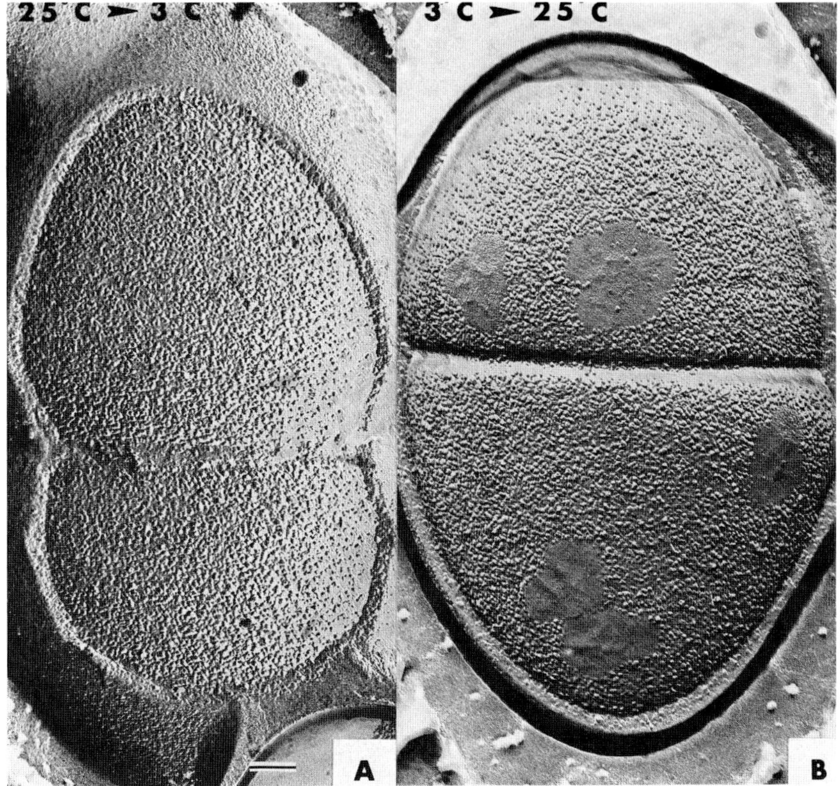

FIG. 21. Effect of temperature on the distribution of particles on the inner leaflet of *Streptococcus faecalis* cell membrane. In (A) the cell was fixed at 25°C and then chilled to 3°C before freeze-fracturing. In (B) the cell was chilled to 3°C fixed and then warmed to 25°C before freeze-fracturing. Chilling before fixing produces areas of the inner leaf which are devoid of particles. (From Tsien and Higgins, 1974; reproduced with permission of the American Society for Microbiology.)

C. FUNCTIONS OF BACTERIAL CELL WALLS AND ENVELOPES

1. PROTECTION OF THE CELL MEMBRANE AGAINST THE ENVIRONMENT

Protection against the environment, particularly against the osmotic shock, was the first important function identified for the cell wall, and evidence supporting this concept was presented by Mitchell and Moyle (1956). Mitchell and Moyle estimated that the interior osmotic pressure of *Staphylococcus aureus* was of the order of 20–30 atm, while that of *Escherichia coli* was 5–6 atm. Such an internal osmotic pressure would exert a force on the cell membrane of *Staphylococcus aereus* of 300 dynes/cm. For a rod of

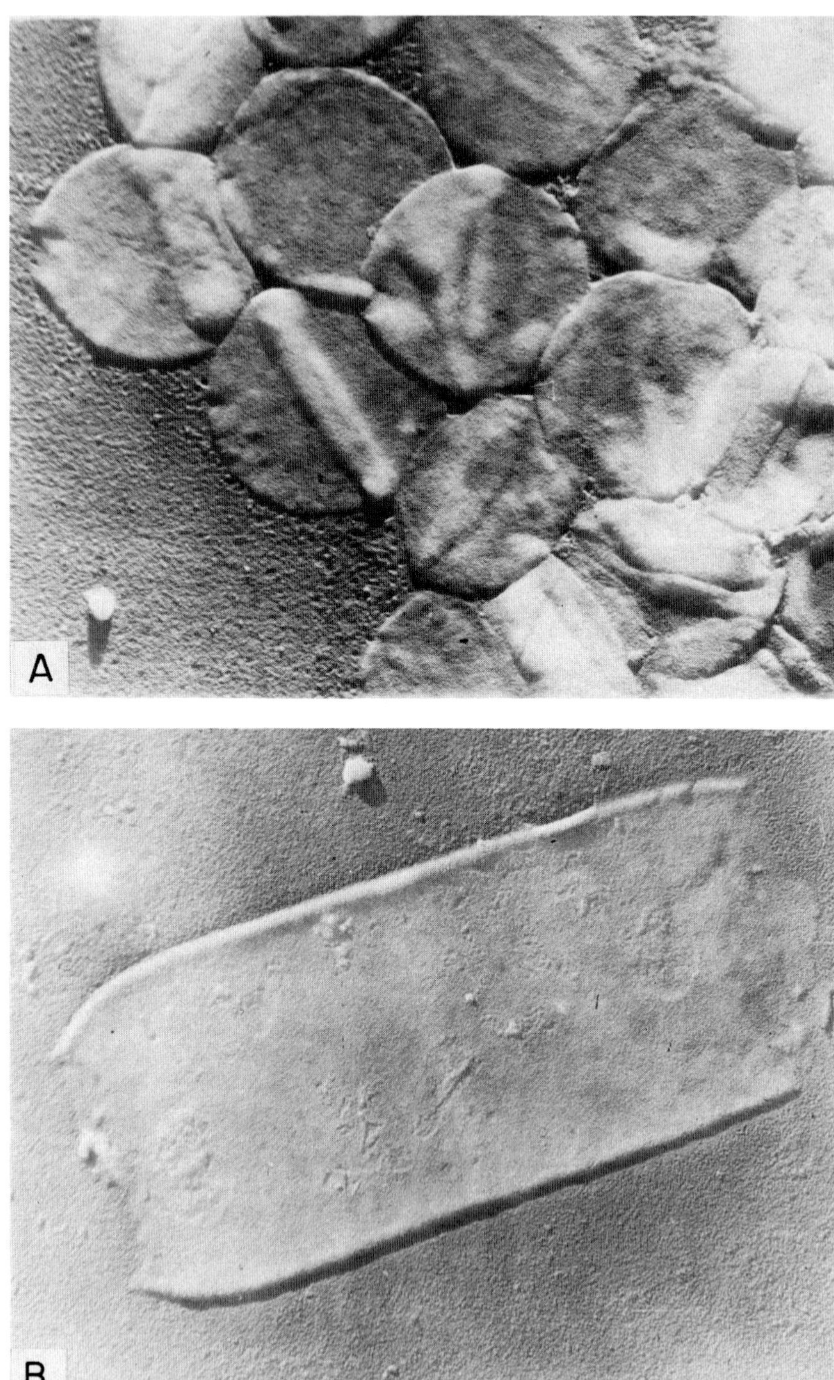

FIG. 22. Isolated peptidoglycan layers retain the shape of the cell. (A) *Streptococcus pyogenes*, (B) *Bacillus cereus*. (From Salton, 1974b.)

the same diameter, the force at right angles to the long axis would be about twice that of the sphere. The forces against the cell membrane are larger in larger organisms, therefore the cell envelope of these organisms must be stronger. Obviously, the cell membrane is not strong enough to protect the protoplast against high internal osmotic pressures relative to the medium. Polar solutes such as phosphate, sodium chloride, and sucrose as well as high molecular weight compounds did not diffuse into the cell (Mitchell and Moyle, 1956). In contrast, glycerol rapidly equilibrated between the medium and the water in the intact cell. Treatment of the cell with trichloroacetic acid or butanol, reagents which destroyed the cell membrane, permitted almost complete equilibration of phosphate between the water of the medium and the water of the cell. The conclusion was that bacteria possessed a semipermeable membrane which allowed exchange of water and glycerol but not of polar or high molecular weight compounds.

The protective role of the cell wall became clearer when it became possible to produce protoplasts of both gram-positive (Weibull, 1953) and gram-negative bacteria (Repaske, 1958) by lysozyme, lysozyme plus EDTA, or by treatment with penicillin. Protoplasts lyse unless they are stabilized by 0.5–1.0 *M* sucrose or polymers such as polyethylene glycol. Even in the presence of a stabilizing agent, protoplasts lyse when aerated because of membrane fragility. Since lysozyme and penicillin affect the peptidoglycan layer of the cell wall, this must be the layer which provides protection to the cell from its environment. According to the calculations of Mitchell and Moyle, the osmotic pressure on the inside of *Staphylococcus aureus* is the result of a cytoplasm with the equivalent of a 12% (w/v) solutes and that of *Escherichia coli* is equivalent to a 2.5% (w/v) solutes. Since both of these solutions would result in osmotic pressures considerably higher than the external medium, the peptidoglycan layer counteracts the forces on the cell membrane caused by the osmotic pressure inside the cell.

2. Role in Determination of Cell Shape

Isolated peptidoglycan layers of both gram-positive and gram-negative bacteria retain the shape of the cell (Fig. 22). When the rigid layer is stripped away from the living cell by either lysozyme or prevented from forming by penicillin, the protoplast assumes a spherical shape (Fig. 23). Under certain conditions, purified outer membranes (i.e., minus peptidoglycan) also retain the shape of the cell, and it is possible that this structure adds some strength to the cell wall. Henning *et al.* (1973) prepared rod-shaped "ghosts" of *Escherichia coli* and other gram-negative bacteria. "Ghosts" are preparations of the outer layer plus some cell membrane material but minus peptidoglycan, and they retain the shape of the cell (see Fig. 26 in Tipper and Wright, this volume, Chapter 6). Others have also reported that treatment of gram-negative bacteria with lysozyme and EDTA in 0.5 *M* sucrose yields

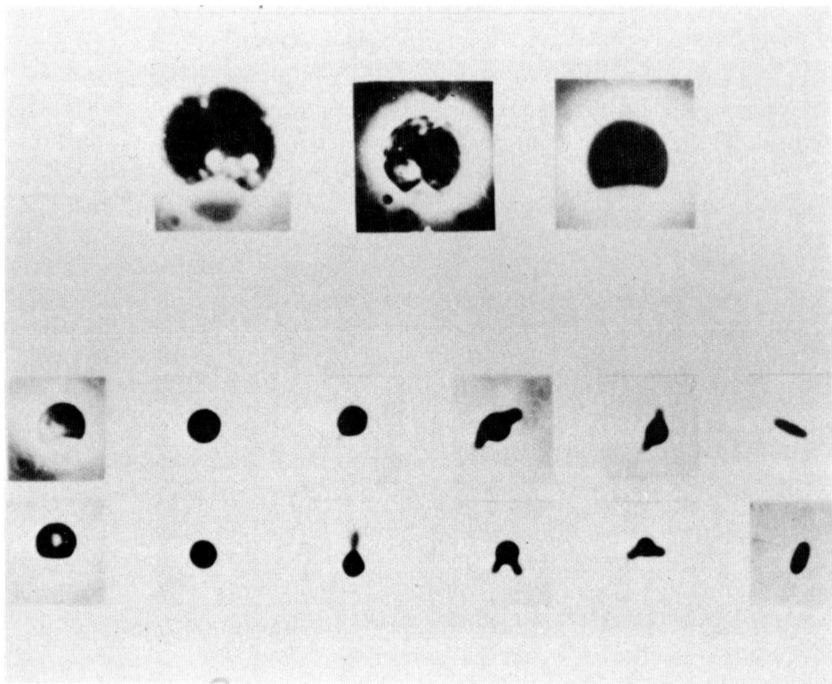

FIG. 23. These figures show the effect of penicillin on growing *Escherichia coli.* Defects occur in the cell wall at the point of new wall synthesis leading eventually to penicillin spheroplasts. (From Lederberg and St. Clair, 1958; reproduced with permission of the American Society for Microbiology.)

preparations lacking peptidoglycan but which retain the shape of the cell (Voss, 1964; Birdsell and Cota-Robles, 1967).

Henning (1975) recently raised a provocative question: What determines the cell shape? Henning's thought was that proteins I and II were responsible for the determination of cell shape (Henning *et al.*, 1973). However, mutants of *Escherichia coli* lacking one or both of these proteins had no obvious defects in either cell shape or physiology (Schnaitman, 1974; Chai and Foulds, 1974; Henning and Haller, 1975), ruling out a role in the determination of cell shape for these proteins. Henning pointed out that there is no proof and very little evidence for any mechanism and that more than one mechanism may be operating. It seems inescapable, however, that the method of synthesis of cell envelope is a factor that should have an important influence on cell wall shape. For example, Schmidt and Stanier (1966) found that stalks of *Caulobacter crescentus* elongated at the base of the stalk (Fig. 24). Swarmer cells were incubated in media with limiting phosphate

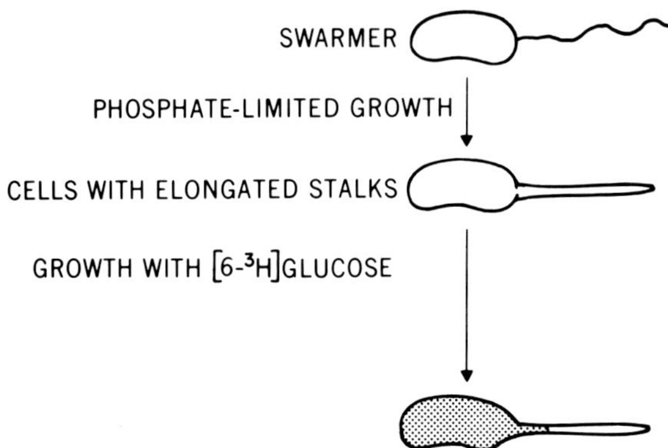

FIG. 24. Mechanism of stalk elongation in *Caulobacter crescentus*. Swarmer (motile) cells were grown in a phosphate-limiting medium until stalks were formed. [6-^{3}H]Glucose was added to the medium, and further stalk growth occurred. The radioactive cells were collected, radioautographed, and the number of silver grains counted in the body and the stalk. Almost all of the radioactivity was found in the cell and the proximal portion of the stalk. (From Schmidt and Stanier, 1966; reproduced with the permission of the Rockefeller Press.)

which causes some elongation of the stalk. Tritiated glucose was added and the incubation continued, resulting in further stalk elongation. Tritiated glucose was incorporated into the cell as well as at the base of the stalk but not into the distal end of the stalk. Therefore, stalk elongation occurred at the base of the stalk rather than at the tip or uniformly over the entire stalk. It is easy to understand how synthesis of the stalk at its base results in a cylindrical stalk, however it is not clear what determines the diameter of the stalk.

The mechanism of cell division in cocci is probably the most important factor responsible for the cell shape of these organisms. Higgins and Shockman (1971) proposed the scheme shown in Fig. 2 in Chapter 6 by Tipper and Wright for cell wall growth and division based on their studies with the electron microscope and on the innunofluorescence studies of Cole and Hahn (1962) and Chung *et al.* (1964). A band on the cell wall marks the boundary between the old wall and new cell wall. As new wall is formed at the midline of the cell, it pushes the old wall outward. A septum forms, closing off the compartment between the two cells, and finally the septum separates into the cell walls of the two daughter cells. The two halves of the streptococcal cell wall are apparently equivalent to the polar caps of bacilli. The new wall is thinnest at the growing point near the notch (Higgins and Shockman, 1976) and becomes rounded due to turgor pressure of the growing cytoplasm (Previc, 1970). After examining these figures, one is led to the conclusion that an important difference between the formation of cocci and rods is that there is no elongation phase of cell wall formation in growth of cocci. The well-known case of rod to sphere morphogenesis of *Arthrobacter crystallopoites* (Ensign and Wolfe, 1964) emphasizes the relationship of cell wall growth in cocci and rods. A little is known about wall formation in other cocci. *Staphylococcus aureus* (Tzagoloff and Novick, 1977) and *Neisseria gonorrhoeae* (Westling-Haggstrom *et al.*, 1977) both divide by forming septa at right angles to the previous plane of division (Fig. 25). When septa between cells are complete, the cells separate, with a snapping motion in *Staphylococcus aureus*, and the walls are forced into their spherical shape by turgor pressure as discussed above.

There are also arguments to explain the shape of bacilli on the basis of cell wall growth. *Escherichia coli* grows by elongation with no change in width during growth in a given medium (Marr *et al.*, 1966). Schwarz *et al.* (1975) believe that new peptidoglycan is laid down in an annulus around the middle of the cell (Fig. 26). As new cell wall is produced, the old cell wall is pushed to the outside. Growth with tritiated diaminopimelic acid added in pulses produced cells that were labeled at the equator. These same authors

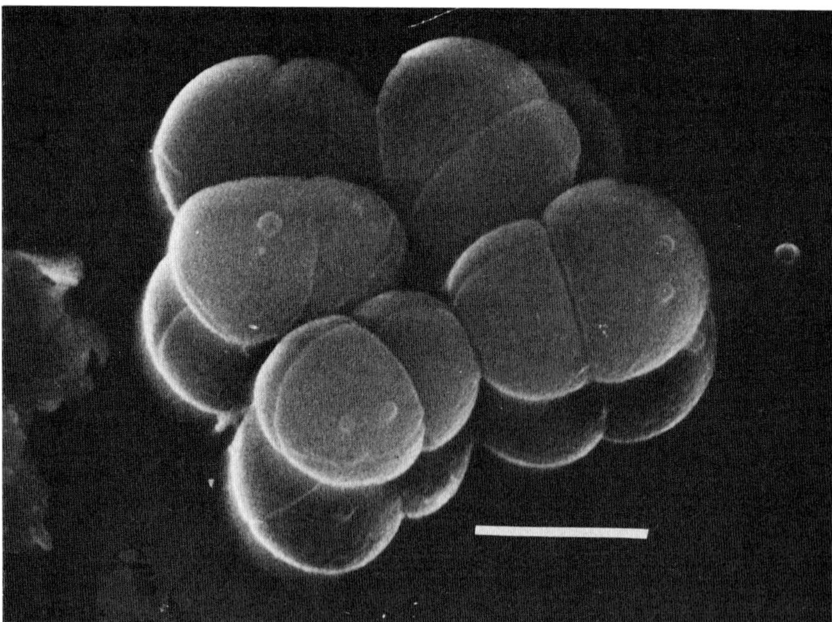

FIG. 25. Scanning electron micrograph of *Staphylococcus aureus* showing planes of division at right angles to each other, middle right, top left. The typical grapelike clusters are formed by postfission movements. (From Koyama *et al.*, 1977; reproduced with permission of the American Society for Microbiology.) Scale, 1 μm.

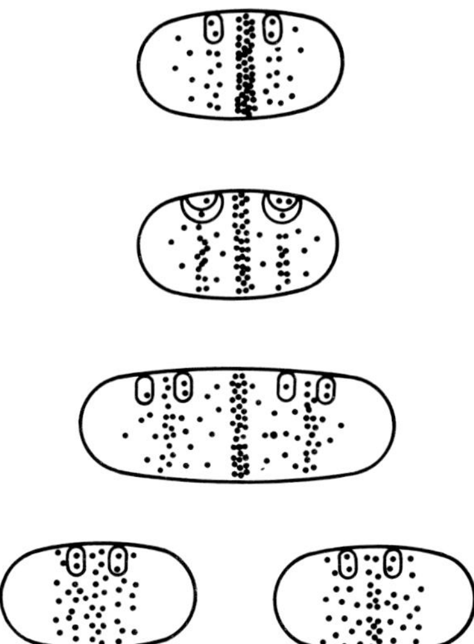

FIG. 26. Zone of new cell wall formation in *Escherichia coli*. The stippled areas represent radioactivity when *Escherichia coli* is grown with tritiated diaminopimelic acid. The most intense zones of radioactivity occur where septum formation will take place. The lighter areas of labeling are probably areas where cell elongation has taken place. The spherical and hemispherical structure are nuclei. (From Schwarz *et al.*, 1975.)

(Schwarz *et al.*, 1975) identified a second type of labeling pattern more diffuse than the first, which may be the result of cell elongation (Fig. 26). These results indicate that new cell wall is produced at the equator where the septum will eventually form. This is similar to the mechanism for synthesis of λ phage receptor in *Escherichia coli* which is also synthesized first at the septum (Ryter *et al.*, 1975). Growth of peptidoglycan at the equator provides a logical explanation for the appearance of *Escherichia coli* grown in the presence of penicillin, in which case bulges caused by weakened walls appear first at the equator (Fig. 23). However, there is much controversy surrounding the mode of envelope growth in rod-shaped bacteria. For example, Begg Donachie (1977) found that receptors for phage T6 were synthesized at both poles of rapidly growing cells, or precisely the converse of where new peptidoglycan appeared according to Schwarz *et al.* (1975). Apparently, once the T6 receptors were formed, they did not move. In contrast, lipopolysaccharide was synthesized at some 200 sites per cell in *Salmonella typhimurium* and then diffused rapidly over the cell surface (Mühlradt *et al.*, 1973, 1974).

A recent proposal by Sargent (1975) for new wall growth in *Bacillus subtilis* looks very much like that for *Escherichia coli* (Fig. 26), where new wall is formed in an annulus at the point where septum formation will occur. Sargent recognized many similarities between his model for wall growth in *Bacillus* and the model of Higgins and Shockman for wall growth in *Streptococcus faecalis*. A principal similarity shared by the two models was the location of new wall formation and the relationship of the daughter nuclei to the area of new wall growth. However, there is at least one other proposal for wall growth in *Bacillus*. Mauck *et al.* (1972) found that tritiated diaminopimelic acid was deposited uniformly over the cell wall and believed that wall growth took place over the entire surface of the *Bacillus megaterium* wall. The differences between the two models for wall growth in *Bacillus* must still be resolved.

3. Ion Exchange by Cell Walls of Gram-Positive Bacteria

The cell wall of gram-positive bacteria is a natural cation exchanger because of the numerous phosphate and carboxylate anions of teichoic acids and carboxylate ions of peptidoglycan. The cation binding capacity of cell walls of *Bacillus subtilis* determined by Beveridge and Murray (1976) in micrograms per milligram dry weight of cell wall was 200 for Mg^{2+}. 200 for Fe^{3+}, 190 for Cu^{2+}. Several other cations were bound but in lesser amounts. Surprisingly, treatment of cell wall preparations with lysozyme destroyed much of the magnesium-binding capacity of the cell, suggesting that magnesium was bound mainly by peptidoglycan rather than by phosphate of teichoic acid. Marquis *et al.* (1976) reported that the affinity of cell walls of *Bacillus megaterium* for ions determined by displacement was H^+, La^{3+}, Cd^{3+}, Sr^{2+}, Ca^{3+}, Mg^{2+}, K^+, Na^+, and Li^+. There have been attempts to correlate cation binding with cell wall layers of gram-positive bacteria, and specifically to identify the location of teichoic acids, however, so far no such correlation has been possible (Garland *et al.*, 1975).

4. Molecular Sieve Functions of the Cell Wall

The concept of a molecular sieve on the cell surface of gram-negative bacteria was proposed by Mitchell (1961). The outer layer was suspected as the diffusion barrier, since there were a number of observations that damage to the outer layer allowed the passage of molecules into the cell that were normally excluded (see review in Costerton *et al.*, 1974). Nakae and Nikaido (1975) made the definitive study on this subject. Stachyose, an oligosaccharide with a mass of 666 daltons, penetrated into the periplasmic space of both plasmolyzed cells and penicillin spheroplasts of *Salmonella typhimurium* with identical kinetics, which shows that it is the outer layer which is the sieve and not the pepeidoglycan. Isolated vesicles of outer layer membranes, however, had a much larger exclusion limit, probably due to damage in preparation.

There have been some interesting speculations on the manner by which molecules pass through the outer layer of gram-negative bacteria. Inouye (1974) proposed that several molecules of Braun's lipoprotein were assembled in such a fashion that they provided a tubular channel through the outer membrane to the periplasmic space. The intriguing part of this concept was that it was possible to arrange the assembly such that hydrophilic groups of the lipoprotein were on the interior of the channel and the hydrophobic groups were on the exterior in the lipid of the outer membrane. The size of the pores would be 1.25 nm for an assembly of six lipoprotein chains. However, Nakae (1976) apparently ruled this mechanism out. Vesicles of lipopolysaccharide and phospholipid from *Salmonella typhimurium* were impermeable to sucrose, but when he added protein from cell envelopes which had been treated with sodium dodecyl sulfate, he found that two protein fractions greatly enhanced the ability of vesicles to pass sucrose. The fraction which contained Braun's lipoprotein was inactive.

Nikaido (1976) obtained evidence for two mechanisms of diffusion through the outer membrane, one for hydrophilic substances and one for hydrophobic substances. Hydrophobic substances, such as actinomycin D, novobiocin, phenol, crystal violet, rifamycin SV, and malachite green, were much more effective against deep rough mutants of *Salmonella typhimurium* than against the wild-type mutants which had lost 80–90% of the lipopolysaccharide. Since many of these substances had molecular weights well below the exclusion limit for oligosaccharides, the conclusion was that their diffusion was selectively restrictrd by the outer membrane. On the other hand, deep rough mutants were no different in their sensitivity to low molecular weight hydrophilic antibiotics than the wild type. Nikaido proposed that hydrophilic antibiotics and oligosaccharides diffused through water-filled pores, while hydrophobic substances diffused through the membrane by dissolving in the lipophilic substances of the outer membrane.

Cell walls of gram-positive bacteria also act as molecular sieves; however, the exclusion limits are much higher than those of gram-negative bacteria. Scherrer and Gerhardt (1971) estimated that the exclusion limits for cell walls of *Bacillus megaterium* were of the order of 100,000 daltons, while the exclusion limit for the protoplast was approximately 600–1000 daltons. They also reestimated their previous data and reported limits of 160,000 daltons for spores of *Bacillus cereus* and 25,000 daltons for cells walls of *Micrococcus lysodeikticus*.

5. Role of Periplasmic Proteins

Substrates for periplasmic enzymes are usually low molecular weight (Table V). In the intestine, which is the natural habitat of *Escherichia coli*, hydrolysis of proteins, polysaccharides, and lipids is done by digestive enzymes. The outer layer allows passage into the periplasmic region of hydrophilic substances, roughly equivalent to tetramers and pentamers of amino acids and dimers and trimers of carbohydrates. Considering these facts, it follows that one function of periplasmic enzymes is to chemically prepare substrates which diffuse through the outer layer for passage through the cell membrane into the cytoplasm. The phosphatases, sulfatases, amidases, and dexoyribonucleoside-catabolizing enzymes are examples of periplasmic proteins with this function. The β-lactamases and neomycin phosphotransferase are special examples of proteins in this class which inactivate antibiotics. A second function of periplasmic proteins is to act as a carrier between membrane-bound enzymes and substrates. The binding proteins and periplasmic cytochromes are examples of proteins which fall into this category. Finally, binding proteins play an important role in chemotaxis and motility, subjects which are discussed in Section I of this chapter and in Chapter 3 by Koshland in this volume.

All of the substrates for periplasmic proteins (Table V) are able to pass through the outer layer with the exception of ribonucleic acid and deoxyribonucleic acid. Roles for ribonuclease I and deoxyribonuclease I have been suggested which are different from the roles proposed for the bulk of the periplasmic enzymes and do not require transport of a high molecular weight substrate past the outer layer or passage of the enzyme into the medium. Kaplan and Apirion (1975) found that degradation of ribosomal RNA was necessary for the survival of starved *Escherichia coli* and that ribonuclease I was one of the enzymes that participated in RNA degradation. They suggested that ribonuclease I might enter the cell through a damaged membrane during starvation and had some circumstantial evidence to support this idea. Ribonuclease I was more active in cells at 45° and 50°C, temperatures which would damage membranes. Smith and Pizer (1968) proposed that deoxyribonuclease I was partially responsible for the resistance of *Escherichia coli* strain W to infection by T2 phage by virtue of its location in the periplasmic space. Cytochrome c_{550} of *Spirillum itersonni* (Garrard, 1972) might also appear to be an exception to the role proposed for periplasmic enzymes in this paper; however, the amount of this enzyme was greatly increased when *Spirillum itersonii* was grown with nitrate. Furthermore, the enzyme was ordinarily membrane-bound, but the amount of periplasmic enzyme greatly increased with nitrate in the medium (Clark-Walker and Lascelles, 1970; Garrard, 1971). These results suggested that cytochrome c_{550} of *Spirillum itersonii* participated in nitrate or nitrite reduction in the periplasmic space, but there were no data to indicate that this occurred.

6. Role of Extracellular Enzymes of Gram-Positive Bacteria

How do gram-positive bacteria process macromolecules prior to transport since they do not have the equivalent of a periplasmic region? There has been the impression for a long time that gram-positive bacteria secrete more degradative enzymes than do gram-negative bacteria. In a recent review of proteolytic enzymes, this impression seems to be borne out (Conrad and Sokatch, 1978). Nineteen of twenty-three well-characterized extracellular proteases were from gram-positive bacteria. Almost all of the extracellular proteases of gram-positive bacteria had molecular weights of 50,000 or less, below the exclusion limits of the cell walls of gram-positive bacteria. Some gram-positive bacteria are especially prone to excreting enzymes, e.g., the spore-forming rods, *Streptococcus pyogenes* and *Staphylococcus aureus*. A consideration of the habitat of these organisms suggests a reason why exoenzymes are effective. *Bacillus* species live in decomposing plant and animal tissues; pathogenic clostridia, *Streptococcus pyogenes*, and *Staphylococcus aureus* live in tissues and cause local necrosis. In these cases, the concentration of nutrients is high, and they are confined as opposed to the dilute environment of ponds and lakes which is the habitat of many gram-negative bacteria.

7. The Effect of Cell Membrane Lipids on Function of Membrane Proteins

Studies of membrane transitions and diffusion of membrane molecules were stimulated by interest in the effect of the physical state of the membrane on membrane function. The overwhelming evidence is that membranes function best when their phospholipids are in the fluid state. This problem has been approached by a study of Arrhenius plots of activity of membrane-bound enzymes, which provide information about the energy of activation of an enzyme-catalyzed reaction. Unsaturated fatty acid mutants of *Escherichia coli* have been valuable in these studies because they can be grown with membrane phospholipids composed of virtually only one type of fatty acid (Silbert and Vagelos, 1974), usually elaidic acid because of its high and sharp transition temperatures.

Esfahani *et al.* (1972) studied the effect of membrane lipids on the activity of a membrane-bound succinate dehydrogenase in an unsaturated fatty acid mutant of *Escherichida coli*. They extracted cell membranes with acetone and added back coenzyme Q_6 and phospholipid. Coenzyme Q_6 alone stimulated activity slightly, but coenzyme Q_6 and phospholipid restored full activity of the dehydrogenase. Furthermore, plots of enzyme activity as a function of the reciprocal of the absolute temperature (Arrhenius plots) were straight for the preparation supplemented with Q_6, but were biphasic for the preparation supplemented by Q_6 and phospholipid. Phospholipid was not only needed for enzyme activity but its presence affected the energy of activation of the enzyme. When the preparation was supplemented with phospholipids high in oleic acid, the change in slope of the Arrhenius plot occurred at 18°C; when the preparation was supplemented with phospholipid rich in elaidic acid, the break occurred at 28°C. Biphasic Arrehnius plots were also obtained for the transport of thiomethyl-β-D-galactoside (Overath *et al.*, 1970), ρ-nitrophenyl-β-D-glucoside (Wilson and Fox, 1971). *o*-nitrophenyl-β-D-galactoside (Linden *et al.*, 1973), and proline, but not for glucose transport (Shechter *et al.*, 1974). Morrisett *et al.* (1975) obtained Arrhenius plots of NADH oxidase and D-lactate oxidase with two distinct changes in slope (Fig. 27). Overath *et al.* (1970) also showed that growth and respiration were affected by the composition of the membrane in an *Escherichia*

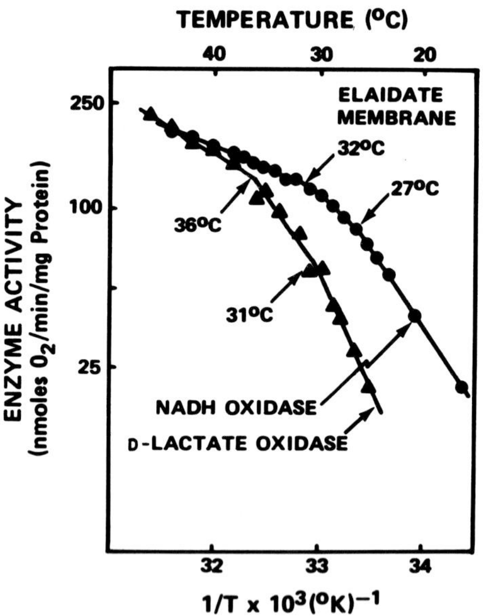

FIG. 27. Arrhenius plots of oxidases for NADH and D-lactate in membranes of *Escherichia coli* strain OL_2^-, an unsaturated fatty acid auxotroph grown in the presence of elaidic acid. There is a transition temperature in both membranes and lipids extracted from this auxotroph that begins at 26°C and extends through 36°C, which is probably due to the presence of dielaidoylphosphatidylethanolamine (DEPE). Therefore, the change in slope at 27°C for NADH oxidase occurs at the beginning of the thermal transition for DEPE, and the change in slope at 31°C for D-lactate oxidase occurs in the middle of the transition for DEPE. (From Morrisett *et al.*, 1975; reproduced with permission of the *Journal of Biological Chemistry*.)

coli unsaturated fatty acid auxotroph and that Arrhenius plots of these functions resembled those for transport of thiomethyl-β-D-galactoside.

In general, the portion of the Arrhenius plot at temperatures above the transition temperature has a shallow slope, meaning that there is a low energy of activation in this region of the curve; the portion of the plot below the transition temperature has a steep slope, reflecting a high energy of activation at temperatures below the transition temperature. The change in slope of the Arrhenius plot, however, occurs at different positions in relation to the transition temperature, sometimes at the start of the transition temperature and sometimes in the middle (Fig. 27). The interpretation of these results is that the ordered arrangement of phospholipid acyl chains in the gel state impairs the movement of proteins in the membrane causing the high energy of activation; conversely, the disordered arrangement of fatty acid side chains in the fluid state produces a lipid membrane where movement of protein molecules is easier, resulting in the lower energy of activation. In support of this concept, both Tsien and Higgins (1974) and Shechter *et al.* (1974) found regions of the inner leaf of the cell membrane that were devoid of particles at low temperatures; in the first case, with *Streptococcus faecalis* (Fig. 21) and, in the second case, with *Escherichia coli.* Morrisett *et al.* (1975) also suggested that lipids normally segregate into regions of different fluidity and composition and this would also affect enzyme activity. The effect of the degree of disorder of membrane lipids is a plausable explanation for the Arrhenius plots obtained with membrane-bound enzymes, particularly for transport involving binding proteins. The lack of a biphasic Arrhenius plot for the transport of glucose in *Escherichia coli* is an obvious exception which needs explanation. Shechter *et al.* (1974) suggested that the difference was due to the fact that glucose was transported by the phosphotransferase system and that only one enzyme of the system, enzyme II, was located in the cell membrane and that perhaps there was no change in configuration of enzyme II during transport.

References

Achtman, M., Willetts, N., and Clark, A. J. (1971). *J. Bacteriol.* **106,** 529.
Altendorf, K. H., and Staehelin, L. A. (1974). *J. Bacteriol.* **117,** 888.
Asakura, S., and Iino, T. (1972). *J. Mol. Biol.* **64,** 251.
Asakura, S., Eguchi, G., and Iino, T. (1968). *J. Mol. Biol.* **35,** 227.
Baumann, P., and Baumann, L. (1977). *Annu. Rev. Microbiol.* **31,** 39.
Bayer, M. E. (1974). *Ann. N.Y. Acad. Sci.* **235,** 6.
Beard, J. P., and Connolly, J. C. (1975). *J. Bacteriol.* **122,** 59.
Beard, J. P., Howe, T. G. B., and Richmond, M. H. (1972). *J. Bacteriol.* **111,** 814.
Begg, K. J., and Donachie, W. D. (1977). *J. Bacteriol.* **129,** 1524.
Bendis, I., and Shapiro, L. (1970). *J. Virol.* **6,** 847.

Berg, H. C. (1974). *Nature* (*London*) **249,** 77.
Berg, H. C. (1975). *Nature* (*London*) **254,** 389.
Berg, H. C. (1976). *J. Theor. Biol.* **56,** 269.
Berg, H. C., and Anderson, R. A. (1973). *Nature* (*London*) **245,** 380.
Berg, H. C., Bromley, D. B., and Charon, N. W. (1978). *Symp. Soc. Gen. Microbiol.* **28,** 285.
Beveridge, T. J., and Murray, R. G. E. (1976). *J. Bacteriol.* **127,** 1502.
Bharier, M. A., and Rittenberg, S. C. (1971). *J. Bacteriol.* **105,** 422.
Bharier, M. A., Eiserling, F. A., and Rittenberg, S. C. (1971). *J. Bacteriol.* **105,** 413.
Birdsell, D. C., and Cota-Robles, E. H. (1967). *J. Bacteriol.* **93,** 427.
Bradley, D. E. (1966). *J. Gen. Microbiol.* **45,** 83.
Bradley, D. E. (1972). *J. Gen. Microbiol.* **72,** 303.
Bradley, D. E. (1975). *Biochem. Biophys. Res. Commun.* **64,** 918.
Bradley, D. E. (1977). *In* "Microbiology—1977" (D. Schlessinger, ed.), p. 127. Am. Soc. Microbiol., Washington, D.C.
Bradley, D. E., and Pitt, T. L. (1974). *J. Gen. Virol.* **24,** 1.
Bradley, D. E., and Rutherford, E. L. (1975). *Can. J. Microbiol.* **11,** 152.
Bragg, P. D., and Hou, D. (1972). *FEBS Lett.* **28,** 309.
Braun, V. (1975). *Biochim. Biophys. Acta* **415,** 335.
Brinton, C. C. (1965). *Trans. N.Y. Acad. Sci.* **27,** 1003.
Brinton, C. C. (1971). *Crit. Rev. Microbiol.* **1,** 105.
Brinton, C. C., Gemski, P., and Carnahan, J. (1964). *Proc. Natl. Acad. Sci. U.S.A.* **52,** 776.
Brodt, P., Leggett, F., and Iyer, R. (1974). *Nature* (*London*) **249,** 856.
Buckmire, F. L. A., and Murray, R. G. E. (1973). *Can. J. Microbiol.* **19,** 59.
Burdett, I. D. J., and Murray, R. G. E. (1974). *J. Bacteriol.* **119,** 1039.
Calladine, C. R. (1976). *J. Theor. Biol.* **57,** 469.
Canale-Parola, E. (1977). *Bacteriol. Rev.* **41,** 181.
Caro, L. G., and Schnös, M. (1966). *Proc. Natl. Acad. Sci. U.S.A.* **56,** 126.
Chai, T., and Foulds, J. (1974). *J. Mol. Biol.* **85,** 465.
Chang, J. Y., Brown, D. M., and Glazer, A. N. (1969). *J. Biol. Chem.* **244,** 5196.
Chung, K. L., Hawirko, R. Z., and Isaac, P. K. (1964). *Can. J. Microbiol.* **10,** 473.
Chwang, A. T., Winet, H., and Wu, T. Y. (1974). *J. Mechanochem. Cell Motil.* **3,** 69.
Clark-Walker, G. D., and Lascelles, J. (1970). *Arch. Biochem. Biophys.* **136,** 153.
Cohen-Bazire, G., and London, J. (1967). *J. Bacteriol.* **94,** 458.
Cole, R. M., and Hahn, J. J. (1962). *Science* **135,** 722.
Conrad, R. S., and Sokatch, J. R. (1978). "Handbook Series in Nutrition and Food." CRC Press, Cleveland, Ohio. In press.
Costerton, J. W., Ingram, J. C., and Cheng, K. J. (1974). *Bacteriol. Rev.* **38,** 87.
Cox, P. J., and Twigg, G. I. (1974). *Nature* (*London*) **250,** 260.
Crawford, E. M., and Gesteland, R. F. (1964). *Virology* **22,** 165.
Cronan, J. E., and Gelmann, E. D. (1975). *Bacteriol. Rev.* **39,** 232.
Curtiss, R. J. (1969). *Annu. Rev. Microbiol.* **23,** 69.
Czajkowski, J., Soltesz, V., and Weibull, C. (1974). *J. Ultrastruct. Res.* **46,** 79.
Datta, N. (1975). *In* "Microbiology—1974" (D. Schlessinger, ed.), p. 9.
Davidson, B. E. (1971). *Eur. J. Biochem.* **18,** 524.
DeLange, R. J., Chang, J. Y., Shaper, J. H., Martinez, R. J., Komatsu, S. K., and Glazer, A. N. (1973). *Proc. Natl. Acad. Sci. U.S.A.* **70,** 3428.
DeLange, R. J., Chang, J. Y., Shaper, J. H., and Glazer, A. N. (1976). *J. Biol. Chem.* **25,** 705.
DePamphilis, M. L., and Adler, J. (1971a). *J. Bacteriol.* **105,** 376.
DePamphilis, M. L., and Adler, J. (1971b). *J. Bacteriol.* **105,** 384.
DePamphilis, M. L., and Adler, J. (1971c). *J. Bacteriol.* **105,** 396.

DePetris, S. (1967). *J. Ultrastruct. Res.* **19,** 45.
Dimmitt, K., and Simon, M. (1971). *J. Bacteriol.* **105,** 369.
Duguid, J. P., and Wilkinson, J. F. (1961). *Symp. Soc. Gen. Microbiol.* **11,** 69.
Duguid, J. P., Smith, I. W., Dempster, G., and Edmunds, P. N. (1955). *J. Pathol. Bacteriol.* **70,** 335.
Duguid, J. P., Anderson, E. S., and Campbell, I. (1966). *J. Pathol. Bacteriol.* **92,** 107.
Ensign, J. C., and Wolfe, R. S. (1964). *J. Bacteriol.* **87,** 925.
Esfahani, M., Crawfoot, P. D., and Wakil, S. F. (1972). *J. Biol. Chem.* **247,** 7251.
Falkinham, J. P., and Curtiss, R. (1976). *J. Bacteriol.* **126,** 1194.
Fitzgerald, J. W., and Laslie, W. W. (1975). *Can. J. Microbiol.* **21,** 59.
Follett, E. A. C., and Gorden, J. (1963). *J. Gen. Microbiol.* **32,** 235.
Frost, L. S., and Paranchych, W. (1977). *J. Bacteriol.* **131,** 259.
Fuerst, J. A., and Hayward, A. C. (1969). *J. Gen. Microbiol.* **58,** 227.
Futai, M. (1974). *J. Membr. Biol.* **15,** 15.
Garland, J. M., Archibald, A. R., and Baddiley, J. (1975). *J. Gen. Microbiol.* **89,** 73.
Garrard, W. T. (1971). *J. Bacteriol.* **105,** 93.
Garrard, W. T. (1972). *J. Biol. Chem.* **247,** 5935.
Ghuysen, J. M., and Shockman, G. D. (1973). *In* "Bacterial Membranes and Walls" (L. Leive, ed.), p. 37. Dekker, New York.
Glauert, A. M. (1962). *Br. Med. Bull.* **18,** 245.
Glauert, A. M., and Thornley, M. J. (1969). *Annu. Rev. Microbiol.* **23,** 159.
Glauert, A. M., Kerridge, D., and Horne, R. W. (1962). *J. Cell Biol.* **18,** 327.
Glossman, H., and Bode, W. (1972). *Hoppe-Seyler's Z. Physiol. Chem.* **353,** 298.
Goldman, P. R., and Northrop, D. B. (1976). *Biochem. Biophys. Res. Commun.* **69,** 230.
Greenawalt, J. W., and Whiteside, T. L. (1975). *Bacteriol. Rev.* **39,** 405.
Haller, I., Hoehn, B., and Henning, U. (1975). *Biochemistry* **14,** 479.
Henning, U. (1975). *Annu. Rev. Microbiol.* **29,** 45.
Henning, U., and Haller, I. (1975). *FEBS Lett.* **55,** 161.
Henning, U., Hohn, B., and Sonntag, I. (1973). *Eur. J. Biochem.* **39,** 27.
Heppel, L. A. (1971). *In* "Structure and Function in Biological Membranes" (L. I. Rothfield, ed.), p. 223. Academic Press, New York.
Heumann, W. (1968). *Mol. Gen. Genet.* **102,** 132.
Heumann, W., and Marx, R. (1964). *Arch. Mikrobiol.* **47,** 325.
Higgins, M. L., and Shockman, G. D. (1970). *J. Bacteriol.* **103,** 244.
Higgins, M. L., and Shockman, G. D. (1971). *Crit. Rev. Microbiol.* **1,** 29.
Higgins, M. L., and Shockman, G. D. (1976). *J. Bacteriol.* **127,** 1346.
Hindennach, I., and Henning, U. (1975). *Eur. J. Biochem.* **59,** 207.
Hofschneider, P. H. (1963). *Z. Naturforsch., Teil B* **18,** 203.
Holt, S. C., and Canale-Parola, E. (1968). *J. Bacteriol.* **96,** 822.
Iino, T. (1969). *Bacteriol. Rev.* **33,** 454.
Iino, T. (1974). *J. Supramol. Struct.* **2,** 372.
Iino, T. (1977). *Annu. Rev. Genet.* **11,** 161.
Inouye, M. (1974). *Proc. Natl. Acad. Sci. U.S.A.* **71,** 2396.
Jahn, J. L., and Landman, M. D. (1965). *Trans. Am. Microsc. Soc.* **84,** 395.
Jephcott, A. E., Reyn, A., and Birch-Anderson, A. (1971). *Acta Pathol. Microbiol. Scand., Sect. B* **79,** 437.
Joseph, R., and Canale-Parola, E. (1972). *Arch. Mikrobiol.* **81,** 146.
Joseph, R., Holt, S. C., and Canale-Parole, E. (1973). *J. Bacteriol.* **115,** 426.
Joys, T. M., and Rankis, V. (1972). *J. Biol. Chem.* **247,** 5180.
Kaback, H. R. (1974). *Science* **186,** 882.

Kaplan, R., and Apirion, D. (1975). *J. Biol. Chem.* **250,** 1854.
Kellogg, D. S., Cohen, I. R., Norins, L. C., Schroeter, A. L., and Reising, G. (1968). *J. Bacteriol.* **96,** 596.
Kondoh, H., and Hotani, H. (1974). *Biochim. Biophys. Acta* **336,** 117.
Kondoh, H., and Yanagida, M. (1975). *J. Mol. Biol.* **96,** 641.
Koyama, T., Yamada, M., and Matsuhashi, M. (1977). *J. Bacteriol.* **129,** 1519.
Kumazawa, N., and Yanagawa, R. (1972). *Infect. Immun.* **5,** 27.
Lagenaur, C., and Agabian, N. (1976). *J. Bacteriol.* **128,** 435.
Lagenaur, C., and Agabian, N. (1977). *J. Bacteriol.* **131,** 340.
Lagenaur, C., Farmer, S., and Agabian, N. (1977). *Virology* **77,** 401.
Larsen, S. H., Reader, R. W., Kort, E. N., Tso, W. W., and Adler, J. (1974). *Nature (London)* **249,** 74.
Lawn, A. M. (1966). *J. Gen. Microbiol.* **45,** 377.
Lawn, A. M., Meynell, G. G., Meynell, E., and Datta, N. (1967). *Nature (London)* **216,** 343.
Lederberg, J., and St. Clair, J. (1958). *J. Bacteriol.* **75,** 143.
Linde, C. D., Wright, K. L., McConnell, H. M., and Fox, C. F. (1973). *Proc. Natl. Acad. Sci. U.S.A.* **70,** 2271.
Listergarten, M. A., and Socransky, S. S. (1964). *J. Bacteriol.* **88,** 1087.
Lowy, J., and Hanson, J. (1965). *J. Mol. Biol.* **11,** 293.
Lüderitz, O., Westphal, O., Staub, A. M., and Nikaido, H. (1971). *In* "Microbial Toxins" (G. Weinbaum, S. Kadis, and S. Ajl, eds.), Vol. 4, p. 145. Academic Press, New York.
MacAlister, T. J., Costerton, J. W., Thompson, L., Thompson, J., and Ingram, J. M. (1972). *J. Bacteriol.* **111,** 827.
McCoy, E. C., Doyle, D., Wiltberger, H., Burda, K., and Winter, A. J. (1975). *J. Bacteriol.* **122,** 307.
McDonough, M. W. (1965). *J. Mol. Biol.* **12,** 342.
MacFarren, A. C., and Clowes, R. C. (1967). *J. Bacteriol.* **94,** 365.
Maier, S., and Murray, R. G. E. (1965). *Can. J. Microbiol.* **11,** 645.
Marquis, R. E., Mayzel, K., and Carstensen, E. L. (1976). *Can. J. Microbiol.* **22,** 975.
Marr, A. G., Harvey, R. J., and Trentini, W. C. (1966). *J. Bacteriol.* **91,** 2388.
Martin, H. H., and Frank, H. (1962). *Z. Naturforsch., Teil B* **17,** 190.
Martinez, R. J. (1963). *J. Gen. Microbiol.* **33,** 115.
Martinez, R. J., Brown, D. M., and Glazer, A. N. (1967). *J. Mol. Biol.* **28,** 45.
Mauck, J., Chan, L., Glaser, L., and Williamson, J. (1972). *J. Bacteriol.* **109,** 373.
Meynell, E., and Datta, N. (1965). *Nature (London)* **207,** 884.
Meynell, E., Meynell, G. G., and Datta, N. (1968). *Bacteriol. Rev.* **32,** 55.
Meynell, G. G., and Lawn, A. M. (1968). *Nature (London)* **217,** 1184.
Mitchell, P. (1961). *In* "Biological Structure and Function" (T. W. Goodwin and O. Lingberg, eds.), Vol. 2, p. 581. Academic Press, New York.
Mitchell, P., and Moyle, J. (1956). *Symp. Soc. Gen. Microbiol.* **6,** 150.
Miura, T., and Mizushima, S. (1968). *Biochim. Biophys. Acta* **150,** 159.
Morrisett, J. D., Pownall, H. J., Plumlee, K. T., Smith, L. C., Zehner, Z. E., Esfahani, M., and Wakil, W. J. (1975). *J. Biol. Chem.* **250,** 6969.
Mühlradt, P. F., and Golecki, J. R. (1975). *Eur. J. Biochem.* **51,** 343.
Mühlradt, P. F., Menzel, J., Golecki, J. R., and Speth, V. (1973). *Eur. J. Biochem.* **35,** 471.
Mühlradt, P. F., Menzel, J., Golecki, J. R., and Speth, V. (1974). *Eur. J. Biochem.* **43,** 533.
Murray, R. G. E., Steed, P., and Elson, H. E. (1965). *Can. J. Microbiol.* **11,** 547.
Nakae, T. (1976). *J. Biol. Chem.* **251,** 2176.
Nakae, T., and Nikaido, H. (1975). *J. Biol. Chem.* **250,** 7359.

Nauman, R. K., Holt, S. C., and Cox, C. D. (1969). *J. Bacteriol.* **98,** 264.
Nikaido, H. (1976). *Biochim. Biophys. Acta* **433,** 118.
Novick, R. P., Clowes, R. C., Cohen, S. N., Curtiss, R., Datta, N., and Falkow, S. (1976). *Bacteriol. Rev.* **40,** 168.
Novotny, C., Carnahan, J., and Brinton, C. C. (1969a). *J. Bacteriol.* **98,** 1294.
Novotony, C., Raizon, E., Knight, W. S., and Brinton, C. C. (1969b). *J. Bacteriol.* **98,** 1307.
Novotny, P., Short, J. A., and Walker, P. D. (1975). *J. Med. Microbiol.* **8,** 413.
O'Brien, E. J., and Bennett, P. M. (1972). *J. Mol. Biol.* **70,** 133.
Oelze, J., and Drews, G. (1972). *Biochim. Biophys. Acta* **265,** 209.
Old, D. C., and Duguid, J. P. (1970). *J. Bacteriol.* **103,** 447.
Olsen, R. J., and Thomas, D. D. (1973). *J. Virol.* **12,** 1560.
Olsen, R. H., Siak, J., and Gray, R. J. (1974). *J. Virol.* **14,** 689.
Oppenheim, J. D., and Salton, M. R. J. (1973). *Biochim. Biophys. Acta* **298,** 297.
Osborn, M. J. (1969). *Annu. Rev. Biochem.* **38,** 501.
Osborn, M. J., Gander, J. E., Parisi, E., and Carson, J. (1972). *J. Biol. Chem.* **247,** 3962.
Osborn, M. J., Rick, P. D., Lehmann, Rupprecht, E., and Singh, M. (1974). *Ann. N.Y. Acad. Sci.* **235,** 52.
Ottow, J. C. G. (1975). *Annu. Rev. Microbiol.* **29,** 79.
Ou, J. T. (1973). *J. Bacteriol.* **114,** 1108.
Ou, J. T., and Anderson, T. F. (1970). *J. Bacteriol.* **102,** 648.
Overath, P., Schairer, H. U., and Stoffer, W. (1970). *Proc. Natl. Acad. Sci. U.S.A.* **67,** 606.
Pemberton, J. M. (1973). *Virology* **55,** 558.
Pijper, A., Neser, M. L., and Abraham, G. (1956). *J. Gen. Microbiol.* **14,** 371.
Pillot, J., and Ryter, A. (1965). *Ann. Inst. Pasteur, Paris* **108,** 791.
Previc, E. P. (1970). *J. Theor. Biol.* **27,** 471.
Prince, R. C., Baccarini-Melandri, A., Hauska, G. A., Melandri, B. A., and Crafts, A. R. (1975). *Biochim. Biophys. Acta* **387,** 212.
Reithmeier, R. A. F., and Bragg, P. D. (1974). *FEBS Lett.* **41,** 195.
Remsen, C. C., Watson, S. W., Waterbury, J. B., and Truper, H. G. (1968). *J. Bacteriol.* **95,** 2374.
Repaske, R. (1958). *Biochim. Biophys. Acta* **30,** 225.
Reyn, A., Birch-Andersen, A., and Berger, U. (1970). *Acta Pathol. Microbiol. Scand., Sect. B* **78,** 375.
Rogers, H. J. (1970). *Bacteriol. Rev.* **34,** 194.
Rogers, H. J. (1974). *Ann. N.Y. Acad. Sci.* **235,** 29.
Rosen, B. P., and Heppel, L. A. (1973). *In* "Bacterial Membranes and Walls" (L. Leive, ed.), p. 209. Dekker, New York.
Rosenbusch, J. P. (1974). *J. Biol. Chem.* **249,** 8019.
Rothman, J. E., and Kennedy, E. P. (1977). *J. Mol. Biol.* **110,** 603.
Ryter, A., Shuman, H., and Schwartz, M. (1975). *J. Bacteriol.* **122,** 295.
Sackmann, E., Trauble, H., Galla, H. J., and Overath, P. (1973). *Biochemistry* **12,** 5360.
Salton, M. R. J. (1974a). *Adv. Microb. Physiol.* **11,** 213.
Salton, M. R. J. (1974b). *In* "Biomembranes," Part A (S. Fleischer and L. Packer, eds.), Methods in Enzymology, Vol. 31, p. 653. Academic Press, New York.
Salton, M. R. J., and Owen, P. (1976). *Annu. Rev. Microbiol.* **30,** 451.
Sargent, M. G. (1975). *J. Bacteriol.* **123,** 7.
Scherrer, R., and Gerhardt, P. (1971). *J. Bacteriol.* **107,** 718.
Schleifer, K. H., and Kandler, O. (1972). *Bacteriol. Rev.* **36,** 407.
Schmidt, J. M. (1966). *J. Gen. Microbiol.* **45,** 347.

Schmidt, J. M., and Stanier, R. Y. (1966). *J. Cell Biol.* **28,** 423.
Schnaitman, C. A. (1970). *J. Bacteriol.* **104,** 882.
Schnaitman, C. A. (1974). *J. Bacteriol.* **118,** 442.
Schwarz, U., Ryter, A., Raimbach, A., Hellio, R., and Hirota, Y. (1975). *J. Mol. Biol.* **98,** 749.
Schweizer, M., and Henning, U. (1977). *J. Bacteriol.* **129,** 1651.
Seidler, R. J., and Starr, M. P. (1968). *J. Bacteriol.* **95,** 1952.
Shahrabadi, M. S., Bryan, L. E., and Van Den Elzen, H. M. (1975). *Can. J. Microbiol.* **21,** 591.
Shapiro, L., Agabian-Keshishian, N., and Bendis, I. (1971). *Science* **173,** 884.
Shechter, E., Letellier, L., and Gulik-Krzywicki, T. (1974). *Eur. J. Biochem.* **49,** 61.
Shimada, K., Kamiya, R., and Asakura, S. (1975). *Nature (London)* **254,** 332.
Shinoda, S., Honda, T., Takeda, Y., and Miwatani, T. (1974). *J. Bacteriol.* **120,** 923.
Silbert, D. F. (1975). *Annu. Rev. Biochem.* **44,** 315.
Silbert, D. F., and Vagelos, P. R. (1974). *In* "Biomembranes," Part B (S. Fleischer and L. Packer, eds.), Methods in Enzymology, Vol. 32, p. 856. Academic Press, New York.
Silverman, M., and Simon, M. I. (1972). *J. Bacteriol.* **112,** 986.
Silverman, M., and Simon, M. I. (1974). *Nature (London)* **249,** 73.
Silverman, M., and Simon, M. I. (1977). *Annu. Rev. Microbiol.* **31,** 397.
Silverman, M., Matsumura, P., Draper, R., Edwards, E., and Simon, M. (1976). *Nature (London)* **261,** 248.
Singer, S. J., and Nicolson, G. L. (1972). *Science* **175,** 720.
Smit, J., Kamio, Y., and Nikaido, H. (1975). *J. Bacteriol.* **124,** 942.
Smith, H. S., and Pizer, L. I. (1968). *J. Mol. Biol.* **37,** 131.
Smith, R. W., and Koffler, H. (1971). *Adv. Microb. Physiol.* **6,** 219.
Sokatch, J. R., and Ferretti, J. J. (1976). "Basic Bacteriology and Genetics." Yearbook Publ., Chicago, Illinois.
Stanish, V. (1974). *J. Gen. Microbiol.* **84,** 332.
Swanson, J. (1973). *J. Exp. Med.* **137,** 571.
Swanson, J., Kraus, S. J., and Gotschlich, E. C. (1971). *J. Exp. Med.* **134,** 886.
Thipayathasana, P., and Valentine, R. C. (1974). *Biochim. Biophys. Acta* **347,** 464.
Tourtellote, M. E., and Zupnik, J. S. (1973). *Science* **179,** 84.
Tsien, H. C., and Higgins, M. L. (1974). *J. Bacteriol.* **118,** 725.
Tweedy, J. N., Park, R. W. A., and Hodgkiss, W. (1968). *J. Gen. Microbiol.* **51,** 235.
Tzagoloff, H., and Novick, R. (1977). *J. Bacteriol.* **129,** 343.
Vaituzis, Z., and Doetsch, R. N. (1969). *J. Bacteriol.* **100,** 512.
Valentine, R. C., and Strand, M. (1965). *Science* **148,** 511.
Valentine, R. C., Silverman, P. M., Ippen, K. A., and Mobach, H. (1969). *Adv. Microb. Physiol.* **3,** 1.
Van Gool, A. P., and Nanninga, N. (1971). *J. Bacteriol.* **108,** 474.
Van Iterson, W., Hoeniger, J. F. M., and Van Zanten, E. N. (1966). *J. Cell Biol.* **31,** 585.
Verkleij, A. J., Ververgaert, P. H. J., Van Deenen, L. L. M., and Elbers, P. F. (1972). *Biochim. Biophys. Acta* **288,** 326.
Völker, H., Schweisfurth, R., and Hirsch, P. (1977). *J. Bacteriol.* **131,** 306.
Voss, J. G. (1964). *J. Gen. Microbiol.* **35,** 313.
Wang, C. Y., and Jahn, T. L. (1972). *J. Theor. Biol.* **36,** 53.
Weibull, C. (1953). *J. Bacteriol.* **66,** 688.
Weidel, W., and Pelzer, H. (1964). *Adv. Enzymol. Relat. Subj. Biochem.* **26,** 193.
Weidel, W., Frank, H., and Martin, H. H. (1960). *J. Gen. Microbiol.* **22,** 159.
Weiner, J. H. (1974). *J. Memb. Biol.* **15,** 1.
Weiss, R. L. (1971). *J. Gen. Microbiol.* **67,** 135.

Westling-Haggstrom, B., Elmros, T., Normark, S., and Winblad, B. (1977). *J. Bacteriol.* **129,** 333.

Wetzel, B. K., Spicer, S. S., Dvorak, H. F., and Heppel, L. A. (1970). *J. Bacteriol.* **104,** 529.

Wilson, G., and Fox, C. F. (1971). *J. Mol. Biol.* **55,** 49.

Wistreich, G. A., and Baker, R. F. (1971). *J. Gen. Microbiol.* **65,** 167.

Yanagawa, R., and Otsuki, K. (1970). *J. Bacteriol.* **101,** 1063.

CHAPTER 6

The Structure and Biosynthesis of Bacterial Cell Walls

D. J. TIPPER AND A. WRIGHT

I. Introduction

The major stimulus to the investigation of bacterial cell walls has been interest in their role in cell growth and survival. Prokaryotic organisms can grow in an unparalleled variety of environments and they, together with eukaryotic microorganisms, are responsible for maintaining the biological cycles essential to all life on this planet. The primal living cells were primitive

ISBN 0-12-307207-7

prokaryotes which antedated primitive eukaryotes by many millenia. As more complex cell types evolved and developed into multicellular organisms, their survival was predicated on some degree of resistance to parasitization by prokaryotes. Although the tissue fluids of warm-blooded animals provide a thermostated, buffered, oxygenated nutrient medium, the tissues of animal hosts may be regarded as environments hostile to most prokaryotic invaders, and it has been important to investigate the role of bacterial cell walls in protection of successful bacterial parasites against constitutive defenses of the host, such as phagocytes. It is equally important to determine the role of cell wall components as targets of adaptive host immune defenses.

Investigation of the biosynthesis of bacterial cell walls has been essential to an understanding of the mode of action of inhibitors of this process, which include a major class of antibacterial agents, the β-lactam antibiotics. The bacterial cell wall is a prime focus for the dynamic interplay of host, bacteria, and drug which determines the initiation and outcome of infectious disease.

Cell wall structures have also been investigated by taxonomists seeking a molecular understanding of the morphological divisions that remain primal in classic bacterial taxonomy (Bergey's Manual of Determinative Bacteriology, 8th edition, 1974). Early investigators of bacterial cell walls observed some of their unique components, such as the D-alanine, D-glutamate, diaminopimelate, and muramic acid constituents of peptidoglycan. It soon became clear that cell wall amino acid composition has considerable taxonomic significance among gram-positive organisms (Cummins and Harris, 1956). Subsequent investigation of peptidoglycan structures has resulted in a much more detailed understanding of the varieties of peptidoglycan structure among gram-positive organisms and have suggested several modifications of existing taxonomic groupings (Schleifer and Kandler, 1972). Serological identification of the superficial antigens of bacteria is also of considerable taxonomic importance, especially among human and animal pathogens, since resistance to infection is frequently dependent upon opsonization by antibodies directed against these specific antigens. For this reason, the investigation of the antigenic components of bacterial cell walls also commenced early and led to the recognition of many unique cell wall components which are principally carbohydrate in nature. These include the polyol phosphate polymers found in many gram-positive bacteria and called teichoic acids by Baddiley (1972) and also the O antigenic polysaccharides of gram-negative bacteria which are part of lipopolysaccharide (LPS), a major component of gram-negative outer membranes. In its role as endotoxin, LPS also plays a major role in the pathogenesis of gram-negative infections, a further stimulus to the investigation of its structure. The structure and biosynthesis of the outer membrane of gram-negative bacteria is discussed in Chapter 7.

Bacterial physiologists are concerned with the role of cell walls in growth, cell division, and morphogenesis of bacteria, and since cell wall synthesis involves a major commitment of the cell's resources, understanding the control of its synthesis is a goal of major importance. Investigations by chemical, enzymological, biophysical, genetic, and electron microscopic techniques have now given us a reasonable understanding of cell wall structure and of some of its functions. Controls involved in the determination of cell shape, size, growth, and division, all of which clearly involve control of cell wall synthesis and turnover, remain poorly understood, although much relevant information already exists in this field (Daneo-Moore and Shockman, 1977).

Bacterial cell walls contain several classes of heteropolymers whose existence helps to differentiate prokaryotic from eukaryotic organisms. The most ubiquitous among these and the most important with respect to cell wall structural functions is peptidoglycan. This polymer has been found in all bacteria which inhabit a hypotonic environment, that is all except the mycoplasma group (mollicutes), the extreme halophiles, and the methanogenic bacteria (see above). *Halococcus morrhuae* cell walls contain a sulfated heteropolysaccharide that appear to be their major structural component (Steber and Schleiffer, 1975). *Methanococcus vannielii* is a motile, methane-producing coccus that is neither osomotically fragile nor salt tolerant. However, its cells are extremely fragile, lysing under minimal mechanical shear. Insensitivity to a variety of inhibitors of peptidoglycan synthesis suggests that the walls of this organism do not contain peptidoglycan. The walls are thin, composed of tetragonally arranged subunits (Jones *et al.*, 1977).

It has recently been demonstrated (Kandler and Hippe, 1977; Kandler and Konig, 1978) that the cell walls of methanogenic bacteria do indeed lack conventional peptidoglycan. One *Methanococcus* species apparently contains a simple repeating polymer of L-Glu, L-Lys, L-Ala, and D-GlcNAc, bearing only a distant relationship to peptidoglycan. Studies of the T1 ribonuclease hydrolysis products of the 16 S rRNA's of methanogenic bacteria (Fox *et al.*, 1977; Woese and Fox, 1977) has shown them to be as different from 16 S rRNA's of "typical bacteria" (which group includes mycoplasma, blue-green algae, chloroplasts, and all the other "eubacteria") as they are from the 16 S rRNA's of eukaryotic cells, both animal and fungal. Indeed, this difference, and the implied age of the evolutionary divergence, is as great as that between eukaryotes and eubacteria. Three primary kingdoms are suggested, and peptidoglycan is found only in one, the eubacteria, and not in all of its members. Nothing is known of the prokaryotic precursors ("urkaryotes") of the eukaryotic cytoplasm, but the third kingdom, encompassing all known methanogenic bacteria and for which the name *Archaebacteria* is suggested (Woese and Fox, 1977), have solved the problem of cell wall strength and growth differently. Because nothing more is currently known about the walls of these organisms, they are tacitly excluded from consideration in the rest of this chapter, even though they occur in various morphological varieties including both gram-positive and gram-negative species.

The differential osmotic pressure between the bacterial cytoplasm and its environment is as high as 20 atm in gram-positive species, and peptidoglycan is the principal cell wall component determining the considerable tensile strength required to contain this pressure. Disruption of the functional integrity of peptidoglycan results in lysis and death of bacterial cells, as internal osmotic pressure disrupts their plasma membrane unless the membranes are protected by an external osmolarity high enough to minimize the stress, in which case spheroplast or protoplast formation results (Guze, 1968).

Disruption of peptidoglycan can be achieved by exposure of sensitive cells to peptidoglycan hydrolases, such as lysozyme, or by exposure of growing cells in inhibitors of peptidoglycan biosynthesis, such as the penicillins. It is remarkable that Alexander Fleming was responsible for the initial observations of both kinds of effect. Because of their obvious relevance to the natural and antibiotic mediated defenses against bacterial pathogens, investigations of these phenomena has been a constant stimulus to the investigation of bacterial cell walls and in particular to the investigation of the structure and biosynthesis of their peptidoglycan component.

Prokaryotic organisms have been evolving for several billion years, yet current knowledge of the structure and biosynthesis of bacterial cell wall peptidoglycans indicates a remarkable fundamental unity. This fundamental structural pattern has survived evolutionary divergence, and must represent an early and highly successful mechanism for solving the problem presented by an hypoosmotic environment. This structural unity will be emphasized in this chapter, although variations will be noted and the more diverse patterns of other cell wall components will be described.

Because prokaryotes (by definition) lack a separate nucleus, their most fundamental differences from eukaryotic organisms lie in the mechanisms of macromolecular synthesis involved in the expression of their genetic information. In prokaryotes, transcription and translation is essentially simultaneous, while eukaryotic organisms may extensively modify primary transcripts before they are excreted to the cytoplasm for translation. The plastids of modern eukaryotic cells are probably remnants of prokaryotic endosymbionts, and modern prokaryotes differ from eukaryotes in that they lack such semiautonomous organelles. In prokaryotes the cytoplasmic membrane, or functionally distinct extrusions of this membrane, perform all of the varied functions of the nuclear, mitochondrial, chloroplast, and plasma membranes of eukaryotic cells, encompassing the basic machineries of respiration and photosynthesis. With the exception of certain members of the mycoplasma group, prokaryotic membranes also differ from eukaryotic membranes in lacking sterols. Since mycoplasma are unique among prokaryotes in lacking cell wall structures, they will receive scant attention in this chapter. However, their existence must not be forgotten in discussing the functions of bacterial cell walls, such as their role in chromosomal segregation and cell division.

As discussed in Chapter 7, the outer membrane of gram-negative bacterial cell walls is fundamentally different, in structure and function, from the cytoplasmic membrane of bacteria.

II. Reviews of Bacterial Cell Walls

Salton's book on bacterial cell walls (Salton, 1964) and the review by Weidel and Pelzer (Weidel and Pelzer, 1964) are now classics within the field. The book by Rogers and Perkins on cell walls and membranes was published in 1968 (Rogers and Perkins, 1968), as were Ghuysen's review of peptidoglycan hydrolases and their use in analyzing peptidoglycan structure (Ghuysen, 1968), and a general review of bacterial cell walls (Ghuysen *et al.*, 1968b). The biosyntheses of peptidoglycan and lipopolysaccharide were reviewed by Osborn (1969) and Strominger (1970), and the structure and function of peptidoglycan was reviewed by Tipper (1970). Bacterial growth and the cell envelope was reviewed by Rogers (1970). One of us published a general review of the field in 1972 (Tipper, 1972) and in the same year, Schleiffer and Kandler published what is still the definitive review of peptidoglycan structure (Schleifer and Kandler, 1972) while Ellwood and Tempest reviewed their work on the effect of environment on cell wall composition (Ellwood and Tempest, 1972).

Two recent collections of reviews and research reprints provide a good summary of the field: "Bacteria Membranes and Walls," edited by Leive (1973) contains excellent reviews on biosynthesis of peptidoglycan (Ghuysen and Shockman, 1973) and lipopolysaccharide (Nikaido, 1973), while volume 235 of the *Annals of the New York Academy of Science* (1976), edited by Salton and Tomasz, contains contributions from most of the laboratories active in the field (cf. Bayer, 1974; Leive, 1974; Osborn *et al.*, 1974; Braun *et al.*, 1974; Rogers, 1974; Strominger *et al.*, 1974; Ghuysen *et al.*, 1974). Glaser (1973) reviewed cell surface polysaccharides and Archibald (1974) reviewed teichoic acids. Braun and Hantke (1974) reviewed cell envelope biochemistry, and Henning's discussion of the determination of cell shape was published in 1975 (Henning, 1975).

Nikaido has summarized the functional properties of the outer membrane of gram-negative bacteria, and especially his own data on the nature of its permeability barrier (Nikaido, personal communication). Chapters on biosynthesis and assembly of cell walls (Ghuysen, 1977) and the bacterial cell surface in growth and division (Daneo-Moore and Shockman, 1977) have recently been published in *Cell Surface Reviews.* Ghuysen's book summarizing the work of his group and their collaborators on the bacterial DD-carboxypeptidase-transpeptidase enzyme system has recently been published (Ghuysen, 1976). Blumberg and Strominger published a review of interaction of penicillin with the bacterial cell in 1974 (Blumberg and Strominger, 1974). It is interesting to note that the data on this topic from Strominger's and Ghuysen's groups, which were apparently in conflict, now appear amenable to a common interpretation (see Section VI,E). Finally, a chapter on cell wall biosynthesis has recently been published (Hussey and Baddiley, 1976) and a new edition of "Cell Walls and Membranes" (Rogers, Perkins, and Ward) is in preparation.

III. Cell Wall Morphology of Gram-Negative and Gram-Positive Bacteria

Bacteria which are large enough to be easily seen by light microscopy and which have a relatively rigid cell wall can be subdivided according to their response to the gram stain. Many bacterial species respond to this stain in an unequivocal manner, but the technique is quantitative rather than qualitative and some species are gram variable (Bartholemew and Mittwer, 1952). A positive gram stain, dependent upon retention of iodine-fixed crystal violet in the presence of alcohol, may depend on the length of alcohol treatment or on the growth phase of such a gram-variable organism.

Among human pathogens, gram negativity is highly correlated with intrinsic resistance to antibiotics, such as penicillin G, the ability to harbor and transmit plasmid determinants of antibiotic resistance (R factors) by conjugation, and the presence of lipopolysaccharide endotoxin. Response to the gram stain thus has considerable predictive value in the investigation of bacterial infections. *Neiserria* are relatively difficult to decolorize and are normally quite sensitive to penicillin G. Nevertheless, they are classified as gram negative, and strains harboring enteric-type R plasmids containing the sequence determining β-lactamase production have recently been isolated (Elwell *et al.*, 1977). Moreover, at least some of the serious consequences of systemic neisserial infection are due to the effects of the cell wall

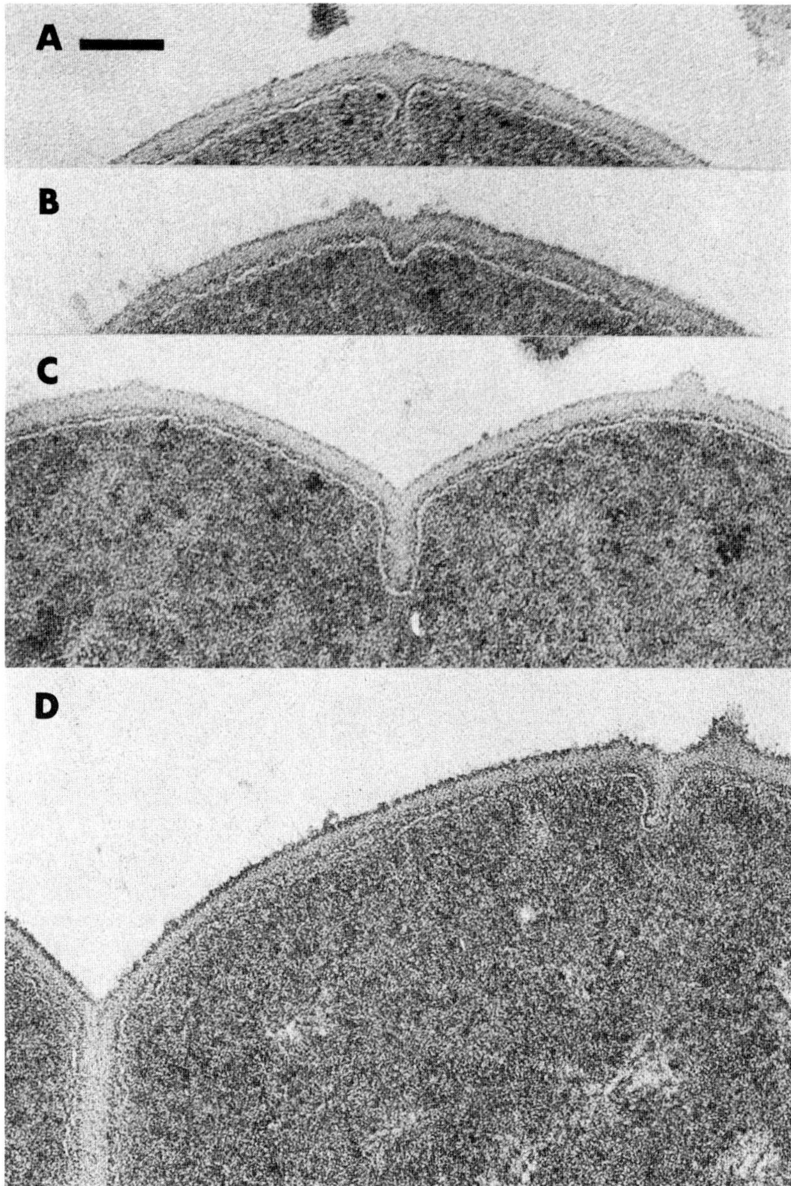

FIG. 1. Thin sections of *Streptococcus faecalis* showing stages in septum development (unpublished observations by Dr. Michael Higgins). The bar in (A) equals 0.1 μm and applies to all four figures in this plate. The culture was doubling in mass every 32 minutes at the time of fixation. Note that invagination of the cytoplasmic membrane, or association with a mesosome (A), occurs prior to visible duplication of the peripheral wall band (see Fig. 2). Duplication of the band (B) indicates insertion of an analus of new wall at the new septum site.

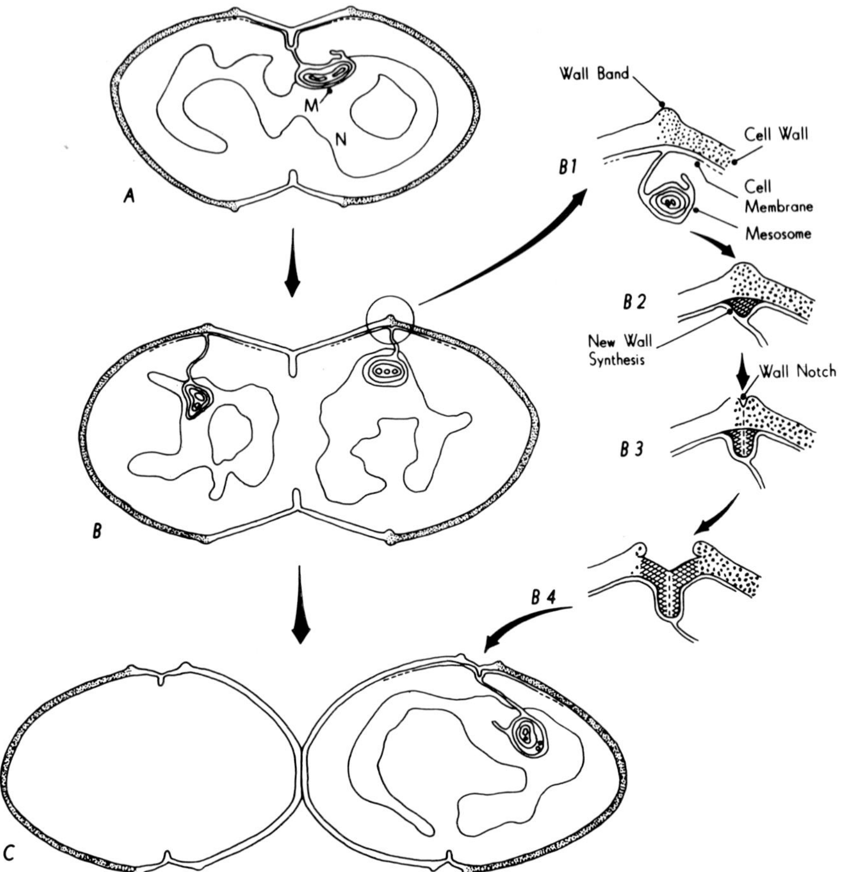

FIG. 2. Schema of the cell division cycle in *S. faecalis*. (From Higgins and Shockman, 1971.) Wall extension is proposed to occur by insertion of new material at the leading edges of the nascent cross wall and at the point where this cross wall meets the peripheral wall. The diplococcus in (A) is in the process of growing a new wall at its crosswall and segregating its nuclear material to the two nascent daughter cocci. In rapidly growing exponential phase cultures, before completion of the central crosswall, new sites of wall elongation are established at the equators of each of the daughter cells at the junction of old polar wall (strippled) and new equatorial wall beneath a band of wall material that encircles the equator (B). Beneath each band a mesosome is formed while the nucleoids separate and the mesosome at the central site is lost. Invagination of the septal membrane appears to be accompanied by centripetal crosswall penetration (B2). A notch is then formed at the base of the nascent crosswall which creates two new wall bands (B3). Wall elongation at the base of the crosswall pushes newly made wall outward. At the base of the crosswall, the new wall peels apart into peripheral wall, pushing the wall bands apart (B4). When sufficient new wall is made so that the wall bands are pushed to a subequatorial position (e.g., from C to A to B) a new crosswall cycle is initiated. Meanwhile the initial crosswall centripetally penetrates into the cell, dividing it into two daughter cocci.

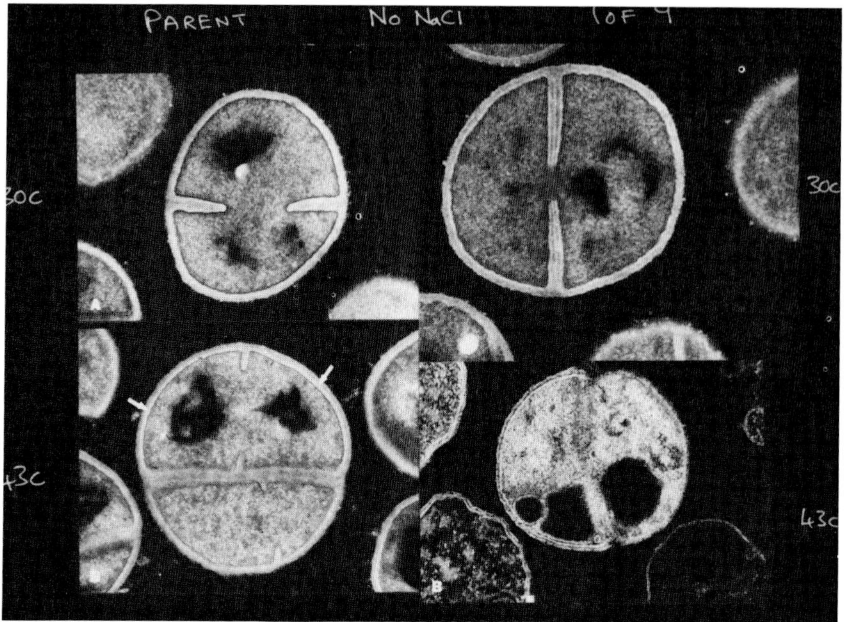

FIG. 3. Thin sections of *Staphylococcus aureus* fixed during growth in low salt at 30°C (top) or 2 hours after a shift to 43°C (bottom). The strains were the temperature-sensitive mutant TOF-9 (right) or its temperature-resistant parent, 655 HT (left). (Courtesy of Dr. Roger M. Cole; see Good and Tipper, 1972.) All are at the same magnification except for the top right, where the magnification is 1.4 times greater. Note that at 43°C (lower left), the parent produces larger cells with more distinct evidence of wall layering than at 30°C. This section also illustrates the orthogonal orientation of sequentially synthesized septa. Strain TOF-9, whose UDP-MurNAc-L-Ala:D-glutamate ligase is temperature-sensitive, rapidly ceases peptidoglycan synthesis on shift from 30° to 43°C, producing fragile cells with defective walls (lower right).

endotoxin. Thus, classification as "gram negative" may take into account other phenotypic correlates of the gram stain response. A further correlation is found with cell wall structure.

Probably the single most significant contribution that electron microscopy has made to the study of bacterial cell walls and the classification of bacteria has been the correlation of response to the gram stain with cell wall profile seen in stained cross sections (Glauert and Thornley, 1969; Bayer, 1974). Gram-positive bacteria normally have a thick, relatively homogeneous cell wall, and although different layers can frequently be distinguished by staining density, they are contiguous and seldom sharply defined. In contrast, gram-negative bacteria normally have a thinner, distinctly layered cell wall which includes an outer membrane resembling the typical trilaminar cyto-

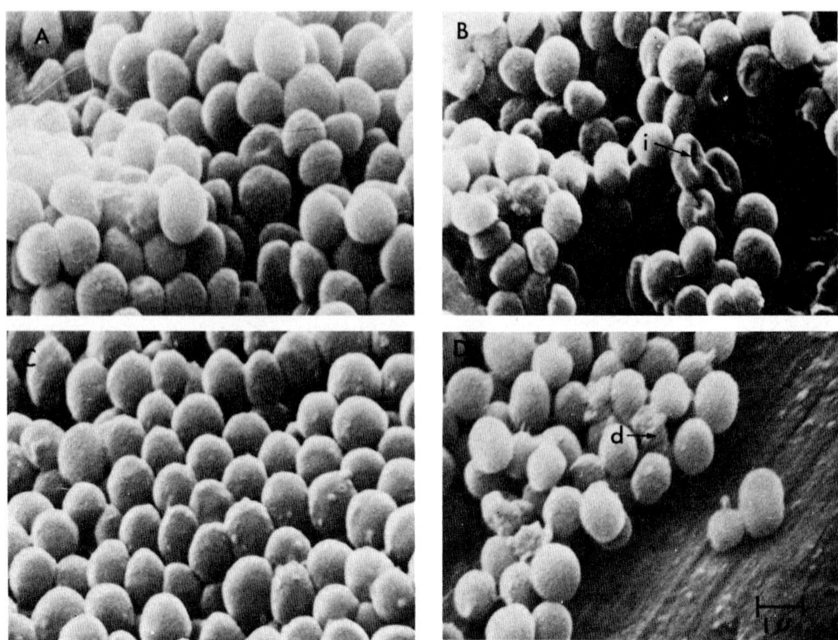

FIG. 4. Scanning electron micrographs of *S. aureus* 655HT(A,C) and TOF-9(B,D) fixed during growth in low salt at 30°C (A,B) or 2 hours after a shift to 43°C (C,D). Several of the mutant cells have lysed (d in D). (From Good and Tipper, 1972.)

plasmic membrane structure in profile. It is usually about 25% thicker and may have a quite different organization (see Chapter 7).

The topology and morphogenesis of the cell wall of the gram-positive coccus *Streptococcus faecalis* has been studied in careful detail by Higgins and Shockman (1971, 1976), and a typical thin section of this organism is shown in Fig. 1 which also shows typical streptococcal septum development. A model of cell wall development in this species is shown in Fig. 2. The wall is essentially homogeneous in cross section, even though it contains polysaccharide, teichoic acid and protein components in addition to peptidoglycan. The wall thickens continuously following extension at the developing septum (Higgins and Shockman, 1976). The cells have a longitudinal axis of symmetry and divide in parallel planes at right angles to this axis, forming typical "streptococcal" chains. *Staphylococcus aureus*, another gram-positive coccus, has a thick wall which does show signs of layering in some stained sections (Fig. 3). Cells of this organism, like those of *Micrococcus luteus*, divide in three planes at right angles to each other in strict sequence

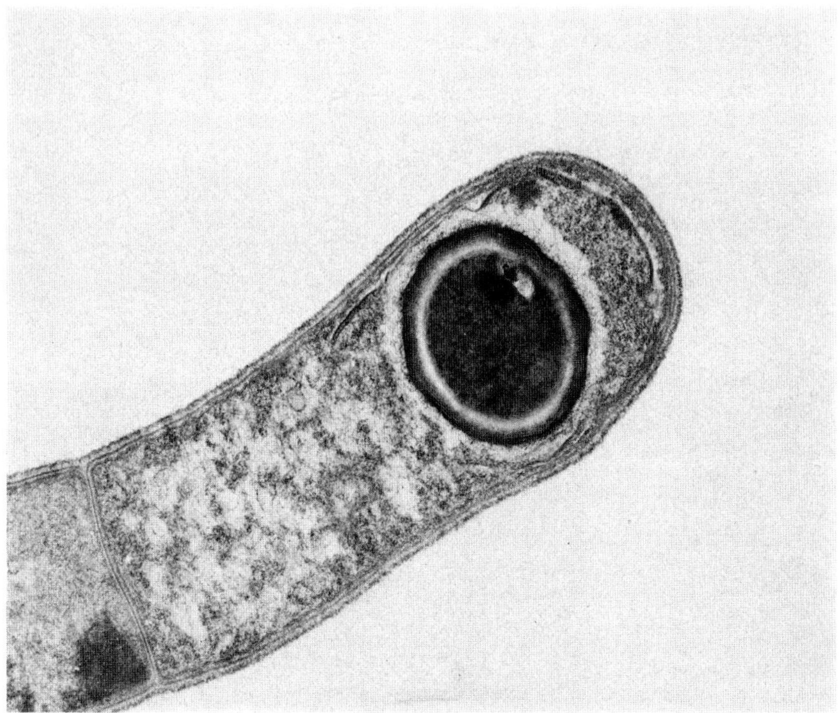

FIG. 5. A cell of *B. sphaericus* 9602 containing an almost mature endospore. The thin darkly stained inner layer of the vegetative cell wall is peptidoglycan and traverses the septum. It is separated from the thick diffusely stained outermost layer (T layer protein) by a lightly stained zone of unknown composition. Neither the T layer nor this lightly stained zone enter the septum except during cell separation. (Courtesy of Dr. S. C. Holt; see Holt *et al.*, 1975.) The mottled spore core is surrounded by the following concentric layers: a barely visible inner forespore membrane, a thin lightly stained primordial cell wall (arrow), an unstained cortex in the first stages of development, the outer forespore membrane, and the multilayered coats. Cross sections of the structured exosporium are visible close to the sporangial cell cytoplasmic membrane.

and so would form regular packets (Koyama *et al.*, 1977), except that asymmetrical cell separation results in the typical irregular "cluster of grapes" form (Fig. 4). See Section VIII,B. While *Bacillus subtilis* walls show little evidence of layering, *B. sphaericus* shows distinct wall layers (Holt *et al.*, 1975) (Fig. 5). In this organism, 70% of the cell wall mass consists of a structured outer "T" layer polypeptide (Fig. 6) while peptidoglycan accounts for only 20% of the wall mass (Hungerer and Tipper, 1969), a low figure for a gram-positive organism. Sporulating cells of *B. sphaericus* are

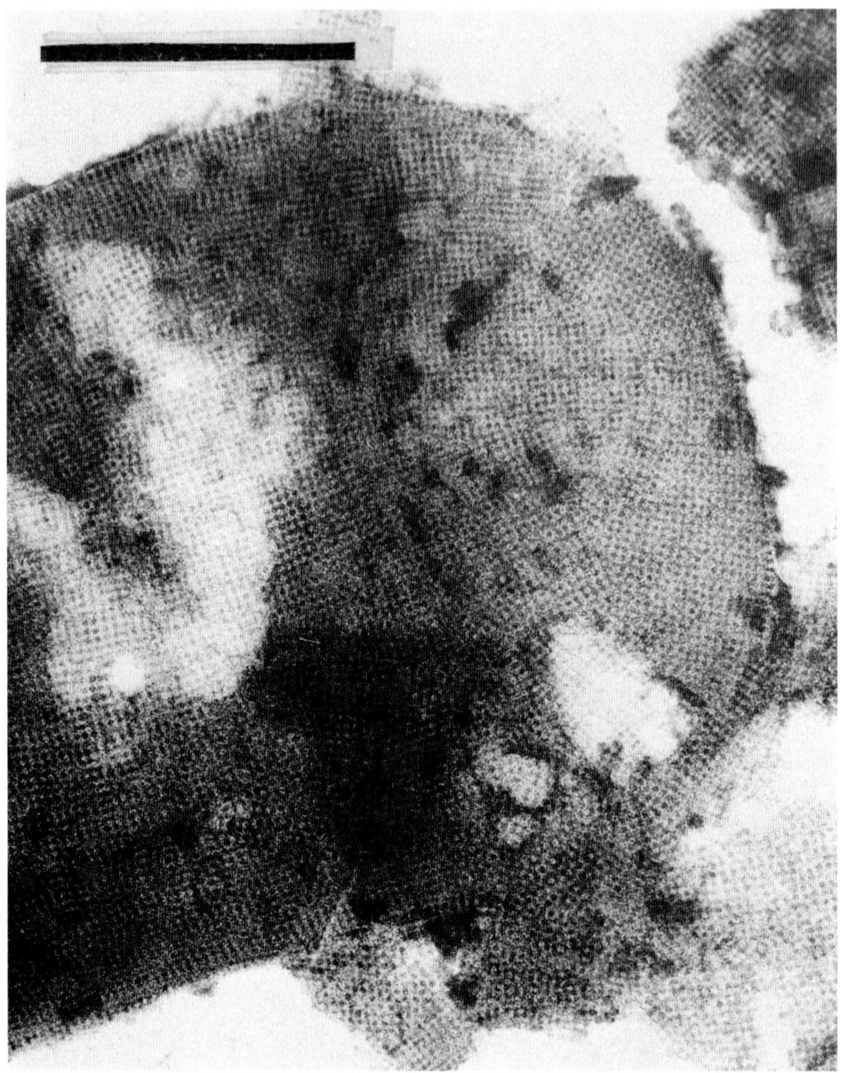

FIG. 6. The terminus of a *B. sphaericus* cell negatively stained with phosphotungstic acid. (Courtesy of Dr. Jack Pate; see Howard and Tipper, 1973.) This micrograph clearly illustrates the tetragonally arrayed subunits of the T layer protein. The bar represents 0.25 μm.

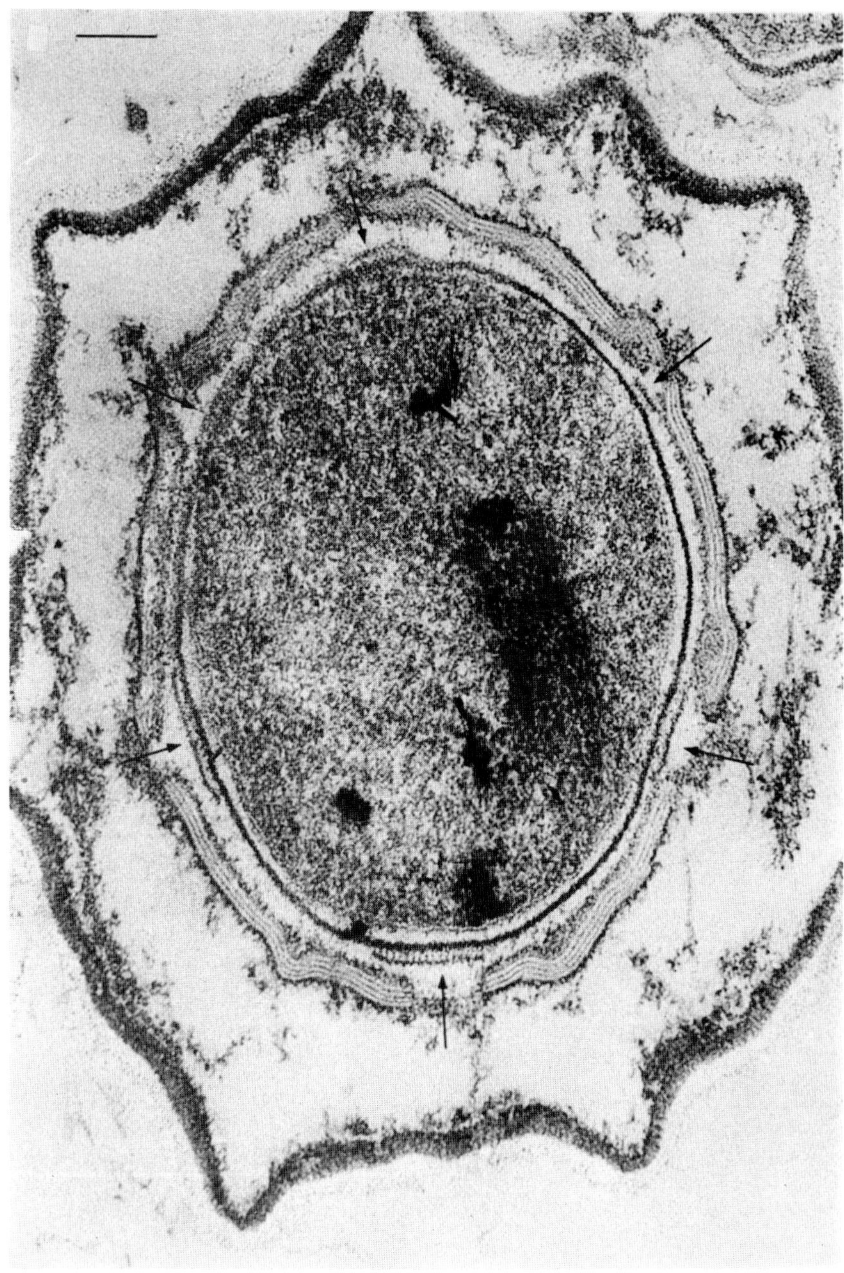

FIG. 7. An early stage in the germination of a spore of *B. polymyxa.* (Courtesy of Dr. R. G. E. Murray; see Murray *et al.*, 1970.) The cortex has disappeared and the primordial cell wall resembles the densely stained peptidoglycan layer seen in *B. sphaericus* cell walls (Fig. 6). The coat is ruptured in several places, and T layer protein is beginning to assemble on the outer surface of the primordial cell wall at these points (arrows). Distinct subunit structure can be seen in the layer at one point.

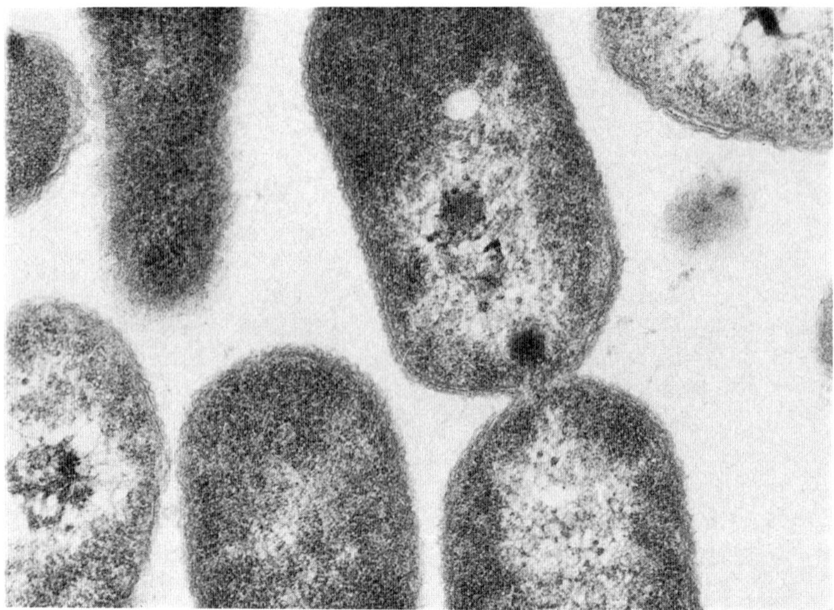

FIG. 8. Thin section of *Pseudomonas aeruginosa* showing the cytoplasmic and outer membranes together with the distinctly separate peptidoglycan layer. (Courtesy of Dr. Lee F. Ellis.)

gram negative. The outer layer seen in thin sections of the wall (Fig. 5) is T layer, while the inner, thinner, and more densely stained layer is peptidoglycan (Howard, Holt, and Tipper, unpublished observations).

The septal wall of streptococcal cells is as undifferentiated in appearance as is the peripheral wall (Fig. 1), and may contain all of the polymeric components of the mature wall. In the diffusely layered *S. aureus* wall (Fig. 3), the septa appear to contain all of the layers. However, the septa of *B. sphaericus* (Fig. 5) are composed only of the peptidoglycan layer. The outer T layer is assembled onto the developing end walls as they separate (Holt *et al.*, 1975). The cell shown in Fig. 5 contains an almost mature spore. Surrounding the central core of this spore is a thin, lightly stained primordial cell wall layer which is surrounded by the almost stain-free cortex (white annulus). This primordial cell wall, like vegetative septal wall, probably consists of "naked" vegetative wall-type peptidoglycan (Tipper and Gauthier, 1972) and develops into mature cell wall following spore germination. *B. polymyxa*, like *B. sphaericus*, has a T layer on its outer cell wall surface (Nermut and Murray, 1967), and its wall looks very similar in cross section.

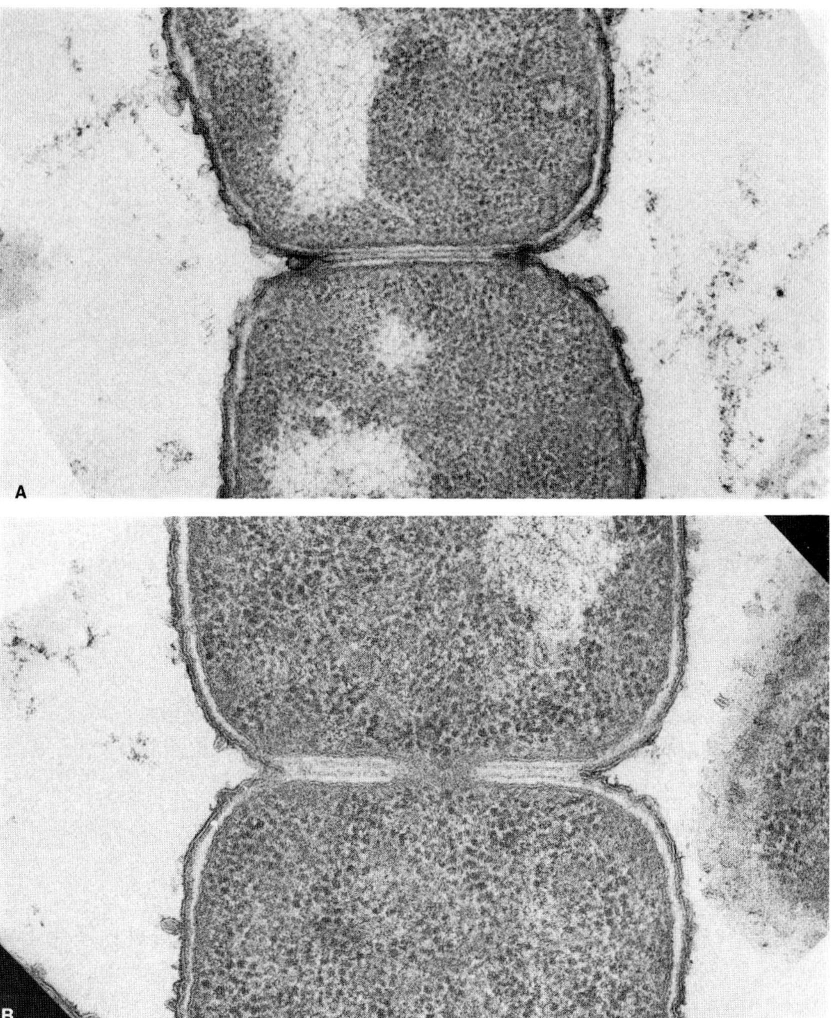

FIG. 9. Thin sections of *E. coli* CRT9F fixed according to the Ryter–Kellenberger technique and illustrating the true morphology of septum development. (Courtesy of Dr. R. G. E. Murray; see Burdett and Murray, 1975.) These micrographs clearly illustrate: (a) the continuity of the cytoplasmic membrane at the cell poles; (b) the thin, homogeneously stained double peptidoglycan layers traversing the septum; (c) the outer trilaminar membrane which enters the septum only to the point of cell separation.

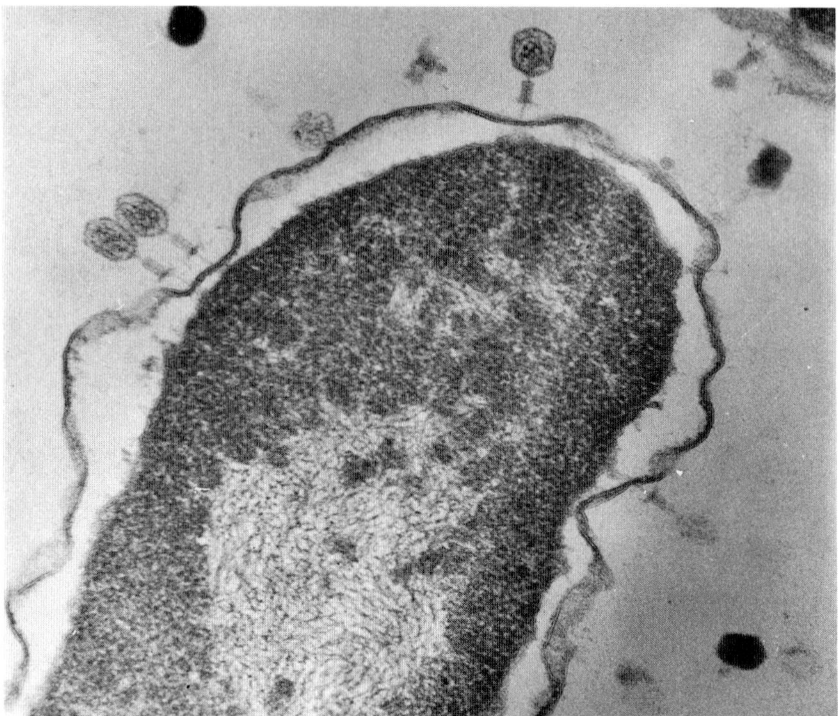

FIG. 10. Phage T2 absorbed to the surface of a plasmolyzed cell of *E. coli* B. (Courtesy of Dr. M. E. Bayer.) Shrinkage of the cytoplasm and its membrane away from the wall clearly reveals the wall profile. The peptidoglycan layer, attached to the inner surface of the outer membrane, is only visible as a diffusely stained border (see Figs. 9 and 11). Extensions of the cytoplasmic membrane contact the wall at the sites of phage attachment.

Assembly of the T layer, which begins on the surface of the primordial cell wall of the germinating spore of *B. polymyxa*, occurs very early in the outgrowth process (Murray *et al.*, 1970) (Fig. 7).

In some gram-negative genera, such as *Pseudomonas*, the peptidoglycan layer, identified by its sensitivity to lysozyme, can be seen as a thin, sharply defined, and continuous layer underlying the outer membrane of the wall (Fig. 8). It is always well separated from the cytoplasmic membrane. The peptidoglycan layer of enterobacterial cell walls can normally be distinguished from the outer membrane only after treatments which denature the outer membrane structure (mild heat, EDTA, etc.) (Bayer, 1975) (See Fig. 9). Lipoprotein, which is covalently linked to the peptidoglycan of Enterobacteriaceae, is an intrinsic component of the outer membrane in its native state and may serve to hold the outer membrane and peptidoglycan layers

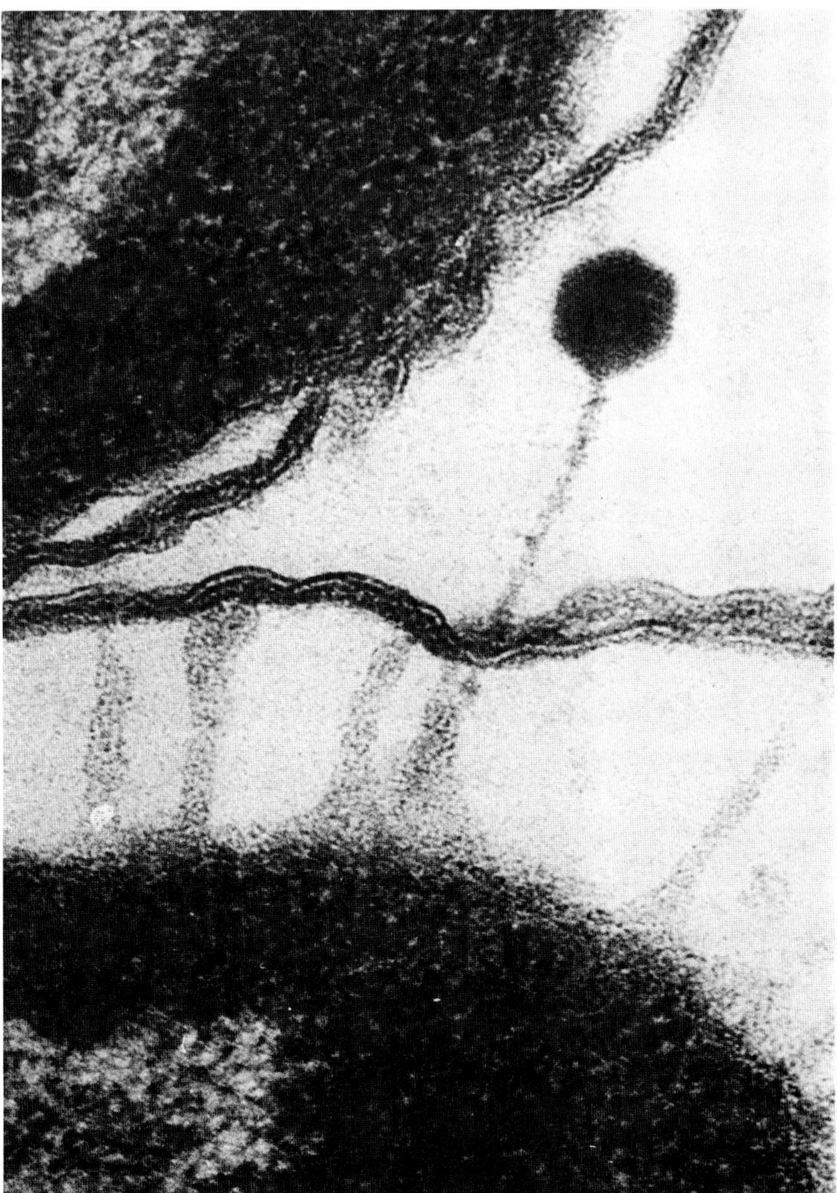

FIG. 11. Phage T5 absorbed to the surface of a plasmolyzed cell of *E. coli* B. (Courtesy of Dr. M. E. Bayer.) This illustrates cell wall profile and wall–membrane contact points at higher magnification than in Fig. 10.

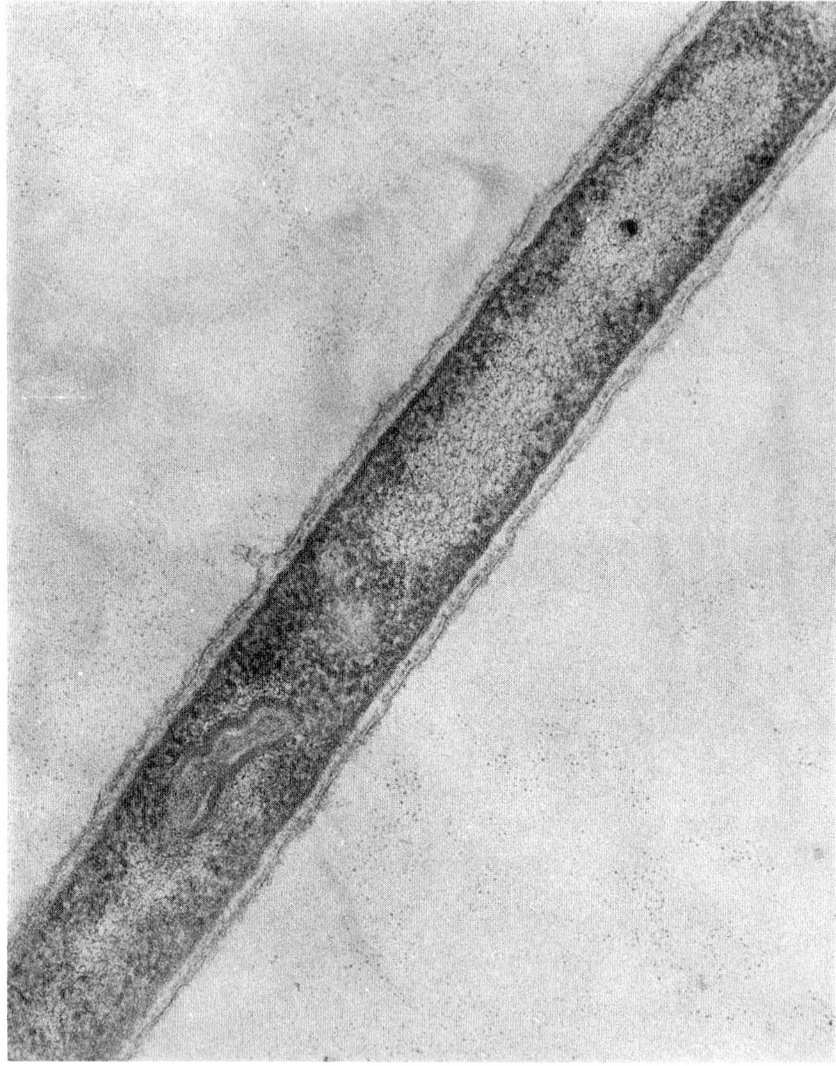

FIG. 12. Thin section of *Sporocytophaga myxococcoides*. (Courtesy of Dr. S. C. Holt.) The wall profile is gram-negative in type.

in intimate contact. The same function has more recently been proposed for the "Rosenbusch" protein and other "porins" (see Chapter 7).

It used to be thought that cell division in gram-negative organisms involved a constriction of the cell, quite at variance with the formation of a planar anulus of septal wall as seen in gram-positive species (Figs. 1 and 3).

Meticulous observations by Burdett and Murray (1975) established the artifactual nature of such constrictions; in properly fixed sections of *E. coli* (Fig. 9), a planar septum is seen. These micrographs clearly demonstrate that the crosswall is bounded by the cytoplasmic membranes of the divided cells and contains two opposed peptidoglycan layers, each of which is contiguous with the layer immediately underlying the outer membrane of the peripheral wall. The outer membrane enters the septum only to the depth of cell separation.

The existence of the outer membrane implies that gram-negative cells employ a fundamentally different strategy toward the environment than do gram-positive cells. The outer membrane has a quite unique composition and serves as an initial osmotic barrier between the environment and the plasma membrane. These two membranes enclose a separate fluid compartment, the periplasm, and most material transported in and out of the cell (with the probable exception of DNA) must pass through this fluid compartment. The periplasm is bridged by small and discrete zones of contact between the outer and cytoplasmic membrane and these zones are the probable sites of phage DNA injection, as shown in Figs. 10 and 11. These micrographs illustrate the sites of attachment of phage to plasmolyzed cells of *E. coli* B. The typical *E. coli* cell wall profile is seen at very high resolution in Fig. 11. The receptor for T2 (Fig. 10) is an outer membrane protein (DePamphilis, 1971), while the receptor for T5 (Fig. 11) is another outer membrane protein that is determined by the *TonA* locus and which also serves as the colicin M, ferrichrome, and albomycin receptors (Braun *et al.*, 1973; Braun, personal communication).

Existence of the outer membrane in gram-negative bacterial cells is, perhaps, a more fundamental difference from gram-positive cells than response to the gram stain, though these properties usually correlate well. It is clear that response to the gram stain is a characteristic of cell wall structure. Thus, for example, it is possible to say that thin sections reveal that *Neisseria*, *Bdellovibrio*, *Myxobacteria* (cf. Fig. 12), *Cyanobacteria* and *Rickettsia* have a "gram-negative type" cell wall. This is also true for budding bacteria, such as *Hyphomicrobium.* Spirochetes (cf. Fig. 13) have a thin flexible wall, and their external sheath and unique mechanisms of motility differentiate them from other groups. As will be demonstrated, gram negativity also correlates with the existence of a single ($A1_\gamma$) chemotype of peptidoglycan, and this may be used as another criterion of gram-negative cell wall type. The Spirochetes are not gram negative by this criterion: at least two members of this species have a peptidoglycan in which the *meso*-DAP of the $A1_\gamma$ type is replaced by L-ornithine (Joseph *et al.*, 1970). The peptidoglycan of *Myxobacter* AL-1 (Harcke *et al.*, 1975), *Caulobacter crescentus* (Goodwin and Schedlarski, 1975), and *Hyphomicrobium* (Jones and Hirsch, 1968) appear to be of the $A1_\gamma$ type.

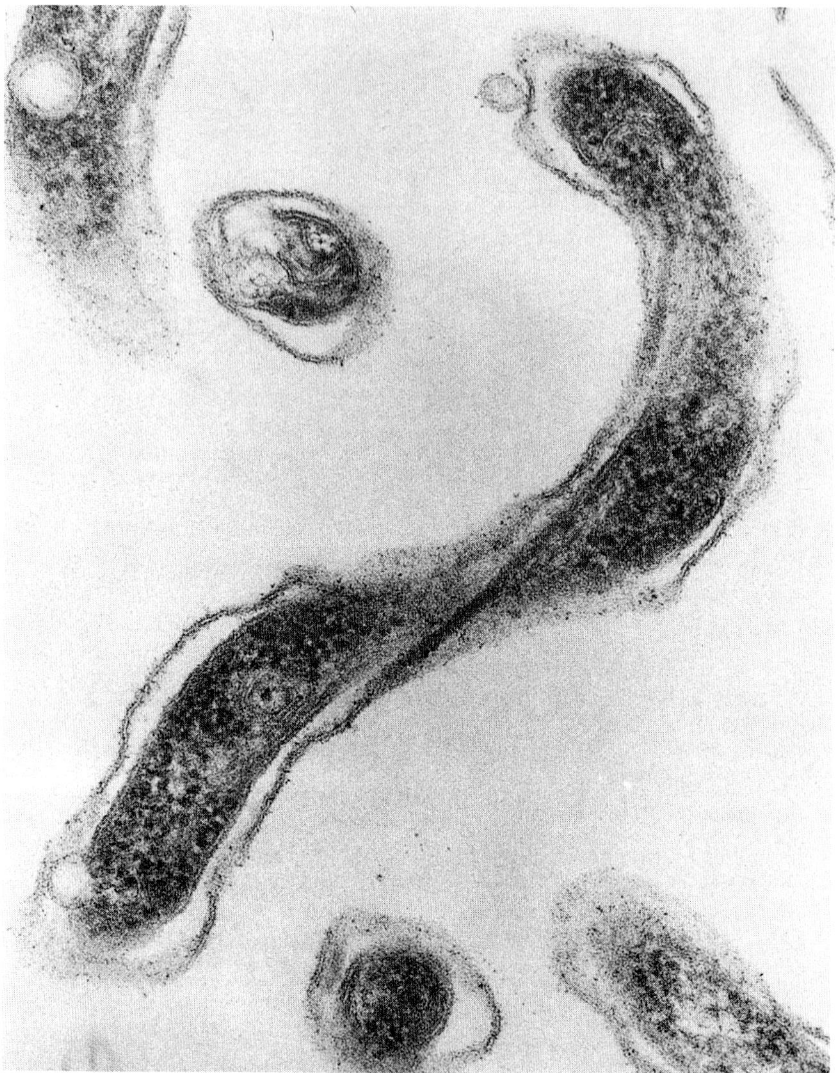

FIG. 13. *Spirochaeta stenostrepta.* This micrograph (courtesy of Dr. S. C. Holt) shows a cell in longitudinal section surrounded by its triple-layered sheath and traversed by the spindle of modified flagellae, which lies between the wall and the sheath. The wall profile, also seen in transverse section, is thin and indistinctly layered.

IV. Bacterial Cell Wall Structure: Definitions of the Cell Wall

There are several possible definitions of cell wall structure (Tipper, 1972). A morphological definition encompassing those stainable structures visible by electron microscopy and which lie outside the plasma membrane is easy to comprehend, but these structures cannot always be correlated with chemically defined components, especially since some commonly used stains fail to reveal neutral polysaccharides. Nevertheless, electron microscopy is indispensable to the study of cell wall structure, and it is important to make these correlations. One might define the bacterial cell wall as all of the insoluble material exterior to the plasma membrane, excluding flagella and capsules. This would include sheaths but would exclude highly hydrated polymers which are not covalently linked to the peptidoglycan in gram-positive bacteria or to outer membrane components in gram-negative bacteria, and which can be removed by gentle washing at neutral pH.

The only reasonable definition of the components of bacterial cell walls is operational and includes those components of the bacterial cell which can be isolated in an insoluble form after mechanical disruption of the cells and centrifugation at moderate *g* forces, e.g., 30 minutes at 25,000 *g*. Because of the lack of precision of this process, judgment must frequently be used to distinguish between true cell wall components and coprecipitated membrane fragments. Moreover, unless appropriate precautions are taken, the breakage of bacterial cells may activate the autolytic hydrolysis of peptidoglycan leading to a loss of cell wall components. This is especially true for species such as *Streptococcus pneumoniae* or *Myxococcus xanthus*, which have highly active autolytic systems (e.g., White *et al.*, 1968). Procedures designed to minimize autolysis, such as treatment with detergents, heat, or other denaturing agents prior to breakage, may lead to their own artifacts. It is, therefore, important in the description of any cell wall preparation to describe the methods used in its isolation. For example, *Staphylococcus aureus* cell walls contain a protein A component covalently linked to their peptidoglycan (Sjoquist *et al.*, 1972). Isolation of cell walls after heat treatment requires subsequent treatment with trypsin to remove precipitated protein, and this would also remove the protein A cell wall component.

Current investigation of the unique components of bacterial cell walls is sufficiently advanced to allow a definition of the major components of many bacterial cell walls on a chemical basis. This is easiest to perform for gram-positive bacteria in which most of the weight consists of peptidoglycan, unique teichoic acid, and polysaccharide components, and sometimes one or two abundant and unique proteins. In the majority of gram-positive organisms, the cell walls do not contain major polypeptide components, and the purity of cell wall preparations can be determined by amino acid analysis.

It is more difficult to assess the purity of preparations of gram-negative cell walls, since the protein composition of outer membranes is quite complex. However, it is unique and should present a characteristic profile after gel electrophoresis.

Since the nonpeptidoglycan components of bacterial cell walls can be altered by mutation, lysogenization, or the acquisition of plasmids, it may also be important to define the genetic composition of the bacterial strain from which cell walls were obtained. Moreover, since the relative abundance of such cell wall components has been shown to depend on the growth conditions as well as on genotype (Elwood and Tempest, 1972), these conditions should also be defined, as should the stage of culture from which the cells were isolated. Finally, it should be pointed out that definitions of cell wall structure based upon those components isolated from broken bacteria fail to distinguish between potential variations of cell wall structure in different locations of the bacterial cell, such as the crosswalls and longitudinal walls.

V. Peptidoglycan Structure

The peptidoglycan of bacterial cell walls is a unique macropolymer of chitin-like glycan chains interlinked into a two- or three-dimensional mesh by cross-linked repeating peptide subunits. The covalent link between glycan and peptide involves the carboxyl group of *N*-acetylmuramic acid (MurNAc), the unique and essential component of this polymer. The chemical structure of the peptidoglycan of *E. coli* cell walls is represented in Fig. 14. This polymer is capable of forming a continuous molecule surrounding the cell, the "bag-shaped macromolecule" postulated by Weidel and Pelzer (1964). A possible organization for such a structure is shown schematically in Fig. 15, which represent a two-dimensional sheet of the *S. aureus* peptidoglycan.

Peptidoglycan has some elasticity (Marquis, 1968; Ou and Marquis, 1970, 1972) and considerable tensile strength and is primarily responsible for the maintenance of bacterial cell shape and the containment of internal osmotic pressure. The outer membrane, and in particular its protein components, probably contribute to these functions in gram-negative bacteria, such as *E. coli* (Henning and Schwarz, 1973; Henning, 1975), but even in these organisms, where the peptidoglycan layer may be only a unimolecular sheet, it provides the major structural strength of the wall.

No macromolecular equivalent of peptidoglycan is found outside of the prokaryotic kingdom: the walls of eukaryotic microorganisms, such as yeasts, perform the same functions with a thick enmeshed, but un-cross-linked array of β-glucans, mannan–protein complexes, and other largely

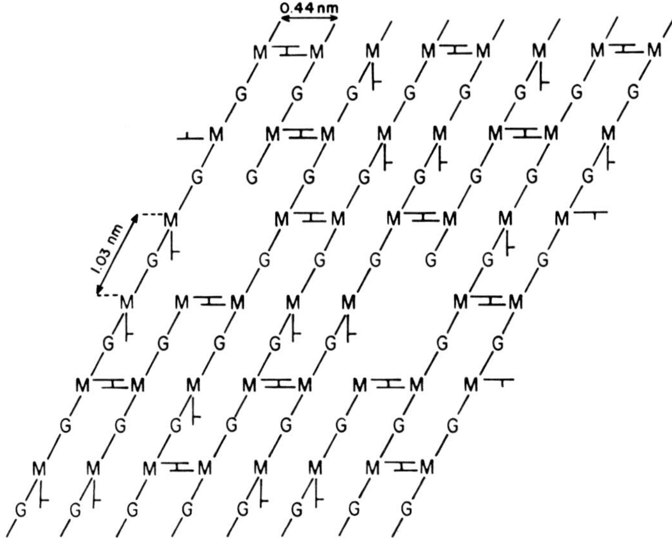

FIG. 14. The unimolecular peptidoglycan sheet of *E. coli* cell walls. (Modified from Ghuysen, 1977.) The glycan chains, polymerized from alternating residues of *N*-acetylglucosamine (G) and *N*-acetylmuramic acid (M), average 50 disaccharide units in length in these walls. In the chitin conformation (see Fig. 16), the intrachain repeat distance is 1.03 nm. If chains are closely stacked in the chitin mode, the interchain repeat distance would be 0.44 nm. This can only occur locally, if it occurs at all. All M residues are substituted by stem peptides of the sequence L-alanyl-D-isoglutamyl-(L)*meso*-diaminopimelyl(L)-D-alanine. These peptides, when present as un-cross-linked monomers, are perfunctorily represented as vertical lines. Horizontal "girders" represent dimers of this peptide, cross-linked via D-alanyl-(D)*meso*-diaminopimelate linkages. (The terminus (D or L) of *meso*-diaminopimelate involved in peptide linkage is indicated in parentheses.). These dimers involve about two-thirds of the peptide units, so that cross-linkage is 33%.

carbohydrate polymers (Nickerson, 1974). These include chitin which forms cross-septa during budding and bud scars after cell separation. The walls of fungal hyphae, the cells of crustaceae, and the cuticles of insects are also rich in chitin, while plant cells employ cellulose and hemicelluloses as structural polymers. Cellulose and chitin have the same preferred chain conformation (Carlström, 1957), and hydrogen bonding between such extended flat chains is essential to the functions of these major structural polymers. This may be true for peptidoglycan in a limited fashion (see Section V,D).

A. THE GLYCAN

1. STRUCTURE

Except for minor modifications, the glycan of all cell wall peptidoglycans studied consists of alternating, β-1,4-linked residues of 2-acetamido-2-deoxy-

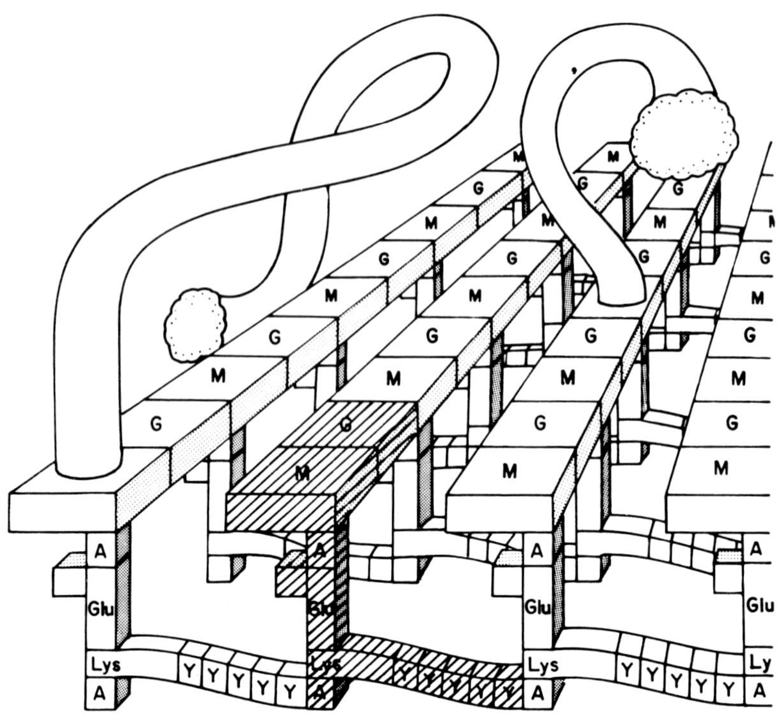

FIG. 15. Schema of the *S. aureus* peptidoglycan. The glycan, formed of polymerized GlcNAc-$\beta\rightarrow$4-MurNAc (G-M) disaccharides, is presented as flat ribbons, to which teichoic acid molecules (flexible "tubes") are occasionally linked as MurNAc-6-phosphodiester substituents. Peptide cross-links hold the glycan chains in a parallel array, forming a sheet. The stem peptides (L-Ala-D-isoGln-L-Lys-D-Ala) are linked to each MurNAc residue and are constrained by the chitin conformation of the glycan chains to project in parallel on one face of the glycan sheet. They are cross-linked by N^{ε}(D-alanyl-pentaglycyl)-L-Lys cross-bridges to form a two-dimensional sheet. These sheets must be stacked and may be cross-linked to form the thickness of the cell wall. Local areas of the peptidoglycan may be highly ordered, as suggested in this schema, but peptide cross-linkage is only 75%, and much of the structure probably forms a more loosely organized and porous net. Average glycan chain length is eight disaccharides. (Tipper *et al.*, 1967b.)

D-glucopyranose (*N*-acetyl-D-glucosamine, GlcNAc) and 2-acetamido-3-*O*-(D-1-carboxyethyl)-2-acetamido-2-deoxy-D-glucopyranose [*N*-acetylmuramic acid (MurNAc)]. The linkages and alternating sequence of this polymer were first chemically established in *S. aureus* (Tipper *et al.*, 1965, 1971; Tipper and Strominger, 1966) and subsequently in a few other species (e.g., Leyh-Bouille *et al.*, 1966), but their generality in peptidoglycans isolated from many other bacteria can be inferred from the consistent results

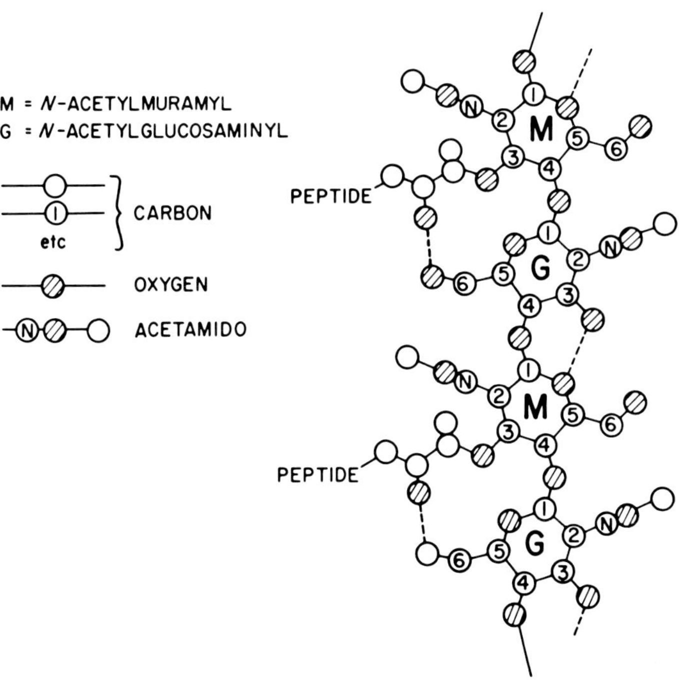

FIG. 16. Linear peptidoglycan in the flat chain conformation of individual cellulose and chitin chains. (From Carlström, 1957.) The hydrogen bond between GlcNAc C-3-OH and the C-5 ring oxygen of adjacent MurNAc residues is typical of chitin. It is proposed (Tipper, 1970) that the glycan strands of peptidoglycan are stabilized in the chitin mode by further hydrogen bonding of these GlcNAc–MurNAc disaccharides between the MurNAc lactyl group and the adjacent GlcNAc C-6-OH. This results in projection of all peptide chains in the same direction. In chitin, such ribbonlike glycan chains are stacked (vertically to the plane of the figure) by hydrogen bonds between acetamido groups.

of hydrolysis of these polymers by endo-*N*-acetylmuramidases, such as lysozyme and the chalaropsis B enzyme (Ghuysen, 1968). Fewer studies have been performed with endo-*N*-acetylglycosaminidases, such as that present in lysostaphin and other preparations of bacterial autolysins (Tipper and Strominger, 1966; Ghuysen, 1968), but they also confirm the uniformity of glycan structure in peptidoglycans. Both types of enzyme cleave alternate linkages in the susceptible portions of the structure to produce disaccharides which can be rigorously compared with those produced from *Staphylococcus aureus*. A more complete analysis of these findings will be found in earlier reviews (Tipper, 1970, 1972).

A survey of more thad 40 species of gram-positive and gram-negative bacteria (Wheat and Ghuysen, 1971) failed to reveal the occurrence of hexosamine configurations other than D-glucosamine and D-glucomuramic acid, although the presence of a small proportion of mannomuramic acid was reported in a different study (Hoshino *et al.*, 1972). The glycan of peptidoglycan may thus be regarded as a modified form of chitin (Tipper, 1970). A diagram of the glycan in the preferred conformation of chitin is shown in Fig. 16. It seems probable that the glycan chains are stabilized in a regular extended conformation of this type. Polymerization and cross-linkage of peptidoglycan in this conformation is shown schematically in Fig. 17, and in a more architectural fashion in Fig. 15.

In gram-positive cell walls, polymers, such as polysaccharides and teichoic acids, may be covalently linked to the C-6 hydroxyl groups of some of the MurNAc residues (Ghuysen *et al.*, 1965a; Knox and Hall, 1965; Knox and Holmwood, 1968; Button *et al.*, 1966; Hughes, 1970). These hydroxyl groups are readily accessible on the face of the glycan sheet not occupied by

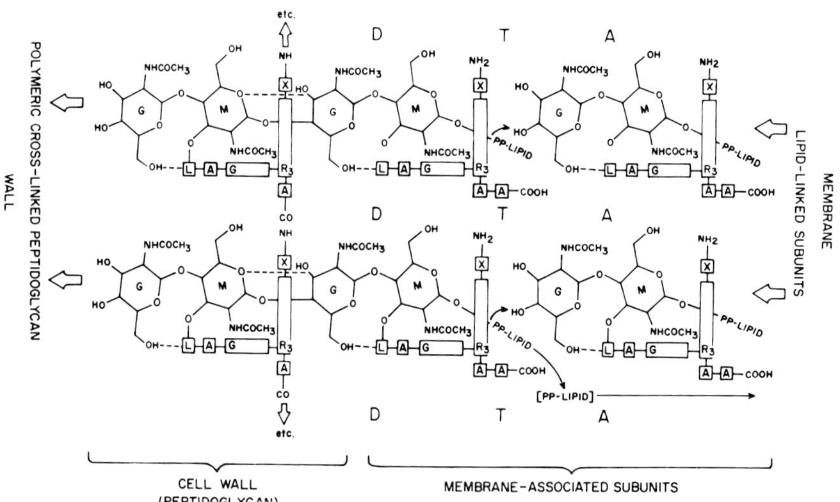

FIG. 17. Polymerization of parallel peptidoglycan chains in chitin conformation. The acceptor for transglycosylation is the disaccharide subunit which is bound to the A site of the polymerase. This disaccharide is pyrophosphate linked by undecaprenol ("lipid"), which is buried in the outer surface of the cytoplasmic membrane. The donor is the polymeric glycan, activated at its most recently added disaccharide subunit, which is at the reducing end (bound at the D site of the polymerase) by pyrophosphate linkage to a membrane-bound carrier, here also presumed to be undecaprenol ("lipid"). The glycan chain grows from its reducing end as a nascent un-cross-linked polymer which becomes cross-linked, as shown at the left, by transpeptidation. This could occur between adjacent nascent chains, as shown, or between nascent chains and preexisting peptidoglycan.

peptide (Fig. 15) and are sometimes acetylated, as in *S. aureus* (Ghuysen and Strominger, 1963). The occurrence of muramic acid-6-phosphate in cell wall hydrolysates suggests covalent polysaccharide attachment, and is widespread (Liu and Gottschlich, 1967; Tipper *et al.*, 1967b).

Minor variations in the glycan structure do occur: in all of the species of mycobacteria and in several species of nocardia which have been examined, the *N*-acetyl group of MurNAc is oxidized to *N*-glycolyl (Azuma *et al.*, 1970). In the peptidoglycans of several spore-forming gram-positive bacilli, 20 to 80% of the glucosamine residues are not *N*-acetylated (Hayashi *et al.*, 1973), probably because of postpolymerization deacetylation.

A major variant of the glycan structure occurs in the spores of gram-positive bacilli. The cortex of these spores consists of peptidoglycan in which about 50% of the *N*-acetylmuramyl peptide subunits have been converted to muramyl lactam (Warth and Strominger, 1969) (Fig. 18). While the cortex glycan could also assume a chitin-like chain conformation, the presence of its lactam residues halves the possible intrachain hydrogen bonding, probably resulting in a much more flexible structure (Tipper and Gauthier, 1972). The cortex is also cross-linked to the extent of only about 20% of the peptides, or 6% of the disaccharide units, giving a very loosely bonded structure when compared to the 25% cross-linked *E. coli* cell wall peptidoglycan structure represented in Fig. 14. The cortex of the spore probably has a rather different function from such a cell wall peptidoglycan. It is probably involved in maintenance of the relatively dehydrated state of the mature spore cytoplasm, which it surrounds. The flexibility of its structure may allow the cortex to adopt a grossly altered conformation when monovalent counterions are exchanged with divalent metal ions. This could result in an expansion

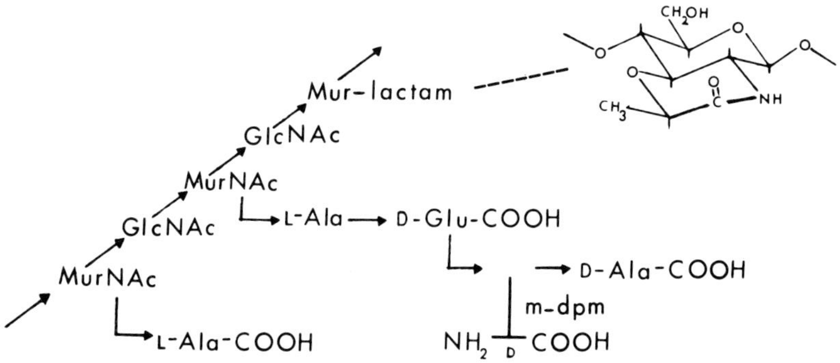

FIG. 18. Spore cortex structure. The spore cortices of *B. subtilis* (Warth and Strominger, 1969) and *B. sphaericus* (Landbeck and Tipper, unpublished observations) are virtually identical polymers of three kinds of subunit; disaccharide–peptide (30%–20% cross-linked), disaccharide-L-Ala-COOH (20%) and disaccharide containing MurNAc lactam (50%). The glycan chains are very long, but disaccharides are only 6% cross-linked, forming a very loosely organized net.

of the cortex which, because it is contained within rigid polypeptide spore coats, compresses and dehydrates the cytoplasmic core of the spore (reviewed in Gould and Dring, 1974; Gould and Measures, 1977).

2. Chain Length of the Glycan

Glycan chain length was first studied in staphylococci (Tipper *et al.*, 1967a; Tipper, 1969a). The glycan was found to be heterodisperse with chains ranging from 3 to 40 disaccharide units (6–80 hexosamines), the number average chain length being 6–8 disaccharides. This may be shorter than the chain length in most bacteria (Ward, 1973), possibly because of the unusually high degree (75%) of peptide cross-linking in staphylococci. It must also be remembered that cleavage of glycan bonds may occur during isolation, in spite of precautions taken against this possibility, so that *in vitro* chain length estimates are always minimal. The only autolytic glycanase activity found in *S. aureus* is an endo-*N*-acetylglucosaminidase (Tipper, 1969a), and most of the reducing end groups found in the glycan of *S. aureus* are GlcNAc residues (Ward, 1973). Because the polymerization of GlcNAc-MurNAc peptide subunits should result in MurNAc reducing end groups (see below), the GlcNAc end groups are probably the consequence of autolysin activity, either *in vivo* or during isolation.

Ward investigated glycan chain lengths in the walls of *S. aureus* and *B. subtilis* and in both normal and autolysin-deficient strains of *B. licheniformis* (Ward, 1973). Walls were obtained from SDS-treated cells so that autolysis during isolation was minimized. The ratio of total muramic acid to total reducing end groups gave number average chain lengths of 9, 54, 45, and 79 disaccharides, respectively, in these bacterial strains. The ratio of total muramic acid to reducing MurNAc residues gave number average chain lengths of 176, 96, 114, and 140 disaccharides, respectively. Assuming that reducing GlcNAc residues resulted from *in vivo* autolysis, while reducing MurNAc residues resulted from chain termini produced during biosynthesis, it appears that all three of these gram-positive species synthesize long glycan chains which are autolytically cleaved to varying extents during subsequent cell growth. It is thus probable that glycan chain length is determined by controlled hydrolysis, even though appropriate autolytic activities have not been identified in all bacterial species investigated. In *E. coli* the glycan chains are terminated by 1,6-anhydro-GlcNAc residues (Fig. 19), probably as a consequence of the action of a different type of autolysin (Holtje *et al.*, 1975; Taylor *et al.*, 1975), giving glycan chains averaging 50 disaccharides in length.

The glycan chain length of *Arthrobacter crystallopoites* peptidoglycan

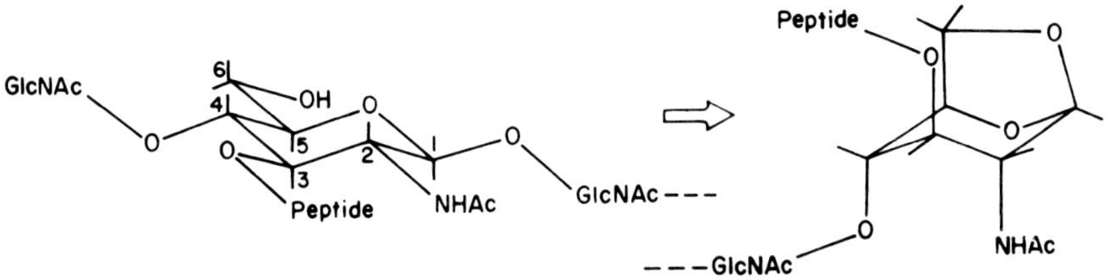

FIG. 19. 1,6-Anhydro-MurNAc structure. This is the product of a glycan hydrolase present in *E. coli* cells.

was found to average 8 disaccharides in spherical cells and 32 disaccharides in rod-shaped cells (Krulwich *et al.*, 1967a). The nutritionally triggered conversion of rod-shaped to spherical cells was accompanied by a transient increase in endo-*N*-acetylmuramidase activity (Krulwich and Ensign, 1968). correlating with the lower chain length in the spheres. However, there seems to be no general correlation between cell shape and glycan chain length, since rod-shaped organisms, such as *L. casei* (Hungerer *et al.*, 1969), have glycans with relatively short average chain lengths of about 6 disaccharide subunits. Peptide variation in *Arthrobacter* peptidoglycan is discussed in Section VIII.

3. SUBSTITUTION OF GLYCAN BY PEPTIDE

Just as muramic acid is the essential structural component of peptidoglycan, its activated derivative, UDP-*N*-acetylmuramyl pentapeptide (Fig. 20), is the key intermediate in peptidoglycan biosynthesis. It is probable, and consistent with all available information, that nascent peptidoglycan in all bacteria is formed by polymerization of disaccharide–peptide subunits. These units are formed from MurNAc pentapeptide, derived from the nucleotide precursor, together with GlcNAc. Further modifications and/or additions to the peptide occur prior to polymerization in many bacterial species. As a consequence, every MurNAc residue in this nascent polymer is substituted by a peptide, and this holds true in the mature peptidoglycans of most bacterial cell walls, although both the peptides and the glycan may be modified by hydrolysis. Major discrepancies are found only in the spore cortex peptidoglycan (see Section V,A) and in the type A2 peptidoglycans found only in certain micrococci such as *M. luteus* (see Section V,B,2). The formation of muramic lactam residues has yet to be demonstrated *in vitro*. However, the coupled activity of *N*-acetylmuramyl-L-alanine amidase and transpeptidase, required to form the peptide cross-links and peptide-free glycan regions in *M. luteus* peptidoglycan, does function *in vitro* (Ghuysen *et al.*, 1973).

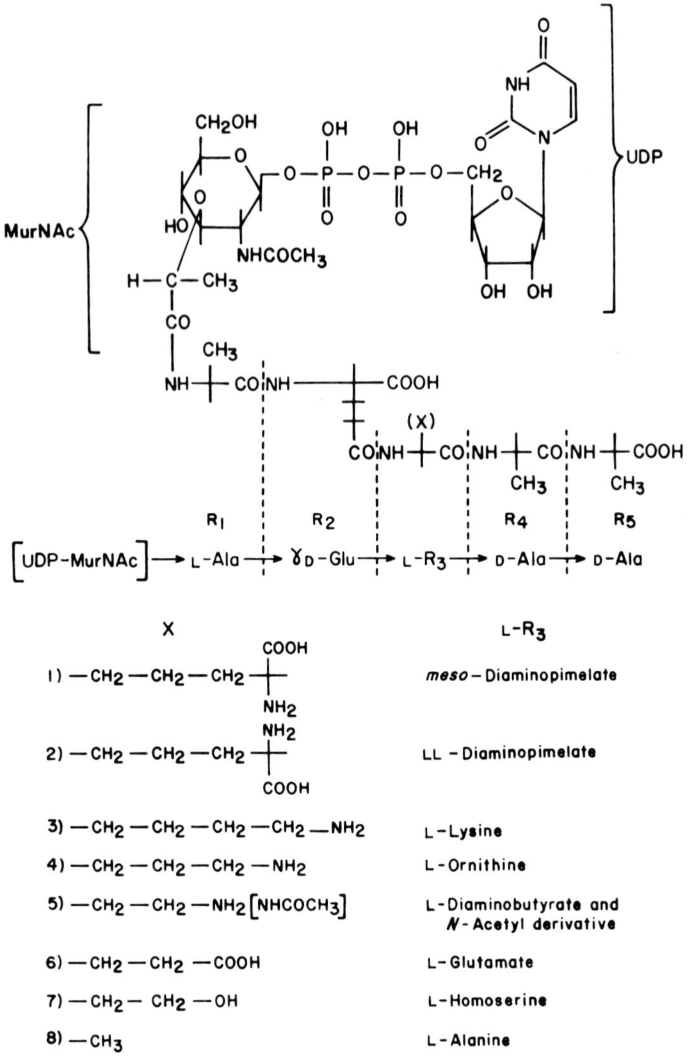

[UDP-MurNAc] → L-Ala → γD-Glu → L-R_3 → D-Ala → D-Ala

	X	L-R_3
1)	$-CH_2-CH_2-CH_2-CH(COOH)(NH_2)$	*meso*-Diaminopimelate
2)	$-CH_2-CH_2-CH_2-CH(NH_2)(COOH)$	LL-Diaminopimelate
3)	$-CH_2-CH_2-CH_2-CH_2-NH_2$	L-Lysine
4)	$-CH_2-CH_2-CH_2-NH_2$	L-Ornithine
5)	$-CH_2-CH_2-NH_2$ [$NHCOCH_3$]	L-Diaminobutyrate and *N*-Acetyl derivative
6)	$-CH_2-CH_2-COOH$	L-Glutamate
7)	$-CH_2-CH_2-OH$	L-Homoserine
8)	$-CH_3$	L-Alanine

FIG. 20. The UDP-MurNAc–pentapeptide precursor of peptidoglycan. The stem pentapeptide, R_1-R_2-R_3-R_4-R_5, has the sequence L-Ala-D-Glu-L-R_3-D-Ala-D-Ala, except in certain corynebacteria where R_1 is L-Ser or Gly. The residue (L-R_3) linking the γ-COOH of D-glutamate to D-alanyl-D-alanine is always an L-α-amino acid. The side chain (X) of residue L-R_3 has eight known variants, as shown. Modification of this peptide structure by addition of further amino acids or oxidation may occur before transfer to undecaprenol phosphate.

B. STRUCTURE OF THE PEPTIDE

Following the pioneering work of Cummins and Harris (1956) which demonstrated both the simplicity of amino acid patterns present in the cell walls of gram-positive organisms and the discrete variations in this pattern among different taxa, sequence investigation has revealed the importance of variations in peptide structure for taxonomy. The limits of variation are presumably imposed by biosynthetic and functional imperatives and indicate the universality of the mechanisms described in Section VI.

The earliest model sequence was that of the UDP-*N*-acetylmuramyl pentapeptide isolated from *S. aureus* treated with penicillin G (Fig. 20). The discovery and characterization of this nucleotide (Park, 1952) was a key event in the structural analysis of peptidoglycan; hypothesized to be a precursor of peptidoglycan on the basis of a similar amino acid composition, this was only proved after several years of investigation on both *S. aureus* peptidoglycan structure and the mechanism of its biosynthesis.

The isolation, characterization and use of specific hydrolytic enzymes (glycanases, amidases and endopeptidases) was essential to the analysis of peptidoglycan structure (Ghuysen, 1968). These enzymes are capable of breaking this enormous structure into its discrete repeating subunits. Sequential use of such enzymes, together with conventional chemical techniques such as N- and C-terminal analysis (Ghuysen *et al.*, 1968b) and Edman degradation (Tipper and Strominger, 1968), allowed complete sequence determination as well as identification of the location of amide groups. When it became clear that all structures investigated followed a common pattern, Kandler and co-workers realized that purely chemical procedures, involving partial acid hydrolysis of intact peptidoglycan and of UDP-MurNAc pentapeptide precursors, could be used for much more rapid analysis of peptidoglycan structures. They used these procedures for the analysis of the cell wall peptidoglycans of many bacterial strains, resulting in the most definitive data currently available (Schleifer and Kandler, 1972). These structures had previously been classified into four types by Ghuysen (1968) and Tipper (1972), and this classification has been revised by Ghuysen (1977) to include the wider variation in structure now known to exist. This review will follow the classification of Schleifer and Kandler (1972).

1. THE STEM PEPTIDE

The pentapeptide sequence (R_1-R_2-R_3-R_4-R_5) found in the nucleotide precursor normally has the sequence L-alanyl-γ-D-glutamyl-L-R_3-D-alanyl-D-alanine (Fig. 20). The only variation is found in certain corynebacteria, in which R_1, the first amino acid (L-alanine), is replaced by L-serine or glycine. The second amino acid (R_2) is always D-glutamate, which is invariably linked to the L-R_3 residue through its γ-carboxyl group. Eight different L-amino acids are known to occur in the R_3 position (Fig. 20), but in every case the amino and carboxyl groups of an L-α-aminoacyl center are linked, respectively, to R_2 (D-glutamate) and R_4 (D-alanine). In structures containing *meso*-DAP, the sites of peptide substitution may be at the D- or L-terminus and are indicated in parentheses. Thus D-Glu-(L)*meso*-DAP(L)-D-Ala is the normal stem peptide sequence. The C-terminal R_4-R_5 D-alanine dimer is invariant. It plays a central role in cross-link formation and is presynthesized as a dimer.

The MurNAc residues of the mature peptidoglycan are substituted by modified forms of this pentapeptide which will be referred to as the stem peptide. The terminal D-alanine of the stem pentapeptide is eliminated in the process of cross-link formation by transpeptidation. The un-cross-linked and C-terminal stem peptides may retain both D-alanine residues (as in *S. aureus*) (Tipper and Strominger, 1965), or may lose one or both of these residues, as in *E. coli*, *B. megaterium* (van Heijenoort *et al.*, 1969) and *B. sphaericus* (Hungerer and Tipper, 1969) as a consequence of carboxypeptidase activity. Besides these events, the major modifications of the stem peptide in forming the peptide subunits of the mature peptidoglycan from the nucleotide pentapeptide are the addition of the cross-bridge amino acids (if any) and amidation.

2. Types of Peptidoglycan Cross-Link and Varieties of Peptide Structure

The structure of the peptide cross-links in all known peptidoglycans is consistent with formation by transpeptidation involving cleavage of the C-terminal R_4–R_5 D-alanyl D-alanine bond of one peptide subunit (donor) and attachment of the penultimate (R_4) D-alanine residue of this subunit to the amino terminus of an amino acid in a second (acceptor) peptide subunit (see Section VI,E). Two major types of cross-link are known. In type A, the amino acceptor is the ω-amino group of a dibasic amino acid in position L-R_3, or the amino terminus of a peptide linked to this residue. In type B structure, the acceptor is an amino group of a C-terminal dibasic amino acid linked to the α-carboxyl group of D-glutamate (R_2) in the stem peptide (Table I).

These two basic types of cross-link can be subdivided according to the nature of the residues involved in the cross-link. Schleiffer and Kandler (1972) recognize four type A and two type B structures, and the relationship between these and the four types recognized in an earlier review by Ghuysen (1968) is shown in Table I.

TABLE I

CLASSIFICATION OF PEPTIDOGLYCAN STRUCTURES[a]

Group	Variant L-R_3	Classification by Ghuysen (1968)	Known types	Cross-link
A: Cross-linked to ω-NH_2 of L-R_3				
A1	$A1_\alpha$ L-Lys	I	1	Direct, e.g.,
	$A1_\beta$ L-Orn		1	D-Ala→(D)-*meso*-DAP
	$A1_\gamma$ *meso*-DAP		4	
A2	L-Lys	III	1	Polymerized stem peptide
A3	$A3_\alpha$ L-Lys	II	23	Glycine and/or monocar-
	$A3_\beta$ L-Orn		3	boxylic L-amino acids, e.g.,
	$A3_\gamma$ L,L-DAP		2	D-Ala→(Gly) 5→L-Lys
A4	$A4_\alpha$ L-Lys	II	9	Contains dicarboxylic amino
	$A4_\beta$ L-Orn		5	acids whose ω-COOH is in
	$A4_\gamma$ *meso*-DAP		2	the peptide chain, e.g., D-Ala→D-isoAspN-β→L-Lys
B: Cross-linked to α-COOH of D-Glu (R_2)[b]				
B1	$B1_\alpha$ L-Lys	IV	1	Contains an L-diamino acid,
	$B1_\beta$ L-Hsr		1	e.g., D-Ala→(ε)
	$B1_\gamma$ L-Glu		1	L-Lys←Gly←Gly←D-Glu
	$B1_\delta$ L-Ala		1	
B2	$B2_\alpha$ L-Orn	IV	1	Contains a D-diamino
	$B2_\beta$ L-Hsr		2	acid, e.g.,
	$B2_\gamma$ L-Dab		1	D-Ala→D-Lys(ε)←αD-Glu

[a] Note that in single clones of some species (e.g., *Bifidobacterium adolescentis*) residue L-R_3 is sometimes L-Lys and sometimes L-Orn.

[b] Schleiffer and Kandler (1972).

In type A1 structures, the cross-link between D-alanine and L-R_3 is direct (Fig. 21). Variants in which L-R_3 is L-lysine, L-ornithine or meso-diaminopimelate are classified as types A1α, A1β and A1γ, respectively (Tables I and II). Type A1γ is by far the most common among type A1 structures and is the most common peptidoglycan type in all bacteria. This type is probably common to all gram-negative bacteria as well as being found in several common gram-positive bacterial species. It is the simplest and the most economical peptidoglycan structure, and its patent success in diverse bacterial species suggests that it is the most highly evolved type (see below). All four possible variations in amidation of type A1γ peptidoglycan occur (Table II).

TABLE II

EXAMPLES OF TYPE A1 PEPTIDOGLYCANS

Residue L-R_3	D-Glu-CO(OH)	*meso*-DAP (D)-CO(OH)	Species
meso-DAP	OH	OH	*E. coli*, cortex of *Bacillus* endospores
meso-DAP	NH_2	OH	*Bacillus licheniformis* (some strains)
meso-DAP	OH	NH_2	*B. subtilis* (wall). *B. licheinformis* (other strains)
meso-DAP	NH_2	NH_2	*C. diptheriae, L. plantarum, M. tuberculosis*
L-Lys	NH_2	—	*Gaffkya homari, Aerococcus viridans*
L-Orn	OH	—	*Spirocheta stenostrepta*

Type A2 peptidoglycans are only known to occur in a few closely related micrococci (Campbell *et al.*, 1969). L-R_3 is L-lysine and the cross-links between D-alanine and L-lysine involve the stem peptide unit itself (Fig. 22). The number of stem peptide subunits in the cross-bridges varies, and some direct D-alanyl-L-lysine cross-links do occur. Formation of the D-Ala-L-Ala linkages between the stem peptide subunits of the cross-bridges requires prior release of N-terminal L-alanine by amidase action on an *N*-acetyl-muramyl-L-alanine linkage (see Section VI,E). A quantity of peptide-free disaccharides equal to the quantity of stem peptides in cross-links must initially exist, and persists in *M. luteus*. They are apparently removed by

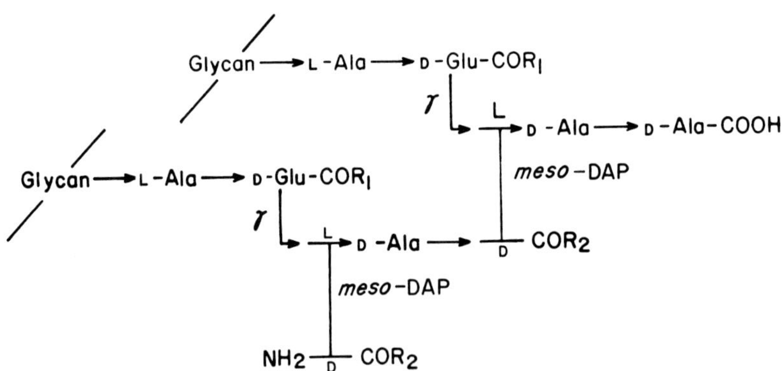

FIG. 21. Peptide structure of type A1γ-peptidoglycans, the most prevalent of the A1 type, with direct cross-linkage of stem peptides. In this and subsequent figures, arrows represent peptide linkages. A horizontal start or finish for such an arrow indicates linkage from an α-COOH or to an α-NH2 group, respectively. A vertical start or finish represents linkage from an ω-COOH or to an ω-NH_2 group, respectively. When L-R_3 is *meso*-DAP, as illustrated here, the structure is that found in *E. coli* and probably all other gram-negative bacteria, and in many gram-positive species. Amidation varies, being absent in *E. coli* ($R_1 = R_2 =$ OH). See Table II.

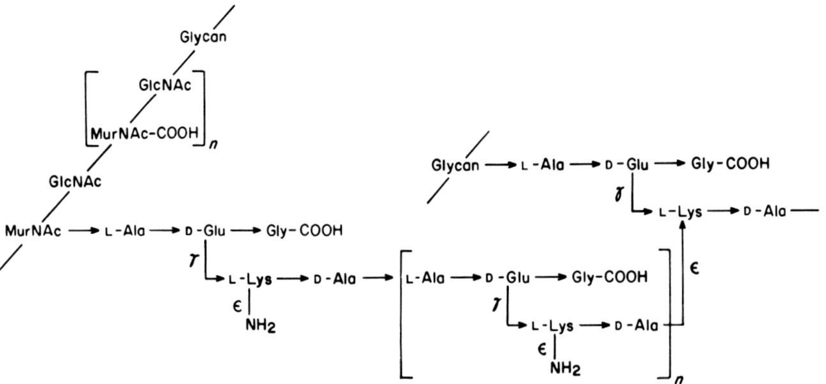

FIG. 22. Peptide structure of type A2 peptidoglycans. This is found only in *Micrococcus luteus* and closely related organisms (*S. lutea*, *M. flavus*, *M. citreus*). Cross-links are formed of D-Ala-L-Ala linked head-to-tail stem peptides of variable number ($n = 3$, average) terminating in a direct N^{ε}(D-alanyl)-L-lysine cross-link, as found in type A1α. Glycine residues are added to each stem pentapeptide subunit at the nucleotide precursor stage, producing hexapeptides. The nascent polymeric peptidoglycan is fully substituted by hexapeptides, and peptide-free disaccharides are formed during head-to-tail cross-link production. These may persist in the mature peptidoglycan, as illustrated, or be removed by glycan hydrolase activity.

subsequent glycanase activity in some related micrococci (Campbell *et al.*, 1969).

In type A3 and A4 peptidoglycans (Fig. 23), a cross-bridge peptide occurs between D-Ala and the L-R_3 residues. Schleifer and Kandler (1972) separate those containing only glycine or monocarboxylic L-amino acids (type A3) from those containing dicarboxylic amino acids (type A4) because synthesis of these two cross-bridge types usually involves different mechanisms (see Section VI,D,4). Four different dibasic amino acids are known to occur in position L-R_3 of these two subgroups, and there is wide variation in the peptide cross-links. Peptidoglycans in these subgroups are found in a wide variety of gram-positive bacteria, and are the most common peptidoglycan

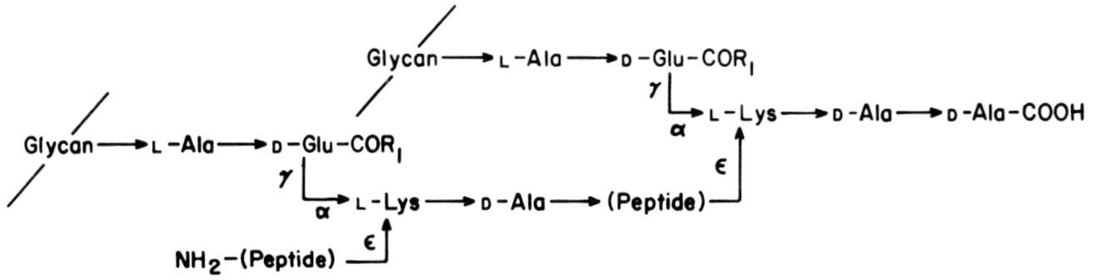

FIG. 23. Peptide structure of types A3 and A4 peptidoglycans. In the most common peptidoglycan structures of this type, L-R_3 is L-lysine (Tables III and IV), though L-R_3 may also be L-ornithine, LL-DAP or *meso*-DAP (Table III). R_1 is usually OH or NH_2, but is occasionally an amino acid or amide, as in certain *Arthrobacter* sp.

TABLE III

EXAMPLES OF TYPES A3 AND A4 PEPTIDOGLYCANS IN WHICH L-R_3 IS NOT L-LYSINE

Residue L-R_3	Cross-bridge	Species
L-Orn	Gly-Gly	*Micrococcus radiodurans*
	L-Ala-L-Ala-L-Ala →	*B. globosum*
	L-Ala-L-Thr-L-Ala-L-Ser-Gly →	*B. longium, Propionibacterium petersonii*
	D-Asp (α-$CONH_2$) $\xrightarrow{\beta}$	*L. cellobiosus*
	D-Asp $\xrightarrow{\beta}$ D-Ser	*B. Bifidum*
	D-Asp $\xrightarrow{\beta}$	*Cellulomonas flavigena*
	D-Glu →	*C. fimi*
	D-Glu (α-$CONH_2$) $\xrightarrow{\gamma}$	*C. biazotea*
L,L-DAP	Gly	*Streptomyces albus*
	Gly-Gly-Gly	*Arthrobacter tumescens*
meso-DAP	D-Glu → D-Glu $\xrightarrow{\gamma}$	*M. conglomeratus*

type, with the exception of type A1γ. Some examples of variation in cross-bridge structure are given in Tables III and IV. Amidation of the D-glutamate α-carboxyl in the stem peptide is common in types of A3 and A4, and in certain *Arthrobacter* species L-alanine amide or glycine amide is found linked at this position (Fiedler *et al.*, 1970). Amidation of the α-carboxyl group of D-aspartate and D-glutamate residues in cross-bridges of type A4 peptidoglycans also occurs frequently (Hungerer *et al.*, 1969).

Peptidoglycans of type B (Figs. 24 and 25), with cross-linkage through the α-carboxyl of D-glutamate (Perkins, 1967, 1968, 1971), are found only among the coryneforme group of bacteria. This group shows the widest variation in peptidoglycan structure of all taxonomic groups and clearly is inadequately

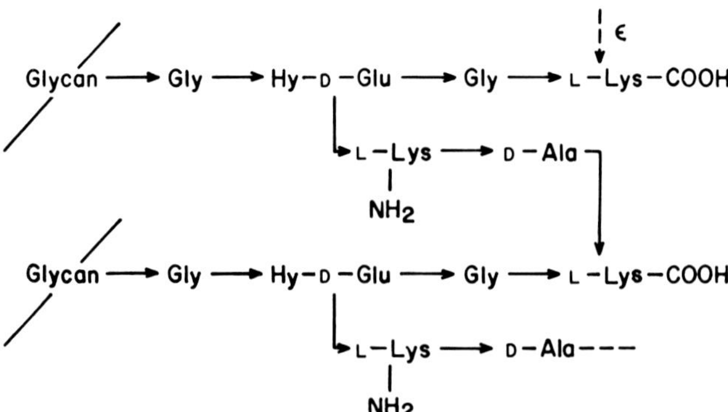

FIG. 24. Peptide structure of a type B1 peptidoglycan. Cross-bridges in all known type B1 peptidoglycans are made up of L-Lys and Gly. L-R_3 varies (Table V). The degree of oxidation of D-Glu depends on the extent of oxygenation of the culture.

TABLE IV

EXAMPLES OF TYPES A3 AND A4 PEPTIDOGLYCANS IN WHICH L-R_3 IS L-LYSINE[a]

Cross-bridge	Species
Type A3	
Gly	*Bifidobacterium sp.*
L-Ala	*A. crystallopoites* (sphere)
L-Ala-Gly-Gly	*A. crystallopoites* (rod)
Gly *or* L-Ala *or* L-Ser	*M. Mucilaginosus* (variable ratio)
L-Ser-L-Ser *and* L-Ser-L-Ala	*Leuconostoc gracile* (constant ratio)
Gly-Gly-Gly-Gly-Gly	*S. aureus*
Gly-L-Ser-Gly-Gly-Gly	*S. aureus*, *S. epidermidis* T-26
L-Ser-Gly-L-Ser-Gly-Gly	*S. epidermidis* T-26
Gly-Gly-Gly-Gly-L-Ala	*S. epidermidis* 66
L-Ala-L-Ala	*L. coprophilus*
L-Ala-L-Ala-L-Ala	*M. roseus* (typical strain)
L-Ala-L-Ala-L-Ala-L-Thr	*M. roseus*
L-Ala-Gly-L-Ala-L-Ala	*Streptococcus* sp.
L-Ala-L-Ser-L-Ala	*Streptococcus* sp.
Gly-L-Thr	*Streptococcus* sp.
L-Ala-L-Thr	*Streptococcus* sp.
Type A4	
D-Isoasparagine	*Bacillus sphaericus*, *L. casei*, *Streptococcus lactis*
D-Asp $\xrightarrow{\beta}$ L-Ala	*B. pasteurii*
D-Glu $\xrightarrow{\gamma}$	*Planococcus* sp.
D-Glu $\xrightarrow{\gamma}$ Gly	*Sporosarcina ureae*
L-Glu $\xrightarrow{\gamma}$ Gly	*M. luteus*
L-Glu $\xrightarrow{\gamma}$ L-Ala	*M. freundreichii*
D-Glu $\xrightarrow{\gamma}$ L-Ser	*M. cyaneus*

[a] Note that the ratio of cross-bridge types in *A. crystallopoites*, *S. aureus*, and *S. epidermidis* depends on both genotype and amino acid supply. In *L. gracile* and *B. rettgeri* (Table V) no such influence of nutrient supply on ratio of cross-bridge type is observed.

defined. Type B structures are found in certain plant pathogenic *Corynebacterium*, in *Brevibacterium* and *Microbacterium* species, and in certain anaerobic *Actinomycetes*, such as *A. israelii*. It is also found in a few *Arthrobacter* species (which are probably misclassified), in *Erysipelothrix rhusopathiae*, and in *Butyribacterium retgerri* (which is now called *Eubacterium limnosum*). The latter is more similar to *Erysipelothrix* than to other *Butyribacterium* (Guinand *et al.*, 1969).

Since residue L-R_3 in group B peptidoglycans is not involved in cross-linkage, it is not necessarily dibasic, and it is in this group that L-homoserine, L-glutamate, L-alanine, and γ-*N*-acetyl-L-diaminobutyrate occur in this position (Tables I and V). On the other hand, the cross-bridge must contain

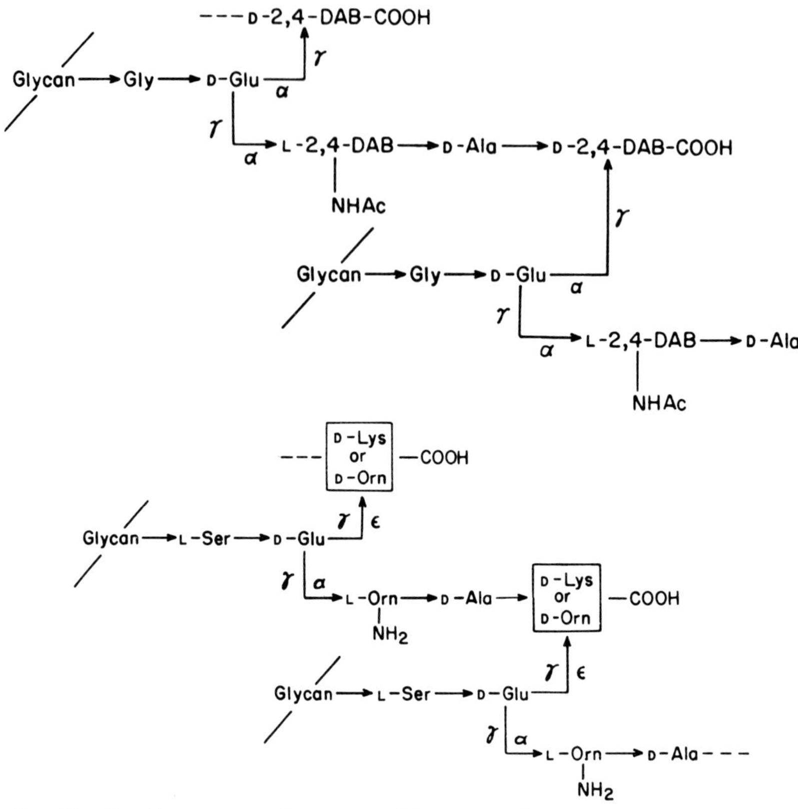

FIG. 25. Peptide structure of two types of B2 peptidoglycans. The *C. insidiosium* peptide (upper) is the only known *N*-acetylated structure. The cross-bridges of *B. rettgeri* (lower) are composed of D-Orn *or* D-Lys in constant ratio. The cross-bridges in *C. poinsettiae* are also D-Orn, the structure being otherwise identical except for the replacement of L-Orn (L-R_3) by L-homoserine.

a dibasic amino acid, and those containing L-diamino acids are classified in group B1 (Fig. 25) (Table V). In all of these structures the L-alanine in position 1 of the stem peptide is replaced by glycine or, less commonly, by L-serine. Another modification sometimes found in group B stem peptides and not found in group A structures, is oxidation of D-glutamate to *threo*-3-hydroxy-D-glutamate. The mesodiaminopimelate occurring in certain aerobic Actinomycetes is sometimes also oxidized to 2,6-diamino-3-hydroxypimelic acid. These strains probably also contain type B peptidoglycans.

C. EVOLUTION OF PEPTIDOGLYCAN STRUCTURE

Schleifer and Kandler (1972) invoked the concept of phylogenetic advancement through simplification and loss of variability in phenotype (reduction)

TABLE V

EXAMPLES OF TYPES B1 AND B2 PEPTIDOGLYCANS[a]

Variant	L-R_3	Cross-link	Species
B1α	L-Lys	(D-Ala)→(ε) L-Lys←Gly←(D-Glu)	*Microbacterium lacticum*
B1β	L-Hsr	(D-Ala)→(ε) L-Lys←Gly←Gly←(D-Glu)	*Brevibacterium imperialis*
B1γ	L-Glu	(D-Ala)→(ε) L-Lys←Gly←Gly←(D-Glu)	*Arthrobacter* sp.
B1δ	L-Ala	(D-Ala)→(ε) L-Lys←Gly←(D-Glu)	*Erysipelothrix rhusopathiae*
B2α	L-Orn	(D-Ala)→D-Lys (ε)←(D-Glu)	*Butyribacterium rettgeri*
		(D-Ala)→D-Orn (δ)←(D-Glu)	
B2β	L-Hsr	(D-Ala)→D-Orn (δ)←(D-Glu)	*C. poinsettiae*
		(D-Ala)→D-Orn (δ)←Gly←(D-Glu)	*M. liquifaciens*
B2γ	L-DAB[b]	(D-Ala)→D-Dab (δ)←(D-Glu)	*C. insidiosium, C. sepedonicum*

[a] Note that in both types the dibasic amino acid is at the N-terminus of the bridge. In type B1, where the cross-bridge dibasic amino acid has the L-configuration, its ω-NH_2 is the acceptor for transpeptidation. In type B2, where the dibasic amino acid has the D-configuration, its α-NH2 is the acceptor for transpeptidation, which therefore produces a DD-linkage, as in type A1$_\gamma$ peptidoglycans. This linkage is susceptible to DD-carboxypeptidases (see Section VI.E).

[b] Diaminobutyrate.

in proposing that the A1γ type of bacterial cell wall peptidoglycan, and in particular the monolayer peptidoglycan structure of gram-negative bacteria, represents the most advanced and sophisticated form of this polymer. The peptidoglycans of gram-positive bacteria would represent a more primitive stage of evolution, the greatest divergence having produced the coryne-bacteria with type B peptidoglycans and perhaps the micrococci with type A2 peptidoglycans. An early separation between the gram-negative and gram-positive groups is also suggested by differences in the structures and properties of the components of the translational machinery, such as the differential response of the ribosomes of these organisms to antibiotics such as erythromycin and lincomycin, determined by RNA methylation (Tanaka and Weisblum, 1975). Evolution of the monolayer A1γ structure may have required concommitant evolution of an outer membrane, with some intrinsic strength, as a barrier to attack by lytic enzymes. Type A1γ structure is also found in the most highly evolved prokaryotes including the myxobacteria (White *et al.*, 1968) and the cyanobacteria.

Spore cortex peptidoglycan has apparently evolved under different functional imperatives and also has the A1γ structure, as does the cell wall peptidoglycan of most gram-positive spore-forming bacilli. However, the cell walls of *B. sphaericus* and *B. pasteurii* have type A4 peptidoglycans, and these organisms, like certain *Arthrobacter* species (Krulwich *et al.*, 1967b), are therefore capable of expressing two types of peptidoglycan structure. This suggests the possibility that exchange of operons capable of determining the difference in these two peptidoglycan structures may have occurred. Rare events of this type may account for the apparent convergence of type A3 and A4 peptidoglycan structures in bacterial species of wide taxonomic distribution. D-Isoasparaginyl cross-links are found, for example, in *Bacillus sphaericus* (Hungerer and Tipper, 1969), *Lactobacillus casei* (Hungerer *et al.*, 1969), and *Streptococcus faecium* (Kandler *et al.*, 1968).

D. The Three-Dimensional Structure of Peptidoglycan

Models for peptidoglycan structures have been proposed by Tipper (1970, 1972), Kelemen and Rogers (1971), Oldmixon *et al.* (1974), and Formanek *et al.* (1974). All acceptable models of general applicability must be capable of producing a continuous two-dimensional sheet and of accommodating the known varieties in cross-link structure, although modifications dependent on cross-bridge length may occur. The model of Keleman and Rogers for *S. aureus* peptidoglycan, in which the long pentaglycine cross-bridges form a β-pleated sheet structure that optimizes hydrogen bonding but gives only a linear macromolecule linking paralled pairs of glycan strands, seems unacceptable.

Tipper (1970) suggested fixation of the glycan in a chitin-like conformation by intrachain hydrogen bonding (Fig. 16) and pointed out that this had the effect of aligning all of the peptide substituents in one direction, while exposing the C-6 hydroxyl of muramic acid residues on the other side of the linear polymer for potential unimpeded linkage to other cell wall polymers. This structure has been assumed by other model builders. The stem peptide, if one includes the D-lactate moiety of muramic acid, has an invariant D-L-D-L-D-D sequence, although the terminal D-alanine may be replaced by glycine or L- or D-aminoacyl centers after transpeptidation. It was suggested (Tipper, 1970) that such a sequence could form a cyclic conformation capable of chelating metal ions, the β-helix proposed for gramicidin (Ascoli *et al.*, 1975), and presenting a relatively hydrophobic surface to the environment, a possible aid in translocation through the membrane. However, binding studies show that chelation of this type probably does not occur in the mature peptidoglycan (Ou and Marquis, 1970), and structures designed to maximize hydrogen bonding are more probable.

A comprehensive model has been suggested by Oldmixon *et al.* (1974) for type A structures, based on the use of space filling models. A schematic representation of peptidoglycan polymerization resulting in this kind of structure is shown in Fig. 15. Peptide subunits, aligned because of the chitin structure of the glycan, are individually folded so as to maximize hydrogen bonding within the peptide and between peptide and disaccharide units. This results in projection of the L-R_3-D-Ala-D-alanine tripeptide at an angle of 90° to the glycan chains, so that transpeptidation will result in cross-linking of a sheet of parallel glycan chains by an orthogonal sheet of parallel peptide chains linked to one face of the glycan sheet (cf. Figs. 15 and 16). Accommodation of cross-bridge amino acids requires only increasing the separation between glycan strands. It is satisfying that formation of this conformation requires γ-linkage of the D-glutamate to the L-R_3 residues and is consistent with the alternation of D and L centers in the stem peptide. The crystalline arrays of chitin have intrachain repeating distances of about 1 nm (Fig. 14), corresponding to the disaccharides. The stacking of such chains through hydrogen bonding of their acetamido groups produces a repeat distance of 0.4 nm (Fig. 14). Until quite recently, X-ray studies (e.g., Balyuzi *et al.*, 1972) had failed to demonstrate repeating distances in peptidoglycans corresponding to the chitin structure. Scattering corresponding to both the 1.03 and 0.44 nm repeat distances of chitin have now been seen, and careful model building studies (Formanek *et al.*, 1974) have shown that it is possible to stack the cross-linked peptide repeating units of both type A and B peptidoglycans into the 0.4 nm interchain separation of the chitin structure only by adopting a specific β-chain conformation. Again it is satisfying that this conformation can only be adopted by a peptide with the alternating D-L-D-L-D sequence of peptidoglycan, and requires the α-D-Glu linkage. It results in a very tightly packed structure that would be impermeable even to water. The composite structure is a thick two-dimensional sheet, and, as pointed out by Braun and Wolff (1975), the peptidoglycan content of *E. coli* would be sufficient to cover only 30% of the cell surface in such a conformation. Moreover, the peptidoglycan layer is permeable to quite large molecules (Scherrer and Gerhardt, 1971) and has considerable elasticity (Marquis, 1968), whereas the chitin structure is rigid. A mixture of chitin-like and of more extended structures is possible, but a structure that is entirely more extended seems more probable. The massive loss of *N*-acetyl groups in the wall peptidoglycan of organisms, such as *B. Cereus* (Hayashi *et al.*, 1973) would prohibit interglycan chain stacking via hydrogen bonding of acetamido groups, so that extensive organization in this fashion cannot be universal. The function of decreased intrachain hydrogen bonding in cortex peptidoglycan has already been discussed (Section V,A).

E. Shape Maintenance in *E. coli*

The peptidoglycan layer in *E. coli* cells has a thickness in stained sections of only 10–15 nm (Figs. 9 and 10), and it has been suggested (Weidel and Pelzer, 1964) that it is a monomolecular sheet. The proposed models of peptidoglycan (e.g., Fig. 15) produce a sheet with glycan chains on one face and peptide chains on the other. Since lipoprotein is attached to *meso*-DAP residues in the peptidoglycan of *E. coli* and is also inserted into the outer membrane (see Chapter 7), the peptide face of such an anisotropic sheet would have to face outward.

In the *E. coli* peptidoglycan, about 30% of the R_4 (D-Ala) residues are involved in cross-link formation (see Fig. 14), and about 10% of the *meso*-DAP residues are covalently linked to lipoprotein. The pattern is probably not random, since trimers of linked peptide subunits do not occur (van Heijenoort *et al.*, 1969), although they are found in the walls of *P. mirabilis*. With an average glycan chain length of 50 disaccharides, a continuous net can be formed, completely encompassing the cell (Fig. 14). The glycan chains average 100 nm in length, and have no necessary orientation with respect to the axis of a rod-shaped organism, such as *E. coli*, whose cell circumference is thirtyfold larger.

When cell envelope fractions of *E. coli* are treated with sodium dodecyl sulfate (SDS), a collapsed, cell-shaped sacculus remains which consists only of the peptidoglycan layer plus covalently attached lipoprotein. Negative

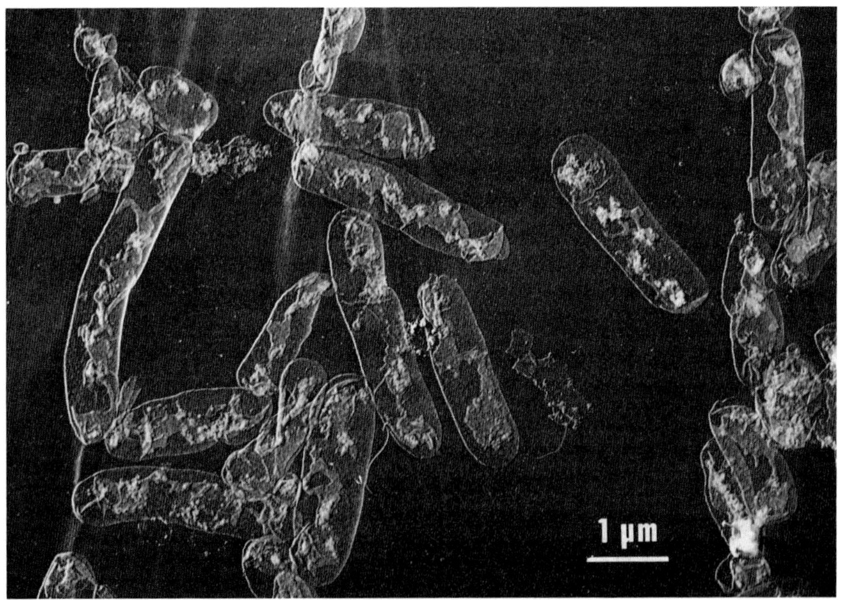

FIG. 26. "Ghost" membranes from *E. coli*. (Courtesy of Dr. U. Henning; see Henning and Schwarz, 1973.) Cell envelopes were prepared by suspending cells in 40% sucrose, 1% Triton X-100 followed by treatment with 4 *M* urea and trypsin. These envelopes consist of outer membrane and peptidoglycan plus remnants of the cytoplasmic membrane. Subsequent treatment with lysozyme in the presence of Mg^{2+} produces these cell-shaped "ghosts" that have lost their peptidoglycan layer. The only proteins retained are from the outer membrane.

staining shows an even distribution of lipoprotein on the surface of this sacculus. After trypsin treatment, which removes the lipoprotein, the sacculus retains its shape, but it is completely disorientated by 6 *M* guanidine-HCl (Leduc and van Heijenoort, 1975), indicating that hydrogen bonds as well as covalent bonds are involved in shape maintenance. It appears that the *E. coli* murein structure has been reduced to the minimum consistent with its structural functions.

Spheroplasts produced by exposure of *E. coli* cells to lysozyme in the presence of 0.5 *M* sucrose have fragmented peptidoglycan and are osmotically fragile. They retain their rod shape but form spherical ghosts on dilution. Ghosts retaining their rod shape are formed if the spheroplasts are treated with Triton X-100, which causes leakage of cytoplasmic constituents, prior to dilution (Henning and Schwarz, 1973) (Fig. 26). They consist largely of outer membrane components. If such ghosts are cross-linked by bifunctional reagents spanning distances as small as 0.3 nm, a stabilizing ghost is produced that resists boiling in 1% SDS (Haller and Henning, 1974). This suggests that the outer membrane proteins are arrayed in a continuous network that may play a role in cell shape maintenance. The phenotype of mutants lacking outer membrane proteins is not consistent with this hypothesis (see Chapter 7).

VI. Biosynthesis of Peptidoglycan

A. Strategy

Because the bacterial cell wall and its components are extracellular, the biosynthesis of its polymeric components is a vectorial process. It can be divided into three separate stages (cf. Fig. 27): the first stage involves cytoplasmic enzymes and soluble substrates and products and results in the formation of UDP-MurNAc pentapeptide. The second stage takes place at the phase boundary formed by the interface between the plasma membrane and cytoplasm and results in formation of the complete subunit of the peptidoglycan. This process is initiated by transfer of phospho-MurNAc pentapeptide to undecaprenol phosphate, a specialized phospholipid which serves to anchor the hydrophilic peptidoglycan intermediates to the membrane and later is presumed to serve in the translocation of the completed peptidoglycan subunits across the membrane to the outer surface where polymerization takes place.

The other components of the peptidoglycan subunit are added, whereas MurNAc pentapeptide is anchored on the cytoplasmic side of the membrane.

In type A1 peptidoglycans, these include only GlcNAc residues, transferred from UDP-GlcNAc, and amide groups added to glutamate and/or *meso*-DAP carboxyl groups. In type A2 peptidoglycans these added constituents include GlcNAc residues and the glycine residues ligated to the α-carboxyl group of D-glutamate. In type A3 and A4 peptidoglycans these substituents also include the cross-bridge amino acids added to the ω-amino group of the L-R_3 residue. In type B1 and B2 peptidoglycans, the cross-bridge amino acids are added to the α-carboxyl group of D-glutamate. All of these reactions

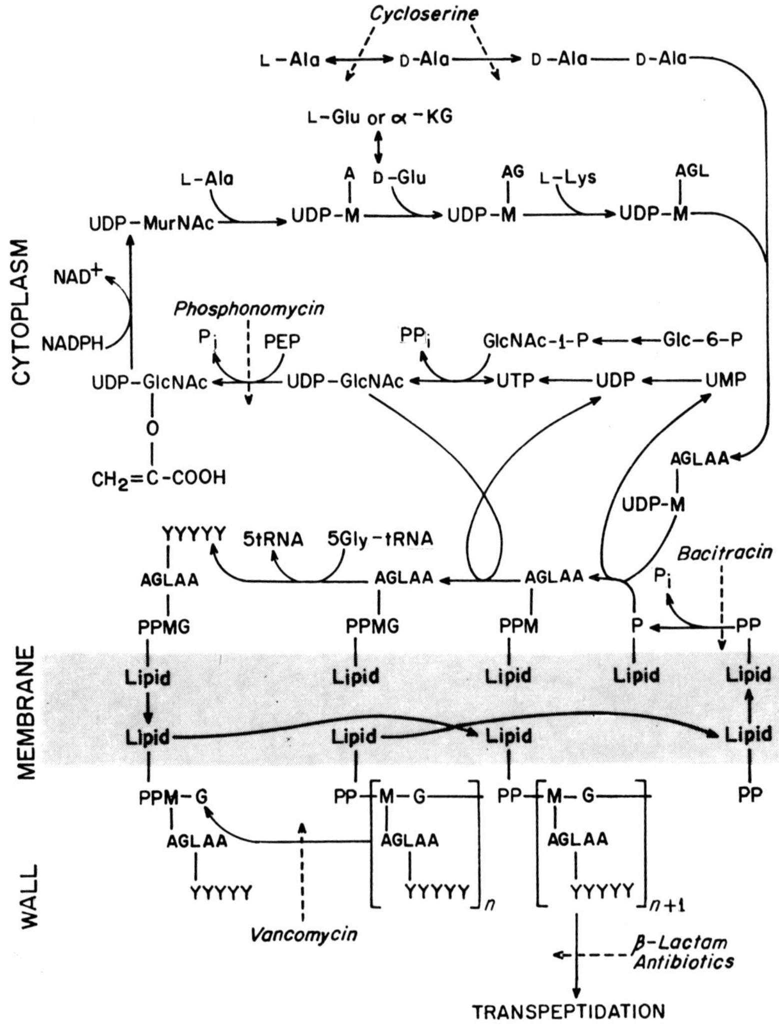

involve membrane-bound enzymes, the membrane-bound undecaprenol-pyrophosphoryl-MurNAc pentapeptide acceptor and soluble substrates, such as UDP-GlcNAc, ammonia, ATP, amino acids and aminoacyl-tRNA's. The final product of stage 2 is a completed disaccharide–peptide subunit, fully modified, containing both D-alanine residues of the pentapeptide and ready for polymerization.

Stage 3 of peptidoglycan biosynthesis is initiated by translocation of the lipid-linked subunit across the membrane to its outer surface where it is polymerized and cross-linked to preexisting peptidoglycan by transpeptidation. Modifications of this nascent polymeric structure by transpeptidation, transglycosylation, and hydrolysis may occur subsequently, and it is believed that controlled events of this type are involved in modulations of the structure which are necessary for cell growth and division.

Similar mechanisms are involved in the synthesis of the lipopolysaccharide (LPS) of gram-negative bacteria. In addition, mechanisms for LPS insertion into the outer membrane must exist (see Chapter 7). The synthesis of the cell wall teichoic acids of gram-positive bacteria probably involves a similar strategy (see Section VII), and is integrated with peptidoglycan synthesis.

The cell wall comprises 20% of the dry weight of gram-negative bacteria and a higher proportion of the dry weight of gram-positive bacteria. The synthesis of peptidoglycan, other glycans, teichoic acids, structural proteins, and the biosynthetic enzymes involved in the synthesis of these wall components requires considerable utilization of cellular resources of energy and metabolites. Careful control of these events is predictable.

The peptide component of the peptidoglycan of gram-positive bacteria comprises 15–40% of the cell wall dry weight and is synthesized by a relatively

FIG. 27. The three phases of peptidoglycan biosynthesis. Synthesis of the *S. aureus* structure is illustrated. In phase 1, cytoplasmic enzymes synthesize UDP-GlcNAc and UDP-MurNAc pentapeptide (UDP-M-AGLAA). Phospho-MurNAc pentapeptide is then transferred to the membrane-bound lipid carrier (undecaprenol phosphate, Lipid-P), initiating the second phase, taking place on the cytoplasmic face of the membrane and leading to synthesis of the complete subunit. Completion of the disaccharide by addition of GlcNAc and addition of the five glycine residues (–YYYYY–) that make up the cross-bridges are illustrated. Amidation of D-Glu-(α)-COOH also occurs at this stage (not shown). The completed subunit is translocated to the outer surface of the membrane where phase 3 is initiated by polymerization of the disaccharide subunits. Subsequent or concomitant events include cross-linkage by transpeptidation, D-alanine carboxypeptidase action (in most organisms), and partial hydrolysis of the completed polymer to form the mature wall structure. This may be subject to slow turnover. A separate set of cytoplasmic enzymes synthesize nucleotide precursors of other cell wall polymers, such as lipopolysaccharides and teichoic acids. Assembly of subunits on undecaprenol phosphate carrier also takes place, and if the polymer is covalently attached to peptidoglycan when found in the wall, as for wall teichoic acids, then ligation of nascent peptidoglycan and teichoic acid apparently takes place at the initiation of phase 3.

economical process. As a structural polymer whose major functions are more dependent on its cross-linked integrity than on its absolute sequence, it can accommodate a much greater variation in structure and a much lower fidelity of synthesis than can the polypeptide of an enzyme. The high fidelity of the ribosome-associated machinery of protein synthesis has a correspondingly high energy requirement and is not involved in peptidoglycan synthesis, even where amino acid components are activated as tRNA derivatives, as in the formation of type A3 cross-bridges. The peptide subunits of peptidoglycan contain 5–10 amino acids and subunit synthesis is dependent upon specific ligases of adequate fidelity whose function is either coupled to the hydrolysis of ATP or to the hydrolysis of aminoacyl-tRNA. Gross variations in the pools of available amino acids can cause variations in peptidoglycan structure (Schleifer *et al.*, 1976).

Finally, it should be emphasized that bacterial cell wall synthesis requires integrated function of the cytoplasmic, membrane-bound, and extracellular enzymes involved in formation and modulation of the final structure. This process must be responsive to the demands of varying cell growth and division rates and is controlled in such a fashion as to produce cells of characteristic shape and size. We know little of how this is achieved (Daneo-Moore and Shockman, 1977).

B. Model Systems Investigated

The basic processes of peptidoglycan biosynthesis were studied by the groups of Strominger, Park, and Neuhaus (Strominger, 1970). The synthesis of UDP-MurNAc pentapeptide was initially investigated in gram-positive cocci, such as *Staphylococcus aureus* and *Streptococcus faecalis*, and the initial studies of the subsequent membrane-associated steps were also performed in *S. aureus*. However, the first demonstration of *in vitro* transpeptidase and carboxypeptidase activity occurred in preparations of *E. coli* membranes (Izaki *et al.*, 1966). Early investigation of the role of carrier lipid in peptidoglycan biosynthesis (Higashi *et al.*, 1967) was closely followed by similar investigations of LPS biosynthesis in *Salmonella anatum* (Wright *et al.*, 1967). Investigation of the related events of D-alanine carboxypeptidase action and transpeptidation has been vigorously pursued by Ghuysen and his collaborators (Ghuysen, 1976), first exploiting the soluble enzymes excreted by *Streptomyces* species. More recently, systems for coupled peptidoglycan polymerization and cross-linking have been investigated in gram-positive organisms by a number of research groups, employing permeabilized cell or "wall–membrane" preparations in which the biosynthetic apparatus is presumably associated in a more or less unchanged manner with the natural cell wall acceptor (see Section VI,E,2).

Penicillin-binding components have been reinvestigated in recent studies of the mode of action of β-lactam antibiotics by Strominger and his colleagues (Blumberg and Strominger, 1974), and these successful studies in gram-positive bacilli have been followed by work in *E. coli* by Pardee and his colleagues, where physiological and genetic studies, related in particular to the mode of action of Mecillinam (Spratt and Pardee, 1975), has led to hypotheses concerning the function of some of these different binding components in cell shape determination and cell division (see Section VI,F).

The organisms involved in these investigations are widely separated taxonomically but have been shown to have similar mechanisms of peptidoglycan synthesis. The generality of these findings can be inferred from data in a much wider range of organisms demonstrating basic similarities in peptidoglycan structure and in sensitivity to inhibitors of peptidoglycan synthesis. Variations in peptidoglycan structure must result from the evolution of specific enzyme systems, within the constraints imposed by the necessity for reproducing a functional cross-linked network. These constraints impose considerable stability on the genes determining peptidoglycan structure, and this is the basis of the utility of peptidoglycan structure determinations in taxonomy.

C. Cytoplasmic Events: Synthesis of UDP-MurNAc Pentapeptide

1. Synthesis of Muramic Acid

The five reactions leading to the synthesis of UDP-GlcNAc from glucose 6-phosphate are illustrated in Fig. 28. Since MurNAc is found only in peptidoglycan, while GlcNAc occurs both in peptidoglycan and in other bacterial polymers, such as teichoic acids and lipopolysaccharides, the addition of phosphoenolpyruvate to GlcNAc, the first step in conversion of UDP-GlcNAc to UDP-MurNAc (Fig. 29), is the first reaction wholly unique to peptidoglycan synthesis. An NADPH-linked dehydrogenase converts the product to UDP-MurNAc (Fig. 29). These two enzymes were first investigated in *Enterobacter cloacae* (Gunetileke and Anwar, 1968), from

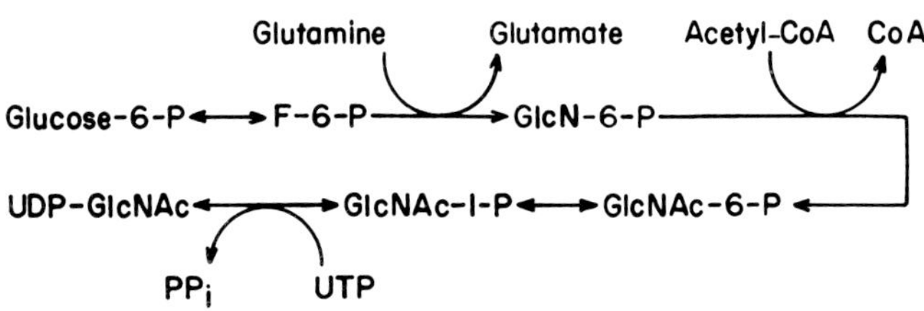

Fig. 28. Synthesis of UDP-GlcNAc.

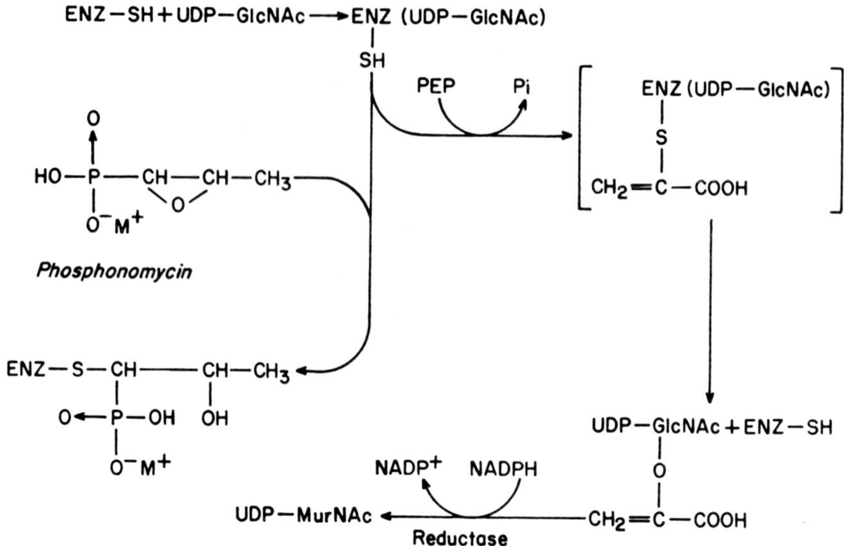

FIG. 29. Synthesis of UDP-MurNAc. Phosphonomycin, acting as an analog of the phosphoenolpyruvate (PEP) substrate of the enolpyruvate transferase, irreversibly inactivates this enzyme by reacting with a cysteine residue. This suggests that this cysteine residue is at the active site and participates in the normal reaction, as shown. A separate reductase converts the product of the transferase to UDP-MurNAc.

which organisms they have now been purified (Taku *et al.*, 1970). They have also been partially purified from *Staphylococcus epidermidis* (Wickus *et al.*, 1973; Wickus and Strominger, 1973).

Phosphonomycin (fosfomycin) competitively inhibits the transfer of enolpyruvate to UDP-GlcNAc, apparently acting as a substrate analog of PEP. It irreversibly inactivates the partially purified transferase of *E. cloacae* at concentrations comparable to those inhibiting growth of this organism. Binding of both PEP and phosphonomycin to the enzyme requires the presence of UDP-GlcNAc, and hydrolysis of the inactivated enzyme gives a residue of cysteine linked through its —SH group to hydroxypropyl phosphonate (Kahan *et al.*, 1974), indicating that the antibiotic becomes covalently linked to this —SH group in the active site, as indicated in Fig. 29.

Since the enolpyruvate transferase is a cytoplasmic enzyme, inhibition by phosphonomycin requires transport of the drug across the plasma membrane. This is achieved via the glucose-repressible L-α-glycerophosphate transport system (Hayashi *et al.*, 1964), and by the inducible hexose monophosphate uptake system (Kadner and Winkler, 1973). Growth conditions repressing these systems and mutations leading to their inactivation result in resistance to phosphonomycin. Unfortunately, these are sufficiently frequent events to severely limit the clinical utility of the drug.

In *Nocardia* and *Mycobacteria*, the oxidation of MurNAc to *N*-glycolylmuramic acid probably occurs at the UDP-MurNAc stage, since *Nocardia asteroides* accumulates UDP-*N*-glycolylmuramic acid in the presence of benzyl penicillin, and extracts of this organism can oxidize UDP-MurNAc in the presence of NADPH.

2. Alanine Racemase

The conversion of L-alanine to D-alanine is catalyzed by alanine racemase, and this reaction has been studied in several species, most extensively in *S. faecalis* by Neuhaus *et al.* (Lynch and Neuhaus, 1966) who identified the cofactor of the enzyme as pyridoxal phosphate. The *S. aureus* alanine racemase also requires a second heat-stable cofactor (Ito and Strominger, 1962).

D-Cycloserine is an antibiotic sometimes used in the treatment of tuberculosis infections refractory to less toxic antibiotics. Its utility is restricted by its severe toxicity, which is (at lèast in part) related to its interaction with pyridoxal phosphate-containing enzymes. D-Cycloserine causes lysis of growing cultures of susceptible bacteria, and causes accumulation of UDP-MurNAc tripeptides (Strominger *et al.*, 1959). Both effects could be reversed by D-alanine, leading Park (1958a) to hypothesize that D-cycloserine inhibited the normal incorporation of D-alanine into the nucleotide precursor of peptidoglycan. This was confirmed and the effects of D-cycloserine explained by demonstration of its competitive inhibitory effect on both alanine racemase and D-alanyl-D-alanine synthetase (Strominger *et al.*, (1960) (Figs. 27 and 30).

The alanine racemases from *S. faecalis*, *S. aureus*, and *Lactobacillus fermenti* are competitively inhibited by D- but not L-cycloserine. Roze and Strominger (1966), investigating the *S. aureus* enzyme, postulated a single binding site for both D- and L-alanine which binds both substrates in a conformation such that the amino and carboxyl groups are located as in the fixed conformation of D-cycloserine. L-Cycloserine, which cannot adopt this conformation, would therefore not be an inhibitor. However, more recent data for the *E. coli* racemase (Lambert and Neuhaus, 1972) is inconsistent with this hypothesis, since both D- and L-cycloserine inhibit this enzyme. Neuhaus *et al.* (1972) have postulated a second model for the *E. coli* racemase, with distinct binding sites for D- and L-alanine. Johnston and Diven (1969), who investigated the *B. subtilis* enzyme, which is also inhibited by both cycloserine isomers, postulated a third model in which a single binding site binds either L- or D-alanine according to the enzyme conformation. This would not explain all of the *E. coli* data, but it is possible that all

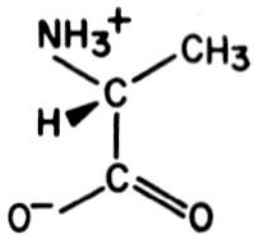

D-Alanine

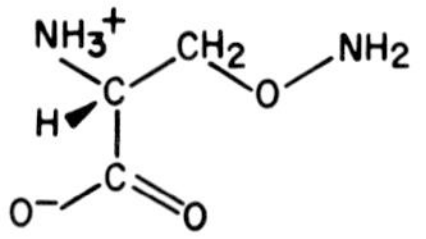

β-Aminoxy-D-alanine

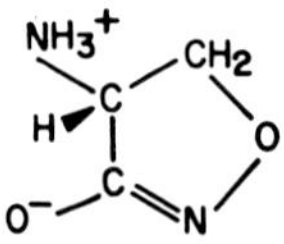

D-Cycloserine

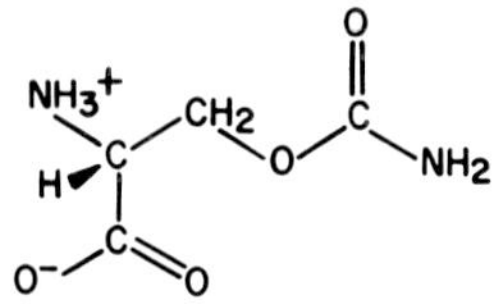

O-Carbamoyl-D-serine

FIG. 30. Inhibitors of alanine racemase and D-alanyl-D-alanyl synthetase. Only D-cycloserine, an analog of D-alanine (as illustrated here), inhibits the synthetase. It also inhibits alanine racemase. L-Cycloserine, β-aminoxy-L- and -D-alanine and *O*-carbamoyl-D-serine also inhibit the alanine racemases of certain bacterial species (see text).

three types of enzyme exist, or that all alanine racemases are two site enzymes in which accessibility of the D- and L-alanine binding sites to D- and L-cycloserine, respectively, varies.

The alanine racemase of *S. faecalis* is also competitively inhibited by *O*-carbamoyl-D-serine, but not by its L isomer (Lynch and Neuhaus, 1966) (Fig. 30). However, both β-aminoxy-D- and L-alanine, formed by cleavage of the rings of D- and L-cycloserine, respectively (see Fig. 30), inhibit the alanine racemase and the growth of *S. faecalis* (Neuhaus *et al.*, 1972). It, therefore, appears that this enzyme may also have separate binding sites for D- and L-alanine and that access of *O*-carbamoyl-L-serine and L-cycloserine to the L-alanine site is sterically hindered.

3. D-ALANYL-D-ALANINE SYNTHETASE

D-Alanyl-D-alanine synthetase is inhibited by D-cycloserine, but not by L-cycloserine, carbamoyl-D- or -L-serine, or β-aminoxy-D- or L-alanine. The *S. faecalis* enzyme has two distinct binding sites for D-alanine which are filled sequentially by donor (N terminal) and receptor (C terminal) amino acids (Neuhaus and Lynch, 1964). Energy for formation of the peptide bond is provided by concomitant hydrolysis of ATP in the presence of potassium ions and either Mg^{2+} or Mn^{2+} ions. This reaction is, therefore, essentially irreversible, and since the alanine racemase has an equilibrium constant of about 1, the entire L-alanine pool of the cell would be converted to D-alanyl-

D-alanine if the synthetase were not regulated. Effective regulation is provided by product inhibition (Neuhaus and Lynch, 1964).

The ultimate effect of inhibition of alanine racemase by cycloserine, carbamoyl-D-serine or β-aminoxyalanine is irreversible inhibition due to covalent binding of the antibiotic to pyridoxal phosphate, the prosthetic group of the enzyme. D-Alanyl-D-alanine synthetase lacks this prosthetic group, and the effects of D-cycloserine are reversible. D-cycloserine binds to both D-alanine binding sites, although the K_i for binding to the acceptor site is fivefold higher than that for binding to the donor site. This K_i is close to the minimal concentration (MIC) causing lysis *in vivo*. Since this MIC is lower than the K_i for alanine racemase, inhibition of the donor site of the synthetase is the probable effective site of D-cycloserine action *in vivo*. D-Cyclothreonine, the *cis*-5-methyl-substituted analog of D-cycloserine, is fivefold less effective *in vivo* and has a fivefold higher K_i for the donor site of the synthetase, substantiating this hypothesis (Lynch and Neuhaus, 1966; Neuhaus *et al.*, 1972).

The synthetase is essentially specific for the utilization of D-alanine as donor, but will utilize glycine, or other D-amino acids, such as D-serine and D-norvaline, as acceptor, producing dipeptides with variable C termini. These dipeptides, such as D-alanyl-D-alanine, competitively inhibit the synthetase, and product binding sites probably exist on the enzyme for this necessary function (Neuhaus *et al.*, 1972).

Interestingly enough, the ligase which normally links D-alanyl-D-alanine to UDP-MurNAc tripeptide has a complementary specificity, having a near absolute requirement for C-terminal D-alanine, but accepting glycine or D-norvaline in the N-terminal position (Neuhaus and Struve, 1965). The combination of the specificities of the ligase and the synthetase thus ensures that only D-alanyl-D-alanine is incorporated into the nucleotide pentapeptide. The effects of D-amino acids on these activities and on transpeptidase (see Section VI,E) appear to be largely responsible for the growth inhibiting action of D-amino acids on bacteria.

To inhibit growth, D-cycloserine must be transported into the cytoplasm, and not unexpectedly it employs a D-alanine transport system for this purpose. Alanine transport in *E. coli* involves separate systems for L-alanine transport and the transport of both D-alanine and glycine (Wargel *et al.*, 1970). The latter system has both high and low affinity segments. Mutants of *E. coli* resistant to D-cycloserine have been obtained with stepwise increases in MIC (Wargel *et al.*, 1970). The initial step is accompanied by loss of the high affinity. D-Alanine transport segment and a second step mutation involves loss of the low-affinity segment. This transport system may primarily be involved in glycine transport, although it seems likely that reutilization of D-alanine released extracellularly by carboxypeptidase and transpeptidase action could have growth advantages under poor growth conditions.

Alanine is transported differently in *S. faecalis* (Reitz *et al.*, 1967). One class of D-cycloserine resistant mutants was unable to transport either D- or L-alanine and presumably was defective in a component common for transport of both amino acids. A second D-cycloserine resistant mutant class had increased levels of both alanine racemase and D-alanyl-D-alanine synthetase activities, suggesting some coordinate control of synthesis of these enzymes, which however, are genetically unlinked (see Section VI) in *E. coli*.

4. Other Unique Amino Acids

The normal pathway for formation of L-lysine from L-aspartate in bacteria involves the production of LL-diaminopimelate which is racemized to *meso*-diaminopimelate and decarboxylated to form L-lysine. L-Homoserine is an intermediate in the production of L-threonine and L-methionine from L-aspartate, and L-ornithine is an intermediate in the formation of L-arginine from L-glutamate. These known reactions account for the synthesis of all the unusual L-amino acids found in the L-R_3 position, with the exception of L-diaminobutyrate.

The mechanisms of formation of D-glutamate has been demonstrated in very few organisms. In *Lactobacillus fermentii* it is produced by a highly specific racemase (Diven, 1969), while in *B. subtilis* it is produced by a relatively specific D-alanine:D-glutamate transaminase (Martinez-Carrion and Jenkins, 1965). The latter enzyme also functions with D-aspartate, which occurs in the cross-bridges of some type A4 peptidoglycans. A similar D-amino acid aminotransferase occurs in *Bacillus sphaericus* (Yonaka *et al.*, 1975). No inhibitors of these enzymes are known, and no known mutants are defective in D-glutamate production. D-Serine occurs in the cross-bridges of certain corynebacteria, and D-lysine, D-ornithine, and D-diaminobutyrate occur in some type B peptidoglycans. The mechanism of their synthesis is unknown. DD-DAP is presumably formed by racemization of *meso*-DAP.

5. Synthesis of the Pentapeptide

The pentapeptide is synthesized by sequential addition of L-alanine, D-glutamate, amino acid L-R_3, and the dipeptide D-alanyl-D-alanine to the carboxyl group of UDP-MurNAc. The addition of each amino acid or dipeptide is catalyzed by a specific ligase which requires either Mg^{2+} or Mn^{2+} as cofactor, and formation of each peptide bond is accompanied by the hydrolysis of one molecule of ATP. Addition of amino acids to the α-carboxyl group of D-glutamate involves peptide chain extension at the C-terminus, and employs the same mechanism. This occurs in type A2 and a few type A3 peptidoglycans and in the addition of the cross-bridge amino acids to the glutamate carboxyl groups in all type B peptidoglycans. The amidation of carboxyl groups is a formally analogous reaction, and the amidation of the α-carboxyl of D-glutamate (stem peptides), D-aspartate (cross-bridges of type A4), and the carboxyl at the D center of *meso*-DAP (stem peptides) is also accompanied by ATP hydrolysis and involves specific enzymes (Siewert and Strominger, 1968).

The sequence of amino acids is determined by the specificity of the ligases for the acceptor amino acid and the donor peptide. For example, the L-lysine ligase of *S. aureus* will not use *meso*-DAP as a substrate for ligation to UDP-MurNAc-L-Ala-D-Glu (Ito and Strominger, 1966), while the converse is true for the *meso*-diaminopimelate ligase of *E. coli* (Mizuno and Ito, 1968). In *B. sphaericus*, where the cell wall peptidoglycan contains L-lysine and the cortex peptidoglycan contains *meso*-diaminopimelate (Powell and Strange, 1957; Hungerer and Tipper, 1969; Tipper, 1969b), the same specificities hold true, and sporulation is accompanied by the synthesis of a *meso*-diaminopimelate ligase activity which is absent in vegetative cells and spores (Tipper and Pratt, 1970). The ligase which adds D-alanyl-D-alanine onto the mucleotide tripeptide in *B. sphaericus* does not distinguish between L-lysine and *meso*-diaminopimelate in the L-R_3 position of the acceptor (Linnett and Tipper, 1974). Because of these specificities, synthesis of diaminopimelate ligase during sporulation in *B. sphaericus* is necessary and sufficient for synthesis of the UDP-MurNAc pentapeptide precursor of cortical peptidoglycan.

Corynebacterium poinsettia has a type B peptidoglycan in which R_1 is glycine and L-R_3 is L-homoserine (Table V,A). The cells of this organism contains a glycine ligase, which has only slight activity with L-alanine as acceptor. The D-glutamate ligase will utilize either UDP-MurNAc-Gly or UDP-MurNAc-L-Ala as donor. However, the L-homoserine ligase will utilize only the UDP-MurNAc-Gly-D-Glu dipeptide as donor, having low activity with the L-Ala-containing nucleotide dipeptide. The L-homoserine ligase also has the usual tight specificity for the acceptor amino acid. It has only low activity with L-lysine, *meso*-DAP, or L-diaminobutyrate. The *C. poinsettia* D-alanyl-D-alanine ligase fails to distinguish between L-homoserine, L-lysine, and *meso*-diaminopimelate (Wyke and Perkins, 1975) in donor tripeptides, so that the final sequence is determined by the specificity of the Gly and L-homoserine ligases only.

Ligation of D-ala-D-ala to UDP-MurNAc tripeptide is reversible in the presence of ADP + P_i in preparations from *S. aureus*, *S. faecalis*, and *B. subtilis* (Egan *et al.*, 1973; Oppenheim and Patchornik, 1974), but the conditions required for this reversal are nonphysiological, and there is no evidence to suggest that significant reversal takes place *in vivo*. Reversibility could otherwise be significant in controlling the accumulation of UDP-MurNAc pentapeptide under conditions of varying demand for peptidoglycan synthesis.

6. Control of UDP-MurNAc Pentapeptide Synthesis

Temperature-sensitive, osmotically fragile (lysis) mutants (designated *tof* or *tsl*), impaired in various steps of UDP-MurNAc pentapeptide synthesis, have been isolated from *E. coli* and *S. aureus*. Symbols have been proposed for the loci in *E. coli;* mutants presumed to reside in these loci have been obtained for UDP-GlcNAc:enolpyruvate reductase (*MurB*) (Matsuzawa *et al.*, 1969), L-alanine ligase (*MurC*), *meso*-diaminopimelate ligase (*MurE*), and D-alanyl-D-alanine ligase (*MurF*) (Lutgenberg, 1971; Lutgenberg and de Haan, 1971; Lugtenberg *et al.*, 1971, 1973; Wijsman, 1972a). Alanine racemase (*alr*) (Wijsman, 1972b) and D-Ala-D-Ala synthetase (*ddl*) mutants are also known (Lugtenberg and van Schijndel van Dam, 1972, 1973). *Staphylococcus aureus* mutants defective in L-alanine, D-glutamate, and L-lysine ligases and in D-alanyl-D-alanine synthetase have been obtained (Good and Tipper, 1972; Gauthier, Good, and Tipper, unpublished observations). UDP-GlcNAc:enolpyrutate ligase (*MurA*) mutants have not been isolated in *E. coli*. The *MurC*, *E* and *F* and *ddl* genes of *E. coli* are closely linked at about 2 minutes on the 100 minutes *E. coli* genetic map (Wijsman, 1972a). This region has been designated "murein A" (*mra*) and also contains linked markers for a D-alanine requirement (*mraA*) and a DD-carboxypeptidase defect (*mraB*) (Miyakawa *et al.*, 1972). The relationship of the later mutation to the better characterized DD-carboxypeptidase defects (see Section VI,F) is not clear. Also in this region are two temperature-sensitive division (*fts*) loci, *ftsA* and *ftsI* (see Section E,8), which may be related to lesions in peptidoglycan synthesis. Genetic evidence is consistent with operon organization of the *murE*, *C*, and *F* genes, and coordinate control of these genes with related functions seem appropriate. J. R. Walker has been able to order the genes in this region precisely, and with respect to another temperature-sensitive cell division mutant he calls *sep*, using defective lambda phage derived from insertion in the adjacent *leuA* gene. *sep* is probably identical with *ftsI*, since function of both is required continuously during septation. *ftsA* function is required only to initiate septation (Walker *et al.*, 1975). The order is (*leuA*) *sep*, *murE*, *F*, *C*, *ddl*, *ftsA*, *envA* (Fletcher *et al.*, 1978). The *murB* gene is apparently part of a second cluster (murein B, *mrb*) at 89 minutes, which includes other less well-characterized defects in peptidoglycan synthesis (Miyakawa *et al.*, 1972). Alanine racemase maps at about 92 minutes. Mutants defective in penicillin-binding proteins have also been mapped, the best characterized being the *rodA* locus at 14 minutes

(Iwaya *et al.*, 1978, see also Section VI,F,4c). New data identifies *sep* with *ftsI* (Walker, Iwaya, and Tipper, unpublished observations). Mutants defective in outer membrane protein synthesis are found in several unlinked loci. Attempts at determining genetic linkage of the *S. aureus* markers by transformation have so far been unsuccessful (Good, Gauthier, and Tipper, unpublished observations).

The *S. aureus* tof mutants, defective in a specific ligase, characteristically accumulate the UDP-MurNAc peptide donor substrate for this ligase at the nonpermissive temperature (Good and Tipper, 1972). Intracellular concentrations easily reach 10 m*M*, and feedback inhibition of these accumulated nucleotides on the enzymes responsible for their synthesis can be detected *in vivo* and demonstrated *in vitro* (Gauthier and Tipper, unpublished observations). For example, UDP-MurNAc-L-Ala and UDP-MurNAc-L-Ala-D-Glu inhibit the L-alanine ligase of *S. aureus in vitro* with K_i of about 0.15 m*M*. The nucleotide tripeptide is a less efficient inhibitor, but the pentapeptide also inhibits both L-Ala and D-Glu ligases. As a consequence, Vancomycin and β-lactam antibiotics, which inhibit late stages in peptidoglycan synthesis and cause accumulation of UDP-MurNAc pentapeptide, also cause accumulation of millimolar concentrations of UDP-MurNAc and UDP-MurNAc-L-Ala. Similarly, mutants defective in D-Glu ligase accumulate almost equal amounts of UDP-MurNAc and UDP-MurNAc-L-Ala at the nonpermissive temperature (Good and Tipper, 1972). Because of the high K_i, inhibition of peptidoglycan synthesis at any stage in *S. aureus* results in accumulation of millimolar concentrations of precursor nucleotides. D-Cycloserine, for example, causes accumulation of large quantities of UDP-MurNAc tripeptide.

D-Cycloserine has similar effects in *E. coli* and *B. subtilis*, and lower level accumulation of acceptor nucleotides is detectable in *E. coli* tsl (temperature-sensitive lysis) mutants at the nonpermissive temperature (Lutgenberg and de Haan, 1971). However, in contrast to gram-positive organisms, such as *S. aureus*, β-lactam antibiotics fail to cause accumulation of UDP-MurNAc pentapeptide in *E. coli.* It seems likely, therefore, that synthesis of this complete precursor is under tight (feedback?) control in *E. coli.* Precursor nucleotides are not present at chemically detectable levels in normal *E. coli* cells, whereas cells of *S. aureus* do contain small but significant quantities. In *B. subtilis* and other gram-positive spore-forming bacilli, β-lactam antibiotics also fail to cause marked pentapeptide accumulation, but Vancomycin (to which *E. coli* is resistant) does cause accumulation. The significance of this is not clear, but suggests differential effects of β-lactam antibiotics on the accumulation of the more proximal lipid-bound intermediates in *B. subtilis* and *S. aureus.*

The most obvious mechanism for feedback control of pentapeptide synthesis by product would be inhibition of the UDP-GlcNAc:enolpyruvate ligase. In fact, the enzymes from *E. cloacae*, *B. cereus* and *E. coli* are all inhibited by both UDP-MurNAc pentapeptide and UDP-MurNAc tripeptide (Taku *et al.*, 1970). In contrast, the enzyme from *S. epidermidis* is not inhibited by these nucleotides, and inhibition *in vitro* was only obtained with UDP-MurNAc (Wickus and Strominger, 1973). It is, therefore, possible that inhibition of the L-alanine ligase by pentapeptide in staphylococci, as described above, is physiologically significant in leading to accumulation of UDP-MurNAc and secondary repression of the entire pathway. The *E. coli* control systems appear to be more direct and efficient, although these observations still fail to explain why tripeptide accumulates in the presence of cycloserine but pentapeptide fails to accumulate in the presence of penicillins. In summary, the role of feedback inhibition in the control of synthesis of UDP-MurNAc pentapeptide remains unclear, but it probably functions at a coarse level in most bacteria. The nucleotide precursors of peptidoglycan and teichoic acid synthesis (e.g., CDP-ribitol, CDP-glycerol) also exert reciprocal feedback inhibitory effects in *B. licheniformis* (Anderson *et al.*, 1973b).

D. Membrane-Associated Events: The Lipid Cycle

The cyclic reutilization of carrier lipid in the membrane-associated events leading to completion of the peptidoglycan subunit and its polymerization is illustrated in Fig. 31.

1. The Membrane-Bound Carrier Lipid

The first evidence for the involvement of a carrier lipid in peptidoglycan biosynthesis derived from the use of a chromatographic assay for *in vitro* peptidoglycan synthesis. Using nucleotide precursors labeled in D-Ala-D-Ala and particulate enzyme preparations, it was observed that, while label was incorporated into a polymeric product which stayed at the origin (and also into free D-alanine), a more rapidly labeled product was formed which migrated near the solvent front. Kinetic studies implied that this was a precursor of the polymeric product, and this was later confirmed when it was demonstrated that membrane preparations could synthesize polymeric peptidoglycan using the purified lipid-bound intermediate as substrate. It was subsequently identified as an undecaprenyl phosphate derivative (Higashi *et al.*, 1967, 1970b).

Shortly after these initial observations, the involvement of a carrier lipid in synthesis of the O antigenic portion of lipopolysaccharide (Wright *et al.*, 1967) and of the membrane-bound phosphomannan of *M. luteus* (Scher

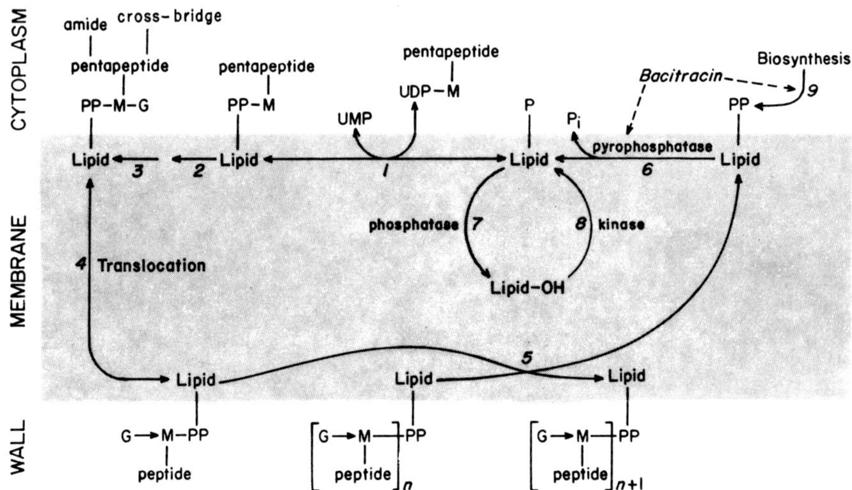

FIG. 31. The lipid cycle in peptidoglycan biosynthesis. The lipid is C_{55}-polyisoprenoid alcohol (see Fig. 32) whose monophosphate derivative is the acceptor for phospho-MurNAc pentapeptide (reaction 1). The product is converted to a lipid pyrophosphate-linked completed peptidoglycan subunit by addition of GlcNAc (from UDP-GlcNAc), amide groups and cross-bridge amino acids (if any) (reactions 2 and 3; not shown in detail). After translocation to the exterior surface of the membrane (reaction 4), the completed subunit acts as the acceptor in peptidoglycan polymerization (reaction 5) which releases lipid PP. The lipid must be converted back to the monophosphate by a pyrophosphatase to complete the cycle (reaction 6). Biosynthesis (reaction 9, see Fig. 32) produces nascent carrier lipid in the pyrophosphorylated form. Bacitracin binds to lipid PP, inhibiting its pyrophosphorolysis and, with less efficiency, the later stages of lipid PP biosynthesis. In at least some organisms, a membrane-bound phosphatase can convert lipid P to the free lipid (reaction 7), while a membrane-bound kinase, best studied in *S. aureus* (see text) (reaction 8), can convert free lipid back to the biosynthetically active monophosphate derivative. This epicycle, by controlling the ratio of lipid OH to lipid P, could control the overall output of the various biosynthetic lipid cycles. These cycles include those involved in synthesis of other extracellular glycans (see Fig. 40).

et al., 1968) was demonstrated and subsequent work showed that undecaprenol, a C_{55}-polyisoprenoid alcohol (Fig. 32), was the carrier in both systems. Recent investigations have demonstrated the involvement of similar carrier lipids in the synthesis of the glycan side chains of glycoproteins in many eukaryotic cell systems (Waechter and Lennarz, 1976). In mammalian systems, the lipid is called dolichol, a C_{80}- to C_{100}-polyisoprenoid alcohol in which the hydroxy-terminal isoprene unit is saturated. Most of the dolichol in tissues is present as a fatty acyl ester, but a small proportion of it is present as the phosphate and is presumed to be involved in glycan synthesis. Glycosylation of asparaginyl residues of glycoproteins involves a glycan core of the type $(\alpha\text{-Man})_n \rightarrow \beta\text{-Man} \rightarrow \beta\text{-GlcNAc} \rightarrow \beta\text{-GlcNAc} \rightarrow \text{AspN}$. Phospho-

GlcNAc is initially transferred from UDP-GlcNAc to dolichol phosphate. The β-linked GlcNAc and mannosyl residues are directly transferred to GlcNAc-PP-dolichol from UDP-GlcNAc and GDP-Man, but the α-Mannosyl residues are first transferred to dolichol-P before transfer to the trisaccharide-PP-dolichol acceptor. Thus, while the growing core remains activated as the dolichol pyrophosphoryl derivatives, these single α-mannosyl residues are activated as the lipid phosphate derivative.

The transfer of oligosaccharide subunits to extracellular polymers (e.g., peptidoglycan and O antigen subunits, teichoic acid link, teichuronic acid subunits, glycoprotein side chain core) seems invariably to involve activation as undecaprenol-PP derivatives. On the other hand, transfer of single glycosyl residues to extracellular polymeric acceptors, as in the synthesis of the membrane-associated mannan in *M. lysodeckticus* (see Section VII,C) or in the glycosylation of the O antigen of *S. anatum* directed by the lysogenic ε15 phage genome (see Section XI) frequently involves activation as the undecaprenol phosphodiester derivative. Transfer releases undecaprenol-P directly. In glycoprotein synthesis, the complete core is transferred from dolichol-PP to the polypeptide. Sialic acid and other residues may be added before or after this transfer (Waechter and Lennarz, 1976).

Tunicamycin inhibits the transfer of phospho-GlcNAc to dolichol-P or undecaprenol-P and so inhibits the glycosylation of asparaginyl residues, the synthesis of teichoic acid (see Section VII), and the P-MurNAc-pentapeptide transferase (G. Tamura *et al.*, 1976; Ward, personal communication). In peptidoglycan synthesis tunicamycin does not inhibit the subsequent transfer of GlcNAc to lipid-PP-MurNAc pentapeptide, in spite of reports to the contrary (Bettinger and Young, 1975).

Undecaprenol has been purified from *M. luteus* (Higashi *et al.*, 1967), *S. aureus* (Higashi *et al.*, 1970b), *S. faecalis* (Umbreit *et al.*, 1972), and *E. coli* (Umbreit and Strominger, 1972a,b), and in each case mass spectroscopy established the structure as C_{55}-polyisoprenoid alcohol. It appears that this single lipid species is involved in synthesis of both O antigens and peptidoglycan in gram-negative bacteria, and in the synthesis of both peptidoglycan and other glycans in gram-positive species (see Sections VII and XI). The lipid carrier may also be involved in the synthesis of capsular polysaccharide in *E. coli* (Troy *et al.*, 1975) and is involved in glycosylation of the cell wall glycoproteins of *Halobacterium* sp. (Mescher and Strominger, 1976). Undecaprenol may thus be a common component in the synthesis of all membrane and extracellular glycans in bacteria. Control of the concentration of the acceptor (monophosphate) form of the lipid could thus serve to control synthesis coordinately of all of these polymers.

Undecaprenol pyrophosphate is synthesized from farnesyl pyrophosphate and 8-isopentenyl pyrophosphate moieties (Fig. 32) in *S. newington* (Chris-

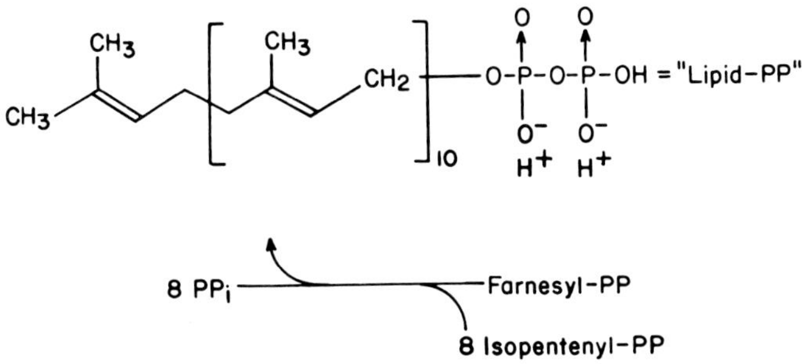

CH3 CH3 CH2 O–P–O–P–OH = "Lipid–PP"
CH3 10 O⁻ O⁻ H+ H+

8 PPi — Farnesyl-PP
8 Isopentenyl-PP

FIG. 32. Biosynthesis of undecaprenol pyrophosphate.

tenson *et al.*, 1969). This nascent pyrophosphate form must be hydrolyzed to the monophosphate by a membrane-bound pyrophosphatase before it can act as an acceptor in glycan synthesis. Transfer of glycan subunits from the lipid intermediate to acceptor in the polymerization of peptidoglycan also produce undecaprenol pyrophosphate. The pyrophosphatase whose action is required to return this lipid to the acceptor state is the primary target of bacitracin action (see Section VI,D,6).

Undecaprenol phosphate can be further converted to the free alcohol in *S. aureus* and *M. luteus* by a membrane-bound phosphatase (Willoughby *et al.*, 1972; Goldman and Strominger, 1972), and it appears that the alcohol may be a depot form of the carrier lipid, since its proportion increases in stationary phase cells of lactobacilli (Thorne and Barker, 1972; Thorne *et al.*, 1974). Membranes of *S. aureus* contain a specific phosphokinase which will catalyze the ATP-dependent phosphorylation of undecaprenol. The enzyme was readily purified from *S. aureus* membrane preparations because of its peculiar property of being soluble in organic solvents but insoluble in water (Higashi *et al.*, 1970a). The solubilized enzyme is lipid-free and inactive, but can be reactivated by the addition of phospholipids (Higashi and Strominger, 1970). The concentration of undecaprenol phosphate in *S. aureus* can thus be controlled by the relative activities of the phosphatase and the kinase, and this ratio may play a central role in controlling the gross rate of peptidoglycan synthesis in this organism. Similar kinases have not yet been described in other bacteria, but the presence of unphosphorylated undecaprenol in all species investigated suggests that the phosphorylation–dephosphorylation system may exist in all bacteria.

2. PHOSPHO-*N*-ACETYLMURAMYL PENTAPEPTIDE TRANSLOCASE

The first reaction in the membrane-associated lipid cycle stage of peptidoglycan biosynthesis is the transfer of phospho-MurNAc pentapeptide from UDP-MurNAc pentapeptide to undecaprenol phosphate (Fig. 31).

The products are undecaprenolpyrophosphoryl-MurNAc-pentapeptide and UMP, so that MurNAc remains activated as its pyrophosphoryl derivative. The reaction has been extensively investigated by Neuhaus *et al.* (see Neuhaus, 1972).

As might be anticipated, the reaction is readily reversible, with an equilibrium constant of about 0.25. Besides translocation to undecaprenol phosphate (the forward reaction), the enzyme also catalyzes an exchange of free uridylic acid and the UMP moiety of UDP-MurNAc pentapeptide. Both reactions require magnesium and potassium ions.

The enzyme has been solubilized from *M. luteus* (Heydanek and Neuhaus, 1969; Umbreit and Strominger, 1972b) and *S. aureus* (Pless and Neuhaus, 1973) membranes by treatment with Triton X-100. The solubilized *M. luteus* enzyme requires the addition of undecaprenol phosphate for the translocation reaction but not for the exchange reaction. Other neutral and polar lipid fractions were required for optimal activity of the translocase and exchange reactions, respectively. On the basis of these data and of similar investigations of the *S. aureus* enzyme, Neuhaus *et al.* (1972) have proposed a five-step reaction sequence (Fig. 33). Translocation requires all five reactions, whereas the exchange reaction involves only steps 2–4, and the presence of sufficient undecaprenol phosphate to form a complex with the enzyme. In the presence of nucleotide pentapeptide and undecaprenol phosphate, a steady state is eventually reached, but the exchange reaction continues and free phospho-MurNAc pentapeptide eventually appears, presumably from breakdown of the enzyme intermediate in reaction 3.

The specificity of the enzyme for its nucleotide substrate partially determines the exclusive incorporation of completed pentapeptides into the

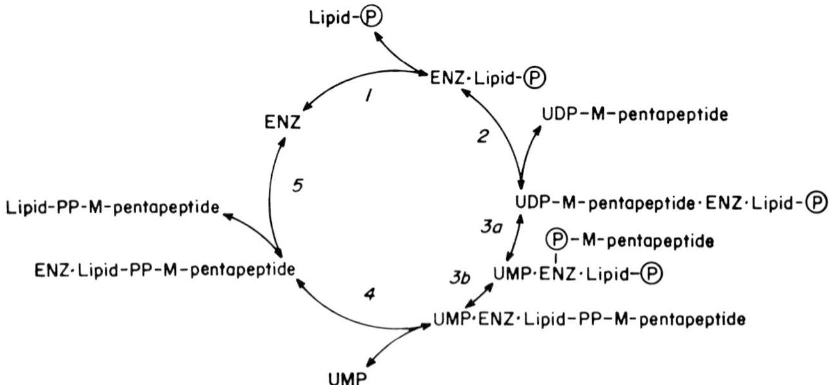

FIG. 33. Mechanism of phospho-MurNAc-pentapeptide translocase. A period represents noncovalent bonding; a dash represents covalent bonding. The intermediate between stages 3a and 3b is hypothetical.

peptidoglycan. The *S. aureus* translocase is only 24% as active with UDP-MurNAc tetrapeptide as with the pentapeptide, and the tripeptide is only 1% as active. The enzyme has relatively high specificity for the L-alanine (R_1) and D-alanine (R_4) residues of the stem peptide, but has a relatively low specificity for the L-R_3 residue and the terminal D-alanine residue (Hammes and Neuhaus, 1974a; Stickgold and Neuhaus, 1967). The enzyme has a high specificity for the UMP moiety of the nucleotide substrate and refuses to accept the 5-fluorouracil analog. Since this analog is accepted by the enzyme involved in synthesis of the nucleotide pentapeptide, addition of 5-fluorouracil results in the accumulation of fluoro-UDP-MurNAc pentapeptide.

3. Completion of the Disaccharide

GlcNAc is transferred from UDP-GlcNAc directly to the MurNAc moiety of lipid-PP-MurNAc pentapeptide. The other product is UDP, and this reaction is essentially irreversible and results in formation of the GlcNAc-MurNAc disaccharide repeating unit of the glycan. Its formation prior to polymerization ensures the alternation of sugar residues in the glycan.

4. Modification of the Stem Peptide

a. Amidation and the Ligation of Amino Acids to D-*Glu-α-COOH.* Amidation of D-glutamate residues was first recognized in *S. aureus* peptidoglycan (Tipper *et al.*, 1967a), and membrane preparations from *S. aureus* catalyze the transfer of NH_2 groups from glutamine or ammonia to lipid-PP-MurNAc pentapeptide or lipid-P,P-disaccharide pentapeptide (Siewert and Strominger, 1968). No transfer to UDP-MurNAc pentapeptide was detectable, and the reaction is accompanied by the hydrolysis of ATP. In *M. luteus*, a similar reaction is involved in the ligation of glycine residues to the α-carboxyl of D-glutamate residues of the stem peptide, and again both the MurNAc- and disaccharide-containing lipid intermediates act as acceptors (Katz *et al.*, 1967). A combination of these ligation and amidation events presumably results in formation of the α-D-glutamylglycine amide and L-alanine amide groups found in the peptidoglycans of certain *Arthrobacter* species (Fiedler *et al.*, 1970). It is also probable that this same mechanism is employed for the sequential addition of the cross-bridge amino acids in type B1 peptidoglycans to the α-carboxyl group of D-glutamate of the stem peptide. All of these reactions, like synthesis of the stem peptide, involve the α-amino group of a free amino carboxylic acid as acceptor. In type B2 cross-bridges, however, the ω-group of D-lys, D-Orn, or D-diamino-butyrate is involved in the cross-bridge peptide chain (Table V), and it is possible that a different mechanism results in formation of this linkage.

b. Formation of Cross-Bridge Peptides of Type A3 *Peptidoglycans.* In contrast to the synthesis of the stem peptide and the type B cross-bridge peptides, the cross-bridges of type A3 and A4 peptidoglycans are synthesized by N-terminal addition to the ω-amino group of the dibasic L-R_3 residue. In type A3 peptidoglycans, where the cross-bridge amino acids are glycine or L-α-aminocarboxylic acids, these amino acids are activated as their tRNA derivatives. This was first suggested by the sensitivity to ribonuclease of the incorporation of glycine into *S. aureus* peptidoglycan *in vitro* (Chatterjee and Park, 1964). The transfer of glycine to the growing pentaglycine cross-bridge of *S. aureus* peptidoglycan apparently occurs sequentially at the N-terminus (the opposite of protein synthesis), and there is no evidence for the presynthesis of peptidyl-tRNA. The reaction involves a membrane-bound enzyme system and is independent of ribosomes and supernatant factors (Matsuhashi *et al.*, 1967; Thorndike and Park, 1969; Kamiryo and Matsuhashi, 1972).

The pentapeptide cross-bridges of *S. aureus* are composed largely of glycine but contain some L-serine. The pentapeptide cross-bridges of *S. epidermidis* T26 contain much more serine in four major sequence variants, some of which occurs as minor components in *S. aureus* (Tipper, 1969a). These sequences indicate differential specificity for L-serine or glycine addition at each stage of cross-bridge elongation, but no evidence for such specificity has been detected *in vitro* (Hilderman and Riggs, 1973). In cells of *S. aureus* grown in medium enriched for L-serine, the peptidoglycan composition resembles that of *S. epidermidis* grown on a serine-deficient medium (Schleifer *et al.*, 1969). Thus, phenotypic variations in cross-bridge sequence can occur in response to medium composition (see Section VII). The probability of addition of a Gly or Ser residue *in vivo* may depend on the relative concentrations of the charged tRNA's, as *in vitro* (Matsuhashi *et al.*, 1967; Hilderman and Riggs, 1973). Nevertheless, clear sequence preference occurs: in either species, an (N^{α})-glycine residue is usually found proximal to the lysine residue. Under conditions of severe glycine deprivation, Gly + Ser pentapeptides or a single *N*-terminal N^{α}-(L-alanyl)-L-lysine residue is found (Schleifer *et al.*, 1969, 1976) (see Section VIII).

Staphylococcus epidermidis was found to contain four species of acceptor tRNA for both glycine and serine, including one species of each which does not participate in protein synthesis. All eight species are active in cross-bridge synthesis *in vitro*, so two appear to have evolved specifically for this function (Bumstead *et al.*, 1968; Petit *et al.*, 1968). The specific glycyl-tRNA was purified and found to contain two isoaccepting species differing in seven bases. Both lack the GTψC sequence characteristic of other tRNA's, and also lack most minor bases, but they are not simply undermodified precursors of other glycyl-tRNA's (Roberts, 1972). Their presence presumably helps the cell to control the flow of glycine residues differentially to peptidiglycan and protein and must allow at least some cross-link formation under conditions of glycine deprivation, where selective utilization of glycine for protein synthesis would be fatal.

The involvement of aminoacyl-tRNA's in cross-bridge formation has also been studied in *M. roseus* (Roberts *et al.*, 1968a), *A. crystallopoites* (Roberts *et al.*, 1968b) and *L. viridescens* (Plapp and Strominger, 1970a,b). In *L. viridescens*, the cross-bridges have the sequence L-Ser-L-Ala. *In vivo* the L-alanine cross-bridge residue is ligated by a soluble enzyme to the UDP-MurNAc pentapaptide precursor, while the L-Ser residue is ligated by a membrane-bound enzyme only to the lipid intermediate. The L-alanyl ligase shows little discrimination *in vitro* between L-seryl-tRNA, L-alanyl-tRNA, or L-cysteinyl-tRNA (Plapp and Strominger, 1970b), so that the *in vivo* sequence specificity is unexplained. In *A. crystallopoites*, by contrast, the membrane-bound ligase that adds the single L-alanyl cross-bridge residue to the lipid intermediate shows tight specificity and does not function with L-seryl-tRNA, L-cysteinyl-tRNA, or L-alanyl-tRNA-Cys, prepared by reduction of L-cysteinyl-tRNA (Roberts *et al.*, 1968b). This ligase thus has specificity for the tRNA as well as for its aminoacyl substituent.

c. Formation of Cross-Bridges of Type A4 *Peptidoglycans. Streptococcus faecium* (frequently referred to as *S. faecalis* by Shockman and his colleagues) had D-isoasparaginyl cross-links. D-Aspartate is activated as the β-D-aspartylphosphate anhydride by a membrane-bound enzyme and is transferred to the lysine εNH_2 of the nucleotide precursor or lipid intermediate by a second membrane-bound activity (Staudenbauer and Strominger, 1972). Kinetic coupling of these two reactions would prevent loss of the rather reactive intermediate. Amidation of the ligated β-D-aspartyl residue occurs last, catalyzed by a soluble enzyme and involving ATP hydrolysis.

The cross-bridges in *Sarcina ureae* are (D-alanyl)-γ-D-glutamylglycyl-(N^{ε}-lysyl) and are synthesized by a hybrid mechanism: glycine, activated as glycyl-tRNA, is added first. D-Glutamate addition requires prior gly addition and is catalyzed by a membrane-bound activity coupled to ATP hydrolysis (Linnett *et al.*, 1974), presumably analogous to the D-aspartate activating activity described above.

d. Formation of Type A2 *Peptidoglycan Cross-Bridges.* This process involves coupled transpeptidase and amidase activities, as shown in Fig. 34 (Ghuysen *et al.*, 1968a). Coordinated interaction of these enzymes with multiple adjacent peptide units is required. In Step 1 (Fig. 34), a conventional type A1 cross-bridge is formed between two adjacent hexapeptides. This triggers amidase action on the muramyl-L-Ala linkage of the (donor) pentapeptide (step 2), producing N-terminal L-Ala which acts as an acceptor for a second transpeptidation reaction (step 3). Repetition of steps 2 and 3

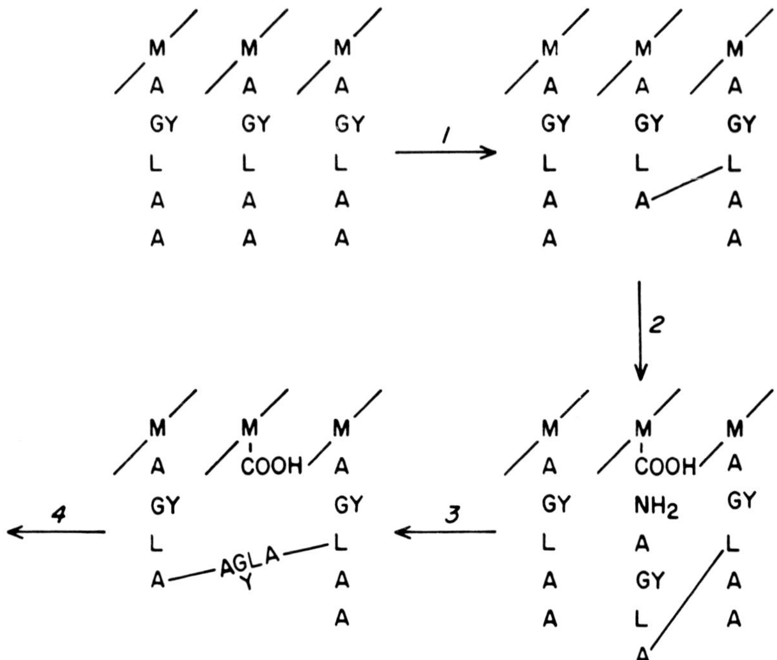

FIG. 34. Synthesis of type A2 cross-bridges. The following events must occur in sequence: (1) formation of a direct D-Ala-(N^{ε})L-Lys cross-link by transpeptidation, (2) amidase action on the N-terminal subunit of the peptide dimer, and (3) transpeptidation to a third peptide unit using the N-terminal L-Ala residue released by amidase action as acceptor. The cross-bridge will now contain one stem pentapeptide. Subsequent repititions of steps 2 and 3 at the N-terminus of the bridge (step 4) will produce bridges with multiple head-to-tail-linked stem peptides (see Fig. 22). The sequence may stop after the first step, and the total number of cross-links is always equal to the number of direct cross-links produced by the first step.

could produce leftward elongation of the cross-bridge. To ensure efficient cross-linkages by this mechanism, the N-terminal lysine of the initial dimer must be as an acceptor for transpeptidation, and the amidase must be unable to act on C-terminal peptides.

5. Polymerization of Completed Disaccharide Peptide Subunits

a. Sequence. The mature cross-linked peptidoglycan is polymerized through both glycosyl and peptide bonds. In principle, these linkages could be formed sequentially, either being formed first, or concomitantly. The scheme shown in Fig. 19 is based on the assumption that glycan elongation (transglycosylation) precedes peptide cross-link formation (transpeptidation), but that the latter reaction must be concomitant and involve nascent donor and acceptor peptide subunits aligned by juxtaposition of their glycan polymerizing sites. More recent data indicates that such a concerted mechanism is not general, if it exists at all: transpeptidation *in vivo* may occur out of phase with and more slowly than incorporation of subunits by glycan chain elongation (Pitel and Gilvarg, 1970) and transpeptidation may be responsible for linking nascent, soluble, polymeric peptidoglycan to preexisting wall (Mirelman *et al.*, 1972) (see below and Section VI,E,3). It remains probable, however, that glycan polymerization precedes transpeptidation, as indicated by the tremendous disparity between glycan and peptide chain lengths. The independence of glycan polymerization from cross-link formation, both *in vitro* and *in vivo*, is direct evidence for this sequence: membrane preparations from many bacterial species are capable of polymerizing un-cross-linked peptidoglycan, and even systems capable of *in vitro* cross-link formation also polymerize un-cross-linked peptidoglycan, and do so exclusively in the presence of penicillins or in the absence of activated cross-bridge amino acids (e.g., Mirelman and Sharon, 1972). In an analogous fashion, intact cells of *M. luteus* (Mirelman *et al.*, 1976), or *B. licheniformis* (Ward and Perkins, 1974) and L forms of *B. licheniformis* (Elliott *et al.*, 1975a,b), growing in the presence of penicillin, produce long soluble chains of un-cross-linked peptidoglycan. Transpeptidation in the absence of glycan chain elongation has only been demonstrated with model peptide substrates (see Section VI,E,2,E,4).

b. Direction of Glycan Elongation. If disaccharide peptide subunits were transferred from lipid-PP to the nonreducing end of preexisting glycan chains, then the GlcNAc residue of newly inserted subunits should be susceptible to periodate oxidation. If, on the other hand, the preexisting polymer, activated as a lipid-PP derivative, were transferred to the GlcNAc residue of lipid-PP-disaccharide peptide subunit acceptors (Figs. 17, 27, and 31), the newly inserted GlcNAc residues will be protected from periodate, but newly inserted peptide subunits will be at the reducing end of the polymer and susceptible to alkali-catalyzed β-elimination (Tipper, 1969a) after removal of the labile lipid-PP moiety by mild acid hydrolysis. By both criteria, elongation of glycan chains by *B. licheniformis* membrane preparations was shown to occur at the reducing end (Ward and Perkins, 1973, 1974). This reducing terminus is blocked by some entity whose susceptibility to acid hydrolysis is consistent with a glycosyl pyrophosphate linkage. It can be labelled with ^{32}P from UDP-MurNAc pentapeptide, however hydrolysis does not yield undecaprenol phosphate and the blocking group has not yet been identified (Ward, personal communication).

Gilvarg *et al.* (Fuchs-Cleveland and Gilvarg, 1976) have recently reported the isolation of a soluble product of peptidoglycan synthesis from intact cells of *B. megaterium*. This is an un-cross-linked peptidoglycan oligomer of discrete size, 11–13 disaccharide units long, with a blocked reducing group.

Lysozyme hydrolysis gives a fragment that may be undecaprenol-PP-disaccharide peptide, although this has yet to be confirmed. No precursor–product relationship has been established for this un-cross-linked oligomer and the mature peptidoglycan, although the oligomer is labeled and chased with essentially the same kinetics as the monomeric lipid intermediate. If it is an intermediate between lipid-linked subunits and polymeric peptidoglycan, then the lack of shorter oligomers is puzzling. It may, alternatively, be a product of cleavage of nascent peptidoglycan chains. The length of the oligomer could then be determined by a hydrolase specificity. If so, this suggests that quite long glycan chains are formed *in vivo* before transpeptidation occurs and that these chains carry reducing terminal undecaprenol-PP. Indirect evidence has recently been obtained suggesting linkage of nascent peptidoglycan to a lipophilic compound whose properties resemble lipoteichoic acid (LTA) (Wong, personal communication). If true, this would implicate LTA as a "carrier" for peptidoglycan synthesis, as previously suggested for wall teichioic acid (Fiedler and Glaser, 1974a,b). The involvement of LTA in wall teichoic acid synthesis is now in doubt (see Section VII).

Elongation of glycan chains at the reducing terminus using lipid-PP-linked substrates as both donor and acceptor, ensures that all reactants are anchored at the membrane surface by their lipid substituents as well as by interaction of the adjacent substrate regions with the membrane-bound glycan polymerase. The process may be analogous to the accèpted model of ribosomal polypeptide synthesis in which tRNA-activated peptide, bound to the donor site, is transferred to aminoacyl-tRNA at the acceptor site, followed by translocation of the elongated peptidyl-tRNA to the donor site in readiness for the next step. Direct translocation of the elongated lipid-PP-glycan to the donor site of a two-site polymerase would ensure efficient glycan polymerization with no necessary requirement for a cross-linked or polymeric acceptor. Such a requirement had been hypothesized by Weidel and Pelzer (1964) and frequently reiterated by other workers. They postulated a role for endo-*N*-acetylmuramidases in providing the acceptor sites on preexisting polymeric peptidoglycan. A requirement for such an acceptor is inconsistent with observations of polymeric glycan synthesis by various bacterial membrane preparations or by reverting protoplasts, e.g., *B. megaterium* (Fitz-James, personal communication) and *B. lichenformis* (Elliott *et al.*, 1975a). It is also inconsistent with the lack of demonstratable muramidase activity in organisms such as *S. aureus* (Tipper and Strominger, 1965).

O antigen biosynthesis occurs in an analogous fashion (see Section XI), while elongation of the polyglycerolteichoic acid of *B. subtilis* occurs in the reverse fashion: glycerol phosphate residues are attached to the terminus of the acceptor distal to the initial CDP-activated glycerol unit which must be linked eventually to the peptidoglycan (see Section VII,B,3).

6. Site of Bacitracin Action

The secondary product of glycan polymerization is undecaprenol pyrophosphate. This must be hydrolyzed to the monophosphate by a membrane-bound pyrophosphatase before it can participate again in glycan synthesis. This enzyme has been purified from *M. luteus* membranes (Goldman and Strominger, 1972). Lipid pyrophosphate must also migrate back to the inner surface of the membrane (Figs. 27 and 31), and this probably occurs prior to hydrolysis, so that phosphate is conserved within the cytoplasm.

Bacitracin, a mixture of cyclic peptide antibiotics produced by sporulating cells of *B. licheniformis* (Johnson *et al.*, 1945), inhibits peptidoglycan synthesis and causes the accumulation of UDP-MurNAc pentapeptide in *S. aureus* (Park, 1958b). In its presence, membrane preparations from *S. aureus* polymerize some peptidoglycan from [^{32}P]UDP-MurNAc pentapeptide plus UDP-GlcNAc, and [^{32}P]undecaprenol pyrophosphate accumulates (Siewert and Strominger, 1967). The [^{32}P]phosphate component of this product is released by membrane preparations from *S. aureus* or *M. luteus*, or by *E. coli* alkaline phosphatase. Bacitracin inhibits hydrolysis by all three preparations, although it failed to inhibit hydrolysis of other substrates by the *E. coli* enzyme (Siewert and Strominger, 1967). Bacitracin, therefore, interacts with the lipid-PP substrate rather than with the enzymes. Formation of a tight complex between bacitracin and lipid-PP, dependent upon divalent metal ions, has been demonstrated (Stone and Strominger, 1971). The association constant increases with the length of the lipid chain, so that while undecaprenol pyrophosphate synthesis is inhibited by bacitracin, the most sensitive event is the pyrophosphorolysis of the complete undecaprenol-PP (Storm and Strominger, 1973).

Surprisingly, protoplasts of *B. megaterium* are killed by bacitracin (Hancock and Fitz-James, 1964), apparently as a consequence of loss of their permeability barrier. Bacitracin was bound to *M. luteus* protoplasts with the same affinity with which it binds to undecaprenol-PP, and it was suggested that formation of a complex between bacitracin and this lipid disrupts membrane function (Storm and Strominger, 1974). The role of this effect in the lethal action of bacitracin in intact cells is unknown. Bacitracin is a bactericidal antibiotic, effective against many gram-positive human pathogens. Most gram-negative species are resistant, but resistance is acquired very rarely among sensitive bacterial species, and presumably only as a consequence of impermeability, since there is no polypeptide target to be mutated and formation of undecaprenol-PP is essential to cell survival. Unfortunately, bacitracin is extremely nephrotoxic, and its use is therefore limited almost exclusively to topical applications. The nephrotoxicity of bacitracin is due to effects on mammalian membrane functions, perhaps due to complex formation with polyprenolpyrophosphates in these membranes.

Bacitracin is also lethal for halophilic bacteria, such as *H. halobium*, whose cell walls contain glycoprotein but are devoid of peptidoglycan (Mescher and Strominger, 1976). It appears that the oligosaccharide substituents linked to the asparaginyl residues of the cell wall polypeptides of this organism are transported and assembled on membrane-bound undecaprenol pyrophosphate. This is also true for synthesis of the reiterated subunits of the O antigenic polysaccharide portion of LPS, and synthesis of this polymer is, therefore, also inhibited by bacitracin (see Section IX). This may also be true for the linkage subunit between peptidoglycan and wall teichoic acid in *S. aureus* (see Section VII). By contrast, assembly of the "core" oligosaccharide of LPS occurs on the lipid A moiety on the inner face of the cytoplasmic membrane, so does not involve undecaprenol-PP and is insensitive to bacitracin. Glycosylation reactions involving sugars activated as undecaprenol-P derivatives (see Section VI,D,1) produce undecaprenol-P directly and are insensitive to bacitracin.

7. Vancomycin and Ristocetin

Vancomycin, ristocetin, and related antibiotics are bactericidal glycopeptide antibiotics whose activity apparently resides in their unusual cyclic phenolic peptide moieties (Perkins and Nieto, 1974). They are highly effective against many gram-positive species, but, like bacitracin, are inactive against most gram-negative organisms, apparently being unable to cross the outer membrane of their cell walls. These antibiotics inhibit peptidoglycan polymerization catalyzed by particulate prepatations from several species, including *S. aureus* (Anderson *et al.*, 1967), *B. megaterium* (Reynolds, 1971), *B. licheniformis* (Ward, 1974), and *Gaffkya* (*Aerococcus*) *homari* (Hammes and Heuhaus, 1974b). Nucleotide and lipid intermediates accumulate in each case. *Gaffkya* has an Alα peptidoglycan, and the cell free system is capable of efficiently polymerizing peptidoglycan from UDP-GlcNAc together with either UDP-MurNAc pentapeptide or UDP-MurNAc tetrapeptide, lacking the terminal D-alanine residue. Only polymerization of the pentapeptide substrate is inhibited by Vancomycin, so Vancomycin must interact with the monomer substrate rather than with the enzyme (Hammes and Neuhaus, 1974b,c). This interaction is the crucial step in inhibition of peptidoglycan synthesis by these antibiotics. Formation of such a complex has recently been demonstrated with a spin-labeled lipid intermediate and *S. aureus* membrane preparations (Johnston and Neuhaus, 1975). Modification of the lipid intermediate by addition of cross-bridge amino acids or amidation is not inhibited.

The formation of tight complexes by Vancomycin with lipid intermediates containing acyl-D-alanyl-D-alanine apparently has secondary effects on membrane function, resulting in permeability changes somewhat analogous to those seen on binding of bacitracia to undecaprenol-PP. For this reason, neither bacitracin nor Vancomycin-like antibiotics can be used to isolate viable L forms. As for bacitracin, resistance to Vancomycin-type antibiotics can only be due to inaccessibility of the biosynthetic intermediate targets, or to inactivation of the antibiotics. For this reason, resistance develops rarely in sensitive gram-positive organisms and resistance in gram-negative organisms is probably due to the outer membrane permeability barrier. Like bacitracin, Vancomycin is highly toxic to man, especially for the auditory system, and it is used only for situations such as endocarditis where a bactericidal antibiotic is required and when less toxic antibiotics have been shown to be ineffective.

Newer antibiotics such as moenomycin, enduracidin and prasinomycin (see Lugtenberg *et al.*, 1971; Linnett and Strominger, 1973) also inhibit peptidoglycan polymerase, but appear to be structural analogs of the undecaprenol-PP-(subunit) enzyme substrate, rather than agents capable of binding tightly to this intermediate. Gardimycin (Somma *et al.*, 1977) is probably similar.

E. The Cross-Linking of Glycan Chains by Transpeptidation

1. The Lethal Target for β-Lactam Antibiotics: Theory

Early observations of the effects of penicillin G on sensitive bacteria indicated a primary effect on cell wall integrity, limited to growing cells, and resulting in lysis unless the cells exposed to penicillin were protected by a medium of high osmolarity. Investigations have followed four major routes: (1) Analysis of the effects of penicillin on growth, viability, cell shape, cell division, and cell integrity. Recent investigations have focused on the role of autolytic enzymes in these effects. (2) Analysis of effects of exposure to penicillin *in vivo* on the structure of the cell wall and of *in vitro* effects on the enzymes involved in peptidoglycan synthesis. (3) Analysis of the cell components which bind penicillin tightly and attempts at correlating this binding with the above biochemical and physiological effects and (4) analysis, by the above techniques, of mutants with altered response to β-lactam antibiotics.

The peptidoglycan formed by polymerization of disaccharide peptide subunits is an ionic water-soluble polymer which must be cross-linked to form an insoluble matrix before it can function as a structural cell wall component. By 1964, all of the reactions leading up to the formation of linear un-cross-linked peptidoglycan had been demonstrated *in vitro*, and none had been found sensitive to β-lactam antibiotics (Anderson *et al.*, 1965). Accumulating information on the structure of intact cross-linked peptidoglycan (e.g., Ghuysen *et al.*, 1965) indicated cross-linkage through penultimate D-alanine residues, and, as pointed out by Wise and Park (1965), formation of such cross-links might involve a transpeptidation reaction in which the D-alanyl-D-alanine bond in one peptide chain is cleaved with formation of a new peptide link to an amino acceptor on an adjacent cross-bridge peptide (Fig. 35). Tipper and Strominger (1965) arrived at the same conclusion independently.

The total bond energy should change little in transpeptidation, so that the reaction might require no energy input in the form of ATP hydrolysis, allowing the reaction to take place outside of the plasma membrane. The extent of reaction would then depend on entropic factors, such as effective removal of the product from the reaction phase due to insolubility, and the reaction should be reversible. The transpeptidase was proposed as the target of β-lactam antibiotic action.

In vivo evidence for the validity of this hypothesis was soon obtained. Wise and Park (1965) demonstrated that cells of *S. aureus*, growing in the presence of sublethal penicillin concentrations, incorporated a higher ratio of alanine to glycine in their walls than did uninhibited cells. Tipper and Strominger (1965) showed, by direct chemical analysis of the peptide subunits, that the C-terminal peptide units of *S. aureus* peptidoglycan retain both of the D-alanine residues found in the nucleotide precursor and that the proportion of un-cross-linked peptide monomer increased in the presence of sublethal concentrations of penicillin. In retrospect, it was fortuitous that an organism was chosen for study in which DD-carboxypeptidase activity (see below) is negligible.

To explain maintenance of the peptide bond energy during a sequential transpeptidation reaction, Tipper and Strominger (1965) hypothesized that the acyl-D-alanyl-D-alanine donor interacts with the enzyme to give an acyl-D-alanyl–enzyme intermediate with the release of the terminal D-alanine, followed by transfer of the acyl-D-alanyl substituent to the amino acceptor (Fig. 35). It was further hypothesized that β-lactam antibiotics, such as the *N*-acyl derivatives of 6-aminopenicillanic and 7-aminocephalosporanic acids, are structural analogs of the particular conformation of acyl-D-alanyl-D-alanine that is bound to sensitive transpeptidases (Fig. 36), and so bind to the donor site of these transpeptidases in a facile fashion. According to this analogy, the carbonyl-nitrogen bond of the highly strained β-lactam ring corresponds to the peptide bond cleaved during transpeptidation, and

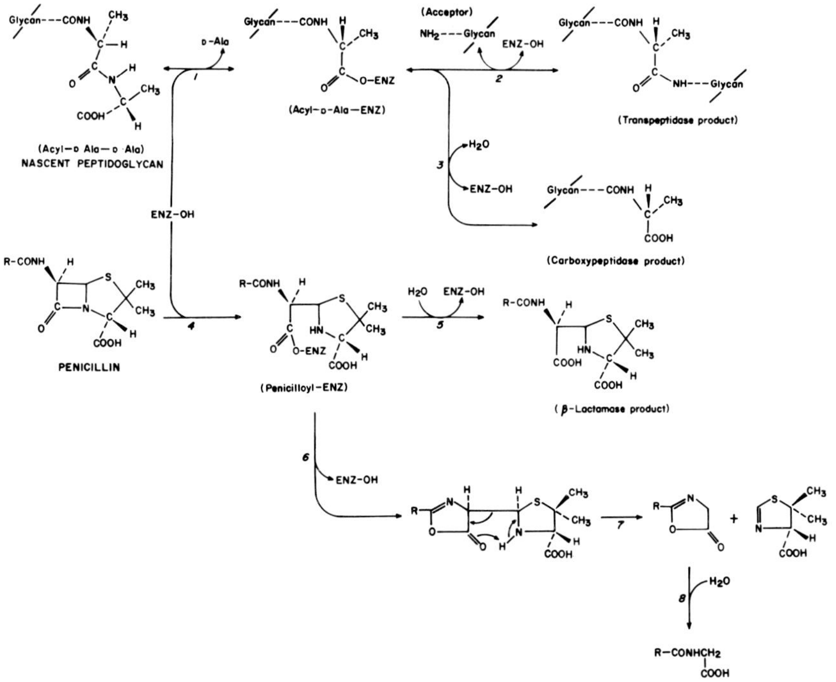

FIG. 35. Transpeptidases, DD-carboxypeptidases, and DD-endopeptidases; their inhibition by penicillins and relationship to β-lactamases. Reaction (1): Acyl-D-alanyl-D-alanine reacts with a functional serine OH group in the donor site of a transpeptidase, endopeptidase, or DD-carboxypeptidase (ENZ-OH) to give an acyl-D-alanyl-ester derivative of the enzyme and free D-alanine. If the amino terminus of a suitable amino acid or peptide is bound to the acceptor site, transpeptidation ensues with regeneration of enzyme (reaction 2). If the acceptor site accepts D-alanine, this reaction is simply a reversal of reaction (1). If the acceptor site binds water, acyl-D-alanine and enzyme are released, resulting in DD-carboxypeptidase action (reaction 3). Enzymes tend to be transpeptidases with weak DD-carboxypeptidase activity, or DD-carboxypeptidases with weak transpeptidase activity (e.g., Table VI). DD-Endopeptidase action results from reversal of reaction 2 followed by regeneration of the enzyme by reaction 3 (see Fig. 37). This only occurs when the hydrolysis product ("acceptor") is a D-α-amino acid with a free (D)-COOH. The cross-linked substrate of DD-endopeptidase action is thus a structural analog of the acyl-D-alanyl-D-alanine substrate for DD-carboxypeptidase action. The fixed configurations of β-lactam antibiotics, such as the penicillins, probably mimic (with variable efficiency) the single-bonded transition state conformation of acyl-D-alanyl-D-alanine, as it is bound to penicillin-sensitive transpeptidases or carboxypeptidases. Penicillins are able to acylate the enzyme very efficiently (reaction 4), forming penicilloyl enzyme by virtue of the reactivity of their β-lactam linkage which occupies the position equivalent to the D-alanyl-D-alanine bond in the normal substrate. If the product is hydrolyzed by water, acting as an acceptor for the penicilloyl substituent, the result is β-lactamase action (reaction 5). Carboxypeptidases and transpeptidases are inefficient β-lactamases, and for some the interaction with a β-lactam antibiotic results in almost irreversible inhibition. Hydroxylamine may release penicilloyl hydroxamate from such inhibited enzymes with much greater efficiency (see text). Certain DD-carboxypeptidases spontaneously release bound penicilloyl derivatives as the two fragments shown. One possible pathway for their formation is illustrated. (From Strominger, 1977.)

could itself be cleaved by the enzyme, forming a penicilloyl enzyme derivative in which the penicilloyl substituent occupies the site normally occupied by acyl-D-alanine (Fig. 35). The β-lactam carboxyl–nitrogen bond in the penicillin and cephalosporin nuclei has much more single- than double-bonded character and the β-lactam antibiotics are closer structural analogs of a single-bonded transition state, formed during cleavage of the D-alanyl-D-alanine peptide link, than they are of the peptide itself (Tipper and Strominger, 1965). A high affinity of the active site for such a transition state analog would facilitate both cleavage of the peptide bond in the natural substrates and binding of the β-lactam antibiotics. While inhibition of D,D-transpeptidases by such substrate analogs need not involve covalent interaction with the enzyme, it was proposed that cleavage of these antibiotics at the β-lactam bond occurs, forming an acyl enzyme.

The penicillin and cephalosporin nuclei are formed from L-cysteinyl-D-valine with the amino group of the L-cysteinyl moiety being the 6-amino group of the penicillins and the 7-amino group of the cephalosporins (Fig. 36), which is thus hypothesized to mimic the α amino group of the penultimate D-alanine residue of the donor substrate for transpeptidation. In both types of β-lactam antibiotic, this amino group is the site of substitution by acyl groups (e.g., "benzyl" in penicillin G) which vary widely in the semisynthetic antibiotic derivatives. If β-lactam antibiotics are analogs of the donor substrate in transpeptidation, inspection of the conformation of this peptide substrate, drawn so as to mimic the solid-state conformation of penicillin G (Fig. 36), suggests that 6 α-methylpenicillin derivatives should be even better substrate analogs than the penicillins themselves. Actually 6 α-methylpenicillin and 7 α-methylcephalosporin derivatives turned out to be almost inactive (Ho *et al.*, 1972). 6 α-methoxypenicillin retained some activity, while 7 α-methoxycephalosporin derivatives are natural antibiotics which were given the generic name cephamycins (Miller *et al.*, 1972). Semisynthetic derivatives of the cephamycins [e.g., Cefoxitin (Wallick and Hendlin, 1974)], are not only highly active but are also resistant to the most common (type IIIA) β-lactamase found in gram-negative bacteria. This β-lactamase is coded by the *bla* gene of the *Tn10* transposon (Elwell *et al.*, 1977).

The inactivity of 6-α-methylpenicillin derivatives apparently conflicts with the substrate analog theory of Tipper and Strominger (1965). This has recently been explained by Virudachalam and Rao (1977), who considered the effect of changes in configuration and of substituents at the C-6 position in the penicillin nucleus, on the conformation of its aminoacyl substituent. They point out that, when C-6 has the natural L configuration, the conformation for its substituent (R—CO—NH—) found in the crystalline state (Fig. 36) cannot be mimicked by acyl-D-alanyl-D-alanine, because this position of the aminoacyl group is energetically disallowed in the dipeptide. In solution, however, the C-6 aminoacyl group of the penicillins could adopt a position equivalent to one allowed for acyl-D-alanyl-D-alanine, but *only* if the C-6 residue of the β-lactam ring has the L-configuration, and only if it does *not* carry an α-methyl group. An α-methoxy group *is* allowed in this conformation. Their conformational analysis of the C-7 residue and its aminoacyl substituents in the cephalosporins leads to parallel conclusions. This refinement of the substrate analog theory thus rationalizes all of the observations, including the fact that totally synthetic penicillin analogs based on D-cysteinyl-D-valine are inactive (Gorman and Ryan, 1972). It further suggests allowable modifications of the penicillin nucleus which might be exploited in investigation of the antibacterial activity of totally synthetic derivatives. An intact β-lactam ring is necessary for the L-cysteinyl moiety to mimic a D-aminoacyl center and makes it impossible for penicillins to mimic L-alanyl-D-glutamate, as hypothesized by Wise and Park (1965).

The β-lactam ring, according to this theory, is essential for activity for conformational reasons, beside the reactivity it confers on the molecule. It had been thought that a nonplanar two-ring system, as found in cephalosporins and penicillins, was essential for reactivity of the β-lactam bond and

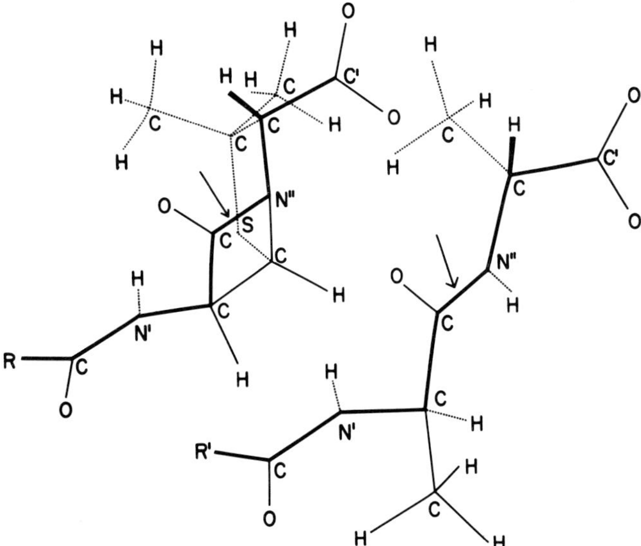

FIG. 36. Two-dimensional projections of three-dimensional models of penicillins (upper structure) and acyl-D-alanyl-D-alanine (lower structure). The conformation of the peptide has been chosen to mimic the fixed conformation of the penicillin nucleus. Arrows indicate the D-Ala-D-Ala bond cleaved by carboxypeptidases and transpeptidases, and the equivalent bond in the penicillin nucleus: the CO–N bond of the β-lactam ring.

therefore for antibiotic activity (Sweet and Dahl, 1970), but new antibiotically active β-lactams (nocardicins) have recently been described which contain only the β-lactam ring (Aoki *et al.*, 1976). Other β-lactam antibiotics have been isolated which have a second five-membered ring in the same configuration as in penicillins, such that the substituents at either end of the β-lactam linkage have the cis configuration, but with different structure. These include olivanic and clavulanic acids and certain of the epithienamycins (Gorman and Huber, 1977). The thienamycins have a very broad spectrum against gram-negative organisms, and are promising new antibiotics. Certain of them have a trans configuration about their β-lactam bond, but since the C_6 substituent is hydroxyethyl rather than acylamino, it may mimic the methyl group of D-alanine rather than its amino group. Like cefoxitin, these thienamycins are resistant to common β-lactamases, apparently due to their 6 α substituents. Clavulanic acid is a potent inhibitor of many β-lactamases, and it may potentiate the action of β-lactamase-sensitive antibiotics against β-lactamase-producing organisms.

If penicillins interact covalently with the donor substrate site of sensitive enzymes and if the penicilloyl enzyme product were stable, the enzyme would be irreversibly inhibited, and the rapidity with which the effects of penicillin on sensitive organisms such as *S. aureus* become irreversible, would be rationalized. If, on the other hand, the enzyme were capable of transferring the penicilloyl moiety to water, it would behave as a β-lactamase (Fig. 35), and it was postulated (Tipper and Strominger, 1965) that enzymes functioning as efficient β-lactamases have evolved from transpeptidases. When an *in vitro* transpeptidation system using *E. coli* membranes was developed and D,D-carboxypeptidase activity was first recognized, it was immediately obvious (Izaki *et al.*, 1966) that such carboxypeptidases could also have evolved from transpeptidases as enzymes which efficiently transfer acyl-D-alanyl intermediates to water instead of to an amino acceptor (Fig. 35). It was later realized that the KM endopeptidase activity observed in culture supernatants of *Streptomyces albus* G (Ghuysen, 1968), and the endopeptidase activity observed much earlier in *E. coli* extracts (Weidel and Pelzer, 1964), were hydrolyzing substrates very similar to acyl-D-alanyl-D-alanine (Fig. 37). Their recognized substrate is the normal acyl-D-alanyl-(D)-*meso*-DAP cross-link product of transpeptidation in the *A1γ* peptidoglycans of organisms such as *E. coli* and *B. megaterium* strain KM, or the acyl-D-alanyl-D-(Lys, etc.) cross-links in the B2 peptidoglycans of organisms such as *B. rettgeri* (Fig. 25). The carboxyl group on the *meso*-DAP or D-Lys residue adjacent to the cleaved peptide bond is free, and the structural analogy to D-alanyl-D-alanine is obvious (Fig. 37). It was shown that the

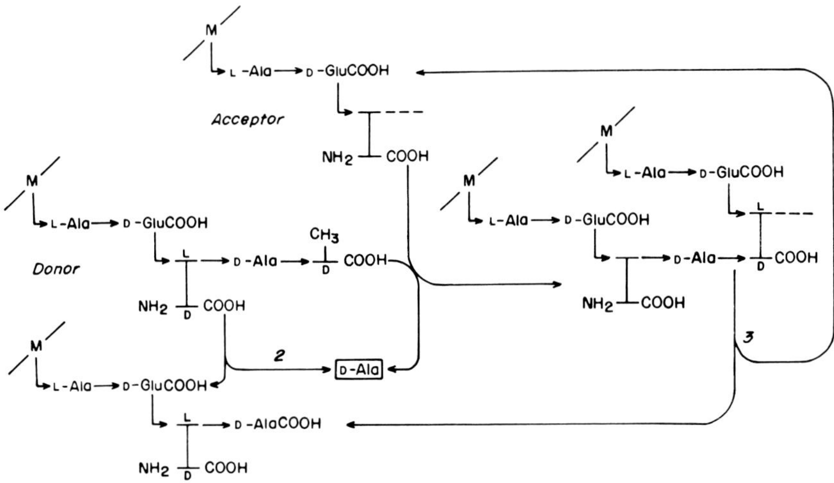

FIG. 37. Relationship of endopeptidase and carboxypeptidase action. The acceptor and donor peptides for transpeptidation (reaction 1) are drawn according to the normal convention for peptidoglycan (cf. Fig. 21), except that the terminal D-Ala residue of the donor is depicted in the same manner as *meso*-DAP. DD-Carboxypeptidase action (reaction 2) on the donor pentapeptide releases tetrapeptide and the terminal D-Ala. Endopeptidase action (reaction 3), hydrolyzes the D-Ala-(D)-*meso*-DAP link in the product of transpeptidation, releasing the acceptor peptide and tetrapeptide. The effect of sequential transpeptidase and endopeptidase action is thus equivalent to DD-carboxypeptidase action.

S. albus G enzyme had acyl-D-alanyl-D-alanine carboxypeptidase activity (Ghuysen *et al.*, 1970), and it was suggested (Bogdanovsky *et al.*, 1969) that the endopeptidase and carboxypeptidase of *E. coli* were identical. It is now clear that the pattern of enzymes in *E. coli* for which acyl-D-alanyl-D-alanine is a substrate is more complex. Several functionally separate transpeptidases may exist with individual patterns of susceptibility to β-lactam antibiotics (see Section VI,F). It is equally clear that enzymes of this type may all have variable degrees of transpeptidase, carboxypeptidase, endopeptidase and β-lactamase activities (Figs. 35 and 37). They differ drastically in the relative efficiencies of these different activities, but probably have a common evolutionary origin. Similarly, available structure-activity data, listed above, is consistent with the hypothesis that β-lactam antibiotics exert their physiological effects by interacting with these enzymes as analogs of acyl-D-alanyl-D-alanine. Thus all penicillin binding proteins may be enzymes of this type.

2. Transpeptidation *in Vivo* and in Particulate *in Vitro* Systems Coupled to Peptidoglycan Synthesis

After the initial success in *S. aureus*, it proved more difficult to obtain *in vivo* evidence of cross linkage inhibition by β-lactam antibiotics in gram-negative species. Martin initially concluded that penicillin-induced unstable L forms of *Proteus mirabilis* had peptidoglycan of markedly reduced cross-linkage (Martin, 1964). Later analysis failed to substantiate this (Katz and Martin, 1970), but recent careful investigation has shown a decrease in cross-linkage, but only from 25 to 23%. Thus concentrations of penicillin G sufficient to induce formation of osmotically fragile L forms in *P. mirabilis* have little effect on their gross cross-linkage frequency. The loss of functional integrity of the peptidoglycan must be due to more subtle effects, probably a consequence of selective loss of activity of a specialized transpeptidase component, responsible for only part of the total activity. Even carboxy-peptidase function is retained in the presence of penicillin (Martin *et al.*, 1976).

Recently, Kamiryo and Strominger (1974) used a simple assay for *in vivo* cross-linkage frequency to characterize the peptidoglycans of *E. coli* mutants selected as temperature sensitive and resistant to low levels of ampicillin. Among several interesting classes, a distinctive class of hypo-cross-linked mutants was found in which the normal 25 to 30% cross-linking was reduced by a maximum of about 30% to 17 to 20%. Using the same assay, it has been found that different β-lactam antibiotics all cause 50 to 60% inhibition of cross-linkage in a given strain of *E. coli* at their minimal inhibitory concentrations, resulting in a reduction of c:oss-linking from 30 to 15% (Curtis *et al.*, 1976). It was suggested that 15% cross-linkage is the minimum required to maintain the integrity of the *E. coli* envelope under the particular growth conditions employed.

In vivo pulse-chase experiments in *S. aureus* (Tipper and Strominger, 1968) showed the [^{14}C] glycine incorporated into the intact insoluble wall fraction first appears in un-cross-linked material. Various β-lactam antibiotics, including ampicillin, methicillin and cephalothin, were found to inhibit this cross-linking reaction, leading to accumulation of label in un-cross-linked peptide monomer attached to the insoluble wall. Soluble un-cross-linked peptidoglycan also accumulated at higher penicillin concentrations, but total peptidoglycan synthesis was partially inhibited. More recent observations on the effects of β-lactam antibiotics on peptidoglycan synthesis *in vitro* and *in vivo* in other gram-positive species, such as *M. luteus* (Mirelman *et al.*, 1972, 1974a,b) and *B. licheniformis* (Ward, 1974; Tynecka and Ward, 1975), indicate that the total rate of peptidoglycan polymerization is unaffected by β-lactam antibiotics. However, as a consequence of inhibition of cross-linking, these antibiotics prevent attachment of nascent peptidoglycan to preexisting wall. Attachment is measured as incorporation of suitably labeled precursors into material insoluble after treatment with boiling 2% SDS. Polymerized un-cross-linked peptidoglycan, which is soluble under these circumstances, can be precipitated with cold 5% TCA.

In the presence of high penicillin concentrations, 100% of the newly synthesized *B. licheniformis* peptidoglycan (Ward, 1974) and 75% of the *M. luteus* (Mirelman *et al.*, 1972) peptidoglycan was soluble. The residual 25% incorporation in *M. luteus* is presumably due to extension of glycan chains already cross-linked to the wall. The *B. licheniformis* strain was autolysin-defective, and the soluble product had a chain length of 100 disaccharides, indicating that glycan chain hydrolysis was minimal or absent. It thus appears that, at least in some gram-positive species, glycan polymerization is independent of transpeptidation and does not involve a preexisting, mature, cross-linked primer (see Section VI,D,5). The self-priming capacity of peptidoglycan polymerizing systems explains the relative ease with which active preparations can be isolated.

Efficient transpeptidation has more complex requirements. While *in vivo* studies in *S. aureus* indicated that peptidoglycan cross-links were formed by a transpeptidation reaction inhibited by penicillin, direct demonstration of this process required the development of *in vitro* systems for transpeptidation.

Initial success was obtained with systems in which cross-linking was coupled to the synthesis of polymeric peptidoglycan from nucleotide precursors, a process requiring several membrane-bound enzymes in functional association. The resemblance of this type of model system to the *in vivo* situation probably depends on the extent to which this process is coupled to linkage of the product to preexisting cell wall material.

Cross-linking was first investigated in the products made by wall-free membrane preparations and is normally observable only in preparations from organisms with type A1 peptidoglycans, where cross-linkage between pentapeptide subunits is direct. Cross-linking in coupled *in vitro* systems derived from organisms with type A2, A3, A4, or B peptidoglycans is inherently more complex and less efficient than in systems from type A1 organisms, since cross-linking will only occur to the N-terminus of the cross-bridge peptide of appropriately modified subunits, whereas active membrane systems will polymerize peptidoglycan whether or not any normally present amide groups or cross-bridge amino acids have been added.

Effective juxtaposition of donor and acceptor peptides for transpeptidation in these more complex systems frequently seems to require a natural association between the membrane polymerizing system and the structurally organized cell wall product. While osmotically fragile wall-free protoplasts of some bacterial species are capable of reversion, a process which must involve *de novo* formation of cross-linked peptidoglycan, this process is relatively slow and inefficient. Indeed, in reversion of lysozyme-induced protoplasts of an autolysin-deficient *B. licheniformis* strain, Elliott *et al.* (1975a,b) found that long glycan chains with little cross-linkage were first formed, and that the extent of cross-linking gradually increased as the cells regained wall organization and osmotic stability. Normal rod shape was regained only after a normal degree of cross-linkage has been achieved.

The first successful *in vitro* transpeptidation system was developed from *E. coli* B membranes (Izaki *et al.*, 1966, 1968; Araki *et al.*, 1966), and similar success was soon obtained with a system from *Salmonella newington* (Izaki *et al.*, 1968). More recently, various gram-positive species, including *B. megaterium* (Wickus and Strominger, 1972, 1974), *B. stearothermophilus* (Linnett and Strominger, 1974), and *Gaffkya homari* (Hammes, 1976; Hammes and Kandler, 1976), all with type A1 peptidoglycans, have been sources of crude membrane fractions capable of *in vitro* peptidoglycan transpeptidation. A successful system was also obtained from *Sporosarcina ureae*, which has a type A4 peptidoglycan containing L-lysyl-D-glutamate cross-bridges (Linnett *et al.*, 1974). Successful polymerization of type A2 peptidoglycan by membranes derived from *M. luteus*, with formation of both types of cross-bridge, has been reported recently (Pellon *et al.*, 1976). Membrane preparations from *B. stearothermophilus* catalyze the polymerization of an amidated A1γ peptidoglycan with a degree of cross-link formation that is unusually high for an *in vitro* system (Linnett and Strominger, 1974).

Bacillus megaterium KM, like *E. coli*, has an unamidated type A1γ peptidoglycan. However, *B. megaterium* walls do contain a proportion of DD-diaminopimelic acid (DAP) and it was found that membrane preparations from this organism would incorporate DD-diaminopimelate in a reaction dependent upon peptidoglycan polymerization and resulting in the production of D-alanyl-DD-DAP linkages. Other D-α-amino acids could also be incorporated, and since DD-DAP has two such centers it could actually cross-link two pentapeptides by transpeptidation at both ends in a head-to-head fashion. This does seem to occur both *in vitro* and *in vivo* (Wickus and Strominger, 1972, 1974). This *in vitro* amino äcid incorporation system was a precursor of the simple model transpeptidation system pioneered by Ghuysen *et al.* (see below). It also produced evidence of unanticipated complexity in the transpeptidation system: differential penicillin sensitivity of transpeptidation involving peptide subunits and DD-DAP as acceptors suggests that different enzymes are involved. Mirelman and Bracha (1974) reported that formation of D-Ala-L-Ala cross-links in *M. luteus in vivo* is some 50-fold less sensitive to penicillin than formation of D-Ala-L-lys cross-links, suggesting that different transpeptidases are involved in these two processes (see Fig. 34).

Organisms such as *S. aureus* and *M. luteus* have complex cross-bridges and, until recently, only preparations derived from such organisms containing cell wall fragments to which membrane still remains associated ("dirty walls") seemed capable of coupled peptidoglycan polymerization and transpeptidation. Permeabilized cells probably retain the closest approximation to the *in vivo* juxtaposition of synthetic enzymes and product and are an extension of the same system. They have been successfully exploited by Schrader and Fan (1974) in *B. megaterium*, who recently demonstrated that reversible inactivation of such preparations by extraction with 1.5 *M* LiCl is due to solubilization of GlcNAc transferase and possibly of the enzyme responsible for DD-DAP incorporation (Taku *et al.*, 1975; Taku and Fan, 1976; Fan, personal communication). Mirelman *et al.* (1976). investigating permeabilized cells of the $pat84_{ts}$ division mutant of *E. coli*, found that the longitudinal wall has a higher degree of cross-linkage than does the cross wall and that inhibition of DD-carboxypeptidase by low concentrations of ampicillin was associated with an increase in the proportion of newly synthesized peptidoglycan attached to preexisting wall, an increase in percentage cross-linkage and an inhibition of septum formation. This mimicked the effects of growth at nonpermissive temperatures in the absence of ampicillin and suggests that, in *E. coli*, the balance of transpeptidase and DD-carboxypeptidase activity is crucial to normal cell division. The behavior of mutants lacking DD-carboxypeptidase activity is inconsistent with this hypothesis (see Section VI,F,3). Dirty wall preparations from *S. aureus* polymerize peptidoglycan from UDP-MurNAc pentapeptide and UDP-GlcNAc and also incorporate glycine if added (Mirelman and Sharon, 1972). Release of the C-terminal D-alanine residue occurs, but only in the presence of glycine, because of tRNA-dependent cross-bridge formation and transpeptidation. The D-alanine release and cross-linking is sensitive to penicillin.

3. The Direction of Transpeptidation

Nascent polymeric peptidoglycan can act as both donor and acceptor for transpeptidation and may form cross-links with itself or with mature cross-linked "murein." Mirelman has suggested that cell wall elongation principally involves extension of preexisting murein, while crosswall synthesis necessarily involves cross-linking between nascent peptidoglycan strands, since it requires centripetal wall extension, away from preexisting murein (see Fig. 2). If different transpeptidases were involved in these two processes (see Section VI,F,4), their relative activity could control cell division and could account for the different cross-link frequency found in longitudinal and crosswalls (Mirelman *et al.*, 1976). Variation in the rate of peptidoglycan synthesis in the *E. coli* cell cycle has been reported by Churchward and Holland (1976).

In organisms, such as *E. coli* and *B. licheniformis*, which have active DD-carboxypeptidases, the C-terminal peptides of mature murein lack the terminal D-alanine residues and cannot act as donors for transpeptidation. Ward and Perkins (1974) found that 40% of the subunits polymerized by "dirty wall" preparations of *B. licheniformis* were cross-linked and that this was reduced by half if the UDP-MurNAc pentapeptide precursor were replaced by its N^{ε}-acetyl-Lys derivative (which cannot act as acceptor). Polymerization of UDP-MurNAc tetrapeptide (which can act only as acceptor) gave a product with only 10% cross-linkage, suggesting that preexisting murein acts more efficiently as acceptor than as donor.

More recent experiments in *Gaffkya homari* (Hammes, 1976; Hammes and Kandler, 1976) have yielded contrary and surprising results. Dirty wall preparations polymerize UDP-MurNAc tetrapeptide and pentapeptide with equal efficiency, and both are incorporated into the wall (SDS-insoluble fraction) by transpeptidation. Penicillin G, at concentrations similar to those inhibiting growth of the organism, inhibits incorporation of the pentapeptide into the wall (but not into soluble polymer), as expected. However, it has no effect on tetrapeptide incorporation into this cross-linked product. Thus, the transpeptidase itself is resistant to penicillin. A penicillin-sensitive DD-carboxypeptidase is present and only tetrapeptides formed by its action can act as acceptors for transpeptidation. Apparently the carboxypeptidase attacks only intermediates in peptidoglycan synthesis, not mature murein, which must retain some of its D-Ala-D-Ala groups to serve as donor. Nascent peptidoglycan can only act as acceptor, not as donor, and the carboxypeptidase is the target of lethal penicillin action in this organism (Hammes and Kandler, 1976). It seems likely that this is an unusual situation, that a transpeptidase is the lethal site of penicillin action in most organisms and that preexisting murein may normally be an acceptor in transpeptidation to nascent peptidoglycan.

4. Soluble DD-Carboxypeptidase/Transpeptidase Activities from Actinomycetes

In their paper which first described an active DD-carboxypeptidase (in particulate preparation from *E. coli* B), Izaki *et al.* (1966) pointed out "that the cross-linking reaction (catalyzed by the same preparations) and the carboxypeptidase could be different manifestations of the activity of one enzyme. Reaction of a MurNAc-tetrapeptidyl–enzyme (acyl-D-alanyl–enzyme) intermediate with another (acceptor) peptide unit would be a transpeptidation, while its reaction with water would result in hydrolysis of the terminal D-alanine residue. However, differential sensitivity of the two reactions to penicillins may suggest that they are catalyzed by different proteins."

Correct positioning of membrane-bound enzyme, the cell wall acceptor, and nascent polymeric substrate is probably essential to most *in vivo* transpeptidation. Reproduction of this system for the assay of a purified soluble transpeptidase would clearly be difficult. The development of model transpeptidation systems, using relatively low molecular weight soluble donors and acceptors, by Jean-Marie Ghuysen, Harold Perkins, and their collaborators, was a major advance which has allowed investigation of solubilized enzyme systems derived from a variety of bacteria. Their data are summarized in three recent reviews (Ghuysen *et al.*, 1974, 1975; Ghuysen, 1976). They found that the simple model peptides used as substrates for the soluble DD-carboxypeptidases from *Streptomyces* species (see later, this section) could also act as donors for transpeptidation, by using amino acids or simple peptides as acceptors. They espoused the theory of identity between the excreted carboxypeptidases and the membrane-bound transpeptidases present in these organisms (Ghuysen *et al.*, 1972). This hypothesis was tenaciously adhered to in the interpretation of much early kinetic data on the activities of the soluble "carboxypeptidase/transpeptidase" enzymes and on the inhibition of these activities by β-lactam antibiotics. These data were also interpreted as demonstrating that β-lactam antibiotics bind to neither the donor nor the acceptor binding sites of these enzymes so are not substrate analogs, but bind to another "allosteric" site (Ghuysen *et al.*, 1974). Their own more recent data, however (Ghuysen *et al.*, 1975; Ghuysen, 1976), shows that interaction between the strain R61 exocellular DD-carboxypeptidase and β-lactam antibiotics gives data "typical of a competitive inhibition of the hydrolysis of peptide donor by antibiotic." These data are, therefore, consistent with the inhibitors being donor substrate analogs (Frère *et al.*, 1975b). Moreover, differences between the excreted and membrane-bound activities have been clearly demonstrated (Table VI), and the presence of multiple penicillin-binding components in *Streptomyces* membranes complicates the interpretation of data derived from such preparations.

Nevertheless, these activities share substrate specificities which demonstrate ontological relatedness (Ghuysen *et al.*, 1973, 1974), and more plausible interpretations of this inherently interesting data can be made. It remains possible that a soluble DD-carboxypeptidase could be derived from a membrane-bound transpeptidase by cleavage at a site that separates a more hydrophobic portion of the molecule (important for membrane binding and efficient transpeptidation action) from a more hydrophilic portion containing the active site. However, it is more probable that carboxypeptidases and transpeptidases are products of different genes derived from a common ancestral gene. In *E. coli* and *B. subtilis*, the killing sites for β-lactam antibiotics have been correlated with specific penicillin-binding components (PBC's), differentiated from the DD-carboxypeptidases, which have been shown (by mutation) to be dispensible for normal growth (see below).

TABLE VI

DD-CARBOXYPEPTIDASE/TRANSPEPTIDASE ACTIVITIES OF ENZYME PREPARATIONS FROM *Streptomyces* STRAIN R61[a]

Preparation	Hydrolysis	Transpeptidation	Ratio
Extracellular	5000	500	0.1
Lysozyme releasable	200	50	0.25
CTAB solubilized	1.3	18	14
Membrane-bound	0.32	11	35

[a] Relative rates for hydrolysis of diacetyl-L-Lys-D-Ala-D-Ala and for transpeptidation using this peptide as donor and glycyl-glycine as acceptor are presented for a single set of reaction conditions (Ghuysen *et al.*, 1975; Ghuysen, 1976) (see text). The extracellular enzyme was purified to protein homogeneity. The "lysozyme releasable" enzyme was released from whole cells by protoplast formation induced by lysozyme in the presence of sucrose. The "CTAB solubilized" enzyme was derived from treatment of whole mycelia with cetyltrimethylammonium bromide (CTAB). The extract had been purified tenfold by Sephadex filtration.

A soluble DD-carboxypeptidase was first purified from *Streptomyces albus* G and was initially recognized and used as the KM endopeptidase (see Section VI,E,1). It later became clear that this endopeptidase was more active against a variety of acyl-D-alanyl-D-alanine substrates as a carboxypeptidase (Ghuysen *et al.*, 1970). Nieto and Perkins (1971) had synthesized diacetyl-L-Lys-D-Ala-D-Ala for use in their studies of the mode of action of Vancomycin, and this tripeptide turned out to be an excellent substrate for the *S. albus* G enzyme (Leyh-Bouille *et al.*, 1970). Following this observation, a variety of synthetic peptides and peptides derived from nucleotide precursors and hydrolysis products of peptidoglycans were used to determine the specificity of the extracellular enzymes produced by various *Actinomycetes*, both as carboxypeptidases and as transpeptidases. Besides the *S. albus* G enzyme, DD-carboxypeptidases have been purified from the culture media of *Streptomyces* strains R61 and R11 and from an organism originally called *Streptomyces* strain R39 but which is probably an *Actinomadura* sp. (Ghuysen *et al.*, 1974).

Streptomyces species, including R61, characteristically have a type A3 peptidoglycan in which L-R_3 is LL-DAP, the cross-bridge is a single glycine residue and the α-COOH of D-Glu is amidated. *Actinomyces* strain R39, in contrast, has a type A1γ peptidoglycan in which L-R_3 is *meso*-diaminopimelate. Its D-Glu α-COOH groups are also amidated. When assayed as carboxypeptidases both enzymes have high specificity for a D-alanyl-D-Ala C terminus, but at least one additional amino terminal (L-R_3) amino acid must be present to make the D-Ala-D-Ala peptide a good substrate and the nature of this L-R_3 substituent is important: the R61 enzyme prefers amino acids with long uncharged side chains, such as the diacetyl lysyl derivative, unless the side chain terminates in Gly-NH_2. The R39 enzyme, in contrast, prefers LL- or *meso*-DAP with a free —NH_2 in the L-R_3 position. Clearly the preference in each case is exactly that expected of an enzyme whose normal substrate is the peptidoglycan or peptidoglycan precursors of its source organism.

Like the *S. albus* G enzyme, both the R61 and R39 enzymes will act as endopeptidases, splitting bisdisaccharide peptide dimers derived from such organisms as *B. megaterium* KM at D-Ala-(D)-*meso*-DAP linkages, but this action is weak compared to their carboxypeptidase activity on diacetyl-L-lysyl-D-Ala-D-Ala.

Preliminary investigations (Pollock *et al.*, 1972; Perkins *et al.*, 1973) showed that diacetyl-L-Lys-D-Ala-D-Ala could participate as donor in transpeptidation reactions, using D-Ala or glycine as acceptor, catalyzed by the R61 and R39 enzymes. Investigation of transpeptidation specificities showed the R11 and R61 enzymes to be very similar, and the R11 enzyme has received little further attention. Because the *S. albus* G enzyme is inactive in model transpeptidation systems and is insensitive to β-lactam antibiotics, major emphasis has been placed on the investigation of the R61 and R39 enzymes. As expected, the specificities of the R39 and R61 enzymes for donors in transpeptidation are the same as those for carboxypeptidase substrates. The specificity for acceptors also reflects the structures of the peptidoglycans in the source organisms.

Using diacetyl-L-lysl-D-ala-D-ala as donor, the R61 enzyme will use a wide variety of acceptors, including glycine and D-amino acids, or peptides in which these are the N-terminal residues, and 6-aminopenicillanic acid. L-Amino acids are not substrates, and the most efficient peptide acceptors have N-terminal glycine. The pattern of functional acceptors clearly resembles the glycyl-LL-DAP acceptor for transpeptidation in the cell wall of strain R61. In contrast, the R39 enzyme could not use many D-amino acids, peptides, or 6-aminopenicillanic acid as acceptors, although glycine, D-Ala and *meso*-diaminopimelate were utilized. In fact, this enzyme has acceptor specificity directed toward the L-alanyl-D-isoglutamyl-*meso*-diaminopimelate sequence of the peptidoglycan of strain R39.

When a simple peptide (e.g., glycylglycine) is used as acceptor by the R61 or R39 enzyme, the ratio of transpeptidation to carboxypeptidase action is dependent on acceptor concentration, but the total rate of donor substrate conversion is independent of acceptor. Transpeptidation is favored over hydrolysis by increasing the pH and decreasing the water content by the addition of ethylene glycol and glycerol (Ghuysen *et al.*, 1974).

The above systems may be regarded as "unnatural" model transpeptidations. A more "natural" model transpeptidation can be catalyzed by the R39 enzyme in which L-alanyl-D-isoglutamyl-*meso*-diaminopimelyl-D-alanyl-D-alanine is used as donor and the same peptide, amidated on the DAP residue, is used as acceptor (Ghuysen *et al.*, 1974). The dimer product, being amidated, is stable in the presence of the enzyme; however, it is an inhibitor of the R39 enzyme. The products are free D-alanine, produced by both carboxypeptidase and transpeptidase activity, and a mixture of monomer, nonapeptide dimer, and the octapeptide dimer in which the terminal D-alanine has been removed by carboxypeptidase action (Fig. 37). The same substrates are utilized by *E. coli* K12 membrane preparations to form the same peptide dimer (Nguyen-Disteche *et al.*, 1974a,b). This is presumably due to the *E. coli* carboxypeptidase IB fraction (Table VII) and not the physiologically functional transpeptidases which are present in much smaller concentrations (see Section VI,F,4). The R61 enzyme will use (N^{α})-acetyl-(N^{ε})-glycyl-L-lysyl-D-alanyl-D-alanine as both acceptor and donor for transpeptidation, an approximation of a natural model system. The products include the heptapeptide dimer, the decapeptide trimer, and the products of DD-carboxypeptidase action on these compounds and on the original tetrapeptide monomer. After incubating 7 m*M* tetrapeptide in 58% ethylene glycol and 25% glycerol, 55% of the products were in the form of dimer and 10% in the form of trimer. Trimer formation is efficient only when the dimer acts as acceptor, rather than as donor (Frère *et al.*, 1976a). It is of interest that Oldmixon *et al.* (1976) concluded, from an analysis of the peptidoglycan of *Streptococcus faecalis*, that peptide cross-linking *in vivo* in this organism also proceeds by a monomer addition mechanism.

Treatment of *Streptomyces* R61 with lysozyme in the presence of sucrose produces protoplasts and releases carboxypeptidase activity that may have been bound to the wall or bound to the outer surface of the cytoplasmic membrane, perhaps in the process of excretion. Like the extracellular enzyme, this activity is primarily a carboxypeptidase with weak transpeptidase activity (Table VI). Use of the model peptide substrates allows detection of a transpeptidase activity in membrane preparations from strain R61 (Dusart *et al.*, 1973). This preparation has only weak carboxypeptidase activity (Table VI). It has recently been demonstrated that strain R61 has multiple penicillin-binding components in its membranes, so that carboxypeptidase and transpeptidase activities may be manifested by different components. The transpeptidase activity assayed is probably a major functional transpeptidase activity of the cell, as its pattern of antibiotic susceptibility reflects that of the intact organism (see Ghuysen *et al.*, 1975). Solubilization of activity from these membranes was only achieved in good yield by treatment with cetyltrimethylammonium bromide (CTAB) (Dusart *et al.*, 1975). Solubilized activity resembles the membrane-bound form, except

TABLE VII

E. coli DD-CARBOXYPEPTIDASE I COMPONENTS[a]

Name	Cell fraction		Solubilized by		PCMB inhibited	50% Inhibition (μg/ml)		Relative activity		Half-life bound Pen G[b]	SDS	PAGE
	Soluble	"Membrane"	0.5*M* LiCl	2% Triton		Pen G[b]	Ampi[b]	Transpep[b]	Endopep[b]		PBC's[b]	MW
IA	No	Yes	No	Yes	Yes	220	465	Moderate	0	{20 min	6	40,000
										{ 5 min	5	42,000
IB	No	Yes	Yes	Yes	No	0.3	0.2	Low	Yes	Long	4	49,000
IC	Yes	No			No	0.3	0.2	Low	Yes	Long	4	49,000

[a] Enzyme IB seems to be identical to the soluble enzyme IC. Both have low transpeptidase activity, are active as endopeptidases, and are inhibited almost irreversibility by low concentrations of penicillin G. Enzyme IA gives two bands on SDS gels; these two components are known to differ only in the rate of loss of bound penicillin G.

[b] Pen G, penicillin G; Ampi, ampillicin; Transpep, transpeptidase; Endopep, endopeptidase; PBC, penicillin-binding components.

for an enhanced carboxypeptidase activity (Table VI). The membrane-bound complex had a specificity profile for acceptors similar to that of the extracellular enzyme. An apparently identical activity can be released directly and in much better yield by treatment of washed mycelia with 2% CTAB in buffer at 37°C for 45 minutes. After concentration by ultrafiltration, tenfold purification was obtained by Sephadex G-100 chromatography. In membrane-bound form the enzyme complex is able to function as a transpeptidase in the frozen state at −20°C, suggesting that the enzyme functions in a liquid (lipid) environment that remains fluid at low temperatures (Dusart *et al.*, 1975). The extracted enzyme complex is inactive below 0°C. All attempts at extracting transpeptidase activity from strain R39 have failed, but other *Streptomyces* strains yield active and stable enzyme complexes on extraction with CTAB. Excellent activity was obtained from *S. rimosus*, whose genome has been partially mapped. If mutants affecting the activities of this enzyme preparation could be obtained and subjected to genetic analysis, this would allow incisive correlation of *in vivo* and *in vitro* effects and analysis of the number of separate gene products with some degree of transpeptidase activity.

As mentioned above, extensive analysis of the kinetics of interaction of β-lactam antibiotics with the R61 and R39 DD-carboxypeptidase/transpeptidase enzymes are consistent with classic competitive inhibition involving competition with the donor peptide (carboxypeptidase substrate) for the same binding site. This is consistent with the substrate analog theory of Tipper and Strominger (1965), although alternative models are not ruled out, since it has yet to be demonstrated that binding of inhibitor and substrate are mutually exclusive (Frère *et al.*, 1975a,b). Benzylpenicillin, bound to the R61 and R39 enzymes, can be slowly released as benzylpenicilloate or as fragments of this molecule (see Section VI,F,2).

F. Penicillin Binding Components: Relationship to Sensitive Enzymes

1. Penicillin Binding Components (PBC's)

Early studies of the binding of radioactive penicillin G to sensitive bacteria, reviewed by Cooper (1956), provided a major impetus to the development of the hypothesis of penicilloylation of the target enzyme as an essential event in the lethal action of β-lactam antibiotics. These early studies showed that a small number of high-affinity binding sites for penicillin occurred in sensitive bacteria; that penicillin, once bound to these sites, could not be released by treatment with β-lactamase, by exchange with excess unlabeled penicillin, or by treatments which effectively denature proteins such as boiling, exposure to 2% SDS, 6 *M* guanidine-HCl, or phenol. This binding appeared to be covalent, and since treatment with alkali released penicilloic acid, binding was hypothesized to involve penicilloyl ester formation. Saturation of these penicillin binding components (PBC's) in different strains of bacteria required penicillin concentrations correlating well with minimal inhibitory concentrations. More recently, Edwards and Park (1969) have investigated competition between various unlabeled β-lactam antibiotics and radioactive benzylpenicillin for PBC's in whole cells of *S. aureus* H. They demonstrated that an excellent correlation exists between the affinity for these sites, and the MIC for each drug, even though *S. aureus* is now known to have multiple PBC's.

Early investigations by Cooper (1956) tentatively identified PBC's as membrane proteins. More recent investigations and in particular the fractionation of membrane proteins by polyacrylamide gel electrophoresis in buffer contains sodium dodecyl sulfate (SDS–PAGE) has resulted in demonstration of the complexity of PBC's. In organisms such as *B. subtilis* and *E. coli*, the major PBC has been identified as a DD-carboxypeptidase whose saturation by penicillin is irrelevant to killing. Other minor PBC's are probable lethal targets. Thus data on binding to the total pool of PBC's in whole cells or membrane fractions cannot be interpreted unequivocally.

Recent investigations of the nature of the interaction between benzylpenicillin and PBC's has concentrated on the readily purifiable DD-carboxypeptidases of *B. subtilis*, *B. stearothermophilus*, *Streptomyces*, and *E. coli*. This work was reviewed in 1974 by Blumberg and Strominger (1974). Spratt (1977) has more recently summarized the properties of the PBC's of *E. coli*. Investigation of the *E. coli* PBC's had been retarded by their relatively low concentration. DD-Carboxypeptidase IA comprises fully two-thirds of the total membrane-bound *E. coli* PBC's and, in its active form, loses bound penicillin so rapidly (Table VII) that its contribution to labeled PBC's is easily underestimated. The remaining PBC's comprise approximately 1000 molecules per cell (Table VIII). The best estimates for gram-positive organisms, such as *B. subtilis*, indicate that they bind five- to tenfold more, or a total of about 10,000 molecules per cell, unequally distributed among the different PBC's.

It was observed that bound penicillin could be released from *B. subtilis* PBC's by treatment with hydroxylamine at neutral pH. This was initially surmized to be a chemical reaction, indicating a highly labile (possibly thiolester) bond between penicillin and the enzyme. This hypothesis was substantiated by the demonstration of an essential thiol group in the enzyme (Lawrence and Strominger, 1970a; Suginaka *et al.*, 1972; Umbreit and Strominger, 1972a,b). It was subsequently found (Blumberg *et al.*, 1974) that the penicillin remained bound when the PBC preparation was denatured

TABLE VIII

E. coli K12 PBC's

Name	MW	Mols/Cell	Map	Symbol	Relative affinity for					Proposed function
					Pen G	Mec	Kex	Lori	Cefox	
IA	95	230	73.5′	ponA	+		+	+ +	+ +	Peripheral wall extension
IB's	90		3.3′	mrc, ponB	+			+	+	Transpeptidase
2	66	20	14.4′	(rod A)	+	+	0	0	0	Rod shape maintenance
3	60	50	1.8′	(ftsI)	+		+ +	+	+	Septum formation
4	44	110	68′	(dacB)	+		+			DD-Carboxypeptidase IB,C
5	42	1800	13.7′	(dacA)	+ +		0			DD-Carboxypeptidase IA
6	40	570			+ +		0			

Molecular weight, inferred from mobility on SDS PAGE, is given in kilodaltons. Map position corresponds to the locus of mutants affecting the quantity of each PBC, as stated in the given reference. Molecules of each PBC per cell (mols/cell) is based on the total amount of (^{14}C) penG bound per cell and comparative binding to individual PBC's, assuming stoichiometric binding (34). Relative binding affinities (negligible, 0; low-moderate, +; high, + +) are determined by direct binding of labelled penicillin G (PenG), mecillinam (Mec) and cefoxitin (Cefox). Binding of other antibiotics, Cefalexin (Kex) and Cephaloridine (Lori), is determined by competition for binding of labelled penicillin G. These data are taken from Tamura *et al.*, 1976; Spratt, 1977; Tamaki *et al.*, 1977; Iwaya and Strominger, 1977; Matsuhashi *et al.*, 1977; Nishimura *et al.*, 1977; Suzuki *et al.*, 1978 and Iwaya *et al.*, 1978.

by boiling, or treatment with SDS, but that hydroxylamine could then no longer cause release of penicilloyl hydroxamate. Hydroxylamine was acting as an acceptor for the penicilloyl moiety in an enzyme-catalyzed transpeptidation, and the data merely demonstrated the superiority of hydroxylamine to water as an acceptor in this system. Besides releasing penicilloyl hydroxamate, hydroxylamine treatment resulted in reactivation of the DD-carboxypeptidase (Lawrence and Strominger, 1970b). Thus, it proved possible to use affinity chromatography, employing penicillins covalently bound to a column matrix, for the purification of PBC's. They can be eluted from the column in active form with buffered hydroxylamine (Blumberg and Strominger, 1972a).

Assays utilizing SDS–PAGE of membrane preparations and bound [^{14}C]penicillin followed by radioautography (see Section VI,F,4) detect only those PBC's which retain their bound penicillin during washing with excess cold penicillin and under the severe denaturing conditions of boiling in the SDS gel electrophoresis buffer. It remains possible that some or all of the organisms investigated contain components which bind penicillins transiently or noncovalently and which could be important in the mode of action of these drugs. However, this is an unnecessary hypothesis, since the behavior of certain of the identifiable PBC's in each organism implicates them as killing sites. Multiple PBC's have been identified in each organism investigated, including five in *B. subtilis;* four in *B. cereus*, *B. stearothermophilis*, and *S. aureus;* several in *Streptomyces* R61; and (initially) six in *E. coli* (Yocum *et al.*, 1974).

a. B. subtilis PBC's. SDS–PAGE of *B. subtilis* membranes pretreated with radioactive benzylpenicillin revealed five bands (Blumberg and Strominger, 1972b). If membranes were pretreated with cephalothin before penicillin G, only PBC's 3 and 5 were detected. PBC 5, which comprises 70% of the total PBC's, was tentatively identified as the previously purified DD-carboxypeptidase of *B. subtilis* by its identical gel migration rate and by competition experiments which demonstrated that it bound various β-lactam antibiotics with the same affinity and kinetics (Umbreit and Strominger, 1973a,b).

The PBC's were purified by affinity chromatography on sepharose containing a bound semisynthetic penicillin, and, after SDS-PAGE, staining for protein revealed the same five components detected as PBC's in gels of whole membrane proteins. It was demonstrated that binding of penicillin to these purified PBC's was stochiometric. PBC 3 was recovered in poor yield, and, if membranes were pretreated with cephalothin, affinity chromatography yielded pure PBC 5 whose identity with the DD-carboxypeptidase was conclusively demonstrated (Blumberg and Strominger, 1972a,b). This protein comprises 0.75% of the total membrane proteins of *B. subtilis* and was isolated in large quantities. The interaction of this protein with penicillins has been extensively investigated.

The competition experiments eliminated PBC 3 and 5 as potential lethal sites, since they fail to bind cephalosporins which kill *B. subtilis*. The investigation of mutants with stepwise increase in resistance to cloxacillin (Buchanan and Strominger, 1976; Buchanan, 1977) has shown that PBC 1 can be deleted without affecting viability. Further acquisition of resistance is associated with a proportional reduction in affinity of PBC 2 for cloxacillin, but not for benzylpenicillin. The mutants remain as sensitive to benzylpenicillin as the parent, and PBC 2 is the probable killing site. This is strong evidence linking binding of β-lactam antibiotics to PBC's to the lethal effects of such antibiotics on the parent organisms. These techniques also demonstrate the multiplicity of potential effective targets; they fail to identify a single PBC as "the" transpeptidase and rather suggests that bacteria have multiple transpeptidase activities. Differential functions may be assigned to these multiple PBC's in *E. coli* (see Section 8,c).

b. S. aureus PBC's. Staphylococcus aureus H has four PBC's of molecular size about 115,000, 110,000, 110,000, and 46,000 daltons (Kozarich and Strominger, 1978; Kozarich, 1977). Blumberg and Strominger (1972b) originally recognized only two, the lowest molecular weight, PBC 4, being undetected because of the rapidity with which it releases bound benzylpenicillin as benzylpenicilloate. This component, besides this β-lactamase activity, also has DD-carboxypeptidase and model transpeptidase activities, although the cell wall peptidoglycan of *S. aureus* shows no signs of carboxypeptidase attack (Tipper and Strominger, 1965, 1968). Affinity chromatography of *S. aureus* H membranes pretreated with penicillin allowed selective adsorption and purification of PBC 4. Its preferred acceptor for transpeptidation with diacetyl-L-Lys-D-Ala-D-Ala is glycine, which at 1 m*M* completely suppresses hydrolysis of the donor, suggesting that carboxypeptidase activity in PBC 4 is not physiologically significant. The kinetics of interaction of the enzyme with this donor are consistent with formation of an acyl-enzyme intermediate (Kozarich and Strominger, 1978). Spontaneous cloxacillin-resistant mutants of *S. aureus* have alterations in the higher molecular weight PBC's, implicating them as lethal targets of cloxacillin action (reported in Kozarich, 1977).

2. Fragmentation of Penicillin G by DD-Carboxypeptidases

When solubilized and purified or when in its native membrane-bound state, *E. coli* DD-carboxypeptidase IA releases bound benzylpenicillin as benzylpenicilloate with a half-life of about 5 minutes (T. Tamura *et al.*,

1976). This enzyme may thus be regarded as a very inefficient β-lactamase (cf. Fig. 35). This is also true of *S. aureus* PBC-4.

The DD-carboxypeptidases of *B. subtilis* and *B. stearothermophilis* were also found to release bound penicillin slowly in aqueous buffer (Blumberg *et al.*, 1974). The rate of reactivation of the enzyme depended upon the particular β-lactam antibiotic bound and the temperature. At 55°C, penicillin G was released from the *B. stearothermophilus* enzyme with a half-life of only 10 minutes, and two fragments, phenacetylglycine and 5,5-dimethyl-Δ^2-thiazoline-4-carboxylic acid were found, the products of breakage of two bonds of the β-lactam ring (see Fig. 34) (Hammarstrom and Strominger, 1975, 1976). Pathways for enzymatic catalysis of the scission of penicilloyl–enzyme to give these products were proposed, and it was suggested that this was mediated by protonation of the thiazoline ring nitrogen in a penicilloyl–enzyme intermediate, the equivalent of protonation of the amino group of the cleaved terminal D-alanine residue of the carboxypeptidase substrate, perhaps a necessary step in release of D-alanine from the active site (Hammarstrom and Strominger, 1975). One plausible mechanism for formation of these products is shown in Fig. 35.

Reactivation of the penicilloylated R61 DD-carboxypeptidase occurs at 37°C in 0.01 *M* $NaPO_4$ buffer with a half-time of 80 minutes, and the products are essentially the same as those released by the *B. subtilis* enzyme: phenacetylglycine and *N*-formyl-D-penicillamine (Frère *et al.*, 1975a, 1976b). The same products are formed from benzylpenicillin bound to the R39 enzyme in buffers in which the released enzyme activity is stable. At low ionic strengths, however, active enzyme is not released, and benzylpenicilloate is produced. The pathway of release of the bound penicillin thus depends on the environmental conditions (Frère *et al.*, 1975b).

Membrane preparations of strain R61 bind a small number of benzylpenicillin molecules with high affinity (to multiple PBC's). Concentrations of other β-lactam antibiotics able to exclude benzylpenicillin from 50% of these binding sites are equal to those concentrations causing 50% inhibition of transpeptidase activity [assayed with $(Ac)_2$-Lys-D-Ala-D-Ala] by these same preparations. They also are close to the 50% lethal dose for these same antibiotics acting on the parent organisms (Marquet *et al.*, 1974). These data indicate that the PBC's detected in these preparations are related both to the lethal target (presumably a physiologically important transpeptidase activity) and to the transpeptidase activity assayed *in vitro* by the model peptide system, helping to establish the validity of this assay in this system. Benzylpenicillin, bound by these membrane preparations, is slowly released as benzylpenicilloate, with recovery of enzyme activity. A minor unknown product is also found. Membranes from *S. rimosus* give the same result (Ghuysen, 1976). The activity solubilized from strain R61 and *S. rimosus* cells by CTAB binds benzylpenicillin, and releases it mostly as phenacetylglycine (Dusart *et al.*, 1975; Ghuysen, 1976). In this respect it resembles the exocellular activity more than the membrane-bound activity, and this may reflect only the effect of the microenvironment on the mechanism of release of bound penicilloate from the same PBC's, as for the R39 exoenzyme.

3. Isolation of Penicilloyl and Acyl-D-alanyl Peptides from DD-Carboxypeptidases

Benzylpenicilloyl peptides have been isolated from the products of interaction of benzylpenicillin with the *B. subtilis* and *Streptomyces* R61 DD-carboxypeptidases, following proteolytic hydrolysis (Georgapapadakou *et al.*, 1977; Frère *et al.*, 1976b). In both instances, a penicilloyl-L-serine link apparently occurs, the sequence in R61 being Val-Gly-Ser (Frère and Ghuysen, personal communications) and in *B. subtilis* being Asn-Ser-Gly (Georgapapadakou and Strominger, personal communication).

If, following interaction of the same enzymes with an acyl-D-alanyl-D-alanine transpeptidation donor, the same peptide sequences were isolated with an acyl-D-alanyl-(serine) substituent, the validity of the transpeptidation substrate analog theory (Tipper and Strominger, 1965) of β-lactam antibiotic action would be established. This requires that an enzyme form a relatively stable intermediate with a simple labeled substrate such as (Ac)2-L-lys-D-Ala-D-Ala, so that quenching of the reaction by enzyme denaturation would yield a covalent intermediate. Both complexes have been isolated by quenching of the reactions with *S. aureus* PBC 4 (Kozarich, 1977) and with *B. subtilis* PBC 5 (DD-carboxypeptidase), and these products have stabilities consistent with a serine ester linkage after denaturation. The *B. subtilis* enzyme has an essential thiol group. The enzyme is inactivated by parachloromercuribenzoate (PCMB) (Umbreit and Strominger, 1973a), but the product still interacts with penicillin and transpeptidase donors to give a relatively stable product (Curtis and Strominger, personal communication). Unfortunately, yields are still low. A depsipeptide substrate analog, which by analogy with the serine proteases should acylate the enzyme more rapidly, has recently been used by Rasmussen and Strominger (1978) to isolate larger quantities of acylated enzyme. Preliminary date suggests homology of an acylated peptide fragment with that acylated by penicillins (Rasmussen, personal communication).

β-lactamase action may also be expected to proceed via a penicilloyl-enzyme intermediate (Fig. 35), and such an intermediate has been identified in the products of interaction of the *S. aureus* β-lactamase with quinacillin (Virden *et al.*, 1975). This enzyme was chosen because of the rapidity with which it is denatured in 5 *M* quanidinium chloride. The particular penicillin derivative chosen has strong UV absorbance, allowing easy detection. Rapid quenching of the hydrolysis of quinacillin with quanidinium chloride gave a covalently bound product with a molar ration to enzyme approaching unity. A penicilloyl peptide was isolated by chymotrypsin hydrolysis, but the amino acid composition was not reported.

4. *Escherichia coli* CARBOXYPEPTIDASES AND PBC'S

Two carboxypeptidase activities were originally observed in an *E. coli* B strain in broken cell preparations (Izaki *et al.*, 1966). Carboxypeptidase I cleaved the terminal D-Ala residue from UDP-MurNAc pentapeptide and was found to be highly sensitive to inhibition by β-lactam antibiotics, which were thought to inhibit it reversibly. It has specificity for D-D linkages. Carboxypeptidase II cleaved the L-D linkage between *meso*-DAP and D-Ala in the UDP-MurNAc tetrapeptide product of carboxypeptiase I action. It was not inhibited by β-lactam antibiotics.

a. D,D-*Carboxypeptidase I Components.* Subsequent investigation in K12 strains has clarified the relationship between the several molecular species found in *E. coli* having carboxypeptidase I activity. All have been purified to electrophoretic homogeneity and their properties are summarized in Table VII (after T. Tamura *et al.*, 1976), which also lists the relationship of these activities to the known PBC's of *E. coli* inner membrane preparations (Table VIII).

The soluble form, IC, comprises about 70% of the total activity and appears to be identical with the membrane-bound fraction IB and the previously recognized endopeptidase that cleaves the D-Ala-(D)*meso*-DAP-(COOH) cross-links in the murein. Fraction IB is equivalent to PBC 4, whose SDS–PAGE mobility corresponds to a MW of 49,000 daltons. By itself it constitutes only 4% of the PBC's (Table VIII). The activities of IB and IC are highly sensitive to inhibition by β-lactam antibiotics, including 6-aminopenicillanic acid, which totally inhibits IC at concentrations well below growth inhibitory concentrations. This suggests that little, if any, of IB and IC activity is essential for growth under laboratory conditions, and this has been confirmed by the isolation, by two independent groups (Iwaya and Strominger, 1977; Matsuhashi *et al.*, 1977), of mutants (*dacB*) totally lacking IB and IC activity and PBC 4. They map near *rodY* at 68 minutes. A second mutant (*dacA*) lacks DD-carboxypeptidase IA activity, although it retains PBC 5 and 6. It maps near *rodA* at 14 minutes (Suzuki *et al.*, 1978). Both mutants and the double mutant, which lacks all carboxypeptidase activity, are physiologically normal (Suzuki *et al.*, 1978) except for a reduced rate of lysis in the *dacB* mutants in the presence of ampicillin, probably a reflection of the loss of endopeptidase activity. A role in control of transpeptidation had been proposed for DD-carboxypeptidases (see Section VI,E,2), but is not supported by this data, which also suggests that none of these enzymes are primary targets for antibiotic action.

Carboxypeptidase IB can be selectively solubilized from membrane preparations by 0.5 *M* LiCl. Carboxypeptidase IA can subsequently be solubilized by treatment with 2% Triton X-100. After further purification, it is found to give two bands on SDS–PAGE corresponding to MW of 42,000 and 40,000 (Spratt, 1977), originally reported as 34,000 and 32,000 (T. Tamura *et al.*, 1976), which are identical to the two major PBC's of *E. coli* membranes, PBC 5 and 6 (Spratt and Strominger, 1976). These proteins are sufficiently abundant to comprise about 0.2% of the total *E. coli* membrane proteins, and can be seen as distinct bands on Coomassie blue-stained SDS gels after electrophoresis (Spratt, 1977). Apart from this small difference in electrophoretic mobility, these two components are known to differ only in the rates at which they release bound penicillin G. In their native state, the half-lives of the penicilloyl group at 30°C are 5 and 19 minutes for PBC 5 and 6, respectively. They can be detected as PBC's only because the release, a slow β-lactamase activity producing benzylpenicilloic acid, is enzymatic and is inhibited by denaturation (and solubilization) of the cytoplasmic membrane preparations with sodium sarcosinate (Spratt, 1977). Carboxypeptidase IA, in contrast to IB and IC, is inhibited by PCMB, indicating the presence of an —SH group important to enzymatic function. The inhibited enzyme no longer functions as a carboxypeptidase, transpeptidase, or β-lactamase, but still binds penicillin G irreversibly (T. Tamura *et al.*, 1976) (see E7). It is worth emphasizing that the β-lactamase coded for by the *E. coli bla* gene is not inhibited by PCMB. The β-lactamase activity of carboxypeptidase IA thus cannot be due to contamination by the *bla* gene product. The original observation of reversible inhibition of DD-carboxypeptidase IB by benzylpenicillin (Izaki *et al.*, 1966) was presumably also due to reactivation following release of bound drug as benzylpenicilloate. The slow but significant rates of release of bound penicilloate and reactivation of inhibited enzyme helps to explain the reversibility of the *in vivo* effects of β-lactam antibiotics under nongrowth conditions.

b. LD-*Carboxypeptidases.* Carboxypeptidase II of *E. coli* has specificity directed against *meso*-DAP(L)-D-Ala-COOH linkages. The *E. coli* murein contains both tripeptide and tetrapeptide C-termini, showing that both carboxypeptidases I and II function *in vivo*. Carboxypeptidase II seems to be periplasmic and an increase in its susceptibility to release by osmotic shock, coupled to cell division (Beck and Park, 1976), suggest that it may play a role in cell cycle control. Beck and Park (1977) suggest that the increase in detectable activity at septation is due to reduction in the effect of an unspecified inhibitor. A role for this enzyme in linkage of murein to lipoprotein by transpeptidation is plausible, (see Section XII), but has not been substantiated. If such a role exists and if lipoprotein serves to anchor murein to the outer membrane, then the rapid association of outer membrane with completed crosswalls during cell separation (cf. Fig. 9) might require a burst of carboxypeptidase II activity at the same time. Certain mycobacteria have been reported to contain some *meso*-DAP–*meso*-DAP cross links of undetermined stereochemistry between their stem peptides (Wietzerbin *et al.*, 1974). A *meso*-DAP(L)-(D)*meso*-DAP(D)-COOH cross-link could conceivably be produced by an LD-carboxypeptidase acting as a transpeptidase.

Sporulating gram-positive bacilli, which usually have an A1γ peptidoglycan structure, such as *E. coli*, also have carboxypeptidase I and II activities and have tripeptide C-termini. This is also true in *B. sphaericus* in which the L-R_3 residue of the cell wall peptidoglycan is L-Lys. The LD-carboxypeptidase has specificity directed against L-Lys-D-Ala-COOH linkages and the C-termini are tripeptides terminating in L-Lys-COOH. The DD-carboxypeptidase of this species hydrolyzes both L-Lys-D-Ala-D-Ala and *meso*-DAP-D-Ala-D-Ala, so that the C-termini of the spore cortex peptidoglycans are *meso*-DAP-D-Ala, which is not a substrate for the LD-carboxypeptidase (Guinand *et al.*, 1974; Arminjon *et al.*, 1977). In *B. subtilis* the LD-carboxypeptidase has a similar specificity. It hydrolyzes the cell wall peptidoglycan in which the (D)-COOH of *meso*-DAP is amidated, but does not hydrolyze the unamidated cortex peptidoglycan (Guinand *et al.*, 1976).

An unusual LD-transpeptidase activity occurs in membrane preparations of *Streptococcus faecalis* (*faceium*) (Coyette *et al.*, 1974). The transpeptidase has specificity for L-R_3-D-Ala-COOH donor and for D-amino acids as acceptors. It does not function as a carboxypeptidase, and its function is unknown. The expected product of its action would be L-lys-D-Asn cross-links. They have not been observed in *S. faecalis*.

c. Escherichia coli PBC's. Envelopes of *E. coli* K12 bind [^{14}C]benzylpenicillin to several components which are solubilized by 1% sodium lauryl sarcosinate, a detergent which selectively solubilizes cytoplasmic membrane proteins in *E. coli* (Filip *et al.*, 1973). The detergent denatures the PBC's, demonstrating that penicillin is covalently bound to them. This denaturation also prevents enzymatic release of the bound penicillin. SDS–PAGE initially showed the presence of 6 PBC's, but the slowest moving component, PBC-1, has now been resolved into 1A and several 1B species. Their properties are outlined in Table VIII. No covalent binding of penicillin to outer membrane or cytoplasmic proteins is detectable (Spratt, 1977). Other labelled β-lactam antibiotics, cefoxitin and mecillinam, bind to a subset of the PBC's (Table VIII), and the PBC's 1a, 1b, 2 and 3 are all targets for lethal antibiotic action.

Although all PBC's bind benzylpenicillin, related penicillin derivatives, and some cephalosporins, cefoxitin fails to bind to PBC2 while cephalexin and cephradine (virtually identical antibiotics) fail to bind to PBC2, 5, or 6. These cephalosporin derivatives have highest affinity for PBC3 (Spratt, 1975, 1977). In contrast, mecillinam, a 6-amidinopenicillanic acid derivative (Lund and Tybring, 1972), binds only to PBC2 (Spratt, 1975; Spratt and Pardee, 1975). The effects of these antibiotics on the morphology of *E. coli* reflect the specificity of these interactions.

Mecillinam at minimal lethal concentrations causes cells of *E. coli* (Lund and Tybring, 1972; James *et al.*, 1975) and of all other gram-negative organisms tested (Iwaya, personal communication) to become enlarged and round. This occurs over a period of several generations. The cells lose viability rapidly, but lyse only very slowly. A proportion of mutants isolated as resistant to mecillinam (*mecR*) after severe mutagenesis have round morphology (*rod* phenotype). *rodX* mutants (Iwaya *et al.*, 1978), also called *rod*A (Matsuzawa *et al.*, 1973) map at 14 minutes and lack PBC2. *rodY* mutants map at 67 minutes and retain PBC2 (Iwaya *et al.*, 1978). Apparently the *rodA* locus is the structural gene for a membrane protein (PBC2) whose function depends upon the *rodY* gene product, and which determines rod cell shape in *E. coli* (Iwaya *et al.*, 1978). It is the lethal target for mecillinam, but is unaffected by cephalosporins. Cells carrying F′ORF4, which are merodiploids for the $14' \rightarrow 10'$ map region, overproduce PBC2 (Iwaya, personal communication). Cephalexin inhibits cell division of *E. coli* at concentrations at which it binds effectively only to PBC3. A mutant (SP63) resistant to low levels of cephalexin has been isolated in which division is temperature sensitive (*fts* phenotype). The ability of PBC3 to bind penicillin in membranes of this mutant is also temperature sensitive, suggesting that the mutation is in the structural gene for PBC3 (Spratt, 1975). The *fts* character of this strain is contransducible with *leu* (Iwaya and Sturgeon, unpublished observations). Another *fts* mutant, called *sep*, has been very accurately located adjacent to *leu* (Fletcher *et al.*, 1978). *Sep* mutants lack PBC 3 and appear to be allelic to *SP63* (Iwaya, Walker and Tipper, unpublished observations). A mutation of similar phenotype, designated *fts* I and also defective in PBC 3, has been reported to map at 1.8 minutes (Nishimura *et al.*, 1977) and is presumably another allele. All presumably lie in the structural gene for PBC 3, a membrane protein whose function is necessary for cell division. Recent models of rod cell growth (Begg and Donachie, 1973, 1977; see also James, 1975) suggest that most elongation occurs at cell termini. It would appear that PBC 3 may be responsible for initiating new zones of centripetal peptidoglycan synthesis in the center of growing cells while PBC 2 is required for the initiation of cylindrical wall growth at the junction of the new crosswall and the older cylindrical wall of the cell. Both PBC 2 and PBC 3 are independent targets of drug action. PBC 3 might be a highly specialized transpeptidase whose function is coupled to the termination of rounds of chromosomal replication. It might be involved in initiating transpeptidation between nascent peptidoglycan chains (see Section VI,E,3), directing growth away from preexisting murein.

Cephaloridine at minimal inhibitory concentrations causes lysis of *E. coli* cells without prior inhibition of cell division or disturbance of cell shape. This β-lactam antibiotic has highest affinity for PBC 1 components and these are presumed to be the major transpeptidases of the cell involved in cell wall elongation (Spratt, 1975, 1977; Spratt *et al.*, 1977) PBC 1 has recently been resolved into a slower moving component, PBC 1A, and a faster moving component comprising at least three related species (PBC 1B) (Spratt, 1977). The individual roles of the 1A and 1B components have now been clarified by genetic studies. Tamaki *et al.* (1977) described a mutant which is simultaneously thermosensitive (lysing at 42°C), hypersensitive to many β-lactam antibiotics, deficient in *in vitro* peptidoglycan synthesis and transpeptidation and deficient in all PBC 1B components. All phenotypes co-revert, including the presence of all of the PBC 1B components, suggesting that all are the consequence of a single point mutation. This was called *mrc* and maps at 3.3 minutes. In certain pseudorevertants, the quantities of PBC 1A and 2 are increased, apparently compensating for the almost total loss of PBC 1B. Suzuki *et al.* (1978) found ten mutants which had lost PBC 1A and one mutant was ts for binding of penicillin to PBC 1A among 500 temperature sensitive (ts) mutant derived from cells of a heavily mutagenized *E. coli* strain. These mutants, called *pon*A, mapped at 73.5 minutes and were not responsible for ts growth, suggesting that 1A is not essential for growth, as previously suggested by Spratt (1977). This is only true in strains containing normal 1B. Suzuki *et al.* (1978) also found 3 mutants retaining 1A but lacking 1B. These mutants all mapped in a locus called *pon*B at 3.3 minutes, but unlike the *mrc* mutant, none were responsible for ts growth. PBC 1B seems to be dispensible in the presence of normal PBC 1A. The multiple PBC 1B bands again cotransduce and corevert. Comparison of the MIC's of cephalosporins for the parent and *pon*B (*mrc*) mutants with the affinity of these antibiotics for PBC 1A and 1B indicates that inactivation of 1A in the presence of 1B is without consequence. Thus, Cephalothin, whose affinity for 1A is high and for 1B is low, kills wild-type cells by inactivating PBC 3 and consequently inhibiting cell division. In the absence of 1B, however, 1A becomes indispensible and the mutants are hypersensitive to cephalothin. Cephaloridine has a high affinity for 1A, and a fairly high affinity for 1B and 3, 1B being slightly more sensitive. This explains why it causes lysis without inhibition of cell division as the major effect on growing *E. coli* cells at its MIC. The *pon*B (*mrc*) mutants are hypersensitive to cephaloridine, corresponding to the higher affinity of the drug for PBC

1A, which becomes the primary target in the absence of 1B. Thus 1A and 1B appear to have redundant functions and both must be inactivated, or absent, to cause cell death. In confirmation, *pon*A *pon*B double mutants could not be constructed and the *pon*A^{ts} *pon*B double mutant is ts for growth, lysing at 42°C (Suzuki *et al.*, 1978). Tamaki *et al.* (35) reported ratios of the MIC's for the parent, relative to their *mrc* (*pon*B) mutant, of 3 for penicillin G and ampicillin, 10 for Cefoxitin and 100 for nocardicin. PBP 1A must become the primary target for each of these antibiotics in this mutant, suggesting a broad and high affinity of PBC 1A for β-lactam antibiotics. *Pon*B (*mrc*) mutants could thus be useful in screening for new antibiotics of this type.

Cloning of the *E. coli* PBC genes is underway in several laboratories. It may thus become possible to purify these proteins and to test their enzymatic properties *in vitro*, perhaps giving direct evidence of the relationship of PBC's to transpeptidases. The extent to which different PBC's continue to function as transpeptidases or carboxypeptidases in crude *in vitro* membrane or wall/membrane preparations appears to be variable, and since such assays measure only the combined surviving activities, the results must be interpreted with caution.

5. Conclusion

In conclusion, β-lactam antibiotics, with the exception of mecillinam, cause gross inhibition of cross-linking of peptidoglycan leading to lysis and death of growing cells. So far, among penicillin-sensitive enzymes, it has been possible to purify and investigate only the DD-carboxypeptidases whose function, while mimicking those of transpeptidase *in vitro*, seem to be entirely dispensible for normal growth, at least in *B. subtilis* and *E. coli*. Nevertheless, the interaction of β-lactam antibiotics with these DD-carboxypeptidases appears to be consistent with penicilloylation of the donor binding site. However, the lethal effects of different β-lactam antibiotics on *E. coli* may result from preferential interaction with different binding components so that multiple, independent lethal targets for these antibiotics exist, at least in gram-negative organisms. The target PBC's are very poorly characterized as enzymes and are only presumed to be transpeptidases. In this respect mecillinam is in a class by itself and its effects help to illustrate the specific role that some of these enzymes may have in cell shape and cell division control in gram-negative organisms. Mecillinam is almost without effect on most gram-positive organisms, but is active against *Streptococcus mutans* (Mychajlonka and Shockman, personal communication). Moreover, gram-positive organisms normally respond to inhibitory concentrations of β-lactam antibiotics, such as cephalexin, by death and lysis with no evidence for selective inhibition of cell division. Again, the oral streptococci are exceptions (see Section VI,G). Thus the most sensitive target for a given β-lactam antibiotic may vary from organism to organism, and data in *E. coli* is not necessarily applicable to *S. aureus*.

The process by which inhibition of these varied targets leads to cell death is discussed in Section VI,G.

G. The Role of Autolysins in Cell Killing by β-Lactam Antibiotics

It was early recognized that β-lactam antibiotics kill only growing organisms. It is for this reason that bacteriostatic antibiotics, such as chloramphenicol, antagonize the bactericidal effects of penicillins (Rogers, 1967). This antagonism understood at this level for 20 years and well established in *in vitro* bacterial cultures, is also demonstrable in certain clinical situations such as bacterial meningitis (e.g., Jawetz, 1967; Mathies *et al.*, 1967). It has frequently been ignored by clinicians. The death of growing cells in the presence of penicillin could simply be due to the continued synthesis of protein and consequent cytoplasmic expansion in cells whose wall is weakened by the absence of adequate cross-linkage. However, autolytic peptidoglycan hydrolases were proposed to play an essential role in cell wall growth (Weidel and Pelzer, 1964), and it was proposed that β-lactam antibiotics lead to an imbalance of peptidoglycan synthesis and autolysis in growing cells, resulting in weakening of the structure and cell lysis.

The most dramatic support for this type of model has come from the studies of Tomasz and his colleagues on the pneumococcus (cf. Tomasz, 1974; Tomasz and Holtje, 1977). The pneumococcal C substance is a complex teichoic acid covalently bound to cell wall peptidoglycan. It contains choline, a required growth factor for this species. The "Forssman antigen" of pneumococci also appears to be a choline-containing teichoic acid, perhaps the "membrane" teichoic acid of this species. None of these polymers has been completely characterized (see Section VIII). In the absence of choline, the cells will utilize ethanolamine for incorporation into its teichoic acids. The normal organism is very fragile, lysing rapidly at the end of vegetative growth or if deoxycholate or β-lactam antibiotics are added during vegetative growth. These normal cells are also competent for transformation and die rapidly in the presence of penicillins. The cells containing ethanolamine fail to autolyze in stationary growth phase, are incompetent and resistant to deoxycholate and fail to lyse or die rapidly in the presence of β-lactam antibiotics, even though cell growth is inhibited. This response to penicillins is termed tolerance (Tomasz *et al.*, 1970). The cells are also tolerant to other inhibitors of peptidoglycan synthesis, such as D-cycloserine. The autolytic enzyme is an *N*-acetylmuramyl-L-alanine amidase which rapidly lyses choline-containing cell walls. The ethanolamine-containing cell walls, however, are resistant, and it appears that choline-containing teichoic acid is necessary for activation of the autolysin (Holtje and Tomasz, 1975a,b). Autolysin activation is stimulated by deoxycholate or β-lactam antibiotics and is the proximal cause of cell death following such exposure.

Mutants can be isolated which are resistant to lysis by deoxycholate because they are deficient in amidase activity. Cells of such mutants are tolerant to penicillin and resistant to deoxycholate even when grown in the presence of choline. When grown in the presence of choline and exposed to autolysin from wild-type cells, however, these mutant cells lyse when penicillin is subsequently added.

The role of modification in wall teichoic acid structure in activating autolysis was first recognized in *Streptococcus zymogenes*. The autolysin in this strain appears to have a primary effect on the cytoplasmic membrane, perhaps activating a "true" autolysin (Basinger and Jackson, 1968). Alanylated teichoic acids inhibit it, but alanine-free teichoic acids have no effect (Davie and Brock, 1966).

Streptococcus pneumoniae Forssman antigen has considerable structural similarity to the C polysaccharide wall teichoic acid. However, whereas the wall teichoic acid activates the amidase, the membrane lipoteichoic acid (LTA) inhibits it (Holtje and Tomasz, 1975a,b). The effects of deoxycholate and β-lactam antibiotics appear to be mediated by loss of this inhibitor; exposure of wild-type cells to deoxycholate or penicillin results in excretion of the membrane teichoic acid (Tomasz and Waks, 1975). The LTA of *S. faecalis* is also an inhibitor of the wall autolysin of this species (Cleveland *et al.*, 1975). LTA's, isolated from *S. faecalis* (*faecium*), *B. subtilis*, and *Lactobacillus fermentum;* the acidic lipomannan of *M. luteus;* and the Forssman antigen of *S. pneumoniae* were tested as inhibitors of the major autolysins produced in these species (Cleveland *et al.*, 1975). Only the Forssman antigen inhibited the *S. pneumoniae* amidase, and it had no effect on the muramidases of *S. faecalis* and *L. acidophilus* or the amidase of *B. subtilis*. In contrast, the LTA's inhibited the latter three enzymes, but not the *S. pneumoniae* amidase. The *S. pneumoniae* system thus demonstrates unique homospecificity. Removal of the fatty acids from the LTA's destroyed their inhibitory activities.

It is possible that cell lysis following exposure to β-lactam antibiotic is a consequence of multiple events. First, inhibition of transpeptidation causes synthesis of un-cross-linked peptidoglycan. Second, this causes, by some unknown mechanism (excretion?), a modification in autolysin inhibitors presumed to be membrane teichoic acids in gram-positive organisms. Third, the activated autolysins find the hypo-cross-linked peptidoglycan formed in the presence of penicillin to be a particularly susceptible substrate. They attack it, further weakening its structure, leading to loss of cell wall integrity, cytoplasmic membrane disruption, and cell death. It must be emphasized, however, that the generality of this type of scheme is unproved, and the bacteriostatic effects of β-lactam antibiotics in tolerant organisms is unexplained.

Organisms highly sensitive to β-lactam antibiotics, such as *Pneumococcus*, *Neisseria*, *B. subtilis*, etc., all have highly active autolytic systems. It is also true that autolysin-defective mutants of *B. licheniformis* are found to give a tolerant response to β-lactam antibiotics (Forsberg and Rogers, 1971) and all gram-positive organisms apparently contain membrane teichoic acids. However, such compounds have not been positively identified in gram-negative organisms. The role of autolysins in the killing of *S. aureus* by β-lactam antibiotics is equally clear. Certain laboratory-derived mutants defective in autolysin activity are no less susceptible to such antibiotics (Chatterjee *et al.*, 1976). However, it has recently been found that a majority of recent clinical *S. aureus* isolates have a tolerant response to β-lactamase-resistant penicillins, even though their growth is inhibited by normally low concentrations. This tolerance frequently extends to Vancomycin, but not cycloserine. The autolysin activity of normal nontolerant strains increases markedly on exposure to penicillins (Best *et al.*, 1974), but this increase is not seen in the tolerant strains, which seem to produce an excess of an uncharacterized (LTA?) autolysin inhibitor (Sabath *et al.*, 1977). Presumably the autolysin-defective strains of Chatterjee *et al.* (1976) can still respond to β-lactam antibiotics with a marked rise in activity. *Streptococcus sanguis* is apparently naturally tolerant to penicillins (Horne and Tomasz, 1977), and so is *S. mutans*. These cariogenic oral streptococci are also unusual in their sensitivity to mecillinam (see Section VI,F,5).

VII. Other Cell Wall Components in Gram-Positive Bacteria

Peptidoglycan seldom comprises more than 50% of cell wall mass in gram-positive bacteria and is frequently a considerably smaller fraction. The rest of the wall usually consists of complex carbohydrate polymers and polypeptides which are covalently linked to the peptidoglycan. An intimate mix of polymers is presumably responsible for the homogeneous cross section of the walls of organisms such as *S. faecalis* (Fig. 1). Even where cell wall profiles show poorly defined layering, as in the walls of *S. aureus* (Fig. 3) and *B. subtilis*, it seems likely that only the relative proportions of polymers may vary from layer to layer. Much more specific wall layers occur in gram-positive species such as *B. sphaericus* (Fig. 5) in which the outer layer consists of a structural polypeptide array, not covalently attached to other wall components and capable of self-assembly (Brinton *et al.*, 1969; Aebi *et al.*, 1973). A similar structured polypeptide layer, assembled on the outside of the outer membrane, can be found in gram-negative species, such as *Spirillum serpens*. Here the basic unit is hexagonal rather than tetragonal, but self-assembly on stripped walls does occur (Murray and Buckmire, 1969). Leduc *et al.* (1977) have recently described an endospore-forming gram-variable bacillus with a multiple-layered cell wall containing large amounts of a high molecular weight polypeptide that is assembled on the wall surface in a hexagonal pattern. This organism is reported to lack lipoteichoic acids as well as wall teichoic acid.

The vast variety of such polymers in bacteria is responsible for the serological specificities used in the identification of strains and species. The widest variation occurs in the actinomycetes whose multiple groups and families, besides containing unique filamentous and sporulating genera such as *Actinomyces*, *Streptomyces*, and *Actinoplanes*, contain genera with lipid-rich walls, such as *Mycobacterium*, *Nocardia*, and typical *Corynebacterium*. The unique wall components of these last three genera play a major role in their interaction with animals as skin flora, pathogens prone to cause chronic infections and modifiers of the adaptive immune system. The ability of some of these organisms to survive in the fatty acid-rich secretions of the skin merely illustrates the importance of nonpeptidoglycan cell wall components in interaction of bacteria with their environment. Bacterial glycolipids and glycophospholipids were recently reviewed by Shaw (1975). Mutants lacking wall components, such as teichoic acid-deficient staphylococci, are easily obtained in the laboratory, where they may grow nicely in the absence of competition, but they are biologically inept, unfitted for survival in their natural environment.

Capsules, exocellular polymers that are usually complex polysaccharides, are not wall components by definition (Section II). However, when present, they also play an important role in bacterial survival. Capsules, for example, are essential determinants of pathogenicity in *S. pneumoniae* and are also the strain-specific antigens for this species. The structures of such bacterial exopolysaccharides have been recently reviewed by Sutherland (1972).

A. Teichoic Acids: Structure

Teichoic acids are polymers containing alditol phosphates which occur in probably all gram-positive bacteria. They are probably absent from most gram-negative species. However 60% of *Butyrivibrio* species examined were found to contain a phenol-extractable polymer reacting with antisera specific for polyglycerol phosphate (Sharpe *et al.*, 1975). This species stains gram negative, but has a gram-positive type of cell wall profile in cross section (see Section III) (Cheng and Costerton, 1977). Also glycerol phosphate occurs in the O antigen of at least one *E. coli* strain (Jann *et al.*, 1970), and ribitol phosphate and uronic acids have been found in the O antigen of *Proteus mirabilis* strains (Gmeiner, 1975). Thus surface polysaccharides of certain gram-negative species strongly resemble teichoic acids. In gram-positive bacteria, teichoic acids are located exclusively in the membrane, wall, or capsular layers of the cell and are the most highly charged of the polyanionic polymers which provide the surface charge to these cells and which probably help to maintain an adequate magnesium ion concentration at the surface of the cytoplasmic membrane. Their structure and biosynthesis was recently reviewed by Archibald (1974), and this section will emphasize recent information on the linkage of wall teichoic acids to peptidoglycan.

1. Polyalditol Phosphates

The first teichoic acids described were simple polymers of 1,5-poly(ribitol phosphate) and 1,3-poly(glycerol phosphate), usually glycosylated and esterified by D-alanyl residues (Fig. 38). Such teichoic acids are found covalently attached to the cell wall peptidoglycans of *staphylococci*, *lactobacilli*, and various other gram-positive bacilli. Some species such as *B. licheniformis* and *L. plantarum*, may contain several species of wall-bound teichoic acid, including examples of more complex types in which a sugar residue forms an integral part of the polymer chain. In contrast to this variety of structures, the teichoic acid polymers found associated with membranes all consist of polymers of 1,3-glycerol phosphate which, in several instances, have been shown to be covalently linked to a glycolipid (Fig. 39). Such lipoteichoic acids (LTA's) may be universal components of gram-positive bacterial envelopes (see Section VII,A,4). Both the wall teichoic acids and the LTA's are glycosylated to variable extents.

2. Ribitol Teichoic Acids

All wild-type *S. aureus* strains contain a polyribitol teichoic acid which makes up about half of the wall mass. The ribitol residues carry GlcNAc residues on the 4-(D) position and alanine ester residues on the 2-(D) position (Mirelman *et al.*, 1970). In some strains, all the GlcNAc residues have all α- or all β-anomeric configuration while others, such as strain Copenhagen (Nathenson *et al.*, 1966), have both anomers. These occur in separate, homogeneously substituted (α or β) chains (Torii *et al.*, 1964). Similar polymers, carrying 4-*O*-β-D-glucopyranosyl-D-ribitol substituents, are found in the walls of some strains of *B. subtilis* and *B. licheniformis*. *Bacillus subtilis* W23 walls contain a mixture of fully glucosylated and unglucosylated teichoic acids (Chin *et al.*, 1966) whose proportions vary with the growth conditions. When grown in continuous culture, only the fully glucosylated form was found (Ellwood and Tempest, 1972). In *L. plantarum* 17-5, the ribitol residues carry either zero, one, or two glucosyl residues, and each

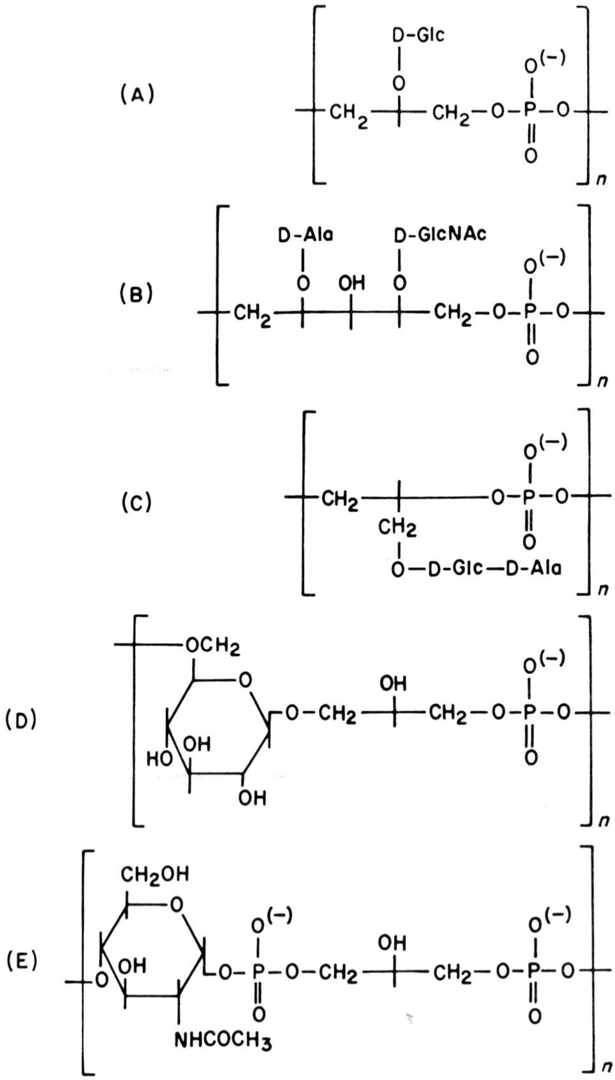

(A) $\left[-CH_2-C(O\text{-D-Glc})-CH_2-O-P(=O)(O^{(-)})-O- \right]_n$

(B) $\left[-CH_2-C(O\text{-D-Ala})-C(OH)-C(O\text{-D-GlcNAc})-CH_2-O-P(=O)(O^{(-)})-O- \right]_n$

(C) $\left[-CH_2-C(CH_2-O\text{-D-Glc-D-Ala})-O-P(=O)(O^{(-)})-O- \right]_n$

(D) $\left[-OCH_2\text{-(glucopyranosyl: HO, OH, OH)}-O-CH_2-C(OH)-CH_2-O-P(=O)(O^{(-)})-O- \right]_n$

(E) $\left[-O\text{-(}CH_2OH\text{, OH, }NHCOCH_3\text{ pyranosyl)}-O-P(=O)(O^{(-)})-O-CH_2-C(OH)-CH_2-O-P(=O)(O^{(-)})-O- \right]_n$

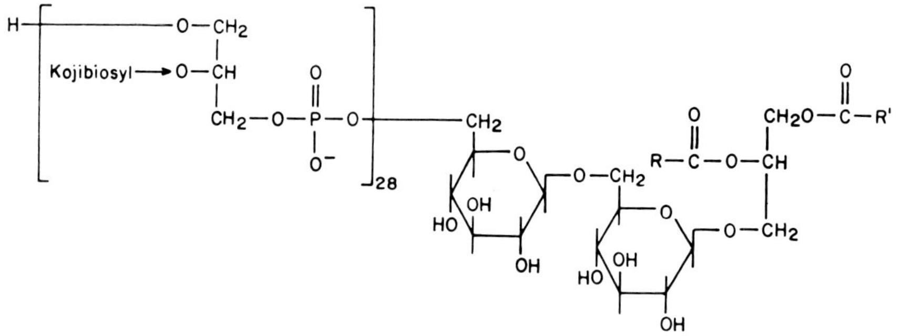

FIG. 39. Structure of the lipoteichoic acid (LTA, membrane teichoic acid) from *S. faecalis* 8191. The disaccharide linking the 1,3-poly(glycerol phosphate) teichoic acid to the diglyceride moiety is kojibiose (6-*O*-β-D-glucosyl-D-glucose) which also substitutes each glycerol residue. R-COOH and R′-COOH are fatty acids.

may be present in a separate homogeneous polymer (Knox and Wicken, 1972).

As these examples illustrate, and as will be further exemplified, the capacity of gram-positive bacteria to synthesize several phosphate-containing cell wall polymers, which may have closely related or unrelated structures, seems to be widespread. The reason for maintenance of this redundant biosynthetic capacity is obscure, as are the mechanisms governing their relative degrees of expression (see Section VIII).

Certain actinomycetes contain ribitol teichoic acids of unusual structure which lack D-alanyl residues. The glucosylated polymer in *A. violaceus* contains *O*-acetyl groups and succinate monoesters are found in the teichoic acids of *A. streptomycini* and *Streptomyces griseus*. The *S. griseus* polymer is unique in having glycerol residues phosphodiester linked to the 3-hydroxyl of some ribitol residues (Bews, 1967).

FIG. 38. Some representative teichoic acid structures. (A) 1,3-Poly(glycerol phosphate). The secondary hydroxyl group may be free, or glycosylated, or esterified by D-alanine. This is the most common type of teichoic acid structure. It is found in all membrane teichoic acids and in the cell wall teichoic acids of many gram-positive organisms, e.g., species of *Staphylococcus*, *Streptococcus*, *Lactobacillus*, and *Bacillus* genera. (B) 1,3-Poly(ribitol phosphate). This structure is also common in the cell wall teichoic acids of the genera mentioned above. Glycosylation usually occurs on the 4-hydroxyl of D-ribitol, and the 2-hydroxyl is esterified by D-alanine, as shown. (C) 2,3-Poly(glycerol phosphate). This polymer, substituted on its primary hydroxyl groups by 6-*O*-D-alanyl-D-glucose as indicated, has been reported to exist in *B. stearothermophilus*. A similar polymer occurs in certain actinomycetes. (D) Poly(6-phospho-α-D-glucosylglycerol). This polymer and its galactose analog have been found in *B. licheniformis* and in *L. plantarum*. Polymers with this type of structure have been found in pneumococcal capsular polysaccharides and pneumococcal C substance. (E) This is a member of the less common type of teichoic acid in which the main chain contains the acid-labile sugar 1-phosphate linkage. This example is from *Staphylococcus lactis* I3.

The complete solubilization of *S. aureus* cell walls by peptidoglycan hydrolases first produced teichoic acid covalently linked to fragments of the peptidoglycan, establishing both the size of the undegraded teichoic acid and the existence of this linkage (Ghuysen *et al.*, 1965). It was demonstrated that the linkage was to the glycan (Tipper *et al.*, 1967a), and a combination of enzymatic hydrolysis and periodate oxidation was used to isolate the linkage region. As reported in 1964 (Tipper, 1964), this was found to contain a periodate-resistant chain of three or four phosphate residues, probably linked to a GlcNAc residue. This linkage structure has recently been characterized in an *S. aureus* H mutant lacking GlcNAc residues on its polyribitol teichoic acid as a trimer of 1,3-glycerol phosphate, phosphodiester linked to the 6-position of a MurNAc residue (Heckels *et al.*, 1975; Coley *et al.*, 1976). The same link has been characterized between the peptidoglycan and the poly(GlcNAc-1-phosphate) polymer in walls of *Micrococcus* sp. 2102 (Coley *et al.*, 1977), and the existence of a small number of glycerol phosphate residues in the walls of *B. subtilis* W23 suggests the presence of a similar "link." This has been confirmed by biosynthetic studies (see Section VII,B,3) which have also demonstrated a requirement for UDP-GlcNAc for *in vitro* linkage of polyribitolteichoic acids to peptidoglycan in several species. Evidence for incorporation of GlcNAc into the linkage region has only been described in preparations from *B. subtilis* W23 (Wyke and Ward, 1977a), and its involvement in the "link" requires substantiation.

3. Glycerol Wall Teichoic Acids

1,3-Poly(glycerol phosphate) teichoic acids are found in the walls of all *Staphylococcus epidermidis* strains and in the walls of many bacilli, streptococci, and lactobacilli. The subunits have a single hydroxyl residue available for glycosylation or esterification by D-alanine. The extent of substitution varies widely. In *B. subtilis* 168, all residues are normally glucosylated (Armstrong *et al.*, 1961), while the walls of several species contain a mixture of fully glucosylated chains and of chains devoid of glycosyl substituents (Burger, 1966). In other wall teichoic acids and in most LTA's, only a few of the glycerol residues are glycosylated, and D-alanyl residues are found on some of the others. For example, the LTA of group D lactobacilli has two D-glucosyl residues per chain of 18 glycerol phosphate residues, while the group E-specific antigen is a wall teichoic acid with about 4 glucosyl residues per chain of 14 glycerol phosphate residues (Sharpe *et al.*, 1964). The D-alanyl residues on these polymers are highly labile to base-catalyzed hydrolysis, because of the adjacent phosphate residues, and the amount of D-alanine remaining on isolated teichoic acid varies markedly with the isolation procedure.

The walls of *B. stearothermophilus* B65 were reported to contain a 2,3-poly(glycerol phosphate) polymer fully substituted by 1-*O*-α-D-glucopyranosyl residues (Wicken, 1966). The D-alanyl residues apparently esterify the glucosyl residues (Fig. 38C). Biosynthetic studies (Kennedy, 1974) were consistent with this structure, and it was shown that the glucosyl residues incorporated *in vitro* could be removed without depolymerization of the synthetic product. A more recent investigation of this strain (Anderson and Arichibald, 1975) failed to detect this polymer, but identified a poly(α-D-glucopyranosylglycerol phosphate)teichoic acid in which the glucosyl residues form an integral part of the chain, a structure identical to that found in *B. licheniformis* (Fig. 38D). Growth conditions were not thought to be responsible for this discrepancy, which remains unexplained. The reverse has occurred in studies of the major wall teichoic acid of *B. subtilis* var. *niger*. This was originally identified as a poly(glucosylglycerol phosphate) (see Ellwood and Tempest, 1972), but more recent studies (DeBoer *et al.*, 1976) have identified it as a 2,3-poly(glycerol phosphate) carrying 1-*O*-β-D-glucosyl substituents. Walls of *Actinomyces antibioticus* have also been reported to contain a 2,3-poly(glycerol phosphate)teichoic acid substituted on the 1-position by a galactosyl-*N*-acetylgalactosamine disaccharide (Zaretskaya *et al.*, 1971).

4. Membrane Teichoic Acids: Lipoteichoic Acids

Whèn various gram-positive bacteria are converted to protoplasts in the presence of 10 m*M* Mg^{2+} ions, a "membrane" 1,3-poly(glycerol phosphate)-teichoic acid remains associated with the protoplast membranes. Protoplasting in low Mg^{2+} ion concentrations leads to solubilization of this teichoic acid (Hughes *et al.*, 1973). In several instances, it has been shown to be covalently linked to lipid (cf. Coley *et al.*, 1972, 1975), and it is probable that most gram-positive bacteria contain such LTA's. The lipid moiety of these LTA's is presumably buried in the outer layer of the cytoplasmic membrane, while the hydrophilic portion projects outwards from this surface. In organisms, such as groups A and F streptococci, where the LTA is the group-specific antigen, a portion of these LTA molecules must penetrate through the wall to be accessible to antibody (Wicken and Knox, 1975). Such intercellation into the wall matrix may also be a regular feature of LTA's, and may play a vital role in the local inhibition of autolysis (see Section VI,G).

In most membrane teichoic acids only a few of the glycerol residues are glycosylated, as in group D lactobacilli (see Section VII,A,3), while some are not glycosylated at all, as in lactobacilli of groups B and C. The LTA of group D streptococci (Fig. 39) is the group specific antigen and is fully glycosylated by kojibiose residues (Ganfield and Pieringer, 1975). The

D-alanyl residues of this polymer esterify some of these kojibiose residues, and do not have the extreme lability to base-catalysed hydrolysis which is characteristic of the D-alanyl–glycerol (phosphate) linkages. The link between lipid and poly(glycerol phosphate) in the LTA of *S. aureus* is gentibiose, but the polymer is only partially substituted by D-glucosyl and D-alanyl residues (Duckworth *et al.*, 1975).

As mentioned in Section F, the Forssman antigen of *S. pneumoniae* is an LTA of unusual structure containing phosphocholine (Briles and Tomasz, 1973, 1975). This species may be an exception to the ubiquitous presence of a poly(glycerol phosphate)-LTA in gram-positive species. The *Micrococcus* species with type A2 peptidoglycan (*M. luteus*, *M. flavus*, *M. sodonensis*) are also exceptions. They contain an LTA-like membrane component containing mannose, glycerol, succinate, and fatty acids (Powell *et al.*, 1975).

5. Teichoic Acids with Sugar Residues Integrated into the Polymer Chain

Certain complex pneumococcal capsular polysaccharides, such as the type 13 and 29 determinants, have (glycosyl)-ribitol phosphate repeating units (Archibald, 1976). The pneumococcal somatic antigen, called "C substance," is a complex wall teichoic acid whose repeating subunits contain choline phosphate, glucose, GalNAc, and a diaminotrideoxyhexose, linked via ribitol phosphate residues (Watson and Baddiley, 1974). Complex teichoic acids of this type are also found in some other streptococci (e.g., Miller, 1969).

Two unexpected products were found during investigation of teichoic acid synthesis by membrane preparations from *B. licheniformis* (Burger and Glaser, 1966). They are poly-(β-D-glucosylglycerol phosphate) (Fig. 38D) and poly(α-D-galactosylglycerol phosphate). They were subsequently found to occur in the wall of the parent organism, together with a 1,3-poly(glycerol phosphate)teichoic acid. *Lactobacillus plantarum* walls also contain three different teichoic acids, all with $(glycosyl)_n$-glycerol phosphate repeating units (Archibald and Coapes, 1971).

Walls of "*Staphylococcus*" *lactis* I3 (*Micrococcus varians*) contain a polymer whose repeating unit contains two phosphate residues and one glycerol residue in the sequence (HPO_4-GlcNAcα → HPO_4-glycerol) (Fig. 38E) (Archibald *et al.*, 1971). This is a representative of the highly acid-labile teichoic acids which have sugar 1-phosphate linkages in their main chain. D-Alanine esterifies the GlcNAc residues, probably at C-6 hydroxyl. The link between this teichoic acid and peptidoglycan apparently involves direct linkage to MurNAc-6-hydroxyl. Acid hydrolysis under conditions breaking only sugar 1-phosphate links of an autolysin-solubilized teichoic acid-peptidoglycan complex gave MurNAc-P-glycerol-P (Button *et al.*, 1966). Other micrococci contain similar teichoic acids. Walls of *Staphylococcus lactis* 2102 contains a simple polymer of 1,6-phosphodiester-linked GlcNAc residues (Archibald and Stafford, 1972). Like the *S. aureus* H polymer, this has been shown to be linked to peptidoglycan by a (glycerol phosphate)$_3$ unit (see Section VII,A,2). Although it shares this link structure with teichoic acids, the polymer lacks alditol, so is not a teichoic acid. *Micrococcus* sp. A1 contains a polymer in which Glc$\alpha \rightarrow$3-GlcNAc disaccharides are join by a phosphodiester between C-1 of GlcNAc and C-6 of Glc (Partridge *et al.*, 1973). *Bacillus subtilis* 168 walls contain a polymer with identical structure, except that the disaccharide is Glc$\beta \rightarrow$3-GlcNAc (Duckworth *et al.*, 1972; Shibaer *et al.*, 1973). These walls also contain a glucosylated 1,3-poly(glycerol phosphate) polymer (Fig. 38A).

B. Teichoic Acid Biosynthesis

The alditol phosphates of wall teichoic acids are derived from CDPribitol and CDPglycerol. These nucleotides are synthesized from CTP and D-glycerol 1-phosphate or D-ribitol 1-phosphate by specific pyrophosphorylases. Ribitol phosphate is produced by reduction of D-ribulose 1-phosphate (Glaser, 1963).

1. Polyalditol phosphate Wall Teichoic Acids

a. Polymerization. Membrane fractions from *L. plantarum* (Glaser, 1964), *B. subtilis* W23 (Chin *et al.*, 1966), and *S. aureus* (Ishimoto and Strominger, 1966) were found capable of transferring ribitol from CDP-[^{14}C]ribitol to an endogenous acceptor to form 1,5-poly(ribitol phosphate). Added poly(ribitol phosphate) or teichoic acid–peptidoglycan complexes would not act as acceptor. Polymerization of glycerol phosphate by particulate membrane fractions has been studied with preparations from *B. licheniformis* and *B. subtilis* (Burger and Glaser, 1964). Again, transfer is from CDPglycerol to an endogenous acceptor. Transfer was found to occur to the polymer terminus susceptible to periodate oxidation, i.e., the glycerol terminus, distal from any carrier molecule (Kennedy and Shaw, 1968). Incorporation involves extension of preexisting chains.

A system capable of using added acceptor was not obtained until a glycerolphosphate polymerase was solubilized from membranes of *B. subtilis* with detergent (Triton X-100) (Mauck and Glaser, 1972a). The solubilized enzyme could be separated from an acceptor, which had also been solubilized, and the acceptor had the properties of lipoteichoic acid (LTA) and was called lipoteichoic acid carrier (LTC) (Fiedler and Glaser,

1974a). A similar system was obtained from *S. aureus*. LTA's, prepared by phenol extraction from a variety of sources were all active as acceptors, although activity was highly variable (Fiedler and Glaser, 1974b). LTC, prepared by Triton X-100 solubilization, is more active than phenol-extracted LTA, and apparently elutes earlier from DEAE columns. This suggests that LTC carries more D-alanyl substituents than LTA (Duckworth *et al*., 1975).

It has not been possible to demonstrate *in vitro* transfer of poly(glycerol phosphate) or poly(ribitol phosphate), polymerized on LTC, to peptidoglycan. Moreover, it has not proven possible to label LTA with CDPglycerol, and it is suspected that the glycerol phosphate subunits of LTA are derived from phosphatidylglycerol (Emdur and Chiu, 1975; Glaser and Lindsay, 1974). In the light of recent data implicating bactoprenol pyrophosphate in the synthesis of the linkage unit of wall teichoic acid, it seems quite possible that *in vitro* demonstrations of polymerization of alditol phosphate on LTC are artifactual (see Section VII,B,3).

b. Glycosylation. Polymerization of ribitol phosphate by *S. aureus* membrane preparations is independent of glycosylation, and these preparations will transfer GlcNAc from UDP-GlcNAc to added polyribitol phosphate acceptor. However, a much greater rate of addition of GlcNAc occurs during simultaneous polymerization of ribitol phosphate (Nathenson and Strominger, 1963), suggesting juxtaposition of the polymerase and GlcNAc transferase enzymes in the membrane. Presumably some of the ribitol-phosphate polymerases of these membranes are associated with an α-GlcNAc transferase and some with β-GlcNAc transferase.

Spores of *B. subtilis* W23 contain no detectable teichoic acid. Synthesis of the normal mixture of fully glycosylated and unglycosylated poly(ribitol phosphates) is initiated early during germination. This is analogous to the early initiation of cell wall T layer protein synthesis during germination of *B. polymyxa* spores (Fig. 7). If spores of *B. subtilis* W23 are germinated in the presence of chloramphenicol, only the fully glucosylated polymer is produced (Chin *et al*., 1968). Apparently protein synthesis is required to produce new teichoic acid polymerase sites free of associated glucosyltransferase. Particulate membrane preparations from *B. subtilis* NCTC 3610 polymerize glycerol phosphate and will form completely glucosylated polymer if given UDP-Glc as well as CDPglycerol (Burger and Glaser, 1964). No evidence was obtained for involvement of carrier lipid in *in vitro* glucosylation of the 1,3-poly(glycerol phosphate)teichoic acid of *B. subtilis* 168 (Brooks *et al*., 1971).

c. Esterification by D-*Alanine.* Several gram-positive bacterial genera contain a cytoplasmic enzyme that produces a D-alanyl–AMP–enzyme complex from D-alanine plus ATP (Baddiley and Neuhaus, 1960). Transfer of D-alanine from this complex to wall teichoic acids has not been demonstrated; however, a second cytoplasmic enzyme in *L. casei* will transfer D-alanine from this complex to a membrane-bound acceptor that is probably LTA (Reusch and Neuhaus, 1971; Neuhaus *et al.*, 1974). A stable L form of *S. pyogenes* contains a D-alanine-deficient membrane teichoic acid, although the cytoplasm of these cells contains both D-alanine-activating enzyme and ligase (Neuhaus *et al.*, 1974). Clearly, alanine addition requires more than the presence of these activities and the polymerization of LTA.

2. Teichoic Acids with Sugar Residues in the Polymer Chain

Incubation of membrane preparations from *B. licheniformis* ATCC9945 with CDPglycerol and UDPglucose or UDPgalactose produces poly(glucosylglycerol phosphate) or poly(galactosylglycerol phosphate). CDPglycerol alone leads to the production of 1,3-poly(glycerol phosphate) (Burger and Glaser, 1966). All of these polymers occur in the wall, and it is not known how their proportions are controlled. Similarly, *B. stearothermophilus* B65 membrane preparations synthesize the glucosylated 2,3-poly(glycerol phosphate) polymer from a mixture of UDPGlc and CDPglycerol. In the absence of UDPGlc, a 1,3-poly(glycerol phosphate)teichoic acid is synthesized. *In vitro*, at least, the proportion of these two polymers formed depends on the UDPGlc concentration (Kennedy, 1974). As mentioned in Section VII,A,3. a more recent study of this strain detected synthesis of a poly(glucosylglycerol phosphate) polymer identical to the *B. licheniformis* teichoic acid (Anderson and Archibald, 1975).

The glycerol phosphate-glucose repeating unit of the *B. licheniformis* polymer is apparently synthesized by sequential transfer of glucose from UDPGlc and glycerol phosphate from CDPglycerol to undecaprenol phosphate to form glycerol-P-glucose-P-Lipid, though direct identification of the lipid is lacking (Hancock and Baddiley, 1972). Transfer of the glycerol-P-glucose subunit to the growing polymer releases P-lipid. Since PP-lipid is not involved in this cycle, the process is insensitive to bacitracin. Addition of UDP-MurNAc pentapeptide reduces the rate of *in vitro* teichoic acid synthesis, presumably due to competition for P-lipid. Teichoic acid synthesis also becomes sensitive to bacitracin, presumably because of trapping of lipid-P as the lipid-PP-bacitracin complex in the peptidoglycan synthetic cycle (Fig. 31) (Anderson *et al.*, 1972). This is indirect evidence for participation of the same lipid carrier pool in synthesis of both teichoic acid and peptidoglycan (see Fig. 40A).

Involvement of a lipid carrier was first demonstrated in synthesis of the *Staphylococcus lactis* 13 polymer (Douglas and Baddiley, 1968; Hussey and Baddiley, 1972), which also proceeds via formation of a lipid-linked precursor, glycerol-P-GlcNAc-PP-lipid. Transfer of the glycerol-P-GlcNAc-P repeating unit occurs at the glycerol end of the chain (Hussey *et al.*, 1969), i.e., to the end distal from the point of attachment to peptidoglycan, as previously demonstrated for polyglycerol wall teichoic acids (Kennedy and Shaw, 1968; Kennedy, 1974). Lipid-P is produced directly by polymeriza-

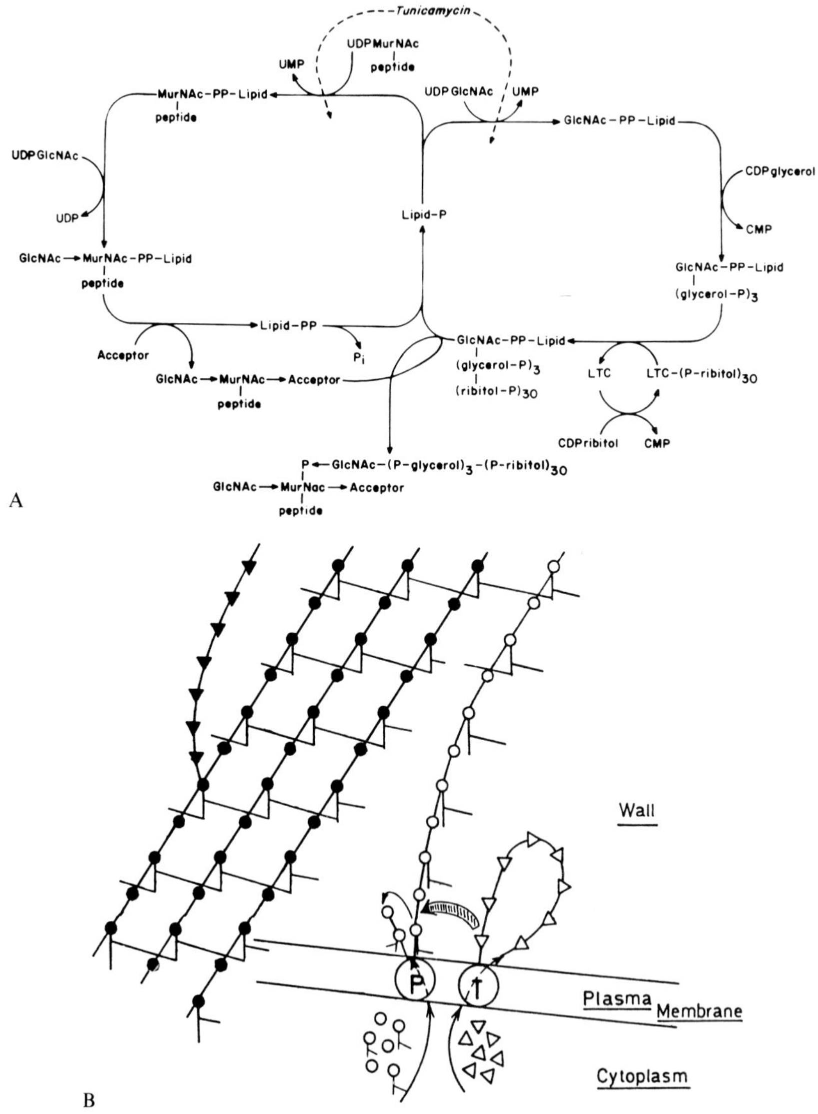

tion in the *S. lactis* I3 system, and the process is insensitive to bacitracin until UDP-MurNAc pentapeptide is added (Watkinson *et al.*, 1971). The *S. lactis* 2102 polymer is also produced by a two-step reaction: transfer of GlcNAc-P to P-lipid produces GlcNAc-P-P-lipid. The GlcNAc-P subunits are then polymerized, also at the nonreducing terminus, to form the cell wall polysaccharide (Brooks and Baddiley, 1969a,b).

Involvement of undecaprenol in all of these processes suggests translocation of the lipid-P-activated subunits to the exterior of the cytoplasmic membrane prior to polymerization. This involves transfer to the preformed "link," itself activated as an undecaprenol-PP-derivative.

3. Transfer of Nascent Polymerized Teichoic Acid to Peptidoglycan

It has recently been found that "dirty wall" preparations from *B. licheniformis* (Wyke and Ward, 1975) *S. aureus* H, *B. subtilis* W23, and *Micrococcus* strain 2103 (Bracha and Glaser, 1976a; Hancock and Baddiley, 1976) will polymerize their cell wall teichoic acids from nucleotide precursors and will transfer part of the polymeric product to peptidoglycan, rendering it insoluble in boiling 2% SDS. Pulse-labeling studies of Mauck and Glaser (1972b) in *B. subtilis* NTCC3610 had previously indicated that teichoic acid becomes linked only to coincidentally synthesized peptidoglycan. When ethanolamine-grown cells of *S. pneumoniae* are pulsed with choline, a thin band of wall at the equatorial growth zone becomes susceptible to autolysin, giving visual demonstration of the exclusive linkage of nascent teichoic acid to nascent peptidoglycan (Tomasz *et al.*, 1975) (see Section VI,F). This has also been demonstrated using phages, for which the wall teichoic acid is the receptor, to detect teichoic acid in phosphate-limited cultures of *B. subtilis*

FIG. 40. Polymerization and linkage to peptidoglycan of a wall teichoic acid. (A) Biochemistry. The link between teichoic acid-like polymers and peptidoglycan has been characterized in two structures as a 1,3-glycerol phosphate trimer, phosphodiester linked to the C-6 hydroxyl of MurNAc (see text). Biosynthetic evidence implicates undecaprenol phosphate (lipid-P) and UDP-GlcNAc in formation of this link, which may thus contain GlcNAc-1-phosphate, as shown. Earlier demonstration of *in vitro* polymerization of ribitol phosphate on a lipoteichoic acid carrier (LTC) may be artifactual. Thus, although an LTC cycle is included in this figure, ribitol phosphate may be transferred directly from CDPribitol to the link. (B) Schema of wall assembly. (From Ghuysen, 1977.) In black: pre-existing cross-linked peptidoglycan (—●—●—) with covalently attached teichoic acid (—▼—▼—). In white: nascent un-cross-linked peptidoglycan (—○—○—) and its precursors; teichoic acid in the process of assembly (—▽—▽—) and its precursors. Peptidoglycan and teichoic acid are polymerized (on a lipophilic carrier, probably undecaprenol phosphate) at their strategically located assembly centers (P and T). Peptidoglycan elongates at its point of attachment to this carrier, while teichoic acid elongates at its other end. When teichoic acid reaches a critical size, it is transferred from the carrier to nascent peptidoglycan.

pulsed with excess phosphate (Archibald and Coapes, 1976). This has also been demonstrated by the use of concanavalin A, which binds exclusively to the α-glucosylteichoic acid substituents of these walls (Anderson *et al.*, 1977).

β-Lactam antibiotics cause excretion of linear un-cross-linked peptidoglycan by autolysin-deficient cells of *B. licheniformis*, and these chains lack attached teichoic acid (Tynecka and Ward, 1975), suggesting that only cross-linked peptidoglycan is an effective acceptor. A rather precise temporal coordination of peptidoglycan synthesis and transpeptidation with teichoic acid synthesis seems to be required, as well as a spatial coordination of the two membrane assembly centers. This is illustrated in Fig. 40B, which also emphasizes the peculiar fact that growing teichoic acid must remain associated with membrane both at its initiation point (lipid-PP-"link") and at its other terminus, the site of polymerization. Wall–membrane preparations from *B. subtilis* W23 polymerize and attach polyribitol phosphate–teichoic acid independently of peptidoglycan synthesis (Ward and Wyke, personal communication).

The identification of the (glycerol phosphate)$_3$ "link" between peptidoglycan and ribitolteichoic acid in *S. aureus* H and *B. subtilis* W23 and between peptidoglycan and poly(GlcNAcl→P) in *Micrococcus* sp. 2102 led to a predicted requirement for CDPglycerol in *in vitro* linkage of these polymers to wall. This also rationalized the presence of CDPglycerol pyrophosphorylase activity in organisms, such as *S. aureus*, lacking glycerol wall teichoic acids, since no role for CDPglycerol in LTA synthesis has been demonstrable. Glycerol is incorporated from CDPglycerol into the wall concomitant with teichoic acid synthesis in "dirty wall" preparations from all three of these species. This incorporation is also dependent on the presence of UDP-GlcNAc and ATP, but does not require CDPribitol. Sequential incubations in each case indicated a requirement for UDP-GlcNAc followed by CDPglycerol. This suggests the link sequence shown in Fig. 40A. However, incorporation of GlcNAc into the link has only been demonstrated in *B. subtilis* W23. In *B. licheniformis*, UDP-GlcNAc seems to be the source of the phosphate of the muramic acid-6-P produced on acid hydrolysis of the *in vitro* product (Wyke and Ward, 1977b and personal communication). Chemical identification of GlcNAc in the link is lacking, apart from preliminary data in *S. aureus* (Tipper, unpublished observations).

The products of incubation without CDPribitol behave like undecaprenol-PP derivatives, being soluble in 70% ethanol, and the first step, which requires UDP-GlcNAc, is inhibited by tunicamycin (Bracha and Glaser, 1976b; Hancock *et al.*, 1976; Wyke and Ward, 1977a). Since this antibiotic specifically inhibits the transfer of P-GlcNAc from UDP-GlcNAc to polyprenol-P, the sequence shown in Fig. 40A is indicated. Presumably, glycosylation of the polyribitol chain occurs during its polymerization while the polymer is attached to lipid-P.

Transfer of ribitol phosphate to LTC is indicated in Fig. 40 as a possible precursor of transfer of the poly(ribitol phosphate) to the undecaprenol-PP link. However, there is no reason to suppose that this occurs *in vivo* or in the dirty wall systems described above, and it seems more likely that LTC is sn amphibolic analog of the normal undecaprenol-PP "link" acceptor that functions *in vitro*. Recent evidence suggesting that peptidoglycan precursors are transferred to LTC may reflect a similar artefact.

C. Teichuronic Acids

Teichuronic acids are cell wall-bound polysaccharides containing uronic acids, which usually alternate with hexosamine, as in the hyaluronate capsule of *S. pyogenes*. The first teichuronic acid characterized was the alternating polymer of GalNAc and glucuronic acid in walls of *B. licheniformis* NCTC6346 (see Hughes and Thurman, 1970, Table 10). This polymer, like the apparently identical teichuronic acid of *B. subtilis* strain W23 (Wright and Heckels, 1975), is produced only during phosphate-limited growth (Section VIII). In *B. subtilis*, teichuronic acid, like teichoic acid, is linked only to simultaneously synthesized peptidoglycan (Mauck and Glaser, 1972a,b). *Micrococcus lysodeikticus* (*M. luteus*) produces a teichuronic polymer of alternating residues of D-glucose and *N*-acetyl-D-mannosaminuronic acid (Perkins, 1963) even under conditions of phosphate excess. Membrane preparations from this organism synthesize this polymer from UDPglucose and UDPmannosaminuronic acid, but only in the presence of UDP-GlcNAc and a heat-stable cofactor that may be an acceptor. The product contains one GlcNAc for every 15 glucose residues, and so may have a "link" related to that identified for teichoic acids. Undecaprenol-P is involved in the formation of a lipid-linked disaccharide intermediate (Page and Anderson, 1972; Anderson *et al.*, 1972). Synthesis of lipoteichoic acid-like succinoylated mannan associated with in *M. luteus* membranes has been studied by Lennarz, who found mannosyl-P-undecaprenol to be an obligatory intermediate (see Scher *et al.*, 1968; Lennarz and Scher, 1972). This lipid intermediate is formed from undecaprenol-P and GDPmannose, and is the donor for transfer of mannosyl residues to the nonreducing terminus of an acceptor (Scher and Lennarz, 1969).

D. Functions of Teichoic Acids

It is reasonable to suppose that LTA's play an essential role in the maintenance of adequate concentrations of metal ions, Mg^{2+} in particular, in the microenvironment of the cytoplasmic membrane's outer surface. Most membrane functions and enzymes require significant Mg^{2+} concentrations. Gram-positive bacteria maintain the synthesis of LTA even under conditions of phosphate starvation which can lead to the cessation of wall teichoic acid synthesis (see Section VIII). Although only certain species of gram-positive bacteria contain wall teichoic acids, most gram-positive species do contain anionic polymers in their walls, such as the uronic acid containing polymers normally found in the walls of micrococci and which predominate in the walls of staphylococci and bacilli grown under conditions of phosphate limitation (Section VIII). It has been suggested that wall teichoic acids provide additional magnesium binding capacity for the surfaces of cells which can grow in the presence of high concentrations of NaCl. *Bacillis subtilis* grown in normal media contains a wall teichoic acid. This is replaced by teichuronic acid when the organisms are grown under conditions of phosphate limitations. If high NaCl concentrations are added to the phosphate-limited medium, however, wall teichoic acid is again produced (Ellwood and Tempest, 1972). This suggests that wall teichoic acid can be functionally replaced by teichuronic acids only when magnesium ions are readily available. While a role for teichoic acids in metal ion binding is plausible and has received considerable experimental support, it should not be forgotten that there are gram-positive species, such as *B. sphaericus*, whose walls contain no detectable phosphate or uronic acid-containing polymers (Hungerer and Tipper, 1969).

Nonpeptidoglycan wall polymers, such as teichoic and teichuronic acids and the M proteins of *S. pyogenes*, play an important role as determinants of pathogenicity. They may protect bacteria against the killing mechanisms of phagocytes or may inhibit phagocytosis (e.g., Steinmetz and Balko, 1973). They may also be the determinants of hypersensitivity reactions. Nonpeptidoglycan polymers may play an architectural role in modifying the three-dimensional structure of the peptidoglycan to which they are attached. Teichoic acids and lipoteichoic acids have been shown to control autolysin activity in several species, and most probably pervade the wall structure to perform this function. The data in *S. pneumoniae* was summarized in Section VI,F. Finally, teichoic acids, when present on cell walls in their native state form important components of certain phage and bacteriocin receptors. To the extent that phage promote genetic exchange and mechanisms for genetic exchange increase adaptability, the presence of phage receptors confers adaptive advantages on bacteria carrying them.

VIII. Effect of Environment on Cell Wall Composition

A. Peptidoglycan Structure

1. Effect of Growth Conditions

Few studies of phenotypic variations in peptidoglycan structure have been made. The most dramatic variations are found in the cross-bridge peptides of staphylococci, whose response to medium composition has been investigated by Schleifer, Hammes, and Kandler, as detailed in their recent review (Schleifer *et al.*, 1976).

As reported by Tipper *et al.* (1967b) and Tipper and Berman (1969), *S. aureus* strain Copenhagen, when grown to stationary growth phase in peptone–yeast extract–glucose medium using Difco products, has almost exclusively pentaglycyl cross-bridges with a few percent of Gly-L-Ser-Gly-Gly-Gly bridges. Cross-linkage is 80%, and N-termini are almost exclusively glycine. Similar medium prepared from Cenovis and Merke products contains much less glycine and proportionally more L-alanine and L-serine. Cells of strain Copenhagen, isolated from this medium, in logarithmic growth phase, have a similar composition to those grown in Difco medium. However, cells from late stationary growth phase in the Cenovis medium have only 60% pentaglycyl cross-bridges, and only 40–50% cross-linking, leaving 10–20% N-terminal glycine. Another 20% of the stem peptides have N-terminal lysine, and so lack any peptide cross-bridge. The residual 20% of the stem peptide carry N-terminal N^{ε}-(L-alanyl)-lysine residues. This gross deficiency in glycine addition is due to growth in limiting glycine concentrations and is abolished by addition of 0.1% glycines to the Cenovis medium (Schleifer *et al.*, 1969, 1976). Swenson and Neuhaus (1976) recently demonstrated that the L-alanyl-(N^{ε}-lysine) derivative of the normal UDP-MurNAc pentapeptide precursor of *S. aureus* cell wall peptidoglycan can be incorporated into peptidoglycan by membrane preparations of *S. aureus.* formation of this derivative in glycine-deficient media may account for the N-terminal alanine residues found in the wall under these conditions.

High glycine concentrations, such as high concentrations of D-amino acids, are lethal for many bacterial species. This is because these amino acids replace D-alanine, especially in the R_4 position of the stem peptide, and the modified stem peptide is a poor substrate for phospho-MurNAc-pentapeptide translocase, peptidoglycan polymerase, or transpeptidase (Hammes *et al.*, 1973; Trippen *et al.*, 1976). The efficiency with which these different enzymes accept the modified substrate varies from species to species, but the general effect is the same: a reduction in cross-linkage resembling the effects of β-lactam antibiotics. Two specific and relatively harmless peptidoglycan amino acid substitutions have been documented: *S. faecalis* will incorporate hydroxylysine when grown in lysine-deficient medium (Shockman *et al.*, 1965) and DAP auxotrophs of *E. coli* will incorporate lanthionine, a sulfur-containing analog, in place of *meso*-DAP (Knusel *et al.*, 1967).

It seems that environmentally induced gross variations in peptidoglycan structure are unusual, unlike the wide variations seen in the ratios of non-peptidoglycan cell wall components. An increase in cross-linkage has been reported in longitudinal walls of *E. coli* (Mirelman *et al.*, 1976) and in stationary growth phase cells (Schwarz and Leutgeb, 1971). A transitory decrease in cross-linkage of newly synthesized peptidoglycan occurs at the time of septation (Mirelman *et al.*, 1976) and correlates with a decreased cross-linkage in septal wall. Apart from this, little variation has been detected in peptidoglycan composition with growth phase.

It remains possible that septa and end walls of rod-shaped organisms differ in other perhaps subtle ways from longitudinal walls, rendering them capable of differential response to morphogenetic stimuli, but there is little evidence to support this concept. The most probable difference between septal and nonseptal walls would be in the concentration of modulators of autolysin activity, such as teichoic acids, which could control cell separation by directing localized autolysis. Such differences might be transitory.

2. Genotypic Variations

As Described in Section IV, some variations in peptidoglycan structure are relatively independent of environmental conditions and appear to be determined by the relative activity of specific synthetic enzymes. *Bifidobacterium adolescentis* and *B. globosum* have L-Lys *or* L-Orn in the L-R_3 position of their stem peptides. The ratio of these amino acids varies from strain to strain, but is relatively constant in a given strain, independent of the growth conditions. *Leuconostoc oenos* has L-Ser-L-Ala or L-Ser-L-Ser cross-bridges, in ratios ranging from strain to strain from 0.1 to 0.8. Again, the ratio is relatively constant in a given strain. *Micrococcus mucilaginosus* has single amino acid cross-bridges that may be Gly, L-Ala, or L-Ser. The relative proportions are highly strain dependent. The reason for the differential degrees of expression of these multiple and apparently redundant gene functions is unknown.

3. Variations Related to Morphogenesis

There is little information concerning variations in the wall composition of the cells and stalks of *Caulobacter* or the cells and buds of *Hyphomicrobium* or during most other cell-cycle-dependent variations in wall structure in bacteria. Changes were noted between rod and sphere forms of *Arthrobacter crystallopoites* ATCC 15481 (Krulwich *et al.*, 1967a,b), but these could be characteristic of the cell types, or a consequence of the different media used to grow the two forms. The rods had longer glycan chains and exclusively L-Ala cross-bridges. The spheres had a few percent of L-Ala-Gly-Gly cross-bridges. Cells growing in media supporting rod-shaped growth revert to spheres in stationary growth phase. Cells from this stage had a wall composition resembling logarithmic-phase rod-shaped cells (Schleifer *et al.*, 1976). Previc and Lowell (1975) reported that a "fast growing mutant" of this same strain of *A. crystallopoites* has diaminopimelate in its rod-shaped walls. This was not observed in earlier studies (Krulwich *et al.*, 1967a,b) and is both unexpected and unexplained.

Myxococcus xanthus cells can be induced to encyst synchronously in suspension culture by the addition of glycerol and other agents. The morphogenesis is accompanied by an increase in peptidoglycan cross-linkage from 50 to 70% (White *et al.*, 1968; Johnson and White, 1972) as well as the synthesis of a new wall polymer rich in GalNAc and glycine.

The synthesis of spore cortex is unique in that it requires expression of several new genes whose functions are not required for vegetative growth (see Section IV) (Tipper and Gauthier, 1972). While initiation of sporulation is controlled by nutrition, these sporulation-specific peptidoglycan synthetic events are late parts of a more complex and linearly dependent chain of transcriptionally controlled events that, once past its early stages, proceeds relatively independent of nutritional modifications. In *B. sphaericus* the sporulation-specific synthesis of UDP-MurNAc-L-Ala-D-Glu-:-*meso*-DAP ligase is a consequence of differential expression of the sporangial genome: the forespore genome expresses the L-Lys ligase required for synthesis of vegetative cell wall peptidoglycan (Tipper and Linnett, 1976). The data on cortex synthesis and related events in *B. sphaericus* has been summarized (Tipper *et al.*, 1977).

B. Nonpeptidoglycan Components

1. Gram-Negative Bacteria

Alteration of the genome of gram-negative bacteria by lysogenization or acquisition of plasmids is frequently associated with alterations in outer membrane composition resulting in modification of phage and bacteriocin receptors. Some outer membrane components, such as the maltase and λ receptor, are inducible. This is also true for components involved in the uptake of citrate and enterochelin (see Section IX). Systematic studies of the effect of environment on cell wall structure in gram-negative organisms have not been described.

2. GRAM-POSITIVE BACTERIA

Changes in the growth environment of gram-positive organisms can cause drastic changes in cell wall structure by grossly altering the ratio of the different cell wall polymers present. These effects are best studied in chemostat cultures and have been intensively investigated by Ellwood and Tempest in gram-positive bacilli and cocci as outlined in their review (Ellwood and Tempest, 1972). The cell wall composition is dependent on growth rate, temperature, pH, and the nature of the growth-limiting substituent. Since most of these parameters vary during standard batch culture growth, the cell wall composition of a bacterial population produced in this way may reflect the consequences of a series of adaptions to changing growth conditions and will be dependent upon the growth phase at the time of harvesting.

In gram-positive bacilli the ratio of cell wall weight to total cell dry weight decreases as the cells grow faster and increase in average cell volume. In contrast, little effect of growth rate is seen in staphylococci or micrococci. The ratio of peptidoglycan to teichoic acids in the bacillus cell walls varies with the growth rate in an unpredictable fashion (Ellwood and Tempest, 1972) (Elwood, 1975).

a. Effects of Specific Nutrient Limitations. Growth rate in chemostat culture can be controlled by dilution rate and by restricting the concentration of one essential nutrient. When growth of *B. subtilis* var. *niger* is limited by glucose, ammonia, sulfate, or K^+ concentration (at 35°C, pH 7), teichoic acids account for 40–50% of the cell wall dry weight and no detectable teichuronic acid is present. In Mg^{2+}-limited culture, the teichoic acid content increases to 65–75%, but in phosphate-limited culture it drops to less than 3%, and large quantities of teichuronic acid are produced. Thus, the most dramatic effects of alteration in nutrient limitation is seen on switching from Mg^{2+} to phosphate-limited growth, and the effects of this switch have been studied most extensively. Similar effects are seen in a variety of other gram-positive organisms that normally contain wall teichoic acids (see Table IX). The organic phosphate contents of walls of *B. subtilis* strains 168, W23, and niger; *B. licheniformis* 6346; and *S. aureus* strain H are all in the range of 2.5–5% in broth-grown cells (indicating 30–60% teichoic acid contents) and somewhat higher in cells grown under Mg^{2+} limitation. This drops to 0.1–0.5% in phosphate-limited growth, when teichuronic acids are synthesized in large amounts. This effect is only marked in *B. subtilis* W23 if it is grown at low dilution rates in phosphate-limiting conditions. At high dilution rates, the phosphate in this medium is not exhausted and teichoic acids are synthesized. In broth culture, teichuronic acids are absent from walls of all these organisms with the exception of *B. licheniformis*. Because of this fact, a teichuronic acid was first isolated and characterized from this orga-

TABLE IX

PRINCIPAL ANIONIC POLYMERS LINKED TO PEPTIDOGLYCAN WHEN GRAM-POSITIVE BACTERIA ARE GROWN UNDER THE CONDITIONS SHOWN

	Growth conditions		
Organisms	Mg^{2+} Limited	Broth	Phosphate Limited
B. subtilis W23	1,5-Poly(ribitol phosphate), glucosylated		Teichuronic acid $\rightarrow$(4-GlcUA-α $\rightarrow$3-GalNAcα)$_n$-
B. subtilis 168	1,3-Poly(glycerol phosphate) (Glc$\beta \rightarrow$3-GalNAc$\rightarrow$P)$_n$		Teichuronic acid (GlcUA-GalNAc)$_n$
B. subtilis var. *niger*	2,3-Poly(glycerol phosphate), glucosylated poly(galactosylglycerol phosphate)		Teichuronic acid (GlcUA-GalNAc)$_n$
B. licheniformis NCTC6346	Poly(glucosylglycerol phosphate), poly(galactosylglycerol phosphate), 1,3-poly(glycerol phosphate), glucosylated		Teichuronic acid $\rightarrow$(4-GlcUA-α $\rightarrow$3-GalNAcα)$_n$-
Staphylococcus aureus	1,5-Poly(ribitol phosphate)-*O*-GlcNAc		(Glucosaminuronic acid-GlcNAc)$_n$
M. luteus (*M. licheniformis*)	1,3-Poly(glycerol phosphate)-2-*O*-GlcNAc	(Glc$\rightarrow$mannosaminuronic acid)$_n$	

nism (Hughes and Thurman, 1970). The characterization of the teichuronic acid synthesized by *B. subtilis* W23 required growth in a chemostat at low dilution rate under phosphate limitation (Wright and Heckels, 1975). The two polymers appear to be identical in structure, so that these two organisms contain similar genes, although their expression is more severely repressed under conditions of phosphate excess in *B. subtilis* W23. Analytical data indicate that the teichuronic acid found in *B. subtilis* var. *niger* is also very similar. *Staphylococcus aureus*, *B. megaterium*, and *M. luteus* contain very different teichuronic acids which contain aminohexuronic acids.

Micrococcus luteus behaves in a different fashion (Ellwood, 1975). It does not synthesize significant amounts of teichoic acid when grown in broth or under phosphate limitations; teichoic acid is only seen in cells grown under Mg^{2+} limitation. Its synthesis appears to be repressed in the presence of adequate amounts of Mg^{2+}.

In general, all of these organisms behave as predicted if one assumes that a major function of wall teichoic acids is to trap Mg^{2+} when its supply is limited. Thus when a culture of *B. subtilis* W23, growing under phosphate limitation at pH 8, was shifted gradually to pH 5, teichoic acid reappeared as a major cell wall component at pH 5.5 and disappeared again when the pH was increased to 6. Possibly ionization and Mg^{2+} binding capacity of the teichuronic acid was suppressed at pH 5.5, requiring teichoic acid synthesis. Increasing the NaCl concentration of phosphate-limited cultures of *B. subtilis* var. *niger* had similar effects: presumably competition by Na^+ with Mg^{2+} for cell-wall-binding sites reduces the effective Mg^{2+} concentration. It is hypothesized (Ellwood and Tempest, 1972) that the precursors of teichoic acids, CDPglycerol and CDPribitol, inhibit synthesis of teichuronic acid presursors (e.g., UDP-GalNAc) in organisms such as *B. subtilis* (but not in *M. luteus*). One would have to hypothesize further that synthesis of CDPglycerol is selectively diminished at low phosphate concentrations, except, perhaps, at low Mg^{2+} concentrations.

b. Turnover. When cells of *B. subtilis* W23 are transferred from phosphate- to potassium-limited growth conditions, teichuronic acid synthesis stops and teichoic acid synthesis starts. Transfer back to phosphate-limited growth has the reverse effect. During the transition, the preexisting polymer is not merely diluted, but is actively excised by autolytic degradation of the peptidoglycan to which it is attached, so that it is excreted in soluble form into the medium (Mauck *et al.*, 1971). Ellwood and Tempest (1972) failed to find similar turnover effects in steady-state chemostat cultures of *B. subtilis* W23, but did find turnover in *B. subtilis* var. *niger* on switch from Mg^{2+} to phosphate-limited growth. Not all bacteria turnover their cell wall, and the half-life of newly incorporated cell wall components in *B. subtilis*, roughly equal to the generation time, is unusually short.

Archibald (1976) has exploited the rapidity of teichoic acid insertion in *B. subtilis* W23 in investigating the kinetics of appearances of newly incorporated teichoic acid at the cell surface. Phage SP50, which employs the teichoic acid as receptor and concanavalin A which also binds to it (Anderson *et al.*, 1977) have both been used for its detection. After addition of a pulse of excess phosphate to a phosphate-limited culture, there was a lag of half a generation before maximum binding developed and before turnover of the incorporated teichoic acid began. It was deduced that insertion of new peptidoglycan and covalently attached wall polymers occurs at the inner surface of the wall, while autolytic turnover occurs at the exterior surface. It takes half a generation in *B. subtilis* for wall components to reach the surface. Phage binding (and therefore incorporation and turnover) occurred with faster kinetics on longitudinal than on end walls. Pulse labeling data indicate that a similar process occurs in *B. subtilis* 168 (Pooley, 1976).

c. Effects of Mutations. Mutants of *M. luteus* that fail to synthesize teichuronic acid (Yamada *et al.*, 1975) do not replace this polymer with teichoic acid. Growth of the mutants requires abnormally high Mg^{2+} concentrations and results in the formation of regular cell packets. The high Mg^{2+} requirement is consistent with a role for teichuronic acid in binding of Mg^{2+}, and packet formation may be a consequence of absence of normal autolysin activation by teichuronic acid. Cells of *S. aureus* with defects in autolysin, induced by detergent treatment or mutation, also grow as regular packets (Koyama *et al.*, 1977) (See Section III).

The lesion in shape determination in certain *rod* mutants in *E. coli*, mutants whose cells are permenantly or conditionally spherical rather than rod shaped, lies in the PBC 3 structural gene. Since PBC 3 binds penicillins, it is probably an enzyme using acyl-D-alanyl-D-alanine as substrate, perhaps a specialized transpeptidase (see Section VI,E). *Rod* mutants have been extensively studied in *B. subtilis* 168 and *B. licheniformis* 6346 by Rogers and coworkers (Rogers, 1977). Genetic studies of the mutants in *B. subtilis* have identified four groups, A, B, C and D, which appear to be point mutations in spite of their pleiotropic effects (Rogers, 1977). Certain of these mutants are temperature-sensitive (Rogers and Thurman, 1978).

Bacillus licheniformis mutants deficient in phosphoglucomutase or UDPglucose pyrophosphorylase cannot make UDPglucose or UDPglucuronic acid, so can neither glucosylate their teichoic acid nor make teichuronic acid. When growth is limited by phosphate concentration, cell wall turnover removes the cell wall teichoic acid, but it is not replaced by teichuronic acid; the cells become round. Addition of glycerol and galactose allows synthesis of poly(galactosylglycerol phosphate), and the morphology is corrected (Forsberg *et al.*, 1973). A *B. subtilis* mutant defective in CDPglycerol pyrophosphorylase (*rodD*) makes walls devoid of teichoic acid and has round morphology (Rogers, 1977).

In temperature-sensitive *rodA* mutants, shift to nonpermissive temperature causes the loss of 80% of the glycerolteichoic acid and 96% of the Glc-GalNAc polymer (Boylan *et al.*, 1972), while peptidoglycan content increases four- to five-fold and the cells become round. A $rodA_{ts}$, phosphoglucomutase double mutant has the *ts rod* phenotype when grown in phosphate excess. "Dirty wall" preparations from this strain, grown at 30°C (rod-shaped cells) will synthesis the Glc-GalNAc polymer from the appropriate nucleotides, while similar preparations from round cells grown at 45°C could not. The *rodA* locus, thus, appears to control polymerization of the Glc-GalNAc polymer (and presumably also glycerolteichoic acid) from its precursors (Hayes *et al.*, 1977).

In summary, a variety of defects causing loss of the bulk of the anionic polymers from *B. subtilis* cell walls results in round morphology. This might be a consequence of reduced Mg^{2+} uptake or of improper autolysin control (activation defect?) because of the absence of the wall anionic polymers (see

Section VIII). While not conclusive, the bulk of the evidence suggests that reduced Mg^{2+} uptake is the common element in *rod* mutants (Rogers, 1977).

The phosphoglucomutase and UDPglucose pyrophosphorylase mutants are grossly defective in MurNAc-L-Ala amidase activity, the major autolysin in *B. subtilis*, and round cells of $rodA_{ts}$ mutants grown at 45°C fail to autolyze. However, a *B. subtilis* mutant producing only 5% of normal amidase activity grows as rods, even under phosphate limitation (Rogers, 1977). Moreover, *rodB* mutants autolyze normally. The *rodB* phenotype is suppressed by high Mg^{2+} concentrations in the presence of specific nutrients, of which L-glutamate appears to be the most effective. High Mg^{2+} concentrations also corrected the rod phenotype of the *rodD* mutant, but not *rodA* mutants (Rogers, 1977).

IX. Conclusion

Current structural investigations of bacterial cell walls are mostly focused on the gram-negative outer membrane, as described in Chapter 7. The highly sophisticated genetics and biochemistry of the Enterobacteriaceae can be exploited in investigating the role of their outer membrane components in such functions as phage adsorption, transport, and chemotaxis, but information in other genera, including important human pathogens such as *Neisseria* and *Pseudomonas*, is largely inferrential. The available data illustrates the fundamental differences in strategy between gram-negative and gram-positive bacteria in their adaption to environmental stress, as mentioned in Section III.

This chapter has emphasized the recent spate of publications on the interaction of β-lactam antibiotics with sensitive enzymes. These investigations should soon result in identifying the polypeptide domains interacting with both substrates and inhibitors in several specific DD-carboxypeptidases and β-lactamases. These data, extrapolated to the transpeptidases inhibited *in vivo*, will give a clear understanding of an important step in the interaction of β-lactam antibiotics with bacteria. The combination of genetic and biochemical investigations of *E. coli* PBC's should result in dissection of the enzymatic functions of the multiple β-lactam antibiotic targets of gram-negative bacteria. Less is known about the functional complexity of gram-positive PBC's. It may be much longer before we understand how these interactions result in autolysin activation and cell death in any species. We are just as far from understanding how β-lactam antibiotics arrest macromolecular synthesis in tolerant bacteria.

General conclusions, relevant to both this and Chapter 7 will be found at the end of Chapter 7.

Acknowledgments

We would like to thank Drs. L. Daneo-Moore, J.-M. Ghuysen, G. D. Shockman, and J. B. Ward for sending us manuscripts of their recent reviews. We would also like to thank Dr. J. B. Ward for helpful discussions and comments. The preparation of this review was supported by Grants AI 10806 and GM 15837, awarded to D. J. T. and A. W., respectively, by the National Institutes of Health.

References

Aebi, U., Smith, P. R., Dubochet, J., Henry, C., and Kellenberger, E. (1973). *J. Supramol. Struct.* **1,** 498–522.

Anderson, A. J., and Archibald, A. R. (1975). *Biochem. J.* **151,** 115–120.

Anderson, A. J., Green, R. S., and Archibald, A. R. (1977). *Proc. Soc. Gen. Microbiol.* p. 85.

Anderson, J. S., Matsuhashi, M., Haskin, M. A., and Strominger, J. L. (1965). *Proc. Natl. Acad. Sci. U.S.A.* **53,** 881–889.

Anderson, J. S., Matsuhashi, M., Hoskin, M. A., and Strominger, J. L. (1967). *J. Biol. Chem.* **242,** 3180.

Anderson, J. S., Page, R. L., and Salo, W. L. (1972). *J. Biol. Chem.* **247,** 2480–2485.

Anderson, R. G., Douglas, L. J., Hussey, H., and Baddiley, J. (1973a). *Biochem. J.* **136,** 871–876.

Anderson, R. G., Hussey, H., and Baddiley, J. (1973b). *Biochem. J.* **127,** 11–25.

Aoki, M., Sakai, H., Kohsaka, M., Konomi, T., Hosoda, J., Kubochi, Y., Iguchi, E., and Imanaka, H. (1976). *J. Antibiot.* **29,** 492.

Araki, Y., Shimada, A., and Ito, E. (1966). *Biochem. Biophys. Res. Commun.* **23,** 518–525.

Archibald, A. R. (1974). *Adv. Microbiol. Physiol.* **10,** 53–95.

Archibald, A. R. (1976). *J. Bacteriol.* **127,** 956–960.

Archibald, A. R., and Coapes, H. E. (1971). *Biochem. J.* **124,** 449.

Archibald, A. R., and Coapes, H. E. (1976). *J. Bacteriol.* **125,** 1195.

Archibald, A. R., and Stafford, G. H. (1972). *Biochem. J.* **130,** 681–690.

Archibald, A. R., Baddiley, J., Heckels, J. E., and Heptinstall, S. (1971). *Biochem. J.* **125,** 353–359.

Arminjon, F., Guinand, M., Vacheron, M. J., and Michel, E. (1977). *Eur. J. Biochem.* **73,** 557–565.

Armstrong, J. J., Baddiley, J., and Buchanan, J. G. (1961). *Biochem. J.* **80,** 254–261.

Ascoli, F., DeAngelis, G., DelBianco, F., and DeSantis, P. (1975). *Biopolymers* **14,** 1109–1114.

Azuma, I., Thomas, D. W., Adam, A., Ghuysen, J. M., Bonaly, R., Petit, J. F., and Lederer, E. (1970). *Biochim. Biophys. Acta* **208,** 444–451.

Baddiley, J. (1972). *Eassays Biochem.* **8,** 35–77.

Baddiley, J., and Neuhaus, F. C. (1960). *Biochem. J.* **75,** 579–587.

Balyuzi, H. H. M., Reavely, D. A., and Burge, R. E. (1972). *Nature* (*London*) *New Biol.* **235,** 252.

Bartholemew, J. W., and Mittwer, T. (1952). *Bacteriol. Rev.* **16,** 1.

Basinger, S. F., and Jackson, R. W. (1968). *J. Bacteriol.* **96,** 1895.

Bayer, M. E. (1974). *Ann. N.Y. Acad. Sci.* **235,** 6–28.

Bayer, M. E. (1975). *In* "Membrane Biogenesis" (A. Tzagoloff, ed.), pp. 393–427. Plenum, New York.

Beck, B. D., and Park, J. T. (1976). *J. Bacteriol.* **126,** 1250–1260.

Beck, B. D., and Park, J. T. (1977). *J. Bacteriol.* **130,** 1292–1302.

Begg, K. J., and Donachie, W. D. (1973). *Nature* (*London*), *New Biol.* **245,** 38–39.

Begg, K. J., and Donachie, W. D. (1977). *J. Bacteriol.* **129,** 1524–1536.

Best, G. K., Best, N. H., and Koral, A. V. (1974). *Antimicrob. Agents Chemother.* **6,** 825.
Bettinger, G. E., and Young, F. E. (1975). *Biochem. Biophys. Res. Commun.* **67,** 16–21.
Bews, B. (1967). Ph.D. Thesis, Univ. of Newcastle upon Tyne, *Newcastle upon Tyne, England.*
Blumberg, P. M., and Strominger, J. L. (1972a). *Proc. Natl. Acad. Sci. U.S.A.* **69,** 3751–3755.
Blumberg, P. M., and Strominger, J. L. (1972b). *J. Biol. Chem.* **247,** 8107–8113.
Blumberg, P. M., and Strominger, J. L. (1974). *Bacteriol. Rev.* **38,** 291–315.
Blumberg, P. M., Yocum, R. R., Willoughby, E., and Strominger, J. L. (1974). *J. Biol. Chem.* **249,** 6828–6835.
Bogdanovsky, D., Bricas, E., and Dezelee, P. (1969). *C. R. Acad. Sci., Ser. D* **269,** 390–393.
Boylan, R. J., Mendelson, N. H., Brooks, D., and Young, F. E. (1972). *J. Bacteriol.* **110,** 281–290.
Bracha, R., and Glaser, L. (1967a). *J. Bacteriol.* **125,** 872–879.
Bracha, R., and Glaser, L. (1967b). *Biochem. Biophys. Res. Commun.* **72,** 1091.
Braun, V., and Hantke, K. (1974). *Annu. Rev. Biochem.* **43,** 89–121.
Braun, V., and Wolff, H. (1975). *J. Bacteriol.* **123,** 888–897.
Braun, V., Schaller, K., and Wolff, H. (1973). *Biochim. Biophys. Acta* **323,** 87–97.
Braun, V., Bosch, V., Hantke, K., and Schaller, K. (1974). *Ann. N.Y. Acad. Sci.* **235,** 66–82.
Briles, E. B., and Tomasz, A. (1973). *J. Biol. Chem.* **248,** 6394–6397.
Briles, E. B., and Tomasz, A. *J. Bacteriol.* **122,** 335–337, 1975.
Brinton, C. C., McNary, J. C., and Carnahan, J. (1969). *Bacteriol. Proc.* p. 48.
Brooks, D., and Baddiley, J. (1969a). *Biochem. J.* **113,** 635–642.
Brooks, D., and Baddiley, J. (1969b). *Biochem. J.* **115,** 307–314.
Brooks, D., Mays, L., Hatefi, Y., and Young, F. E. (1971). *J. Bacteriol.* **107,** 223–229.
Buchanan, C. E. (1977). *In* "Microbiology—1977" (D. Schlessinger, ed.), pp. 191–194. Am. Soc. Microbiol., Washington, D.C.
Buchanan, C. E., and Strominger, J. L. (1976). *Proc. Natl. Acad. Sci., U.S.A.* **73,** 1816–1820.
Bumstead, R. M., Dahl, J. L., Soll, D., and Strominger, J. L. (1968). *J. Biol. Chem.* **243,** 779–782.
Burdett, I. D. J., and Murray, R. G. E. (1975). *J. Bacteriol.* **119,** 1039–1056.
Burger, M. M. (1966). *Proc. Natl. Acad. Sci. U.S.A.* **56,** 910–917.
Burger, M. M., and Glaser, L. (1964). *J. Biol. Chem.* **239,** 3168.
Burger, M. M., and Glaser, L. (1966). *J. Biol. Chem.* **241,** 494–506.
Button, D., Archibald, A. R., and Baddiley, J. (1966). *Biochem. J.* **99,** 11c–14c.
Campbell, J. N., Leyh-Bouille, M., and Ghuysen, J. M. (1969). *Biochemistry* **8,** 193–200.
Carlström, D. (1957). *J. Biophys. Biochem. Cytol.* **3,** 669.
Chatterjee, A. N., and Park, J. T. (1964). *Proc. Natl. Acad. Sci. U.S.A.* **51,** 9–16.
Chatterjee, A. N., Wong, W., Young, F. F., and Gilpin, P. W. (1976). *J. Bacteriol.* **125,** 961–967.
Cheng, K. J., and Costerton, W. (1977). *J. Bacteriol.* **129,** 1506–1512.
Chin, T., Burger, M. M., and Glaser, L. (1966). *Arch. Biochem. Biophys.* **116,** 358–367.
Chin, T., Younger, J., and Glaser, L. (1968). *J. Bacteriol.* **95,** 2044.
Christenson, J. G., Gross, S. K., and Robbins, P. W. (1969). *J. Biol. Chem.* **244,** 5436–5439.
Churchward, G., and Holland, J. B. (1976). *J. Mol. Biol.* **105,** 245–261.
Cleveland, R. F., Holtje, J. V., Wicken, A. J., Tomasz, A., Daneo-Moore, L., and Shockman, G. D. (1975). *Biochem. Biophys. Res. Commun.* **67,** 1128–1135.
Coley, J., Duckworth, M., and Baddiley, J. (1972). *J. Gen. Microbiol.* **73,** 587–591.
Coley, J., Duckworth, M., and Baddiley, J. (1975). *Carbohydr. Res.* **40,** 41–52.
Coley, J., Archibald, A. R., and Baddiley, J. (1976). *FEBS Lett.* **61,** 240–243.
Coley, J., Archibald, A. R., and Baddiley, J. (1977). *Proc. Soc. Gen. Microbiol.* p. 84.
Cooper, P. D. (1956). *Bacteriol. Rev.* **20,** 28–48.
Coyette, J., Perkins, H. R., Polacheck, I., Shockman, G. D., and Ghuysen, J. M. (1974). *Eur. J. Biochem.* **44,** 459–468.

Cummins, C. S., and Harris, H. (1956). *J. Gen. Microbiol.* **14,** 583–600.
Curtis, N. A., Hughes, J. M., and Ross, G. W. (1976). *Antimicrob. Agents Chemother.* **9,** 208–213.
Daneo-Moore, L., and Shockman, G. D. (1977). *In* "Membrane Assembly and Turnover" (G. Poste, ed.), Cell Surface Reviews, Vol. 4, pp. 597–716. Elsevier, Amsterdam.
Davie, J. M., and Brock, T. (1966). *J. Bacteriol.* **92,** 1623.
DeBoer, W. R., Kruyssen, F. J., and Wouters, T. M. (1976). *Eur. J. Biochem.* **62,** 1–6.
DePamphilis, M. (1971). *J. Virol.* **1,** 683–686.
Diven, W. F. (1969). *Biochim. Biophys. Acta* **191,** 702–706.
Douglas, L. J., and Baddiley, J. (1968). *FEBS Lett.* **1,** 114–116.
Duckworth, M., Archibald, A. R., and Baddiley, J. (1972). *Biochem. J.* **130,** 691–696.
Duckworth, M., Archibald, A. R., and Baddiley, J. (1975). *FEBS Lett.* **53,** 176–179.
Dusart, J., Marquet, A., Ghuysen, J. M., Frere, J. M., Moreno, R., Leyh-Bouille, M., Johnson, K. Lucchi, C. H., Perkins, H. R., and Nieto, M. (1973). *Antimicrob. Agents Chemother.* **3,** 181–187.
Dusart, J., Marquet, A., Ghuysen, J. M., and Perkins, H. R. (1975). *Eur. J. Biochem.* **56,** 57–65.
Edwards, J. R., and Park, J. T. (1969). *J. Bacteriol.* **99,** 459–462.
Egan, A., Lawrence, P., and Strominger, J. L. (1973). *J. Biol. Chem.* **248,** 3122–3130.
Elliott, T. S. J., Ward, J. B., and Rogers, H. J. (1975a). *J. Bacteriol.* **124,** 623–632.
Elliott, T. S. J., Ward, J. B., Wyrick, P. B., and Rogers, H. J. (1975b). *J. Bacteriol.* **124,** 905–917.
Ellwood, D. C. (1975). *Proc. Soc. Gen. Microbiol.* **3,** 16–17.
Ellwood, D. C., and Tempest, D. W. (1972). *Adv. Microb. Physiol.* **7,** 83–116.
Elwell, L. P., Roberts, M., Mayer, L. W., and Falkow, S. (1977). *Antimicrob. Agents Chemother.* **11,** 528–533.
Emdur, L. I., and Chiu, T. M. (1975). *FEBS Lett.* **55,** 216–219.
Friedler, F., and Glaser, L. (1974a). *J. Biol. Chem.* **249,** 2684–2689.
Fiedler, F., and Glaser, L. (1974b). *J. Biol. Chem.* **249,** 2690–2695.
Fiedler, F., Schleifer, K. H., Cziharz, B., Interschick, E., and Kandler, O. (1970). *Spisy Prirodoved. Fak. Univ. J. E. Purkyne Brne* **47,** 111–122.
Filip, C., Fletcher, G., Wilff, J. L., and Earhart, C. F. (1973). *J. Bacteriol.* **115,** 717–722.
Fitz-James, P. (1974). *Ann. N.Y. Acad. Sci.* **235,** 345–346.
Fletcher, G., Irwin, C. A., Henson, J. M., Fillingim, C., Malone, M. M., Walker, J. R. *J. Bacteriol.* **133,** 91–100, 1978.
Formanek, H., Formanek, S., and Wawra, H. (1974). *Eur. J. Biochem.* **46,** 279–294.
Forsberg, C. W., and Rogers, H. J. (1971). *Nature* (*London*) **229,** 227.
Forsberg, C. W., Wyrick, P. B., Ward, J. B., and Rogers, H. J. (1973). *J. Bacteriol.* **113,** 969–984.
Fox, G. E., Magrum, L. J., Balch, W. E., Wolfe, R. S., and Woese, C. R. (1977). *Proc. Natl. Acad. Sci. U.S.A.* **74,** 4537.
Frère, J.-M., and Ghuysen, J.-M. (1976). *FEBS Lett.* **63,** 112–116.
Frère, J.-M., Ghuysen, J.-M., Degelaen, J., Loffet, A., and Perkins, H. R. (1975). *Nature* (*London*) **258,** 168–170.
Frère, J.-M., Ghuysen, J.-M., and Perkins, H. P. (1975b). *Eur. J. Biochem.* **57,** 353–359.
Frère, J.-M., Duez, C., Ghuysen, J.-M., and Vandekerhove, J. (1976a). *FEBS Lett.* **70,** 257–260.
Frère, J.-M., Ghuysen, J.-M., Vanderhaeghe, H., Adriaens, P., Degelaen, J. and DeGraeve, J. (1976b). *Nature* (*London*) **260,** 451–454.
Fuchs-Cleveland, E., and Gilvarg, C. (1976). *Proc. Natl. Acad. Sci. U.S.A.* **73,** 6200–6204.
Ganfield, M. C. W., and Pieringer, R. A. (1975). *Eur. J. Biochem.* **24,** 116–122.
Georgapapadakou, N., Hammarstrom, S., and Strominger, J. L. (1977). *Proc. Natl. Acad. Sci. U.S.A.* **74,** 1009–1012.
Ghuysen, J.-M., and Strominger, J. L. (1963). *Biochemistry* **2,** 1119–1125.

Ghuysen, J.-M. (1968). *Bacteriol. Rev.* **33,** 425–464.
Ghuysen, J.-M. (1976). "E. R. Squibb Lectures on Chemistry and Microbial Products" (W. E. Brown, ed.), pp. 1–164. Univ. of Tokyo Press, Tokyo.
Ghuysen, J.-M. (1977). *In* "Membrane Assembly and Turnover" (G. Poste, ed.), Cell Surface Reviews, Vol. 4, pp. 463–596. Elsevier, Amsterdam.
Ghuysen, J.-M., and Strominger, J. L. (1963).
Ghuysen, J.-M., and Shockman, G. D. (1973). *In* "Bacterial Membranes and Walls" (L. Leive, ed.), Vol. 1, pp. 37–130. Dekker, New York.
Ghuysen, J.-M., Tipper, D. J., and Strominger, J. L. (1965a). *Biochemistry* **4,** 474–485.
Ghuysen, J.-M., Tipper, D. J., Birge, C. H., and Strominger, J. L. (1965b). *Biochemistry* **4,** 2244.
Ghuysen, J.-M., Bricas, E., Lache, M., and Leyh-Bouille, M. (1968a). *Biochemistry* **7,** 1450–1460.
Ghuysen, J.-M., Strominger, J. L., and Tipper, D. J. (1968b). *In* "Comprehensive Biochemistry" (M. Florkin and E. H. Stotz, eds.), Vol. 26A, pp. 53–104. Am. Elsevier, New York.
Ghuysen, J.-M., Leyh-Bouille, M., Bonaly, R., Nieto, M., Perkins, H. R., Schleifer, K. H., and Kandler, O. (1970). *Biochemistry* **9,** 2955–2960.
Ghuysen, J.-M., Leyh-Bouille, M., Frère, J.-M., Dusart, J., Johnson, K., Nakel, M., Coyette, J. Perkins, H. R., and Nieto, M. (1972). *In* "Molecular Mechanisms of Antibiotic Action on Protein Biosynthesis and Membranes" (E. Munoz, F. Garcia-Ferrandiz, and D. Vazquez, eds.), pp. 406–426. Elsevier, Amsterdam.
Ghuysen, J.-M., Leyh-Bouville, M., Campbell, J. N., Moreno, R., Frère, J. M., Duez, C., Nieto, M., and Perkins, H. R. (1973). *Biochemistry* **12,** 1243–1251.
Ghuysen, J.-M., Leyh-Bouille, M., Frère, J.-M., Dusart, J., Marquet, A., Perkins, H. R., and Nieto, M. (1974). *Ann. N.Y. Acad. Sci.* **235,** 236–266.
Ghuysen, J.-M., Frère, J.-M., Leyh-Bouille, M., Dusart, J., Nguyen-Disteche, M., Coyette, J., Marquet, A., Perkins, H. R., and Nieto, M. (1975). *Bull. Inst. Pasteur, Paris* **73,** 101–140.
Glaser, L. (1963). *Biochim. Biophys. Acta* **67,** 525.
Glaser, L. (1964). *J. Biol. Chem.* **239,** 3178–3186.
Glaser, L. (1973). *Annu. Rev. Biochem.* **42,** 91–112.
Glaser, L., and Lindsay, B. (1974). *Biochem. Biophys. Res. Commun.* **59,** 1131–1136.
Glauert, A. M., and Thornley, M. J. (1969). *Annu. Rev. Microbiol.* **23,** 1525.
Gmeiner, J. (1975). *Eur. J. Biochem.* **58,** 627–629.
Goldman, R., and Strominger, J. L. (1972). *J. Biol. Chem.* **247,** 5116–5122.
Good, C. M., and Tipper, D. J. (1972). *J. Bacteriol.* **111,** 231–241.
Goodwin, S. D., and Schedlarski, J. G. (1975). *Arch. Biochem. Biophys.* **170,** 23–36.
Gorman, M., and Huber, F. (1977). *Annu. Rep. Ferment. Processes* **1,** 327–346.
Gorman, M., and Ryan, C. W. (1972). *In* "Cephalosporins and Penicillins: Chemistry and Biology" (E. M. Flynn, ed.), p. 540. Academic Press, New York.
Gould, G. W., and Dring, G. J. (1974). *Adv. Microb. Physiol.* **11,** 137–164.
Gould, G. W., and Measures, J. C. (1977). *Philos. Trans. R. Soc. London, Ser. B* **278,** 151–166.
Guinand, M., Ghuysen, J.-M., Schleifer, K. H., and Kandler, O. (1969). *Biochemistry* **8,** 200–206.
Guinand, M., Michel, G., and Tipper, D. J. (1974). *J. Bacteriol.* **120,** 173–184.
Guinand, M., Michel, G., and Balassa, G. (1976). *Biochem. Biophys. Res. Commun.* **68,** 1287–1293.
Gunetileke, K. C., and Anwar, R. A. (1968). *J. Biol. Chem.* **243,** 5770–5778.
Guze, L. B. (1968). "Microbial Protoplasts, Spheroplasts and L-Forms." Williams & Williams, Baltimore, Maryland.
Haller, I., and Henning, U. (1974). *Proc. Natl. Acad. Sci. U.S.A.* **71,** 2018–2021.

Hammarstrom, S., and Strominger, J. L. (1975). *Proc. Natl. Acad. Sci. U.S.A.* **72,** 3463–3467.
Hammarstrom, S., and Strominger, J. L. (1976). *J. Biol. Chem.* **251,** 7947–7949.
Hammes, W. P. (1976). *Eur. J. Biochem.* **70,** 107–113.
Hammes, W. P., and Kandler, O. (1976). *Eur. J. Biochem.* **70,** 97–106.
Hammes, W. P., and Neuhaus, F. C. (1974a). *J. Biol. Chem.* **249,** 3140–3150.
Hammes, W. P., and Neuhaus, F. C. (1974b). *Antimicrob. Agents Chemother.* **6,** 722–728.
Hammes, W. P., and Neuhaus, F. C. (1974c). *J. Bacteriol.* **120,** 210–218.
Hammes, W., Schleiffer, K. H., and Kandler, O. (1973). *J. Bacteriol.* **116,** 1029.
Hancock, I. C., and Baddiley, J. (1972). *Biochem. J.* **127,** 27–37.
Hancock, I. C., and Baddiley, J. (1976). *J. Bacteriol.* **125,** 880–886.
Hancock, I. C., Wiseman, G., and Baddiley, J. (1976). *FEBS Lett.* **69,** 75.
Hancock, R., and Fitz-James, P. C. (1964). *J. Bacteriol.* **87,** 1044–1050.
Harcke, E., Van Massow, F., and Kuhlwein, H. (1975). *Arch. Microbiol.* **103,** 251–257.
Hayashi, H., Araki, Y., and Ito, E. (1973). *J. Bacteriol.* **113,** 592–598.
Hayashi, S., Koch, J. P., and Linn, E. C. C. (1964). *J. Biol. Chem.* **239,** 3098–3105.
Hayes, M. V., Ward, J. B., and Rogers, H. J. (1977). *Proc. Soc. Gen. Microbiol.* p. 85.
Heckels, J. E., Archibald, A. R., and Baddiley, J. (1975). *Biochem. J.* **149,** 637–647.
Henning, U. (1975). *Annu. Rev. Microbiol.* **29,** 45–60.
Henning, U., and Schwarz, U. (1973). *In* "Bacterial Membranes and Walls" (L. Leive, ed.), pp. 413–438. Dekker, New York.
Heydanek, M. G., and Neuhaus, F. C. (1969). *Biochemistry* **4,** 1474–1481.
Higashi, Y., and Strominger, J. L. (1970). *J. Biol. Chem.* **245,** 3691–3696.
Higashi, Y., Strominger, J. L., and Sweeley, C. C. (1967). *Proc. Natl. Acad. Sci. U.S.A.* **57,** 1878–1884.
Higashi, Y., Siewert, G., and Strominger, J. L. (1970a). *J. Biol. Chem.* **245,** 3683–3690.
Higashi, Y., Strominger, J. L., and Sweeley, C. C. (1970b). *J. Biol. Chem.* **245,** 3697–3702.
Higgins, M. L., and Shockman, G. D. (1971). *Crit. Rev. Microbiol.* **1,** 29–72.
Higgins, M. L., and Shockman, G. D. (1976). *J. Bacteriol.* **127,** 1346–1358.
Hilderman, R. H., and Riggs, H. G. (1973). *Biochem. Biophys. Res. Commun.* **50,** 1095–1103.
Ho, P. P. K., Towner, R. D., Indelicato, J. M., Spitzer, W. A., and Koppel, G. A. (1972). *J. Antibiot.* **25,** 627–628.
Holt, S. C., and Leadbetter, E. R. (1969). *Bacteriol. Rev.* **33,** 346.
Holt, S. C., Gauthier, J. J., and Tipper, D. J. (1975). *J. Bacteriol.* **122,** 1322–1338.
Holtje, J. V., and Tomasz, A. (1975a). *Proc. Natl. Acad. Sci. U.S.A.* **72,** 1690–1694.
Holtje, J. V., and Tomasz, A. (1975b). *J. Biol. Chem.* **250,** 6072–6076.
Holtje, J. V., Mirelman, D., Sharon, N., and Schwarz, U. (1975). *J. Bacteriol.* **124,** 1067–1076.
Horne, D., and Tomasz, A. (1977). *Antimicrob. Agents Chemother.* **11,** 888–896.
Hoshino, O., Zehari, U., Smay, P., and Jeanloz, R. W. (1972). *J. Biol. Chem.* **247,** 381–390.
Howard, L., and Tipper, D. J. (1973). *J. Bacteriol.* **113,** 1491–1504.
Hughes, A. H., Hancock, I. C., and Baddiley, J. (1973). *Biochem. J.* **132,** 83.
Hughes, R. C. (1970). *Biochem. J.* **117,** 431–439.
Hughes, R. C., and Thurman, P. F. (1970). *Biochem. J.* **117,** 441–449.
Hungerer, K. D., and Tipper, D. J. (1969). *Biochemistry* **8,** 3577–3587.
Hungerer, K. D., Fleck, J., and Tipper, D. J. (1969). *Biochemistry* **8,** 3567–3587.
Hussey, H., and Baddiley, J. (1972). *Biochem. J.* **127,** 39–50.
Hussey, H., and Baddiley, J. (1976). *In* "The Enzymes of Biological Membranes" (A. Martonosi, ed.), Vol. 2, pp. 227–326. Plenum, New York.
Hussey, H., Brooks, D., and Baddiley, J. (1969). *Nature (London)* **221,** 665–666.
Ishimoto, N., and Strominger, J. L. (1966). *J. Biol. Chem.* **241,** 639.
Ito, E., and Strominger, J. L. (1962). *J. Biol. Chem.* **237,** 2696–2703.

Ito, E., and Strominger, J. L. (1966). *J. Biol. Chem.* **234,** 210–216.
Iwaya, M., and Strominger, J. L. (1977). *Proc. Natl. Acad. Sci. U.S.A.* **74,** 2980–2984.
Iwaya, M., Goldman, R., Tipper, D. J., Feingold, R., and Strominger, J. L. Morphology of temperature-dependent morphological mutant of *E. coli. J. Bacteriol.* **136,** In press.
Iwaya, M., Weldon Jones, C., Khorana, J., and Strominger, J. L. (1978). *J. Bacteriol.* **133,** 196–202.
Izaka, K., Matsuhashi, M., and Strominger, J. L. (1966). *Proc. Natl. Acad. Sci. U.S.A.* **55,** 656–663.
Izaki, K., Matsuhashi, M., and Strominger, J. L. (1968). *J. Biol. Chem.* **243,** 3180–3192.
James, R. (1975). *J. Bacteriol.* **124,** 918–929.
James, R., Haga, J. Y., and Pardee, A. R. (1975). *J. Bacteriol.* **122,** 1283–1292.
Jann, B., Jann, K., Schmidt, G., Orskov, I., and Orskov, F. (1970). *Eur. J. Biochem.* **15,** 29–39.
Jawetz, E. (1967). *Antimicrob. Agents Chemother.* pp. 203–209.
Johnson, B. A., Anker, H., and Meleney, F. L. (1945). *Science* **102,** 376–377.
Johnson, P. Y., and White, D. (1972). *J. Bacteriol.* **112,** 849.
Johnston, L. S., and Neuhaus, F. C. (1975). *Biochemistry* **4,** 2754–2760.
Johnston, M. M., and Diven, W. F. (1969). *J. Biol. Chem.* **244,** 5414–5420.
Jones, H. E., and Hirsch, P. (1968). *J. Bacteriol.* **96,** 1037.
Jones, J. B., Bowers, B., and Stadtman, T. (1977). *J. Bacteriol.* **130,** 1357.
Joseph, R., Holt, S. C., and Canale-Parola, E. (1970). *Bacteriol. Pro;.* p. 57.
Kadner, R. J., and Winkler, H. H. (1973). *J. Bacteriol.* **113,** 895–900.
Kahan, F. M., Kahan, J. S., Cassidy, P. J., and Kropp, H. (1974). *Ann. N.Y. Acad. Sci.* **235,** 364–386.
Kamiryo, T., and Matsuhashi, M. (1972). *J. Biol. Chem.* **247,** 6306–6311.
Kamiryo, T., and Strominger, J. L. (1974). *J. Bacteriol.* **117,** 568–577.
Kandler, O., and Hippe, H. (1977). *Arch. Microbiol.* **113,** 57–60.
Kandler, O., and Konig, H. (1978). *Hoppe-Seyler's Z. Physiol. Chem.* **359,** 282–283.
Kandler, O., Schleifer, K. H., and Dandl, R. (1968). *J. Bacteriol.* **96,** 1935–1939.
Katz, W., and Martin, H. H. (1970). *Biochem. Biophys. Res. Commun.* **39,** 744–749.
Katz, W., Matsuhashi, M., Deitrich, C. P., and Strominger, J. L. (1967). *J. Biol. Chem.* **242,** 3207–3217.
Kelemen, M. V., and Rogers, H. J. (1971). *Proc. Natl. Acad. Sci. U.S.A.* **68,** 992–996.
Kennedy, L. D. (1974). *Biochem. J.* **138,** 525–535.
Kennedy, L. D., and Shaw, D. R. D. (1968). *Biochem. Biophys. Res. Commun.* **32,** 861–865.
Knox, K. W., and Hall, E. A. (1965). *Biochem. J.* **96,** 302–309.
Knox, K. W., and Holmwood, K. J. (1968). *Biochem. J.* **108,** 363–368.
Knox, K. W., and Wicken, A. J. (1972). *Infect. Immun.* **6,** 43.
Knusel, F., Neusch, J., Scherrer, M., and Schmid, K. (1967). *Schweiz. Z. Pathol. Bakteriol.* **30,** 87.
Koyama, T., Yamada, M., and Matsuhashi, M. (1977). *J. Bacteriol.* **129,** 1518–1523.
Kozarich, J. W. (1977). *In* "Microbiology—1977" (D. Schlessinger, ed.), pp. 203–208. Am. Soc. Microbiol. Washington, D.C.
Kozarich, J. W., and Strominger, J. L. (1978). *J. Biol. Chem.* **253,** 1272–1278.
Krulwich, T. A., and Ensign, J. C. (1968). *J. Bacteriol.* **96,** 857.
Krulwich, T. A., Ensign, J. C., Tipper, D. J., and Strominger, J. L. (1967a). *J. Bacteriol.* **94,** 734–740.
Krulwich, T. A., Ensign, J. C., Tipper, D. J., and Strominger, J. L. (1967b). *J. Bacteriol.* **94,** 741–750.
Lambert, M. P., and Neuhaus, F. C. (1972). *J. Bacteriol.* **110,** 978–987.
Lawrence, P. J., and Strominger, J. L. (1970a). *J. Biol. Chem.* **245,** 3653–3659.

Lawrence, P. J., and Strominger, J. L. (1970b). *J. Biol. Chem.* **245,** 3660–3666.

Leduc, M., and Van Heijenoort, J. (1975). *FEBS Meet., 10th, Paris* Abstr. No. 1055.

Leduc, M., Rousseau, M., and Van Heijenoort, J. (1977). *Eur. J. Biochem.* **80,** 153–163.

Leive, L. (1973). *In* "Bacterial Walls and Membranes" (L. Leive, ed., pp. xi–xv. Dekker, New York.

Leive, L. (1974). *Ann. N.Y. Acad. Sci.* **238,** 109–129.

Lennarz, W. J., and Scher, M. G. (1972). *Biochim. Biophys. Acta* **265,** 417–442.

Leyh-Bouille, M., Ghuysen, J.-M., Tipper, D. J., and Strominger, J. L. (1966). *Biochemistry* **5,** 3079–3090.

Leyh-Bouille, M., Ghuysen, J.-M., Bonaly, R., Nieto, M., Perkins, H. R., Schleifer, K. H., and Kandler, O. (1970). *Biochemistry* **9,** 2962–2970.

Linnett, P. E., and Strominger, J. L. (1973). *Antimicrob. Agents Chemother.* **4,** 231–236.

Linnett, P. E., and Strominger, J. L. (1974). *J. Biol. Chem.* **249,** 2489–2496.

Linnett, P. E., and Tipper, D. J. (1974). *J. Bacteriol.* **120,** 342–354.

Linnett, P. E., Roberts, R. J., and Strominger, J. L. (1974). *J. Biol. Chem.* **249,** 2497–2506.

Liu, T. Y., and Gottschlich, E. C. (1967). *J. Biol. Chem.* **242,** 471.

Lugtenberg, E. J. J. (1971). Ph.D. Thesis, Pijksuniversiteit te Utrecht, Utrecht.

Lugtenberg, E. J. J., and de Haan, P. G. (1971). *Antonie van Leeuwenhoek; J. Microbiol. Serol.* **37,** 537–552.

Lugtenberg, E. J. J., and van Schijndel van Dam, A. (1972). *J. Bacteriol.* **110,** 35–40.

Lugtenberg, E. J. J., and van Schijndel van Dam, A. (1973). *J. Bacteriol.* **113,** 96–104.

Lugtenberg, E. J. J., van Schijndel van Dam, A., and van Bellegem, M. H. (1971). *J. Bacteriol.* **108,** 20–29.

Lugtenberg, E. J. J., de Haas-Menger, L., and Ruyters, W. H. M. (1973). *J. Bacteriol.* **109,** 326–335.

Lund, F., and Tybring, L. (1972). *Nature (London), New Biol.* **236,** 135–137.

Lynch, J., and Neuhaus, F. C. (1966). *J. Bacteriol.* **91,** 449–460.

Marquet, A., Dusart, J., Ghuysen, J.-M., and Perkins, H. R. (1974). *Eur. J. Biochem.* **46,** 515–523.

Marquis, R. E. (1968). *J. Bacteriol.* **95,** 775–781.

Martin, H. H. (1964). *J. Gen. Microbiol.* **136,** 641–650.

Martin, H. H., Schilf, W., and Maskos, C. (1976). *Eur. J. Biochem.* **71,** 585–593.

Martinez-Carrion, M., and Jenkins, W. T. (1965). *J. Biol. Chem.* **240,** 3538–3546.

Mathies, A. W., Leedom, J. M., Ivler, D., Wehrle, P. F., and Portnoy, B. (1967). *Antimicrob. Agents Chemother.* pp. 218–223.

Matsuhashi, M., Dietrich, C. P., and Strominger, J. L. (1967). *J. Biol. Chem.* **242,** 3191–3206.

Matsuhashi, M., Takagaki, Y., Marayama, I. N., Tamaki, S., Nishimura, Y., Suzuki, H., Ogino, U., and Hirota, Y. (1977). *Proc. Natl. Acad. Sci. U.S.A.* **74,** 2976–2979.

Matsuzawa, H., Matsuhashi, M., Oka, A., and Sugino, Y. (1969). *Biochem. Biophys. Res. Commun.* **36,** 682–689.

Matsuzawa, H., Sato, T., and Imahori, K. (1973). *J. Bacteriol.* **115,** 436–442.

Mauck, J., and Glaser,L. (1972a). *Proc. Natl. Acad. Sci. U.S.A.* **69,** 2386–2390.

Mauck, J., and Glaser, L. (1972b). *J. Biol. Chem.* **247,** 1180–1187.

Mauck, J., Chan, L., and Glaser, L. (1971). *J. Biol. Chem.* **246,** 1820–1827.

Mescher, M. F., and Strominger, J. L. (1976). *Proc. Natl. Acad. Sci. U.S.A.* **73,** 2687–2691.

Miller, F. (1969). Ph.D. Thesis, Univ. of Newcastle upon Tyne, Newcastle upon Tyne, England.

Miller, T. W., Goegelman, P. T., Weston, R. G., Putker, I., and Wolf, F. J. (1972). *Antimicrob. Agents Chemother.* **2,** 132–135.

Mirelman, D., and Bracha, R. (1974). *Antimicrob. Agents Chemother.* **5,** 663–666.

Mirelman, D., and Sharon, N. (1972). *Biochem. Biophys. Res. Commun.* **46,** 1909–1917.

Mirelman, D., Beck, B. D., and Shaw, D. P. D. (1970). *Biochem. Biophys. Res. Commun.* **39,** 712.
Mirelman, D., Bracha, P., and Sharon, N. (1972). *Proc. Natl. Acad. Sci. U.S.A.* **69,** 3355–3359.
Mirelman, D., Bracha, R., and Sharon, N. (1974a). *FEBS Lett.* **39,** 105–110.
Mirelman, D., Bracha, R., and Sharon, N. (1974b). *Ann. N.Y. Acad. Sci.* **235,** 326–347.
Mirelman, D., Yashouv-Gan, Y., and Schwarz, U. (1976). *Biochemistry* **15,** 1781–1788.
Miyakawa, T., Matsuzawa, H., Matsuhashi, M., and Sugino, Y. (1972). *J. Bacteriol.* **112,** 950–958.
Mizuno, Y., and Ito, E. (1968). *J. Biol. Chem.* **243,** 2665–2672.
Murray, R. G. E., and Buckmire, F. L. A. (1969). *J. Gen. Microbiol.* **57,** xxiii.
Murray, R. G. E., Hall, M. M., and Marak, J. (1970). *Can. J. Microbiol.* **16,** 883–888.
Nathenson, S. G., and Strominger, J. L. (1963). *J. Biol. Chem.* **238,** 3161.
Nathenson, S. G., Ishimoto, N., Anderson, J. S., and Strominger, J. L. (1966). *J. Biol. Chem.* **241,** 651.
Nermut, M. V., and Murray, R. G. E. (1967). *J. Bacteriol.* **93,** 1949–1965.
Neuhaus, F. C. (1972). *Acc. Chem. Res.* **4,** 297–303.
Neuhaus, F. C., and Lynch, J. L. (1964). *Biochemistry* **3,** 471–480.
Neuhaus, F. C., and Struve, W. G. (1965). *Biochemistry* **4,** 120–131.
Neuhaus, F. C., Carpenter, C. V., Lambert, M. P., and Wargel, R. J. (1972). *In* "Molecular Mechanisms of Antibiotic Action on Protein Biosynthesis and Membranes" (E. Munoz, F. Ferrandiz, and D. Vazquez, eds.), pp. 339–362. Elsevier, Amsterdam.
Neuhaus, F. C., Linzer, R., and Reusch, V. M., Jr. (1974). *Ann. N.Y. Acad. Sci.* **235,** 502–518.
Nguyen-Distèche, M., Pollock, J. J., Ghuysen, J.-M., Puig, J., Reynolds, P., Perkins, H. R., Coyette, J., and Salton, M. (1974a). *Eur. J. Biochem.* **41,** 439–446.
Nguyen-Disteche, M., Ghuysen, J.-M., Pollock, J. J., Reynolds, P. E., Perkins, H. R., Coyette, J., and Salton, M. R. J. (1974b). *Eur. J. Biochem.* **41,** 447–455.
Nguyen-Disteche, M., Pollock, J. J., Ghuysen, J.-M., Puig, J., Reynolds, P. E., Perkins, H. R., Coyette, J., and Salton, M. R. J. (1974c). *Eur. J. Biochem.* **41,** 457–463.
Nickerson, W. J. (1974). *Ann. N.Y. Acad. Sci.* **235,** 105.
Nieto, M., and Perkins, H. R. (1971). *Biochem. J.* **124,** 845–852.
Nikaido, H. (1973). *In* "Bacterial Membranes and Walls" (L. Leive, ed.), pp. 131–209. Dekker, New York.
Nishimura, Y., Takeda, Y., Nishimura, A., Suzuki, H., Inouye, M., and Hirota, Y. *Plasmid* **1,** 67–77, 1977.
Oldmixon, E. H., Glauser, S., and Higgins, M. L. (1974). *Biopolymers* **13,** 2037–2060.
Oldmixon, E. H., Dezelee, P., Ziskin, M. C., and Shockman, G. D. (1976). *Eur. J. Biochem.* **68,** 271–280.
Oppenheim, B., and Patchornik, A. (1974). *FEBS Lett.* **48,** 172–175.
Osborn, M. J. (1969). *Annu. Rev. Biochem.* **38,** 501–538.
Osborn, M. J., Rick, P. D., Lehmann, V., Rupprecht, E., and Singh, M. (1974). *Ann. N.Y. Acad. Sci.* **235,** 52–65.
Ou, L. T., and Marquis, R. E. (1970). *J. Bacteriol.* **101,** 92–101.
Ou, L. T., and Marquis, R. E. (1972). *Can. J. Bacteriol.* **18,** 623–629.
Page, R. L., and Anderson, J. S. (1972). *J. Biol. Chem.* **247,** 2471–2479.
Park, J. T. (1952). *J. Biol. Chem.* **194,** 897.
Park, J. T. (1958a). *Biochem. J.* **70,** 2P.
Park, J. T. (1958b). *Symp. Soc. Gen. Microbiol.* **8,** 49–61.
Partridge, M. D., Davison, A. L., and Baddiley, J. (1973). *J. Gen. Microbiol.* **74,** 169.
Pellon, G., Bondet, C., and Michel, G. (1976). *J. Bacteriol.* **125,** 509–517.
Perkins, H. R. (1963). *Biochem. J.* **86,** 475–483.
Perkins, H. R. (1967). *Biochem. J.* **102,** 29c–32c.

Perkins, H. R. (1968). *Biochem. J.* 47p.

Perkins, H. R. (1971). *Biochem. J.* **121,** 417–423.

Perkins, H. R., and Nieto, M. (1974). *Ann. N.Y. Acad. Sci.* **235,** 348–363.

Perkins, H. R., Nieto, M., Frère, J.-M., Leyh-Bouille, M., and Ghuysen, J.-M. (1973). *Biochem. J.* **131,** 707–718.

Petit, J. F., Strominger, J. L., and Soll, D. (1968). *J. Biol. Chem.* **243,** 757–767.

Pitel, P. W., and Gilvarg, C. (1970). *J. Biol. Chem.* **245,** 6711–6717.

Plapp, R., and Strominger, J. L. (1970a). *J. Biol. Chem.* **245,** 3667–3674.

Plapp, R., and Strominger, J. L. (1970b). *J. Biol. Chem.* **245,** 3675–3682.

Pless, D. D., and Neuhaus, F. C. (1973). *J. Biol. Chem.* **248,** 1568–1576.

Pollock, J. J., Ghuysen, J.-M., Linder, R., Salton, M. R. J., Perkins, H. R., Nieto, M., Leyh-Bouille, M., Frère, J.-M., and Johnson, K. (1972). *Proc. Natl. Acad. Sci. U.S.A.* **69,** 662–666.

Pooley, H. M. (1976). *J. Bacteriol.* **125,** 1127–1139.

Powell, D. A., *et al.* (1975). *Proc. Soc. Gen. Microbiol.* **3,** 28.

Powell, J. F., and Strange, R. E. (1957). *Biochem. J.* **65,** 700–708.

Previc, E. P., and Lowell, A. (1975). *Biochim. Biophys. Acta* **411,** 377–385.

Rasmussen, J. R., Strominger, J. L. (1978). *Proc. Natl. Acad. Sci. U.S.A.* **75,** 84–88.

Reitz, R. H., Slade, H. D., and Neuhaus, F. C. (1967). *Biochemistry* **6,** 2561–2570.

Reusch, V. M., Jr., and Neuhaus, F. C. (1971). *J. Biol. Chem.* **246,** 6136–6143.

Reynolds, P. E. (1971). *Biochim. Biophys. Acta* **237,** 255–272.

Roberts, R. J. (1972). *Nature (London), New Biol.* **237,** 44–45.

Roberts, W. S. L., Strominger, J. L., and Soll. D. (1968a). *J. Biol. Chem.* **243,** 749–756.

Roberts, W. S. L., Petit, J. F., and Strominger, J. L. (1968b). *J. Biol. Chem.* **243,** 768–772.

Rogers, H. J. (1967). *Biochem. J.* **103,** 90–102.

Rogers, H. J. (1970). *Bacteriol. Rev.* **34,** 194–214.

Rogers, H. J. (1974). *Ann. N.Y. Acad. Sci.* **235,** 29–51.

Rogers, H. J. (1977). *In* "Microbiology—1977" (D. Schlessinger, ed.), pp. 25–34. Am. Microbiol. Soc., Washington, D.C.

Rogers, H. J., and Perkins, H. R. (1968). "Cell Walls and Membranes." Spon, London.

Rogers, H. J., and Thurman, P. F. (1977). *J. Bacteriol.* **133,** 298–305.

Roze, U., and Strominger, J. L. (1966). *Mol. Pharmacol.* **2,** 92–94.

Sabath, L. D., Wheeler, N., Laverdiere, M., Blazevic, D., and Wilkinson, B. J. (1977). *Lancet* **i,** 443–447.

Salton, M. R. J. (1964). "The Bacterial Cell Wall." Am. Elsevier, New York.

Scher, M., and Lennarz, W. J. (1969). *J. Biol. Chem.* **244,** 2777–2789.

Scher, M., Lennarz, W. J., and Sweeley, C. C. (1968). *Proc. Natl. Acad. Sci. U.S.A.* **59,** 1313–1320.

Scherrer, P., and Gerhardt, P. (1971). *J. Bacteriol.* **107,** 718–735.

Schleifer, K. H., and Kandler, O. (1972). *Bacteriol. Rev.* **36,** 407–477.

Schleifer, K. H., Huss, L., and Kandler, O. (1969). *Arck. Mikrobiol.* **68,** 387.

Schleifer, K. H., Hammes, W. P., and Kandler, O. (1976). *Adv. Microb.* **13,** 245–292.

Schrader, W. P., and Fan, D. P. (1974). *J. Biol. Chem.* **249,** 4815–4818.

Schwarz, U., and Leutgeb, W. (1971). *J. Bacteriol.* **106,** 558.

Sharpe, M. E., Davison, A. L., and Baddiley, J. (1964). *J. Gen. Microbiol.* **34,** 333.

Sharpe, M. E., Brock, J. H., Wicken, A. J., and Knox, K. W. (1975). *Proc. Am. Soc. Microbiol. Annu. Meet., 75th. New York* Abstr. K118.

Shaw, N. (1975). *Adv. Microbiol. Physiol.* **12,** 141–164.

Shibaer, V. N., Duckworth, M., Archibald, A. R., and Baddiley, J. (1973). *Biol. J.* **135,** 383.

Shockman, G. D., Thompson, J. S., and Canova, M. J. (1965). *J. Bacteriol.* **90,** 575.

Siewert, G., and Strominger, J. L. (1967). *Proc. Natl. Acad. Sci. U.S.A.* **57,** 767–773.
Siewert, G., and Strominger, J. L. (1968). *J. Biol. Chem.* **243,** 783–791.
Sjoquist, J., Movitz, J., Johansson, I. B., and Hjelm, H. (1972). *Eur. J. Biochem.* **30,** 190–194.
Somma, S., Merati, W., and Parent, F. (1977). *Antimicrob. Agents Chemother.* **11,** 396–401.
Spratt, B. G. (1975). *Proc. Natl. Acad. Sci. U.S.A.* **72,** 3117–3127.
Spratt, B. G. (1977). *Eur. J. Biochem.* **72,** 341–352.
Spratt, B. G., and Pardee, A. B. (1975). *Nature (London)* **254,** 516–517.
Spratt, B. G., and Strominger, J. L. (1976). *J. Bacteriol.* **127,** 660–663.
Spratt, B. G., Jobanputra, V., and Schwarz, U. (1977). *FEBS Lett.* **79,** 374–378.
Staudenbauer, W., and Strominger, J. L. (1972). *J. Biol. Chem.* **247,** 5095–5102.
Steber, J., and Schleiffer, K. M. (1975). *Arch. Microbiol.* **105,** 175–177.
Steinmetz, P. R., and Balko, C. (1973). *N. Engl. J. Med.* **289,** 846–852.
Stickgold, R. A., and Neuhaus, F. C. (1967). *J. Biol. Chem.* **242,** 1331–1337.
Stone, K. J., and Strominger, J. L. (1971). *Proc. Natl. Acad. Sci. U.S.A.* **68,** 3223–3227.
Storm, D. R., and Strominger, J. L. (1973). *J. Biol. Chem.* **248,** 3940–3945.
Storm, D. R., and Strominger, J. L. (1974). *J. Biol. Chem.* **249,** 1823–1827.
Strominger, J. L. (1970). *Harvey Lect.* **64,** 171–213.
Strominger, J. L. (1977). *In* D. Schlessinger (ed.). "Microbiology—1977." American Society for Microbiology, p. 177–181.
Strominger, J. L., Threnn, R. H., and Scott, S. S. (1959). *J. Am. Chem. Soc.* **81,** 3803–3804.
Strominger, J. L., Ito, E., and Threnn, R. H. (1960). *J. Am. Chem. Soc.* **82,** 998–999.
Strominger, J. L., Willoughby, E., Kamiryo, T., Blumberg, P. M., and Yocum, R. R. (1974). *Ann. N.Y. Acad. Sci.* **235,** 210–224.
Suginaka, H., Blumberg, P. M., and Strominger, J. L. (1972). *J. Biol. Chem.* **247,** 5279–5288.
Sutherland, I. W. (1972). *Adv. Microb. Physiol.* **8,** 143–208.
Suzuki, H., Nishimura, Y., and Hirota, Y. (1978). *Proc. Natl. Acad. Sci. U.S.A.* **75,** 664–668.
Sweet, R. M., and Dahl, L. F. (1970). *J. Am. Chem. Soc.* **92,** 5489–5507.
Swenson, J. C., and Neuhaus, F. C. (1976). *J. Bacteriol.* **125,** 626–634.
Taku, A., and Fan, D. P. (1976). *J. Biol. Chem.* **251,** 6154–6156.
Taku, A., Gunetileke, K. G., and Anwar, R. A. (1970). *J. Biol. Chem.* **245,** 5012–5016.
Taku, A., Gardner, H. L., and Fan, D. P. (1975). *J. Biol. Chem.* **250,** 3375–3380.
Tamaki, S., Nakajima, S., and Matsuhashi, M. (1977). *Proc. Natl. Acad. Sci. U.S.A.* **74,** 5472–5476.
Tamura, G., Sasaki, T., Matsuhashi, M., Takatsuki, A., and Yamosaki, M. (1976). *Agric. Biol. Chem.* **40,** 447–449.
Tamura, T., Imae, Y., and Strominger, J. L. (1976). *J. Biol. Chem.* **251,** 414–423.
Tanaka, T., and Weisblum, B. (1975). *J. Bacteriol.* **123,** 771–774.
Taylor, A., Das, C. B., and Van Heijenoort, J. (1975). *Eur. J. Biochem.* **53,** 47–54.
Thorndike, J., and Park, J. T. (1969). *Biochem. Biophys. Res. Commun.* **35,** 642–647.
Thorne, K. J. I., and Barker, D. C. (1972). *J. Gen. Microbiol.* **70,** 87–98.
Thorne, K. J. I., Swales, L. S., and Barker, D. C. (1974). *J. Gen. Microbiol.* **80,** 467–473.
Tipper, D. J. (1964). *Fed. Proc. Fed. Am. Soc. Exp. Biol.* **23,** 379.
Tipper, D. J. (1969a). *Biochemistry* **8,** 2192–2212.
Tipper, D. J. (1969b). *Bacteriol. Proc.* p. 24.
Tipper, D. J. (1970). *Int. J. Syst. Bacteriol.* **26,** 361–377.
Tipper, D. J. (1972). *In* "Subunits in Biological Systems" (G. D. Fasman and S. N. Timasheff, eds.), Biological Macromolecules Series, Vol. VIB, pp. 121–205, 331–345. Dekker, New York.
Tipper, D. J., and Berman, M. F. (1969). *Biochemistry* **8,** 2183–2192.

Tipper, D. J., and Gauthier, J. J. (1972). *In* "Spores V" (H. O. Halvorson, P. Hanson, and L. L. Campbell, eds.), pp. 3–12. Am. Soc. Microbiol., Washington, D.C.
Tipper, D. J., and Linnett, P. E. (1976). *J. Bacteriol.* **126,** 213–221.
Tipper, D. J., and Pratt, I. (1970). *J. Bacteriol.* **103,** 305–317.
Tipper, D. J., and Strominger, J. L. (1965). *Proc. Natl. Acad. Sci. U.S.A.* **54,** 1133–1141.
Tipper, D. J., and Strominger, J. L. (1966). *Biochem. Biophys. Res. Commun.* **22,** 48.
Tipper, D. J., and Strominger, J. L. (1968). *J. Biol. Chem.* **243,** 3169–3179.
Tipper, D. J., Ghuysen, J.-M., and Strominger, J. L. (1965). *Biochemistry* **4,** 468–473.
Tipper, D. J., Katz, W., Strominger, J. L., and Ghuysen, J.-M. (1967a). *Biochemistry* **6,** 921–929.
Tipper, D. J., Strominger, J. L., and Ensign, J. C. (1967b). *Biochemistry* **6,** 906–920.
Tipper, D. J., Tomoeda, M., and Strominger, J. L. (1971). *Biochemistry* **10,** 4683.
Tipper, D. J., Pratt, I., Guinand, M., Holt, S. C., and Linnett, P. E. (1977). *In* "Microbiology—1977" (D. Schlessinger, ed.), pp. 50–68. Am. Soc. Microbiol., Washington, D.C.
Tomasz, A. (1974). *Ann. N.Y. Acad. Sci.* **235,** 439–447.
Tomasz, A., and Holtje, J. V. (1977). *In* "Microbiology—1977" (D. Schlessinger, ed.), pp. 209–215. Am. Soc. Microbiol., Washington, D. C.
Tomasz, A., and Waks, S. (1975). *Proc. Natl. Acad. Sci. U.S.A.* **72,** 4162–4166.
Tomasz, A., Albino, A., and Zanati, E. (1970). *Nature (London)* **227,** 138–141.
Tomasz, A., McDonnell, M., Wesphal, M., and Zande, E. (1975). *J. Biol. Chem.* **250,** 341–337.
Torii, M., Kabat, E. A., and Bezer, A. E. (1964). *J. Exp. Med.* **120,** 13.
Trippen, B., Hammes, W. P., Schleiffer, K. M., and Kandler, O. (1976). *Arch. Microbiol.* **109,** 247–261.
Troy, F. A., Vijay, I. K., and Tesche, N. (1975). *J. Biol. Chem.* **250,** 156–163.
Tynecka, Z., and Ward, J. B. (1975). *Biochem. J.* **146,** 253–267.
Umbreit, J. N., and Strominger, J. L. (1972a). *J. Bacteriol.* **112,** 1306–1309.
Umbreit, J. N., and Strominger, J. L. (1972b). *Proc. Natl. Acad. Sci. U.S.A.* **69,** 1972–1974.
Umbreit, J. N., and Strominger, J. L. (1973a). *J. Biol. Chem.* **248,** 6759–6766.
Umbreit, J. N., and Strominger, J. L. (1973b). *J. Biol. Chem.* **248,** 6767–6771.
Umbreit, J. N., Stone, K. J., and Strominger, J. L. (1972). *J. Bacteriol.* **112,** 1302–1305.
van Heijenoort, J., Elbaz, L., Dezelee, P., Petit, J. F., Bricas, E., and Ghuysen, J.-M. (1969). *Biochemistry* **8,** 207–213.
Virden, R., Bristow, A. F., and Pain, R. H. (1975). *Biochem. J.* **149,** 397–407.
Virudachalam, R., and Rao, V. S. R. (1977). *Int. J. Pept. Protein Res.* **10,** 51–59.
Waechter, S., and Lennarz, W. J. (1976). *Annu. Rev. Biochem.* **45,** 95–112.
Walker, J. R., Koranck, R., Allen, J. S., and Gustafson, R. A. (1975). *J. Bacteriol.* **123,** 693–703.
Wallick, H., and Hendlin, D. (1976). *Antimicrob. Agents Chemother.* **5,** 25–32.
Ward, J. B. (1973). *Biochem. J.* **133,** 395–398.
Ward, J. B. (1974). *Biochem. J.* **141,** 227–241.
Ward, J. B., and Perkins, H. R. (1973). *Biochem. J.* **135,** 721–728.
Ward, J. B., and Perkins, H. R. (1974). *Biochem. J.* **139,** 781–784.
Wargel, R. J., Shadur, C. A., and Neuhaus, F. C. (1970). *J. Bacteriol.* **103,** 778–788.
Warth, A. D., and Strominger, J. L. (1969). *Proc. Natl. Acad. Sci. U.S.A.* **64,** 528–535.
Watkinson, R. J., Hussey, H., and Baddiley, J. (1971). *Nature (London), New Biol.* **229,** 57–59.
Watson, M. J., and Baddiley, J. (1974). *Biochem. J.* **137,** 399–404.
Weidel, W., and Pelzer, H. (1964). *Adv. Enzymol.* **26,** 193–232.
Wheat, R. W., and Ghuysen, J.-M. (1971). *J. Bacteriol.* **105,** 1219–1221.
White, D., Dworkin, M., and Tipper, D. J. (1968). *J. Bacteriol.* **95,** 2186–2197.
Wicken, A. J. (1966). *Biochem. J.* **99,** 108.
Wicken, A. J., and Knox, K. W. (1975). *Science* **187,** 1161–1167.

Wickus, G. G., and Strominger, J. L. (1972). *J. Biol. Chem.* **247,** 52.
Wickus, G. G., and Strominger, J. L. (1973). *J. Bacteriol.* **113,** 287–290.
Wickus, G. G., and Strominger, J. L. (1974). *J. Biol. Chem.* **247,** 53.
Wickus, G. G., Rubenstein, P. A., Warth, A. D., and Strominger, J. L. (1973). *J. Bacteriol.* **113,** 291–294.
Wietzerbin, J., Das, B. C., Petit, J. F., Lederer, E., Leyh-Bouille, M., and Ghuysen, J.-M. (1974). *Biochemistry* **13,** 3471–3476.
Wijsman, H. J. W. (1972a). *Genet. Res.* **20,** 65–74.
Wijsman, H. J. W. (1972b). *Genet. Res.* **20,** 269–277.
Willoughby, E., Higashi, Y., and Strominger, J. L. (1972). *J. Biol. Chem.* **247,** 5113–5115.
Wise, E. M., and Park, J. T. (1965). *Proc. Natl. Acad. Sci. U.S.A.* **54,** 75–81.
Woese, C. R., and Fox, G. E. (1977). *Proc. Natl. Acad. Sci. U.S.A.* **74,** 5088.
Wright, A., and Heckels, J. E. (1975). *Biochem. J.* **167,** 187–189.
Wright, A., Dankert, M., Fennessey, P., and Robbins, P. W. (1967). *Proc. Natl. Acad. Sci. U.S.A.* **57,** 1798–1803.
Wyke, A. W., and Perkins, H. R. (1975). *J. Gen. Microbiol.* **88,** 159–168.
Wyke, A. W., and Ward, J. B. (1975). *Biochem. Biophys. Res. Commun.* **65,** 877–885.
Wyke, A. W., and Ward, J. B. (1977a). *J. Bacteriol.* **130,** 1055.
Wyke, A. W., and Ward, J. B. (1977b). *FEBS Lett.* **73,** 159–163.
Yamada, M., Hirose, A., and Matsuhashi, M. (1975). *J. Bacteriol.* **123,** 678–686.
Yocum, R. R., Blumberg, P. M., and Strominger, J. L. (1974). *J. Biol. Chem.* **249,** 4863–4871.
Yonaka, K., Misono, H., Yamanoto, T., and Soda, K. (1975). *J. Biol. Chem.* **250,** 6983–6989.
Zaretskaya, M. Z., Naumova, I. B., and Shabarova, Z. A. (1971). *Biochimia* **36,** 97.

CHAPTER 7

The Outer Membrane of Gram-Negative Bacteria

A. WRIGHT AND D. J. TIPPER

I. Introduction

As discussed in Chapter 6 by Tipper & Wright, the gram-negative cell has a fundamentally different strategy toward the external environment than the gram-positive cell. In the gram-negative cell a membrane is present, external to the peptidoglycan layer, termed the outer membrane, that acts as a permeability barrier between the external environment and the cytoplasmic membrane. It is an essential component of all gram-negative cells and apparently cannot be dispensed with, even under laboratory conditions. Interactions between the cell and other agents, such as viruses, proteins,

ISBN 0-12-307207-7

and a large variety of small molecules, are initiated in the outer membrane, and most of its components have some receptor function of this type.

Biological membranes consist of a lipid phase, in the form of a bimolecular leaflet, plus associated proteins some of which traverse the membrane and others which are associated with one or the other surface of the leaflet. The two halves of the bimolecular leaflet appear, in general, to be asymmetric with respect to both lipid and protein species. For example, the membrane of the erythrocyte contains mainly phosphatidylcholine only in the outer half of the bimolecular leaflet, while phosphatidylethanolamine and phosphatidylserine are present on its inner half (Bretscher, 1973; Zwaal *et al.*, 1975). In *Bacillus megaterium* the phospholipid distribution is also asymmetric with twice as much phosphatidylethanolamine in the inner half of the cell membrane as in the outer half (Rothman and Kennedy, 1977).

The outer membrane of the gram-negative bacteria is also asymmetric and in addition has a unique composition. It contains many fewer polypeptide species than are present in the cytoplasmic membrane. Nevertheless this membrane acts as a selective permeability barrier to both hydrophilic and hydrophobic substances, and it performs numerous functions for the cell. Its relative simplicity combined with the ability to manipulate its structure by genetic means make it ideal for the study of both its assembly and function. The structure, biosynthesis, function, and assembly of the major components of the outer membrane are the subject of this chapter.

II. General Properties of the Outer Membrane

A. Isolation

Electron microscopy combined with antibody and phage typing of cells gave early indications that a number of cellular components were present or accessible at the cell surface and were therefore potential components of the outer membrane. More recently, methods have been developed for fractionating cell envelope fragments into cytoplasmic and outer membrane components. Such fractionation is possible because the outer membrane has a higher density than the cytoplasmic membrane (Miura and Mizushima, 1968) due to the presence of lipopolysaccharide in the outer membrane. Cells can be broken either mechanically using a French press (Schnaitman, 1970) or by lysis with lysozyme and EDTA (Osborn *et al.*, 1972a) to give envelope fragments which are fractionated on sucrose gradients to yield purified inner and outer membranes. Mechanical breakage yields envelope fragments containing peptidoglycan, whereas lysozyme–EDTA treatment removes most of the peptidoglycan. The peptidoglycan, when present, remains associated with outer membrane fragments, increasing their density. Thus, depending upon the method of cell disruption, the density of the outer membrane can vary.

Membrane fractions, separated on sucrose density gradients after mechanical breakage of cells, have been shown to be at least threefold enriched based on the distribution of lipopolysaccharide (outer membrane) and oxidative enzymes (cytoplasmic membrane) (Schnaitman, 1970; Koplow and Goldfine, 1974). The lysozyme–EDTA lysis method (Osborn *et al.*, 1972a) is applicable, with minor modifications, to many gram-negative organisms. In the case of *Salmonella typhimurium* LT2, fractionated membranes produced by this procedure are relatively pure (e.g., 2% inner membrane in outer membrane and about 10% outer membrane in inner membrane), their densities being 1.18 (cytoplasmic) and 1.24 (outer). The density of the outer membrane can vary from strain to strain, being dependent upon lipopolysaccharide structure (see Section IV). Since separation of the two membranes is relatively efficient, it appears that there are no strong interactions holding them together. This may not be strictly true, since it is known that there are regions where the outer and cytoplasmic membranes appear to form junctions (see Fig. 10 in Tipper and Wright, this volume, Chapter 6). If these junctions involve fairly strong interactions between the two membranes then hybrid membrane fragments, albeit a minority, would be produced upon cell disruption. They would probably have a hybrid density and might be separable from both the outer and cytoplasmic membranes. The separated outer and cytoplasmic membranes have relatively broad density distributions in sucrose gradients and could easily include such hybrid membrane fragments. Interestingly, artifactual association of outer membrane fragments with cytoplasmic membrane fragments to form hybrid membranes appears to be minimal, which is probably due to the very different structures of the two membranes. This is in contrast to cytoplasmic membrane fragments which can be induced to form hybrids by mild sonication (Tsukagoshi and Fox, 1971).

The cytoplasmic membrane contains a large variety of enzymatic activities, including succinic and lactic dehydrogenases, enzymes involved in phospholipid biosynthesis and in lipopolysaccharide biosynthesis, ATPase, and various cytochromes. In contrast, the outer membrane is almost devoid of enzymatic activity. One exception is phospholipase A_1 which appears to be an outer membrane component.

B. Composition

In the isolation procedure of Osborn *et al.* (1972a), the fraction of total protein recovered in the total cell envelope of *S. typhimurium* was found to vary from about 13 to 18%. Upon fractionation of this envelope material, 40–65% of the envelope protein was present in the outer membrane and 25–35% was present in the cytoplasmic membrane. These numbers may vary from strain to strain and may also depend upon the method of isolation, but they are probably fairly typical. The protein content of the outer membrane is clearly higher than that of the cytoplasmic membrane. The major components of the outer membrane are protein, phospholipid, and lipopolysaccharide. Lipopolysaccharide is found only in the outer membrane, and, as will be seen in Section XII, the protein species present in the outer membrane are unique and different from these in the cytoplasmic membrane. The weight ratio of protein–lipopolysaccharide–phospholipid in the outer membrane of wild-type *S. typhimurium* is 1:1:0.3 (Osborn *et al.*, 1972a) but in a UDPgalactose-4-epimerase-minus mutant of *S. typhimurium* which makes an incomplete lipopolysaccharide, it is 1:0.3:0.3. Thus, the weight contribution of the lipopolysaccharide can vary depending upon the lipopolysaccharide structure. Comparable values for the protein–lipopolysaccharide–phospholipid ratios in the cytoplasmic membrane of the same organism are 1.0:0.1:0.5–0.6. The presence of lipopolysaccharide in this case is probably due to contamination of the cytoplasmic membrane fraction by outer membrane material. The phospholipid–protein ratio is significantly higher (1.5–2 times) in the cytoplasmic membrane than in the outer membrane. This is an important feature of the outer membrane, which is discussed in Section XIII on outer membrane organization.

III. Phospholipids

The major phospholipids of *E. coli* and *Salmonella* are phosphatidylethanolamine, phosphatidylglycerol, and cardiolipin. They are all present in the outer membrane (White *et al.*, 1972), although the relative amounts of phosphatidylglycerol and cardiolipin are markedly lower than in the cytoplasmic membrane (Osborn *et al.*, 1972a). In *S. typhimurium* the phosphatidylglycerol content of the outer membrane is about one-half that in the cytoplasmic membrane and the cardiolipin content is reduced to about one-fourth. The significance of these differences is not known but might reflect the assembly or functional requirements of the outer membrane components. Exchange of phospholipids between the outer and cytoplasmic membranes has been reported, but it must be somewhat restricted otherwise a completely homogeneous composition would be expected.

IV. Lipopolysaccharides

Lipopolysaccharides are ubiquitous among the gram-negative bacteria. They have been of clinical interest because they are involved in pathogenicity and they are responsible for the endotoxin activity of the cell. They are heat-stable molecules which can be readily extracted from cells or cell envelopes (Luderitz *et al.*, 1966; Galanos *et al.*, 1969), and they can be purified with relative ease. Details of lipopolysaccharide structure are available in many reviews and textbooks. Only the more recent information is discussed in detail in this section.

A. Structure

The structure of the lipopolysaccharide of *S. typhimurium* is shown in Fig. 1. It consists of a lipid, called lipid A, which is covalently joined to an oligosaccharide termed the R core, which is in turn linked to a polysaccharide. The structure of lipid A is relatively invariable and the R core sugars to which it is attached are similar in different species. In contrast, the polysaccharide has both a variable composition and structure; it carries the determinants of O antigenic specificity and is therefore referred to as the O polysaccharide or O antigen.

Mutants of *S. typhimurium* defective at almost every stage of lipopolysaccharide synthesis have been isolated and their lipopolysaccharide structures analyzed. The O antigen and various R core sugars can be lost by mutation leaving lipid A linked to an incomplete R core. The chemotypes of the various incomplete structures are indicated in Fig. 1. The S (smooth)* structure is complete, the various R (rough) structures lack O polysaccharide and in most cases also have an incomplete R core.

The analysis of structures of this kind indicated that the molecule is synthesized in a stepwise manner with R core sugars and O polysaccharide being added to lipid A in sequence (Nikaido, 1973). "Wild-type" bacteria, such as *E. coli* K12 and *E. coli* B, have an incomplete R core structure with no O polysaccharide component. They appear to be naturally occurring rough mutants. The cell can lose most of the lipopolysaccharide structure by mutation without significant effect on its growth properties, but the lipid

* Organisms that have an S-type lipopolysaccharide form colonies with a smooth appearance whereas those with an R-type lipopolysaccharide form colonies with a rough appearance.

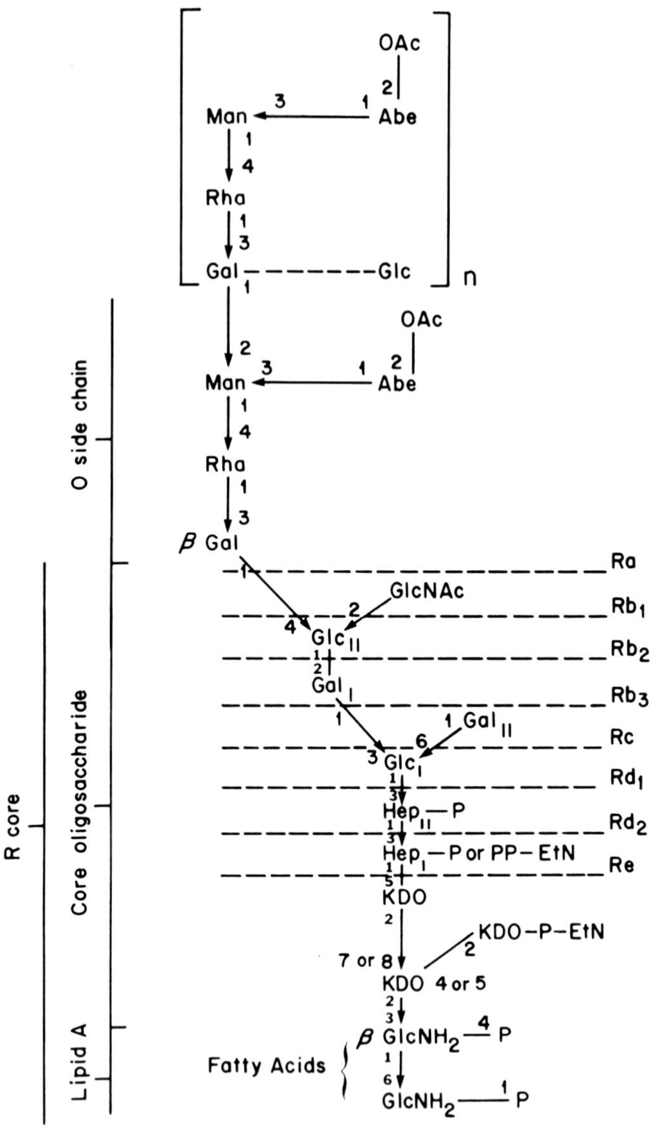

FIG. 1. Structure of the *Salmonella typhimurium* lipopolysaccharide. Abe, abequose; EtN, ethanolamine; Gal, D-galactose; Glc, D-glucose; $GlcNH_2$, glucosamine; Hep, L-glycero-D-mannoheptose; KDO, ketodeoxyoctonate; Man, D-mannose; OAc, acetyl; P, phosphate; PP, pyrophosphate; Rha, L-rhamnose. (Modified from Hussey and Baddiley, 1976.)

A and ketodeoxyoctonate (KDO) components appear to be essential for cell viability.

1. LIPID A

The basic structural unit present in the lipid A of *S. minnesota* is a β-1,6-linked D-glucosaminyl-D-glucosamine disaccharide which carries phosphate groups in glycosidic linkage (at C-1) and in ester linkage (at C-4′) and which is fully acylated by a series of fatty acids (Luderitz *et al.*, 1973) (Fig. 2). This unit is linked to the terminal KDO linkage of the R core by a ketosidic bond through its 3′-hydroxyl group. The fatty acid substituents are usually medium length saturated and 3-D-hydroxy fatty acids. Lipid A is the only cellular lipid which contains hydroxy fatty acids. In *S. minnesota* the major fatty acids are C_{12},C_{16}-saturated acids and β-hydroxymyristic acid. The amino groups are exclusively acylated with β-hydroxymyristic acid. Lauric (C_{12}), myristic (C_{14}), and 3-myristoxymyristic acid are present in ester linkage. Similar lipid A structures are present in a wide range of organisms including *E. coli* (M. Rosner and H. G. Khorana, personal communication), *Shigella* (Hase and Reitschel, 1976), *Pseudomonas* (Drewry *et al.*, 1973), *Fusobacterium nucleatum* (Hase *et al.*, 1977), and an anaerobe, *Selenomonas ruminantium* (Kamio *et al.*, 1971), although the latter contains no phosphate. In all cases examined the 3-hydroxy fatty acids have the D configuration. Some organisms contain no glucosamine in lipid A; in two *Rhodopseudomonas*

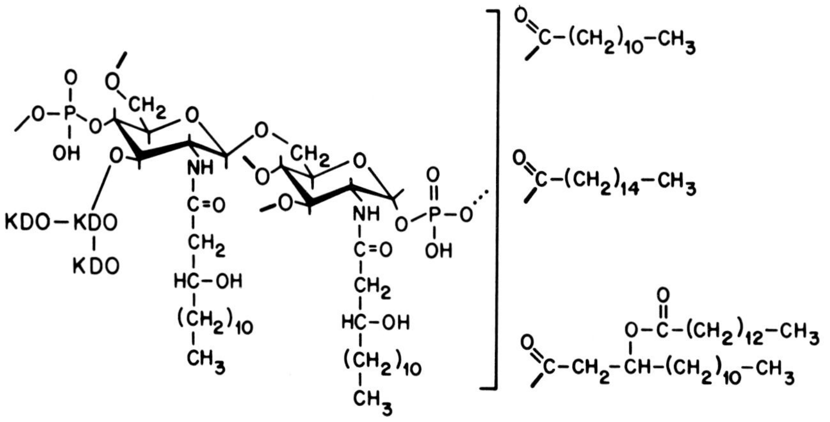

$$\left.\begin{array}{l} \text{O}\!=\!\text{C}-(\text{CH}_2)_{10}-\text{CH}_3 \\ \text{O}\!=\!\text{C}-(\text{CH}_2)_{14}-\text{CH}_3 \\ \text{O}\!=\!\text{C}-\text{CH}_2-\text{CH}(\text{O}-\text{CO}-(\text{CH}_2)_{12}-\text{CH}_3)-(\text{CH}_2)_{10}-\text{CH}_3 \end{array}\right]$$

KDO—KDO, KDO; NH—C=O—CH₂—HC—OH—(CH₂)₁₀—CH₃ (two amide chains); O—P(=O)(OH)—O phosphate groups

FIG. 2. Structure of lipid A from *Salmonella minnesota* R595. (From Luderitz *et al.*, 1973.)

strains, 2,3-diamino-2,3-dideoxy-D-glucose is the amino sugar component of the lipid (Reitschel *et al.*, 1977).

The ester-linked fatty acid components of lipid A can vary widely depending upon its source, but only one type of fatty acid, 3-hydroxy fatty acid, is amide linked. The chain length of the hydroxy fatty acid differs in individual bacterial groups. Usually in organisms such as the Enterobacteriaceae only one type of hydroxy fatty acid is present in amide linkage but in some organisms two or three different hydroxy acids are amide bound. The 3-hydroxy acids, both ester and amide bound, represent about 55–75% of the total fatty acids in most lipopolysaccharides; however, there are exceptions. In *Aerobacter aerogenes* lipopolysaccharide hydroxy fatty acids are only 6% of the total and in *Brucella* they are absent (Reitschel *et al.*, 1977). Lipid A contains all of the 3-hydroxy fatty acids of the cell but contains no unsaturated or cyclopropane fatty acids (Reitschel *et al.*, 1977). Its fatty acid composition is thus markedly different from that of the phospholipids. There is some indication that lipid A monomers (Fig. 2) are joined together through pyrophosphate linkages in *Salmonella minnesota* (Luderitz *et al.*, 1973) and in *E. coli* (Rosner and Khorana, personal communication). Presumably the glycosidic and ester-linked phosphate groups are involved in these linkages. The extent of such cross-linking is uncertain, and its role in lipid A function is at present unclear.

The sugar, 4-aminoarabinose, is sometimes present in lipid A of *Salmonella* as a substituent on about 20–30% of the ester-bound phosphate groups. A *Salmonella* mutant with a temperature-sensitive defect in KDO synthesis accumulates an acidic lipid A precursor containing nonsubstituted phosphate groups at nonpermissive temperature under certain growth conditions. If the growth conditions are changed, a neutral precursor is formed; the glycosidic phosphate is substituted with ethanolamine, the ester phosphate with 4-aminoarabinose (Reitschel *et al.*, 1977). Thus, growth conditions may lead to variation in lipid A structure in normal cells.

Lipid A represents a family of bacterial lipids. Like phospholipids, it contains a variety of fatty acid constituents, and its basic structure may show a good deal of microheterogeneity, for example, in its phosphate content and in the degree to which it is substituted by other minor constituents.

2. R Core

The R core structure has been studied in detail in *S. typhimurium* (Fig. 1) and in *S. minnesota*. It is a complex oligosaccharide containing *N*-acetyl-D-glucosamine, D-glucose, D-galactose, a 7-carbon sugar L-glycero-D-manno-heptose and an 8-carbon sugar acid, ketodeoxyoctonate (KDO). It also contains phosphate groups, *O*-phosphorylethanolamine and *O*-pyrophos-

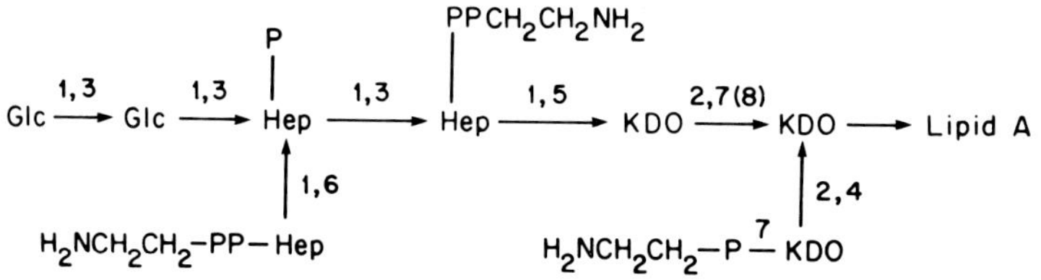

$$
\begin{array}{ccccccccccccc}
 & & & & P & & PPCH_2CH_2NH_2 & & & & & & \\
 & & & & | & & | & & & & & & \\
Glc & \xrightarrow{1,3} & Glc & \xrightarrow{1,3} & Hep & \xrightarrow{1,3} & Hep & \xrightarrow{1,5} & KDO & \xrightarrow{2,7(8)} & KDO & \longrightarrow & \text{Lipid A} \\
 & & & & \uparrow{\scriptstyle 1,6} & & & & & & \uparrow{\scriptstyle 2,4} & & \\
 & & & & H_2NCH_2CH_2{-}PP{-}Hep & & & & & & H_2NCH_2CH_2{-}P\overset{7}{-}KDO & &
\end{array}
$$

FIG. 3. Structure of *E. coli* BB lipopolysaccharide. (From Prehm *et al.*, 1975.)

phorylethanolamine (Lehmann *et al.*, 1971). An additional heptose unit is present in the R core of some *Salmonella* lipopolysaccharides (Hämmerling *et al.*, 1973), and there is also some variation in the degree of substitution by phosphate and by pyrophosphorylethanolamine. In certain *S. typhimurium* R forms L-glycero-D-mannoheptose is replaced by D-glycero-D-mannoheptose (Lehmann *et al.*, 1973).

The structure of the R core of *E. coli* B has recently been determined. It contains only glucose, heptose, and KDO (Prehm *et al.*, 1975) (Fig. 3). Presumably *E. coli* B is lacking one or more of the genes required for complete R core synthesis. Its structure is not unlike that of the R core of *Salmonella typhimurium* in both composition and linkage specificity.

3. O POLYSACCHARIDE

The O polysaccharide of *S. typhimurium* (Fig. 1) consists of tetrasaccharide repeating units containing the hexoses D-galactose and D-mannose, a deoxyhexose, L-rhamnose and a dideoxyhexose, abequose. Dideoxyhexoses are relatively rare sugars in nature, but are found quite frequently as components of O polysaccharides. The repeating units may be further substituted with additional glycosyl groups or in some instances by acetyl groups. More subtle structural variations also occur. For example, the O polysaccharide of *Salmonella anatum* contains galactose, mannose, and rhamnose in the same sequence as that of *S. typhimurium* but the glycosidic linkages are different; thus the two O polysaccharides, as expected, have no antigenic determinants in common.

$$
\begin{array}{ll}
\textit{S. typhimurium} & \nrightarrow \text{Man } 1 \xrightarrow{\alpha} 4 \text{ Rha } 1 \xrightarrow{\beta} 3 \text{ Gal } 1 \underset{\eta}{\overset{\alpha}{\nrightarrow}} 2 \text{ Man} \rightarrow \\
 & \quad\ \uparrow\alpha \qquad\qquad\qquad\qquad\qquad\qquad\qquad \uparrow\alpha \\
 & \quad \text{Abe} \qquad\qquad\qquad\qquad\qquad\qquad\quad \text{Abe} \\
\textit{S. anatum} & \nrightarrow \text{Man } 1 \xrightarrow{\beta} 4 \text{ Rha } 1 \xrightarrow{\alpha} 3 \text{ Gal } 1 \underset{\eta}{\overset{\alpha}{\nrightarrow}} 6 \text{ Man} \rightarrow
\end{array}
$$

All O polysaccharide structures have a basic structural simplicity in that they consist of repeating sequences usually containing two, three, or four monosaccharide components. Variability in structure is the basis for the serological classification of gram-negative bacteria. The *Salmonella* alone have been grouped into more than 40 serological groups based on O polysaccharide structure. Examples of other types of O polysaccharide structure are reviewed elsewhere (Luderitz *et al.*, 1966; Nikaido, 1973).

B. Biosynthesis

Two completely independent pathways are used for the biosynthesis of lipopolysaccharide. In one of these the lipid A–R core structure is formed and in the other the O polysaccharide is produced. These two components are then joined together to form the complete molecule.

1. Lipid A

Little direct information is available on the synthesis of this complex lipid; however recent studies (Rick and Osborn, 1977; Rick *et al.*, 1977) have provided information on some of the steps in its assembly. These workers have isolated a conditional mutant of *S. typhimurium* that is temperature sensitive for both KDO synthesis and cell growth which accumulates a precursor of lipid A at nonpermissive temperature. KDO is formed by the following series of reactions.

$$\text{D-Ribulose-5-P} \rightleftarrows \text{D-arabinose-5-P} \quad (1)$$

$$\text{D-Arabinose-5-P} + \text{PEP} \rightleftarrows \text{KDO-8-P} + P_i \quad (2)$$

$$\text{KDO-8-P} \rightarrow \text{KDO} + P_i \quad (3)$$

Three enzymes catalyze these reactions which are, respectively, D-ribulose-5-phosphate isomerase, KDO-8-phosphate synthetase, and KDO-8-phosphate phosphatase. The active donor in lipopolysaccharide synthesis is CMP–KDO, which is formed by the interaction of CTP and KDO (Ghalambor and Heath, 1966). The mutant isolated by Rick and Osborn (1977) has a defect in KDO-8-phosphate synthetase; at a nonpermissive temperature (42°C) the enzyme has a markedly reduced ability to catalyze reaction (2) due to an increased apparent K_m for D-arabinose-5-phosphate. Under nonpermissive conditions lipopolysaccharide synthesis in the mutant stops and a precursor of lipid A accumulates in the cell (Rick *et al.*, 1977). The precursor resembles the phosphorylated diglucosamine lipid A backbone shown in Fig. 2, but differs in that it contains one ester and two amide-linked β-hydroxymyristic acid residues but lacks ester-linked saturated fatty acids. Two phosphomonester groups are present, one of which was shown to be glycosidically linked. The structure suggests that it is a direct precursor of lipid A and that the attachment of KDO residues to lipid A occurs at an intermediate stage of lipid A synthesis prior to addition of the saturated fatty acid components (Rick *et al.*, 1977).

Earlier observations by Heath and co-workers (1966) had suggested that KDO might be transferred to lipid A prior to complete acylation because partial deacylation of lipid A was required to convert it to an efficient acceptor of KDO units from CMP–KDO *in vitro*. Transferase activities have been solubilized from *S. typhimurium* membranes that can transfer KDO from CMP–KDO to the underacylated lipid A precursor *in vitro* (Munson and Osborn, 1975). A definite relationship between the underacylated precursor and intact lipid A has been demonstrated *in vivo* by Osborn *et al.* (1978), who have shown that the precursor which accumulates in the KDO$^-$ mutant at high temperature, is incorporated into lipopolysaccharide when KDO synthesis is restored by returning cells to permissive conditions. The sequence of reactions shown in Fig. 4 has been proposed for the pathway of lipid A biosynthesis (Rick *et al.*, 1977) based on the above results. In this scheme a phosphorylated diglucosamine unit is N- and O-acylated with β-hydroxymyristic acid. KDO residues are added; then either saturated fatty acids are added to the structure which is finally polymerized or polymerization occurs first, followed by fatty acid addition. No information is available at present on the formation of the phosphorylated glucosamine backbone, the synthesis of β-hydroxymyristic acid, or the reactions which join lipid A units to one another through pyrophosphate bridges (polymerization).

The defect in KDO-8-phosphate synthetase described above causes growth stasis in mutant cells under nonpermissive conditions (Rick and

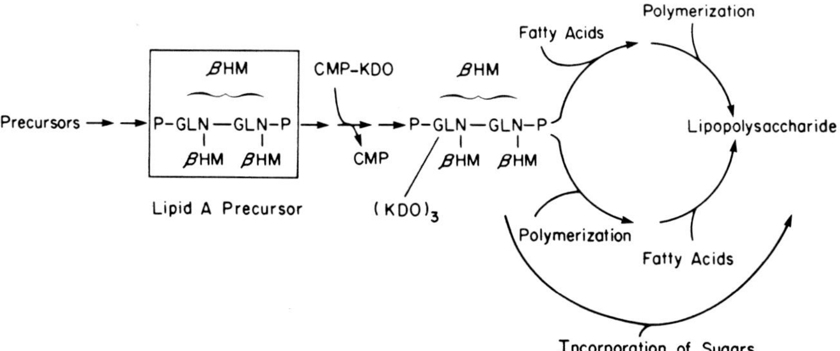

FIG. 4. Scheme for the biosynthesis of the lipid A–KDO region of *S. typhimurium* lipopolysaccharide. Abbreviations are given in the legend to Fig. 1, except: GLN, glucosamine; βHM, β-hydroxymyristic acid.

Osborn, 1977). A failure to produce lipopolysaccharide would have a dramatic effect on outer membrane structure which might be responsible for the effect (see Section V). Alternatively, the accumulation of an abnormal lipid in the cell, presumably in the cytoplasmic membrane, could be responsible for the observed inhibition of growth.

2. R Core

R core synthesis has been studied in detail in cell-free systems only in *Salmonella*, but the general principles of R core assembly (Nikaido, 1973; Hussey and Baddiley, 1976) are probably similar in most gram-negative bacteria. The R core is assembled by the stepwise transfer of sugars and other components, such as phosphate and *O*-phosphorylethanolamine, to lipid A or a lipid A precursor (see Section IV,B,1) (Fig. 5). Mutants that are defective in the production of any of the nucleotide sugar precursors or in specific transferase activities produce incomplete R core structures which lack all sugars distal to the lesion. For example, mutants of *S. typhimurium* defective in UDPglucose synthesis produce an R core–lipid A structure lacking galactose and *N*-acetylglucosamine in addition to glucose. In fact, studies of mutants with defects in either UDPglucose or UDPgalactose synthesis gave the first indications that R core assembly was a stepwise process (Nikaido, 1973).

As stated in Section IV,B,1 complete acylation of lipid A probably occurs at some stage after the KDO units are transferred to it. The metabolic precursor of the heptose units which are added to the terminal KDO acceptor is still not known, and nothing is known about the heptose transfer reactions. In intact lipopolysaccharide, the heptose units (specifically heptose I) are phosphorylated. The phosphate groups are derived from ATP and are probably added to the R core as it is growing. Phosphate transfer to carbon 4 of heptose I has been demonstrated *in vitro* using as acceptor a *Salmonella* cell envelope fraction containing phosphate-deficient lipopolysaccharide [derived from a phosphate-minus (P^-) mutant] (Mühlradt, 1969, 1971). A P^- Rd_1 lipopolysaccharide structure, lacking glucose I, could act as an acceptor, but the presence of glucose I increased the acceptor efficiency. For this reason, it was suggested that glucose I is added to heptose II prior to phosphorylation (Mühlradt, 1971). The heptose II unit in *Salmonella* lipopolysaccharide is found to be phosphorylated also (Hämmerling *et al.*, 1973); this phosphate unit appears to be added after addition of glucose I. *O*-Pyrophosphorylethanolamine units are found on C-4 of heptose I. The role of the phosphate units is unclear. It has been claimed that the heptose units of different R core chains may be cross-linked by

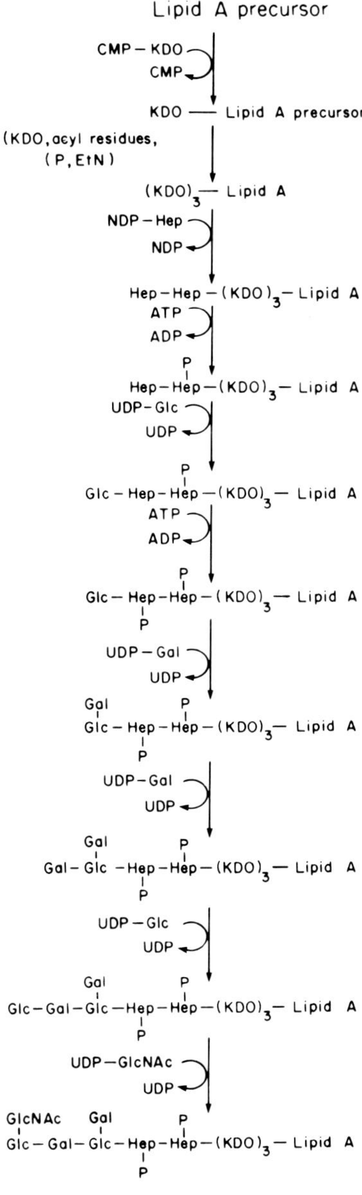

FIG. 5. Biosynthesis of the R core portion of the *Salmonella typhimurium* lipopolysaccharide. (Modified from Hussey and Baddiley, 1976.) Abbreviations are given in the legend to Fig. 1.

phosphodiester bridges, but there is no convincing evidence for such structures. The high density of negative charges in this region of the lipopolysaccharide may be important for the structural integrity of the outer membrane, since P^- mutants have significantly altered permeability properties.

The addition of glucose, galactose, and *N*-acetylglucosamine to the R core (Fig. 5) has been studied in detail, especially the reactions involving transfer of glucose and galactose (reviewed in Osborn, 1969; Nikaido, 1973; Hussey and Baddiley, 1976). Transfer of glucose I and galactose I to R core occurs *in vitro* in cell envelope fractions derived from mutants defective in the synthesis of UDPglucose or UDPgalactose. These preparations contain the glycosyltransferases and defective lipopolysaccharide which acts as an endogenous acceptor (Rothfield *et al.*, 1964). Rothfield and co-workers in an elegant series of experiments have purified the soluble forms of both the glucosyl- and galactosyltransferase enzymes and have reconstituted an active system which contains enzyme, lipopolysaccharide acceptor, and phospholipid. This system at least partially recreates a membrane environment in which enzyme, substrate, and acceptor can interact. Phospholipid is an essential component of the system, phosphatidylethanolamines from a variety of sources being active (Rothfield and Pearlman, 1966). Heating and cooling of a mixture of phospholipid and lipopolysaccharide forms a binary complex to which the transferase enzymes bind forming a stable ternary complex (Rothfield and Takeshita, 1965). The galactosyltransferase complex has been reconstituted as a mixed monolayer on a water surface (Romeo *et al.*, 1970a,b). The order of addition of the components is important; lipopolysaccharide and phospholipid must first interact to form a complex to which enzyme will bind. The ternary complexes interact with substrate sugar nucleotides, resulting in transfer of the sugar to the lipopolysaccharide acceptor. The properties of the reconstituted systems are quite similar to those of the enzymes in cell envelope preparations, although in the monolayer system transfer efficiency is low (Rothfield and Romeo, 1971). This suggests that the mobility of the enzyme in the monolayer system is low, whereas in an aqueous suspension of the complex it has more freedom of movement. The apparent failure of the galactosyltransferase to dissociate from the complex in the monolayer system could be due to nonspecific interactions between the enzyme and phospholipid in the *in vitro* system (Rothfield *et al.*, 1972). Rothfield and Romeo (1971) proposed a model for addition of monosaccharides to the R core in which incomplete R core units interact sequentially with the transferase enzymes. Each transferase enzyme would bind to the appropriate incomplete R core acceptor molecule and would dissociate from it once transfer had occurred.

Transfer of the first galactose unit to the R core appears to require the presence of phosphate groups on the heptose units. Cell envelope fractions from a P^- mutant of an Rc strain of *S. minnesota*, which contains galactose-deficient lipopolysaccharide, do not incorporate galactose efficiently when incubated with UDPgalactose (Mühlradt *et al.*, 1968). Galactose incorporation is stimulated by a prior incubation of cell envelope fraction with ATP and a soluble phosphate transferase from a P^+ strain, presumably because of phosphorylation of heptose units in the P^- lipopolysaccharide. Thus, the structure of the acceptor probably determines which transferase enzyme it will interact with.

3. O Polysaccharide

The O polysaccharide is formed in a series of reactions that are analogous to those used for synthesis of the glycan chains of the peptidoglycan (see Section VI,D, in Tipper and Wright, this volume, Chapter 6). All of these reactions have been studied in *in vitro* systems in which particulate or membrane fractions from cells have been used as the source of enzymes. All of the enzymes involved in O polysaccharide synthesis are membrane bound, although some of them can be relatively easily solubilized.

a. Formation of the O Polysaccharide Repeating Unit. The sugar components of the O polysaccharide are transferred from their nucleoside diphosphate derivatives to undecaprenol phosphate, a C_{55}-polyisoprenoid lipid present in the cytoplasmic membrane, the same lipid which is involved in peptidoglycan synthesis (Wright *et al.*, 1967). In this initial set of transfer reactions from the cytoplasmic sugar donors to the lipid–phosphate acceptor, the oligosaccharide repeating unit of the O polysaccharide is produced. In the case of *S. anatum* three transfer reactions are involved (Wright *et al.*, 1965; Weiner *et al.*, 1965), as shown in Fig. 6.

In the first reaction Gal-1-P is transferred from UDPGal to undecaprenol-P forming undecaprenol-PP-Gal and UMP. The reaction is inhibited by UMP and can also be reversed by it. A UDPGal–UMP exchange reaction can also occur that forms the basis of a sensitive assay of the reaction (Rundell and Shuster, 1973). This reaction, catalyzed by undecaprenol-PP-Gal synthetase, is analogous to the first reaction in the synthesis of the glycan chain of peptidoglycan (see Section VI in Tipper and Wright, this volume, Chapter 6). The transferase has an apparent K_m of 2.5×10^{-5} M for UDPgalactose and requires Mg^{2+} for activity (Rundell and Shuster, 1973). It has been solubilized, and in its soluble form can catalyze the exchange reaction between UDP galactose and UMP. It has been postulated that since both reactions utilize the same C_{55}-polyisoprenoid lipid, the availability of undecaprenol-P might coordinate the synthesis of peptidoglycan and lipopolysaccharide (Nikaido, 1973). The *in vitro* exchange reaction catalyzed by phospho-*N*-acetylmuramylpentapeptide translocase

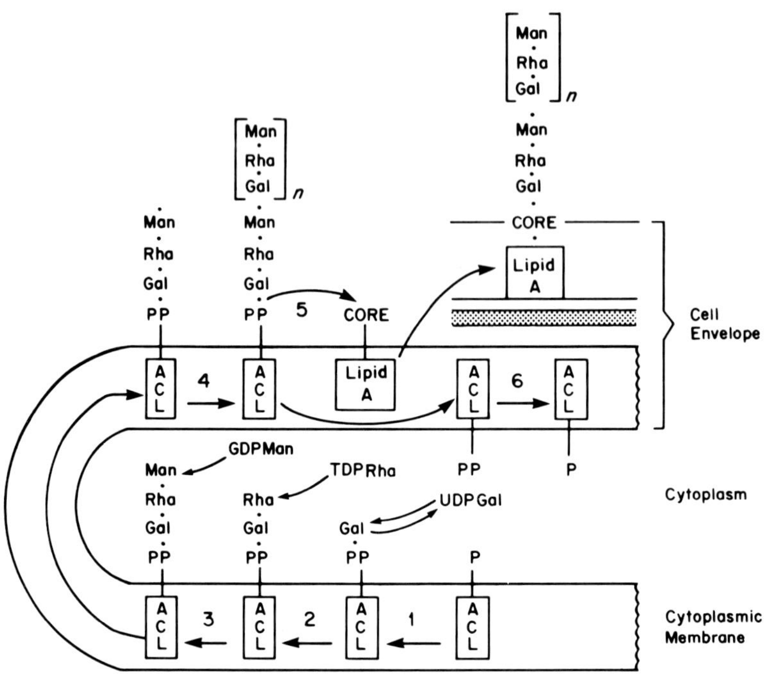

FIG. 6. O Antigen biosynthesis. Synthesis of the *S. anatum* O antigen is illustrated. The trisaccharide subunit, mannosylrhamnosylgalactose is assembled by sequential transfer of monosaccharides to membrane-bound C_{55}-polyisoprenoid phosphate (ACL-P) (reactions 1–3). Trisaccharides are polymerized (reaction 4) and the lipid-linked O antigen polymer is transferred to lipid A–core acceptor (reaction 5) forming the completed lipopolysaccharide molecule. The ACL-PP produced in reaction 5 is dephosphorylated (reaction 6) to regenerate ACL-P and complete the cycle.

is not affected by the accumulation of O polysaccharide precursors on undecaprenol (Rundell and Shuster, 1975), which led to the suggestion that different pools of undecaprenol-P are available for peptidoglycan synthesis and O polysaccharide synthesis. However, *in vitro* experiments indicate that the lipid acceptor has a high degree of mobility or freedom of movement in the cell membrane (Kanegasaki and Wright, 1970). Thus, it seems more likely that factors other than separate lipid pools prevent competition between the two systems.

Subsequent reactions in the formation of the oligosaccharide repeating unit are the transfer of rhamnose from TDPrhamnose to undecaprenol-PP-Gal to form undecaprenol-PP-Gal-Rha and transfer of mannose from GDPmannose to undecaprenol-PP-Gal-Rha to form undecaprenol-PP-Gal-Rha-Man.

All reactions are catalyzed by enzymes of the cytoplasmic membrane (Osborn *et al.*, 1972b) and all require Mg^{2+}, but none have been studied in detail. In all reactions but the first, the sugar is transferred to an acceptor with the release of the corresponding nucleoside diphosphate. The lipid intermediates involved in these reactions have been isolated and characterized. They can be added to particulate enzyme preparations *in vitro* (Osborn *et al.*, 1972b; Kanegasaki and Wright, 1970) and can participate in all of the subsequent reactions of O antigen synthesis. For example, isolated undecaprenol-PP-Gal can act as an acceptor of rhamnose, mannose, and abequose in an *in vitro* system using a particulate fraction derived from the cell envelope as enzyme. Since the nucleoside diphosphate sugars are present in the cytoplasm, it is likely that the lipid-linked oligosaccharide repeating sequence is formed on the cytoplasmic face of the membrane.

b. Polymerization of O Polysaccharide Repeating Units. The mechanism of polymerization of lipid-bound oligosaccharides has been demonstrated in *S. anatum* by Bray and Robbins (1967). The lipid intermediate in this case is undecaprenol-PP-Gal-Rha-Man. In a first step the galactosyl phosphate bond of one trisaccharide–lipid (donor) is cleaved and the galactosyl bond is transferred to the terminal mannosyl residue of a trisaccharide–lipid (acceptor) forming hexasaccharide–lipid (Fig. 7). In a second step, the galactosyl phosphate bond of the newly formed hexasaccharide lipid is cleaved and the galactosyl linkage is transferred to the terminal mannose

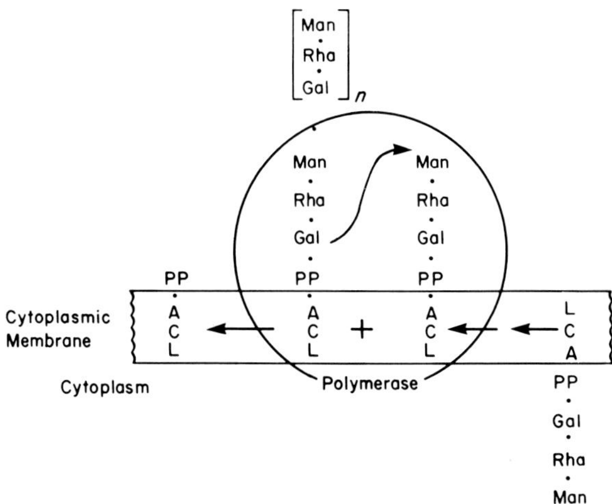

FIG. 7. Direction of chain growth during polymerization of *Salmonella* O antigen. ACL-linked intermediates are pictured as being closely associated with the polymerase in the cytoplasmic membrane. The growing chain is transferred to the new ACL-linked monomeric unit; thus chains grow at their reducing ends.

residue of a trisaccharide–lipid forming a chain containing three repeating units or nine monosaccharides. When the process is repeated, polysaccharide chains of increasing length are formed. The growing chain is always added to the new trisaccharide subunit; thus chain growth occurs at the reducing end of the growing chain. This is analagous to the growth of polypeptide chains. Growth of the glycan chains of peptidoglycan also occur in this way (see Section VI in Tipper and Wright, this volume, Chapter 6). The purified lipid-linked precursor of the O polysaccharide of *S. anatum* undecaprenol-PP-Gal-Rha-Man is polymerized efficiently when added to a particulate enzyme fraction containing O polysaccharide polymerase (Kanegasaki and Wright, 1970). Addition of the lipid intermediate to cell membrane preparations requires special conditions, such as the presence of nonionic detergent or repeated freezing and thawing of a mixture of membrane fraction and a homogeneous aqueous suspension of the intermediate. No soluble cofactors are required for the polymerization reaction apart from divalent cations. The lipid intermediates are freely mobile in the membrane.

In the O polysaccharide of *S. typhimurium* the abequose units are not in the main polymer chain but are present as side groups. *In vitro*, the polymerase is as active with undecaprenol-PP-Gal-Rha-Man as substrate as it is with the normal undecaprenol-PP-Gal-Rha-Man-Abe (Osborn and Weiner, 1968). In the group D *Salmonella* the dideoxyhexose tyvelose is present as an O polysaccharide side group, as abequose in *S. typhimurium*. Mutants which are blocked in the conversion of CDP paratose to CDP tyvelose, a reaction catalyzed by an epimerase in these cells, produce an O polysaccharide containing paratose instead of tyvelose side groups (Sasaki and Uchida, 1976). Thus, even *in vivo*, the polymerase is not highly specific for the side group sugars.

In the E group *Salmonella* the galactosyl linkage in the undecaprenol-PP-Gal-Rha-Man precursor has the α-configuration. Polymerization in group E1 organisms produces galactosylmannose linkages with the α-configuration, but polymerization in group E2 organisms produces galactosylmannose linkages with the β-configuration (Wright *et al.*, 1965). Lysogenization of a group E1 organism by a temperature virus called ε15 produces a group E2 organism. The virus carries two genes which bring about this change; one gene gives rise to an inhibitor of the E1 host O polysaccharide polymerase (α-polymerase), the other is the structural gene for a new phage-specified O polysaccharide polymerase (β-polymerase). Both polymerases use the same substrate. By replacing one polymerase with another O polysaccharide with different structures can be produced. Both polymerases are membrane bound and have similar properties. Cells lysogenic for a mutant of ε15 with a defect in the inhibition of host α-polymerase activity produce O polysaccharide chains having either α- or β-galactosylmannose linkages;

single polysaccharide chains produced in these cells have either all α- or all β-galactosyl linkages. This implies that O polysaccharide chains are polymerized by one enzyme during chain growth, rather than moving from one enzyme to another reminiscent of the mechanism of glycosylation of teichoic acids (Section VIII, in Chapter 6).

In order to have continued polymerization of O polysaccharide subunits, undecaprenol-PP must be cleaved to regenerate undecaprenol-P. A specific phosphatase must be present in the gram-negative cell membrane for this purpose, presumably with properties like those of the phosphatase present in *S. aureus* (see Section VI,D in Tipper and Wright, this volume, Chapter 6). Bacitracin inhibits O polysaccharide polymerization by binding to undecaprenol-PP preventing its reutilization in the reaction sequence.

c. Modification of O Polysaccharides. O Polysaccharides often have side groups attached to the main polymer which are added to the nascent polysaccharide chain during growth. Although abequose is a side group on the *S. typhimurium* O polysaccharide, it is incorporated into the lipid-linked oligosaccharide prior to polymerization and is, therefore, not considered a modification of the chain. Modification is often incomplete, whereas the main chain sugar components are always present in equimolar amounts.

i. Acetylation. Acetyl groups are present in a wide range of O polysaccharides and as terminal or side groups are responsible for specific antigenic characteristics of these polysaccharides. In *S. anatum*, which has been studied in most detail, acetyl groups are present in ester linkage on C-6 of the galactosyl units of the O polysaccharide (Keller, 1966). The acetylation reaction has been demonstrated *in vitro* using cell envelope fractions of *S. anatum* as source of enzyme and endogenous polysaccharide acceptor; acetyl-CoA is the acetyl donor. Various oligosaccharides are also active acceptors of acetyl groups in the cell-free system, especially those that contain the rhamnosylgalactose sequence of the O polysaccharide. The *in vitro* experiments give little indication of the stage at which acetylation occurs *in vivo*. In mutants of *S. anatum* that are defective in R core synthesis, an undecaprenol-linked O polysaccharide accumulates which is substituted with *O*-acetyl groups. This suggests that acetylation occurs during growth of polysaccharide chains and certainly prior to transfer of the polysaccharide to the R core.

Strains of *S. typhimurium* defective in polymerization of lipid-linked oligosaccharides carry antigenic determinant 05 which is specified by acetyl substituents on the abequose units (Mäkelä, 1966). Thus, it seems likely that acetyl groups can be transferred to single repeating sequences. The acetylation reaction most likely involves those sequences of the growing O polysaccharide chain that are closest to the membrane.

ii. Glucosylation. The *Salmonella* group E3 O polysaccharide repeating sequence is —Man—Rha—Gal(Glu)—, the glucose substituents being linked to C-4 of the galactosyl groups. The glucose donor is UDPglucose, but transfer of glucose is a two-step process involving formation of a glucosyl–lipid intermediate (Wright, 1971) as follows.

UDPglucose + undecaprenol-P ⇄ undecaprenol-P-glucose + UDP

Undecaprenol-P-glucose + O polysaccharide →
glucosyl-*O*-polysaccharide + undecaprenylphosphate

The glucosyl–lipid intermediate which is involved only in the transfer of single sugar units has a phosphodiester group in contrast to the oligosaccharide–lipid intermediates involved in polymerization which have pyrophosphate groups. Transfer of mannose units to mannan in *Micrococcus lysodeikticus* occurs in a manner analagous to that described above (Scher *et al.*, 1968; Scher and Lennarz, 1969).

It is unclear how glucosylation is coordinated with O polysaccharide chain growth. Some polymerization of O polysaccharide subunits must occur prior to glucosylation, since the transglucosylase is specific for β-galactosyl units which are produced by the O polysaccharide polymerase. Using a cell-free system, Sasaki *et al.* (1974) have shown that O polysaccharide chains polymerized prior to addition of UDPglucose but still attached to the undecaprenol lipid can act as acceptors of glucose units, and it has also been shown that endogenous lipopolysaccharide itself can act as an acceptor in cell-free systems. Wright and Kanegasaki (1971) proposed that glucose units are transferred during O polysaccharide polymerization to the β-galactosyl units nearest the undecaprenol–lipid, since both substrate and acceptor are linked to the membrane. Glucosylation by this mechanism, shown in Fig. 8 ensures a systematic addition of glucose to the growing chain. This model is not incompatible with the above *in vitro* results, which could easily be due to an altered relationship between enzyme and substrate caused by disruption of the cell envelope.

Glucosylation of O polysaccharide in *S. typhimurium* occurs by a mechanism identical to that described above (Nikaido *et al.*, 1971; Nikaido and Nikaido, 1971; Takeshita and Mäkelä, 1971).

4. The Ligase Reaction: Joining of O-Polysaccharide and R Core–Lipid A

All of the reactions involved in O polysaccharide and R core–lipid A synthesis occur on the cytoplasmic membrane. The joining of these com-

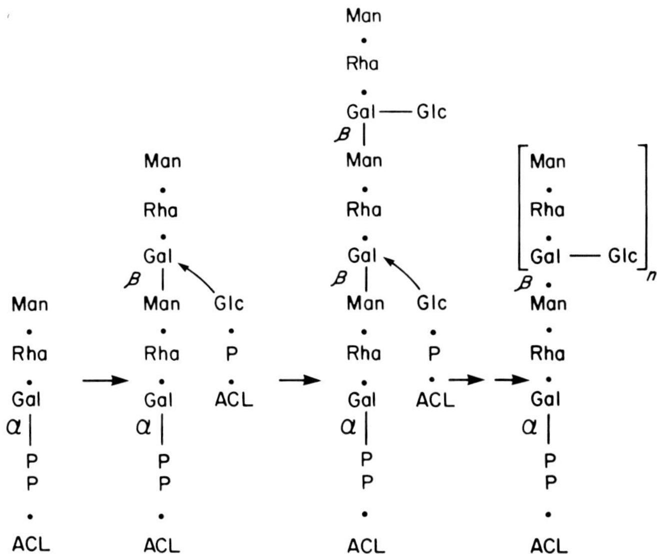

FIG. 8. Glucosylation of *S. newington* of O antigen. Glucosylation occurs during polymerization of the O antigen. The glucosyl units are transferred from a lipid carrier (ACL-P = C_{55}-polyisoprenoid phosphate) to the β-galactosyl unit nearest the reducing end of the polymer. The galactosyl unit at the reducing end is not glucosylated, as it has the α-configuration.

ponents, in which O polysaccharide is transferred from the undecaprenol carrier lipid to glucose II of the R core (Fig. 1), is catalyzed by O antigen-lipopolysaccharide ligase also occurs at this site. The reaction does not depend upon the length of the O polysaccharide chain, since mutants that lack O polysaccharide polymerase produce lipopolysaccharide with single repeating units attached to the R core. It is, thus, unlikely that the ligase recognizes the length of chains that it transfers. The reaction has been demonstrated *in vitro* (Cynkin and Osborn, 1968), but it is not understood in detail. Once the biosynthesis of the lipopolysaccharide is complete, it is transported to the outer membrane and is no longer accessible to the enzymes involved in its assembly. This has been demonstrated by Osborne *et al.* (1972b) using a conditional mutant of *S. typhimurium* in which O polysaccharide synthesis is dependent upon the addition of exogenous mannose. In the absence of mannose, the mutant produces an Ra-type lipopolysaccharide; when mannose is added to mutant cells, an S-type lipopolysaccharide is produced. Newly synthesized O polysaccharide is added only to newly synthesized R core–lipid A acceptor and not to the Ra lipopolysaccharide formed prior to its synthesis. Once in the outer membrane the Ra lipopolysaccharide is no longer available to the ligase.

C. Mutants with Defects in Lipopolysaccharide Synthesis

1. R Core

In *S. typhimurium* there are a number of genes, termed *rfa*, in the *str–xyl–metA* region of the chromosome which are largely responsible for R core biosynthesis; a number of these are in a group termed the *rfa* cluster (Stocker and Mäkelä, 1971; Nikaido, 1973). R core mutants with lipopolysaccharide chemotypes Ra-Re (Fig. 1) have been mapped in this region. For example, glucosyltransferase I (*rfa G*) and galactosyltransferase I (*rfa H*) have their structural genes in this region (Osborn, 1968; Wilkinson and Stocker, 1968). Many, if not all, of the transferase activities involved in R core biosynthesis are likely to have their genes located at this locus. The *rfa* mutants are not usually defective in O polysaccharide synthesis, but since there is no complete R core acceptor in such mutants, O polysaccharide remains linked to the undecaprenol lipid carrier. It is present in this form in relatively small amounts.

Loss of the O antigen and much of the R core by mutation does not affect cellular properties; however, *rfa* mutants containing lipopolysaccharide lacking glucose, or heptose (Rd_1, Rd_2, Re chemotypes—see Fig. 1) have drastically altered outer membrane properties (see Section XII).

2. O Polysaccharide

Many *Salmonella* rough mutants that form a complete R core but no O polysaccharide have been characterized and most fall into a single class called *rfb*. The *rfb* cluster is located between *met G* and *his* on the *S. typhimurium* chromosome. The *rfb* genes specify enzymes that participate in the synthesis of nucleotide derivatives of the sugars abequose, mannose, and rhamnose, all of which are present in the O polysaccharide. In addition, the *rfb* region probably contains genes determining the transferases required to assemble the O repeating units (Nikaido *et al.*, 1966). Many of the transferases have not actually been mapped though a gene *rfb P*, required for production of UDPgalactose: undecaprenol-phosphate; galactose-1-phosphatetransferase has been identified (Levinthal and Nikaido, unpublished data, cited in Mäkelä *et al.*, 1976). Mutations at a locus termed *rfc* located between the *his* and *trp* genes in *S. typhimurium* affect O polysaccharide polymerization. Mutants of this type produce a lipopolysaccharide with single O polysaccharide subunits attached to the R core.

V. Outer Membrane Proteins

A. MAJOR SPECIES

More than 20 species of polypeptide, identified as protein bands on SDS-polyacrylamide gels, are present in the outer membrane of *E. coli* and *Salmonella* and probably in all other gram-negative bacteria. As protein fractionation methods have improved, the number of identifiable protein components has been increasing. However, a very few species of protein, termed major outer membrane proteins, make up 70% or more of the mass of the protein in the outer membrane (Schnaitman, 1970). Most gram-negative organisms have between one and four major proteins in the molecular weight range 33,000–38,000 which total about 10^5 copies each per cell. The other major protein, present in most cells, is a small molecular weight lipoprotein which has been studied in detail. Since much more is known about all aspects of this latter component its properties will be discussed before those of the other proteins mentioned above.

1. THE BRAUN LIPOPROTEIN

The most abundant protein in the *E. coli* cell envelope in terms of numbers of molecules is a small lipoprotein of MW 7200, which is present in about 3×10^5 to 6×10^5 copies per cell and thus represents about 5–7% of total cellular protein. This protein, termed the Braun lipoprotein, is present in two forms in the *E. coli* cell, one being covalently linked to peptidoglycan, the other being free in the outer membrane. One-third of the lipoprotein is peptidoglycan-bound, while two-thirds is free. A similar lipoprotein is found in several other organisms; in some cases it may not be present in a peptidoglycan-bound form (Braun, 1975). The lipoprotein is one of the most intensively studied macromolecules in *E. coli;* its entire amino acid sequence is known, and furthermore its *in vitro* biosynthesis from purified messenger RNA has recently been demonstrated. It represents a unique model system for understanding the biosynthesis of a membrane protein and the mechanism by which it is inserted into the envelope of the bacterial cell. Reviews on lipoprotein by Braun (1975) and by Inouye (1975) are recommended for more detailed information.

a. Structure. The primary structure of the mature lipoprotein, as elucidated by Braun and his co-workers, is shown in Fig. 9. It has 58 amino acid residues with cysteine at the N-terminus and lysine at the C-terminus. It contains only 15 different amino acids lacking glycine, histidine, phenylalanine, proline, and tryptophan, a property which has been very useful for studies of its *in vivo* synthesis. [Although the protein portion of the mature

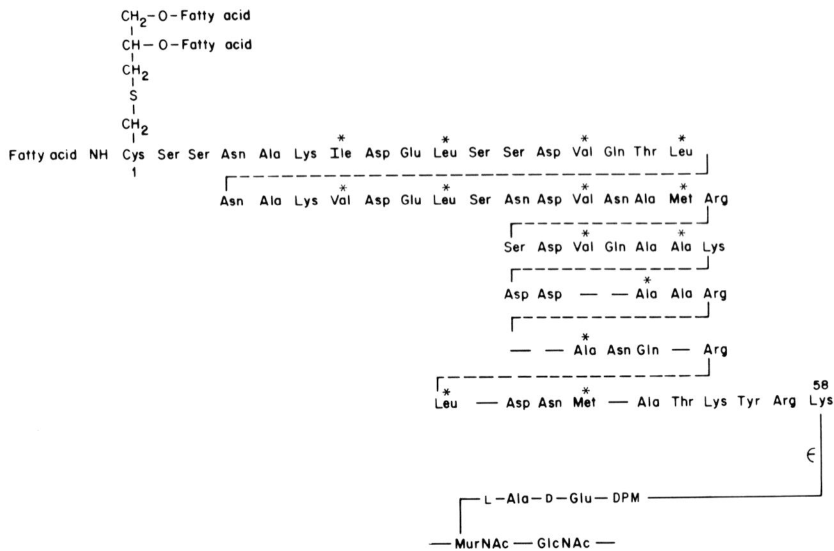

FIG. 9. Structure of the Braun lipoprotein (Braun, 1975). The asterisks indicate the hydrophobic amino acids at every 3.5 position. The terminal cysteine residue is linked to a diglyceride moiety by a thioether bond, and its amino group is blocked by a fatty acid. The C-terminal lysine is linked to a diaminopimelic acid unit in the peptidoglycan. DPM, diaminopimelic acid.

lipoprotein contains no glycine, glycine is essential for its synthesis because lipoprotein is formed from a longer polypeptide (prolipoprotein) which contains glycine near its N-terminus (Section V,A,c).] One striking feature of the structure is the repetitive amino acid sequence. The sequence from amino acid 4 (asparagine) to amino acid 17 (leucine) is almost identically repeated in the next sequence of 15 amino acids, and is followed by partial repetitions as shown in the diagram. Braun has suggested that the lipoprotein gene evolved by gene duplication and deletion from an original gene coding for a short stretch of amino acids.

The N-terminal cysteine residue is covalently joined to a diacylglycerol by a thioether bond (Hantke and Braun, 1973); the amino group of the cysteine is also acylated. The fatty acid residues do not seem to be unique, having a composition similar to that found in phospholipids from the same cells. As mentioned above, the lipoprotein was first isolated in its peptidglycan-bound form; chemical analysis of the lipoprotein–peptidoglycan linkage indicated that the C-terminal lysine of the lipoprotein was joined covalently through its ε-amino group to the L-carboxyl group of diaminopimelic acid (Braun and Bosch, 1972). More recently, analysis of the free form of the lipoprotein has indicated that it has an identical structure to the bound form.

No modifications of the lipoprotein appear to occur when it is joined to peptidoglycan.

Optical measurements on preparations of lipoprotein liberated from peptidoglycan, either by trypsin (see following paragraph) or lysozyme treatment indicate that lipoprotein has a very high α-helical content (Braun, 1973). The lipoprotein preparations used for this purpose were derived from cells which had been boiled in 4% sodium dodecyl sulfate (SDS). Very little helical structure was observed in the presence of SDS after such drastic treatment; however, the α-helical structure was reestablished upon removal of SDS. Two three-dimensional structures have been proposed (Inouye, 1974; Braun, 1975) for the lipoprotein, based on the α-helical content and are shown in Fig. 10. In the first model (Inouye, 1974), the molecule is depicted as a rod which is all α-helix apart from small segments at the N and C termini. In the other, the molecule is largely α-helix, but it is folded at a region of β-structure at amino acids 25–29; in this model, as in the first, the N and C termini are not α-helical (Braun, 1975). Recent work by Lee *et al.* (1977) indicates that the α-helical content of lipoprotein is about 88% in the presence of 0.015% SDS, supporting the former model.

The only region of the lipoprotein molecule which is sensitive to tryptic cleavage is the C-terminus. Trypsin cleaves between lysine (55) and tyrosine (56) and between arginine (57) and lysine (58) liberating bound lipoprotein from peptidoglycan. There are several other potential trypsin cleavage sites, but these are resistant, suggesting a highly ordered structure such as is postulated in either model in Fig. 10.

Whether the conformations established in the above way represent the *in vivo* state of lipoprotein is unclear, although it seems likely that they do. Further physical studies will hopefully show whether the lipoprotein is a helical rod (model 1) or a folded helical rod (model 2). An interesting property of both models is the distribution of hydrophobic fatty acids within the structure. They are present every three or four amino acids in a strictly alternating sequence; in a helical structure which has 3.6 amino acids per turn they line up on one face of the helix (Fig. 10). This type of organization within the molecule led Inouye (1974) to suggest that several helical rods might interact in the membrane to form organized groups of lipoprotein molecules.

b. Location. As mentioned in Section V,A,1, two-thirds of the lipoprotein in the cell is present in a free form. Analyses of separated inner and outer membranes indicated that all of the free lipoprotein was present in the outer membrane. In order to determine the orientation of the peptidoglycan-bound lipoprotein, cells were treated with lysozyme, which liberates lipoprotein linked to two murein repeating units, then fractionated to yield inner and outer membranes. Greater than 90% of the bound form of lipoprotein was present in the outer membrane fraction (Bosch and Braun, 1973), suggesting that the lipoprotein extends from the peptidoglycan outwards into the outer membrane.

There is approximately one lipoprotein molecule bound to peptidoglycan per 10 muramic acid residues (Braun and Wolff, 1970). It has been suggested that the bound lipoprotein molecules extend outward from the peptidoglycan, their free lipid-carrying ends being imbedded in the outer membrane.

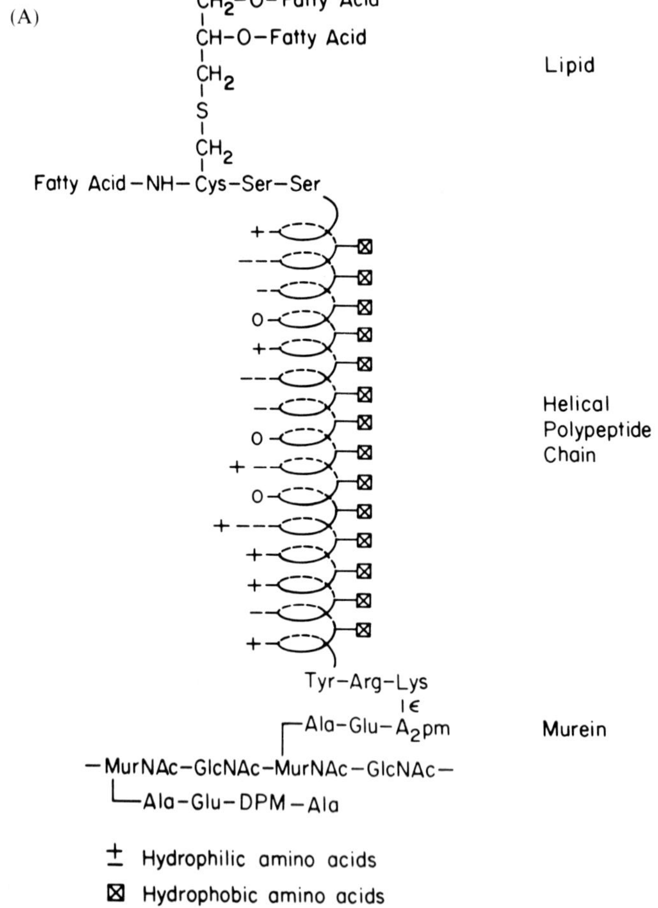

FIG. 10. Two possible models of the Braun lipoprotein. (From Braun, 1975.) In (A), the structure is shown as a single helical rod (Inouye, 1974). All of the hydrophobic side chains are arrayed on one face of the α-helix, and all of the hydrophilic side chains are on the other face. It has been proposed that groups of such molecules could aggregate in the membrane forming

At first sight this would seem an ideal way to anchor the outer membrane to the cell. Although this may be a function of the lipoprotein, it is not an essential one because mutants have now been isolated in which there is little, if any, bound form of the lipoprotein; other mutants that completely lack lipoprotein have also been isolated (see Section C).

The disposition of free lipoprotein in the outer membrane is more difficult to determine. Antibody directed against lipoprotein can interact with whole

(B)

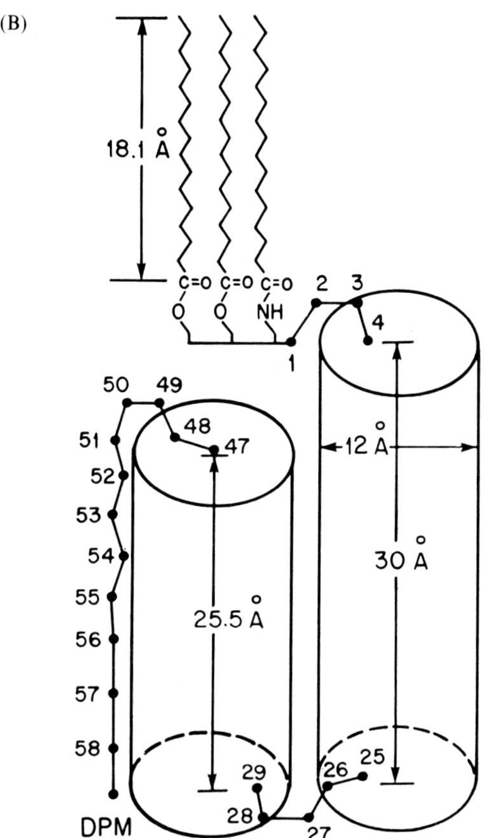

hydrophilic channels. In the second structure (B), the molecule is folded as shown, the cylinders representing the α-helical regions of the polypeptide (Braun, 1975). The extended structure could span the outer membrane. The folded structure could span half the thickness of the outer membrane. DPM or A_2pm, diaminopimelic acid.

cells, although it is adsorbed most efficiently by rough mutants that have defects in lipopolysaccharide biosynthesis (Braun, 1975). Lipopolysaccharide acts as a barrier to the penetration of many different types of molecules including antibodies, thus removal of its outer portions by mutation might be expected to allow the antibody to reach the cell surface more easily. However, lipopolysaccharide mutants can have drastically altered outer membrane structure and composition (see Section V). Despite such alterations, it seems reasonable to conclude that lipoprotein is exposed at the surface of the outer membrane. The outer membrane is known to be highly asymmetric in that lipopolysaccharide molecules are present exclusively in the outer half of the bilayer and phospholipids are present mainly in its inner half. Whether or not there is an asymmetric distribution of free lipoprotein remains to be seen.

c. Biosynthesis. i. Synthesis of the Polypeptide. One of the most useful properties of the Braun lipoprotein for biosynthetic studies is its lack of certain amino acids. Cells starved for three of these, proline, histidine, and tryptophan, continue to synthesize lipoprotein at a time when the synthesis of most other proteins is drastically reduced (Hirashima and Inouye, 1973). The product, radioactively labeled with one of its constituent amino acids, e.g., arginine, can be readily fractionated using standard chromatographic procedures. Using this technique, it has been shown that the lipoprotein messenger RNA has a half-life of about 11 minutes (Hirashima *et al.*, 1973), which is considerably longer than the half-life of cytoplasmic protein messenger RNA molecules. The lipoprotein is one of several membrane proteins synthesized in *E. coli* minicells (Levy, 1975). In this system, the messenger half-life is at least 1 hour. Minicells are produced by aberrant cell division that occurs close to a cell pole; they lack the bacterial chromosome; thus protein synthesis must be occurring from "trapped" messenger RNA. The reason for the extraordinary messenger stability in this system is not known.

The biosynthesis of lipoprotein is also unusual in that it is markedly resistant to the antibiotic puromycin both *in vivo* and *in vitro* using *E. coli* polyribosomes (Halegoua *et al.*, 1974; Hirashima and Inouye, 1975). Lipoprotein messenger RNA has recently been isolated from *E. coli* and purified (Takeishi *et al.*, 1976). It is about 360 ± 10 nucleotides long, which is sufficient RNA to code for two lipoprotein molecules; however analysis of oligonucleotide fragments indicates that the molecule codes for only one molecule of lipoprotein. The RNA does not have a polyadenylate sequence at its 3′-end. Interestingly, when this purified mRNA was used to direct protein synthesis in a cell-free system from *E. coli*, a single protein of higher molecular weight than lipoprotein was produced (Inouye *et al.*, 1977b). This protein, termed prolipoprotein, contains the complete lipoprotein amino acid sequence plus 20 additional amino acids at the N-terminus.

Met–Lys–Ala–Thr–Lys(5)–Leu–Val–Leu–Gly–Ala(10)–

Val–Ile–Leu–Gly–Ser(15)–Thr–Leu–Leu–Ala–Gly(20)

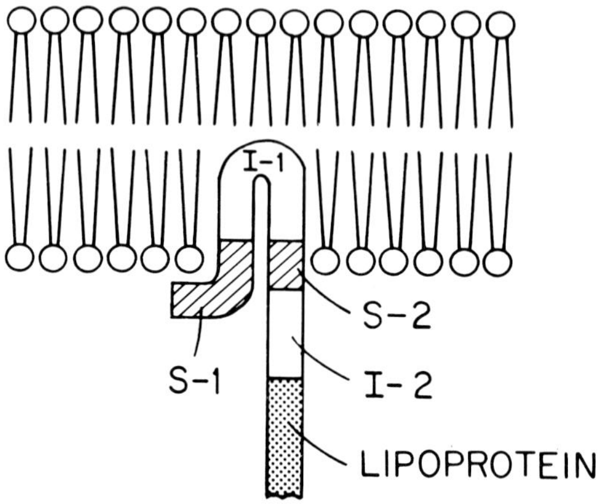

I-1

S-2

S-1

I-2

LIPOPROTEIN

FIG. 11. Amino acid sequence of the N-terminal peptide extension of the Braun lipoprotein. (From Inouye *et al.*, 1977b.) The amino acid sequence of the lipoprotein peptide extension and a possible arrangement in the lipid bilayer is shown. The S-1, I-1, S-2, and I-2 regions are alternately hydrophilic and hydrophobic containing amino acids 1-5, 6-14, 15, 16, and 17-20, respectively.

This peptide extension has a high content of hydrophobic amino acids which are grouped together in a nonrandom manner (Fig. 11). Inouye *et al.* (1977b) have suggested that the structure may fold as shown in Fig. 11. It is an attractive possibility (suggested in Inouye *et al.*, 1977b) that the peptide extension plays a key role in the transport of the lipoprotein to its eventual location in the cell. Lipoprotein is actively synthesized in toluene-treated cells of *E. coli* in the presence of added amino acids. If the cells are briefly pretreated with rifampicin, prior to toluenization, lipoprotein is almost the sole product, presumably due to the stability of its message. Two species of lipoprotein product have been identified, one being the prolipoprotein the other being lipoprotein. The presence of both precursor molecule and product suggests that toluenization of cells interferes with the processing enzyme which converts prolipoprotein to lipoprotein (Inouye *et al.*, 1977b).

The synthesis of *E. coli* alkaline phosphatase in a cell-free system gives rise to a polypeptide product that appears to have a higher molecular weight than the mature alkaline phosphatase found in the periplasm (Inouye and Beckwith, 1977), and the product is relatively hydrophobic. The alkaline phosphatase precursor might have a peptide extension analogous to that

present in the lipoprotein. The proalkaline phosphatase polypeptide is processed to mature size by an outer membrane fraction of *E. coli* (Inouye and Beckwith, 1977). Alkaline phosphatase and the Braun lipoprotein have the common feature that they are exported to locations outside the cytoplasmic membrane. Another exported protein, the penicillinase of *Bacillus licheniformis* (Yamamoto and Lampen, 1975, 1976), is also synthesized in precursor form. Blobel and co-workers (Blobel and Sabatini, 1971; Blobel and Dobberstein, 1975) have postulated from an analysis of the structure of secretory proteins that proteins destined for export into membrane-bound compartments have short-lived amino terminal peptide extensions that determine their transport across the membrane ("signal sequence") by establishing functional ribosome–membrane junctions. Peptide extensions that have been examined have a high content of hydrophobic amino acids (Devillers-Thiery *et al.*, 1975; Schechter *et al.*, 1975).

Secreted proteins in higher cells, particularly those from liver and pancreas, are synthesized on membrane-bound polyribosomes. This is probably also true for "secreted" proteins in bacteria. Some convincing evidence that it is the protein product rather than the ribosomes or mRNA which is responsible for the specificity of attachment of polysomes to membranes has been obtained by Wirth *et al.* (1977) from studies of Sindbis virus envelope protein synthesis. Ribosomes engaged in outer membrane protein synthesis in *E. coli* do not appear to differ from those engaged in the synthesis of cytoplasmic proteins (Randall and Hardy, 1975) which tends to support this idea. It seems likely that the N-terminal leader sequence of the Braun lipoprotein precursor will perform this same membrane attachment function in bacteria and that other outer membrane proteins will be synthesized and exported in the same way.

ii. Posttranstational Modification. If the apoprotein precursor of lipoprotein is produced as described above, it will be present in the cytoplasmic membrane upon completion of the peptide chain. It must then be cleaved to produce the mature protein, converted to lipoprotein by addition of the lipid moiety, and transported outside the cell (not necessarily in the given order). The proteolytic cleavage reaction can occur in toluenized cells, but the mechanism of cleavage is not understood. Addition of lipid to the protein presumably occurs in the cytoplasmic membrane. Glycerol is probably first transferred to the protein, the glycerol donor being phosphatidylglycerol (H. Wu, personal communication); the glycerol hydroxyl groups and the terminal cysteine amino group must then be acylated.

iii. Linking of Lipoprotein to Peptidoglycan. The mechanism of this reaction is unknown, however, it has been suggested that the enzyme LD-carboxypeptidase which is also a transpeptidase (Section VI,E) is responsible for linkage of lipoprotein to peptidoglycan. The bond energy of the diamino-

pimelyl-D-alanine link could be used for formation of the peptide bond diaminopimelyl(ε-lysyl)-lipoprotein (Braun, 1975). Attachment of lipoprotein to murein is insensitive to penicillin G (Braun *et al.*, 1974) and mecillinam (Braun and Wolff, 1975). Thus, transfer of lipoprotein to peptidoglycan may be the *in vivo* function of the penicillin insensitive LD-carboxypeptidase (Nguyen-Disteche *et al.*, 1974).

Pulse chase experiments indicate that a newly made free form of the lipoprotein is the precursor of the peptidoglycan-bound form (Inouye *et al.*, 1972). However, during a chase period the free form of lipoprotein appears to become sequestered and is no longer available for transfer to peptidoglycan (Inouye *et al.*, 1974). Translocation of the free form of the lipoprotein into the outer membrane presumably removes it as a substrate for joining to peptidoglycan.

d. Mutants with Defects in Lipoprotein Synthesis. The formation of lipoprotein involves a series of posttranslational modifications which include cleavage of apoprotein, addition of glycerol to the cysteine sulfhydryl group of the protein, fatty acid transfer to the glycerol moiety to form diglyceride, *N*-acylation of the cysteine amino group, and finally attachment of free lipoprotein to peptidoglycan. The sequence of these reactions is not necessarily in the given order, and none of the reactions have been described in detail. The isolation and characterization of mutants defective in lipoprotein synthesis has been initiated to define the steps involved and the order in which they occur and also to determine the function of lipoprotein in the cell.

i. Mutants with Lipoprotein of Altered Structure. Wu and Lin (1976) isolated mutants of *E. coli.* K12 with defects in lipoprotein synthesis, one of which had a very low content of peptidoglycan-bound lipoprotein. The free from of lipoprotein, which is located in the outer membrane of this mutant, was subsequently shown to lack covalently linked diglyceride (Wu *et al.*, 1977) and as expected contained a cysteine residue with a free sulfhydryl group. The mutant protein contained isoleucine, leucine, glutamic acid, lysine, and possibly glycine as extra amino acids and had a blocked N-terminus. These properties suggest it is the unprocessed apoprotein that accumulates in this mutant. Its structure must prevent it from being linked to peptidoglycan. The mutant protein forms dimers containing disulfide bridges which can be converted to monomers by reduction.

Suzuki *et al.* (1976) described an apparently similar mutant of *E. coli* K12 which produces a lipoprotein also able to form disulfide-linked dimers; however, the lipoprotein has the normal diglyceride component. The presence of an internal cysteine residue, the result of an amino acid replacement, is responsible for the dimerization of the mutant lipoprotein in this case (Inouye *et al.*, 1977a). The mutation termed *lpp* is located between *aro D* and *man* on the *E. coli* chromosome (Suzuki *et al.*, 1976). It almost

certaintly defines the structural gene for lipoprotein. The mutation described by Wu and co-workers, termed *mlp A*, is located in the same region of the chromosome between *aro D* and *man* (Yem and Wu, 1977). Merodiploid derivatives of the *mlp* A^- mutant containing F′*mlp* A^+ produce both normal and mutant forms of the lipoprotein, suggesting that the *mlp A* mutation is in the structural gene for lipoprotein rather than in a gene for a processing enzyme (Yem and Wu, 1977). The *mlp A* mutation is interesting in that it appears to prevent processing of the apoprotein. Primary amino acid sequence determination should show which region of the mutant protein is altered. The lack of glycerol in the accumulated precursor suggests that glycerol addition occurs after cleavage of the apoprotein.

Cells with the *lpp* (Suzuki *et al.*, 1976) and *mlp A* mutations (Wu and Lin, 1976) are viable and grow normally, except that the *mlp* A^- mutant cannot form colonies at the extreme growth temperature of 45°C (Yem and Wu, 1977).

ii. Mutants Lacking Lipoprotein. One mutant of this type has been described (Hirota *et al.*, 1977). It produces neither bound nor free lipoprotein, nor does it produce lipoprotein mRNA. The mutation giving rise to this phenotype is located in the same region of the chromosome as the mutants described above, between *aro D* and *man*. It is believed to be due to a small deletion, thus the simplest explanation for the properties of the mutant is that it has lost the lipoprotein structural gene.

The lipoprotein negative mutant (*lpo*) grows and divides normally but has an increased sensitivity to cationic dyes, EDTA, and detergents and leaks periplasmic enzymes. Thus, lipoprotein does not appear to be essential to the cell, although it clearly plays a role in maintaining the integrity of the outer membrane.

iii. Mutants defective in lipoprotein synthesis. A conditional amber mutant of *E. coli* K12 (ST 715) which produces peptidoglycan-linked lipoprotein at a reduced rate under nonpermissive conditions was isolated in a strain carrying a temperature-sensitive amber suppressor (Torti and Park, 1976). At nonpermissive temperature the mutant stops dividing and forms filaments which eventually lose viability. It is unclear if this effect on cell physiology is due to reduced synthesis of lipoprotein or to a pleiotropic effect of the ST 715 mutation on both lipoprotein biosynthesis and a second process essential for cell growth. The properties of the ST 715 mutant are due to a single mutation at *min 74* on the *E. coli* map near *mal A*. Thus it is clearly not a lipoprotein structural gene mutant.

2. MAJOR OUTER MEMBRANE PROTEINS (33–38K)

a. General Properties. The number of major outer membrane proteins in the molecular weight range 33,000 to 38,000 varies from one to four in different bacterial strains. At present they can be defined by molecular size

on gels (as denatured polypeptides), by their solubility properties in detergent solutions, and by their interaction with phages and bacteriocins. Table I indicates the number and size of these components in several strains of *E. coli* and in *S. typhimurium*. Unfortunately, the nomenclature is quite complex, each strain having its own set of letters and/or numbers to designate the polypeptide components. This system will persist until a more functional nomenclature can be adopted. The nomenclature in Table I is used in this section without repeated reference to the Table.

There are two main types of proteins within this group (called the 33–38K group for simplicity), those which can be extracted from cell envelope fractions with 2% SDS at 60°C, and those which cannot. This might be an oversimplification, and there are probably some exceptions, but at present it is a convenient classification. The proteins which cannot be extracted under the above conditions can be solubilized with 2% SDS at 100°C. A protein of this type has been characterized in detail by Rosenbusch (1974). It is the major envelope protein of *E. coli* BE, has a polypeptide MW of 36,500 and is present in 10^5 copies per cell. The protein is very tightly associated with the peptidoglycan layer in cell envelope fragments and is arranged on the outer face

TABLE I

MAJOR OUTER MEMBRANE PROTEINS OF SEVERAL GRAM-NEGATIVE BACTERIA[a]

MW	*E. coli* K12[b,c,d]	*E. coli* 0111[e]	*E. coli* B/r[c,f]	*E. coli* MX74[i]	*S. typhimurium*[g]
40,000	a				
38,500 (36,500)	b, Ia	1	Ia,A_1 matrix protein	*e*	36,000
38,000 (36,000)	c, Ib				
		2			35,000
					34,000
33,000	d, II*	3a + 3b	II*, B	*g*, *h*	33,000
17,000	III		III		[h]
7,500	IV	[h]	IV	*j*	[h]

[a] Modified from Nikaido (1978).
[b] Lugtenberg *et al.* (1975).
[c] Garten *et al.* (1975).
[d] Rosenbusch (1974).
[e] Schnaitman (1970).
[f] Bragg and Hou (1972).
[g] Ames *et al.* (1974).
[h] Torti (1977).
[i] Inouye and Yee (1973).
[j] Boman *et al.* (1974).
[k] In cases where two systems of nomenclature have been established for the same protein in the same organism, these are given in the vertical columns. For example b and Ia are alternative names for the 38,500 (36,500) protein of *E. coli* K12. The proteins of the different organisms listed in the Table that correspond to one another are grouped in the same horizontal line. Thus protein b (Ia) of *E. coli* K12 corresponds to protein 1 of *E. coli* 0111 and to protein Ia (A1) of *E. coli* B/r. Molecular weights of *S. typhimurium* proteins are listed separately since they do not correspond exactly with the *E. coli* molecular weights. Protein 2 of *E. coli* 0111 is indicated as having a molecular weight between proteins Ib and II* of *E. coli* K12.

of the peptidoglycan in a lattice structure which has hexagonal symmetry. This symmetrical structure persists even if peptidoglycan is removed by lysozyme treatment (Steven *et al.*, 1977). Because of these properties this protein has been termed the "matrix" protein. While in the lattice the protein is insensitive to trypsin but on heating at 100°C in 2% SDS it becomes sensitive. A large fraction of the protein in the lattice is in the β-configuration, and in this form it binds SDS poorly. The stability of the lattice structure even in the absence of peptidoglycan is quite remarkable and indicates strong protein–protein interactions as well as strong protein–peptidoglycan interactions. A covalent linkage of at least some of the matrix protein units to peptidoglycan could account for the observed stability. Since the Braun lipoprotein extends upward from the peptidoglycan surface, it may interact with the matrix protein. It would be of interest to study the matrix protein–peptidoglycan interaction in a lipoprotein-minus mutant. The hexagonal symmetry of the lattice is based on a unit of threefold symmetry (Steven *et al.*, 1977), and if it is assumed that there are three protein monomers per unit the protein could cover a large part of the cell surface. Whether or not such lattice structures really exist in the untreated cell remains to be seen.

Proteins with similar properties to the *E. coli* BE matrix protein are present in a number of *E. coli* K12 strains (Rosenbusch, 1974). Proteins Ia and Ib of Hindenach and Henning (1975), which are equivalent to proteins b and c of Lugtenberg *et al.* (1975), are peptidoglycan associated (Lugtenberg *et al.*, 1976). This is probably also the case for protein 1 of *E. coli* 0111 (Schnaitman, 1974a). The SDS solubility properties of the 33–38K proteins of the *Salmonella* outer membrane suggest that they too have properties like the matrix protein (Ames *et al.*, 1974).

A protein of the class that is easily extractable from the cell envelope with SDS is protein II* of *E. coli* K12 (Henning *et al.*, 1973), also called protein d (Lugtenberg *et al.*, 1975). Its relationship to the protein matrix is not known. Proteins in other strains that correspond to II* are 3a in *E. coli* 0111 (Schnaitman, 1974a) and B in *E. coli* B/r (Bragg and Hou, 1972). This class of proteins has been termed "heat-modifiable" because heating during sample preparation for gel analysis gives a form with a reduced mobility in acrylamide gels; the basis for this alteration is unknown.

Mutants lacking protein II* (d or 3A) are defective as recipients in conjugation with donor cells carrying F-type pili (Skurray *et al.*, 1974; Havekes *et al.*, 1976; Manning and Reeves, 1976a). The isolated protein complexed with lipopolysaccharide can block F pilus-mediated conjugation (Van Alphen *et al.*, 1977). Additional properties of this protein are discussed in the Section V,A,2,c.

Protein 2 of *E. coli* 0111 (Schnaitman *et al.*, 1975) is unique in that it is present as a consequence of lysogenization of this organism with phage Pa-2. It is absent from most of the *E. coli* strains that have been examined.

b. Biosynthesis. The mRNA for the 33–38K proteins, like that of Braun lipoprotein mRNA, is considerably more stable than normal (3 minutes), ranging from 5.5–11.5 minutes. (Hirashima *et al.*, 1973; Lee and Inouye, 1974). In *E. coli* minicells the mRNA for proteins I and II* is even more stable with a half-life of 40–80 minutes) (Levy, 1975). Membrane binding of polysomes producing these proteins may decrease the rate of mRNA degradation. No details of the biosynthesis of these proteins are known.

c. Factors that Alter the 33–38K Protein Composition. i. Growth Conditions. In *E. coli* 0111 the 33–38K proteins are affected by growth conditions (Schnaitman, 1974b). Protein 1 is the major species in exponential growth, while protein 3 is the major species in stationary phase. Under conditions of catabolite repression, protein 2 decreases and protein 1 increases. Fluctuations in the levels of the matrix proteins b and c of *E. coli* K12 with growth temperature and medium have been observed (Lugtenberg *et al.*, 1976). When one decreases, the other increases maintaining a fairly constant level of total matrix protein.

In *E. coli* BE, which has only one matrix protein (protein b), effects of growth conditions and temperature are much less pronounced (Lugtenberg *et al.*, 1976). Therefore, under a variety of growth conditions, the cell maintains a fairly constant level of matrix protein; a decrease in the level of one results in a compensatory increase in the level of another. This could suggest that these proteins are essential for cell growth but as will be seen below, this does not appear to be the case.

ii. Lipopolysaccharide mutations. "Deep rough" lipopolysaccharide mutants of both *Salmonella typhimurium* (Ames *et al.*, 1974) and *E. coli* (Koplow and Goldfine, 1974; Van Alphen *et al.*, 1976) are markedly deficient in the 33–38K proteins of the outer membrane. It is not apparent why the "deep rough" lipopolysaccharide defects (structures Rd_1, Rd_2, Re) result in reduced amounts of protein in the outer membrane. The proteins are not excreted nor are they present elsewhere in the cell. They must, therefore, either be produced normally, then degraded, or their synthesis must be regulated. If these proteins are normally transported to the outer membrane in a protein–lipopolysaccharide complex, failure to form such a complex could lead to an accumulation of protein in the cytoplasmic membrane which might regulate its own synthesis. Mutants containing galactose-deficient lipopolysaccharides (Ames *et al.*, 1974) with Rc structures (Fig. 1) do not show the same drastic reduction in outer membrane proteins.

The most important difference between the Rc and Rd structure is the reduced amount of heptose-linked phosphate in the latter. The low levels of outer membrane proteins in deep rough mutants have a marked effect on overall membrane structure which increases their sensitivity to ionic detergents, dyes and certain drugs.

iii. Protein mutations. Bacteriophage-resistant and colicin-tolerant mutants of *E. coli* K12 have been isolated that lack one or more of the 33–38K major proteins. *tolF* mutants of *E. coli* K12 have been shown to lack protein Ia (Chai and Foulds, 1977). Certain phage-resistant mutants of *tolF* strains lack proteins Ia and Ib; independent mutations lead to the loss of these proteins.

Colicin-tolerant mutants of *E. coli* K12, designated P530 and P692, isolated by Davies and Reeves (1975b) lack protein I (Ia and Ib) (Henning and Haller, 1975). Mutants of *E. coli* K12 tolerant to colicin G (Chai and Foulds, 1974), resistant to bacteriophage K3 (Skurray *et al.*, 1975), or resistant to bacteriophage Tu II* (Henning and Haller, 1975) lack protein II*. Bacteriophage Tu II*-resistant mutants of strains P530 and P692 (protein I-minus) are missing proteins Ia, Ib, and II* (Henning and Haller, 1975). The double mutants, although lacking all of the 33–38K major proteins, grow and divide normally and do not have increased sensitivity to detergents, EDTA, or antibiotics. Lipopolysaccharide structure in these mutants is normal. Mutants of *E. coli* K12 lacking proteins b, c, and d have been described by Van Alphen *et al.* (1976); however, these also have a defective lipopolysaccharide structure. The mutants had lost about 75% of their outer membrane protein with no effect on growth and division. Nurminen *et al.* (1976) isolated a group of phage-resistant mutants of *S. typhimurium* lacking 33 and 36K major outer membrane proteins. Like the *E. coli* mutants, they have normal growth properties.

Why is the loss of a major portion of the outer membrane protein not lethal? It is likely that one role of the "matrix" protein is to anchor the outer membrane to peptidoglycan. The Braun lipoprotein probably also has an anchoring function, therefore, loss of matrix protein may not destabilize the outer membrane. If a stable, anchored, outer membrane is required for cell viability, loss of both the matrix protein and the Braun lipoprotein might be expected to be lethal. Surprisingly, this is not the case, since Henning and co-workers (1977, personal communication), have constructed mutants lacking both protein I and lipoprotein (*lpo* mutation) which have normal growth properties. Either an anchoring function is unnecessary or proteins other than protein I and lipoprotein also have this function. There may well be a significant amount of functional redundancy among outer membrane proteins. Protein II* and lipoprotein may have an essential function in common, since mutants lacking both proteins grow very poorly (U. Henning, personal communication).

B. Other Proteins of the Outer Membrane

A major protein of MW 15,000 present in *E. coli* B/r has been implicated in the coupling of DNA replication and cell elongation (James, 1975). This protein increases in amount when cells are induced to form filaments with nalidixic acid, and it decreases in amount in cells in which elongation is inhibited by mecillinam. The assignment of a function to this protein is quite tentative, since effects of the drug treatments on cells are poorly understood.

A variety of proteins that act as bacteriophage receptors have been identified in *E. coli.* The T6 receptor is a 25,000 MW polypeptide which makes up about 8% of the outer membrane (Section VII,B,4). The receptor protein for phages T1, T5, and ϕ80 is a minor protein of 78,000 MW (Section VII,B,2). A number of outer membrane proteins are inducible. For example, the protein receptor for bacteriophage λ, the product of the *lamB* gene of the maltose operon, is induced by maltose from a level of 60 molecules/cell to 6000 molecules/cell (Section VII,B,3). Growth of *E. coli* K12 in iron-deficient medium leads to induction of three outer membrane proteins of MW 74,000, 81,000, and 84,000 (Braun *et al.*, 1976; Hancock and Braun, 1976b; Hancock *et al.*, 1976). In *S. typhimurium* four outer membrane proteins are induced by growth of cells in iron-deficient minimal medium which together make up 20% of the outer membrane protein (Bennett and Rothfield, 1976). Three of these proteins of MW 82,000, 79,000, and 77,000 are regulated by iron concentration, while the fourth (MW 45,000) is not. The iron-regulated proteins in the two organisms are involved in iron uptake (see Section VII,B,2). Mutants of both *E. coli* K12 (Davies and Reeves, 1975a; Braun *et al.*, 1976) and *S. typhimurium* (Bennett and Rothfield, 1976) have been isolated which produce the iron regulated proteins constitutively. Additional outer membrane proteins are discussed in Section VII.

C. Variation in Outer Membrane Protein Composition

The outer membrane is clearly not just a structural layer of invariable composition. Its protein composition varies significantly with growth conditions, suggesting that it has a functional role in the physiology of the cell.

Loss of a major fraction of the outer membrane protein does not appear to affect the cell growth and division. Thus, either the proteins that are lost are unnecessary and can be dispensed with or several proteins share the same function (functional redundancy) and can compensate for one another.

VI. Orientation of Components of the Outer Membrane

Most membranes are freely permeable to small molecules that dissolve in the lipid bilayer, such as detergents, hydrophobic dyes, and water-insoluble drugs. Gram-negative bacteria are insensitive to such agents unless the outer membrane is at least partially removed, for example, by treating with EDTA which releases up to 50% of the lipopolysaccharide plus protein and phospholipid (Leive, 1974). EDTA-treated cells are sensitive to detergents, dyes (such as methylene blue), and drugs (such as actinomycin D) (Leive, 1974). This indicates that the outer membrane acts as a permeability barrier to hydrophobic substances, thus it must have an atypical membrane structure.

What features of the outer membrane could give it such properties? To answer this question, it is necessary to understand how the various components of the outer membrane are arranged in the bilayer.

A. Lipopolysaccharide Asymmetry

Despite early findings to the contrary, lipopolysaccharide is present only in the outer half of the bilayer. This was demonstrated by Mühlradt and Golecki (1975) who used ferritin-labeled anti-lipopolysaccharide antibody to locate lipopolysaccharide in the outer membrane of spheroplasts of *S. typhimurium.* All of the ferritin label was localized on the outer surface of the outer membrane. A critical parameter in this experiment was temperature, which had to be maintained at 0°C; at higher temperatures rearrangement of the outer membrane occurred and lipopolysaccharide was detected on both the inner and outer surfaces of the membrane.

B. Phospholipid Asymmetry

Early studies of deep rough mutants (Rd_1, Rd_2, Re) of both *E. coli* and *Salmonella* indicated dramatic alterations in their sensitivity to detergents, dyes, drugs (Roantree *et al.*, 1969; Schlecht and Westphal, 1970; Tamaki and Matsuhashi, 1971; Leive, 1974), and polycyclic mutagens (Ames *et al.*, 1973) compared to organisms with an S- or R-type lipopolysaccharide structure up to chemotype Rc. The deep rough mutations cause a marked decrease in the amount of outer membrane protein in both *E. coli* (Koplow and Goldfine, 1974) and *S. typhimurium* (Ames *et al.*, 1974). A careful analysis of protein, phospholipid and lipopolysaccharide in the outer membrane of deep rough mutants of *S. typhimurium* indicated that along with the decrease in protein, there was a compensatory increase in phospholipid of about 70% (Smit *et al.*, 1975). It was calculated by Nikaido and his co-workers that in both smooth and rough strains of *S. typhimurium* the amount of phospholipid in the outer membrane is slightly less than is required to cover one side of the membrane as a monolayer. In deep rough

Rd and Re mutants in which there is a 70% increase in phospholipids, they must be distributed between the two halves of the outer membrane.

Smit *et al.* (1975) proposed a model for the outer membrane in which all of the phospholipid is normally present in the inner half of the leaflet, and all of the lipopolysaccharide is normally present in the outer half of the leaflet. It was proposed that the absence of a lipid bilayer in a normal outer membrane explained its impermeability. Only in deep rough mutants is there a true phospholipid bilayer which explains the increased sensitivity of deep rough mutants to hydrophobic molecules.

A direct test for the presence of phospholipids in the outer leaflet of the outer membrane was performed by treating intact cells of *S. typhimurium* with phospholipase C (Kamio and Nikaido, 1976). Phospholipid degradation was observed only in the case of deep rough mutants, supporting the above model of membrane organization. An alternative procedure using cyanogen bromide-activated dextran, a nonpenetrating reagent, to label amino groups resulted in 1.5–3% of total phospholipid being linked in the case of a deep rough mutant, but no linkage in the case of an S strain or its Rc mutant. Results obtained from electron spin resonance studies of an Rc mutant of *S. typhimurium* indicate that lipopolysaccharide and phospholipid are segregated in the outer membrane, which also supports the asymmetric distribution model (Nikaido *et al.*, 1977a).

C. Protein Asymmetry

Freeze-fracture studies of both *E. coli* (Van Gool and Nanninga, 1971) and *S. typhimurium* (Smit *et al.*, 1975) show the outer half of the outer membrane to be densely covered with particles, 8–10 nm in diameter (presumed to be protein), whereas the inner half is quite smooth. Because fractures through the outer membrane bilayer are quite rare, the structures that are observed may not be truly representative of the whole outer membrane. It seems likely, in fact, that many proteins must span the two leaflets. Regions of this type, where proteins span the membrane, may be much more resistant to freeze cleavages and thus would not be seen using the freeze cleavage technique. Deep rough mutants (Smit *et al.*, 1975) contain decreased numbers of particles in the outer half of the outer membrane, which correlates well with the reduced amount of protein. Cleavage through the outer membrane bilayer occurs more frequently in the case of deep rough mutants, which could be due to the presence of phospholipids in both halves.

The exact arrangement of the proteins in the outer half of the outer membrane is still uncertain. Much of the outer membrane protein appears to be exposed at the cell surface because treatment of intact cells of *S. typhimurium* with CNBr-activated dextran results in about 10 out of 13 outer membrane polypeptides becoming dextran-linked (Kamio and Nikaido, 1977). Out of the four major proteins of molecular weight around 35,000, three are exposed on the cell surface. Proteins, such as the matrix protein of *E. coli* BE (Section V,A,2,a), may exist *in vivo* as the same kind of lattice structure that has been characterized in cell envelope fractions. In the form of a lattice, this protein could cover a large part of the cell surface. Other major proteins (Braun lipoprotein and the heat-modifiable major protein) and minor proteins might interact with the matrix protein (Henning *et al.*, 1973). Since the matrix protein remains bound to peptidoglycan after treatment with SDS at 60°C (Section V,A,2,a), it is likely that some of the matrix protein penetrates through the bilayer to interact with peptidoglycan and is perhaps covalently joined to it. However strong protein–protein interactions appear to be the major force holding the lattice together.

The whole outer membrane must also interact with peptidoglycan-bound Braun lipoprotein, which projects from the peptidoglycan into the outer membrane. The extent to which it penetrates the outer membrane is not known, but it is large enough to reach to the outer surface. The proteins of the outer membrane are tightly packed because most of them can be cross-linked by bifunctional imidoester reagents with short spacer arms (Haller and Henning, 1974). The densely packed particles present in the outer leaflet, seen in electron micrographs of freeze-fractured cells, are also indicative of tight packing of proteins (Smit *et al.*, 1975), at least in some regions of the outer membrane.

D. Dynamic Aspects of the Outer Membrane

Newly made lipopolysaccharide first appears on the *Salmonella* cell surface at a few hundred discrete sites as shown by specific labeling with ferritin-labeled antibody (Mühlradt *et al.*, 1973, 1974). This was demonstrated using a mutant with a conditional defect in lipopolysaccharide synthesis (galactose-minus) which when placed under permissive conditions (supplied with galactose) produced lipopolysaccharide of normal structure and antigenicity. The appearance of lipopolysaccharide at discrete sites is followed by a rapid lateral diffusion resulting in a uniform distribution over the cell surface. The diffusion rate of 3×10^{-13} cm^2/second is much slower than the lateral diffusion rate of phospholipids of 5×10^{-9} cm^2/second (Lee *et al.*, 1973). The slow diffusion could be due to the structural properties of the lipopolysaccharide or alternatively could be due to the unusual makeup of the outer leaflet of the outer membrane. It is also possible that lipopolysaccharide diffuses in association with other membrane components.

The diffusion properties of two outer membrane proteins have been examined. One of these is the receptor for bacteriophage λ, which can be induced by addition of maltose to growing cells and can be detected by phage attachment using the electron microscope. Newly made receptor first appears on the cell surface in the region of the growth zone and does not diffuse away (Ryter *et al.*, 1975). The receptor for bacteriophage T6 behaves in a similar manner (Begg and Donachie, 1973). The combination of an amber mutation in the T6 receptor gene and a temperature-sensitive amber suppressor makes it possible to study conditional synthesis of the receptor protein which can be detected in the same way as the λ receptor. Receptor production in cells devoid of T6 receptor results in a uniform distribution of newly formed receptor over the cell surface. However, if receptor synthesis is switched off in cells that have a normal amount of receptor, preexisting receptor remains where it is and does not diffuse into the new growth zone of the cells (Begg and Donachie, 1973). Protein movement appears to be more restricted than lipopolysaccharide movement, as might be expected if much of the outer membrane protein is in the form of a structural matrix. Covalent linkage of proteins to the murein layer would be expected to prevent diffusion, but there is no evidence to indicate that either the λ or T6 receptor proteins are associated with the murein in this way.

VII. Functions of the Outer Membrane

A. General Permeability Properties

Outer membrane permeability is anomalous in two respects compared to other membranes, including the cytoplasmic membrane. It is relatively impermeable to small hydrophobic molecules, such as detergents, dyes, and drugs, and it is freely permeable to small hydrophilic molecules. The hydrophobic barrier properties of the outer membrane of both *E. coli* and *Salmonella* are altered by treating cells with EDTA, which releases lipopolysaccharide and proteins from the outer membrane. Deep rough lipopolysaccharide mutations also affect the permeability of the outer membrane to hydrophobic compounds. Both EDTA-treated cells and deep rough mutants (Gustafsson *et al.*, 1973) are sensitive to dyes (such as crystal violet), to drugs (such as novobiocin and actinomycin), and to detergents (such as deoxycholate and SDS). EDTA treatment may strip off regions of the outer membrane (Bayer and Leive, 1977) and allow hydrophobic molecules direct access to the cytoplasmic membrane. The outer membranes of deep rough mutants, unlike normal outer membranes contain areas of phospholipid bilayer (Section VI,B) which can be penetrated by these reagents.

Nikaido (1976) studied the rate of diffusion of a semisynthetic penicillin, nafcillin, into deep rough mutants and found it to be ten-fold less than its rate of diffusion into spheroplasts. Thus, in intact cells the rate-limiting step in the uptake of these drugs is its passage across the outer membrane. Diffusion rates were shown to be extremely sensitive to temperature and were also influenced by the hydrophobicity of the diffusing molecule. These characteristics suggest that the diffusing molecule dissolves in the phospholipid phase of the membrane prior to its appearance within the cell (Sha'afi *et al.*, 1971).

The permeability of the outer membrane to small hydrophilic molecules does not seem to be related to the type of permeant molecule but rather to its size. Solutes above a certain size class which varies from 600 to 1000 MW, depending upon the type of solute, do not penetrate the outer membrane. The exclusion limit seems to depend upon the molecular dimensions of the solute, which are influenced by the presence or absence of charged groups. Uptake of most monosaccharides and amino acids is not retarded by the outer membrane to any significant degree.

Early observations by Payne and Gilvarg (1968) on the uptake of peptides by *E. coli* indicated that peptides greater than 562 MW (tetralysine) could not penetrate the cell. Similar observations in *S. typhimurium* were made by Decad and Nikaido (1976) using a series of sucrose–raffinose oligosaccharides. This oligosaccharide series does not penetrate the cytoplasmic membrane, either by passive diffusion or by active transport. Sucrose (MW 342), and raffinose (MW 504) penetrate the outer membrane, stachyose (MW 666) penetrates it less well, and verbascose (MW 828) does not penetrate it at all. The outer membrane seiving properties are unchanged even if the peptidoglycan layer is degraded with lysozyme, indicating that it is the outer membrane and not the peptidoglycan which is the permeability barrier (Nakae and Nikaido, 1975). Isolated outer membrane vesicles, which can be prepared by heating outer membrane fragments in the presence of Mg^{2+}, show the same seiving properties as intact cells (Nakae and Nikaido, 1975).

Nikaido and his co-workers have proposed that penetration of the outer membrane by hydrophilic compounds is likely to occur by one of two mechanisms, diffusion through water-filled pores or carrier-mediated uptake. To date the experimental evidence appears to favor diffusion through water-filled pores because the process is nonspecific with regard to the type of molecule that can penetrate the membrane, there is a sharp size cut-off, and the temperature coefficient is low, which is expected for a process which involves diffusion through water.

The properties of liposomes (closed vesicle structures) prepared *in vitro* using various components of the outer membrane of *S. typhimurium* shed further light on penetration of the outer membrane by hydrophilic substances. Liposomes containing only phospholipids and lipopolysaccharide are relatively impermeable to glucose, sucrose, and higher oligosaccharides (Nikaido and Nakae, 1973). However, if outer membrane proteins are also present the liposomes have permeability properties like the outer membrane (Nakae, 1975). From such results it was concluded that outer membrane proteins are essential for forming water-filled pores.

Nakae (1976a,b) has shown that the SDS-insoluble major outer membrane proteins (matrix proteins) from both *S. typhimurium* and *E. coli* B are highly active in pore production in the liposome system. Protein preparations containing mainly the Braun lipoprotein showed little activity in pore formation. The artificial system allowed uptake of a variety of small molecules and had a molecular weight cut-off similar to that of the outer membrane itself. These experiments suggest that one role of the matrix protein might be to form water-filled pores which span the outer membrane.

If the matrix proteins of the outer membrane are mainly responsible for pore formation, mutants lacking these proteins might be expected to have decreased outer membrane permeability to small hydrophilic molecules. This has indeed been found to be the case in mutant strains (Nurminen *et al.*, 1976) of *S. typhimurium* lacking two of the major outer membrane proteins known to be active in pore formation *in vitro* (Nakae, 1976a). Nikaido *et al.* (1977b) have shown that the permeability coefficient of the outer membrane of such mutants for cephaloridine, a β-lactam antibiotic, is about one-tenth that of the wild-type outer membrane. In contrast, loss of the 33,000 MW major outer membrane protein did not alter the permeability coefficient. Reduced outer membrane permeability does not seem to have a drastic effect on either *E. coli* or *S. typhimurium* because mutants lacking the major 33,000 to 38,000 MW proteins appear to have normal growth properties.

B. Transport Systems of the Outer Membrane

The outer membrane, although acting as a molecular sieve allowing entry of only relatively small hydrophilic molecules, has specific systems associated with it which facilitate the uptake of certain nutrients of large molecular weight. Specific transport systems for vitamin B_{12}, maltose and maltodextrins, ferric iron, and nucleosides have been identified. Some of these systems are also involved in the uptake of bacteriocins by cells and they also act as bacteriophage receptors.

1. VITAMIN B_{12}

The first well-characterized physiological function described for the outer membrane was the uptake of vitamin B_{12} which occurs in cells of *E. coli* in two distinct steps, an initial rapid binding to an outer membrane receptor followed by an energy-dependent transfer of vitamin B_{12} into the cytoplasm (DiGirolamo and Bradbeer, 1971; White *et al.*, 1973). The involvement of a specific receptor in the outer membrane is consistent with the idea that substances larger than 500–600 MW (B_{12} is 1355 MW) cannot enter the cell through the water-filled pores of the membrane. The vitamin B_{12} receptor is also a receptor for the E colicins and for bacteriophage BF23 (DiMasi *et al.*, 1973; Bradbeer *et al.*, 1976; Buxton, 1971; Jasper *et al.*, 1972). Vitamin B_{12}, the E colicins, and bacteriophage BF23 all compete for the outer membrane receptor; cells can be protected from either E colicin killing or BF23 infection by addition of excess vitamin B_{12}.

The outer membrane receptor for vitamin B_{12}, the E colicins and phage BF23 is a protein specified by the *bfe* gene (also called *btu B*) in *E. coli*. Binding of vitamin B_{12}, colicins, and phage can be eliminated by a single mutation at the *bfe* locus; there appears to be a single binding site on the protein to which all three attach. The receptor protein is normally present in about 200–250 copies per cell. It has been extracted from the membrane in soluble form and has a molecular weight of about 60,000 (Sabet and Schnaitman, 1973). The isolated protein has full receptor activity only for colicin E3.

The transfer of vitamin B_{12} from the outer membrane into the cell requires energy derived from the proton motive force (Bradbeer and Woodrow, 1976), and it occurs against a concentration gradient. Energy-dependent transport is sensitive to osmotic shock which releases a periplasmic B_{12} binding protein (White *et al.*, 1973); however the role of this periplasmic protein in vitamin B_{12} transport is unclear. Transport could occur by direct transfer of vitamin B_{12} from the outer receptor to the cytoplasmic membrane, perhaps at adhesion sites between the outer and cytoplasmic membranes, or it could occur via the periplasm. The energy-dependent uptake of B_{12} is dependent on a functional *ton B* product (Bassford *et al.*, 1976), which is also required for normal function of a number of other transport systems including siderophore-mediated ferric iron transport (see Section VII,B,2). *ton B* mutants are resistant to bacteriophages T1 and ϕ80 and to all of the group B colicins which adsorb to the outer membrane (Davies and Reeves,

1975a; Hancock and Braun, 1976a). Thus *ton B* mutations have pleiotropic effects on a number of uptake systems. *ton B* might, therefore, be regarded as a common "gate" which is involved in the entry into the cell of both small molecules, such as vitamin B_{12}, and macromolecules, such as colicins and phage nucleic acids. Although *ton B* mutations affect vitamin B_{12} uptake, they do not alter the sensitivity of the cell to the E colicins or to bacteriophage BF23 (Davies and Reeves, 1975b). Cells which are tolerant to the E colicins because of mutations in the E colicin uptake system (but not the *bfe* locus) are still sensitive to phage BF23 and can transport vitamin B_{12} (Kadner and Bassford, 1977). Thus, although a common surface receptor is used by vitamin B_{12}, the E colicins, and phage BF23, different uptake mechanisms are involved in their subsequent transport into the cell. Specific shut-off of the synthesis of the *bfe* gene product in *E. coli* results in a rapid loss of sensitivity to colicins E2 and E3 followed, subsequently, by a loss of sensitivity to phage BF23. Vitamin B_{12} uptake is not affected, suggesting that the *bfe* receptor protein is still present and can remain active in vitamin B_{12} uptake for some time after its synthesis (Bassford *et al.*, 1977a). Only newly synthesized *bfe* protein appears to be functional as a receptor for the uptake of the E colicins and for successful infection by phage BF23 (Bassford *et al.*, 1977b). The properties of the *bfe*-dependent uptake systems led Bassford *et al.*, 1977a,b) to suggest there are different functional states of the outer membrane receptor protein. They make the interesting speculation that only the newly synthesized *bfe* protein is properly oriented, perhaps with respect to the inner membrane, for lethal uptake of E colicins and phage BF23. Subsequent growth of the cell would lead to loss of the proper orientation, first with respect to colicin uptake and subsequently with respect to phage uptake. Vitamin B_{12} uptake is not affected by the changing orientation and requires only the continued synthesis of a functional *ton B* product (Bassford *et al.*, 1977a). Perhaps the *bfe* protein first emerges on the cell surface at adhesion sites and only at these sites is it an effective receptor for colicin uptake and phage infection.

2. Iron Transport

Under conditions of iron starvation, *E. coli* and other Enterobacteriaceae excrete a siderophore (iron chelator) called enterochelin or enterobactin, which solubilizes iron that would otherwise be unavailable to the cells (O'Brien and Gibson, 1970; Pollack and Neilands, 1970). The ferric enterochelin complex is too large to diffuse through the outer membrane and thus is transported into the cell via a specific high affinity outer membrane–cytoplasmic membrane system similar to that used for vitamin B_{12}. There are two other high-affinity systems for uptake of iron complexed with either citrate or ferrichrome.

Guterman (1973) suggested, from the observation that hyperexcretion of enterochelin by certain mutants of *E. coli* protected them from the lethal action of colicin B, that colicin B and ferric enterochelin compete for a common site. It was later demonstrated directly that colicin B (and D) and enterochelin compete for binding to the cell surface and for uptake (Wayne *et al.*, 1976; Pugsley and Reeves, 1976).

Growth of *E. coli* or *S. typhimurium* in iron-deficient medium results in the overproduction of certain outer membrane proteins (Braun *et al.*, 1976; Hancock and Braun, 1976b; McIntosh and Earhart, 1976; Bennett and Rothfield, 1976) which appear to be regulated by iron concentration and also appear to be involved in uptake of ferric chelates. *Escherichia coli* mutants with defects in the *ton B* locus are defective in all high-affinity iron transport systems (Frost and Rosenberg, 1975). Such mutants overproduce the same outer membrane proteins that are induced by iron starvation (Davies and Reeves, 1975a; Braun *et al.*, 1976). In *E. coli* there are three iron-regulated outer membrane proteins with MW 74,000, 81,000, and 83,000. Mutants of *E. coli* defective in uptake of ferrienterochelin (*feu*) have been isolated (Hantke and Braun, 1975a; Pugsley and Reeves, 1976). One class of mutants, *feu B*, lack the 81,000 MW outer membrane protein and are resistant to colicin B (Hancock *et al.*, 1976). It seems likely that the 81,000 MW protein is a receptor required for ferrienterochelin uptake (Wayne *et al.*, 1976).

The 74,000 MW protein, which is regulated by iron levels, is probably the receptor for colicin Ia, since it is absent in colicin Ia-resistant cells (Hancock *et al.*, 1976). Regulation of the colicin Ia receptor by iron has also been demonstrated directly (Konisky *et al.*, 1976). Although it has been claimed that the 74,000 MW colicin Ia receptor is necessary for ferrienterochelin uptake (Hancock and Braun, 1976b), this does not seem to be the case because mutants lacking the Ia receptor protein have a normal capacity for enterochelin-mediated uptake (Soucek and Konisky, 1977). The 74,000 MW protein may have a role in iron uptake, since its synthesis is regulated by iron levels but at present its role is unclear.

The second high-affinity iron uptake system mediates transport of iron complexed with ferrichrome. Mutants of *E. coli* defective in the *ton A* gene are defective in this transport system and are missing a 78,000 MW outer membrane protein (Braun *et al.*, 1973). Such mutants are also resistant to phages T1, T5, ϕ80, and colicin M; the purified 78,000 MW protein product of the *ton A* gene acts *in vitro* as a receptor for both phages and colicin (Braun *et al.*, 1973). Ferrichrome specifically blocks adsorption of ϕ80 and colicin

M to sensitive *E. coli* cells, thus preventing killing by these agents (Wayne and Neilands, 1975; Hantke and Braun, 1975b). Ferrichrome and bacteriophage T5 compete for partially purified receptor (Luckey *et al.*, 1975).

Albomycin-resistant mutants of *S. typhimurium* appear to be analogous to *ton A* mutants of *E. coli*, since they are defective in ferrichrome-mediated uptake (Luckey *et al.*, 1972). In *S. typhimurium*, bacteriophage ES18 uses the ferrichrome receptor protein as its receptor; its adsorption can be blocked by ferrichrome (Luckey and Neilands, 1976). The amount of the *ton A* product in the cell does not appear to be regulated by iron levels (Hancock *et al.*, 1976). The *ton A* uptake system for iron is quite independent of the system used for uptake of ferrienterochelin.

The third high-affinity iron uptake system mediates transport of iron complexed with citrate (Frost and Rosenberg, 1973). Growth of *E. coli* cells in medium containing citrate leads to induction of an outer membrane protein of MW 81,000, which is produced in relatively large amounts (Hancock *et al.*, 1976). This protein may be the outer membrane receptor which recognizes the iron–citrate complex.

These systems for siderophore iron transport are highly specific, a different protein or proteins acting as a surface receptor in each case. The uptake process, subsequent to binding of complexes on the cell surface, is not understood. Reduction of the metal ion appears to be a prominent feature of the process (Luckey and Neilands, 1976). Mutations in the *ton B* locus abolish all high-affinity transport systems for iron (Frost and Rosenberg, 1975) and in addition prevent infection by colicins B, D, Ia, Ib, M, and V and by bacteriophages T1 and ϕ80 (Hancock and Braun, 1976a). Receptor levels in *ton B* mutants are elevated for all of these agents, thus a step subsequent to binding must be blocked. It has been suggested that the *ton B* mutation uncouples an energy-dependent step required for uptake.

3. Maltose and Maltodextrins

The outer membrane receptor for bacteriophage λ is involved in the uptake of maltose by *E. coli*. The receptor is specified by a gene, *lam B*, which is part of the maltose operon, and its synthesis is controlled by the maltose regulatory gene *mal T* (Hofnung, 1974; Hofnung *et al.*, 1974; Randall-Hazelbauer and Schwartz, 1973). The λ receptor is a protein of MW 55,000 located exclusively in the outer membrane (Randall-Hazelbauer and Schwartz, 1973). Mutants with defects in the *lam B* protein show reduced maltose uptake when the external maltose concentration is low (Szmelcmann *et al.*, 1976; Szmelcmann and Hofnung, 1975); maltotriose uptake is even more reduced in such mutants. Although loss of the *lam B* product affects maltose uptake, maltose does not show any affinity for the purified *lam B* protein nor does it protect cells against λ infection. Phage λ normally interacts with the wild-type *lam B* protein in a reversible manner; however, if the receptor is treated with certain solvents, the interaction becomes irreversible (Schwartz, 1975), suggesting that a specific conformation of the receptor is required for irreversible adsorption. Maltose and maltodextrin uptake and phage adsorption do not involve the same sites on the *lam B* protein, since not all phage-resistant *lam B* mutants are defective in maltose uptake (Szmelcmann and Hofnung, 1975). The primary function of the *lam B* protein is probably the uptake of maltose and maltodextrins. In agreement with this idea, most Mal^+ strains of *E. coli* and *Shigella*, even though they do not support the growth of phage λ and often do not adsorb the phage, still have λ receptor activity which can be detected in cell extracts (Schwartz and LeMinor, 1975). Maltose itself is small enough to pass through the water-filled pores of the outer membrane; thus, the specific transport system may play its major role in the transport of maltodextrins which are too large to enter by this route.

Maltose uptake in *E. coli* is absolutely dependent on a periplasmic maltose-binding protein, which indicates that maltose passes from the outer membrane receptor to the periplasm (Kellerman and Szmelcmann, 1974). The *lam B* protein appears to be inserted only in specific regions of outer membrane as detected by phage adsorption (Ryter *et al.*, 1975). It will be very interesting to know what its relationship is to other proteins of the outer membrane and also to know how it accomplishes the translocation of substrates from the outer surface of the cell into the periplasm.

4. Nucleosides

Mutants of *E. coli* defective in the *tsx* locus which are resistant to bacteriophage T6 and colicin K have been shown to be defective in nucleoside uptake (Hantke, 1976). The T6 receptor specified by the *tsx* gene is a protein (Weltzien and Jesaitis, 1971) of MW 25,000 to 32,000 which is present in the outer membrane of *E. coli* (Hantke, 1976; Manning and Reeves, 1976b).

Mutants defective in nucleoside uptake (*nup* mutants) that have mutations located in the same region of the chromosome as the *tsx* gene have been characterized (McKeown *et al.*, 1976). The *nup* and *tsx* loci are probably the same. A direct role for the *tsx* protein in nucleoside uptake has not yet been determined, but if this is indeed its function it will be another example of a specific uptake mechanism in the outer membrane. The reason for a specific system in this case is not obvious, since the molecular weights of nucleosides are well below the cut-off point for diffusion through the outer membrane pores.

VIII. Overall Structure of the Outer Membrane

Smit *et al.* (1975) have proposed a model for the structure of the outer membrane (Fig. 12) based on the known asymmetry of its components. There is evidence that lipopolysaccharide and protein make up a good portion of the outer leaflet and that most of the phospholipid is present only in the inner leaflet. Some proteins must clearly span the membrane, but the extent to which this occurs is not known. Clearly many of the proteins are exposed at the cell surface (Section VI). Many of these function as receptors for uptake of nutrients, and they are also used as receptor sites by bacterio-

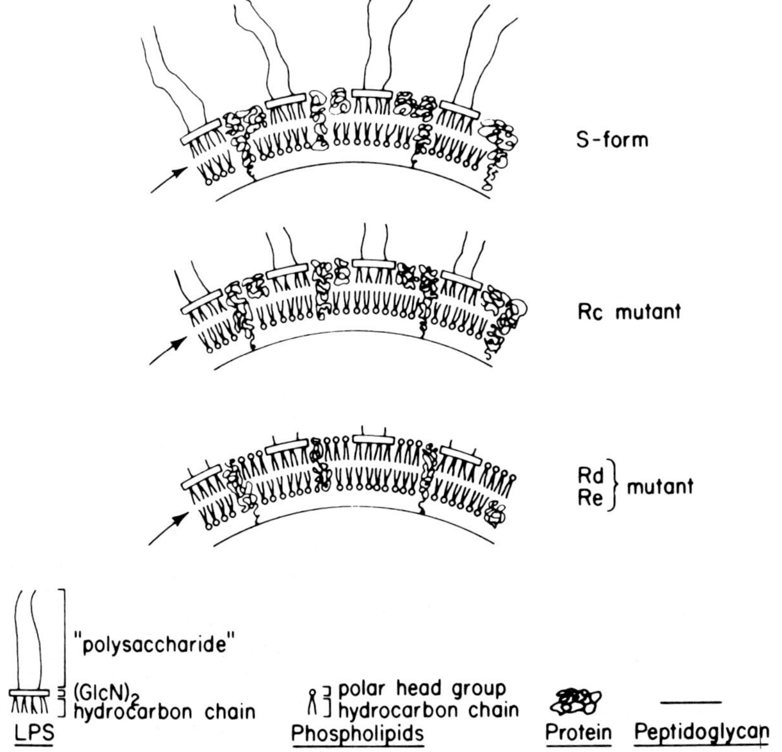

FIG. 12. Schematic representation of outer membrane structure of *S. typhimurium*. The S-form and Rc mutant outer membranes contain lipopolysaccharide and protein in the outer leaflet, and all of the phospholipid is present in the inner leaflet along with some protein. The Rd and Re mutants contain reduced amounts of protein in the outer membrane; phospholipid is present in both leaflets (see text). The arrows show the presumed plane of fracture. (From Smit *et al.*, 1975.)

phages and colicins. Receptors can have different functional states, probably depending upon whether they are newly synthesized or "older" (Section VII). Now that most of the components of the outer membrane have been defined, it should be possible to obtain a better understanding of its detailed architecture, in particular the spatial relationships between its protein components and how they interact with the underlying layers of the envelope.

IX. Biosynthesis of the Outer Membrane

As discussed in previous sections, all of the components outside the cytoplasmic membrane are formed on this membrane or at least in partial association with it. The problem beyond biosynthesis is one of export. It is clear that proteins such as the Braun lipoprotein associate with the cytoplasmic membrane and can perhaps even pass from its inner surface to its outer surface, but how it is further translocated is not clear. The problem is the same for all of the outer membrane components.

Some outer membrane components may be exported via junctions between the cytoplasmic and outer membranes. Such junctions or adhesion sites were first identified in electron micrographs of thin sections of plasmolyzed cells of *E. coli* (Bayer, 1968a; Bayer and Starkey, 1972) and shown to be present at a level of 200–400 per cell. The nature of the junctions has remained elusive; assuming they are stable, they must exist in various subfractions of the cell envelope, but they have not yet been isolated. Interestingly, adsorbed bacteriophages are preferentially located at adhesion sites, suggesting that release of viral DNA and infection of the cell takes place in these areas (Bayer, 1968b; Bayer and Starkey, 1972). Mühlradt *et al.* (1973) have shown using ferritin-labeled antibody that newly synthesized lipopolysaccharide first appears on the cell surface of *S. typhimurium* at a limited number of localized sites (see Section VI) which are coincident with adhesion sites. The total number of adhesion sites in *S. typhimurium* was also in the range of 200–400 per cell (Mühlradt *et al.*, 1973).

Newly synthesized lipopolysaccharide, detected by its ability to adsorb a specific bacteriophage, also appears at localized regions coincident with adhesion sites on the surface of *S. anatum* (Bayer, 1974, 1975). Bayer and Starkey (1972) have calculated that the adhesion sites in *E. coli* comprise about 6% of the total surface.

Kulpa and Leive (1976) used a biochemical approach to study the insertion of newly made lipopolysaccharide into the outer membrane of a *gal E* mutant of *E. coli* 0111, which lacks UDPgalactose-4-epimerase and hence makes a galactose-deficient lipopolysaccharide. The *gal E* mutant can synthesize an S-type lipopolysaccharide if it is provided with exogenous galactose. Cells of the gal E mutant grown briefly in the presence of galactose and then disrupted and fractionated on sucrose density gradients give rise to outer membrane fragments, a fraction of which have a higher than normal density. In a parallel experiment in which the brief period of exposure to galactose was followed by a period of growth in galactose-free medium, the high-density outer membrane fraction is no longer detectable. Kulpa and Leive (1976) interpret these results to indicate that the newly made lipopolysaccharide enters the outer membrane at a limited number of discrete sites and subsequently diffuses away to give an even distribution over the cell surface. The lipopolysaccharide produced in the presence of galactose has a higher density than the galactose-deficient lipopolysaccharide (see Section II). Thus, the membrane regions in which it first appears show an increased density. The number of growth sites calculated from this experiment was in the range of 10–50, which is in good agreement with the estimates made by Mühlradt *et al.* (1973) using the method described above. This appears to be a promising approach for isolation of regions of the membrane containing adhesion sites; however, the lateral diffusion of the lipopolysaccharide may limit its usefulness.

Several outer membrane proteins and periplasmic proteins have been shown to be synthesized as precursor molecules that carry a hydrophobic sequence (signal sequence) of amino acids at their *N*-termini (Section V,A,1,c). This sequence is thought to signal the initiation of the secretion process by imbedding itself in the cytoplasmic membrane. Different signal sequence structures could determine the ultimate location of the exported proteins, in the outer membrane or periplasm, but it seems more likely that other features of the polypeptide structure are involved in determining their localization. Deep rough mutants, which produce incomplete lipopolysaccharide, are deficient in major outer membrane proteins. Perhaps lipopolysaccharide–protein complex formation is required for efficient export of at least some of the outer membrane proteins. Besides the absence of additional core sugars, a major difference between deep rough and Rc mutant lipopolysaccharides is the absence of heptose-linked phosphate residues. These highly polar residues may be important in determining interaction of the core with outer membrane proteins. Many of the receptor activities of the outer membrane proteins are expressed *in vitro* only if they are complexed with lipopolysaccharide. Interactions between lipopolysaccharide and protein may be absolutely necessary to produce the correct or active protein conformation. The assembly of organized complexes of protein, lipopolysaccharide, and phospholipid on the cytoplasmic membrane, prior to export, would allow a coordinated insertion of all of the components into the outer membrane. Lipopolysaccharide, assembled on the inner surface of the cytoplasmic membrane, may form a nucleus for the assembly of outer membrane proteins;

transport of the entire complex, from cytoplasmic membrane to outer membrane, could occur perhaps with concomitant eversion to properly orient the components with respect to the cell surface. The process might be analogous to the formation of enveloped viruses. It remains to be seen whether the areas of adhesion between the outer and cytoplasmic membranes are essential sites of export of macromolecules to the cell surface or whether there are other ways in which this process can occur. The membrane fractionation methods and genetic techniques for manipulating membrane structure that are currently available should allow a better understanding of this important process.

The mode of insertion of the major proteins (porins) of the outer membrane of *S. typhimurium* has been examined by Smit and Nikaido (1978) using ferritin-labeled antibody to detect emergence of the proteins on the cell surface. All three major proteins MWT 34,000, 35,000, and 36,000 appear on the cell surface at a limited number of uniformly distributed discrete sites located at or near zones of adhesion between the inner and outer membranes. The protein, once deposited on the cell surface, does not diffuse away rapidly but forms a patch in the region of the insertion. These results suggest that the major protein framework of the outer membrane grows by a diffuse intercalation mechanism. For continued insertion over the entire cell surface by this mechanism, Smit and Nikaido (1978) suggest the formation of new adhesions and possibly the lateral displacement of existing adhesions. As described earlier in this section, lipopolysaccharide emerges on the cell surface at similar sites. These observations are consistent with the export to the cell surface of at least some of the protein in the form of a protein–lipopolysaccharide complex. A similar diffuse intercalation mechanism for insertion of matrix protein into the outer membrane of *E. coli* has been proposed by Begg (1978) from autoradiographic studies of cell envelope ghosts.

The growth of the outer membrane has been examined by Begg and Donachie (1977) using the phage T6 outer membrane receptor as a marker (see also Section VI,D). Newly synthesized receptor appears on the cell surface at many sites randomly distributed over the cell surface; however, if T6 receptor synthesis is specifically switched off, the preexisting receptors appear to remain within a conserved unit. Further growth under these conditions results in cells devoid of T6 receptor at either one or both poles. A T6 receptor distribution of this type is incompatible with a model of outer membrane growth by diffuse random intercalation and suggests a unique insertion mechanism for this particular protein.

Another receptor protein, the bacteriophage λ receptor, emerges on the cell surface in a fairly localized region near the potential division site of the cell (Ryter *et al.*, 1975). Whether or not it can subsequently diffuse laterally on the cell surface is unclear (see Section VI,D). From the examples given, it seems likely that the major proteins of the outer membrane are inserted by random intercalation into the existing membrane structure. This is not the case for all proteins as indicated by the behavior of the λ receptor protein. More detailed information on the mechanism of export of the outer membrane proteins and the association of these proteins with their neighbors in the cell envelope will be necessary for a better understanding of the overall growth process.

X. Conclusion

The complexity of gram-negative outer membrane structure and function as discussed in this chapter is only now being described in some detail. These data illustrate the fundamental differences between gram-negative and gram-positive bacteria in adaptation to variations in their environment, as mentioned in Chapter 6. The complexity of wall architecture in gram-negative organisms has been obvious for some time and poses obvious questions concerning mechanisms of insertion of outer membrane components and control of extension of both outer membrane and peptidoglycan components in the growing and dividing cell. The dearth of information concerning the distribution of different wall polymers in the gram-positive cell wall suggests that the relative simplicity of these structures, compared to those of the gram-negative cells, may be more apparent than real. We are still a long way from understanding the macroscopic events of cell wall growth, cell shape, and cell division control in any bacterium. Control of wall polymer synthesis is not even understood at the biochemical level.

Synthesis of the unique polymers of both gram-negative and gram-positive organisms is a complex process requiring the integrated function of several membrane assembly sites. Some of the complexities of lipopolysaccharide biosynthesis, such as the presynthesis of polymeric O antigen and its transfer to synthesize separately lipid A–core, have been recognized for some time. Similar complexities in the biosynthesis of teichoic acids have only recently been described, in large part because of the relative intractability of teichoic acid structure to sequential analysis. This has left the structure of the teichoic acid–peptidoglycan link unclear and the steps in its synthesis a matter of conjecture. The relationship between polymerization and cross-linking of peptidoglycan is understood at a rudimentary level, and it is possible that gram-negative cells, with a thin, minimal peptidoglycan structure, must control this process more stringently than gram-positive organisms. The apparent dependence on transpeptidation, of linkage of polymers such as teichoic acids to peptidoglycan, is not understood. In fact, understanding of the coordination of synthesis of the various polymers present in the walls and envelopes of bacteria of any kind is almost totally lacking. The controls of and reasons for cell wall turnover are similarly obscure: organisms like *E. coli* B, *S. aureus*, and *B. subtilis* 168 turnover their peptidoglycans at less than 2, 12, and 60% per generation, respectively. The expression of the synthetic and degradation capacities of bacterial cells may be severely dependent on environmental influences, and yet mutations causing loss of capacity for synthesis of nonpeptidoglycan components may be well tolerated in gram-positive organisms. Moreover, a gram-negative organism such as *E. coli* can lose an outer membrane protein comprising 5% of the total cell protein without marked growth impairment. In general, cell growth rates, division frequency, and shape seem to be little affected by gross changes in cell wall composition, except in an indirect fashion, possibly related to the uptake of divalent metal ions.

Different bacterial species have evolved widely divergent strategies for ensuring their own survival, and it is certainly presumptuous to suppose that any universally valid generalization can be made. Prokaryotes were present for billions of years prior to the evolution of the first primitive multicellular organisms and will probably survive the death of the last such organism on this planet. Our ultimate progenitors will be our ultimate progeny.

Acknowledgments

We would like to thank Drs. U. Henning, M. Inouye, H. Nikaido, M. J. Osborn, J. T. Park, P. D. Rick, M. Rosner, J. B. Ward, and H. Wu for helpful discussion and comments. The preparation of the review was supported by Grants AI 10806 and GM 15837, awarded to D. J. T. and A. W., respectively, by the National Institutes of Health.

References

Ames, B. N., Lee, F. D., and Durston, W. E. (1973). *Proc. Natl. Acad. Sci. U.S.A.* **70,** 782–786.

Ames, G. F., Spudich, E. N., and Nikaido, H. (1974). *J. Bacteriol.* **117,** 406–416.

Bassford, P. J., Bradbeer, C., Kadner, R. J., and Schnaitman, C. A. (1976). *J. Bacteriol.* **128,** 242–247.

Bassford, P. J., Schnaitman, C. A., and Kadner, R. J. (1977a). *J. Bacteriol.* **130,** 750–758.

Bassford, P. J., Kadner, R. J., and Schnaitman, C. A. (1977b). *J. Bacteriol.* **129,** 265–275.

Bayer, M. E. (1968a). *J. Gen. Microbiol.* **53,** 395–404.

Bayer, M. E. (1968b). *J. Virol.* **2,** 346–356.

Bayer, M. E. (1974). *Ann. N.Y. Acad. Sci.* **215,** 6–28.

Bayer, M. E. (1975). *In* "Membrane Biogenesis" (A. Tzagoloff, ed.), pp. 393–427. Plenum, New York.

Bayer, M. E., and Leive, L. (1977). *J. Bacteriol.* **130,** 1364–1381.

Bayer, M. E., and Starkey, T. W. (1972). *Virology* **49,** 236–256.

Begg, K. J. (1978). *J. Bacteriol.* **135,** 307–310.

Begg, K. J., and Donachie, W. D. (1973). *Nature (London), New Biol.* **245,** 38–39.
Begg, K. J., and Donachie, W. D. (1977). *J. Bacteriol.* **129,** 1524–1536.
Bennett, R. L., and Rothfield, L. I. (1976). *J. Bacteriol.* **127,** 498–504.
Blobel, G., and Dobberstein, D. (1975). *J. Cell. Biol.* **67,** 835–851.
Blobel, G., and Sabatini, D. D. (1971). *In* "Biomembranes" (L. E. Manson, ed.), Vol. 2, pp. 193–195. Plenum, New York.
Boman, H. C., Nordstrom, K., and Normark, S. (1974). *Ann. N.Y. Acad. Sci.* **235,** 569–586.
Bosch, V., and Braun, V. (1973). *FEBS Lett.* **34,** 307–310.
Bradbeer, C., and Woodrow, M. L. (1976). *J. Bacteriol.* **128,** 99–104.
Bradbeer, C., Woodrow, M. L., and Kalifah, L. I. (1976). *J. Bacteriol.* **125,** 1032–1039.
Bragg, P. D., and Hou, C. (1972). *Biochim. Biophys. Acta* **272,** 478–488.
Braun, V. (1973). *J. Infect. Dis.* **128,** Suppl., 9–15.
Braun, V. (1975). *Biochim. Biophys. Acta* **415,** 335–377.
Braun, V., and Bosch, V. (1972). *Proc. Natl. Acad. Sci. U.S.A.* **69,** 970–974.
Braun, V., and Wolff, H. (1970). *Eur. J. Biochem.* **14,** 387–391.
Braun, V., and Wolff, H. (1975). *J. Bacteriol.* **123,** 888–897.
Braun, V., Schaller, K., and Wolff, H. (1973). *Biochim. Biophys. Acta* **323,** 87–97.
Braun, V., Bosch, V., Hantke, K., and Schaller, K. (1974). *Ann. N.Y. Acad. Sci.* **235,** 66–82.
Braun, V., Hancock, R. E. W., Hantke, K., and Hartman, A. (1976). *J. Supramol. Struct.* **5,** 37–58.
Bray, D. C. H., and Robbins, P. W. (1967). *Biochem. Biophys. Res. Commun.* **28,** 334–339.
Bretscher, M. S. (1973). *Science* **181,** 622–629.
Buxton, R. S. (1971). *Mol. Gen. Genet.* **113,** 154–156.
Chai, T. J., and Foulds, J. (1974). *J. Mol. Biol.* **85,** 465–474.
Chai, T. J., and Foulds, J. (1977). *J. Bacteriol.* **130,** 781–786.
Cynkin, M. A., and Osborn, M. J. (1968). *Fed. Proc., Fed. Am. Soc. Exp. Biol.* **27,** 293.
Davies, J. K., and Reeves, P. (1975a). *J. Bacteriol.* **123,** 96–101.
Davies, J. K., and Reeves, P. (1975b). *J. Bacteriol.* **123,** 102–117.
Decad, G., and Nikaido, H. (1976). *J. Bacteriol.* **128,** 325–336.
Devillers-Thiery, A., Kindt, T., Scheele, G., and Blobel, G. (1975). *Proc. Natl. Acad. Sci. U.S.A.* **72,** 5016–5020.
DiGirolamo, P. M., and Bradbeer, C. (1971). *J. Bacteriol.* **106,** 745–750.
DiMasi, D. R., White, J. C., Schnaitman, C. A., and Bradbeer, C. (1973). *J. Bacteriol.* **115,** 506–513.
Drewry, D. T., Lomax, J. A., Gray, G. W., and Wilkinson, S. G. (1973). *Biochem. J.* **133,** 563–572.
Frost, G. E., and Rosenberg, H. (1973). *Biochim. Biophys. Acta* **330,** 90–101.
Frost, G. E., and Rosenberg, H. (1975). *J. Bacteriol.* **124,** 704–712.
Galanos, C., Luderitz, O., and Westphal, O. (1969). *Eur. J. Biochem.* **9,** 245–249.
Garten, W., Hindennach, I., and Henning, U. (1975). *Eur. J. Biochem.* **59,** 215–221.
Ghalambor, M. A., and Heath, E. C. (1966). *J. Biol. Chem.* **241,** 3222–3227.
Gustafsson, P., Nordstrom, K., and Normark, S. (1973). *J. Bacteriol.* **116,** 893–900.
Guterman, S. K. (1973). *J. Bacteriol.* **114,** 1217–1224.
Hämmerling, G., Lehmann, V., and Luderitz, O. (1973). *Eur. J. Biochem.* **38,** 453–458.
Halegoua, S., Hirashima, A., and Inouye, M. (1974). *J. Bacteriol.* **120,** 1204–1208.
Haller, I., and Henning, U. (1974). *Proc. Natl. Acad. Sci. U.S.A.* **71,** 2018–2021.
Hancock, R. E. W., and Braun, V. (1976a). *J. Bacteriol.* **125,** 409–415.
Hancock, R. E. W., and Braun, V. (1976b). *FEBS Lett.* **65,** 208–210.
Hancock, R. E. W., Hantke, K., and Braun, V. (1976). *J. Bacteriol.* **127,** 1370–1375.
Hantke, K. (1976). *FEBS Lett.* **70,** 109–112.

Hantke, K., and Braun, V. (1973). *Eur. J. Biochem.* **34,** 284–296.
Hantke, K., and Braun, V. (1975a). *FEBS Lett.* **59,** 277–281.
Hantke, K., and Braun, V. (1975b). *FEBS Lett.* **49,** 301–305.
Hase, S., and Reitschel, E. T. (1976). *Eur. J. Biochem.* **63,** 101–107.
Hase, S., Hofstad, T., and Reitschel, E. T. (1977). *J. Bacteriol.* **129,** 9–14.
Havekes, L. M., Lugtenberg, B., and Hoekstra, W. P. M. (1976). *Mol. Gen. Genet.* **146,** 43–50.
Heath, E. C., Mayer, R. M., Edstrom, R. D., and Beaudreau, C. A. (1966). *Ann. N.Y. Acad. Sci.* **133,** 315–333.
Henning, U., and Haller, I. (1975). *FEBS Lett.* **55,** 161–164.
Henning, U., Hohn, B., and Sonntag, I. (1973). *Eur. J. Biochem.* **39,** 27–36.
Hindenach, I., and Henning, U. (1975). *Eur. J. Biochem.* **59,** 207–213.
Hirashima, A., and Inouye, M. (1973). *Nature (London)* **242,** 405–407.
Hirashima, A., and Inouye, M. (1975). *Eur. J. Biochem.* **60,** 395–398.
Hirashima, A., Childs, G., and Inouye, M. (1973). *J. Mol. Biol.* **79,** 373–389.
Hirota, Y., Suzuki, H., Nishimura, Y., and Yasuda, S. (1977). *Proc. Natl. Acad. Sci. U.S.A.* **74,** 1417–1420.
Hofnung, M. (1974). *Genetics* **76,** 169–184.
Hofnung, M., Hatfield, D., and Schwartz, M. (1974). *J. Bacteriol.* **117,** 40–47.
Hussey, H., and Baddiley, J. (1976). *In* "The Enzymes of Biological Membranes" (A. Martonosi, ed.), Vol. 2, pp. 227–326. Plenum, New York.
Inouye, H., and Beckwith, J. (1977). *Proc. Natl. Acad. Sci. U.S.A.* **74,** 1440–1444.
Inouye, M. (1974). *Proc. Natl. Acad. Sci. U.S.A.* **71,** 2396–2400.
Inouye, M. (1975). *In* "Membrane Biogenesis" (A. Tzagoloff, ed.), pp. 351–391. Plenum, New York.
Inouye, M., and Yee, M. L. (1973). *J. Bacteriol.* **113,** 304–312.
Inouye, M., Shaw, J., and Shen, C. (1972). *J. Biol. Chem.* **247,** 8154–8159.
Inouye, M., Hirashima, A., and Lee, N. (1974). *Ann. N.Y. Acad. Sci.* **235,** 83–90.
Inouye, S., Lee, N., Inouye, M., Wu, H. C., Suzuki, H., Nishimura, Y., Iketani, H., and Hirota, Y. (1977a). *J. Bacteriol.* **132,** 308–313.
Inouye, S., Wang, S., Sekizawa, J., Halegoua, S., and Inouye, M. (1977b). *Proc. Natl. Acad. Sci. U.S.A.* **74,** 1004–1008.
James, R. (1975). *J. Bacteriol.* **124,** 918–929.
Jasper, P. E., Whitney, E., and Silver, S. (1972). *Genet. Res.* **19,** 305–312.
Kadner, R. J., and Bassford, P. J. (1977). *J. Bacteriol.* **129,** 254–264.
Kamio, Y., and Nikaido, H. (1976). *Biochemistry* **15,** 2561–2570.
Kamio, Y., and Nikaido, H. (1977). *Biochim. Biophys. Acta* **464,** 589–601.
Kamio, Y., Kim, K. C., and Takahashi, H. (1971). *J. Biochem. (Tokyo)* **70,** 187–191.
Kanegasaki, S., and Wright, A. (1970). *Proc. Natl. Acad. Sci. U.S.A.* **67,** 951–958.
Keller, J. M. (1966). Ph.D. Thesis, Massachusetts Inst. Technol., Cambridge, Massachusetts.
Kellerman, O., and Szmelcmann, S. (1974). *Eur. J. Biochem.* **47,** 139–149.
Konisky, J., Soucek, S., Frick, K., Davies, J. K., and Hammond, C. (1976). *J. Bacteriol.* **127,** 249–257.
Koplow, J., and Goldfine, H. (1974). *J. Bacteriol.* **117,** 525–543.
Kulpa, C. F., and Leive, L. (1976). *J. Bacteriol.* **126,** 467–477.
Lee, A. G., Birdsell, J. M., and Metcalfe, J. C. (1973). *Biochemistry* **12,** 1650–1659.
Lee, N., and Inouye, M. (1974). *FEBS Lett.* **39,** 167–170.
Lee, N., Cheng, E., and Inouye, M. (1977). *Biochim. Biophys. Acta* **465,** 650–656.
Lehmann, V., Luderitz, O., and Westphal, O. (1971). *Eur. J. Biochem.* **21,** 339–347.
Lehmann, V., Hammerling, G., Nurminen, M., Minner, I., Ruschmann, E., Luderitz, O., Kuo, T. T., and Stocker, B. A. D. (1973). *Eur. J. Biochem.* **32,** 268–275.

Leive, L. (1974). *Ann. N.Y. Acad. Sci.* **238,** 109–129.
Levy, S. B. (1975). *Proc. Natl. Acad. Sci. U.S.A.* **72,** 2900–2904.
Luckey, M., and Neilands, J. B. (1976). *J. Bacteriol.* **127,** 1036–1037.
Luckey, M., Pollack, J. R., Wayne, R., Ames, B. N., and Neilands, J. B. (1972). *J. Bacteriol.* **111,** 731–738.
Luckey, M., Wayne, R., and Neilands, J. B. (1975). *Biochem. Biophys. Res. Commun.* **64,** 687–693.
Luderitz, O., Staub, A. M., and Westphal, O. (1966). *Bacteriol. Rev.* **30,** 192–255.
Luderitz, O., Galanos, C., Lehmann, V., Nurminen, M., Rietschel, E. T., Rosenfelder, G., Simon, M., and Westphal, O. (1973). *J. Infect. Dis.* **128,** S17–S29.
Lugtenberg, B., Meijers, J., Peters, R., van der Hoek, P., and van Alphen, L. (1975). *FEBS Lett.* **58,** 254–258.
Lugtenberg, B., Peters, R., Bernheimer, H., and Berrendsen, W. (1976). *Mol. Gen. Genet.* **147,** 251–262.
McIntosh, M. A., and Earhart, C. F. (1976). *Biochem. Biophys. Res. Commun.* **70,** 315–322.
McKeown, M., Kahn, M., and Hanawalt, P. (1976). *J. Bacteriol.* **126,** 814–822.
Mäkelä, P. H. (1966). *J. Bacteriol.* **91,** 1115–1125.
Mäkelä, P. H., Schmidt, G., Mayer, H., Nikaido, H., Whang, H. Y., and Neter, E. (1976). *J. Bacteriol.* **127,** 1141–1149.
Manning, P. A., and Reeves, P. (1976a). *J. Bacteriol.* **127,** 1070–1079.
Manning, P. A., and Reeves, P. (1976b). *Biochem. Biophys. Res. Commun.* **71,** 466–471.
Miura, T., and Mizushima, S. (1968). *Biochim. Biophys. Acta* **150,** 159–161.
Mühlradt, P. F. (1969). *Eur. J. Biochem.* **11,** 241–248.
Mühlradt, P. F. (1971). *Eur. J. Biochem.* **18,** 20–27.
Mühlradt, P. F., and Golecki, J. R. (1975). *Eur. J. Biochem.* **51,** 343–353.
Mühlradt, P. F., Risse, H. J., Luderitz, O., and Westphal, O. (1968). *Eur. J. Biochem.* **4,** 139–145.
Mühlradt, P. F., Menzel, J., Golecki, J. R., and Speth, V. (1973). *Eur. J. Biochem.* **35,** 471–481.
Mühlradt, P. F., Menzel, J., Golecki, J. R., and Speth, V. (1974). *Eur. J. Biochem.* **43,** 533–539.
Munson, R. S., and Osborn, M. J. (1975). *Fed. Proc., Fed. Am. Soc. Exp. Biol.* **34,** 669.
Nakae, T. (1975). *Biochem. Biophys. Res. Commun.* **64,** 1224–1230.
Nakae, T. (1976a). *J. Biol. Chem.* **251,** 2176–2178.
Nakae, T. (1976b). *Biochem. Biophys. Res. Commun.* **71,** 877–884.
Nakae, T., and Nikaido, H. (1975). *J. Biol. Chem.* **250,** 7359–7365.
Nguyen-Disteche, M., Pollock, J. J., Ghuysen, J. M., Puig, J., Reynolds, P., Perkins, H. R., Coyette, J., and Salton, M. (1974). *Eur. J. Biochem.* **41,** 439–446.
Nikaido, H. (1973). *In* "Bacterial Membranes and Walls" (L. Leive, ed.), pp. 131–209. Dekker, New York.
Nikaido, H. (1976). *Biochim. Biophys. Acta* **433,** 118–132.
Nikaido, H. (1978). To be published.
Nikaido, H., and Nakae, T. (1973). *J. Infect. Dis.* **128,** Suppl., 30–34.
Nikaido, H., Nikaido, K., and Mäkelä, P. H. (1966). *J. Bacteriol.* **91,** 1126–1135.
Nikaido, H., Nikaido, K., Nakae, T., and Mäkelä, P. H. (1971). *J. Biol. Chem.* **246,** 3902–3911.
Nikaido, H., Song, S. A., Shaltiel, L., and Nurminen, M. (1977a). *Biochem. Biophys. Res. Commun.* **76,** 324–330.
Nikaido, H., Takeuchi, Y., Ohnishi, S. I., and Nakae, T. (1977b). *Biochim. Biophys. Acta* **465,** 152–164.
Nikaido, K., and Nikaido, H. (1971). *J. Biol. Chem.* **246,** 3912–3919.
Nurminen, M., Lounatmaa, K., Sarvas, M., Mäkelä, P. H., and Nakae, T. (1976). *J. Bacteriol.* **127,** 941–955.
O'Brien, I. G., and Gibson, F. (1970). *Biochim. Biophys. Acta* **215,** 393–402.

Osborn, M. J. (1968). *Nature* (*London*) **217,** 957–960.
Osborn, M. J. (1969). *Annu. Rev. Biochem.* **38,** 501–538.
Osborn, M. J., and Weiner, I. M. (1968). *J. Biol. Chem.* **243,** 2631–2639.
Osborn, M. J., Gander, J. E., Parisi, E., and Carson, J. (1972a). *J. Biol. Chem.* **247,** 3962–3972.
Osborn, M. J., Gander, J. E., and Parisi, E. (1972b). *J. Biol. Chem.* **247,** 3973–3986.
Osborn, M. J., Rick, P. D., and Rasmussen, N. S. (1978). *J. Biol. Chem.* (to be published).
Payne, J. W., and Gilvarg, C. (1968). *J. Biol. Chem.* **243,** 6291–6299.
Pollack, J. R., and Neilands, J. B., (1970). *Biochem. Biophys. Res. Commun.* **38,** 989–992.
Prehm, P., Stirm, S., Jann, B., and Jann, K. (1975). *Eur. J. Biochem.* **56,** 41–55.
Pugsley, A. P., and Reeves, P. (1976). *J. Bacteriol.* **127,** 218–228.
Randall, L. L., and Hardy, S. J. S. (1975). *Mol. Gen. Genet.* **137,** 151–160.
Randall-Hazelbauer, L. L., and Schwartz, M. (1973). *J. Bacteriol.* **116,** 1436–1446.
Reitschel, E. T., Hase, S., King, M.-T., Redmond, J., and Lehmann, V. (1977). *In* "Microbiology 1977" (*Amer. Soc. Microbiol.*, ed.), pp. 262–268.
Rick, P. D., and Osborn, M. J. (1977). *J. Biol. Chem.* **252,** 4895–4903.
Rick, P. D., Fung, L. W. M., Ho, C., and Osborn, M. J. (1977). *J. Biol. Chem.* **252,** 4904–4912.
Roantree, R. J., Kuo, T., MacPhee, D. G., and Stocker, B. A. D. (1969). *Clin. Res.* **17,** 157.
Romeo, D., Girard, A., and Rothfield, L. (1970a). *J. Mol. Biol.* **53,** 475–490.
Romeo, D., Hinckley, A., and Rothfield, L. (1970b). *J. Mol. Biol.* **5,** 491–501.
Rosenbusch, J. P. (1974). *J. Biol. Chem.* **249,** 8019–8029.
Rothfield, L., and Pearlman, M. (1966). *J. Biol. Chem.* **247,** 1386–1392.
Rothfield, L., and Romeo, D. (1971). *Bacteriol. Rev.* **35,** 14–38.
Rothfield, L., and Takeshita, M. (1965). *Biochem. Biophys. Res. Commun.* **20,** 521–527.
Rothfield, L., Osborn, M. J., and Horecker, B. L. (1964). *J. Biol. Chem.* **239,** 2788–2795.
Rothfield, L., Romeo, D., and Hinckley, A. (1972). *Fed. Proc., Fed. Am. Soc. Exp. Biol.* **31,** 12–17.
Rothman, J. E., and Kennedy, E. P. (1977). *Proc. Natl. Acad. Sci. U.S.A.* **74,** 1821–1825.
Rundell, K., and Shuster, C. W. (1973). *J. Biol. Chem.* **248,** 5436–5442.
Rundell, K., and Shuster, C. W. (1975). *J. Bacteriol.* **123,** 928–936.
Ryter, A., Shuman, H., and Schwartz, M. (1975). *J. Bacteriol.* **122,** 295–301.
Sabet, S. F., and Schnaitman, C. A. (1973). *J. Biol. Chem.* **248,** 1797–1806.
Sasaki, T., and Uchida, T. (1976). *J. Bacteriol.* **117,** 13–18.
Sasaki, T., Uchida, T., and Kurahashi, K. (1974). *J. Biol. Chem.* **249,** 761–772.
Schechter, I., McKean, D. J., Guyer, R., and Terry, W. (1975). *Science* **188,** 160–162.
Scher, M., and Lennarz, W. J. (1969). *J. Biol. Chem.* **244,** 2777–2789.
Scher, M., Lennarz, W. J., and Sweeley, C. C. (1968). *Proc. Natl. Acad. Sci. U.S.A.* **59,** 1313–1320.
Schlecht, S., and Westphal, O. (1970). *Zentralbl. Bakteriol., Parasitenkd. Infektionskr. Hyg., Abt. 1: Orig.* **213,** 354–364.
Schnaitman, C. A. (1970). *J. Bacteriol.* **104,** 882–889.
Schnaitman, C. A. (1974a). *J. Bacteriol.* **118,** 442–453.
Schnaitman, C. A. (1974b). *J. Bacteriol.* **118,** 454–464.
Schnaitman, C. A., Smith, D., and DeSalas, F. M. (1975). *J. Virol.* **15,** 1121–1130.
Schwartz, M. (1975). *J. Mol. Biol.* **99,** 185–201.
Schwartz, M., and LeMinor, L. (1975). *J. Virol.* **15,** 679–685.
Sha'afi, R. I., Gary-Bobo, G. M., and Solomon, A. K. (1971). *J. Gen. Physiol.* **58,** 238–258.
Skurray, R. A., Hancock, R. E. W., and Reeves, P. (1974). *J. Bacteriol.* **119,** 726–735.
Smit, J., and Nikaido, H. (1978). *J. Bacteriol.* **135,** 687–702.
Smit, J., Kamio, Y., and Nikaido, H. (1975). *J. Bacteriol.* **124,** 942–958.
Soucek, S., and Konisky, J. (1977). *J. Bacteriol.* **130,** 1399–1401.

Steven, A. C., Ten Heggeler, B., Muller, R., Kistler, J., and Rosenbusch, J. P. (1977). *J. Cell. Biol.* **72,** 292–301.
Stocker, B. A. D., and Mäkelä, P. H. (1971). *In* "Microbial Toxins" (S. J. Ajl, G. Weinbaum, and S. Kadis, eds.), Vol. 3, pp. 369–386. Academic Press, New York.
Suzuki, H., Nishimura, Y., Idetani, H., Campesi, J., Hirashima, A., Inouye, M., and Hirota, Y. (1976). *J. Bacteriol.* **127,** 1494–1501.
Szmelcmann, S., and Hofnung, M. (1975). *J. Bacteriol.* **124,** 112–118.
Szmelcmann, S., Schwartz, M., Silhavy, T. J., and Boos, W. (1976). *Eur. J. Biochem.* **65,** 13–19.
Takeishi, K., Yasumura, M., Pirtle, R., and Inouye, M. (1976). *J. Biol. Chem.* **251,** 6259–6266.
Takeshita, M., and Mäkelä, P. H. (1971). *J. Biol. Chem.* **246,** 3920–3927.
Tamaki, S., and Matsuhashi, M. (1971). *J. Bacteriol.* **105,** 968–975.
Torti, S. V. (1977). Ph.D. Thesis, Tufts Univ., Boston, Massachusetts.
Torti, S. V., and Park, J. T. (1976). *Nature* (*London*) **263,** 323–326.
Tsukagoshi, N., and Fox, C. F. (1971). *Biochemistry* **10,** 3309–3313.
Van Alphen, W., Lugtenberg, B., and Berendsen, W. (1976). *Mol. Gen. Genet.* **147,** 263–269.
Van Alphen, L., Havekes, L., and Lugtenberg, B. (1977). *FEBS Lett.* **75,** 285–290.
Van Gool, A. P., and Nanninga, N. (1971). *J. Bacteriol.* **108,** 474–481.
Wayne, R., and Neilands, J. B. (1975). *J. Bacteriol.* **121,** 497–503.
Wayne, R., Frick, K., and Neilands, J. B. (1976). *J. Bacteriol.* **126,** 7–12.
Weiner, I. M., Higuchi, T., Rothfield, L., Saltmarsh-Andrew, M., Osborn, M. J., and Horecker, B. L. (1965). *Proc. Natl. Acad. Sci. U.S.A.* **54,** 228–235.
Weltzien, H. U., and Jesaitis, M. A. (1971). *J. Exp. Med.* **133,** 534–553.
White, D., Lennarz, W., and Schnaitman, C. (1972). *J. Bacteriol.* **109,** 686–690.
White, J. C., DiGirolamo, P. M., Fu, M. L., Preston, Y. A., and Bradbeer, C. (1973). *J. Biol. Chem.* **248,** 3978–3986.
Wilkinson, R. G., and Stocker, B. A. D. (1968). *Nature* (*London*) **217,** 955–957.
Wirth, D. F., Katz, F., Small, B., and Lodish, H. (1977). *Cell* **10,** 253–263.
Wright, A. (1971). *J. Bacteriol.* **105,** 927–936.
Wright, A., and Kanegasaki, S. (1971). *Physiol. Rev.* **51,** 748–784.
Wright, A., Dankert, M., and Robbins, P. W. (1965). *Proc. Natl. Acad. Sci. U.S.A.* **54,** 235–241.
Wright, A., Dankert, M., Fennessey, P., and Robbins, P. W. (1967). *Proc. Natl. Acad. Sci. U.S.A.* **57,** 1798–1803.
Wu, H. C., and Lin, J. J. C. (1976). *J. Bacteriol.* **126,** 147–156.
Wu, H. C., Hou, C., Lin, J. J. C., and Yem, D. W. (1977). *Proc. Natl. Acad. Sci. U.S.A.* **74,** 1388–1392.
Yamamoto, S., and Lampen, J. O. (1975). *J. Biol. Chem.* **250,** 3212–3213.
Yamamoto, S., and Lampen, J. O. (1976). *Proc. Natl. Acad. Sci. U.S.A.* **73,** 1457–1461.
Yem, D. W., and Wu, H. C. (1977). *J. Bacteriol.* **131,** 759–764.
Zwaal, R. F. A., Roelofsen, B., Comfurius, P., and van Deenen, L. L. (1975). *Biochim. Biophys. Acta* **406,** 83–96.

CHAPTER 8

Bacteriophage and Bacteria: Friend and Foe

HARRISON ECHOLS

I. Introduction

Every reader of this chapter is undoubtedly aware of the central role of bacteriophage in the recent rapid progress in molecular biology. The relatively small and easily isolated genome of the phage has made these creatures ideal for the combined genetic and biochemical study that has been so successful in revealing mechanisms of gene expression, replication, and regulation utilized by the phage and their bacterial hosts. One biological feature that has been critical for these endeavors is the remarkable diversity of the phage population. Thus, the small RNA viruses have provided a packaged messenger RNA for the analysis of protein synthesis, the small DNA viruses an ideal substrate for the study of DNA replication, and the large DNA viruses a wealth of control systems optimal for the study of gene activation and repression.

Despite the detailed analysis of the life cycle of a number of different phage, the broader biological questions remain: How has this diverse collection of phage species evolved? What evolutionary pressures retain them (out of the hands of the friendly molecular biologist)? I will attempt

ISBN 0-12-307207-7

to summarize some of the available information relevant to these questions by reviewing ways in which phage and their hosts interact. The chapter will begin with some evolutionary comments, review briefiy general properties of phage growth, and then cover some specific examples that indicate some of the diversity of the phage–host relationship: bacteriophages λ, M13, T7, and T4. Since phage have featured prominently in a large number of reviews in recent years and in the series of books "Comprehensive Virology," the reader who wishes to learn more about phage growth can find an abundance of sources. Because of the wide scope of this chapter, I will not attempt to give complete literature references, but will refer mainly to more specific review articles or to papers with general discussions of the topic under consideration.

II. Comments on the Existence of Phage

A. Why Are There Bacterial Viruses?

Many whose occupation involves the growth of bacterial cultures have asked this question in anguish. However, several previous authors have thought more charitably about the ecological niche of the humble phage; the reader may find particularly interesting the discussions of Lwoff (1953), Adams (1959), Campbell (1961), and Bradley (1967). The following discussion depends especially on the analysis of Campbell.

The general possibilities can be resolved to three: (1) Phage have no ecological niche, but are continually created and lost. (2) Phage and their hosts are in a predator-prey equilibrium. (3) Phage are a net benefit to their hosts. I find the first of these possibilities to be extremely unlikely. Campbell (1961) points out the abundance and complexity of phage as contraindicators, to which I would add the enormous diversity of phage species that has become evident in the last fifteen years. The latter two possibilities are both reasonable and not mutually exclusive. I suspect that each approximates most closely the biological situation for different viruses and environmental conditions.

B. Predator–Prey

In the absence of an intervening higher force, such as a molecular biologist, one might think that the virus would wipe out the host population and disappear for lack of new worlds to conquer. With a little reflection, however, it becomes clear that an equilibrium should be established because the finite rate constant for adsorption will eventually match the infection rate to the bacterial birth rate; as the phage has to search longer and longer for a host, the hunted bacterium will sometimes divide before the dread phage arrives [for quantitative treatments, see Campbell (1961) and Levin *et al.* (1977)]. The existence of such an equilibrium in an experimental system has been demonstrated by Levin *et al.* (1977).

Although this simplified equilibrium analysis may apply in some cases, the real situation is probably determined more often by the capacity of the host population to survive infection by phage by other mechanisms: resistance to virus through host mutation or lysogeny. In the former, the host becomes incapable of supporting viral growth, generally through mutational alteration of the specific adsorption site for the phage; in the latter, the host acquires the viral genome in a relatively stable association (prophage) maintained by a repression mechanism that turns off the lytic potential of the prophage and of any of its relatives which infect at a later time. In the lysogenic state, the viral DNA is generally inserted into the host DNA and replicated along with it (alternatively, the viral DNA may remain as an extrachromosomal plasmid).

Both of these resistance mechanisms can be reversed, leading to a new population of free viruses. The phage can overcome host resistance through the acquisition of new genetic potential by mutation or recombination (such as the capacity to adsorb to a different site on the bacterial cell membrane). In the lysogenic case, the prophage can be induced to lytic growth, generating a population of free phage. This induction mechanism usually occurs at a low level in a lysogenic bacterial population, but can happen to the entire bacterial population in response to a potentially disastrous environmental perturbation—damage to the host DNA that blocks its capacity for replication.

In summary, there are three general types of host–virus interactions that can characterize the population dynamics: for convenience, I will term then simple equilibrium, mutational variation, and environmental variation.

C. Phage as a Friend

There can obviously be evolutionary pressure for the maintenance of a bacteriophage population if the phage is a net benefit to the host population. Two related ways in which this might occur are the processes generally termed transduction and lysogenic conversion. In transduction, the phage facilitates bacterial genetic recombination by serving as a vector carrying bacterial DNA from one cell to another. Transduction may occur by two mechanisms: replacement, in which the phage-carried bacterial DNA replaces a segment of host DNA by homologous general recombination, or addition, in which the transducing bacterial DNA is added to the preexisting complement of host DNA, generally when the viral DNA and some associated bacterial DNA are inserted into the host genome in a lysogenic response to infection.

In the case of lysogenic conversion, one or more prophage genes act to change the properties of the bacterial host in some discernible manner; for example, the bacterial cell wall might be altered, conferring resistance to infection by certain phage. The difficulties of biological definition become apparent at this point because transduction by addition differs from lysogenic conversion only in that known bacterial genes are added in the former case (see Section III). Independent of the details of these processes, the potential advantages of transduction and lysogenic conversion should be apparent: new genetic material can be acquired that may provide immediate salvation in a difficult situation or serve as an extra piece of DNA for further mutational adaptation later.

D. Where Do Bacterial Viruses Come from?

Having argued that phage have an ecological niche and are not continuously created and lost, I should add that phage are probably less unique a creation than I have so far implied. The capacity of phage DNA to enter and exit from the bacterial chromosome or to exist as an autonomous plasmid suggests a relationship to other DNA elements capable of this behavior (e.g., the F plasmid or the transposable "insertion elements").

From work with experimental systems, it is clear that phage can become plasmids or inserted chromosomal elements incapable of lytic growth; in the simplest case for a plasmid, the phage DNA maintains its replication and some regulation genes but loses the genes for encapsulation of its nucleic acid. The reverse route is certainly more complicated because the genetic potential for encapsulation must be acquired. This capacity might be borrowed from another phage by genetic recombination, creating a "hybrid" phage; there is a considerable body of evidence for the existence of this process (see Campbell, 1977). One might argue that the more dramatic evolution of plasmid to phage without genes from a parent phage is exceedingly unlikely because structures such as phage capsids are not found in bacterial cells, leaving open the question of which bacterial genes might be borrowed to create the new phage. However, multiple-subunit structures are very common among bacterial proteins. Given that proteins typically recognize their own kind, the mutational transition from a multimeric protein that binds to DNA to a much larger structure that surrounds it seems to me to be a plausible, although rare, event (see Section V). The foregoing discussion implies that the bacterium came before the phage, an order of events that I think is correct, in spite of the possible extrapolation to the statement that "elephants gave rise to fleas."

III. Some General Characteristics of Bacteriophage Growth

A. Classification by Nucleic Acid Structure and Growth Pattern

Since I assume that the general types of phage and their growth will already be familiar to most readers, I will summarize only those features that will be pertinent to the further discussion of the phage–host relationship. For extensive reviews, the articles by Bradley (1967), Calendar (1970), and the "Comprehensive Virology" series edited by Fraenkel-Conrat and Wagner (1974, 1975, 1976, 1977) will be helpful.

By the results of the infection process, phage may be classified as "temperate" or "virulent." For a virulent virus, the genetic program is restricted to the production of new viral particles and their release from the host cell, generally by lysis. A temperate virus has the potential for two developmental pathways: the productive pathway, which normally culminates in the production of new viral particles and lysis of the host cell, and the lysogenic pathway, in which the infected cell survives with the genes for the productive pathway turned off and the viral DNA replicating along with the host chromosome, usually inserted into it. As noted above, the lysogenic response is usually reversible, "spontaneously" at a low frequency (10^{-2}–10^{-5} per cell generation), but "induced" in every cell in which the capacity for chromosomal DNA replication has been inhibited.

By the properties of their genome, phage may be classified as single-stranded RNA, single-stranded DNA, or double-stranded DNA. The viruses with single-stranded nucleic acid have relatively small genomes (3–10 genes), are capable only of a productive infection, and have a simple structure. The viruses with double-stranded DNA are typically larger (30–150 genes), are often temperate, and have a complex particle structure with heads, tails, and miscellaneous adornments facilitating attachment to the host cell and injection of DNA. The double-stranded phages, virulent and temperate, have an ordered, temporally regulated pattern of intracellular events that provides for efficient productive or lysogenic development.

B. Features of Productive Growth

1. Large DNA Phage: The Efficient Assembly Line

Viewed from the phage point of view, the basic developmental problem of the productive or lytic pathway is to maximize viral yield. This optimization principle is most evident for the large, double-stranded DNA phage, which divide productive development into a replication-oriented early phase and an encapsulation-oriented late phase. This sequential pattern provides for a maximal concentration of cell energy and resources on DNA replication and genetic recombination, followed by a similar concentration on virion formation and cell lysis. The principal advantage of this temporal progression is probably the delay of encapsulation and lysis until a large number of new viral genomes have been produced. The molecular mechanisms for the early-late shift in viral development have two major features: sequential regulation of transcription of the phage DNA and a transition from an early to late form of the DNA itself.

Although the biochemical details vary, the overall transcription pattern seems to be similar from one phage to another (see Fig. 1) (see also Rabussay and Geiduschek, 1977). The genes for replication and recombination proteins are typically transcribed during the early stage, and then transcription of genes for head, tail, and lysis proteins is turned on through the action of one or more proteins specified by the early region of the phage

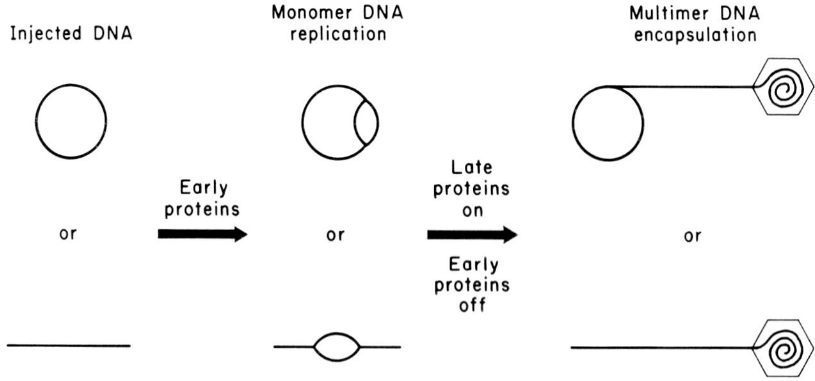

Fig. 1. Productive development by large DNA phage. The injected linear DNA from the phage particle may circularize (above) or remain linear (below). Early proteins are used for replication, recombination, and regulation of the late stage of viral development, in which synthesis of encapsulation and lysis proteins is turned on and synthesis of early proteins is turned off. Multimeric DNA serves as the substrate for encapsulation.

DNA. The transcription of early genes is generally reduced during the late stage of lytic development, often (but perhaps not always) through the action of another early gene protein. The early period may sometimes itself be divided into two temporal transcriptional periods: an "immediate–early" stage utilizing the host RNA polymerase and a "delayed early" stage activated by a phage protein produced during the earliest stage. Thus, temporal control of transcription involves sequential modification of the regions of DNA transcribed. The biochemical mechanisms are highly varied: for example, a new RNA polymerase for phage T7, new polymerase subunits for SP01 and T4, and elimination of an RNA chain termination event for λ (Rabussay and Geiduschek, 1977).

The late transition in the DNA itself has been much less studied than the transcription switch, but appears to be equally ubiquitous. Encapsulation typically proceeds from a very long multimeric piece of DNA and probably normally will occur only from such a concatemeric structure. In contrast, the early replicative structure is a monomer (either circular or linear) in every case so far analyzed. The replicative switch probably serves as an additional mechanism to prevent phage DNA from being packaged before a large replicative pool has accumulated. As for transcription, the biochemical mechanisms are varied: probably recombination at the ends for phages T4 and T7 and a late rolling-circle mode of replication for λ (see Fig. 1) (see also Matthews, 1977; Skalka, 1977).

2. Small Phage: The Simple Life

The large double-stranded DNA phage have used their large genome to optimize their capacity for infection, intracellular development, and the potential for the alternative life styles of productive infection or lysogenization. Such phage are clearly the products of a rather lengthy evolutionary process; their complexity and sophisticated growth pattern contrasts sharply with the apparent simplicity of the small viruses.

The small RNA viruses (e.g., f1, R17, Qβ) carry a genome that codes only for part of a replication protein (other subunits are borrowed from the host), and two encapsulation proteins required for the icosahedral shell (Eoyang and August, 1974; Zinder, 1974). Even for this simplest case, however, there is developmental regulation. The phage gene for the replication protein is translated efficiently early after infection and then the expression of this gene is shut off by the major coat protein. The replication complex is highly specific for the phage RNA and generates principally viral (+) strand RNA in the replication process. Thus, the RNA virus is a considerably more sophisticated creature than an "escaped messenger RNA."

The small DNA viruses have more genes than the RNA viruses, at least three of which are involved in viral DNA replication. These viruses appear in two different morphologies, icosahedral (e.g., ϕX174, S13, G4) or filamentous (e.g., fl, M13) (Denhardt, 1977; Ray, 1977). The principal regulatory feature involved in growth of the small DNA viruses appears to be the replicative switch from the single-stranded infecting DNA to the double-stranded form used for further replication and transcription back to more single-strands for encapsulation. The initial double-stranded molecule is produced by the host replication machinery, but initiation of progeny double-stranded replication and the formation of progeny single strands depend on phage genes (Kornberg, 1974). The replicative switch serves the developmental function noted already for the large DNA phages: encapsulation does not begin until many progeny DNA molecules have accumulated. In addition, the encapsulation of single-stranded DNA can probably proceed by a less exact mechanism than that of a double-stranded virus because there is no "competing" single-stranded host DNA that must be avoided (some other possible advantages of single-stranded DNA are noted in III,B,3). A unique aspect of the filamentous viruses is their uptake into and release from an intact cell as complete virions, thus avoiding the need for a complex injection or lysis mechanism (for this reason, I have used the term productive infection rather than the more usual lytic infection).

3. Problems of Productive Growth: Where Have All the Host Cells Gone?

The spectacular efficiency of the highly regulated productive infection by a large DNA phage, typically 100 progeny particles in 10–20 minutes, raises an obvious question: Can viral evolution lead to self-destruction? The environmental variation provided by lysogeny is one obvious answer to the chancy oscillation provided by mutational variation; the capacity of the temperate phage to conserve its natural resources has led to the suggestion that all phage are temperate for some hosts. However, there seems to be a clear distinction between the temperate and virulent DNA phages isolated from nature that argues for an ecological niche for virulent phage as such. A complex virulent phage, such as T4, devotes a substantial portion of its genome to the shut-off of host gene expression, degradation of host DNA, and new synthetic enzymes for efficient reutilization of the host nucleotides. Such a brutal treatment of the typical host of course precludes lysogenization. In addition, the simple phages lack the genome size to afford the luxury of a lysogenic pathway.

Lacking a lysogenic hideout, the virulent phage has two alternative routes: rapid genome variation through mutation and recombination or an infection route that is not readily eliminated by bacterial mutation. The large virulent DNA phage are well equipped for rapid variation: they supply their own replication and recombination enzymes, and their recombination frequencies are extremely high. In the case of "related" virulent phage studied in the laboratory, the observed alterations in the tail fiber adsorption structure are most easily explained by recombinational exchange which grossly alters the region of the tail fiber that interacts with the lipopolysaccharide of the bacterial outer membrane (Beckendorf, 1973; Davis and Hyman, 1971).

One might guess that the small virulent phage choose an infection route that may be less efficient than the syringe mechanism used by the large phage but less subject to mutational variation. At least the RNA phage and filamentous DNA phage do seem to follow this prescription: adsorption requires specialized bacterial structures, the F or R pili, that are needed for sexual transfer of bacterial episomes (and whose structure is specified by the episomal DNA). Since it is unlikely that a mutation will alter the repeating subunit structure of the pilus to phage resistance and retain sexual function, a phage-resistant mutation will not keep pace with the rapid sexual transfer of the wild-type plasmid that confers phage sensitivity. In addition, the sex pilus probably specifies in some way a recognition element for the single-stranded DNA that is transferred during conjugation, and thus these specialized structures may facilitate transfer of single-stranded nucleic acid into and/or protection inside of the host cell.

So far I have framed the discussion of adaptation for survival mainly in terms of capacity for adsorption. However, intracellular mechanisms exist that are also important. Double-stranded phage face two clearly defined "natural enemies" in the nuclease realm: restriction endonucleases and general exonucleases. Presumably the presence of glucosylated hydroxymethylcytosine in phage T4 DNA is one example of mutational variation to escape restriction nucleases; another is the capacity of phage T3 to destroy *S*-adenosylmethionine, a cofactor for one common type of restriction nuclease (Matthews, 1977).

The bacterial exonuclease most likely to trouble infecting phage is the ubiquitous ATP-dependent nuclease that functions in bacterial general recombination (RecBC DNase) (Clark, 1971). This enzyme is highly active on double-stranded DNA *in vitro* and *in vivo*. The large double-stranded phages (e.g., T4, T7, and λ) specify an inhibitor for the RecBC enzyme (Tanner and Oishi, 1971; Sakaki *et al.*, 1973; Sakaki, 1974). Apparently evolution of the large double-stranded phage has bypassed most of the widespread intracellular defense mechanisms.

The small DNA phages, lacking the adaptive capacity of the large phage, probably rely on their single-stranded and circular genome. Single-stranded DNA is not sensitive to restriction nucleases and by the time it becomes double-stranded has presumably been modified to the protected status of host DNA. Although sensitive to RecBC DNase *in vitro*, single-stranded DNA seems to escape *in vivo*, as does single-stranded DNA transferred by conjugation.

The "gap" in genome size and structural complexity between the "large" and "small" phage may reflect the two adaptive pathways for phage indicated in the foregoing discussion: a complex, highly organized developmental pattern that can overwhelm most host defense mechanisms or a simple subtle one that can sneak by. In the general terms of Section II, the former is probably best described by mutational variation, and the latter by simple equilibrium.

One form of host defense (or phage attack) mechanism that is so far purely hypothetical is a "population" mechanism, in which there exists a bacterial cell-to-cell signaling system that has as one of its purposes a "warning" system to other uninfected cells of the population. The infected cell might release a negatively chemotactic substance driving other cells away. Alternatively the phage might subvert such a signaling system by causing release of a positively chemotactant substance that attracts other potential victims. Thus one can imagine mutational variation at the level of population as well as cellular responses to viral infection (Echols, 1974).

C. Features of Lysogenic Development

Choosing again the phage perspective, the fundamental developmental problem of the lysogenic pathway is to provide for three separable stages: establishment, maintenance, and induction (Echols, 1972; Weisberg *et al.*, 1977). The establishment of lysogeny in an infected cell generally requires repression of the genes needed for productive infection and integration of the viral DNA into the host DNA (Fig. 2). Lysogeny is then maintained through continued repression of the phage genome. Induction involves the release of repression, excision of the viral DNA, and sequential expression of the genes for DNA replication and recombination, head and tail formation, and cell lysis. For phage P1, lysogeny involves controlled replication as a plasmid, rather than the passive replication of the integrated prophage.

As for the productive development just described in Section III, B, the lysogenic pathway is subject to an impressive degree of regulation. This has been worked out most completely with phage λ, and the following discussion relys heavily on this work; however, I think that the statements made below are a reasonably general description for most temperate phage so far studied. The specific routes of phage λ are indicated in Section IV.

The efficient establishment of lysogeny requires a coordination mechanism

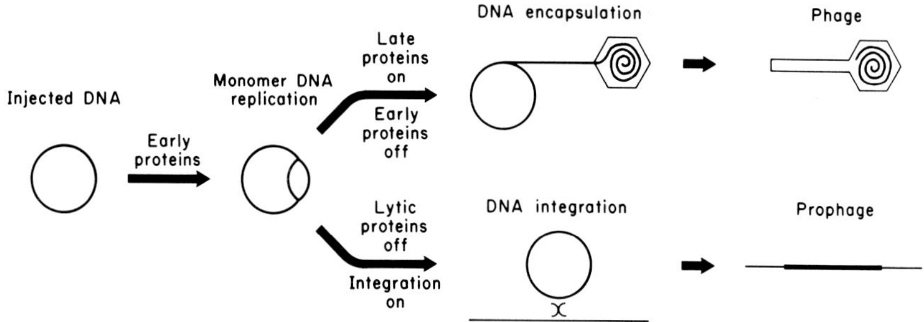

FIG. 2. Developmental pathways for temperate phage. After an early phase common to both pathways, viral development may follow the productive pathway (see Fig. 1) or the lysogenic pathway, in which synthesis of lytic proteins is turned off and the viral DNA is inserted into the host genome by a specific recombination event.

between two functionally distinct events: the turnoff of gene expression and site-specific recombination. Since integrative recombination requires a phage-specified protein, repression cannot be established immediately, but must occur before there is an irreversible commitment to productive growth. In the cases so far defined, the phage accomplishes this by an early stage common to both productive and lysogenic development. The lysogenic pathway then goes to repression and integration, and the productive pathway to encapsulation and lysis. The distinction between the two pathways is probably often aided by a different DNA substrate for each; covalently closed circular DNA produced by early replication is used for integrative recombination and multimeric DNA produced by later rolling circle replication is required for encapsulation (Fig. 2). The integration pattern varies greatly for different temperate phage. Many have a single site that is strongly favored (e.g., λ, 434, ϕ80). However, phage P2 has several potential insertion sites (Bertani and Bertani, 1971), and phage mu appears to integrate randomly in the host genome (Howe and Bade, 1975).

The maintenance of lysogeny for an integrated prophage is the simplest stage of the lysogenic pathway; the requirement is to shut off all of the phage genes except those needed to provide the repression (some genes may be spared for lysogenic conversion events—see below). The repression process is aided by the sequential positive regulation of productive genes noted above; only the early genes must be turned off because an early regulatory protein is required to activate the late genes. As a consequence, a single phage protein typically suffices to maintain repression of an integrated prophage. Maintenance of plasmid lysogeny is intrinsically more complicated and may require more regulatory genes (Scott, 1975). Because the prophage repression mechanism also works on another infecting phage, a lysogenic cell is "immune" to productive infection by additional phage.

Industion involves a reversal of the establishment process in response to a signal that the host cell is in serious trouble. Repression is released, the viral DNA is excised from the host genome, and productive development generally ensues. An important by-product of prophage excision is the rare production (frequency about 10^{-6}) of transducing variants that have acquired nearby bacterial genes (and generally lost phage genes) through an aberrant excision event (Fig. 2) (see also Campbell, 1977). The initiating environmental event is an inhibition of the capacity of the host genome to replicate. The typical inducing agents are classical inhibitors of DNA replication: ultraviolet light, mitomycin C, thymine deprivation. However, the inhibition of chromosomal replication by the use of a temperature-sensitive mutation affecting a DNA replication protein also serves as an inducing event. The onset of phage development is triggered by the inactivation of the repressor protein that maintains lysogeny. For phage λ, the inactivation involves a proteolytic eleavage (Roberts and Roberts, 1975).

The sequence of events between the inhibition of DNA replication and the inactivation of the repressor has not been clearly established. This "induction pathway" is at least in part identical to the pathways for "error-prone" repair for damaged DNA and regulation of cell division because there are bacterial mutations that affect all three [e.g., certain mutations in the *recA* and *lexA* genes turn off all three and other mutations in (or near) these two genes activate all three] (Witkin, 1976). Thus the phage presumably steals a host distress signal to initiate its escape (Fig. 3). Recent experiments indicate that the process involves an activation of the RecA protein to turn on its own synthesis, possibly by proteolytic cleavage of the Lex repressor that represses the *recA* gene (and perhaps other repair and cell division genes) (Mount, 1977; Roberts *et al.*, 1977; McEntee, 1977). In this

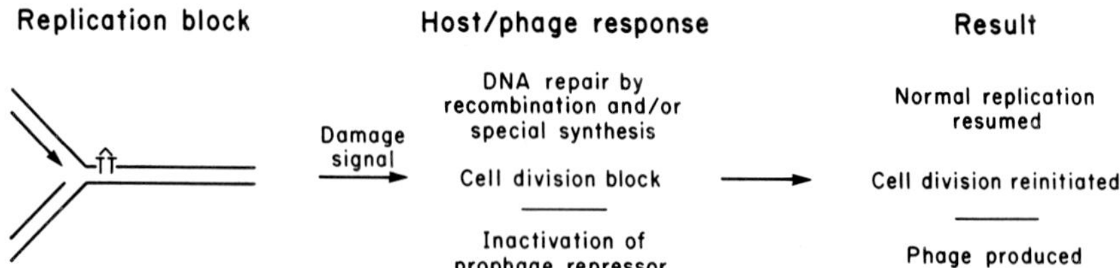

FIG. 3. Prophage induction and related events. An inhibition of host DNA replication (indicated here as the thymine dimer produced by uv irradiation) triggers a "damage signal" that turns on a pathway for DNA repair and turns off cell division. In nonlysogenic cells, normal replication, and cell division are resumed after the DNA is repaired. In cells harboring an "inducible" prophage, the prophage repressor is inactivated and productive development ensues.

model, the prophage repressor has become a target of a general host derepression system aimed at bacterial repair genes.

Although not all temperate phage are inducible by the DNA inhibition pathway, the process does appear to be a very general one. Some of the "noninducible" phage may respond to different signals. The advantage of induction for the phage is obvious. For the bacterial host induction will serve to increase recombination through the production of transducing particles (see Section III,D) and may sometimes provide for "cured" cells free of prophage, but not killed by productive growth. When "times are bad," new genetic potential is particularly useful, and the metabolic load of the prophage may become a survival factor.

The last statement raises the question of why the excess 1% DNA provided by the prophage is tolerated at all in a bacterial population. Transduction and lysogenic conversion probably provide part of the answer. However, the nature of site-specific recombination and its regulation make the prophage hard to shake off. The recombination event that inserts the prophage involves very little DNA homology; thus the prophage is not lost through general recombination at an appreciable frequency. In addition, the integration–excision reaction of Fig. 2 appears to be far toward insertion once repression is established, and the integration but not excision reaction occurs at a low constitutive level in a prophage (Weisberg *et al.*, 1977).

Another general question about the lysogenic pathway is why those mutants able to grow on lysogenic cells ("virulent" mutants) do not replace the original temperate phage population. If selection for lysogenization is an advantage, then the phage has presumably evolved a mechanism to prevent this occurrence. In fact, there are probably several (exemplified by λ in Section IV): (1) Two separate transcription units must escape repression for effective productive growth in a lysogenic cell. (2) The binding (operator) sites for the repressor that maintains lysogeny can be reiterated, so that more than one mutation per operator region is required for complete insensitivity to the prophage repressor (Ptashne *et al.*, 1976). (3) Another repressor protein needed for successful lytic growth can bind to the same operator region as the maintenance repressor, indicating that mutations leading to escape from the prophage repressor may usually suffer in productive growth (Folkmanis *et al.*, 1976).

D. Transduction and Lysogenic Conversion

The capacities for transduction and lysogenic conversion are widespread among large DNA viruses. By the nature of the bacterial DNA chosen, transducing phage are often classified as "generalized" or "specialized."

A generalized transducing phage will carry any segment of bacterial DNA with approximately equal frequency, whereas a specialized transducing phage bears only a limited portion of the bacterial genome.

Generalized transduction presumably occurs because of a mistake in a continuous ("headful") encapsulation mechanism in which host DNA is chosen instead of the concatemeric phage DNA that is the proper substrate (Fig. 4) (Ikeda and Tomizawa, 1965; Streisinger *et al.*, 1967; Tye *et al.*, 1974). The resultant phage particle carries only bacterial DNA; the DNA is injected into the recipient cell and normally is incorporated into the host genome by homologous general recombination, leading to transduction by replacement.

Specialized transduction typically occurs because an aberrant excision event during prophage induction results in the addition of adjacent bacterial DNA to the phage DNA (Fig. 4). The resultant phage particle carries both bacterial and viral DNA; because of limitations in the amount of DNA that can be encapsulated, the transducing phage has typically lost some phage genes and may be incapable of normal productive or lysogenic development (a "defective" phage) (Campbell, 1977). When the transducing phage DNA enters the recipient cell, the normal lysogenic pathway will lead to transduction by "addition," in which the phage-borne bacterial genes are added to the preexisting complement of host genes by insertion of the DNA through integrative recombination at the prophage site. Transduction may also occur by general recombination in the region of the bacterial DNA homologous to that carried by the virus. In this case, a single cross-over

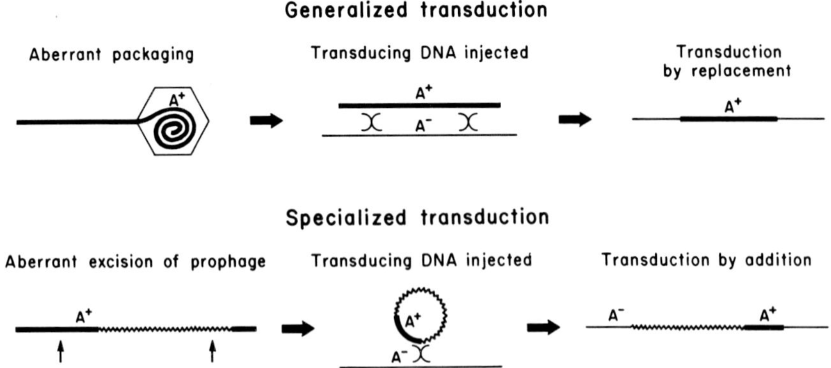

FIG. 4. Transduction mechanisms. In generalized transduction, bacterial DNA is collected from a donor host by an aberrant encapsulation event and acquired by a recipient cell through homologous general recombination. In specialized transduction, bacterial DNA adjacent to a prophage site is collected from a donor host by an aberrant excision event and acquired by a recipient cell through integrative or general recombination.

event will insert the complete transducing phage and produce transduction by addition; a multiple cross-over event can produce transduction by replacement, in which only a single copy of the bacterial DNA remains and the phage genes are not inserted. Specialized transduction might also involve the incorporation of bacterial DNA that is not adjacent to the prophage site, through transposable elements capable of nonhomologous recombination; so far this has been found mainly for the case of transposable drug resistance and for the IS insertion elements (Bukhari *et al.*; 1977).

Lysogenic conversion is a phenotypic consequence of certain phage genes that are expressed in the prophage state (Barksdale and Arden, 1974). The consequences may include changes in the bacterial cell membrane (ε15, P2), the capacity of other phage to grow intracellularly (λ, P1, P2, P22), or extracellular proteins such as diptheria toxin. As noted earlier, lysogenic conversion does not differ in mechanism from transduction by addition. The genes responsible for lysogenic conversion are presumed to be genes not normally present in the host, although one can certainly imagine "conversion" events deriving from elevated expression of bacterial genes by virtue of their association with a phage promoter. Lysogenic conversion clearly provides a mechanism by which a temperate phage may be of net benefit to a host. In addition, both transduction by addition and lysogenic conversion provide a source of "extra" DNA for bacterial or viral evolution.

IV. Specific Examples of Bacteriophage Growth

A. Phage λ—Variety in Life Style

1. General Outline of Phage λ Development

The temperate phage λ has been the most extensively studied of all bacteriophage, characterized fondly in a biography (Hershey, 1971) and numerous recent review articles (Echols, 1971a, 1972, 1973, 1974; Herskowitz, 1973; Weisberg *et al.*, 1977). In this section, I will use λ to show how the general framework described in Section III is executed in a particular case. For more details and a more extensive reference list, the reader should consult the review articles noted above. In Section IV,B I will outline the productive growth of the virulent phages M13, T7, and T4.

Bacteriophage λ is a phage of moderate structural complexity with an icosahedral head about 54 nm in diameter and a tail of 150 nm, terminating in a short tail fiber; the genome is a double-stranded DNA molecule of MW 30×10^6 with a 12-base complementary single-stranded region at each end that provides for intracellular circularization of the DNA by pairing of these "cohesive sites."

When λ DNA is injected into a cell, the first events are formation of a molecule and limited transcription of a small region of the λ DNA, principally the *N* and *cro* genes (Fig. 5). Following this "immediate early" period the N protein (product of the *N* gene) activates the "delayed early" stage of viral development in which "leftward" transcription extends through the recombination genes and "rightward" transcription is enhanced from the replication genes and extends through gene *Q* (Echols, 1971b; Thomas, 1971). The N-activated stage of transcription is accompanied by an association of the λ DNA with the cell membrane; the mechanism and importance of this association is unclear.

At this juncture a "decision" is made between productive and lysogenic development. Productive growth requires the activity of two additional regulatory proteins, Q and Cro. Q acts to turn on transcription of the genes for head, tail, and lysis proteins, and Cro functions to repress RNA synthesis from the early genes. Because λ genes for related functions are clustered, the regulatory pattern is quite simple, involving two positive and one negative signal for the host RNA polymerase (Echols, 1971b; Thomas, 1971). N tells polymerase to ignore the termination signals at the end of the *N*, *cro*, and *P* genes; *Q* advises the enzyme to transcribe the late region of DNA as a single transcription unit, and *Cro* prevents continued initiation events at the early promoter sites, p_L and p_R (Fig. 5) (Weisberg *et al.*, 1977).

Productive growth also utilizes a replication switch from a simple circle to a rolling circle replicative form (Fig. 6) (Skalka, 1977). The early

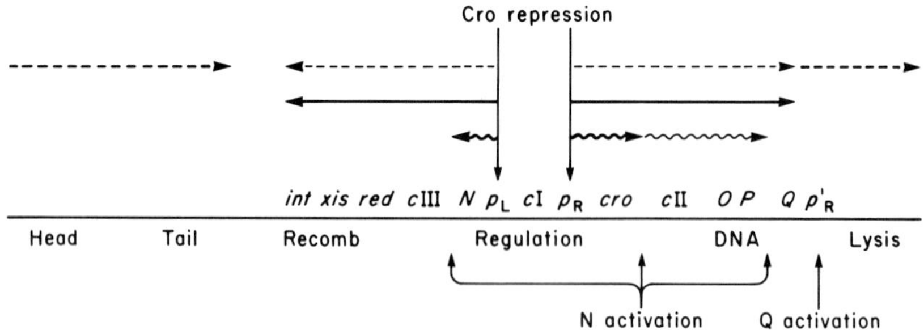

FIG. 5. The transcription pattern of λ DNA during productive development. During the immediate early phase (wavy arrow), RNA chains initiated at the early promoter sites p_L and p_R are mainly terminated at the end of the *N* and *cro* genes (termination sites t_L and t_{R1}, respectively); some rightward RNA chains continue through the replication genes (to t_{R2}). During the delayed early phase (solid arrow), the N protein provides for extension of the immediate early transcripts into the remainder of the early gene region. During the late phase (dashed arrow) the Cro protein reduces initiation of the early RNA and Q protein provides for transcription of the late gene region, initiated at the late promoter site p_R'.

Early replication

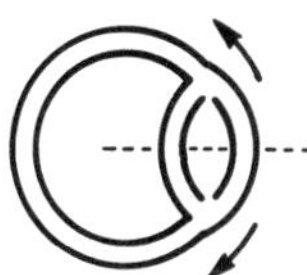

Initiation by phage O/P

Symmetric elongation
by host enzymes

Late replication

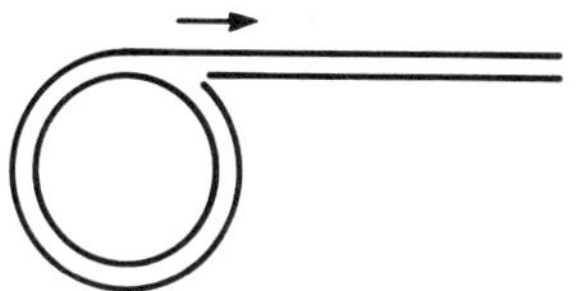

Initiation by O/P off

Asymmetric elongation
by host enzymes

FIG. 6. The replication pattern of λ DNA during productive development. Early replication occurs by a "simple circle" mode, involving symmetric, bidirectional chain growth from a unique origin specified by the phage O and P proteins. Late replication occurs by a "rolling circle" mode, involving asymmetric continuous chain growth along parental circle and linear tail.

stage involves bidirectional replication from a unique origin; initiation is specified by the phage *O* and *P* gene products and chain propagation requires most (if not all) of the same enzymatic machinery used by the host. The late stage of replication proceeds by a rolling circle mechanism in which initiation from the origin has terminated. The regulation of this switch is still obscure, although the Cro protein is probably involved directly or indirectly (Folkmanis *et al.*, 1977).

The alternate route of lysogenic development depends on the action of two additional regulatory proteins, the products of the *cII* and *cIII* genes. The cII/cIII proteins potentiate and coordinate the lysogenic pathway through three distinct regulatory functions: activation of RNA synthesis for the *cI* and *int* genes and inhibition of transcription for late lytic genes (see Fig. 7) (Echols, 1972; Weisberg *et al.*, 1977). The positive regulatory activities turn on the lysogenic pathway because the cI protein maintains the repression essential for lysogeny and the Int protein catalyzes the insertion of λ DNA into the host genome; the negative regulatory activity prevents productive development in a cell destined for lysogenization. The "coordination capacity" of cII/cIII presumably prevents some potential disasters: repression without integration or simultaneous productive and lysogenic development. In addition, cII/cIII exert differential regulation on integrative versus excisive recombination, presumably by turning on transcription of the *int* gene, but not the *xis* gene required only for excision of λ DNA (Fig. 7); thus, the site-specific recombination reaction is far toward insertion during lysogenic development.

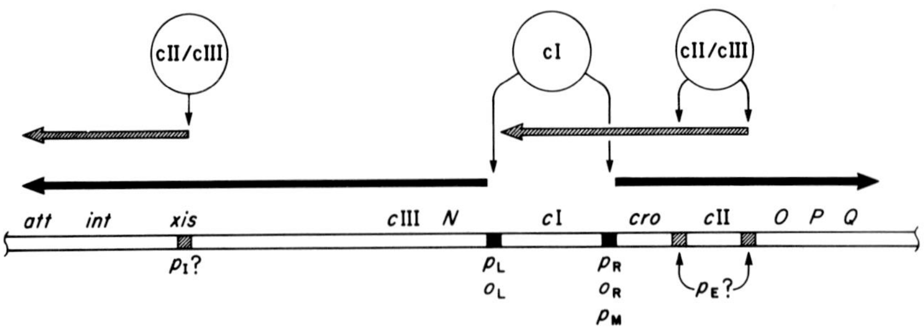

FIG. 7. The transcription pattern of λ DNA during lysogenic development. During the establishment phase, the cII and cIII proteins activate leftward transcription of the *cI* and *int* genes, initiated at the promoter sites p_E and p_I, and inhibit rightward transcription of lytic genes (Fig. 5). During the maintenance phase, the cI protein represses nearly all λ transcription by preventing synthesis of early gene RNA from the early promoter sites p_L and p_R; the cI protein also regulates its own further synthesis by controlling *cI* gene transcription from the maintenance promoter p_M.

Once established, lysogeny is maintained by the capacity of cI protein to block viral development by preventing the initiation of transcription from the early promoter sites p_L and p_R. The cI protein also regulates its own further synthesis by activating transcription of the *cI* gene initiated at the maintenance promoter p_M (Fig. 7) (and repressing transcription of *cI* when the concentration of cI protein becomes high) (Ptashne *et al.*, 1976).

Coordinately regulated along with the *cI* gene is the *rex* gene to its left. The biochemical role of the Rex protein is obscure. Growth of certain T4 mutants (T4rII) is blocked by Rex, possibly indicating that Rex functions normally to exclude certain classes of phage (Howard, 1967; Gussin and Peterson, 1972). Rex has also been implicated in the capacity of cells lysogenic for λ to outgrow nonlysogenic cells under "slow growth" conditions (glucose-limited growth in a chemostat) (Lin *et al.*, 1977). Thus the *rex* gene appears to be an example of a lysogenic conversion event in which the presence of a temperate phage is a net benefit to the host under certain "stress" conditions.

As noted in Section III.C., induction provides an escape route for the prophage. For λ, the cI protein is inactivated, and transcription is initiated at the early promoter sites. The early N-activated transcription unit from p_L includes the *int* and *xis* genes for the proteins that catalyze excisive recombination. Thus, the release of repression will be followed by prophage excision, and productive growth can occur. Since newly synthesized cI protein will also be inactivated, a return down the lysogenic pathway will not normally occur.

Although most of the regulatory proteins that catalyze the developmental pathways for λ and their functional role are becoming clear, a good bit of the biology and most of the biochemistry are obscure for even this "most-studied" phage. There are still biochemical functions in search of genes, and vice versa. I will describe a few of these to point out that a capacity to enumerate regulatory genes and pathways does not constitute a very high level of understanding.

2. Partition between Productive Growth and Lysogenization

The mechanism for "choice" of pathway is a major mystery. The only well-defined influence is the number of infecting phage per cell; at high multiplicity, lysogenization is favored, a reasonable conservation mechanism for a limited host population. This multiplicity effect probably results mainly from more efficient action of cII/cIII at high multiplicity (Kourilsky, 1974; Court *et al.*, 1975). Another potential environmental influence, the bacterial growth rate, seems to have surprisingly little effect. Apparently, as long as the cells can grow exponentially, the partition between productive and lysogenic development (about 2:1 at 5 phage/cell) is not markedly altered (Echols *et al.*, 1975). There is no information that I know of as to whether those cells chosen for lysogenization are picked on a statistical basis or by virtue of some special characteristic, such as stage in replication or cell division cycle.

Although cyclic AMP has been proposed as a possible direct regulator of the lysogenic pathway (Hong *et al.*, 1971; Grodzicker *et al.*, 1972), efforts to find a direct, postinfection effect have failed (Jordan *et al.*, 1973). The lack of marked variation with growth rate of the host cells also argues against a major influence of cyclic AMP because the concentration of this compound varies with growth rate. Nevertheless, there is probably an indirect effect of the cyclic AMP system because mutants defective in the synthesis or activity of cyclic AMP give somewhat reduced lysogenization by wild-type λ (and the residual lysogenization by cIII$^-$ is greatly reduced) (Grodzicker *et al.*, 1972; Belfort and Wulff, 1974).

3. Biochemistry of Phage λ Development

At a biochemical level, only the cI and Cro proteins among RNA regulators have been analyzed in depth with purified components. Both cI and Cro are "classical" repressors that bind to the same operator regions of λ DNA (o_L and o_R) and block initiation by RNA polymerase at the early promoters p_L and p_R. Both cI and Cro also regulate the *c*I gene; regulation by *c*I can be positive or negative, by Cro only negative. The binding sites are reiterated for cI and for Cro (three each for o_L and o_R). As judged by binding affinity, cI is a "strong repressor" that should maintain repression at a low intracellular concentration, whereas Cro is a "weak repressor" that probably requires a high concentration and thus has a delayed developmental action (for more detailed discussions, see Ptashne *et al.*, 1976; Takeda *et al.*, 1977). The N protein probably antagonizes in some way the bacterial protein (ρ) used to terminate RNA synthesis (Roberts, 1970; Franklin, 1974; Adhya *et al.*, 1974). Even the general mechanism by which Q and cII/cIII work is unclear at present; their capacity to activate transcription might result from new initiation of RNA synthesis or from antitermination of short ("leader") RNA chains produced by the unmodified host RNA polymerase (Echols, 1971b; Roberts, 1975; Honigman *et al.*, 1976).

Our understanding of the regulation of DNA events is considerably worse off than the RNA picture. Most aspects of the replicative switch in DNA replication are not understood at all (Echols, 1974; Skalka, 1977). The regulatory role of γ-protein has been defined as an inhibitor of the host RecBc DNase, thereby allowing for the linear rolling-circle "tail" (Enquist and Skalka, 1973). However, otherwise the replicative switch is a case of biochemical functions in search of genes. As noted above, there is a turn off of early initiation and a turn on of the rolling-circle. These events might be phage-regulated, host-regulated, or left to chance. Given the extreme precision with which λ regulates its transcription pattern, I suspect that a similar attention to detail will eventually be found for DNA replication.

The mechanism for site-specific recombination and its regulation is just entering the stage of biochemical analysis. An *in vitro* recombination system has been worked out (Mizuuchi and Nash, 1976), and an *in vitro* DNA-binding assay for the Int protein has led to the partial purification and characterization of Int (Kotewicz *et al.*, 1977). Superhelical DNA and host protein "factors" are required besides Int for the complete reaction (Gellert *et al.*, 1976; Williams *et al.*, 1977; Miller and Friedman, 1977). The requirement for superhelical DNA has one obvious potential regulatory function: the substrate for integration will be the (presumably repressed) closed circle and not a rolling-circle, a potential disaster for the cell.

4. Genes in Search of Functions

There are also λ genes in search of biochemical functions, which may ultimately have a lot to tell us about the murky area of host–phage interactions. One (or possibly two) genes in the recombination region act to shut off bacterial operons sensitive to the cyclic AMP activation system [termed Hin or Car function (Cohen and Chang, 1970; Echols, 1974)]. Another λ gene of the recombination region, termed *kil*, contributes to loss of host cell viability (Greer, 1975); the only well-defined phenotype associated with the Kil function is filament formation by the host. Whether the Car and Kil phenotypes so far observed represent the "real" functions of these genes or are adventitious effects of some other aspect of viral development is unclear. One possibility I find interesting is that these mysterious effects of certain phage genes might represent an example of the population response involving cell-to-cell signaling mentioned earlier.

B. The Virulent Phage T4, T7, and M13

1. M13—The Subtle Approach

Specific events during the life cycle of phage T4, T7, and M13 have been reviewed in detail recently, and I will consider only general features here, mainly in an effort to point out the rather dramatic differences in the ways virulent phage execute the productive cycle. For much more complete discussions the reader should consult the articles by Matthews (1977) and Rabussay and Geiduschek (1977) for phages T4 and T7; Studier (1972) for T7; Kornberg (1974), Denhardt (1975), and Ray (1977) for M13.

M13 is a filamentous phage, about 870 nm long, that contains a single-stranded circular DNA genome of MW 1.9×10^6. The infection process by M13 appears to involve the formation of a complex with the cytoplasmic membrane (perhaps by insertion of the coat protein). The production of this complex requires the presence of the F pilus; however it is not clear whether the extended F pilus is required as an adsorption site or whether the pilus provides accessibility to a membrane site. Complex formation permits the conversion of the single-stranded parental DNA to a double-stranded form by a reaction in which the host RNA polymerase synthesizes a primer RNA that is elongated by the host DNA polymerase III (Wickner *et al.*, 1973). The initial stages of the replication process seem to involve the participation of a minor virion protein (the gene 3 protein) (Jazwinski *et al.*, 1973). Because different single-stranded DNA phages "select" different components of the host replication apparatus, the minor virion protein may function as a "pilot protein" to direct the formation of the replication complexes used for single-stranded (and double-stranded) DNA replication (Kornberg, 1974). Thus the initial stage of M13 infection may be thought of as an insertion of viral DNA into the membrane and a transfer reaction in which a replication complex and a single-strand DNA-binding protein replace the coat protein.

Once the double-stranded DNA (RF) has been synthesized, the host RNA polymerase transcribes this template into a number of RNA chains; these RNA transcripts start from a number of promoter sites but all appear to terminate at a single site. One of the viral proteins (gene 2 protein) is required to initiate the double-strand stage of DNA replication (RF to RF), catalyzed by a number of host proteins [at least the *rep*, *dnaB*, *dnaG*, and *dnaE* (polymerase III) gene products] (Ray, 1977). The shift to progeny single-strand replication (RF to SS) requires the phage-specified gene 5 protein, a single-strand DNA-binding protein that may catalyze the shift by coating one strand of a rolling circle replicative intermediate, preventing its further use as a template (Salstrom and Pratt, 1971; Mazur and Zinder, 1975). The protein 5–DNA complex is somehow cleaved and circularized, and the DNA is transferred into a coat assembled on the cell membrane. Finally, the newly synthesized virion is released from the membrane of the intact host cell. The intracellular sequence of events is shown diagrammatically in Fig. 8.

The life cycle of M13 is notable for its conservation of natural resources. By a clever escape scheme, the host cells are not killed and can continue to produce more M13. As befits a simple phage, M13 also makes maximal use of the host enzymatic machinery, using the host RNA polymerase and replication enzymes, suitably directed by one protein to organize the replication complex, another to tell it where to begin double-strand replication, and a third to produce the switch to progeny single-strand replication. The transcription apparatus seems principally organized to grab as many RNA polymerase enzymes as possible; the variation in the level of proteins produced (proteins 5 and 8 are produced in very large amounts) principally represents differences in the rate of translation.

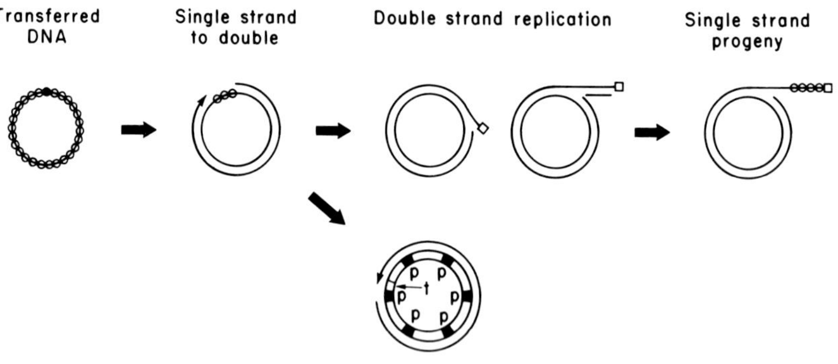

FIG. 8. Phage M13 development. The single-strand DNA is transferred from the virion into a protein–DNA aggregate consisting of a replication complex (closed circle) and the host single-strand DNA-binding protein (open circle). After the single strand is converted to double, the viral DNA is transcribed from a number of promoter sites (*p*), terminating at a single site (*t*). The phage gene 2 protein (open square) initiates double-strand replication by a specific nick, and replication probably proceeds by a rolling circle mechanism. The shift to progeny single-strand replication utilizes the phage single-strand DNA-binding protein (open circle), which probably prevents double-strand synthesis on the tail of the rolling circle.

2. Phage T7—Organized Virulence on a Small Scale

Phage T7 represents the next level of complexity—a "minimal" phage for sequential development at the transcription level. T7 has an icosahedral head about 60 nm in diameter, a stubby tail of 15 nm, and a double-stranded DNA genome of MW 25×10^6; the DNA has a terminal redundancy (repeat) of about 250 nucleotides, and has a unique sequence in each phage particle

As one might expect from its more sophisticated structure and greater genome size compared to M13, T7 has an injection mechanism for its DNA and a regulated transcription pattern (Studier, 1972; Matthews, 1977; Rabussay and Geiduschek, 1977). The earliest stage of viral RNA synthesis involves transcription of 20% of the phage DNA by the host RNA polymerase, starting at three promoter sites near one end of the DNA (the "left end" of Fig. 9). This lengthy early transcript is processed into five RNA chains that serve as messages for the early proteins (Dunn and Studier, 1973).

One of these early proteins (the gene 1 product) is a new RNA polymerase that transcribes the rest of T7 DNA, using several promoter sites specific for this enzyme (Golomb and Chamberlin, 1974; Niles and Condit, 1975). The other early proteins (proteins 0.3, 0.7, 1.1, and 1.3) are not essential for productive growth under typical conditions, but three have known functions: abolition of host restriction (0.3), DNA ligase (1.3), and a protein kinase (0.7) that is likely to participate in shutoff of host DNA transcription. The host turnoff mechanism probably is also responsible for shutting off the earliest stage of T7 transcription at later times.

One class of T7 proteins produced as a consequence of transcription by T7 polymerase are those required for host DNA degradation and T7 DNA replication (including a DNA polymerase). These replication proteins ("early or delayed early" for most phage) are distinguished from the encapsulated proteins (typical "lates") only by continued synthesis of late proteins after a turnoff of production of the replication enzymes (Studier, 1972). The mechanism for this switch is not known; it may involve a dif-

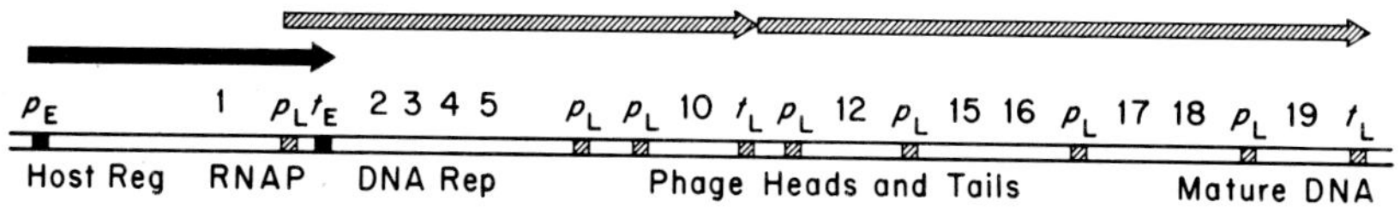

FIG. 9. The transcription pattern of phage T7. During the early phase, RNA chains initiated at three adjacent promoter sites (p_E) terminate at a single site (t_E). The rest of the T7 DNA is transcribed by the T7 RNA polymerase specified by gene 1, starting at a number of late promoter sites (p_L), with at least two termination sites (t_L).

ference in the rate of RNA synthesis at late times, a greater functional stability of late protein RNA, or a greater ability of the RNA chains for late proteins to initiate protein synthesis (and thus compete for synthetic machinery at high RNA concentration).

DNA replication by phage T7 proceeds bidirectionally from an origin about 17% from the "left" end (Fig. 9) (Dressler *et al.*, 1972). The early replicating structure is probably a linear monomer, but later in viral development linear concatemers accumulate, probably by recombination at the redundant ends. The packaging of this multimeric structure into virions as unique, double-stranded, and redundant molecules must involve a single-strand cutting and gap-filling reaction, but the details are not known. Replication and perhaps packaging occur in association with the cell membrane.

Phage T7 represents a classical "virulent" phage in the sense that host functions are taken over completely by the virus. Host transcription is shut off, the host DNA is degraded and used for precursors to make viral DNA, and the ravished cell lyses at the end of the productive cycle to yield new phage particles. Although less tightly organized into replicative and maturation phases than λ, T4, or other large DNA phage, T7 also exhibits a temporally organized developmental cycle and a replicative switch.

3. Phage T4—Rococo Elegance

Phage T4 represents a marvel of viral evolution—a genome of such size (130×10^6) and a virion of such complexity that its very existence seems surprising. T4 has an icosahedral head of diameter 95 nm and an elegant contractile tail of 110 nm to which is attached tail fibers (used for initial adsorption) as long as the tail itself. T4 DNA has a terminal redundancy of about 3000 bases, and a population of phage particles carry DNA molecules that begin and end in different regions of the T4 genome (the viral DNA is "circularly permuted").

Although the number and type of transcription switches that regulate T4 RNA synthesis is incompletely understood, there is a clear separation between an early phase for genes involved in host shutoff and DNA replication and a late phase for genes that specify encapsulation proteins (Rabussay and Geiduschek, 1977). Efficient transcription of late genes requires a modification of the host RNA polymerase by at least three phage proteins produced during the early period (products of genes *33*, *55*, and *45*) and also requires viral DNA replication. During the late stage of viral development, there is a shutoff of production of early proteins that involves control at the level of RNA synthesis, probably modulated by translational effects.

In addition to the separation between early and late transcription, there is probably a transition from the "immediate-early" transcription of the injected DNA to a "delayed-early" period, requiring at least viral RNA synthesis; this may involve the elimination of termination events for immediate-early RNA chains. There appear to be also new initiation events that require immediate-early RNA and protein synthesis (termed "quasi-late" transcription). The study of T4 transcription is complicated by the enormous genome size and the existence of multiple transcription units (shown schematically in Fig. 10).

A special feature of T4 regulation is the existence of self-regulating proteins whose normal biochemical function is replicative. The single-strand DNA-binding protein (gene *32* product) and the T4 DNA polymerase (gene *43* product) are the clearly defined examples (Russel, 1973; Krisch *et al.*, 1974; Gold *et al.*, 1976). For 32 protein, the mechanism probably involves binding of the protein to gene 32 RNA in a way that blocks initiation of protein synthesis.

T4 DNA replication and recombination requires a multitude of phage proteins (at least 15), many of which serve to provide precursors from the degraded host genome. Six of these proteins (32, 41, 43, 44, 45, and 62) (Alberts *et al.*, 1975) are explicitly involved in the replication process itself, and others are probably involved in initiation of replication. T4 also supplies its own pathways of DNA repair that parallel those of the host, an excision repair system and a "postreplicative" repair system. The replication process probably involves bidirectional chain growth from several origins. Linear concatemers are produced late in viral development, most likely by recombination at the redundant ends, and these oligomeric molecules are presumably packaged by a "headful" mechanism in which DNA is inserted sequentially into phage heads one unit at a time, providing the terminal redundancy and the circularly permuted population (Doermann, 1973).

The assembly of the T4 virion is also an enormously complex event, involving sequential addition of proteins along three pathways: heads, tails, and tail fibers (Casjens and King, 1975). Some fifty T4 genes are involved. Both replication and the final assembly stage seem to occur in association with the cell membrane.

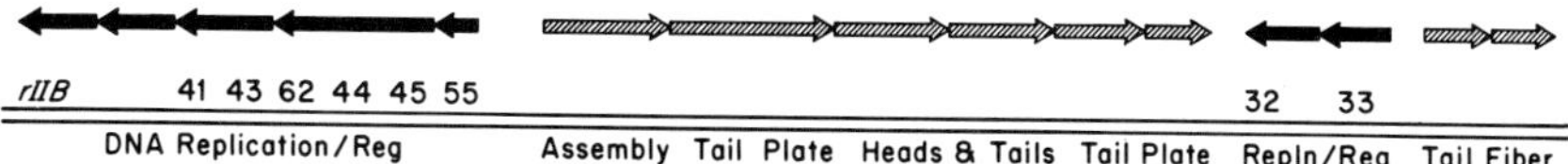

FIG. 10. Highly simplified transcription pattern of phage T4. Early genes are transmitted mainly in the opposite direction from late genes. Both early and late genes are expressed in several transcription units. The ends of the linear DNA differ in the viral DNA population, but all genes are probably expressed in each case because of the lengthy terminal redundancy.

Phage T4 represents the virulent phage in its most sophisticated form. The virus provides for its own DNA replication, recombination, and repair, utilizing only a highly modified RNA polymerase and the protein synthetic machinery of the host for new macromolecular synthesis (phage T4 helps this out with a few extra tRNA's). Host protein and RNA synthesis are shut off very early after infection, and the host DNA is degraded and used for viral DNA synthesis. The virus provides for a new base (hydroxymethylcytosine) to replace the host cytosine and a glucosylation mechanism utilizing this special base, presumably to provide protection against restriction nucleases. Transcription is tightly organized to separate the replicative and encapsulation phases, and even some individual gene products regulate their own synthesis. For all its grandeur, however, one wonders how phage T4 evolved to such complexity when T7 does quite well as a virulent phage with a vastly simpler life style and λ has two separate lives with one-quarter the genes. Perhaps the extreme independence of phage T4 allows it to cross more easily from one bacterial host to another with a minimum of mutational variation.

V. Conclusions

This chapter began with two general questions: How have phage evolved? What evolutionary pressures retain the diverse population of phage? Neither of these questions has been answered, but I hope that the intervening pages have served to focus some of the facets of the problem and at least indicate the enormous diversity of phage–host relationships. To summarize my own opinion, I believe that the diversity of the phage universe is characterized mainly by three population relationships: small phage–host plasmid mutational transitions, large phage–host plasmid recombination, and phage–phage recombination.

There is a remarkable (although possibly superficial) relationship between the bacterial plasmid and the small DNA phage. For phage M13, the double-stranded replicative form reels off a single-stranded DNA that is inserted into a coat and transferred to another cell rather than the direct conjugative transfer of single-strand DNA typical of a plasmid—not a vast difference. Like plasmid replication, double-strand M13 replication uses bacterial replication proteins, but provides its own initiation protein.

The capacity of large temperate phage to exist as plasmids suggests that a continuing relationship might also exist between these phage species and nonviral plasmids. Because of the more complex virion structures of the larger phages, such transitions most likely involve the union of a plasmid replication "module" and a segment of DNA specifying head and tail structures; these unions would be facilitated by the clustering of genes for specific functions and their site of action. The extensive regions of homology and modular nonhomology found for temperate phage are consistent with this possibility (for an extensive discussion, see Campbell, 1977). Finally, the rapid variation needed for adaptation of large virulent phage seems most likely to result from frequent recombination to generate new adsorption structures and new capabilities for intracellular development.

The foregoing discussion makes some proposals about the adaptive maintenance of various classes of phage, but does not say anything about how M13 grew into T4. The "missing link," to my way of thinking, is the gap in genome size and complexity between phages such as M13 and T7. Once T7 has been achieved, the evolution of T4 does not seem to be such a major jump in virion structure and developmental pattern. My guess is that a single-strand phage (such as ϕX174) grew into T7 by purloining a region of DNA from a bacterium that carries out temporal regulation of a specialized structure (e.g., a pilus), thus joining a replication module with a head to a regulation module with a tail. This act of creation can be extremely infrequent because it is not required for maintenance of the phage population.

One of the pleasures derived from writing a chapter such as this is the realization of the diversity found even among the humble phage. Extrapolated to the entire biological universe, this realization becomes awe.

Acknowledgments

I thank the community of phage workers for providing the information and concepts on which this article has been based, with a special thanks to Allan Campbell for his biological insight and encouragement.

The preparation of this article has been aided by the National Institute of General Medical Sciences, Grant GM 17078.

References

Adams, M. H. (1959). "Bacteriophages." Wiley (Interscience), New York.

Adhya, S., Gottesman, M., and deCrombrugghe, B. (1974). *Proc. Natl. Acad. Sci. U.S.A.* **71,** 2534–2538.

Alberts, B. M., Morris, C. F., Mace, D., Sinha, N., Bittner, M., and Moran, L. (1975). *In* "DNA Synthesis and its Regulation" (M. M. Goulran, P. C. Hanawalt, and C. F. Fox, eds.), pp. 241–269. Benjamin, New York.

Barksdale, L., and Arden, S. B. (1974). *Annu. Rev. Microbiol.* **28,** 265–299.

Beckendorf, S. K. (1973). *J. Mol. Biol.* **73,** 37–53.

Belfort, M., and Wulff, D. (1974). *Proc. Natl. Acad. Sci. U.S.A.* **71,** 779–783.

Bertani, L. E., and Bertani, G. (1971). *Adv. Genet.* **16,** 200–237.

Bradley, D. E. (1967). *Bacteriol. Rev.* **31,** 230–314.

Bukhari, A. I., Shapiro, J. A., and Adhya, S. L., eds. (1977). "DNA Insertion Elements, Plasmids, and Episomes," Cold Spring Harbor Laboratory, Cold Spring Harbor, New York.
Calendar, R. C. (1970). *Annu. Rev. Microbiol.* **24,** 241–296.
Campbell, A. (1961). *Evolution* **15,** 153–165.
Campbell, A. (1977). *In* "Comprehensive Virology" (R. R. Wagner and H. Fraenkel-Conrat, eds.), Vol. 8, pp. 259–328. Plenum, New York.
Casjens, S., and King, J. (1975). *Annu. Rev. Biochem.* **44,** 555–611.
Clark, A. J. (1971). *Annu. Rev. Microbiol.* **25,** 437–464.
Cohen, S. N., and Chang, A. C. Y. (1970). *J. Mol. Biol.* **49,** 557–575.
Court, D., Green, L., and Echols, H. (1975). *Virology* **63,** 484–491.
Davis, R. W., and Hyman, R. W. (1971). *J. Mol. Biol.* **62,** 287–301.
Denhardt, D. T. (1975). *Crit. Rev. Microbiol.* , 161–223.
Denhardt, D. T. (1977). *In* "Comprehensive Virology" (R. R. Wagner and H. Fraenkel-Conrat, eds.), Vol. 7, pp. 1–104. Plenum, New York.
Doermann, A. H. (1973). *Annu. Rev. Genet.* **7,** 325–341.
Dressler, D., Wolfson, J., and Magazin, M. (1972). *Proc. Natl. Acad. Sci. U.S.A.* **69,** 998–1002.
Dunn, J. J., and Studier, F. W. (1973). *Proc. Natl. Acad. Sci. U.S.A.* **70,** 3296–3300.
Echols, H. (1971a). *Annu. Rev. Biochem.* **40,** 827–854.
Echols, H. (1971b). *In* "The Bacteriophage λ" (A. D. Hershey, ed.), pp. 247–270. Cold Spring Harbor Lab., Cold Spring Harbor, New York.
Echols, H. (1972). *Annu. Rev. Genet.* **6,** 157–190.
Echols, H. (1973). *In* "Genetic Mechanisms of Development" (F. Ruddle, ed.), pp. 1–14. Academic Press, New York.
Echols, H. (1974). *Biochimie* **56,** 1491–1496.
Echols, H., Green, L., Kudrna, R., and Edlin, G. (1975). *Virology* **66,** 344–346.
Enquist, L. W., and Skalka, A. (1973). *J. Mol. Biol.* **75,** 185–212.
Eoyang, L., and August, L. T. (1974). *In* "Comprehensive Virology" (R. R. Wagner and H. Fraenkel-Conrat, eds.), Vol. 2, pp. 1–60. Plenum, New York.
Folkmanis, A., Takeda, Y., Simuth, J., Gussin, G., and Echols, H. (1976). *Proc. Natl. Acad. Sci. U.S.A.* **73,** 2249–2253.
Folkmanis, A., Maltzman, W., Mellon, P., Skalka, A., and Echols, H. (1977). *Virology* **81,** 352–362.
Fraenkel-Conrat, H., and Wagner, R. R., eds. (1974). "Comprehensive Virology," Vol. 1. Plenum, New York.
Fraenkel-Conrat, H., and Wagner, R. R., eds. (1975). "Comprehensive Virology," Vol. 2. Plenum, New York.
Fraenkel-Conrat, H., and Wagner, R. R., eds. (1976). "Comprehensive Virology," Vol. 7. Plenum, New York.
Fraenkel-Conrat, H., and Wagner, R. R., eds. (1977). "Comprehensive Virology," Vol. 8. Plenum, New York.
Franklin, N. C. (1974). *J. Mol. Biol.* **89,** 33–48.
Gellert, M., Mizuuchi, K., O'Dea, M. H., and Nash, H. A. (1976). *Proc. Natl. Acad. Sci. U.S.A.* **73,** 3872–3876.
Gold, L., O'Farrell, P. Z., and Russel, M. (1976). *J. Biol. Chem.* **251,** 7251–7262.
Golomb, M., and Chamberlin, M. (1974). *Proc. Natl. Acad. Sci. U.S.A.* **71,** 760–764.
Greer, H. (1975). *Virology* **66,** 589–604.
Grodzicker, T., Arditti, R. R., and Eisen, H. (1972). *Proc. Natl. Acad. Sci. U.S.A.* **69,** 366–370.
Gussin, G. N., and Peterson, V. (1972). *J. Virol.* **10,** 760–765.

Hershey, A. D., ed. (1971). "The Bacteriophage λ." Cold Spring Harbor Lab., Cold Spring Harbor, New York.
Herskowitz, I. (1973). *Annu. Rev. Genet.* **7,** 289–324.
Hong, J.-S., Smith, G. R., and Ames, B. N. (1971). *Proc. Natl. Acad. Sci. U.S.A.* **68,** 2258–2262.
Honigman, A., Hu, S.-L., Chase, R., and Szybalski, W. (1976). *Nature* (*London*) **262,** 112–116.
Howard, B. D. (1967). *Science* **158,** 1588–1589.
Howe, M., and Bade, E. (1975). *Science* **190,** 624–632.
Ikeda, H., and Tomizawa, J. (1965). *J. Mol. Biol.* **14,** 85–109.
Jazwinski, S. M., Marco, R., and Kornberg, A. (1973). *Proc. Natl. Acad. Sci. U.S.A.* **70,** 205–209.
Jordan, E., Green, L., and Echols, H. (1973). *Virology* **55,** 521–523.
Kornberg, A. (1974). "DNA Synthesis." Freeman, San Francisco, California.
Kotewicz, M., Chung, S., Takeda, Y., and Echols, H. (1977). *Proc. Natl. Acad. Sci. U.S.A.* **74,** 1511–1515.
Kourilsky, P. (1974). *Biochimie* **56,** 1511–1516.
Krisch, H. M., Bolle, A., and Epstein, R. H. (1974). *J. Mol. Biol.* **88,** 89–104.
Levin, B. R., Stewart, F. M., and Chao, L. (1977). *Amer. Nat.* **111,** 3–24.
Lin, L., Bitner, R., and Edlin, G. (1977). *J. Virol.* **21,** 554–559.
Lwoff, A. (1953). *Bacteriol. Rev.* **17,** 269–337.
McEntee, K. (1977). *Proc. Natl. Acad. Sci. U.S.A.* **74,** 5275–5279.
Matthews, C. K. (1977). *In* "Comprehensive Virology" (R. R. Wagner and H. Fraenkel-Conrat, eds.), Vol. 7, pp. 179–294. Plenum, New York.
Mazur, B. J., and Zinder, N. D. (1975). *Virology* **68,** 490–502.
Miller, H. I., and Friedman, D. I. (1977). *In* "DNA Insertion Elements, Plasmids, and Episomes" (A. I. Bukhari, J. A. Shapiro, S. L. Adhya, eds.), pp. 349–356, Cold Spring Laboratory, New York.
Mizuuchi, K., and Nash, H. A. (1977). *Proc. Natl. Acad. Sci. U.S.A.* **73,** 3524–3528.
Mount, D. W. (1977). *Proc. Natl. Acad. Sci. U.S.A.* **74,** 300–304.
Niles, E. G., and Condit, R. C. (1975). *J. Mol. Biol.* **98,** 57–67.
Ptashne, M., Backman, K., Humayun, M. Z., Jeffrey, A., Maurer, R., Meyer, B., and Sauer, R. T. (1976). *Science* **194,** 156–161.
Rabussay, D., and Geiduschek, E. P. (1977). *In* "Comprehensive Virology" (R. R. Wagner and H. Fraenkel-Conrat, eds.), Vol. 8, pp. 1–196. Plenum, New York.
Ray, D. S. (1977). *In* "Comprehensive Virology" (R. R. Wagner and H. Fraenkel-Conrat, eds.), Vol. 7, pp. 105–178. Plenum, New York.
Roberts, J. W. (1970). *Cold Spring Harbor Symp. Quant. Biol.* **35,** 121–126.
Roberts, J. W. (1975). *Proc. Natl. Acad. Sci. U.S.A.* **72,** 3300–3304.
Roberts, J. W., and Roberts, C. W. (1975). *Proc. Natl. Acad. Sci. U.S.A.* **72,** 147–151.
Roberts, J. W., Roberts, C. W., and Mount, D. W. (1977). *Proc. Natl. Acad. Sci. U.S.A.* **74,** 2283–2287.
Russel, M. (1973). *J. Mol. Biol.* **79,** 83–94.
Sakaki, Y. (1974). *J. Virol.* **14,** 1611–1612.
Sakaki, Y., Karu, A. E., Linn, S., and Echols, H. (1973). *Proc. Natl. Acad. Sci. U.S.A.* **70,** 2215–2219.
Salstrom, J. S., and Pratt, D. (1971). *J. Mol. Biol.* **61,** 489–501.
Scott, J. R. (1975). *Virology* **65,** 173–178.
Skalka, A. (1977). *Curr. Top. Microbiol. Immunol.* **78,** 202–237.
Streisinger, G., Emrich, J., and Stahl, M. M. (1967). *Proc. Natl. Acad. Sci. U.S.A.* **57,** 292–295.
Studier, F. W. (1972). *Science* **176,** 367–376.

Takeda, Y., Folkmanis, A., and Echols, H. (1977). *J. Biol. Chem.* **252,** 6157–6183.
Tanner, D., and Oishi, M. (1971). *Biochim. Biophys. Acta* **228,** 767–769.
Thomas, R. (1971). *In* "Bacteriophage Lambda" (A. D. Hershey, ed.), pp. 211–220, Cold Spring Harbor Laboratory, Cold Spring Harbor, New York.
Tye, B.-K., Chan, R. K., and Botstein, D. (1974). *J. Mol. Biol.* **85,** 485–500.
Weisberg, R. A., Gottesman, S., and Gottesman, M. E. (1977). *In* "Comprehensive Virology" (R. R. Wagner and H. Fraenkel-Conrat, eds.), Vol. 8, pp. 197–258. Plenum, New York.
Wickner, W., Schekman, R., Geider, K., and Kornberg, A. (1973). *Proc. Natl. Acad. Sci. U.S.A.* **70,** 1764–1768.
Williams, J. G. K., Wulff, D. L., and Nash, H. A. (1977). *In* "DNA Insertion Elements, Plasmids, and Episomes" (A. I. Bukhari, J. A. Shapiro, and S. L. Adhya, eds.), pp. 357–361, Cold Spring Harbor Laboratory, New York.
Witkin, E. (1976). *Bact. Revs.* **40,** 869–907.
Zinder, N. D., ed. (1974). "RNA Phages." Cold Spring Harbor Lab., Cold Spring Harbor, New York.

CHAPTER 9

Control of Cell Division in *Escherichia coli*

CHARLES E. HELMSTETTER, OLGA PIERUCCI, MARTIN WEINBERGER, MARGARET HOLMES, AND MOON-SHONG TANG

I. Introduction

This chapter is concerned with the regulation of, and coordination between the multitude of events which must take place for *Escherichia coli* to divide. When does a cell begin preparing for division, what are the synthetic processes which must take place, and how are these processes interrelated? One might think that cells begin to prepare for division at the moment they have completed the previous division. This is not the case. Let us consider a dividing cell and look back in time to guess what steps had to be completed

ISBN 0-12-307207-7

prior to the process of cell fission. Just before separation of the daughter cells at division, the parent cell must have formed a septum–crosswall, or equivalent division process, to separate the cellular components destined to become the two daughter cells. Prior to completion of the steptum–crosswall, the cell must have replicated its genome, if both daughters are to be genetically identical. Completion of replication and segregation of the genome could have taken place before the beginning of septum–crosswall formation, after it started, or perhaps at the same time. Looking back farther in time, prior to replication of the chromosome, the cell had to prepare for initiation of replication by forming the necessary biosynthetic and structural components. Thus, one group of events required for division can be identified: (a) preparation for initiation of chromosome replication, (b) replication and segregation of the chromosome, and (c) initiation and completion of septum–crosswall formation. Event (b) must follow event (a), but event (c) might not be coupled to completion of (b) or to any point in (a) or (b). In addition, there must be other synthetic processes required for division, which ensure that the daughter cells mirror the mother cell in composition and structure. Our discussion of bacterial duplication will be centered around the easily identifiable chromosome replication, septum–crosswall formation sequence(s).

In tracing the sequence of events leading to division, mention was not made of a previous division. The steps we identified leading to one division may or may not have started at the previous division. The "starting" point would depend on (i) when cells begin preparing for initiation of chromosome replication, (ii) when cells begin the other processes which lead specifically to division, and (iii) how these preparations interact, if at all. Thus, it should not be assumed that the duplication process is cyclic and that the cycles begin and end with division. The duplication process need not be cyclic, since the sequences need not repeat one after the other, but instead, they could overlap. We will develop this concept in detail in Section II and show that reasonable estimates of the relationship between chromosome replication and the division cycle of *E. coli* can be determined with a minimal amount of experimental data. The ideas presented are intended to serve as a conceptual framework for analyzing the cell division process. Subsequent sections deal with experimental evidence on specific details of the general scheme.

II. ($I + C + D$) Concept

A. Chromosome Replication–Cell Division Relationships

To begin analyzing the relationship between chromosome replication and cell division in *E. coli*, three time intervals must be defined. The time required for a cell to prepare for initiation of a round of chromosome replication is *I*, the time for a round of chromosome replication is *C*, and the time between completion of chromosome replication and the subsequent cell division is *D*. The duplication of *E. coli* can be described in terms of $(I + C + D)$ (Helmstetter *et al.*, 1968). Cells first prepare the biosynthetic and structural components needed for initiation of chromosome replication. When the preparation is completed in *I* minutes, chromosome replication begins. Chromosome replication takes *C* minutes, and separation of the daughter cells by division is completed *D* minutes later. It should be emphasized that the *D* period is an interval of time which exists between two sepecific events, and it should not be assumed that the *D* period is synonymous with initiation and completion of septum–crosswall formation. The time between initiation of visible septum–crosswall formation and cell division is *T* (Clark, 1968a; Gudas and Pardee, 1974; Woldringh, 1976). The relationship between *D* and *T* is discussed in Section V.

The relationship between chromosome replication and cell division can be determined for cells growing at any rate by applying the $(I + C + D)$ concept to a hypothetical cell containing a single nonreplicating chromosome. An example of the construction of chromosome replication and division patterns for a case in which $I = 70$ minutes, $C = 40$ minutes, and $D = 20$ minutes is shown in Fig. 1. The numerical values for *I*, *C*, and *D* were chosen arbitrarily to simplify the analysis, but these are realistic values, as will be shown in Section V. The construction begins at the far left of the figure with a hypothetical cell. The saw-toothed line shows the time, *I*, required for this cell to progress through the steps necessary for initiation of chromosome replication. Once preparation is completed, chromosome replication takes place as indicated by the solid horizontal line. The cell divides *D* minutes after completion of chromosome replication, as is indicated by the dotted line. The second $(I + C + D)$ duplication sequence is shown below the first sequence, with the cell beginning to prepare for the second initiation the moment the first initiation takes place. The same sequence is then repeated a third time. Note that the sequences do not interact. The $(I + C + D)$ concept simply states that once a cell initiates a round of chromosome replication, a new $(I + C + D)$ sequence is inaugurated. Since the construction begins with a single hypothetical cell at the far left of the diagram, two cells are formed at the first division, four at the second division, and so on. Therefore, if the number of cells were plotted versus time, this construction results in an idealized, synchronously dividing population.

From the construction shown in Fig. 1, the pattern of chromosome replication during the division cycle can be determined. The chromosome replication pattern is shown in the upper portion of the figure. This pattern is obtained by following the construction from left to right and drawing the

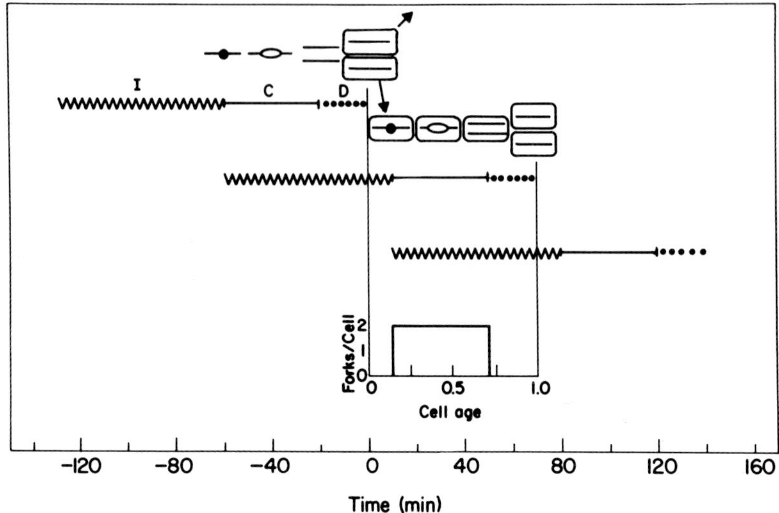

FIG. 1. Construction of chromosome replication and cell division patterns for *E. coli* with $I = 70$ minutes, $C = 40$ minutes, and $D = 20$ minutes. The construction is based on the $I + C + D$ sequence beginning at the left (-130 min) with an hypothetical cell containing a single chromosome. Preparation for chromosome replication requires I minutes (saw-toothed lines), chromosome replication occupies C minutes (solid line), and the cell divides D minutes later (dotted line). At the time chromosome replication begins (-60 minutes), preparation for the next $(I + C + D)$ sequence commences below the first. The vertical lines indicate divisions. The chromosome configurations are shown at the top as linear structures, rather than circular molecules, for convenience. A closed circle at the center of the chromosome indicates initiation of chromosome replication. The next chromosome configuration (-40 minutes) shows a half-replicated chromosome as a result of bidirectional replication from the point of initiation. The configurations to the left of zero time are hypothetical. The configurations between 0 and 70 minutes, i.e., between the first two divisions, indicate the chromosome replication patterns during the division cycle of cells growing under these conditions. The arrows indicate that only one cell is followed through the division cycle. In the overall development of a culture from a single cell, the number of cells formed at successive divisions would be two, four, etc. The rate of chromosome replication during the division cycle is shown in the lower portion of the figure in terms of the number of replication forks per cell.

replicating chromosomal structures as indicated. At the far left, the hypothetical cell contains a single chromosome. After preparation, replication is initiated, as indicated by the closed circle on the chromosome. The chromosome then replicates, presumably bidirectionally and symmetrically as indicated by the movement of the two replication forks along the chromosome. Chromosome replication terminates, the cell divides D minutes later, and each daughter cell receives a single chromosome. Ten minutes after division, the cells have completed preparation for the next initiation event in the second duplication sequence, a new round of replication is initiated, and the $(C + D)$ sequence repeats. The relationship between chromosome replication and the cell division cycle is obtained by observing the chromosome replication pattern between the first two divisions. In the example shown, chromosome replication initiates 10 minutes after division, the chromosome replicates during the next 40 minutes, and then there is a period devoid of chromosome replication during the last 20 minutes of the division cycle. The rate of chromosome replication during the cycle can also be estimated from this construction and is shown in the lower portion of the figure.

A key point to note in the construction in Fig. 1 is that the relationship between chromosome replication and cell division is best described as a repeating series of linear sequences which overlap. As shown, the reproduction of the bacterial cell is not cyclic since the duplication sequences do not necessarily repeat one after the other. On the other hand, the preparation for initiation of chromosome replication is assumed to be cyclic. Since I is defined as the time required for a cell to prepare for initiation, it is reasonable to assume that I is also the interinitiation time in cells in a culture undergoing balanced growth (Campbell, 1957). (I intervals could overlap, but at present there is no evidence to suggest this possibility.) Thus, the duplication of the *E. coli* cell can be visualized as the cyclic achievement of the capacity for initiation of chromosome replication, followed by cell division $(C + D)$ minutes later. The synthetic events required for division begin before the previous division, and the frequency of cell division is determined by the frequency of initiation of chromosome replication, i.e., the interdivision time (τ) equals I.

In the construction shown in Fig. 1, it is assumed that the rate of DNA chain elongation is constant during C. However, the $(I + C + D)$ construction simply stipulates a time for a round of chromosome replication, and does not require specification of the rate of progression of DNA replication during the C period.

The relationship between chromosome replication and cell division in cells growing at any rate can be determined by applying the $(I + C + D)$ rule in the manner described above. Figure 2 shows an example in which $I = C = 40$ minutes and $D = 20$ minutes. The construction is performed exactly as described for Fig. 1, beginning with a hypothetical cell containing a single chromosome. Preparation for initiation of chromosome replication requires I minutes, and once this is completed, cell division takes place $(C + D)$ minutes later. Again, the chromosome replication pattern and the rate of chromosome replication during the division cycle are shown at the top and bottom of the figure, respectively.

As a final example, a construction is shown in Fig. 3 for the case in which $I = D = 20$ minutes and $C = 40$ minutes. Since I is less than C, a new round of chromosome replication begins before the previous round has terminated.

This is another important aspect of the $(I + C + D)$ concept. Each time cells have completed preparing for initiation of chromosome replication, a new initiation event takes place, irrespective of the status of the cell with respect to ongoing chromosome replication or septum–crosswall formation. There may be additional rules which govern the frequency of initiation, but this subject will be discussed in later sections.

B. Mathematical Relationships

A number of equations have been derived which permit calculations of reasonable estimates of various cellular parameters based on average values of C, D, and τ in exponentially growing cultures. These mathematical relationships have been discussed in detail by Sueoka and Yoshikawa (1965), Cooper and Helmstetter (1968), Donachie (1968), Bleecken (1969), and Pritchard and Zaritsky (1970).

Based on the reasoning in Section II,A [i.e., the $(I + C + D)$ concept], and assuming that replication forks proceed bidirectionally and symmetrically along the chromosome with uniform velocity, the average chro-

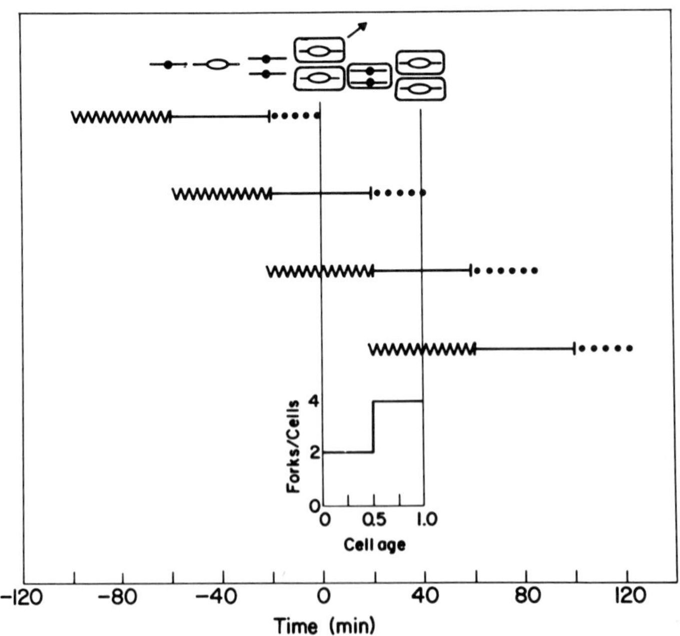

FIG. 2. Construction of chromosome replication and cell division patterns for $I = C = 40$ minutes and $D = 20$ minutes (see legend to Fig. 1 for details).

mosomal DNA content per cell in an exponentially growing population is given by

$$\bar{G} = (\tau/C \ln 2)(2^{(C+D)/\tau} - 2^{D/\tau})$$

where $\bar{G}$ is the average number of chromosome equivalents of DNA per cell; a chromosome equivalent being the mass of DNA corresponding to a single nonreplicating chromosome.

If it is assumed that chromosome replication initiates when the cell mass per chromosome origin reaches a fixed value (see Section IV.A.), the mean cell mass in an exponential-phase culture is given by

$$\bar{M} = k2^{(C+D)/\tau}$$

where k is a constant. Thus, the ratio $\bar{G}/\bar{M}$ is a function of C and τ:

$$\bar{G}/\bar{M} = (\tau/kC \ln 2)(1 - 2^{-C/\tau})$$

If an exponential-phase culture is treated such that ongoing rounds of

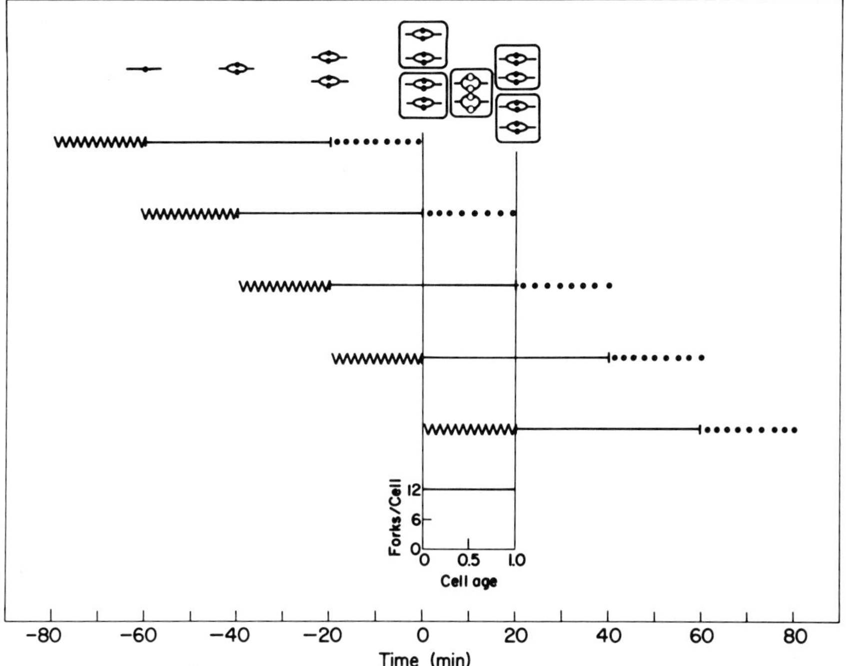

FIG. 3. Construction of chromosome replication and cell division patterns for $I = D = 20$ minutes, and $C = 40$ minutes (see legend to Fig. 1 for details).

replication can be completed but new rounds cannot start (see Section III,A and III,B), the fractional increase in chromosomal DNA content is

$$\Delta\bar{G} = \frac{C\,(\ln 2)(2^{C/\tau})}{\tau(2^{C/\tau} - 1)} - 1$$

The cell age, in fractions of the division cycle, at replication of a specific gene on the chromosome is given by

$$a_x = (n + 1)\tau - [(1 - x)C + D]$$

where n is the smallest integer so that $(n + 1)\tau \geq [(1 - x)C + D]$ and x is the fraction of the C period at which the gene replicates. Consequently, cell age at initiation of replication is

$$a_I = (n + 1)\tau - (C + D)$$

and cell age at termination of replication is

$$a_T = (n + 1)\tau - D$$

The mean number of copies of a gene per cell in an exponential-phase culture is

$$\bar{X} = 2^{[(1 - X)C + D]/\tau}$$

Consequently, the mean number of origins per cell is

$$\bar{X}_0 = 2^{(C + D)/\tau}$$

and the mean number of termini per cell is

$$\bar{X}_T = 2^{D/\tau}$$

C. General Aspects of the Concept

Three important aspects of this conceptualization of bacterial reproduction will be emphasized. First, valuable experimental approaches, which led to many of the ideas presented, were developed in Maaløe's laboratory during the late 1950's and early 1960's (see Maaløe and Kjeldgaard, 1966). As an example, one important contribution was the finding by Schaechter *et al.* (1958) that average cellular mass and RNA and DNA content of *Salmonella typhimurium* increased when steady-state cultures were grown in a variety of culture media which supported progressively faster growth rates. These and similar studies emphasized the value of performing comparative analyses on a single cell line growing under different, definable conditions. The basic idea is that observations of the changes in the cellular content or rate of formation of a macromolecule which take place during gradual changes in growth conditions can be used to identify regulatory aspects of the synthesis of that macromolecule. It was on this basis that the $(I + C + D)$ concept was developed, i.e., by observation of changes in the relationship between chromosome replication and the division cycle as the growth rate was varied (Cooper and Helmstetter, 1968).

Second, the $(I + C + D)$ concept is based primarily on the results of studies performed on strain B/r of *E. coli*. Specific conclusions on cell division control developed with a specific strain should not be applied directly to other bacterial species or even to other strains of the same species. During the course of this chapter, it will be suggested that the biosynthetic events which must be completed for duplication of the bacterial cell may be the same in different strains, but that the time required for completion of individual events may vary in different strains growing under identical conditions. Furthermore, it will be suggested that the preparative steps required for initiation of certain of these events, such as chromosome replication, may also be the same in different strains, but that the specific step which limits the frequency of initiation of the event may be strain-dependent. Thus, strain-dependent aspects of the bacterial duplication process should not be considered unusual.

Third, the $(I + C + D)$ concept described in Section II,A was applied to an idealized or "average" population of cells in which variations in lengths of individual steps in individual cells in the population were not considered. The lengths of the steps can vary considerably in individual cells, e.g., the standard deviation of interdivision times for *E. coli* is about 25% [see the extensive discussion of this point by Koch (1977)]. Furthermore, we do not intend to imply that the duplication process is entirely deterministic, i.e., that all cells proceed in a fixed step-by-step sequence toward division. One or more steps could involve probabilistic components (Cattaneo *et al.*, 1961; Burns and Tannock, 1970; Smith and Martin, 1973, 1974; Minor and Smith, 1974; Koch, 1977). The recent ideas of Smith and Martin with regard to probabilities in cell division were based on earlier observations that the S, G_2, and M phases of the mitotic cycle of eukaryotes varied very little in a population of cells, whereas the G_1 phase could be highly variable. Furthermore, the distribution of intermitotic times was not normal, but was invariably skewed toward longer time intervals. To account for these observations, Smith and Martin suggested that cellular proliferation is composed of a probabilistic phase which occurs before initiation of chromosome replication (within G_1 in eukaryotes) and then a deterministic phase (comprising chromosome replication and division). In the probabilistic state the cell is not progressing toward division, and it can remain in the state for any length of time with a constant probability of entering the deterministic phase. The theory accounts satisfactorily for the distribution of interdivision times in a wide variety of bacterial, yeast, and animal cells (Minor and Smith, 1974; Shilo *et al.*, 1976).

With regard to bacterial cells, Koch (1977) has discussed some related ideas regarding initiation of chromosome replication, although he does not suggest that a deterministic phase leading to division follows initiation of replication. He proposes that chromosome replication is not accurately phased with the growth of individual cells, based primarily on an autoradiographic analysis of slow-growing *E. coli* cells by Chai and Lark (1970) which suggested a very broad distribution in the cell size or age at initiation of chromosome replication. Since the distribution of cell sizes at division was much narrower, Koch concluded that initiation is less well controlled than cell division and, therefore, that initiation cannot time cell division. The concepts of probabilistic factors in initiation of chromosome replication, and independent pathways for chromosome replication and cell division, will be discussed in detail in subsequent sections.

III. Initiation of Chromosome Replication

Aspects of the molecular biology of bacterial duplication are discussed in the remainder of this chapter, the information presented being selective rather than comprehensive. Little emphasis is placed on the bacterial envelope and its involvement in cell division, as the subject has been covered in an excellent recent review by Daneo-Moore and Shockman (1977). Additional information on cellular shape determination has been discussed by Henning (1975), and the involvement of penicillin-binding proteins in cell division has been reviewed by Spratt (1977). In this section, the requirements for initiation of chromosome replication are discussed first (Sections III,A and III,B), followed by some general conclusions on the regulation of initiation (Section III,C).

A. Macromolecular Requirements

1. Cell Mass

What macromolecular synthetic processes are required for initiation of a round of chromosome replication, and which of these processes determines the time at which initiation takes place? One observation bearing on this question was the discovery that protein synthesis is required for initiation. It was suggested by Maaløe and Hanawalt (1961) and by Lark *et al.* (1963) that inhibition of protein synthesis in *E. coli* might not interfere with the completion of rounds of replication in progress, but might prevent initiation of new rounds. A second major accomplishment was the identification of an apparent relationship between initiation and cellular mass in *E. coli* (Hanawalt *et al.*, 1961; Maaløe and Kjeldgaard, 1966). It was observed that when amino acids were restored to a previously starved culture, the fraction of cells synthesizing DNA gradually increased, and the larger cells started first. However, when DNA synthesis was inhibited for about one doubling time after amino acids were restored, all the cells began synthesizing DNA simultaneously when the inhibition was removed. Thus, the capacity for initiation of chromosome replication was achieved in the absence of DNA synthesis, and the results suggested that chromosome replication was not initiated until a critical cellular mass was achieved. These early ideas were refined and extended in subsequent reports (Pritchard and Lark, 1964; Hewitt and Billen, 1964; Boyle *et al.*, 1967), solidifying the notions that synthesis of the components needed for initiation can continue in the absence of DNA synthesis and that completion of one round of replication does not trigger initiation of the next round. The latter conclusion was further supported by findings that new rounds of replication can begin before termination of the previous round in rapidly growing cells (Yoshikawa *et al.*, 1964; Helmstetter and Cooper, 1968) and that rounds are separated by periods devoid of DNA synthesis in slowly growing cells (see Section V).

There is now a widely held belief that there may be a relationship between cellular mass, or a property related to mass, and initiation of chromosome replication in *E. coli*. It has been suggested that the mean cellular mass or size per chromosomal origin at initiation of chromosome replication in an *E. coli* population is constant (Donachie, 1968; Helmstetter *et al.*, 1968; Pritchard *et al.*, 1969). Although accurate determinations of cell size distributions at initiation have not been reported because of experimental difficulties, it has been observed that average cell size per origin at initiation is reasonably constant in cells growing with interdivision times between about 20 and 70 minutes at 37°C. This relationship does not hold in cells growing with longer interdivision times (Helmstetter, 1974b). In addition, the relationship between size and initiation is not a general property of all bacterial species, e.g., it does not appear to exist in *S. faecalis* (Daneo-Moore and Shockman, 1977). The processes capable of determining the frequency of initiation may be the same in all species, but the rate-limiting process in a particular species growing in a given environment need not be reflected in a constant cell mass per origin at initiation.

2. Protein Synthesis

The manner in which cell size might be related to initiation control remains unknown, but it is clear that extensive macromolecular synthesis is required in preparation for initiation. Protein synthesis is required, but not necessarily up to the moment of initiation itself. In general, inhibition of protein synthesis blocks initiation in rapidly growing cells ($\tau \leq 60$ minutes at 37°C) (for a review, see Helmstetter, 1974a,b). However, *E. coli* strain $15T^-$, and certain slowly growing cultures of strain B/r, can continue to initiate chromosome replication for several minutes in the presence of high concentrations of chloramphenicol, or during starvation for required amino acids (Lark and Renger, 1969; Lark, 1972, 1973; Helmstetter, 1974a,c).

Under certain conditions, initiation of chromosome replication can continue for extended periods in the absence of protein synthesis. Rosenberg *et al.* (1969) and Kogoma and Lark (1970) found that starvation of *E. coli* $15T^-$ for thymine resulted in subsequent DNA replication in the absence of protein synthesis at a reduced rate for at least 5 hours. Kogoma and Lark referred to this phenomenon as stable chromosome replication and have proposed that under these conditions the replication complex which is normally destroyed at completion of a round of replication remains usable for subsequent rounds of replication. Stable replication can be induced by a variety of treatments which inhibit DNA synthesis without inhibiting RNA and protein synthesis, or during a nutritional shift upward which stimulates RNA and protein synthesis prior to stimulation of DNA replication (Kogoma and Lark, 1975). Recently, Kogoma (1978) has reported the isolation of *E. coli* $15T^-$ mutants capable of semiconservative chromosome replication in the absence of protein synthesis without any prior inductive treatments.

3. RNA Synthesis

There are numerous examples of the direct involvement of RNA, or a functional RNA polymerase, in initiation of chromosome replication (Lark, 1972; Messer, 1972; Helmstetter, 1974a,b; Hiraga and Saitoh, 1974; Murakami *et al.*, 1976; Zyskind *et al.*, 1977), but the involvement of RNA synthesis in the timing of initiation is uncertain. Cells of *E. coli* B/r growing at rates such that they initiate chromosome replication before or after the *D* period do not initiate in the presence of high concentrations of rifampicin, an inhibitor of initiation of RNA synthesis (Messer, 1972; Helmstetter, 1974a). On the other hand, slow-growing cells of strain B/r which normally initiate during the *D* period and glucose-grown *E. coli* $15T^-$ can progress to initiation for several minutes in rifampicin (Silberstein and Billen, 1971; Lark, 1972; Helmstetter, 1974a). One likely role for the RNA would be as a primer for initiation of chromosomal DNA synthesis. Messer *et al.* (1975) have isolated an RNA–DNA copolymer which appears to possess the properties of an RNA primer linked to origin DNA. This RNA (called oRNA) consists of approximately 30 to 40 nucleotides, and seems to be quite stable (Messer, personal communication). After a 60-minute chase (1.5τ) practically all labeled oRNA can be recovered.

4. Cell Envelope

The involvement of the cell envelope in initiation and replication of the chromosome is suggested indirectly by the apparent attachment of the chromosome to the envelope, probably at the site of replication and/or the origin and terminus of replication (Leibowitz and Schaechter, 1975; Thilo and Vielmetter, 1976). Marvin (1968) proposed that initiation of chromosome replication involves the formation of a site for DNA synthesis on the cytoplasmic membrane. He suggested that the presence of a membrane site for DNA replication prevents the formation of a new site within a fixed distance. A new round of chromosome replication would begin when the membrane area around an existing site had increased enough to permit formation of a new replication site. Pardee *et al.* (1973), Gudas and Pardee (1974), and Helmstetter (1974b) have also suggested that the timing of initiation of chromosome replication in *E. coli* can be controlled by the cell envelope and/or processes leading to cell division. Particularly convincing in this regard was the finding that certain slow-growing cells of *E. coli* B/r continued to initiate chromosome replication in the absence of protein and RNA synthesis, as long as septum–crosswall formation continued (Helmstetter, 1974a).

An involvement of lipids in initiation has also been implied (Barbu *et al.*, 1970; Nunn and Tropp, 1972; Fralick and Lark, 1973). Fralick and Lark (1973) suggested that initiation takes place at a membrane site and that the functioning of this site depends on availability of unsaturated fatty acids. They observed that initiation was inhibited in the presence of 3-decynoyl-*N*-acetylcysteamine (DAC), an inhibitor of unsaturated fatty acid synthesis. However, Thilo and Vielmetter (1976) have questioned this interpretation. These authors found that replication continued to initiate when an unsaturated fatty acid auxotroph of *E. coli* was starved for oleate. They suggest that DAC used by Fralick and Lark might not be specific for inhibition of unsaturated fatty acid synthesis and may have affected initiation due to other metabolic disturbances.

B. Temperature-Sensitive Initiation Mutants

1. Response to Temperature Shifts

Some of the more significant recent studies on the control of initiation of chromosome replication have involved analyses of *E. coli* mutants which are temperature sensitive for chromosome replication. Mutants with temperature-sensitive defects in two genes, *dnaA* and *dnaC*, have been studied most extensively and our comments will focus on these.

dnaA$_{ts}$ and *dnaC*$_{ts}$ mutants are defective in initiation of chromosome replication at nonpermissive temperature (about 40°C), although some *dnaC*$_{ts}$ mutant alleles have been reported to be defective in elongation as well (Wechsler, 1975; and see brief descriptions of these mutants in Table IV, in Section VI,D). When exponential-phase cultures of *E. coli* possessing either *dnaA*$_{ts}$ or *dnaC*$_{ts}$ initiation-defective mutations are grown at permissive temperature (25°–30°C) and shifted to nonpermissive temperature, new rounds of chromosome replication are not initiated, but ongoing rounds continue to completion. The cells divide following completion of the ongoing rounds. Two allele-specific phenomena are observed when such mutants are then returned to permissive temperature. With some, the activity of the gene product is thermoreversible, such that the product synthesized at nonpermissive temperature becomes active upon shift to permissive temperature (Abe and Tomizawa, 1971; Beyersmann *et al.*, 1971; Wolf, 1972; Schubach *et al.*, 1973; Hiraga and Saitoh, 1974; Evans and Eberle, 1975; Hanna and Carl, 1975; Messer *et al.*, 1975; Zyskind *et al.*, 1977). With other alleles, new synthesis of the product at permissive temperature is required for reinitiation of chromosome replication (Hirota *et al.*, 1968, 1970; Kuempel, 1969; Bagdasarian *et al.*, 1972; Schubach *et al.*, 1973). The alternative phenotypes were detected by incubating the mutants for a period at nonpermissive temperature, and then reducing the temperature to a permissive level in the presence of an inhibitor of protein synthesis such as chloramphenicol. In the case of the thermoreversible mutants, reinitiation of chromosome replication took place at permissive temperature in the presence of chloramphenicol, indicating that all proteins required for initiation were produced at high temperature and could be used later at permissive temperature.

2. oRNA Synthesis in *dnaA*$_{ts}$ and *dnaC*$_{ts}$ Mutants

Experiments of the type described above, i.e., growth at permissive temperature, shift to the nonpermissive for an extended period (e.g., 1 to 2 hours), and then return to permissive conditions, are being used extensively to investigate the properties of temperature-sensitive initiation mutants. One question concerns whether oRNA synthesis requires the presence of active *dnaA* or *dnaC* gene product. The experiments involve transferring thermoreversible mutants from nonpermissive to permissive temperature in the presence of an inhibitor of RNA synthesis, such as rifampicin or streptolydigin, and determining whether reinitiation occurs. If reinitiation takes place, either all necessary RNA synthesis can be completed at nonpermissive temperature, or RNA polymerase action becomes insensitive to the inhibitor before the product functions. It has been shown that thermoreversible *dnaC*$_{ts}$ mutants can reinitiate in rifampicin (Evans and Eberle, 1975; Hanna and Carl, 1975; Messer *et al.*, 1975; Saitoh and Hiraga, 1975; Zyskind *et al.*, 1977). The results of such experiments with *dnaA*$_{ts}$ mutants are not as clear. It has been reported that thermoreversible *dnaA*$_{ts}$ mutants can reinitiate when rifampicin is added at, or 5 minutes before, return to permissive temperature (Evans and Eberle, 1975; Hanna and Carl, 1975; Messer *et al.*, 1975), but not when the inhibitor is added 10 minutes before the temperature shift, or long enough to obtain 99% inhibition of RNA synthesis (Zyskind *et al.*, 1977).

In a further analysis of this question, Messer *et al.* (1975) investigated specifically whether oRNA is formed at nonpermissive temperature in *dnaA*$_{ts}$ and *dnaC*$_{ts}$ mutants. The mutants were incubated at 42°C and pulse-labeled with tritiated uridine. Labeled oRNA was recovered from a *dnaC*$_{ts}$ mutant, as would be expected if it were synthesized at 42°C, but recovery was much lower in the *dnaA*$_{ts}$ mutants. They concluded, as have Zyskind *et al.* (1977), that the *dnaA* product acts before or during synthesis of oRNA (RNA polymerase synthesizes this oRNA) and *dnaC* is involved in later steps (probably in the first deoxyribonucleotide polymerization event). Kung and Glaser (1978) also conclude that the *dnaA* product acts before the *dnaC* product based on temperature shift experiments with a strain possessing a cold-sensitive *dnaC* mutation and a heat-sensitive *dnaA* mutation.

3. Initiation Regulation in *ts*-Initiation Mutants

To investigate the controlling elements in initiation frequency, Messer *et al.* (1975) determined the effect of chloramphenicol on DNA synthesis in *dna*$_{ts}$ mutants growing at temperatures between 30° and 37°C. At 33°C, the rate of DNA synthesis increased immediately upon addition of the drug (30 μg/ml) to a thermoreversible *dnaA*$_{ts}$ mutant, apparently due to initiation of new rounds of replication as opposed to an increase in the rate of fork movement. This response was only observed in two *dnaA*$_{ts}$ alleles and not in *dnaC*$_{ts}$, or four other *dnaA*$_{ts}$ mutant alleles which were not thermoreversible. They interpreted the stimulation of initiation by chloramphenicol to mean that a repressor protein is involved in the regulation of initiation. Presumably, the *dnaA*$_{ts}$ mutants growing at an intermediate temperature were limited in initiation by the availability of sufficient active *dnaA* gene product. They proposed that an initiation repressor, which binds to the operator of the oRNA operon, is in equilibrium with free repressor, and that repressor concentration is regulated by matching synthesis and decay rates. The *dnaA* gene product is required for the release of the repressor, and oRNA synthesis would occur when sufficient *dnaA* product (antirepressor) is formed. Addition of chloramphenicol at intermediate temperatures would result in a blockage of synthesis of repressor such that it would decay and release the repression, without further need for the antirepressor.

Blau and Mordoh (1972), Hansen and Rasmussen (1977) and Eberle (personal communication) have also concluded that the *dnaA* gene product is a possible candidate, at least under certain growth conditions, for a rate-limiting element in initiation, possibly acting as an activator (or antirepressor). Alternatively, the behavior of $dnaA_{ts}$ mutants grown at intermediate temperatures is consistent with the *dnaA* gene product functioning as an essential cofactor for RNA polymerase activity in oRNA synthesis (Orr, Meacock and Pritchard, personal communication). The continued initiation in chloramphenicol could be explained, in this case, by more efficient utilization of the limiting quantity of the *dnaA* protein due to increased availability of RNA polymerase molecules for stable RNA synthesis. The *dnaA* gene product could be in excess during balanced growth and act as the rate-limiting element only during growth at intermediate temperatures. There would then be other factors needed for initiation which might function coincident with, or before, the *dnaA* gene product.

4. Accumulation of Initiation Potential at Nonpermissive Temperature

Another question concerns the number of initiations which can occur at permissive temperature in the absence of protein synthesis after long-term preincubation at nonpermissive temperature. Earlier experiments, in which residual DNA synthesis was measured, were interpreted as indicating that temperature-sensitive initiation mutants initiated one round of chromosome replication after being held at nonpermissive temperature and shifted to permissive temperature in the presence of chloramphenicol (Abe and Tomizawa, 1971; Beyersmann *et al.*, 1971). However, it is now clear that thermoreversible $dnaA_{ts}$ and $dnaC_{ts}$ mutants are capable of initiating at least two rounds at permissive temperature in the presence of chloramphenicol after preincubation for 90 minutes or longer at nonpermissive temperature (Schubach *et al.*, 1973; Evans and Eberle, 1975; Hanna and Carl, 1975; Hansen and Rasmussen, 1977; Helmstetter, 1978). The maximal amount of initiation potential that can accumulate at nonpermissive temperature is allele specific, and seems to be less than the cell mass increase and either equal to or less than the increase in cell concentration at the nonpermissive temperature (Hanna and Carl, 1975). This finding supports the idea of a coupling between initiation and the processes leading to cell division, as suggested earlier (see Section III,A,4). Continued cell division in the absence of initiation at the nonpermissive temperature could create potential (e.g., envelope sites) for initiation.

The second initiation after the temperature shift downward occurs synchronously at 30 to 50 minutes postshift, depending on the temperature, with or without continued protein synthesis (Evans and Eberle, 1975; Helmstetter, 1978). What is the explanation for the long interval between the two initiations? Considering the delay and the synchronous second initiation, factors other than the total initiation potential accumulated at nonpermissive temperature must govern the interinitiation time in this situation. Clearly, the timing of the second initiation is not determined by protein synthesis. The length of the interval between the first two initiations in the absence of protein synthesis could be determined by a requirement for the first set of replication forks to progress a minimum distance along the chromosome. During rapid balanced growth of *E. coli*, new rounds of chromosome replication are not initiated until the previous round has progressed at least $0.5C$, i.e., about 20 minutes at 37°C (Helmstetter and Pierucci, 1976). More frequent initiation has not been detected during balanced growth at 37°C, and in fact, the maximum frequency of initiation could be determined by the minimum permissible distance between replication forks. In addition, when *E. coli* thy^- strains were thymine-starved to permit accumulation of potential for initiation of at least two rounds of replication, the interval between the first two initiations subsequent to restoration of thymine was 12 to 20 minutes (Helmstetter, 1971; Zaritsky, 1975a). Also consistent with this possibility is the finding that balanced growth cannot be achieved when C is elongated by thymine limitation in cultures with mass doubling times at 37°C of 50 minutes or shorter (Zaritsky and Pritchard, 1973). If the mass doubling time were shorter than the minimum fractional C time required between successive initiations, balanced growth would not be achieved.

C. Conclusions Regarding the Regulation of Initiation Frequency

The final point for discussion in this section concerns the factor(s) which determine(s) the interinitiation time during steady state growth of *E. coli*. Before analyzing this point specifically, let us first consider the properties of positive and negative control mechanisms for initiation frequency regulation.

The basic assumption behind a positive control mechanism is that it involves the synthesis of a structure or a substance necessary for chromosome replication, at a rate determined by the growth rate (Jacob *et al.*, 1963). An activator could be formed at a specific time during the cell cycle, or initiation could take place when an activator (probably, but not necessarily, a protein) has accumulated to a threshold level (Maaløe and Kjeldgaard, 1966; Tsanev and Sendov, 1966; Donachie and Masters, 1968; Helmstetter *et al.*, 1968; Bleecken, 1971; Sompayrac and Maaløe, 1973). The relationship between cell size and initiation could be a reflection of this accumulation. Activation might involve the formation of the protein and RNA components of a chromosome replication apparatus, duplication of an existing apparatus, and/or growth of the cell envelope.

The basic principles of a negative control scheme for initiation of chromosome replication were described by Pritchard *et al.* (1969). An inhibitor of initiation is assumed to be produced by a gene located near the chromosomal origin which is transcribed only when the gene replicates. From the fixed number of messenger RNA molecules produced, a fixed amount of inhibitor protein is translated at all growth rates. An increase in cell volume by growth would lead to a gradual dilution of the inhibitor until a critical concentration was reached such that initiation could take place. Initiation would then result in the production of a new quantity of inhibitor, and further growth would then be necessary for the next initiation. The frequency of initiation would thus be determined by the dilution rate of the inhibitor which would be the reciprocal of the growth rate. The same basic concept for negative control was proposed by Rosenberg *et al.* (1969), based on the continuation of chromosome replication in the absence of protein synthesis after thymine starvation (Section III,A,2). They suggested that an anti-inhibitor is synthesized continuously, and initiation takes place when sufficient levels of antiinhibitor have accumulated to inactivate the inhibitor.

A slightly different negative control scheme would assume that an unstable inhibitor is synthesized at a constant rate proportional to gene dosage (Ycas *et al.*, 1965). Its concentration would depend not only on cell volume but also on the relative rates of its synthesis and breakdown. The concentration would fall as the cell volume increases. At a critical volume, replication would initiate, resulting in a doubling of the rate of synthesis of the inhibitor, assuming the gene producing the substance were near the chromosomal origin.

One of the interesting aspects of the negative control model proposed by Pritchard *et al.* (1969) concerns its predictions regarding the replication of large transmissable plasmids such as F. Initiation of F replication could be controlled in the same manner as initiation of chromosome replication, with F producing an inhibitor of its own initiation, and initiation taking place when its concentration decreases to a critical value. The number of F particles relative to the number of chromosome origins in an F^+ cell would be a function of the ratio of the critical cellular volumes for their respective initiations. Upon integration of F into chromosome, initiation of replication from the F replicon might be prevented if the cell volume required for replication of F in the autonomous state were larger than the volume required for initiation of chromosome replication.

Both positive and negative control mechanisms can be self-regulating, i.e., an early initiation would tend to be followed by a late initiation, and vice versa. As examples of self-regulating systems, if initiation were normally triggered when a fixed concentration of activator (positive) or antirepressor (negative) per origin was reached, and initiation was abnormally early on one occasion, the concentration per origin would fall to a lower than normal level at the moment of initiation. Consequently, a longer interval of growth would be required for the next initiation.

Returning to the specific question of initiation regulation in *E. coli*, it should be evident, from the observations described in Sections III,A and III,B, that extensive preparations are required before chromosome replication can be initiated. Among the possible requirements are (1) formation of a replication apparatus which includes the necessary enzymatic activities required for replication, (2) synthesis of necessary envelope components, which may involve a division between, or segregation of, completed chromosomes, (3) release of any repressors or inhibitors which regulate expression of essential initiation genes, such as the oRNA cistron, (4) progression of previously initiated replication forks a minimum distance along the chromosome, and (5) growth to a critical cell size which could be a reflection of other requirements in this list. Which of these factors, or others which have been neglected, regulates the timing of initiation? It seems likely that both activators and repressors are involved in initiation of chromosome replication, and that there might normally be tight coupling between completion of formation of the necessary positive elements and release of the repression, as Messer *et al.* (1975) have emphasized. We can go a step further and suggest that it is not likely that there is only one factor, whether it be inactivation of a repressor or synthesis of an activator, which determines the timing of initiation in *E. coli.* The molecular activities required for initiation may be the same in all strains of *E. coli*, but the particular activity which determines the timing of initiation at a particular temperature in a given carbon source could be strain dependent. Conditions can be obtained in which *E. coli* cells may control the frequency of initiation by envelope growth (Section III,A,4), by activity of a *dna* gene product (Section III,B,3) or by the position of a replication fork (Section III,B,4). It would not be surprising if it is eventually shown that, upon transfer of a strain of *E. coli* from one growth condition to another, the factor which determines the timing of initiation during the transition is strain dependent.

IV. Plasmids and Chromosome Replication

A. Replication of Chromosomes Possessing Integrated Plasmids

A good deal of information on the regulation of chromosome replication and cell division may come from studies on the duplication of cells possessing plasmids. Plasmids are extrachromosomal genetic elements capable of stable autonomous replication (Novick *et al.*, 1976). Certain large plasmids, such as the *E. coli* sex factor F and composite antibiotic resistance R factors, are present in a small number of copies per cell. The average number of autonomous F or R plasmids per chromosomal origin in exponential-phase *E. coli* cultures increases when the growth rate is decreased by nutritional means (Pritchard *et al.*, 1975; Engberg and Nordstrom, 1975; Finkelstein and Helmstetter, 1977). As a consequence, the average cell mass/plasmid ratio at initiation of F plasmid replication decreases with decreasing growth rate and approaches the cell mass/origin ratio at initiation of chromosome replication. With these observations in mind, and considering the possibility that F (or R) replication could be controlled by a mechanism similar to that which controls chromosome replication (Section III,C), we will now describe some studies on the replication of chromosomes which possess integrated plasmids.

Escherichia coli cells which are not known to possess integrated plasmids (e.g., F^- and F^+ cells), initiate replication at about 82 minutes on the genetic map, and from this point two replication forks proceed in opposite directions around the circular chromosome (Masters and Broda, 1971; Yahara, 1971; McKenna and Masters, 1972; Prescott and Kuempel, 1972; Bharnik *et al.*, 1973; Fujisawa and Eisenstark, 1973; Hohlfeld and Vielmetter, 1973; Jonasson, 1973; Louarn *et al.*, 1974; Bachmann *et al.*, 1976, and see Section IV,B). The chromosome appears to replicate bidirectionally at all growth rates (Rodriguez and Davern, 1976). When plasmids become integrated into the chromosome, the plasmid DNA usually replicates passively under control of the chromosomal replication system, although some early studies were consistent with initiation of replication at an integrated plasmid at slower growth rates or after amino acid starvation (Nagata, 1963; Wolf *et al.*, 1968; Eberle, 1970). However, the possibility that an integrated plasmid could direct chromosome replication was most clearly demonstrated by the observation that the temperature-sensitive phenotypes of *dnaA*$_{ts}$ mutants could be suppressed by integration of F, R, or P2 replicons (Lindahl *et al.*, 1971; Nishimura *et al.*, 1971; Moody and Runge, 1972; Bagdasarian *et al.*, 1975; Tresguerres *et al.*, 1975; Datta and Barth, 1976). *dnaA*$_{ts}$ mutants with plasmids integrated at a variety of chromosomal loci are capable of replication and division at nonpermissive temperature. Furthermore, Chandler *et al.* (1977) have shown that a *dnaA*$_{ts}$ mutant suppressed by the F-like R plasmid R100.1 initiates replication at the location of the integrated plasmid at nonpermissive temperature.

From the foregoing, it is evident that chromosome replication initiates at the site of an integrated plasmid under certain special conditions. What then are the conditions, in general, in which any Hfr strain might be capable to initiating replication from an integrated plasmid? Pritchard and coworkers (1969, 1975) suggest that the ability of an integrated plasmid to initiate chromosome replication might depend on the growth rate, or the cell mass/plasmid ratio, if it is assumed that initiation of plasmid replication in the autonomous state is controlled by a factor related to cell mass. Upon integration, initiation of replication from the plasmid replicon might be prevented if the cell mass required for replication of the plasmid in the autonomous state were larger than the mass required for initiation of the chromosome. At slower growth rates, when the cell mass at chromosome initiation and at autonomous plasmid replication become very similar, the probability of initiation from an integrated plasmid might increase. Consistent with this idea, Chandler *et al.* (1976) found that replication initiated from the normal chromosomal origin in a majority of Hfr cells during exponential growth, but from both the normal origin and the site of the integrated F after synchronization of chromosome replication by amino acid starvation followed by thymine starvation in the presence of required amino acids. A similar result was observed when a $dnaA_{ts}$ mutant containing an integrated R plasmid was subjected to the same synchronization regime at permissive temperature (Bird *et al.*, 1976). A consequence of the synchronization regime in these experiments is that the cell mass increased without an increase in the number of F plasmids during the thymine-starvation period. Thus, the probability of replication initiating at the integrated F might have increased because the number of copies of F was now below the number a cell of that mass could support if F were autonomous.

A second interesting observation is that chromosome replication continues for a longer time during amino acid starvation of Hfr strains relative to F^- strains (e.g., Caro and Berg, 1968). Initiation of replication at the integrated F is a plausible explanation, particularly since F-like plasmids can replicate at least once in the autonomous state during inhibition of protein synthesis (Bazaral and Helinski, 1970; Spratt, 1972; Goebel, 1974; Kline, 1974). The findings of Rosenberg *et al.* (1969) and of Kogoma and Lark (1970, 1975) that chromosome replication continued for an extended period during inhibition of protein synthesis, subsequent to a period of thymine starvation (Section III,A,2), could also be due to initiation from an integrated replicon.

Considering the above, it would appear that chromosome replication can initiate at the site of an integrated plasmid in Hfr strains with a probability which is related to the growth rate, the cell mass/plasmid ratio, and the location of the plasmid in the chromosome. The probability apparently increases as the growth rate decreases or when the cell mass/plasmid ratio increases. It is also clear from these studies, particularly on integrative suppression, that *E. coli* can grow and divide when initiation of replication takes place at sites other than the normal origin.

The next question concerns the site of termination of chromosome replication. Normally, the two replication forks terminate near 32 minutes on the genetic map, i.e., 180° from the origin at 82 minutes. When replication starts from an integrated plasmid, it might terminate at the normal site or when the two replication forks meet, assuming bidirectional replication from the integrated replicon. Louarn *et al.* (1977) investigated this question with *dnaA*$_{ts}$ mutants suppressed by R plasmids integrated at different locations. They concluded that replication was bidirectional from the R integration site at nonpermissive temperature and that replication stopped between 27 and 43 minutes, presumably at the normal terminus. It is, therefore, possible that the terminus is a defined genetic region on the chromosome which relates in some way to the subsequent cell division.

An interesting recent finding with regard to the termination site concerns the mode of chromosome replication upon induction of prophage P2*sig*$_5$, which has a temperature-sensitive repressor. *dnaA*$_{ts}$ mutants with P2*sig*$_5$ integrated at *att*P2II near 85 minutes are capable of chromosome replication and cell division at 42°C (Lindahl *et al.*, 1971; Kuempel *et al.*, 1978). On the other hand, when the prophage is integrated at 47 minutes (i.e., near the terminus), *dnaA*$_{ts}$ mutants are not capable of extensive growth at 42°C, but some chromosome replication can take place upon shift to the high temperature (Kuempel *et al.*, 1977). Replication starts at the prophage site and proceeds initially counterclockwise to *aroD* (37 minutes). Then the replication fork apparently slows as it passes through the subsequent region, particularly between 31 and 27 minutes. At later times, one-half of the replication forks also travel clockwise such that overall replication is bidirectional. The authors concluded that the terminus is near 31 minutes and that the terminus inhibits replication fork movement, but does not necessarily stop it.

B. Growth of Cells Possessing Autonomous Plasmids

The presence of autonomously replicating plasmids in a cell can profoundly affect cell division and cell size. For instance, an R factor, R*ts*1, is not maintained at 42°C and interferes with ability of its host cell (*Proteus mirabilis* or *E. coli*) to grow at the high temperature (Terawaki *et al.*, 1968; Terawaki and Rownd, 1972; DiJoseph *et al.*, 1973; DiJoseph and Kaji,

1974). At 42°C, *E. coli* cells containing R*ts*1 produce membrane invaginations and filamentation is observed. R*ts*1 DNA is synthesized, but it is not converted from linear molecules to covalently closed circles. When R*ts*1 is integrated into the chromosome, thermosensitive cell division is suppressed (Yoshimoto and Yoshikawa, 1975). Therefore, upon integration, some plasmid genes related to replication are phenotypically suppressed. Perhaps R*ts*1 causes growth defects at 42°C only when it is producing its own replication machinery. If both the plasmid and the chromosome specify products with similar functions for replication, the formation of an inactive product by the plasmid could competitively interfere with the action of the product encoded by the chromosome, thereby affecting chromosome replication and/or processes leading to cell division.

Collins and Pritchard (1973) and Engberg *et al.* (1975) have shown that the average size of *E. coli* is increased by the presence of F′*lac* or R plasmids. Engberg *et al.* investigated the effects of an R plasmid, R1*drd*19, on cell division in *E. coli* cultures growing at different rates. It was found that at slower growth rates, plasmid-containing cells were larger, on the average, than cells lacking the plasmid, but were essentially the same size at faster growth rates. The increased average cell size was due, in part, to the formation of filaments in a portion of the population. The authors suggest that the chromosome and plasmid replicons compete for some factors involved in DNA replication. A delay in chromosome initiation and/or elongation as a consequence of this competition could increase the probability of not forming a septum, thereby resulting in filamentation.

Finally, there is the special case of cells which contain a plasmid with a region which is homologous to the chromosomal replication origin. Such strains might prove useful for investigating DNA replication controls if some of the origin genes on the plasmid remained functional. Analysis of competition between the plasmid and chromosomal origin might enable identification of the properties of the regulating molecules and their sites of action. With these considerations in mind, Masters (1975) isolated F′ strains of *E. coli* which were merodiploid for various regions near the origin. It was predicted that cells containing an F′ plasmid with a functional chromosomal origin would grow slower and have higher mass/DNA ratios. This would be so if there were competition between the chromosomal origin and the origin located on the F′ for the substances or structures required for initiation of chromosome replication. Masters found that merodiploid strains which contained F′'s carrying the chromosomal region between the genes *bgl* and *mtl* (between 81 and 83 minutes on the genetic map) possessed these unusual properties. Some of the F′ plasmids containing the presumptive chromosomal origin were also temperature-sensitive for F replication, but at high temperature, when only the chromosomal origin system could function, the plasmids were not stably maintained (personal communication). von Meyenburg *et al.* (1977) reported similar findings with an F′ carrying genes between *aro*E and *ilv*. Slow growth and heterogenous size distributions were observed, but again all deletion forms of the plasmid contained the vegetative origin of F, suggesting that the chromosomal origin (*ori*C) was not functional. It remains possible that the slow, heterogenous growth of cells possessing such plasmids is due to the diploid configuration of genes located near the origin rather than to a functional activity of the origin itself. In fact, it has been suggested recently that the altered cell growth properties might be consequences, at least in part, of a site in the *uncA*, *uncB* region of the chromosome, which is located near the origin (von Meyenburg *et al.*, 1978).

Another approach to the isolation of plasmids containing a functional *oriC* has involved direct selection for plasmids replicating under direction of the chromosomal origin. Hiraga (1976) predicted that if an F′-containing cell were mated with an Hfr, the only incoming F′ which could become stably replicating in the Hfr would be one which contained a chromosomal origin, since the F replication system would be blocked by the incompatibility mechanism. He found that only plasmids carrying the *dnaA* to *ilv* segment (82–83 minutes) of the chromosome were capable of autonomous replication in the Hfr. The results suggested that a site within this region is required for these F′ plasmids to replicate, and Hiraga has designated this site *poh* for "permissive on *Hfr*." He suggests that the *poh* site could represent the origin of chromosome replication. Recently, Hiraga (personal communication) found that cultures harboring the F′-poh^+ plasmid formed filaments, as described by Masters, with the frequency of filamentation being higher in poor growth media than in rich media. In addition, F′-poh^+ plasmids are capable of replicating in cells possessing mutations in a chromosomal gene (*mafA*), the product of which is required for autonomous replication of F (Wada *et al.*, 1977).

Yasuda and Hirota (1977) have cloned a 6 megadalton DNA fragment containing the replication origin of *E. coli* on an *Eco*R1-generated fragment coding for ampicillin resistance. This plasmid, which replicates under direction of the chromosomal origin, contains the region between *uncA* and *rsbK* of the chromosome. A 1.2 megadalton DNA fragment of the plasmid, excised by *Bam*H1 digestion, has been shown to contain the replication origin. Finally, the chromosomal origin has also been isolated on λ*asn* specialized transducing phages, and subsequently as *oriC* plasmids deleted of λDNA, with a multiplicity of different plasmid copy numbers per cell (von Meyenburg *et al.*, 1978; Messer *et al.*, 1978). The origin was located within 1.5 megadaltons to the left of *asn* on the genetic map. For the future, continued analysis of the effects of these and other similar plasmids on DNA replication and cell division should yield significant new information on the control mechanisms for the processes.

V. Chromosome Replication and Segregation

A. The *C* Period

The time required for a round of chromosome replication (the *C* period) depends on the bacterial strain, the composition of the growth medium, and the incubation temperature. Early studies in Maaløe's laboratory indicated that chromosome replication might occupy most of the interdivision interval of rapidly growing *E. coli* (Maaløe and Kjeldgaard, 1966). These and other findings eventually led to the concept that *C* might not vary much from a value of about 40 minutes in *E. coli* cells growing with interdivision times between 20 and 60 minutes at 37°C. Numerous studies on a variety of strains of *E. coli* have supported this concept. Some experimental values are given in Table I.

In cells growing with longer interdivision times at 37°C, there can be variation in the length of the *C* period. Our interpretation of current information is that the length of *C* in slow-growing cells is strain dependent. As seen in Table I, a variety of measurements indicate that the *C* period increases in some strains of *E. coli* with increasing interdivision time, but the extent of these increases is not uniform. The *C* period in steady-state cultures of different strains growing with interdivision times greater than about 70 minutes has been reported to be as short as 40 minutes or as long as two-thirds of the interdivision time. Even substrains of *E. coli* B/r maintained in different laboratories (which we have designated substrains A, F, and K) vary in *C* period duration during slow growth, i.e., at $\tau = 120$ minutes, *C* was 80 minutes in *E. coli* B/r A, but less than 60 minutes in *E. coli* B/r F (Table I). Availability of precursors may be a rate-limiting factor during slow growth, since elongation of *C* can be achieved by thymine limitation of thymine auxotrophs (Pritchard and Zaritsky, 1970; Zaritsky and Pritchard, 1971; see also Section VI,C).

The location of chromosome replication in the division cycle of slow-growing *E. coli* ($\tau \geq 60$ minutes at 37°C) is also strain dependent. All slowly growing *E. coli* cells examined so far contained a period devoid of chromosome replication at the end of the division cycle, but some strains possessed a period devoid of replication at the beginning of the cycle as well (Lark, 1966; Kubitschek *et al.*, 1967; Helmstetter and Pierucci, 1976).

Once a round of chromosome replication is initiated, replication can continue in the absence of protein synthesis, although the rate of fork movement

TABLE I

AVERAGE VALUES FOR *C* AND *D* IN *E. coli* GROWING AT 37°C

	C (minutes)		*D* (minutes)		
Strain[a]	For $\tau < 60$ minute[b]	For τ_{max}[c]	For $\tau < 60$ minute[b]	For τ_{max}[c]	Reference
B/r A	42 (25–60)	80 (120)	22 (25–60)	40 (120)	Helmstetter and Pierucci (1976)
B/r A	38 (29)				Churchward and Bremer (1977)
	48 (47)				
	52 (51)	64 (120)			
B/r A		83 (135)		48 (135)	Koppes *et al.* (1978)
B/r A				53 (220)	Woldringh *et al.* (1977)
B/r A		71 (119)		11 (119)	Kubitschek and Newman (1978)
B/r K	42 (25–60)	70 (150)	14 (25–60)	20 (150)	Helmstetter and Pierucci (1976)
B/r K	47 (20–60)	47 (1200)			Kubitschek and Freedman (1971)
B/r K		106 (210)		17 (210)	Koppes *et al.* (1978)
B/r K			23 (38)	26 (300)	Kubitschek (1974)
B/r K		66 (120)		14 (120)	Kubitschek and Newman (1978)
B/r K				20 (200)	Woldringh *et al.* (1977)
B/r F	42 (25–60)	60 (120)	16 (25–60)	20 (120)	Helmstetter and Pierucci (1976)
B/r F		132 (240)		25 (240)	Koppes *et al.* (1978)
B/r		62 (95)		33 (95)	Gudas and Pardee (1974)
K12	39 (22–58)	47 (220)			Chandler *et al.* (1975)
K12	39 (27)				Lane and Denhardt (1975)
	42 (54)				
K12	53 (52)				Zaritsky and Pritchard (1971)
K12		62 (103)		41 (103)	Gudas and Pardee (1974)
K12	40[d]	60			Rodriguez and Davern (1976)
K12				24 (270)	Kubitschek (1974)
15		47 (75)		28 (75)	Gudas and Pardee (1974)
$15T^-$		120 (160)		40 (160)	Gudas and Pardee (1974)
$15T^-$	48 (48)				Zaritsky and Pritchard (1971)
$15T^-$	40 (40)	80 (120)			Bird and Lark (1970)

[a] In some experiments thymine-requiring strains were used, in which case the values shown are for experiments employing cultures grown in high thymine concentrations or in thymine plus a deoxynucleoside.

[b] Values for *C* and *D* are given for generation times less than 60 minutes, with the specific (or range) of generation times shown in parentheses.

[c] Values for *C* and *D* are given for representative longer generation times, which are shown in parentheses.

[d] Values for *C* at 42°C in a reversible temperature-sensitive chromosome initiation mutant (CT28) grown in glucose-minimal medium (for $\tau < 60$ minutes) or aspartate minimal medium (for τ_{max}).

may be slower (Pato, 1975). Maaløe and Hanawalt (1961) originally suggested that ongoing rounds progressed to completion. This suggestion has received considerable support, but there has been some question as to whether rounds of replication actually finish in the absence of protein synthesis, or simply reach a point very close to the terminus of the chromosome (Marunouchi and Messer, 1973). Marunouchi and Messer observed that if cells were exposed to chloramphenicol for a time sufficient to complete ongoing rounds of chromosome replication, cell division did not take place when chloramphenicol was removed and nalidixic acid, an inhibitor of DNA synthesis, was added. A similar observation was reported by Dix and Helmstetter (1973). One interpretation of these experimental results is that rounds did not terminate in chloramphenicol. Alternatively, rounds may have terminated, and nalidixic acid blocked some other aspect of cell division. Indeed, when DNA replication was prevented by thymine starvation after long-term exposure to chloramphenicol, the majority of cells divided (Dix and Helmstetter, 1973). A criticism of this latter experiment is that thymine starvation might not have prevented replication of a very short terminal region of the chromosome (Dix and Helmstetter, 1973). Thus, there is evidence to support both points of view, either that rounds of replication are actually completed in the absence of protein synthesis, or that they proceed up to a point very close to the terminus.

B. The *D* Period

During the *D* period, cell division is essentially insensitive to inhibitors of DNA, RNA, and protein synthesis in *E. coli* B/r, although experiments to date are not sufficiently sensitive to determine whether there might be a short interval (about 5 minutes) between completion of chromosome replication and the development of insensitivity to the inhibitors (Clark, 1968b; Helmstetter and Pierucci, 1968; Dix and Helmstetter, 1973). As was the case for the *C* period, the length of *D* is essentially constant in a given strain of *E. coli* growing exponentially at 37°C with interdivision times between about 20 and 70 minutes (Table I). The precise length of the *D* period is strain dependent and ranges from approximately 14 minutes in one substrain of *E. coli* B/r to about 30 minutes in an *E. coli* K12. If the length of the *D* period is partly determined by the time required for septum–crosswall formation, it would not be surprising if it varied in different strains, since cell widths are strain dependent (Woldringh *et al.*, 1977). Whether fortuitous or not, B/r substrains F and K are thinner than substrain A and their *D* periods are shorter. It is conceivable that the apparent constancy of *D* in a given strain over a range of increasing growth rates is due, in part, to a balance between an increase in size of the septum–crosswall (i.e., cell width) and an increase in the rate of its synthesis (Woldringh *et al.*, 1977). The observations of Sloan and Urban (1976) that the rate of cell division increases immediately upon a nutritional shift upward of slow-growing *E. coli* ($\tau \geqq 120$ minutes) is consistent with this possibility.

In more slowly growing cells, the length of *D* appears to depend upon the strain and the growth rate, as was the case for *C* (Table I). The *D* period increases with increasing interdivision times in some strains, but the magnitude of this increase is variable.

C. THE *T* PERIOD

The *T* period is the time from appearance of visible septal constrictions in cells (i.e., the experimentally determined time of initiation of septum–crosswall formation) to cell fission (Clark, 1968a; Gudas and Pardee, 1974; Woldringh, 1976). Woldringh (1976) and Woldringh *et al.* (1977) have reported an extensive electron microscopic analysis of the *T* period and its relationship to nucleoplasm separation in substrains of *E. coli* B/r. Nucleoplasm separation appeared to take place very close to termination of chromosome replication in cells growing with doubling times of 32 and 60 minutes. The *T* period was a constant of 10 minutes in cells growing with doubling times of 60 minutes or less. In more slowly growing cells, the *T* period increased with the doubling time, more so in *E. coli* B/r A than in B/r K. The *D* period was always longer than the *T* period, whereas the nucleoplasm did not separate at the beginning of *D* during slow growth, as it did during rapid growth, but rather in the *T* period or about 10 minutes before fission.

D. SEGREGATION

During balanced growth, there is a partitioning of cellular components at cell division. The process can be random or nonrandom. In the case of a random distribution, cellular components (e.g., soluble protein molecules) are partitioned passively by an equal division of the cytoplasm. Alternatively, nonrandom segregation of cellular components implies the conservation of subcellular aggregations which are transmitted into progeny cells as discrete units. A model for nonrandom, oriented chromosome segregation was proposed by Jacob *et al.* (1963). They suggested that DNA strands were anchored to an envelope component and separated passively into daughter cells by the growth of envelope between attachment sites. If all DNA strands were permanently attached in this fashion, chromosome segregation would be oriented. The results of studies on the extent of orientation in chromosome segregation, which will be summarized here, have been contradictory.

The process of chromosome segregation has been examined by labeling cells with radioactive thymine, removing the label, and continuing cultivation in its absence. If postlabeling incubation is under conditions which permit determination of ancestral relationships [e.g., by growing single cells into chains of cells in medium made viscous by the addition of Methocel (Lin *et al.*, 1971)], the mode of segregation can be identified, in principle. During rapid growth, chromosome segregation in *E. coli* and *B. subtilis* has been reported to be random (Ryter *et al.*, 1968; Chai and Lark, 1970; Lin *et al.*, 1971). At slower growth rates, oriented segregation has been reported in *Lactobacillus acidophilus* (Chai and Lark, 1967), *B. subtilis* (Eberle and Lark, 1966), *E. coli* (Pierucci and Zuchowski, 1973; Cooper and Weinberger, 1977; Cooper *et al.*, 1978), and *Clostridium botulinum* (Kang and Grecz, 1975). Thus, the observed degree of nonrandomness in chromosome segregation is variable, and, in general, is related to the growth rate (Pierucci and Helmstetter, 1976; Cooper and Weinberger, 1977).

An observation made by Lin *et al.* (1971) illustrates a potential pitfall in conventional segregational analyses. After a brief pulse of [^{3}H]-thymidine, the number of labeled cells in chains of rapidly growing recombination-proficient *E. coli* was threefold larger than in recombination-deficient mutants (*recB* or *recC*). This suggested that postlabeling rearrangements could introduce significant variations in observed segregation patterns. Pierucci and Zuchowski (1973) were able to circumvent this problem by examining only chains of cells which were descendent from a cell containing a single replicating chromosome at the time of pulse labeling. Such chains have two labeled cells. The distribution of radioactivity in these chains obtained after two, three, and four divisions was found to be consistent with the idea that labeled DNA strands segregated as if only one DNA strand of the chromosome were attached to the cell envelope. This attachment presumably took place at the time the strand was first used as a template (Pierucci and Zuchowski, 1973).

Oriented nonrandom segregation has also been reported by Cooper and Weinberger (1977). Cells were pulse labeled with [^{3}H]-thymidine, washed, grown in nonlabeled medium, until each cell contained at most one radioactive strand, and seeded in Methocel. The ratio of the numbers of labeled cells in terminal and internal positions was taken as a measure of the degree of orientation in the segregation process. It was found that this ratio was strain and growth rate dependent. In other experiments (Cooper *et al.*, 1978), *E. coli* B/r K cells were eluted from [^{3}H]-thymidine pulse-labeled membrane-bound populations. The pattern of chromosome segregation in cells eluted after•1, 2, . . . , 10 generations of growth was time independent. This result supported the probabilistic strand inertia model proposed by Cooper and Weinberger (1977). The model stipulates that each DNA strand tends to segregate toward the cell pole to which it had previously segregated.

Care should be taken in examining the available chromosome segregation data. The conclusion that chromosomal segregation is nonrandom is probably correct. However, the apparent growth rate and strain-specific components of the process could be consequences of experimental artifacts and not representative of primary physiological events.

VI. Cell Division

A. Mode of Envelope Growth

Any adequate explanation for the control of bacterial cell division must be capable of predicting cellular properties under various conditions of growth. How are properties, such as cell shape and frequency of cell division, related to envelope synthetic patterns and chromosomal DNA content, and how do environmental changes, such as nutritional shifts, affect these cell properties? Before discussing possible answers to these questions, it is important to note that current ideas on cell division control are based to a large extent on certain growth concepts which are probably correct in principle, but not completely clear in specific details. These concepts will be surveyed in this section.

Escherichia coli cells are presumed to grow during the division cycle by increasing in length only (Marr *et al.*, 1966). Qualitatively this is certainly true, but small quantitative changes in cell width during the division cycle cannot be ruled out. It is generally considered that this increase in length is achieved by elongation at a small, fixed number of growth zones in the cell. However, in spite of extensive studies using a variety of approaches to determine the growth patterns of envelope in both gram-positive and gram-negative organisms, the details of the synthetic processes are not completely established. Spherical gram-positive cells clearly possess equatorial cell wall growth zones responsible for both elongation and crosswall formation (Daneo-Moore and Shockman, 1977). On the other hand, rod-shaped gram-positive cells may elongate either by diffuse intercalation of newly synthesized murein into the wall (Archibald and Coapes, 1976; Highton and Hobbs, 1972; Fan *et al.*, 1975; de Chastellier *et al.*, 1975; Mauck *et al.*, 1972) or by zonal growth (Burdett and Higgins, 1978). The high level of turnover of wall in bacillary forms could mask zonal growth, if it existed (Mauck *et al.*, 1972; Mendelson and Reeve, 1973; Pooley, 1976).

In gram-negative *E. coli*, membrane phospholipids and most membrane proteins are either introduced at a large number of sites in the envelope or introduced at a few sites and rapidly dispersed (van Tubergen and Setlow, 1961; Lin *et al.*, 1971; Tsukagoshi *et al.*, 1971; Wilson and Fox, 1971; Green and Schaechter, 1972). The murein layer is apparently synthesized at a discrete number of loci, based on autoradiographic evidence of zones of radioactive diaminopimelic acid incorporation into *E. coli* sacculi (Ryter *et al.*, 1973; Schwarz *et al.*, 1975). In these experiments, radioactivity was located preferentially in the equatorial region in cells growing with generation times of 50 and 80 minutes at 37°C, and additionally at subterminal positions in more rapidly growing cells. Zones of high murein synthetic activity, at least during the process of septum–crosswall formation, are also suggested by the appearance of penicillin-induced bulges in these same general regions of the cell or saculli (Mathison, 1968; Schwarz *et al.*, 1969; Hoffman *et al.*, 1972). The bulges were observed with concentrations of penicillin (about 50 U/ml) which prevented cell division but not elongation, and they appeared at times when the cells would have been in the process of crosswall formation had the drug not been present. Recently, an increase in murein synthetic activity has been reported to occur at the time of septation (Mirelman *et al.*, 1978).

Ryter *et al.* (1975) have reported evidence for zonal growth of the outer membrane of *E. coli*. They measured the locations of λ receptor sites in cells induced for receptor synthesis for short periods of time. The λ receptors were more frequent near the septum in the larger, presumably older cells, and near the poles in smaller cells. This distribution of the receptors was consistent with equatorial growth of the outer membrane, provided the smaller cells were daughters of cells induced during the time of septum formation and division. Begg and Donachie (1973, 1977) investigated outer membrane formation by analysis of T6 attachment site synthesis in a strain of *E. coli* carrying an amber mutation in the *tsx* gene, responsible for the production of the T6 receptor, and a temperature-sensitive amber suppressor. These cells were first grown at nonpermissive temperature and then at permissive temperature in the presence of T6 and a low concentration of penicillin to prevent division. The distribution of T6 attachment sites was asymmetrical. More phage were located at one of the poles, suggesting polar growth of the outer membrane. Thus, both murein and outer membrane appear to possess zonal growth properties, at least during septum–crosswall formation and perhaps during elongation as well. The extent of spatial coincidence of these zones in the different layers, as well as the organization of zones in specific layers, have yet to be resolved.

TABLE II

FORMATION, ACTIVITY, OR TURNOVER OF ENVELOPE COMPONENTS DURING THE DIVISION CYCLE OF *E. coli*

Component	Strain	τ (minutes)	Change in formation[a] activity or turnover during cell cycle	Stage in cycle at change	Experimental[b]	References
Murein synthesis	B/r	43	Step	Midcycle	1, 2	Hoffman *et al* (1972)
	B	45	Burst	Division	3	Mirelman *et al.* (1978)
Envelope protein formation or activity						
Total envelope proteins	B/r	43	Step (or Burst)	Midcycle	1	Hakenbeck and Messer (1977a)
Total envelope proteins	B/r	65	Step	First half-cycle	1	Churchward and Holland (1976)
Envelope proteins of MW						
36,500	B/r	65	Step	First half-cycle	1	Churchward and Holland (1976)
56,000	B/r	65	Step	Near division	1	Churchward and Holland (1976)
76,000	B/r	65	Burst	Near division	1	Churchward and Holland (1976)
Free lipoprotein	B/r	45	Step (or Burst)	Midcycle	2	James and Gudas (1976)
Bound lipoprotein	B/r	45	Continuous		2	James and Gudas (1976)
Wall-bound murein hydrolases	B/r	43	Step (or Burst)	Midcycle	1	Hakenbeck and Messer (1977b)
Endopeptidase	B/r	43	Step	—	1	Hakenbeck and Messer (1977b)
Transglycosylase	B/r	43	Continuous		1	Hackenbeck and Messer (1977b)
Carboxypeptidase I	K12	—	Continuous		2	Beck and Park (1976)
Carboxypeptidase II	K12	—	Step	During septation	2	Beck and Park (1976)
Amidase	K12	—	Continuous		2	Beck and Park (1976)
Cytochrome *b1*	K12	—	Step	—	4, 5	Ohki (1972)

Phospholipid synthesis						
Total phospholipids	B/r	65	Continuous		1	Churchward and Holland (1976)
Total phospholipids	B/r	43	Step (or Burst)	Midcycle	1	Hakenbeck and Messer (1977a)
Phosphatidylethanolamine	K12	60	Continuous		4	Ohki (1972)
Phosphatidylethanolamine	B/r	25 to 60	Step	At initiation of chromosome	1	Pierucci (unpublished observations)
Phosphatidylglycerol	K12	60	Continuous		4	Ohki (1972)
Cardiolipin	K12	60	Continuous		4	Ohki (1972)
Phospholipid turnover						
Phosphatidylglycerol	K12	50	Step	Near division	4, 5	Ohki (1972)
Phosphatidylglycerol	B/r	50	Continuous		1	Zuchowski and Pierucci (1978)
Cardiolipin	B/r	50	Continuous		1	Zuchowski and Pierucci (1978)

[a] The rate of formation or activity of an envelope component increases during the division cycle either continuously without apparent fluctuations or discontinuously, as indicated by "step" or "burst." A step increase implies a single abrupt change in rate of formation or activity during a short interval in the division cycle. A burst indicates an abrupt increase followed by a decrease in formation or activity. A step (or burst) indicates an abrupt increase, followed by a slight decrease in rate of formation or activity several minutes later. Phospholipid turnover refers to either continuous or stepwise loss of radioactivity from previously labeled phospholipid.

[b] (1) Cell cycle studies performed on newborn cells released from membrane-bound populations; (2) separation of the smallest cells from an exponential-phase population by centrifugation through sucrose gradients; (3) synchronization by amino acid starvation; (4) synchronization of chromosome replication and cell division by incubation of a temperature-sensitive chromosome initiation mutant at nonpermissive temperature, followed by a shift to permissive temperature; (5) synchronization by exposure to 45°C for 16 minutes.

B. Synthesis, Activity, and Turnover of Envelope Components during the Division Cycle

If some envelope precursors are incorporated at zones, the rate of incorporation of these envelope components might be expected to increase when new zones become active, or decrease when old zones cease to function. Some cell cycle relationships in the incorporation, activity, and turnover of envelope components are summarized in Table II. As shown in the table, there are stepwise increases or maxima in the rates of formation (or activities) of murein (Hoffman *et al.*, 1972), wall-bound murein hydrolases (Hakenbeck and Messer, 1977b; Mirelman *et al.*, 1978), free lipoprotein (James and Gudas, 1976), envelope proteins (Ohki, 1972; Beck and Park, 1976; Churchward and Holland, 1976; Hakenbeck and Messer, 1977a) and phospholipids (Hakenbeck and Messer, 1977a; Pierucci, unpublished observations) during the division cycle of *E. coli.* In cells growing with an interdivision time of about 45 minutes at 37°C, the increases in rates of formation or activities of all of these envelope components coincided with the increase in penicillin sensitivity, suggesting a coupling to initiation of septum formation. However, at this growth rate a round of chromosome replication terminates and new rounds initiate at about the same time. Thus, some oscillations in formation or activity of envelope components are probably related to the process of septum–crosswall formation itself, while others could be specifically coordinated with initiation or termination of chromosome replication. Information on the extent of coupling to chromosome replication is very limited. In support of a possible coordination, phosphatidylethanolamine synthesis increases abruptly at a time which coincides with initiation of chromosome replication in cells growing with a variety of interdivision times shorter than 50 minutes (Pierucci, unpublished observations). On the other hand, Holland (personal communication), suggests that envelope protein formation is not controlled by chromosome replication, since alterations in DNA replication by thymine limitation do not alter the differential rate of envelope protein synthesis.

Whether coupled to replication of a specific chromosomal locus or not, there must be some interaction between chromosome replication and envelope formation. Inhibition of DNA synthesis eliminates cyclic variations in sensitivity to penicillin (Hoffman *et al.*, 1972) or in activities of murein hydrolases (Hakenbeck and Messer, 1977b). In addition, the synthesis of some proteins found in the envelope are altered by inhibition of DNA synthesis. Synthesis of outer membrane proteins G (MW = 15,000) and Y (MW = 35,000) decrease after inhibition of DNA synthesis, whereas synthesis of protein X (MW = 40,000), located in both cytoplasm and membrane, increases and soon constitutes 3 to 4% of the total protein of the cell

(Inouye and Pardee, 1970; Gudas and Pardee, 1976; Gudas, 1976). These proteins could be involved in cell growth processes. James (1975) suggests that protein G might be a structural protein, based on its appearance at a reduced rate associated with loss of rod shape caused by FL1060, an amidino-penicillanic acid, and at an increased rate during filament formation caused by inhibition of DNA synthesis.

Protein X is the product of the *recA* gene (McEntee *et al.*, 1976; Gudas and Mount, 1977). It has been suggested that protein X is a specific negative effector of cell septation (Howe and Mount, 1978; Satta and Pardee, 1978). When DNA synthesis was blocked (or DNA was damaged) and protein X was produced, cell division was inhibited. Under conditions when DNA synthesis was blocked and protein X was not overproduced, division continued. When protein X production was selectively inhibited with rifampicin, septation was enhanced. It was suggested that protein X may bind to single stranded DNA and displace it from the septum site on the cell membrane, thereby affecting both DNA repair and cell division pleiotropically (Satta and Pardee, 1978), or that it is a protease which attacks a positive effector for division (Howe and Mount, 1978). Thus, a relationship between chromosome replication and the synthesis of proteins possibly involved in cell division appears to exist, but further studies will be necessary to clarify this relationship. For instance, Gudas *et al.*, (1976) identified an 80,000 molecular weight outer membrane protein (protein D) which appeared to be formed during a brief period of the cell cycle. Boyd and Holland (1977) have argued that this protein is the product of the *feuB* locus, a binding protein essential for iron transport, whose synthesis is oscillatory after filtration of cells.

In summary, most current ideas on the control of *E. coli* cell division adhere to the basic concept that cells increase in length during the division cycle by elongation at a small, fixed number of growth zones. The observations described above on the changes in individual envelope components during the division cycle are consistent with these conclusions. We will now focus our discussion of cell division on a few plausible models for the control of division in *E. coli*, recognizing that new alternatives and modifications will continue to be developed.

C. Models for Cell Division Control

Current ideas on the relationship between chromosome replication and cell division in *E. coli* can be divided into two major categories. Chromosome replication and the processes leading to cell division could follow either *interdependent* or *independent* pathways. Each of these categories can be subdivided: changes in cell widths with growth rate could be consequences of either *passive* responses to changes in internal hydrostatic pressure or *active* synthetic processes involved in total surface area formation. The basic properties of one example of each of these four fundamental cell division control possibilities are shown in Table III. Each will be described in turn.

1. Interdependent–Passive

Pritchard and co-workers investigated the interrelationship between chromosome replication and cell division by examining the relationship between *C* and *D*. As one approach, they limited the thymine concentration in cultures of thymine-requiring *E. coli*, and attempted to determine values for *C* and *D* by measuring $\overline{M}$, $\overline{G}$, and $\overline{G}/\overline{M}$ (see equations in Section II,B; see also Zaritsky and Pritchard, 1973; Meacock and Pritchard, 1975). With decreasing thymine concentration, *C* increased, $\overline{M}$ increased, $\overline{G}/\overline{M}$ decreased, and $\overline{G}$ changed little. The data best fit the idea that *D* decreased with decreasing thymine concentration, i.e., with increasing *C*. As a second approach for investigating the relationship between *C* and *D*, the kinetics of cell division following an abrupt decrease (step-down) or increase (step-up) in thymine concentration were measured. In a step-down experiment, the rate of cell division continued at the preshift value for about 20 minutes, which is consistent with a 20-minute *D* period, and then it decreased abruptly. This result corresponds to what would be expected if a terminal event of chromosome replication contributes to the regulation of the timing of cell division. In the complementary step-up experiment, the rate of cell division accelerated rapidly, with a slight delay of at most 10 minutes. If the delay period represents the length of *D* in the prestep conditions, this result is consistent with their other findings that a lengthening of *C* is accompanied by a decrease in *D*.

Pritchard and co-workers generally concluded that termination of chromosome replication can determine the time of the ensuing cell division, but that the length of the period between termination and division (*D*) is not invariant. The apparent decrease in *D* with an increase in *C* is consistent with an earlier proposal of Jones and Donachie (1973) that separate division-specific processes might begin coincident with a very early event in chromosome replication, perhaps initiation. This implies that there is a minimum time of $(C + D)$ minutes between initiation of chromosome replication and cell division, with the value of *D* depending on the length of *C*.

Another important aspect of these studies was the finding that a decrease in thymine concentration caused an increase in cell volume due to an increase in width, with a slight decrease in length (Zaritsky and Pritchard, 1973). Recently, Meacock *et al.* (1978) reaffirmed this conclusion by dimensional analysis of sacculi isolated from cultures grown in different thymine concentrations.

TABLE III

MODELS FOR CELL GROWTH AND DIVISION

	Interdependent-passive[a] *E. coli*	Independent-passive[b] *E. coli*
Stage in cell cycle at doubling in rate of envelope elongation	Completion of chromosome replication	At length 2Λ or 20 minutes before division
Rate of envelope elongation prior to doubling[c]	α/τ	Λ/τ
L_0[d]	$\alpha(1 + D/\tau)$	$\Lambda(1 + 20/\tau)$
$\bar{L}$[d]	$\alpha 2^{D/\tau}/\ln 2$	$\Lambda 2^{20/\tau}/\ln 2$
W[d]	Proportional to $2^{C/2\tau}$	—

		Independent-active	
	Interdependent-active[e] *S. faecalis*	*S. typhimurium*[f]	*E. coli* for $(C + D)/\tau > 1.0$[g]
Rate of envelope surface growth	Function of $1/\tau$	Function of $1/\tau$	Function of $1/\tau$
Number of envelope growth sites	Function of $1/\tau$	Function of $1/\tau$	Function of $(C + D)/\tau$
Rate of envelope growth per site[c]	Constant, independent of τ	Function of $1/\tau$	$\beta_0/(C + D)$
Stage at site activation	About $\frac{3}{4}C$	About $\frac{1}{2}C$	Initiation of chromosome replication
Stage at site inactivation	At completion of septum formation	About $(C/2 + D)$ minutes after activation	$(C + D)$ minutes after activation

[a] Zaritzky and Pritchard (1973).

[b] Donachie *et al.* (1976).

[c] α, Λ and β_0 are proportionality constants, α and Λ have dimensions of length, β_0 has dimensions of area. α and Λ are equivalent to one-half the length of a cell D or 20 minutes prior to division, respectively. Λ is also the length of a newborn cell at $\mu = 0$.

[d] L_0, length of newborn cell; $\bar{L}$ average cell length; W, cell width.

[e] Shockman *et al.* (1974).

[f] Case and Marr (1976).

[g] Pierucci (1978).

Zaritsky and Pritchard (1973) developed a model for cell growth and division, which has been described in most detail by Pritchard (1974) and which embodies many of the earlier ideas of Previc (1970). It was suggested that the cell extends its length at a constant rate which doubles once during the cycle at the end of a round of chromosome replication (Zaritsky and

Pritchard, 1973; Pritchard, 1974; Zaritsky, 1975b). The rate of linear extension would be proportional to the growth rate. The relationships between envelope growth, cell length, and cell width and C, D, and τ are shown in Table III. Cell width is a function of C, and the increase in width caused by a step-down (an increase in C) is explained by a delay in the increase in rate of cell length extension due to a delay in termination of chromosome replication. Mass would continue to increase, resulting in an increase in hydrostatic pressure which would then cause an increase in cell width.

According to the model, new envelope subunits are added to existing envelope in *E. coli* in a manner similar to that in *Streptococcus faecalis* (Higgins and Shockman, 1971; Shockman *et al.*, 1974), i.e., new envelope is laid down primarily at the leading edge of a nascent septum–crosswall (Pritchard, 1974). It is supposed that new envelope is added as if the cell were always attempting to form a septum–crosswall, but as the cell increases in mass, the increasing internal pressure causes extension in length. When the rate of envelope synthesis doubles, the hydrostatic pressure falls rapidly until it is no longer sufficient to draw out the newly added envelope units, and a septum–crosswall develops. Faster-growing cells would be wider than slower-growing cells due to the delay in the response of envelope synthesis relative to mass following a nutritional shift upward (Previc, 1970; Pritchard, 1974; Zaritsky, 1975b). The essence of the model is that there is no specific division signal and no fundamental difference between crosswall formation and cell elongation.

2. Independent–Passive

Donachie *et al.* (1976) have proposed a model which is an example of an independent–passive scheme. The basic properties of the model (shown in Table III) are that cells become committed to division, and increase their rate of elongation, at a critical cell length which is independent of the growth rate. The critical length is twice the minimum cell length, which is defined as the length of a newborn cell as the growth rate approaches zero. Attainment of this critical length is not necessarily coincident with, or coupled to, the end of a round of chromosome replication. It could be coupled to formation of a series of division-specific proteins whose synthesis might begin concurrent with initiation of chromosome replication (Pierucci and Helmstetter, 1969; Jones and Donachie, 1973). The cell presumably extends in length at a fixed number of growth sites, with the rate of extension per site being proportional to the growth rate. The number of growth sites doubles when the cell reaches the critical length and that signals the start of septation which takes 20 minutes at 37°C at all growth rates. It is further assumed that following a change in nutritional conditions, cells do not change their rate of elongation until they reach the critical length.

Analysis of both interdependent- and independent-type models has involved measurements of $\bar{L}$ or L_0 versus μ during steady-state growth at 37°C. Donachie *et al.* (1976) determined $\bar{L}$ versus growth rate for *E. coli* B/r A and a K12 strain. They found that $\bar{L}$ increased from about 2.0 to 4.0 μm when μ increased from 0 to 3.0 doublings/hour. This result was interpreted to indicate that cell length at about 20 minutes before division was the same at all growth rates, and if it is assumed that elongation is linear and increases twofold at this time, this length is 2.77 μm and the minimum length (length of a newborn cell at $\mu = 0$) is 1.37 μm. Grover *et al.* (1977) and Cullum and Vicente (1978) have also reported stepwise increases in rate of elongation during the division cycle of B/r strains, although the cell ages (or sizes) at the steps were somewhat different. Thus, the findings in these studies on cell length versus growth rate are consistent with the basic concepts of either model.

Woldringh *et al.* (1977) have shown that *E. coli* B/r A and B/r K differ significantly in length. The change in length of B/r K with growth rate was found to be very similar to that reported previously for B/r H266, B/r A, and K12 (Donachie *et al.*, 1976; Grover *et al.*, 1977), whereas the change in B/r A was not. If it is assumed that the linear rate of cell elongation doubles at a fixed time in the division cycle, the data indicate that this time would be 13 minutes before division in B/r K and 44 minutes before division in B/r A. The value of 13 minutes is in reasonable agreement with previous findings of 15 to 20 minutes for B/r H266, and with the length of the *D* period in this organism (Table I), but the value of 44 minutes for B/r A is longer than the *D* period and is not consistent with either model. More recently, similar studies from the same laboratory (Koppes *et al.*, 1978) indicate that substrains B/r K and B/r A increase in length exponentially during the division cycle. It seems clear that the cellular dimensional analyses performed to date on *E. coli* cannot be used to distinguish between models, either because of difficulties in performing accurate measurements or the similar dimensional predictions of different models.

An alternative approach, employed by Donachie *et al.* (1976), involved analysis of cell growth during a nutritional shift upward of newborn cells. The rate of cell elongation remained unchanged until it increased abruptly at 20 minutes before division. The same results were obtained when DNA synthesis was inhibited simultaneously with the shift upward. They concluded that termination of chromosome replication is not the trigger for initiation of cell division.

The results of the nutritional shift experiments reported by Donachie *et al.* (1976) are not consistent with earlier findings of Kubitschek (1971) that cell volume increased linearly in *E. coli* in the absence of DNA synthesis, which would suggest that an increase in the rate of volume growth requires chromosome replication. They are also at odds with the results of similar experiments performed by Sargent on *B. subtilis* (Sargent, 1975a). Sargent's findings, and resultant model for cell length control in *B. subtilis*, were very similar to the Zaritsky-Pritchard conclusions, i.e., growth zones function at a constant rate proportional to μ, the number of zones doubles at nuclear segregation, and the doubling of zones is dependent on chromosome replication. Nevertheless, the findings of Donachie *et al.* (1976) raise serious questions as to an obligatory coupling between chromosome replication and processes directly affecting the control of cell elongation. Supporting this conjecture is the recent report that cell shape changes during thymine limitation are strain- and growth media-dependent (Begg and Donachie, 1978). For instance, during growth of thymine-requiring *E. coli* K12 in limiting thymine in poorer media, cell volume increased as a consequence of an increase in length but not width. Similarly, it has been shown that reduction in replication velocity by introduction of a rep^- mutation results in longer, but not wider, *E. coli* cells (Zaritsky and Woldringh, personal communication), although thymine limitation results in an increase in width in the same cells. The effects of thymine limitation on cell shape may not reflect a coupling between chromosome replication and cell shape determination, but rather a reduced differential rate of synthesis of an envelope component whose synthesis involves thymine (Begg and Donachie, 1978; Meacock *et al.*, 1978).

3. Interdependent–Active and Independent–Active

A series of alternative ideas on cell growth, chromosome replication, and cell division suggest that synthesis of the entire envelope surface is regulated deterministically, not just cell elongation (Previc, 1970; Pritchard, 1974; Case and Marr, 1976; Pierucci, 1978). With regard to an interdependent–active scheme, one view suggests that the rate of envelope synthesis doubles near termination of chromosome replication, with the rate of synthesis being proportional to the growth rate (Previc, 1970). The most detailed interdependent–active model was described by Shockman *et al.* (1974) for *Streptococcus faecalis*, and is summarized in Table III. The basic idea is that surface growth sites would be activated at about $\frac{3}{4}$ C, and the activity would terminate at the division which followed completion of the round of replication (Shockman *et al.*, 1974; Case *et al.*, 1977). In their recent review article, Daneo-Moore and Shockman (1977) suggest that the surface biosynthetic sequence could activate closer to the time of initiation of chromosome replication such that replication and envelope growth follow a separate series of timed events which might start at about the same time, i.e., an independent–active scheme.

Detailed independent–active models for cell division (Table III) have been proposed for both *S. typhimurium* (Case and Marr, 1976) and *E. coli* (Pierucci, 1978). For *E. coli*, it is proposed that (1) new sites of envelope surface growth are inaugurated at initiation of chromosome replication, (2) envelope growth sites are potential division positions, (3) the rate of elongation per site is independent of the growth rate, and (4) each site remains active in growth for $(C + D)$ minutes. This model for *E. coli* satisfactorily accounts for cell surface area measurements with B/r A and B/r F and the values calculated from data reported previously for *E. coli* B/r H266 (Grover *et al.*, 1977).[1]

In this section, we have presented a limited number of examples of some basic ideas on cell division control. Various modifications of these ideas will undoubtedly be forthcoming, particularly with regard to the timing element for inauguration of new sites of envelope growth. In principle, if initiation of new sites is "triggered" by an event which takes place a short time before division, then the average number of sites per cell would increase slowly with increasing growth rate, and the rate of elongation per site would have to increase more rapidly, e.g., proportional to μ, for a model to be consistent with existing information. Conversely, if site initiation is triggered well before division, the number of sites per cell would increase more rapidly with growth rate and the rate of elongation per site would have to increase less rapidly, i.e., approach a constant. The problem now is to develop the key experiments which can differentiate between these models for the regulation of cell division. Additional analyses of cell dimensions in strains growing at various rates should be valuable, particularly with regard to the extent of coupling between chromosome replication and cell growth. Another approach would involve analysis of the growth and division properties of mutants which are defective in cell division. In the next section, we will make some general comments concerning such mutants, summarize some of their more interesting properties, and then attempt to reach a general conclusion on cell division regulation.

D. Mutants Defective in Cell Division

Analysis of growth patterns and cell division properties of mutants which are defective in cell division should provide information to complement that which has been obtained by biochemical and physical analysis. Descriptions of some representative mutants with interesting division properties are

[1] An extensive analysis of models for surface growth regulation has been presented recently by Rosenberger *et al.* (1978).

TABLE IV

MUTANTS OF *E. coli* WITH UNUSUAL DIVISION PROPERTIES

Entry	Mutant	Map[a] position (minutes)	Defect	Division,[b] response	Other[b]	References
1	*dnaA*$_{ts}$	82	Initiation of chromosome replication	Some continued division at RT to form chromosomeless cells, with or without thymine starvation, with or without a secondary mutation	Can be integratively suppressed; product of some alleles thermoreversible	Beyersmann *et al.* (1971), 1974); Carl (1970); Gross (1972); Hirota *et al.* (1968)
2	*dnaB*$_{ts}$	91	Elongation of chromosome; *dnaB252* is initiation defective	Some continued division at RT to form chromosomeless cells, with or without thymine starvation, with or without a secondary mutation	Product of some alleles thermoreversible	Kuempel (1971); Lindahl *et al.* (1971); Nishimura *et al.* (1971); Schubach *et al.* (1973); Wechsler and Gross (1972)
3	*dnaC*$_{ts}$	99	Initiation of chromosome replication	Some continued division at RT to form chromosomeless cells, with or without thymine starvation	Suppressible or nonsuppressible. Product of some alleles thermoreversible	Wolf (1972)
4	*dnaI*$_{ts}$	39	Chromosome replication	Some evidence for chromosomeless cell formation		Beyersmann *et al.* (1974)
5	*dnaP*$_{ts}$	84	Initiation of chromosome replication	No evidence for chromosomeless cell formation	Possible suppression. Primary defect in envelope	Wada and Yura, (1974)

6	*ftsA*	2	Division defective; formation of multinucleate filaments at RT	Does not require protein synthesis after shift RT to PT for resumption of division.	PAT84 osmotically reversible, other alleles not reversible. Increased PEN resistance at RT and PT	Pages *et al.* (1975); Ricard and Hirota (1973)
7	*ftsB*	32–34	Division defective; formation of multinucleate filaments at RT	Chromosomeless cell formation at RT	Osmotically reversible. Increased PEN resistance at PT and RT	Pages *et al.* (1975); Ricard and Hirota (1973)
8	*ftsC*	4–9	Division defective; formation of multinucleate filaments at RT	Chromosomless cell formation at RT	Osmotically reversible	
9	*ftsD*	86	Division defective; formation of multinucleate filaments at RT	Requires protein synthesis after shift RT to PT for resumption of division	One of two alleles osmotically reversible. Increased PEN resistance at PT and RT	Pages *et al.* (1975); Ricard and Hirota (1973)
10	*ftsE*	73–76	Division defective; formation of multinucleate filaments at RT	Does not require protein synthesis after shift RT to PT for resumption of division	Two of three alleles osmotically reversible	Pages *et al.* (1975); Ricard and Hirota (1973)
11	*ftsF*	82	Division defective; formation of multinucleate filaments at RT		Osmotically reversible. Increased PEN resistance at RT and PT	Pages *et al.* (1975); Ricard and Hirota (1973)
12	*ftsG*	29–30	Division defective; formation of multinucleate filaments at RT		Not osmotically reversible	Pages *et al.* (1975); Ricard and Hirota (1973)

(*continued*)

TABLE IV (*Continued*)

Entry	Mutant	Map[a] position (minutes)	Defect	Division.[b] response	Other[b]	References
13	*ftsH(y-16)*	68	Division defective; formation of multinucleate filaments at RT	Does not require protein synthesis after shift RT to PT for resumption of division	Not osmotically reversible. Loss of viability at RT prevented by adenine and accelerated by guanosine and cytosine	Santos and de Almeida (1975)
14	*ftsH(ASH124)*	89	Division defective; formation of multinucleate filaments at RT	Does not require protein synthesis after shift RT to PT for resumption of division; no residual filaments due to series of sequential divisions	Osmotically reversible. Cell density dependent. Reduction in growth rate at PT	Holland and Darby (1976)
15	*BUG6*	3	Division defective; formation of multinucleate filaments at RT	Does not require protein synthesis after shift RT to PT for resumption of division. No residual filaments due to rapid nonsequential divisions	Osmotically reversible. Cell density dependent	Reeve *et al.* (1970)
16	Fil_{ts}	88	Division defective; formation of multinucleate filaments at RT	Immediate halt in division with shift to RT. Cell division [only from pole(s)] resumes after shift from RT	Osmotically reversible. Cell density dependent. Medium dependent	Stone (1973)

				resulting in residual filaments. Septum formation stimulated by CAM at PT and RT		
17	*ts20*	1	Division defective; formation of multinucleate filaments at RT	Immediate halt in division with shift to RT. Requires protein synthesis after shift RT to PT for resumption of division. After shift from RT, residual filaments plus normal-size cells	Not osmotically reversible. Cell density dependent. RNA turnover at RT	Nagai and Tamura (1972)
18	*sep*	2	Division defective; formation of multinucleate filaments with incomplete septa at RT	Immediate halt in division with shift to RT. Requires protein synthesis after shift RT to PT for resumption of division. No residual filaments after shift from RT due to central and polar divisions	Not osmotically reversible	Allen *et al.* (1974) Fletcher *et al.* (1978)
19	*ts52*	35	Division defective; formation of multinucleate filaments at RT	After shift to RT, 70% increase in cell number. CAM-stimulated cell division at RT. Residual filaments; one terminal division per filament after shift to PT	Not osmotically reversible. Medium dependent. CAM effect due to inhibition of peptide bond formation	Zusman *et al.* (1972)

(*continued*)

TABLE IV (*Continued*)

Entry	Mutant	Map[a] position (minutes)	Defect	Division,[b] response	Other[b]	References
20	*MAC*-1(*divA*)	3 ± 1	Division defective	Cells induced to divide at RT with addition of CAM or RIF, independent of completion of chromosome replication		de Pedro *et al.* (1975); de Pedro and Canovas (1978)
21	*minA minB*	10 29	Placement of division septum	Frequent formation of small, chromosomeless cells	Septa are structurally normal. Nucleated cells are elongated. When normal cell division is inhibited (e.g., by addition of NAL) minicell divisions are also inhibited.	Adler *et al.* (1967, 1969); Adler and Hardigree (1972); Clark (1968b); Frazer and Curtiss (1975); Teather *et al.* (1974)
22	*minA minB capR*	10 (*capR*)	Placement of division septum plus *capR*-mediated radiation sensitivity and filamentation (see entry 25)	Inhibition of both normal and minicell division after radiation exposure		Adler *et al.* (1969)
23	*minA minB BUG6*		Placement of division septum plus thermosensitive cell division (see entry 15)	Inhibition of both normal and minicell divisions at RT. After shift RT to PT burst of cell division results		Khachatorians *et al.* (1973)

				in production of minicells as well as normal-sized cells.		
24	*minA minB ts52*		Placement of division septum plus thermosensitive cell division (see entry 19)	Inhibition of both normal and minicell divisions at RT. At PT, fraction of minicells is half that of parental minicell strain. Burst of cell divisions resulting from addition of CAM includes minicells as well as normal-size cells		Zusman and Krotoski (1974)
25	*capR* (*lon*)	10	Sensitive to radiation and agents which damage DNA	Filamentation after exposure to radiation inhibition of chromosome synthesis or nutritional shift	Overproduction of capsular polysaccharide. Filamentation can be suppressed by *lexA* or *recA* mutations (but not capsular polysaccharide overproduction	Donch *et al.* (1968); Gayda *et al.* (1976); Green *et al.* (1969); Howard-Flanders *et al.* (1964); Hua and Markovitz (1972); Leighton and Donachie (1970)
26	*capR sulA* *capR sulB*	22 (*sulA*) 2 (*sulB*)	*sulA* and *sulB* suppress *capR* specified sensitivity to radiation and to agents which damage DNA	Suppression of *capR* directed filamentation	*sulA* and *sulB* suppress overproduction of of capsular polysaccharide in *E. coli* B, but not *E. coli* K-12	Donch *et al.* (1971); Gayda *et al.* (1976); James and Gillies (1973); Johnson and Greenberg (1975)

(continued)

TABLE IV (*Continued*)

Entry	Mutant	Map[a] position (minutes)	Defect	Division,[b] response	Other[b]	References
27	*recA*	58	Recombination deficient, and very sensitive to radiation and agents which damage DNA	Continued division in the absence of chromosome replication resulting in the formation of chromosomeless cells	Reduced spontaneous mutation frequency. Spontaneous DNA degradation	Clark and Margulies (1965); Howard-Flanders and Theriot (1966); Inouye (1971); Jenkins and Bennett (1976); Rupp and Howard-Flanders (1968); Smith and Meun (1970); Willetts *et al.* (1969)
28	tif_{ts}	58	Increased capacity to repair uv irradiated λ DNA at RT, but accompanied by mutagenesis	Filamentation at RT	Induction of λ at RT. Does not affect DNA recombination capacity	Castellazzi *et al.* (1972a); Kirby *et al.* (1967); Witkin (1975)
29	tif_{s} *sfiA* tif_{ts} *sfiB*	22 (*sfiA*) 2 (*sfiB*)	Increased capacity to repair uv irradiated λ DNA at RT, but accompanied by mutagenesis	Normal cell division at RT	Induction of λ at RT. Does not affect DNA recombination capacity	George *et al.* (1975)
30	tif_{ts} *zab*	58 (*zab*)	Very radiation sensitive; *zab* suppresses the increased repair capacity of tif_{ts}	Normal cell division at RT	Recombination proficient. No spontaneous DNA degradation. Reduction of spontaneous λ induction and mutagenesis	Castellazzi *et al.* (1972b); Castellazzi (1976)

31	tif_{ts} *recA*		Very radiation sensitive. *recA* suppresses the increased repair capacity of tif_{ts}	Normal cell division at RT	Recombination and DNA repair deficient. Reduced spontaneous mutation frequency. Spontaneous DNA degradation	Castellazzi *et al.* (1972b); Castellazzi (1976)
32	*lexA*	90	Sensitive to radiation and agents which damage DNA	Continued division in the absence of chromosome replication resulting in the formation of chromosomeless cells	Recombination proficient. Very low rate of spontaneous mutation	Howard-Flanders and Boyce (1966); Howe and Mount (1975, 1978); Mount *et al.* (1972)
33	tif_{ts} *lexA*		Radiation sensitive; *lexA* suppresses the increased repair capacity of tif_{ts}	Normal cell division at RT	Very similar to tif_{ts} *zab*	Castellazzi *et al.* (1972b); Castellazzi (1976)
34	*lexA* tsl_{ts}	90 (*tsl*)	Decreased uv sensitivity at PT	Normal division at PT. Filamentation at RT	Low λ induction at RT. Low spontaneous mutation frequency. DNA degradation after exposure to uv	Mount *et al.* (1973)
35	*lexA* tif_{ts} *sfiA spr*	90 (*spr*)	Increased capacity to repair uv irradiated λ DNA at both PT and RT. Increased uv resistance	Normal cell division		Mount (1977)

[a] The map positions cited are from Bachmann *et al.* (1976) or extrapolated from primary sources to the recalibrated linkage map.

[b] Key to abbreviations: RT, nonpermissive temperature; PT, permissive temperature; PEN, penicillin; CAM, chloramphenicol; RIF, rifampicin; NAL, nalidixic acid.

presented in Table IV. Before commenting on the behavior of specific mutants, it is necessary to note that studies on cell division mutants are often complicated by the need to distinguish between direct and indirect effects the genetic lesion on the division parameter of interest (Slater and Schaechter, 1974). For example, the phenotype of conditional cell division mutations depends on the genetic background in which the mutation is placed (Walker and Kovarik, 1975). Furthermore, genetic analysis of cell division mutants has not progressed very far. In some cases, the complex phenotypes associated with these mutations are pleiotropic effects of a single mutation and not the result of closely linked multiple mutations (Slater and Schaechter, 1974). Unfortunately, few analyses have progressed to the point of complementation tests which are required to determine if two mutations with similar phenotypes are allelic (e.g., *BUG6* and *MAC1*). The potential that these (and other) mutants have for providing answers to the complex questions posed by the cell division process has not been fully realized.

1. Conditional Cell Division Mutants

Some mutants are temperature-sensitive for cell division (Table IV, entries 6–20). At the permissive temperature, these mutants grow and divide normally. At nonpermissive temperature, macromolecular synthesis continues, but cell division is blocked resulting in the formation of multinucleate filaments. The residual increase in cell number upon shift of an exponential-phase culture to nonpermissive conditions (5–20%) is consistent with the possibility that not all cells in the *D* period are capable of division at nonpermissive temperature. Viability is generally not affected by a brief (1–2 hour) exposure to nonpermissive conditions, and upon return to permissive conditions cell division resumes. In some cases, there is a rapid increase in cell number subsequent to the temperature shift downward (*BUG6*, *ftsA*, *ftsH*, *AX655*), demonstrating an accumulation of divisional potential at nonpermissive temperature. The expression of this division potential is in some instances (e.g., *ftsA*, *BUG6* and fil_{ts}) independent of concurrent protein synthesis. When *E. coli BUG6* was held at nonpermissive temperature for at least 12 minutes and then returned to permissive temperature, the concentration of cells doubled synchronously at 20 minutes postshift, with or without continuing protein and DNA synthesis (Reeve and Clark, 1972). Apparently, division potential accumulated in cells which were blocked *D* minutes before division.

MAC1 is a particularly interesting new temperature-sensitive cell division mutant (de Pedro *et al.*, 1975; de Pedro and Canovas, 1978). The mutation maps at 3 minutes and could be allelic with *ftsA* or *BUG6*. When synchronous cultures of the mutant were heat-shocked for 10 minutes at various times in the division cycle, all cells required 55 to 60 minutes for the next division at 30°C. Thus, all cells were set back to the same state in progression to division, suggesting that the product of this gene could control the timing of cell division. When synchronous cultures were shifted to 42°C and rifampicin or chloramphenicol was added at 60 minutes postshift, the cells divided once at the nonpermissive temperature, even when DNA synthesis was inhibited at the time of the shift. Thus, termination of chromosome replication was not required for the division. Induction of cell division by chloramphenicol has also been reported by Zusman *et al.* (1972) in *ts52*, but DNA replication was required. The possible significance of the findings with *MAC1* will be discussed later.

2. Minicell Mutants

Another group of mutants which have supplied information on cell division are mutants which produce minicells (Table IV, entries 21–24). Minicells are small chromosomeless cells formed by a septation unusually close to one end of a rod-shaped bacterium. Minicell-producing mutants have been described in *E. coli* (Adler *et al.*, 1967, 1969), *B. subtilis* (Reeve *et al.*, 1973), and in other species (Frazer and Curtiss, 1975). Having no DNA other than perhaps plasmid DNA which segregates independently of chromosomal DNA, they are incapable of dividing, but do contain RNA and protein (Frazer and Curtiss, 1975).

Explanations for the appearance of minicells have generally been based on the idea that the cell poles in such mutants are active division sites, either due to premature utilization of normal division positions (Adler and Hardigree, 1972) or the absence of normal shut-off of division sites located at the cell poles after division (Teather *et al.*, 1974). Consistent with these ideas, Khachatorians *et al.* (1973) estimated the total number of divisions, including minicell divisions, in an *E. coli* minicell mutant and found that the total division potential of the mutant was the same as for the parental strain. Similarly, as a result of studying cell length distributions, Teather *et al.* (1974) concluded that the frequency of production of minicells can account for the elongation of chromosome-containing cells. They proposed that only enough division potential accumulates at each mass doubling for one division, and that this division can occur with an equal probability at normal division positions or at polar minicell-producing positions. Based on these assumptions, the predicted mean length of chromosome-containing cells was calculated and found to be indistinguishable from the observed value. This model may not hold for minicell-producing mutants of *B. subtilis*. Mendelson (1975) and Mendelson and Coyne (1975) reported that the total number of divisions in these strains is less than expected relative to parental.

However, the latter studies involved light microscopy and Donachie (personal communication) suggests that the results in *B. subtilis* are identical to those in *E. coli* if an electron microscopy is used to enable detection of all polar minicell septa.

In an effort to deduce the possible control of septum placement, mutants have been constructed which contain the minicell-producing mutations and another mutation affecting cell division (Table IV, entries 22–24). First, Adler *et al.* (1969) described *min capR* hybrid mutants. The cessation of cell division upon radiation, typical of *capR* mutants, also inhibited divisions which would have produced minicells. Therefore, *min* shares a common regulatory step with *capR*.

Second, *min* was combined with the temperature-sensitive cell division mutation in *E. coli BUG6* (Khachatourians *et al.*, 1973). At 30°C, this mutant formed minicells, and the average volume of chromosome-containing cells was twice that of the parental *BUG6* strain. At 42°C, all cell division, including minicell division, ceased and filaments were formed. The filaments had some septumlike constrictions including minicell-type constrictions. After a temperature shift downward, there was a burst of cell division resulting in the formation of normal cells and minicells. These results show that minicell production shares a common regulatory mechanism with normal division in *E. coli BUG6* also.

Finally, an *E. coli* minicell-producing mutant was combined with *ts52* (Zusman and Krotoski, 1974). This double mutant produced both minicells and normal cells at the permissive temperature, except the minicell yield was half that of the minicell-forming parental strain. At the nonpermissive temperature, division stopped after a residual division characteristic of the *ts52* mutant (Table IV, entry 19). Fifteen minutes after the addition of chloramphenicol to the medium at 41°C, there was a synchronous division at both normal and minicell sites. In the presence of chloramphenicol, minicells were produced only at the poles of the filamentous cells. Therefore, both at the permissive temperature, and at nonpermissive temperature in the presence of the drug, the *ts52* mutation affects the expression of the *min* phenotype in some way, in either case reducing the number of minicells.

3. Effects of Repair-Associated Genes on Cell Division

DNA metabolism other than chromosome replication, e.g., DNA repair, may affect cell division (Table IV, entries 25–35). First, enhancement of repair capacity can impair cell division, such as in tif_{ts} mutants or when tsl^- mutations are introduced into $lexA^-$ strains (Table IV, entries 28 and 34).

The thermoinducible repair and thermosensitive cell division in tif_{ts} mutants can be suppressed coordinately by *recA*, *zab*, or *lexA* mutations (Table IV). Introduction of *sfiA* or *B* mutations into tif_{ts} cells suppresses only thermosensitive cell division, while maintaining thermoinducible repair capacity. Thus, the repair function which is capable of impairing cell division is not an essential element in the progression of a cell through its division cycle.

Protein X, the product of *recA*, which is synthesized when DNA synthesis is blocked (Section VI,B), is synthesized constitutively under conditions of *tif* gene expression (Gudas, 1976), supporting the notion that *tif* regulates the synthesis of specific proteins (Gudas and Pardee, 1975). The study of protein X in tif_{ts} sfi^- double mutants may give some insight as to whether this protein, besides being involved in repair, takes part in the regulation of cell division.

In a second apparent relationship between repair capacity and cell division, some mutations can increase repair capacity and simultaneously reverse a defect in cell division. An example is a $capR^-$ mutant possessing a sul^- mutation (Table IV, entry 26). $capR^-$ mutants overproduce capsular polysaccharide resulting in the formation of mucoid colonies in minimal medium. In addition, $capR^-$ mutants are very radiation sensitive. The mutants form nonseptate filaments when incubated in complex medium after exposure to radiation. Introduction of a sul^- mutation in these cells decreases radiation sensitivity and prevents filament formation, with or without affecting the mucoid phenotype. $capR^-$ mutations have been found to affect synthesis of enzymes located in at least four spatially separate operons involved in capsular polysaccharide biosynthesis. Gayda *et al.* (1976) proposed that the *capR* gene specifies a protein which functions as a repressor. In $capR^-$ mutants, this *capR* protein is inactive and the operons are derepressed. The mucoid phenotype then appears as a result of overproduction of capsular polysaccharide. They proposed that the *capR* gene controls cell division and radiation sensitivity in the same way that it controls capsular polysaccharide synthesis, i.e., that the *capR* gene product controls an operon(s) whose structural gene (*sul*) is involved in radiation sensitivity and cell division. A sul^- mutation will, therefore, prevent the synthesis of active products which can cause radiation sensitivity and/or abnormal cell division.

It is clear that repair capacity and cell division are related, but the details of this relationship are unknown due to the differential effects of alterations in repair capacity on cell division. Witkin (1976) has proposed that cell division control genes may have the same repressor as repair genes but that the extent of derepression in different genes may vary in different mutants. Repair genes and genes controlling cell division could be subject to differential repression by the same repressor.

4. Chromosomeless Cell Production

Perhaps the most significant information contained in Table IV is that some mutants of *E. coli* are capable of division for extended periods of time in the absence of DNA replication, resulting in the formation of chromosomeless cells. Consider first the temperature-sensitive chromosome replication mutants. Upon shift to nonpermissive temperature, normal cell division continues for $(C + D)$ minutes in the case of initiation defective mutants (e.g., $dnaA_{ts}$) or for D minutes in the case of elongation defective mutants (e.g., $dnaB_{ts}$). After this time, filaments begin to form. Chromosomeless cells of near normal size can be produced at the extremeties of the filaments for several hours in both classes of mutants. This phenomenon has been observed upon shift to nonpermissive temperature in certain $dnaA_{ts}$, $dnaB_{ts}$, or $dnaC_{ts}$ mutants of *E. coli* (Hirota *et al.*, 1968; Inouye, 1969; Hirota and Ricard, 1972; Helmstetter, unpublished observations), a $dnaC_{ts}$ mutant of *S. typhimurium* (Spratt and Rowbury, 1971), and a temperature-sensitive mutant of *B. subtilis* (Sargent, 1975b). A secondary mutation (designated *divA*, *B*, or *C*) is required for chromosomeless cell formation in some mutants (Hirota *et al.*, 1968; Hirota and Ricard, 1972), but not in other mutants (Inouye, 1969; Spratt and Rowbury, 1971). The requirement for the secondary mutation was demonstrated by the finding that when dna_{ts} genes were transferred to different strains, not all of the temperature-sensitive recipients could form chromosomeless cells at nonpermissive temperature. In addition, in matings between the dna_{ts} mutants and different Hfr strains, some recipients which remained temperature-sensitive were incapable of chromosomeless cell formation. Finally, some thymine-requiring, temperature-sensitive mutants continue to divide at nonpermissive temperature in the absence of thymine. Thus, completion of ongoing rounds of chromosome replication is not essential for chromosomeless cell formation (Inouye, 1969; Sargent, 1975b). [Division may not continue when chromosome replication is inhibited with nalidixic acid, but in view of the differing effects of thymine starvation and nalidixic acid on normal cell division (Dix and Helmstetter, 1973; Sargent, 1975b), the effects of nalidixic acid cannot be ascribed to inhibition of chromosome replication alone.]

As stated above, chromosomeless cell formation begins some time after shift to nonpermissive temperature, subsequent to an interval of normal cell division. In *S. typhimurium*, where this subject has been studied most extensively, the second phase of division starts at about 80 minutes, even if rounds of replication were terminated prior to the temperature shift [by amino acid starvation at 25°C (Spratt and Rowbury, 1971; Shannon *et al.*, 1972)]. The results suggest that division is triggered at initiation and that it takes $(C + D)$ minutes to prepare for division along a pathway which is independent of chromosome replication. Consistent with the concept of independently timed pathways, when newborn cells of *E. coli* B/r F $dnaC_{ts}$ cells growing in glucose minimal medium were shifted from 26° to 41°C, a synchronous cell division took place at 50 minutes, producing daughters possessing one nonreplicating chromosome, and then a second synchronous division took place about 50 minutes later to produce a chromosomeless daughter cell (Helmstetter, unpublished observations).

$recA^-$ and $lexA^-$ mutants of *E. coli* can also divide to form chromosomeless cells in the absence of DNA synthesis (Inouye, 1971; Howe and Mount, 1975, 1978), as does at least one *B. subtilis* strain (Donachie *et al.*, 1971). The appearance of chromosomeless cells of normal size at regular intervals during inhibition of DNA synthesis in these mutants, and in dna_{ts} mutants, may hold one of the keys to understanding the control of cell division. Previously, it was suggested that the products of these genes, such as *recA* and *dnaA*, could be repressors of cell division and that mutation in these genes caused derepression of some processes leading to cell division, enabling division to take place in the absence of DNA replication (e.g., Inouye, 1971). An alternative explanation for the data should be considered. One obvious similarity between *recA* mutants and temperature-sensitive chromosome replication mutants, with regard to the formation of chromosomeless cells, is that they are relatively insensitive to inactivation due to selective inhibition of DNA synthesis (thymineless death) compared to their parental strains (Inouye, 1971; Bouvier and Sicard, 1975; Sicard and Bouvier, 1977). Particularly important in this regard are the recent findings that chromosomeless cell formation occurs when cell mass increase is capable of continuing extensively in the absence of DNA replication and segregation (Howe and Mount, 1975, 1978; Sargent, 1977). Furthermore, the $divC^-$ mutation, which confers the ability for chromosomeless cell formation on temperature-sensitive mutants defective in chromosome replication, was selected on the basis of mitomycin C resistance (Hirota and Ricard, 1972). It also confers resistance to inactivation due to incubation at nonpermissive temperature. We suggest that the products of the mutated genes may not be directly involved in cell division, and that cell division continues in the mutants during inhibition of DNA synthesis because they remain viable.

If this analysis is correct, then initiation of normal cell division is not necessarily coupled to replication of the chromosome. However, chromosome replication and cell division might interact in the sense that if a round of replication is not completed and the chromosomes do not segregate, the division which would normally take place between the replicated chromosomes is prevented (Woldringh *et al.*, 1977). This would explain why chromosomeless cells are usually formed in filaments, because the cells must first grow long enough so that division can take place at a position in the cell which is not blocked by the physical presence of the chromosome.

E. Conclusions Regarding the Regulation of Cell Division

Analysis of the bulk of information presented in this section leads to the conclusion that chromosome replication and the processes leading to cell division likely follow independent pathways, as suggested earlier by Donachie and co-workers (Section VI,C) and more recently by Koch (Section II,C). Certain events of the division-specific pathway, such as the changes in rates of synthesis of envelope components, might take place concurrent with initiation of chromosome replication, but this coupling is not obligatory for cell division. The timing of cell division is probably determined by events in the cell division pathway and not by chromosome replication, except that the region of the cell at which septum–crosswall formation is initiated must eventually become free of chromosomal material in order for that division to be completed. The observed relationships between chromosome replication, cell growth, and cell division could be consequences of this "veto" power of the chromosome. The timing of normal cell division between replicated chromosomes would be determined solely by the division-specific pathway as long as chromosome replication is completed prior to initiation of the septum–crosswall. On the other hand, cell division, and associated biosynthetic processes, could be timed by chromosome replication if termination of replication were delayed relative to the initiation of septum–crosswall formation in the division-specific pathway. This hypothesis could explain the interactions between replication and division which were described in this section and also why certain strains of *E. coli* divide at a smaller than normal size subsequent to completion of chromosome replication in the absence of protein synthesis (Ron *et al.*, 1975, 1977). Finally, if the proposed absence of an obligatory coupling between chromosome replication and cell division is correct, then it would be predicted that if chromosome replication could be inhibited without affecting any other structural or transcription–translation function of the chromosome, cell division might continue indefinitely.

Acknowledgments

We are grateful to all those who sent us information and manuscripts prior to publication. Special thanks are due Florence Florkowska and Carol Helmstetter for their efforts in preparing the manuscript. Work performed in the authors' laboratories was supported, in part, by United States Public Health Service Research Grants CA 08232 and AI 13761 to C. H., GM 21006, and AI 13453 to O. P., and Postdoctoral Fellowship GM 05327 to M. W.

References

Abe, M., and Tomizawa, J. (1971). *Genetics* **69,** 1–15.

Adler, H. I., and Hardigree, A. A. (1972). *In* "Biology and Radiobiology of Anucleate Systems" (S. Bonotto, R. Kirchman, R. Goutier, and J. R. Maisin, eds.), Vol. 1, pp. 51–56. Academic Press, New York.

Adler, H. I., Fisher, W. D., Cohen, A., and Hardigree, A. A. (1967). *Proc. Natl. Acad. Sci. U.S.A.* **57,** 321–326.

Adler, H. I., Fisher, W. D., and Hardigree, A. A. (1969). *Trans. N.Y. Acad. Sci.* **31,** 1059–1070.

Allen, J. S., Filip, C. C., Gustafson, R. A., Allen, R. G., and Walker, J. R. (1974). *J. Bacteriol.* **117,** 978–986.

Archibald, A. R., and Coapes, H. E. (1976). *J. Bacteriol.* **125,** 1195–1206.

Bachmann, B. J., Low, K. B., and Taylor, A. L. (1976). *Bacteriol. Rev.* **40,** 116–167.

Bagdasarin, M., Zdzienicka, M., and Bagdasarian, M. (1972). *Mol. Gen. Genet.* **117,** 129–142.

Bagdasarian, M., Hryniewicz, M., Zdzienicka, M., and Bagdasarian, M. (1975). *Mol. Gen. Genet.* **139,** 213–231.

Barbu, E., Polonovski, J., Rampini, C., and Lux, M. (1970). *C. R. Acad. Sci., Ser. D* **270,** 2596–2599.

Bazaral, M., and Helinski, D. R. (1970). *Biochemistry* **9,** 399–406.

Beck, B. D., and Park, J. J. (1976). *J. Bacteriol.* **126,** 1250–1260.

Begg, K. J., and Donachie, W. D. (1973). *Nature (London), New Biol.* **245,** 38–39.

Begg, K. J., snd Donachie, W. D. (1977). *J. Bacteriol.* **129,** 1524–1536.

Begg, K. J., and Donachie, W. D. (1978). *J. Bacteriol.* **133,** 452–458.

Beyersmann, D., Schlicht, M., and Schuster, H. (1971). *Mol. Gen. Genet.* **111,** 145–158.

Beyersmann, D., Messer, W., and Schlicht, M. (1974). *J. Bacteriol.* **118,** 783–789.

Bharnik, G., Cummings, D. J., and Taylor, A. L. (1973). *Biochim. Biophys. Acta* **312,** 793–799.

Bird, R. E., and Lark, K. G. (1970). *J. Mol. Biol.* **49,** 343–366.

Bird, R. E., Chandler, M., and Caro, L. (1976). *J. Bacteriol.* **126,** 1215–1223.

Blau, S., and Mordoh, J. (1972). *Proc. Natl. Acad. Sci. U.S.A.* **69,** 2895–2898.

Bleecken, S. (1969). *J. Theor. Biol.* **25,** 137–158.

Bleecken, S. (1971). *J. Theor. Biol.* **32,** 81–92.

Bouvier, F., and Sicard, N. (1975). *J. Bacteriol.* **124,** 1198–1204.

Boyd, A., and Holland, I. B. (1977). *FEBS Lett.* **76,** 20–24.

Boyle, J. V., Goss, W. A., and Cook, T. M. (1967). *J. Bacteriol.* **94,** 1664–1671.

Burdett, I. D., and Higgins, M. L. (1978). *J. Bacteriol.* **133,** 959–971.

Burns, F. J., and Tannock, I. F. (1970). *Cell Tissue Kinet.* **3,** 321–334.

Campbell, A. (1957). *Bacteriol. Rev.* **21,** 263–272.

Carl, P. L. (1970). *Mol. Gen. Genet.* **109,** 107–122.

Caro, L., and Berg, C. M. (1968). *Cold Spring Harbor Symp. Quant. Biol.* **33,** 559–573.

Case, M. L., and Marr, A. G. (1976). *Abstr., Am. Soc. Microbiol.* p. 127.

Case, M. L., Daneo-Moore, L., Carson, D. D., and Lancy, P. L. (1977). *Abstr., Am. Soc. Microbiol.* p. 157.

Castellazzi, M. (1976). *J. Bacteriol.* **127,** 1150–1156.

Castellazzi, M., George, J., and Buttin, G. (1972a). *Mol. Gen. Genet.* **119,** 139–152.

Castellazzi, M., George, J., and Buttin, G. (1972b). *Mol. Gen. Genet.* **119,** 153–174.

Cattaneo, S. M., Quastler, H., and Sherman, F. G. (1961). *Nature (London)* **190,** 923–924.

Chai, N. C., and Lark, K. G. (1967). *J. Bacteriol.* **94,** 415–421.

Chai, N. C., and Lark, K. G. (1970). *J. Bacteriol.* **104,** 401–409.

Chandler, M., Bird, R. E., and Caro, L. (1975). *J. Mol. Biol.* **94,** 127–132.

Chandler, M., Silver, L., Roth, Y., and Caro, L. (1976). *J. Mol. Biol.* **104,** 517–523.

Chandler, M., Silver, L., and Caro, L. (1977). *J. Bacteriol.* **131,** 421–430.

Churchward, G. G., and Bremer, H. (1977). *J. Bacteriol.* **130,** 1206–1213.
Churchward, G. G., and Holland, I. B. (1976). *J. Mol. Biol.* **105,** 245–261.
Clark, A. J., and Margulies, A. D. (1965). *Genetics* **53,** 451–459.
Clark, D. J. (1968a). *Cold Spring Harbor Symp. Quant. Biol.* **33,** 823–838.
Clark, D. J. (1968b). *J. Bacteriol.* **96,** 1214–1224.
Collins, J., and Pritchard, R. H. (1973). *J. Mol. Biol.* **78,** 143–155.
Cooper, S., and Helmstetter, C. E. (1968). *J. Mol. Biol.* **31,** 519–540.
Cooper, S., and Weinberger, M. (1977). *J. Bacteriol.* **130,** 118–127.
Cooper, S., Schwimmer, M., and Scanlon, S. (1978). *J. Bacteriol.* **134,** 60–65.
Cullum, J., and Vincente, M. (1978). *J. Bacteriol.* **134,** 330–337.
Daneo-Moore, L., and Shockman, G. D. (1977). *In* "Cell Surface Reviews" (G. Poste and G. L. Nicolson, eds.), Vol. 4, pp. 597–715. Elsevier/North-Holland, New York.
Datta, N., and Barth, P. T. (1976). *J. Bacteriol.* **125,** 811–817.
de Chastellier, C., Hellio, R., and Ryter, A. (1975). *J. Bacteriol.* **123,** 1184–1196.
de Pedro, M. A., and Canovas, J. (1978). (Submitted for publication.)
de Pedro, M. A., Llamas, J. E., and Canovas, J. L. (1975). *J. Gen. Microbiol.* **91,** 307–314.
DiJoseph, C. G., and Kaji, A. (1974). *Proc. Natl. Acad. Sci. U.S.A.* **71,** 2515–2519.
DiJoseph, C. G., Bayer, M. E., and Kaji, A. (1973). *J. Bacteriol.* **115,** 399–410.
Dix, D. E., and Helmstetter, C. E. (1973). *J. Bacteriol.* **115,** 786–795.
Donachie, W. D. (1968). *Nature (London)* **219,** 1077–1079.
Donachie, W. D., and Masters, M. (1968). *In* "The Cell Cycle: Gene-Enzyme Interactions" (G. M. Padilla, G. L. Whitson, and I. L. Cameron, eds.), pp. 37–76. Academic Press, New York.
Donachie, W. D., Martin, D. T., and Begg, K. J. (1971). *Nature (London), New Biol.* **227,** 1220–1225.
Donachie, W. D., Begg, K. J., and Vicente, M. (1976). *Nature (London)* **264,** 328–333.
Donch, J. J., Green, M. H. L., and Greenberg, J. (1968). *J. Bacteriol.* **96,** 1704–1710.
Donch, J. J., Chung, Y. S., Green, M. H. L., Greenberg, J., and Warren, G. (1971). *Genet. Res.* **17,** 185–193.
Eberle, H. (1970). *Proc. Natl. Acad. Sci. U.S.A.* **65,** 467–474.
Eberle, H., and Lark, K. G. (1966). *J. Mol. Biol.* **22,** 183–186.
Engberg, B., and Nordstrom, K. (1975). *J. Bacteriol.* **123,** 179–186.
Engberg, B., Hjalmarsson, K., and Nordstrom, K. (1975). *J. Bacteriol.* **124,** 633–640.
Evans, I. M., and Eberle, H. (1975). *J. Bacteriol.* **121,** 883–891.
Fan, D. P., Beckman, B. E., and Gardner-Eckstrom, H. I. (1975). *J. Bacteriol.* **123,** 1157–1162.
Finkelstein, M., and Helmstetter, C. E. (1977). *J. Bacteriol.* **132,** 884–895.
Fletcher, G., Irwin, C. A., Henson, J. M., Fillingim, C., Malone, M. M., and Walker, J. R. (1978). *J. Bacteriol.* **133,** 91–100.
Fralick, J. A., and Lark, K. G. (1973). *J. Mol. Biol.* **80,** 459–475.
Frazer, A. C., and Curtiss, R., III (1975). *Curr. Top. Microbiol. Immunol.* **69,** 1–84.
Fujisawa, T., and Eisenstark, A. (1973). *J. Bacteriol.* **115,** 168–176.
Gayda, R. C., Yamamoto, L. T., and Markovitz, A. (1976). *J. Bacteriol.* **127,** 1208–1216.
George, J., Castellazzi, and Buttin, G. (1975). *Mol. Gen. Genet.* **140,** 309–332.
Goebel, W. (1974). *Eur. J. Biochem.* **41,** 51–62.
Green, E., and Schaechter, M. (1972). *Proc. Natl. Acad. Sci. U.S.A.* **69,** 2312–2316.
Green, M. H. L., Greenberg, J., and Donch, J. (1969). *Genet. Res.* **14,** 159–162.
Gross, J. D. (1972). *Curr. Top. Microbiol. Immunol.* **57,** 39–74.
Grover, N. B., Woldringh, C. L., Zaritsky, A., and Rosenberger, R. F. (1977). *J. Theor. Biol.* **67,** 181–194.
Gudas, L. J. (1976). *J. Mol. Biol.* **104,** 567–587.
Gudas, L. J., and Mount, D. W. (1977). *Proc. Natl. Acad. Sci. U.S.A.* **74,** 5280–5284.

Gudas, L. J., and Pardee, A. B. (1974). *J. Bacteriol.* **117,** 1216–1223.

Gudas, L. J., and Pardee, A. B. (1975). *Proc. Natl. Acad. Sci. U.S.A.* **72,** 2330–2334.

Gudas, L. J., and Pardee, A. B. (1976). *J. Mol. Biol.* **101,** 459–477.

Gudas, L. J., James, R., and Pardee, A. B. (1976). *J. Biol. Chem.* **251,** 3470–3479.

Hakenbeck, R., and Messer, W. (1977a). *J. Bacteriol.* **129,** 1234–1238.

Hakenbeck, R., and Messer, W. (1977b). *J. Bacteriol.* **129,** 1239–1244.

Hanawalt, P. C., Maaløe, O., Cummings, D. J., and Schechter, M. (1961). *J. Mol. Biol.* **3,** 156–165.

Hanna, M. H., and Carl, P. L. (1975). *J. Bacteriol.* **121,** 219–226.

Hansen, F. G., and Rasmussen, K. V. (1977). *Mol. Gen. Genet.* **155,** 219–225.

Helmstetter, C. E. (1971). *In* "Drugs and Cell Regulation: Organizational and Pharmacological Aspects on the Molecular Level" (E. Mihich, ed.), pp. 1–20. Academic Press, New York.

Helmstetter, C. E. (1974a). *J. Mol. Biol.* **84,** 1–20.

Helmstetter, C. E. (1974b). *J. Mol. Biol.* **84,** 21–36.

Helmstetter, C. E. (1974c). *J. Bacteriol.* **120,** 565–567.

Helmstetter, C. E. (1978). Submitted for publication.

Helmstetter, C. E., and Cooper, S. (1968). *J. Mol. Biol.* **31,** 507–518.

Helmstetter, C. E., and Pierucci, O. (1968). *J. Bacteriol.* **95,** 1627–1633.

Helmstetter, C. E., and Pierucci, O. (1976). *J. Mol. Biol.* **102,** 477–486.

Helmstetter, C. C., Cooper, S., Pierucci, O., and Revelas, E. (1968). *Cold Spring Harbor Symp. Quant. Biol.* **33,** 809–822.

Henning, U. (1975). *Annu. Rev. Microbiol.* **29,** 45–60.

Hewitt, R., and Billen, D. (1964). *Biochem. Biophys. Res. Commun.* **15,** 588–592.

Higgins, M. L., and Shockman, G. D. (1971). *Crit. Rev. Microbiol.* **1,** 29–72.

Highton, P. J., and Hobbs, D. G. (1972). *J. Bacteriol.* **109,** 1181–1190.

Hiraga, S. (1976). *Proc. Natl. Acad. Sci. U.S.A.* **73,** 198–202.

Hiraga, S., and Saitoh, T. (1974). *Mol. Gen. Genet.* **132,** 49–62.

Hirota, Y., and Ricard, M. (1972). *In* "Biology and Radiobiology of Anucleate Systems" (S. Bonotto, R. Kirchman, R. Goutier, and J. R. Maisin, eds.), Vol. 1, pp. 29–50. Academic Press, New York.

Hirota, Y., Jacob, F., Ryter, A., Buttin, G., and Nakai, T. (1968). *J. Mol. Biol.* **35,** 175–192.

Hirota, Y., Mordoh, J., and Jacob, F. (1970). *J. Mol. Biol.* **53,** 369–387.

Hoffman, B., Messer, W., and Schwarz, U. (1972). *J. Supramol. Struct.* **1,** 29–37.

Hohlfeld, R., and Vielmetter, W. (1973). *Nature (London), New Biol.* **242,** 130–132.

Holland, I. B., and Darby, V. (1976). *J. Gen. Microbiol.* **92,** 156–166.

Howard-Flanders, P., and Boyce, R. P. (1966). *Radiat. Res., Suppl.* **6,** 156–184.

Howard-Flanders, P., and Theriot, L. (1966). *Genetics* **53,** 1137–1150.

Howard-Flanders, P., Simson, E., and Theriot, L. (1964). *Genetics* **49,** 237–246.

Howe, W. E., and Mount, D. W. (1975). *J. Bacteriol.* **124,** 1113–1121.

Howe, W. E., and Mount, D. W. (1978). *J. Bacteriol.* **133,** 1278–1281.

Hua, S. A., and Markovitz, A. (1972). *J. Bacteriol.* **110,** 1089–1099.

Inouye, M. (1969). *J. Bacteriol.* **99,** 842–850.

Inouye, M. (1971). *J. Bacteriol.* **106,** 539–542.

Inouye, M., and Pardee, A. B. (1970). *J. Biol. Chem.* **245,** 5813–5819.

Jacob, F., Brenner, S., and Cuzin, F. (1963). *Cold Spring Harbor Symp. Quant Biol.* **28,** 329–348.

James, R. (1975). *J. Bacteriol.* **124,** 918–929.

James, R., and Gillies, N. E. (1973). *J. Gen. Microbiol.* **76,** 429–436.

James, R., and Gudas, L. J. (1976). *J. Bacteriol.* **125,** 374–375.

Jenkins, S. T., and Bennett, P. M. (1976). *J. Bacteriol.* **125,** 1214–1216.

Johnson, B. F., and Greenberg, J. (1975). *J. Bacteriol.* **122,** 570–574.

Jonasson, J. (1973). *Mol. Gen. Genet.* **120,** 69–90.

Jones, N. C., and Donachie, W. D. (1973). *Nature* (*London*), *New Biol.* **243,** 100–103.
Kang, T. W., and Grecz, W. (1975). *In* "Spores VI" (P. Gerhardt, R. N. Costilow, and H. L. Sadoff, eds.), pp. 513–519. Am. Soc. Microbiol., Washington, D.C.
Khachatorians, G. G., Clark, D. J., Adler, H. I., and Hardigree, A. A. (1973). *J. Bacteriol.* **116,** 226–229.
Kirby, E. P., Jacob, F., and Goldthwait, D. A. (1967). *Proc. Natl. Acad. Sci. U.S.A.* **58,** 1903–1910.
Kline, B. C. (1974). *Biochemistry* **13,** 129–146.
Koch, A. (1977). *Adv. Microb. Physiol.* **16,** 49–98.
Kogoma, T. (1978). *J. Mol. Biol.* **121,** 55–69.
Kogoma, T., and Lark, K. G. (1970). *J. Mol. Biol.* **52,** 143–164.
Kogoma, T., and Lark, K. G. (1975). *J. Mol. Biol.* **94,** 243–256.
Koppes, L. J. H., Woldringh, C. L., and Nanninga, N. (1978). *J. Bacteriol.* **134,** 423–433.
Kubitschek, H. E. (1971). *J. Bacteriol.* **105,** 472–476.
Kubitschek, H. E. (1974). *Mol. Gen. Genet.* **135,** 123–130.
Kubitschek, H. E., and Freedman, M. L. (1971). *J. Bacteriol.* **107,** 95–99.
Kubitschek, H. E., and Newman, C. N. (1978). *J. Bacteriol.* **136,** 179–190.
Kubitschek, H. E., Bendigkeit, H. E., and Loken, M. R. (1967). *Proc. Natl. Acad. Sci. U.S.A.* **57,** 1611–1617.
Kuempel, P. (1969). *J. Bacteriol.* **100,** 1302–1310.
Kuempel, P. (1971). *Adv. Cell Biol.* **1,** 3–56.
Kuempel, P. L., Duerr, S. A., and Seeley, N. R. (1977). *Proc. Natl. Acad. Sci. U.S.A.* **74,** 3927–3931.
Kuempel, P. L., Duerr, S. A., and Maglothin, P. D. (1978). *J. Bacteriol.* **134,** 902–912.
Kung, F. C., and Glaser, D. A. (1978). *J. Bacteriol.* **133,** 755–762.
Lane, H. E. D., and Denhardt, D. T. (1975). *J. Mol. Biol.* **97,** 99–112.
Lark, C. (1966). *Biochim. Biophys. Acta* **119,** 517–525.
Lark, K. G. (1972). *J. Mol. Biol.* **64,** 47–60.
Lark, K. G. (1973). *J. Bacteriol.* **113,** 1066–1069.
Lark, K. G., and Renger, H. (1969). *J. Mol. Biol.* **42,** 221–235.
Lark, K. G., Repko, T., and Hoffman, E. J. (1963). *Biochim. Biophys. Acta* **76,** 9–24.
Leibowitz, P. J., and Schaechter, M. (1975). *Int. Rev. Cytol.* **41,** 1–26.
Leighton, P. M., and Donachie, W. D. (1970). *J. Bacteriol.* **102,** 810–814.
Lin, E. C. C., Hirota, Y., and Jacob, F. (1971). *J. Bacteriol.* **108,** 375–385.
Lindahl, G., Hirota, Y., and Jacob, F. (1971). *Proc. Natl. Acad. Sci. U.S.A.* **68,** 2407–2411.
Louarn, J., Funderburgh, M., and Bird, R. E. (1974). *J. Bacteriol.* **120,** 1–5.
Louarn, J., Patte, J., and Louarn, J.-M. (1978). *J. Mol. Biol.* **115,** 295–314.
Maaløe, O., and Hanawalt, P. C. (1961). *J. Mol. Biol.* **3,** 144–155.
Maaløe, O., and Kjeldgaard, N. O. (1966). "Control of Macromolecular Synthesis." Benjamin, New York.
McKenna, W. G., and Masters, M. (1972). *Nature* (*London*) **240,** 536–539.
Marr, A. G., Harvey, R. J., and Trentini, W. C. (1966). *J. Bacteriol.* **91,** 2388–2389.
Marunouchi, T., and Messer, W. (1973). *J. Mol. Biol.* **78,** 211–228.
Marvin, D. A. (1968). *Nature* (*London*) **219,** 485–486.
Masters, M. (1975). *Mol. Gen. Genet.* **143,** 105–111.
Masters, M., and Broda, P. (1971). *Nature* (*London*), *New Biol.* **232,** 137–140.
Mathison, G. E. (1968). *Nature* (*London*) **219,** 405–407.
Mauck, J., Chan, L., Glaser, L., and Williamson, J. (1972). *J. Bacteriol.* **109,** 373–378.
McEntee, K., Hesse, J. E., and Epstein, W. (1976). *Proc. Natl. Acad. Sci. U.S.A.* **73,** 3979–3983.
Meacock, P. A., and Pritchard, R. H. (1975). *J. Bacteriol.* **122,** 931–942.

Meacock, P. A., Pritchard, R. H., and Roberts, E. M. (1978). *J. Bacteriol.* **133,** 320–328.

Mendelson, N. H. (1975). *J. Bacteriol.* **121,** 1166–1172.

Mendelson, N. H., and Coyne, S. I. (1975). *J. Bacteriol.* **121,** 1200–1202.

Mendelson, N. H., and Reeve, J. N. (1973). *Nature* (*London*), *New Biol.* **243,** 62–64.

Messer, W. (1972). *J. Bacteriol.* **112,** 7–12.

Messer, W., Dankwarth, L., Tippe-Schindler, R., Womack, J. E., and Zahn, G. (1975). *In* "DNA Synthesis and its Regulation" (M. Goulian and P. Hanawalt, eds.), Vol. 3, pp. 602–617. Benjamin, New York.

Messer, W., Bergmans, H. E. N., Meijer, M., Womack, J. E., Hansen, F. G., and Meyenburg, K. von (1978). *Mol. Gen. Genet.* **162,** 269–275.

Meyenburg, K. von, Hansen, F. G., Nielsen, L. D., and Jørgensen, P. (1977). *Mol. Gen. Genet.* **158,** 101–109.

Meyenburg, K. von, Hansen, F. G., Nielsen, L. D., and Riise, E. (1978). *Mol. Gen. Genet.* **160,** 287–295.

Minor, P. D., and Smith, J. A. (1974). *Nature* (*London*) **248,** 241–243.

Mirelman, D., Yashouv-Gan, Y., Nuchamovitz, Y., Rozenhak, S., and Ron, E. Z. (1978). *J. Bacteriol.* **134,** 458–461.

Moody, E. E. M., and Runge, R. (1972). *Genet. Res.* **19,** 181–186.

Mount, D. W. (1977). *Proc. Natl. Acad. Sci. U.S.A.* **74,** 300–304.

Mount, D. W., Low, K. B., and Edmiston, S. J. (1972). *J. Bacteriol.* **112,** 886–893.

Mount, D. W., Walker, A. C., and Kosel, C. (1973). *J. Bacteriol.* **116,** 950–956.

Murakami, S., Inuzuka, N., Yamaguchi, M., Yamaguchi, K., and Yoshikawa, H. (1976). *J. Mol. Biol.* **108,** 683–704.

Nagai, K., and Tamura, G. (1972). *J. Bacteriol.* **112,** 959–966.

Nagata, T. (1963). *Proc. Natl. Acad. Sci. U.S.A.* **49,** 551–558.

Nishimura, Y., Caro, L., Berg, C. M., and Hirota, Y. (1971). *J. Mol. Biol.* **55,** 441–456.

Novick, R. P., Clowes, R. C., Cohen, S. N., Curtiss, R., III, Datta, N., and Falkow, S. (1976). *Bacteriol. Rev.* **40,** 168–189.

Nunn, W. D., and Tropp, B. E. (1972). *J. Bacteriol.* **109,** 162–168.

Ohki, M. (1972). *J. Mol. Biol.* **68,** 249–264.

Pages, J. M., Piovant, M., Lazdunski, A., and Lazdunski, C. (1975). *Biochimie* **57,** 303–313.

Pardee, A. B., Wu, P. C., and Zusman, D. R. (1973). *In* "Bacterial Membranes and Wall" (L. Leive, ed.), pp. 357–412. Dekker, New York.

Pato, M. (1975). *J. Bacteriol.* **123,** 272–277.

Pierucci, O. (1978). *J. Bacteriol.* **125,** 559–574.

Pierucci, O., and Helmstetter, C. E. (1969). *Fed. Proc., Fed. Am. Soc. Exp. Biol.* **28,** 1755–1760.

Pierucci, O., and Helmstetter, C. E. (1976). *J. Bacteriol.* **128,** 708–716.

Pierucci, O., and Zuchowski, C. (1973). *J. Mol. Biol.* **80,** 477–503.

Pooley, H. M. (1976). *J. Bacteriol.* **125,** 1127–1138.

Prescott, D. M., and Kuempel, P. L. (1972). *Proc. Natl. Acad. Sci. U.S.A.* **69,** 2842–2845.

Previc, E. P. (1970). *J. Theor. Biol.* **27,** 471–497.

Pritchard, R. H. (1974). *Philos. Trans. R. Soc. London, Ser. B* **267,** 303–336.

Pritchard, R. H., and Lark, K. G. (1964). *J. Mol. Biol.* **9,** 288–307.

Pritchard, R. H., and Zaritsky, A. (1970). *Nature* (*London*) **226,** 126–131.

Pritchard, R. H., Barth, P. T., and Collins, J. (1969). *Symp. Soc. Gen. Microbiol.* **19,** 263–297.

Pritchard, R. H., Chandler, M. G., and Collins, J. (1975). *Mol. Gen. Genet.* **138,** 143–155.

Reeve, J. N., and Clark, D. J. (1972). *J. Bacteriol.* **110,** 117–121.

Reeve, J. N., Groves, D. J., and Clark, D. J. (1970). *J. Bacteriol.* **104,** 1052–1064.

Reeve, J. N., Mendelson, N. H., Coyne, S. I., Hallock, L. L., and Cole, R. M. (1973). *J. Bacteriol.* **114,** 860–873.

Ricard, M., and Hirota, Y. (1973). *J. Bacteriol.* **116,** 314–322.
Rodriguez, R. L., and Davern, C. I. (1976). *J. Bacteriol.* **125,** 346–352.
Ron, E. Z., Rozenhak, S., and Grossman, N. (1975). *J. Bacteriol.* **123,** 374–376.
Ron, E. Z., Grossman, N., and Helmstetter, C. E. (1977). *J. Bacteriol.* **129,** 569–573.
Rosenberg, B. H., Cavalieri, L. F., and Ungers, G. (1969). *Proc. Natl. Acad. Sci. U.S.A.* **63,** 1410–1417.
Rosenberger, R. F., Grover, N. B., Zaritsky, A., and Woldringh, C. L. (1978). *J. Theor. Biol.* **73,** 711–721.
Rupp, D., and Howard-Flanders, P. (1968). *J. Mol. Biol.* **31,** 291–304.
Ryter, A., Hirota, Y., and Jacob, F. (1968). *Cold Spring Harbor Symp. Quant. Biol.* **33,** 669–676.
Ryter, A., Hirota, Y., and Schwarz, U. (1973). *J. Mol. Biol.* **78,** 185–195.
Ryter, A., Shuman, H., and Schwartz, M. (1975). *J. Bacteriol.* **122,** 295–301.
Saitoh, T., and Hiraga, S. (1975). *Mol. Gen. Genet.* **137,** 249–261.
Santos, D., and de Almeida, D. F. (1975). *J. Bacteriol.* **124,** 1502–1507.
Sargent, M. G. (1975a). *J. Bacteriol.* **123,** 7–19.
Sargent, M. G. (1975b). *J. Bacteriol.* **123,** 1218–1234.
Sargent, M. G. (1977). *Mol. Gen. Genet.* **155,** 329–338.
Satta, G., and Pardee, A. B. (1978). *J. Bacteriol.* **133,** 1492–1500.
Schaechter, M., Maaløe, O., and Kjeldgaard, N. O. (1958). *J. Gen. Microbiol.* **19,** 592–606.
Schubach, W. H., Whitmer, J. D., and Davern, C. I. (1973). *J. Mol. Biol.* **74,** 205–221.
Schwarz, U., Asmus, A., and Frank, H. (1969). *J. Mol. Biol.* **41,** 419–429.
Schwarz, U., Ryter, A., Rambach, A., Hellio, R., and Hirota, Y. (1975). *J. Mol. Biol.* **98,** 749–759.
Shannon, K. P., Spratt, B. G., and Rowbury, R. J. (1972). *Mol. Gen. Genet.* **118,** 185–197.
Shilo, B., Shilo, V., and Simchen, G. (1976). *Nature (London)* **264,** 767–769.
Shockman, G. D., Daneo-Moore, L., and Higgins, M. L. (1974). *Ann. N.Y. Acad. Sci.* **235,** 161–197.
Sicard, N., and Bouvier, F. (1977). *J. Bacteriol.* **132,** 779–783.
Silberstein, S., and Billen, D. (1971). *Biochim. Biophys. Acta* **247,** 383–390.
Slater, M., and Schaechter, M. (1974). *Bacteriol. Rev.* **38,** 199–221.
Sloan, J. B., and Urban, J. E. (1976). *J. Bacteriol.* **128,** 302–308.
Smith, J. A., and Martin, L. (1973). *Proc. Natl. Acad. Sci. U.S.A.* **70,** 1263–1267.
Smith, J. A., and Martin, L. (1974). *In* "Cell Cycle Controls" (G. M. Padilla, I. L. Cameron, and A. Zimmerman, eds.), pp. 43–60. Academic Press, New York.
Smith, K. C., and Meun, D. H. C. (1970). *J. Mol. Biol.* **51,** 459–472.
Sompayrac, L., and Maaløe, O. (1973). *Nature (London), New Biol.* **241,** 133–135.
Spratt, B. G. (1972). *Biochem. Biophys. Res. Commun.* **48,** 496–503.
Spratt, B. G. (1977). *In* "Microbiology–1977" (D. Schlessinger, ed.), pp. 182–190. Am. Soc. Microbiol., Washiagton, D.C.
Spratt, B. G., and Rowbury, R. J. (1971). *Mol. Gen. Genet.* **114,** 35–49.
Stone, A. B. (1973). *J. Bacteriol.* **116,** 741–750.
Sueoka, N., and Yoshikawa, H. (1965). *Genetics* **52,** 747–757.
Teather, R. M., Collins, J. F., and Donachie, W. D. (1974). *J. Bacteriol.* **118,** 407–413.
Terawaki, Y., and Rownd, R. (1972). *J. Bacteriol.* **109,** 492–498.
Terawaki, Y., Kakizawa, H., Takayasu, H., and Yoshikawa, M. (1968). *Nature (London)* **219,** 284–285.
Thilo, L., and Vielmetter, W. (1976). *J. Bacteriol.* **128,** 130–143.
Tresguerres, E. F., Nandadasa, H. G., and Pritchard, R. H. (1975). *J. Bacteriol.* **121,** 554–561.
Tsanev, R., and Sendov, B. (1966). *J. Theor. Biol.* **12,** 327–341.
Tsukagoshi, N., Fielding, F., and Fox, C. F. (1971). *Biochem. Biophys. Res. Commun.* **44,** 497–502.

van Tubergen, R. P., and Setlow, R. B. (1961). *Biophys. J.* **1,** 589–625.
Wada, C., and Yura, T. (1974). *Genetics* **77,** 199–220.
Wada, C., Yura, T., and Hiraga, S. (1977). *Mol. Gen. Genet.* **152,** 211–217.
Walker, J. R., and Kovarik, A. (1975). *J. Bacteriol.* **123,** 752–754.
Wechsler, J. A. (1975). *J. Bacteriol.* **121,** 594–599.
Wechsler, J. A., and Gross, J. D. (1971). *Mol. Gen. Genet.* **113,** 273–284.
Willetts, M. S., Clark, A. J., and Low, B. (1969). *J. Bacteriol.* **97,** 244–249.
Wilson, C., and Fox, C. F. (1971). *Biochem. Biophys. Res. Commun.* **44,** 503–509.
Witkin, E. M. (1975). *Mol. Gen. Genet.* **142,** 87–103.
Witkin, E. M. (1976). *Bacteriol. Rev.* **40,** 869–907.
Woldringh, C. L. (1976). *J. Bacteriol.* **125,** 248–257.
Woldringh, C. L., de Jong, M. A., van den Berg, W., and Koppes, L. (1977). *J. Bacteriol.* **131,** 270–279.
Wolf, B. (1972). *Genetics* **72,** 569–593.
Wolf, B., Newman, A., and Glaser, D. A. (1968). *J. Mol. Biol.* **32,** 611–629.
Yahara, I. (1971). *J. Mol. Biol.* **57,** 373–376.
Yasuda, S., and Hirota, Y. (1977). *Proc. Natl. Acad. Sci. U.S.A.* **74,** 5458–5462.
Ycas, M., Sugita, M., and Bensam, A. (1965). *J. Theor. Biol.* **9,** 444–470.
Yoshikawa, H., O'Sullivan, A., and Sueoka, N. (1964). *Proc. Natl. Acad. Sci. U.S.A.* **52,** 973–980.
Yoshimoto, H., and Yoshikawa, M. (1975). *J. Bacteriol.* **124,** 661–667.
Zaritsky, A. (1975a). *J. Bacteriol.* **122,** 841–846.
Zaritsky, A. (1975b). *J. Theor. Biol.* **54,** 243–248.
Zaritsky, A., and Pritchard, R. H. (1971). *J. Mol. Biol.* **60,** 65–74.
Zaritsky, A., and Pritchard, R. H. (1973). *J. Bacteriol.* **114,** 824–837.
Zuchowski, C., and Pierucci, O. (1978). *J. Bacteriol.* **133,** 1533–1535.
Zusman, D. R., and Krotoski, D. M. (1974). *J. Bacteriol.* **120,** 1427–1433.
Zusman, D. R., Inouye, M., and Pardee, A. B. (1972). *J. Mol. Biol.* **69,** 119–136.
Zyskind, J. W., Deen, L. T., and Smith, D. W. (1977). *J. Bacteriol.* **129,** 1466–1475.

Index

C

F

G

T

U

V

W

X